B. C. W9-BTS-848

ANATOMY OF SEED PLANTS

2nd Edition

ANATOMY OF SEED PLANTS

KATHERINE ESAU
Professor of Botany, Emeritus
University of California
Santa Barbara, California

John Wiley and Sons
New York/Santa Barbara/London/Sydney/Toronto

BOOKS BY ESAU (K.)

Plant Anatomy

Anatomy of Seed Plants

Copyright © 1960, 1977, by John Wiley & Sons, Inc.

All rights reserved. Published simultaneously in Canada.

No part of this book may be reproduced by any means, nor transmitted, nor translated into a machine language without the written permission of the publisher.

Library of Congress Cataloging in Publication Data:

Esau, Katherine, 1898-
Anatomy of seed plants.

Includes bibliographies.
1. Botany—Anatomy. 2. Phanerogams.
I. Title.
QK641.E8 1977 582'.04 76-41191
ISBN 0-471-24520-8

Printed in the United States of America

10 9 8 7 6 5 4 3 2 1

Preface

The first edition of the *Anatomy of Seed Plants* was completed 16 years ago—a long span of time if one considers the great expansion of biological knowledge in the recent past, made possible by the technological progress in all areas of study of living matter. The emendations of the interpretations of structure and development and the increase in understanding of the relation between structure and function provide much new material that must be brought to the attention of the student. Yet, in a course of plant anatomy that may be a student's first introduction to this field of botany, the fundamental aspects of structure cannot be omitted. Thus, a writer of a plant anatomy textbook faces the problem of achieving a balance between providing adequate basic information and reviewing the exciting new discoveries and the resulting interpretations.

Such balance could hardly be achieved within the physical limits of the first edition. Hence, the revision required an increase in length of the text and an addition of new illustrations, as well as some changes in the distribution of the material. With the growth of information on subcellular structure, it was no longer adequate to include the description of the cell in the chapter on parenchyma and that of the cell wall in the chapter on sclerenchyma. The cell and the cell wall are now treated in separate chapters. The spectacular increase in volume of research on the reproductive parts of the plant has caused an expansion of the exposition on the flower and a division of this topic into two chapters. The greater emphasis on the reproductive cycle made it logical to move the chapter on the embryo to the end of the book.

Because of some comments by users of the first edition, the key for the identification of woods has been removed but the characters used in such identifications are assembled in a systematic way, so that the instructor or the student can construct a key if an exercise in wood identification is part of the course. The summary is based on commercially available microscope slides. The introduction of new concepts and terms has required additions to the glossary. A change was made in a taxonomic matter: adoption of family names according to Takhtajan.*

* Takhtajan, Armen. *Flowering plants.* Washington, D. C., Smithsonian Institution Press. 1969.

Among the specific topics, those dealing with the meristems and their activity, the origin and development of leaves, and the ontogenetic and phylogenic origin of the vascular system have been conceptually broadened. The information gained from ultrastructural research was used to elucidate the ontogeny and functional specialization of the conducting cells in the xylem and the phloem.

One of the sources of satisfaction in the work on this book was the awareness that research in plant anatomy has been greatly invigorated by the remarkable development of interest in structure in various fields of botany. Ultrastructural research has made the understanding of plant structure increasingly more important to the plant physiologist and the plant biochemist. Plant anatomy is of major concern to the plant ecologist studying the relation between the plant and the environment and to the plant pathologist seeking to understand the relation between the host plant and the parasite.

Acknowledgments

Since the new edition retains parts of the first, the acknowledgments in the preface to the first edition are still pertinent. Numerous new illustrations have been added and the accompanying legends indicate the sources and give the names of the persons who were kind enough to lend the negatives or supply the prints. Mr. Paul L. Conant again lent microscope slides for photography and for the compilation of characters used for the identification of woods. Mr. Robert H. Gill was especially helpful in taking a number of the electron micrographs.

The reviews of the entire manuscript by Professors Ray F. Evert and Thomas L. Rost are gratefully acknowledged. These reviews and the personal discussions with Professor Evert were particularly helpful in that they conveyed the viewpoints of persons notably interested in teaching.

December 1976

Katherine Esau

Preface to the First Edition

The writing of a second book in plant anatomy by the same author needs, perhaps, a brief comment. The idea of the present book began to develop as soon as my first book, *Plant Anatomy,* John Wiley and Sons, 1953, appeared. Even during the writing of that book the publishers expressed a preference for a relatively short text designed for a one-semester course. My idea, however, was to write a comprehensive treatise in which I could develop the concepts fully and include much detail on ontogeny of plant structure. Very considerately, my publishers raised no objections and cooperated to the fullest extent. Appreciating their attitude, I offered to write a short text later. After the first book underwent classroom tests and when reviews by teachers and investigators became available, the idea to write a new, shorter text became increasingly more appealing to me. Furthermore, the intensive research in plant anatomy of the last decade made me wish to bring the subject matter up to date.

The book may best be characterized by comparing it with my first book. The new book and the old are essentially alike in their basic approach: both follow the "classic" method of discussing the cells and tissues first, then the plant parts composed of these units. The new book, however, contains fewer developmental details, the concepts are defined with less of the historical background, and the terms are not projected back to their Greek and Latin origins. On the other hand, the present book contains a glossary, which was prepared with especial regard to the latest developments of concepts and terminologies. The bibliographies at the ends of chapters are short; moreover, they consist mostly of references published since the appearance of *Plant Anatomy.* This limitation is pointed out in the first chapter, with the thought that the larger book may be consulted for the older and the classic works in plant anatomy. This treatment seemed to be the best solution of the problem of dealing with the vast increase of botanical literature during the last decade.

In the organization of the present book an effort was made to achieve brevity not by deleting topics but by combining closely related subject matter. Thus, the protoplast and the cell wall are reviewed not in separate chapters as in the old book but jointly with parenchyma, collenchyma, and sclerenchyma. The vascular cambium is described after the xylem tissue—a sequence that facilitates the explanation of the arrangement of cells in the cambium. Finally, the apical meristems are taken up in the chapters on root and stem instead of separately. Thus the activity of these meristems may be studied in close relation to the plant parts they produce. This approach is especially suitable for introducing the modern researches on apical meristems, which emphasize the causal aspects of the specific organization of plant organs. Two other features in the arrangement of material should be mentioned. First, the chapters dealing with embryo development are placed before the discussion of the individual tissues and organs. These chapters are used for introducing the concepts of growth, differentiation, and organization of the plant. Second, the root is treated before the stem because it is least difficult to introduce the concepts pertaining to the division into vascular and nonvascular regions in the plant by referring to the relatively simple root structure.

Since this book is intended primarily for students who have had a relatively limited experience in the study of plants, a few words about the significance of plant anatomy among the plant sciences should be helpful. Whether we deal with plants as horticulturists, agronomists, foresters, plant pathologists, or ecologists, we must know what the plant is like and how it functions. We gain this knowledge by studying the structure of the plant, its development, and its various activities. It would be ideal to pursue the study of the plant from all these aspects jointly, but we find it more efficient to concentrate on one aspect at a time. Therefore we divide the study of organisms into separate areas: first the two broad ones, morphology and physiology; then each of these two into more circumscribed areas, such as cytology, anatomy, taxonomy, ecology, and others. Obviously the various areas are not sharply separated; moreover, the study of one area invariably raises questions that are properly answerable only by reference to some other area.

In this interrelationship of sciences, anatomy plays a major role. A realistic interpretation of plant function by the physiologist must rest on a thorough knowledge of the structure of cells and tissues associated with that function. Notable examples of functions, the understanding of which has been materially enhanced by studies of structures of the parts concerned, are photosynthesis, movement of water, translocation of food, and absorption by roots. Knowledge of plant anatomy is also indispensable for the advancement of research in plant pathology. The effect of a parasite cannot be fully understood unless the normal structure of the invaded plant part is known. Furthermore, the warding off of the effects of the parasite, or even the resistance of the parasite itself, may be revealed by structural changes or structural peculiarities of the host. The explanation of the successes or failures of many horticultural practices, such as grafting, pruning, vegetative propagation, and the associated phenomena of callus formation, wound healing, regeneration, and development of adventitious roots and buds, are more meaningful if the structural features underlying these phenomena are properly understood. Fruitful results are obtained by the ecologist when he relates the behavior of plants in different environments with structural peculiarities of plants growing in those environments. It seems significant that one of the most active fields in modern plant research—the study of development of form and organization—is often carried out on the basis of constant correlation between biochemical and structural changes in the experimental plant. By this correlation a much more complete picture of development is obtained than would result from considering the biochemical changes separately from those in the number, sizes, and structure of cells. Finally, plant anatomy is interesting for its own sake. It is a gratifying experience to follow the ontogenetic and evolutionary development of structural features and gain the realization of the high degree of complexity and the remarkable orderliness in the organization of the plant.

Acknowledgments

Various persons mentioned in the legends for the illustrations kindly provided material of one kind or another and made possible to include so many new illustrations in the book. Professor John E. Sass was particularly generous in lending numerous negatives and slides. Mr. Paul L. Conant loaned the slides used for the key for identification of woods. Mrs. Margery P. Mann ably assisted with the finishing of the photomicrographs.

I also wish to acknowledge the helpful reviews of the manuscript by Professor Vernon I. Cheadle, Professor Charles LaMotte, and Mr. Charles H. Lamoureux. Dr. James J. Dunning has lightened the task of preparing the glossary by making the initial collection of words and definitions. I am very grateful to Mrs. Mary M. Brinton who typed the manuscript with devotion and exceptional accuracy.

December 1959

Katherine Esau

Contents

ANATOMY OF SEED PLANTS

1 Introduction

This book deals with the internal structure of extant seed plants. Angiosperms are emphasized, but some features of the vegetative parts of gymnosperms are also reviewed. The anatomy of the flower, fruit, and seed of angiosperms is described in the concluding chapters.

The seed plant has a highly evolved body that bears signs of structural and functional specialization expressed in the differentiation of this body, externally, into organs and, internally, into various categories of cells, tissues, and tissue systems. Three vegetative organs, the root, the stem, and the leaf, are commonly recognized. The flower is interpreted as an assemblage of organs, some of which are reproductive (stamens and carpels), others sterile (sepals and petals). With regard to the internal structure, one stresses the distinctive features of cells and tissues and establishes types on the basis of these distinctions.

This particularizing of the plant body and the associated categorizing of its parts are logical and convenient approaches to the study of the plant because they bring into focus the structural and the functional specialization of its parts, but they must not be emphasized to the degree that they might obscure the essential unity of the plant. This unity is clearly perceived if the plant is studied developmentally, an approach that reveals the gradual emergence of organs and tissues from a relatively undifferentiated body of the young embryo. A similar change, from less differentiated to more differentiated, from less particularized to more particularized, has occurred in the evolution of seed plants, so that one commonly conceives of the root, the stem, the leaf, and the floral organs as parts phylogenetically interrelated and views the distinctive cells and tissues as derivatives of unspecialized cells of the type now called parenchyma cells. Even a static view of the parts of an adult plant reveals their unity and interdependence; the same tissue systems are common to all of them.

Thus, the separation of the plant into organs can be made only approximately. It is impossible, for example, to draw a clear demarkation between shoot and root and between stem and leaf, and the flower in many ways resembles the vegetative shoot. The internal structures, similarly, are not sharply delimited, and the categories of cells and tissues show much intergrading.

INTERNAL ORGANIZATION OF THE PLANT BODY

The plant body consists of morphologically recognizable units, the *cells,* each enclosed in its own cell wall and united with other cells by means of a cementing intercellular substance. Within this united mass certain groupings of cells are distinct from others structurally or functionally or both. These groupings are referred to as *tissues.* The structural variations of tissues are based on differences in the component cells and their type of attachment to each other. Some tissues are structurally relatively simple in that they consist of one cell type; others, containing more than one cell type, are complex.

The arrangement of tissues in the plant as a whole and in its major organs reveals a definite structural and functional organization. Tissues concerned with conduction of food and water—the vascular tissues—form a coherent system extending continuously through each organ and the entire plant. These tissues connect places of water intake and food synthesis with regions of growth, development, and storage. The nonvascular tissues are similarly continuous, and their arrangements are indicative of specific interrelations (such as between storage and vascular tissues) and of specialized functions (such as support or storage). To emphasize the organization of tissues into large entities, revealing the basic unity of the plant body, the expression *tissue system* has been adopted.

Although the classification of cells and tissues is a somewhat arbitrary matter, for purposes of an orderly description of plant structure the establishment of categories is necessary. Moreover, if the classifications issue from broad comparative studies, in which the variability and the intergrading of characters are clearly revealed and properly interpreted, they not only are descriptively useful but also reflect the natural relation of the entities classified.

In agreement with Sachs (1875), in this book the principal tissues of a vascular plant are grouped on the basis of topographic continuity into three tissue systems, the *dermal,* the *vascular,* and the *fundamental* (or *ground*) system. The dermal system comprises the *epidermis,* that is, the primary outer protective covering of the plant body, and the *periderm,* the protective tissue that supplants the epidermis, mainly in plant parts that undergo a secondary increase in thickness. The vascular system contains two kinds of conducting tissues, the *phloem* (food conduction) and the *xylem* (water conduction).

The fundamental system includes tissues that, in a sense, form the ground substance of the plant but at the same time show various degrees of specialization. The main ground tissues are *parenchyma* in all its varieties, *collenchyma,* the thick-walled supporting tissue related to parenchyma, and *sclerenchyma,* the main supporting tissue with thick, hard, often lignified walls.

Within the plant body, the various tissues are distributed in characteristic patterns depending on plant part or plant taxon or both. Basically the patterns are alike in that the vascular tissue is embedded in ground tissue and the dermal tissue forms the outer covering. The principal variations in patterns depend on the relative distribution of the vascular and ground tissues. In a dicotyledon, for example, the vascular tissue of the stem forms a hollow cylinder, with some ground tissue enclosed by the cylinder (*pith*) and some located between the vascular and the dermal tissues (*cortex*). In the leaf, the vascular tissue forms an anastomosing system embedded in the ground tissue, here differentiated as mesophyll. In the root, the vascular cylinder may enclose no pith, but a cortex is present.

The cells and tissues of the plant are usually derived from the zygote (fertilized egg) through intermediate stages of development embodied in the embryo. The embryonic stage, however, is not completely abandoned after the embryo develops into the adult plant. Plants have the unique property of open growth resulting from presence of embryonic tissue zones, the *meristems,* in which the addition of new cells continues while other plant parts reach maturity. Meristems at apices of roots and shoots, the *apical meristems,* produce cells the derivatives of which differentiate into new parts of root and shoot. This growth is called *primary* and the resultant plant body, the primary body. In many plants, stems and roots are increased in thickness by addition of vascular tissues to the primary body. This thickening growth is produced by the *vascular cambium* and is called *secondary* growth. Commonly secondary growth also involves the formation of periderm by the meristem *phellogen.* The vascular cambium and the phellogen are referred to as *lateral meristems* because of their position parallel with the sides of stem and root.

SUMMARY OF TYPES OF CELLS AND TISSUES

As was implied at the beginning of this chapter, separation of cells and tissues into categories is, in a sense, contrary to the facts that structural features vary and intergrade with each other and that they are capable of changing from one to the other. Cells and tissues do, however, acquire differential properties in relation to their positions in the plant body. Classifications of cells and tissues serve to deal with the phenomena of differentiation—and the resultant diversification of plant parts—in a manner that allows making generalizations about common and divergent features among related and unrelated taxa. They make possible treating the phenomena of ontogenetic and phylogenetic specialization in a comparative and systematic way. In the following is a summary of information on the generally recognized categories of cells and tissues of seed plants.

EPIDERMIS

Epidermal cells form a continuous layer on the surface of the plant body in the primary state. They show various special characteristics related to their superficial position. The main mass of epidermal cells, the epidermal cells proper, vary in shape but are often tabular. Other epidermal cells are guard cells of the stomata and various trichomes, including root hairs. The epidermis may contain secretory and sclerenchymatic cells. The principal distinctive features of the epidermal cells of the aerial parts of the plant are the cuticle on the outer wall and the cutinization of the outer and of some or all of the other walls. The epidermis gives mechanical protection and is concerned with restriction of transpiration and with aeration. In stems and roots having secondary growth, the epidermis is commonly replaced by the periderm.

PERIDERM

The periderm comprises cork tissue or *phellem,* cork cambium or *phellogen,* and *phelloderm.* The phellogen occurs near the surface of axial organs having secondary growth and is itself secondary in origin. It arises in epidermis, cortex, phloem, or pericycle and produces phellem toward the outside, phelloderm toward the inside. Phelloderm may be small in amount or absent. The cork cells are commonly tabular in form, are compactly arranged, lack protoplasts at maturity, and have suberized walls. The phelloderm usually consists of parenchyma cells.

PARENCHYMA

Parenchyma cells form continuous tissues in the cortex of stem and root and in the leaf mesophyll. They also occur as vertical strands and rays in the vascular tissues. They are primary in origin in cortex, pith, and leaf and primary or secondary in the vascular tissues. Parenchyma cells are characteristically living cells, capable of growth and division. The cells vary in shape, are often polyhedral, but may be stellate or much elongated. Their walls are often primary, but secondary walls are not uncommon. Parenchyma is concerned with photosynthesis, storage of various materials, wound healing, and origin of adventitious structures. Parenchyma cells may be specialized as secretory or excretory structures.

COLLENCHYMA

Collenchyma cells occur in strands or continuous cylinders near the surface of the cortex in stems and petioles and along the veins of foliage leaves. It is uncommon in roots. Collenchyma is a living tissue closely related to parenchyma; in fact, it is usually regarded as a form of parenchyma specialized as supporting tissue in young organs. The shape of cells varies from short prismatic to much elongated. The most distinctive feature is the unevenly thickened primary wall.

SCLERENCHYMA

Sclerenchyma cells may form continuous masses, or they may occur in small groups or individually among other cells. They may develop in any or all parts of the plant body, primary or secondary. They are strengthening elements of mature plant parts. Sclerenchyma cells have thick, secondary, often lignified walls and may lack protoplasts at maturity. Two forms of cells are distinguished, sclereids and fibers. The sclereids vary in shape from polyhedral to elongated and may be much branched. Fibers are generally long, slender cells.

XYLEM

The xylem is a structurally and functionally complex tissue which, in association with the phloem, is continuous throughout the plant body. It consists of several kinds of cells and is concerned with water conduction, storage, and support. The xylem may be primary or secondary in origin. The principal water-conducting cells are tracheids and vessel members. The vessel members are joined end to end into vessels. Storage occurs in parenchyma cells, which are arranged in vertical files and, in the secondary xylem, also as rays. Mechanical cells are fibers and sclereids.

PHLOEM

The phloem is a complex tissue composed of several kinds of cells. The phloem tissue occurs throughout the plant body, together with the xylem, and may be primary or secondary in origin. It is concerned with conduction and storage of food and with support. The principal conducting cells are sieve cells and sieve-tube members, both typically enucleate at maturity. Sieve-tube members are joined end to end into sieve tubes and are associated with companion cells, which are special parenchyma cells. Other phloem parenchyma cells occur in vertical files. Secondary phloem also contains parenchyma in the form of rays. Supporting cells are fibers and sclereids.

SECRETORY STRUCTURES

Secretory cells—cells producing a variety of secretions—do not form clearly delimited tissues but occur within other tissues, primary and secondary, as single cells or as groups or series of cells, and also in more or less definitely organized formations on the surface of the plant. The principal secretory structures on plant surfaces are glandular epidermal cells and hairs and various glands, such as, floral and extrafloral nectaries, certain hydathodes, and digestive glands. The glands are usually differentiated into secretory cells on their surfaces and nonsecretory cells supporting the secretory. Internal secretory structures are secretory cells, intercellular cavities or canals lined with secretory cells (resin ducts, oil ducts), and secretory cavities resulting from disintegration of secretory cells (oil cavities). Laticifers may be placed among the internal secretory structures. They are either single cells (nonarticulated laticifers), usually much branched, or series of cells united through partial dissolution of walls (articulated laticifers). Laticifers contain a fluid called latex, which may be rich in rubber. They are commonly multinucleate.

GENERAL REFERENCES

Most of the bibliographic citations appearing at the ends of chapters 2 to 24 were selected from the recent literature, but the extended lists of references given in Esau's *Plant Anatomy* (1965) were also used for the presentation and the interpretation of the subject matter. This introductory chapter gives a selected list of domestic and foreign books in plant anatomy and some in plant morphology. Most of the books pertain to seed plants, but some works dealing with the structure of lower vascular plants are included.

Aleksandrov, V. G. *Anatomiya rastenij* [Anatomy of plants.] 2nd ed. Moskva, Vysshaya Shkola. 1966.

Bailey, I. W. *Contributions to plant anatomy.* Waltham, Massachusetts, Chronica Botanica Company. 1954.

Biebl, R., and H. Germ. *Praktikum der Pflanzenanatomie.* 2nd ed. Wien, Springer. 1967.

Bierhorst, D. W. Morphology of vascular plants. New York, Macmillan. 1971.

Bold, H. C. *Morphology of plants.* 3rd ed. New York, Harper and Row. 1973.

Boureau, E. *Anatomie végétale.* 3 vols. Paris, Presses Universitaires de France. 1954, 1956, 1957.

Braune, W., A. Leman, and H. Taubert. *Pflanzenanatomisches Praktikum.* 2nd ed. Jena, Gustav Fischer. 1971.

Carlquist, S. *Comparative plant anatomy.* New York, Holt, Rinehart and Winston. 1961.

Clowes, F. A. L., and B. E. Juniper. *Plant cells.* Oxford, Blackwell Scientific Publications. 1968.

Cutler, D. F. *Anatomy of the monocotyledons. IV. Juncales.* Oxford, Clarendon Press. 1969.

Cutter, E. G. *Plant anatomy: experiment and interpretation.* Part I. *Cells and tissues.* Part 2. *Organs.* London, Edward Arnold. 1969 and 1971.

De Bary, A. *Comparative anatomy of the*

vegetative organs of the phanerogams and ferns. (English translation by F. O. Bower and D. H. Scott.) Oxford, Clarendon Press. 1884.

Eames, A. J. *Morphology of vascular plants. Lower groups.* New York, McGraw-Hill. 1936.

Eames, A. J. *Morphology of the angiosperms.* New York, McGraw-Hill. 1961.

Eames, A. J., and L. H. MacDaniels. *An introduction to plant anatomy.* 2nd ed. New York, McGraw-Hill. 1947.

Esau, K. *Plant anatomy.* 2nd ed. New York, John Wiley and Sons. 1965.

Fahn, A. *Plant anatomy.* 2nd ed. Oxford, Pergamon Press. 1974.

Foster, A. S., and E. M. Gifford, Jr. *Comparative morphology of vascular plants.* 2nd ed. San Francisco, W. H. Freeman. 1974.

Haberlandt, G. *Physiological plant anatomy.* London, Macmillan and Company. 1914.

Hayward, H. E. *The structure of economic plants.* New York, Macmillan. 1938.

Jackson, B. D. *A glossary of botanic terms.* 4th ed. New York, Hafner Publishing Co. 1953.

Jane, F. W. *The structure of wood.* 2nd ed. Revised by K. Wilson and D. J. B. White. London, Adam and Charles Black. 1970.

Jeffrey, E. C. *The anatomy of woody plants.* Chicago, University of Chicago Press. 1917.

Kaussmann, B. *Pflanzenanatomie.* Jena, Gustav Fischer. 1963.

Linsbauer, K. *Handbuch der Pflanzenanatomie.* Vol. 1 and following. Berlin, Gebrüder Borntraeger. 1922–1943.

Mansfield, W. *Histology of medicinal plants.* New York, John Wiley and Sons. 1916.

Metcalfe, C. R. *Anatomy of the monocotyledons. I. Gramineae. V. Cyperaceae.* Oxford, Clarendon Press. 1960 and 1971.

Metcalfe, C. R., and L. Chalk. *Anatomy of the dicotyledons.* 2 vols. Oxford, Clarendon Press. 1950.

Rauh, W. *Morphologie der Nutzpflanzen.* Heidelberg, Quelle und Meyer. 1950.

Sachs, J. *Textbook of botany.* Oxford, Clarendon Press. 1875.

Sinnott, E. W. *Plant morphogenesis.* New York, McGraw-Hill. 1960.

Solereder, H. *Systematic anatomy of the dicotyledons.* 2 vols. (English translation by L. A. Boodle and F. E. Fritsch.) Oxford, Clarendon Press. 1908.

Solereder, H., and F. J. Meyer. *Systematische Anatomie der Monokotyledonen.* Berlin, Gebrüder Borntraeger. No 1, 1933; No. 3, 1928; No. 4, 1929; No. 6, 1930.

Tomlinson, P. B. *Anatomy of the monocotyledons. II. Palmae. III. Commelinales—Zingiberales.* Oxford, Clarendon Press. 1961 and 1969.

Troll, W. *Praktische Einführung in die Pflanzenmorphologie.* Part 1: *Der vegetative Aufbau.* Part 2: *Die blühende Pflanze.* Jena, Gustav Fischer. 1954 and 1957.

Wardlaw, C. W. *Organization and evolution in plants.* London, Longmans, Green and Co. 1965.

Zimmermann, W., P. Ozenda, and H. D. Wulff, eds. *Encyclopedia of plant anatomy* (*Handbuch der Pflanzenanatomie*). Vol. 2 and following. Berlin and Stuttgart, Borntraeger. 1951 and following.

2

Development of the Seed Plant

THE EMBRYO

Gymnosperms and angiosperms, which constitute the main mass of vegetation on the surface of the earth, occur in a great variety of forms, some of which seem hardly interrelated. Yet, when viewed developmentally, seed plants reveal the same basic plan of structure and are remarkably similar in early stages of growth. The highly organized plant body of a seed plant represents the sporophyte phase of the life cycle. It begins its existence usually with the fertilized egg, the *zygote,* which develops into the embryo by characteristic steps prefiguring the adult organization (chapter 24).

The cell divisions that transform the unicellular zygote into a multicellular plant occur in preferred orientations from the early stages of embryo development, often beginning with the first division. Thus, patterns are established in the distribution of cells, and the embryo as a whole (fig. 2.1) assumes a specific form in which an axis and one or more leaflike appendages, the *cotyledons,* can be recognized. Because of its location below the cotyledons, the stemlike axis is called *hypocotyl.* At its lower end (the *root pole*), the hypocotyl bears the incipient root, at its upper end (the *shoot pole*), above the cotyledons, the incipient shoot. The root may be represented by its meristem (apical meristem of the root) or by a primordial root, the *radicle.* Similarly, the apical meristem of the shoot located at the shoot pole may or may not initiate the development of a shoot above the cotyledons. If a primordial shoot is present, it is called *epicotyl* or *plumule.*

Cell divisions in the embryo and the concomitant differential growth and vacuolation of the resulting cells initiate the organization of tissue systems.[13] The component tissues are still meristematic but their position and cytologic characteristics indicate a relation to mature tissues appearing in the subsequently developing seedling. The future epidermis is represented by a meristematic surface layer, the *protoderm* (or *dermatogen*). Beneath it, the *ground meristem* of the future cortex is distinguishable by cell vacuolation, which is

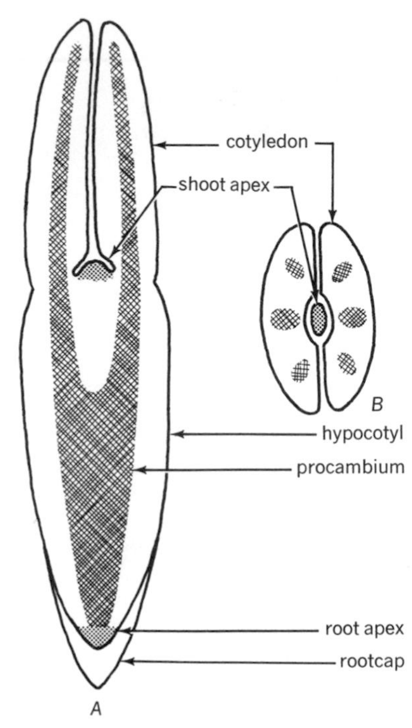

Figure 2.1 Diagram of embryo of a dicotyledon. *A*, longitudinal section. The apical meristem of the shoot (shoot apex) is located between the cotyledons, that of the root (root apex) is covered by the rootcap. The procambium extends through the hypocotyl and the cotyledons. *B*, transection through the cotyledons and the apical meristem of the shoot. The procambium appears as three bundles in each cotyledon.

more pronounced here than it is in contiguous tissues. The centrally located, less vacuolated tissue extending through the hypocotyl-root axis and the cotyledons is the meristem of the future primary vascular system. This meristem is the *procambium* (fig. 2.1). It may also be called *provascular tissue* or *provascular meristem* if its future destiny needs to be emphasized. Longitudinal divisions and elongation of cells impart a narrow, elongated form to the procambial cells. The conformation of the procambium as a whole varies in embryos of different plant taxa, but it is consistently that of a coordinated, orderly system continuous between the cotyledons and the hypocotyl-root axis. The vascular system of the seedling developing from the embryo is an enlarged and differentiated replica of the embryonic procambial system (chapter 24).

FROM EMBRYO TO THE ADULT PLANT

After the seed germinates, the apical meristem of the shoot forms, in regular sequence, leaves and nodes and internodes (fig. 2.2). Apical meristems in the axils of leaves produce axillary shoots which, in turn, have other axillary shoots. As a result of such activity, the plant bears a system of branches on the main stem. If the axillary meristems remain inactive, the shoot fails to branch as, for example, in many palms. The apical meristem of the root located at the tip of the hypocotyl—or of the radicle, as the case may be—forms the taproot (the primary root). In many plants the taproot produces branch roots (secondary roots) from new apical meristems originating deep in the taproot (endogenous origin). The branch roots produce further branches in their turn. Thus, a much branched root system results. In some plants, notably in monocotyledons, the root system of the adult plant develops from adventitious roots arising on the stem. Such a root system may appear brushlike (fibrous root) because the individual roots are similar to one another in length and form.

The growth outlined above comprises the vegetative stage in the life of a seed plant. At an appropriate time, determined in part by an endogenous rhythm of growth and in part by environmental conditions, especially light and temperature,[11] the vegetative apical meristem of the shoot is changed into a reproductive apical meristem, that is, in angio-

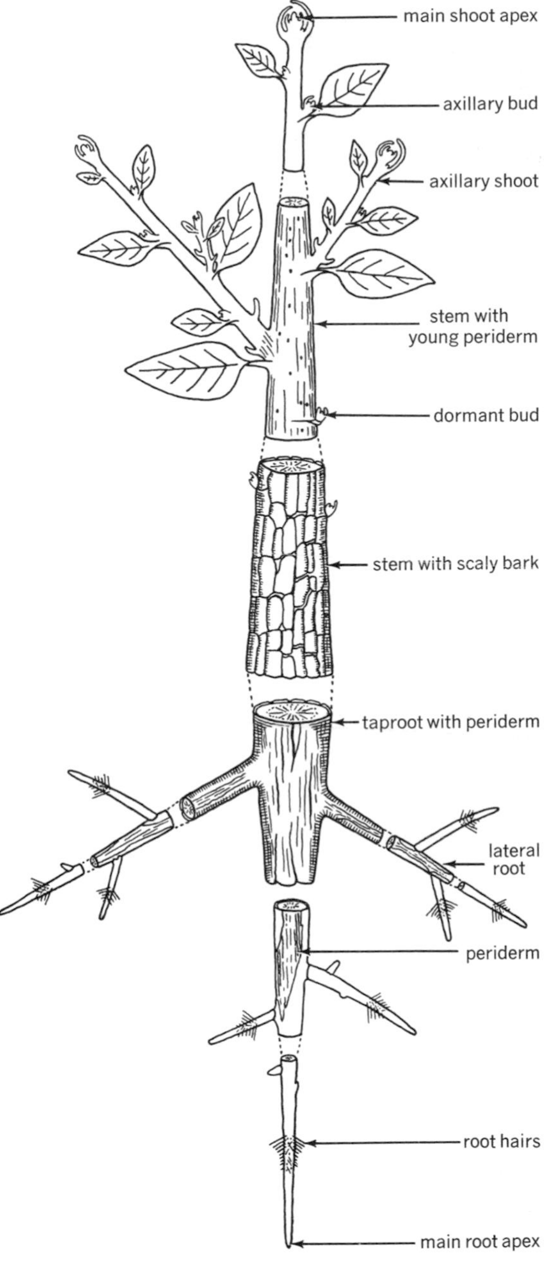

Figure 2.2. Diagram of a perennial dicotyledonous plant illustrating branching of shoot and root, increase in thickness of stem and root by secondary growth, and development of periderm and bark on the thickened axis. The apices of main and lateral (axillary) shoots bear leaf primordia of various sizes. Root hairs occur some distance from the apices of the main (taproot) and lateral roots. (Adapted from Rauh, W., *Morphologie der Nutzpflanzen,* Quelle und Meyer, Heidelberg, 1950.)

sperms, into a floral apical meristem, which produces a flower or an inflorescence. The vegetative stage in the life cycle of the plant is thus succeeded by the reproductive stage.

The stage of development that ends with the maturation of the more or less direct derivatives of the apical meristems is referred to as primary growth. A complete plant body with roots, stems, leaves, flowers, fruits, and seeds and with its dermal system (epidermis), the ground-tissue system, and the vascular system is produced by primary growth. Small dicotyledonous annuals and most of the monocotyledons complete their life cycle by primary growth. The majority of the dicotyledons and gymnosperms, however, show a secondary stage of growth resulting from the activity of the vascular cambium. This meristem increases the amount of vascular tissues and causes thereby the thickening of the axis (stem and root; fig. 2.2). The formation of the protective tissue, periderm, which replaces the epidermis, is also regarded as part of secondary growth. The secondary addition of vascular tissues and protective covering makes possible the development of large, much branched plant bodies, such as are characteristic of trees.

Although it is appropriate to think of a plant as becoming "adult" or "mature," in that it develops from a single cell into a complex but integrated structure capable of reproducing its own kind, an adult seed plant is a constantly changing organism. It maintains the capacity to add new increments to its body through the activity of apical meristems of shoots and roots (fig. 2.3) and to increase the volume of its secondary tissues through

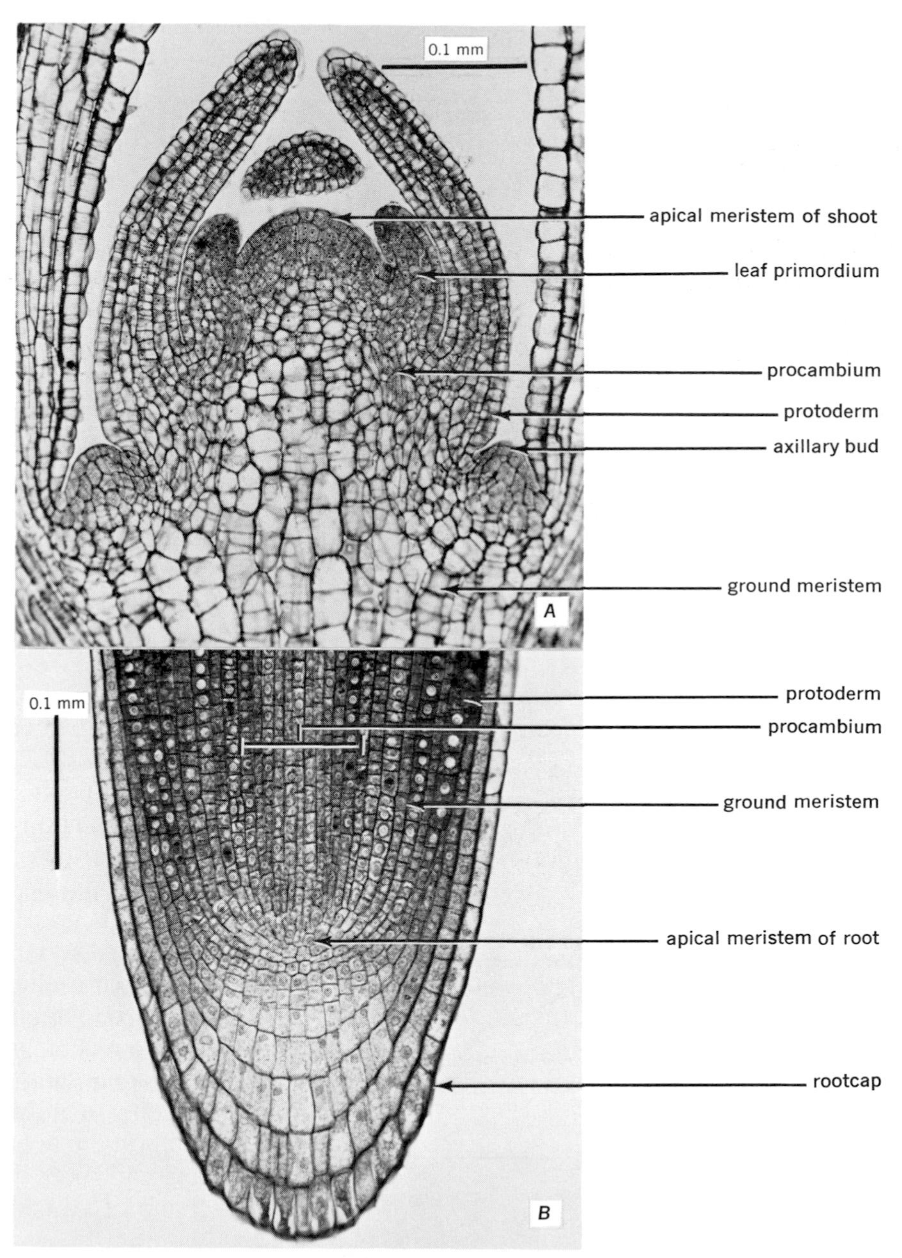

Figure 2.3 Shoot tip (*A*) and root tip (*B*) of seedling of flax (*Linum usitatissimum*) in longitudinal sections. Both illustrate apical meristems and derivative primary meristematic tissues. *A*, primodia of leaves and axillary buds are present. *B*, rootcap covers apical meristem. (*A*, from J. E. Sass, *Botanical Microtechnique*, 3rd ed. The Iowa State College Press 1958.)

the activity of lateral meristems. The outward manifestations of growth are associated with corresponding activities at the cellular level. Growth and differentiation require synthesis and degradation of protoplasmic and cell wall materials and involve an exchange of organic and inorganic substances circulating by way of the conducting tissues and diffusing from cell to cell to their ultimate destinations. A variety of processes take place in specialized organs and tissue systems in providing organic substrates for metabolic activities. An outstanding feature of the living state of a plant is that its perpetual changes are highly coordinated and occur in orderly sequences.[22] Moreover, as other living organisms, plants exhibit rhythmic phenomena, some of which clearly match environmental periodicities and indicate an ability to measure time.[21]

APICAL MERISTEMS AND THEIR DERIVATIVES

The formation of new cells, tissues, and organs through the activity of apical meristems involves division of cells. Certain cells of the meristems undergo divisions in such a way that one product of a division becomes a new body cell, the other remains in the meristem. The cells remaining in the meristem may be called the initials, their products by division contributed to the body, the derivatives. The concept of initials and derivatives should, however, include the qualification that the initials are not inherently different from their derivatives and may become supplanted by their derivatives. The concept of initials and derivatives is taken up from various aspects in connection with the descriptions of the vascular cambium (chapter 10) and the apical meristems of root (chapter 14) and shoot (chapter 16). Suffice it to point out here that, according to a common view, certain cells in the meristems act as initials mainly because they occupy the proper position for such an activity and that the apical meristems of roots and shoots of the higher vascular plants (gymnosperms and angiosperms) contain groups of initials.

The term apical meristem refers to a complex of cells composed of the initials and their immediate derivatives (fig. 2.3). The derivatives usually divide also and produce one or more generations of cells before the cytologic changes, denoting differentiation of specific types of cells and tissues, occur near the tip of root or shoot. Furthermore, divisions continue at levels where such changes are already discernible. In other words, growth, in the sense of cell division, is not limited to the very tip of root or shoot but extends to levels considerably removed from the region usually called the apical meristem. In fact, the divisions some distance from the apex are more abundant than at the apex.[5] In the shoot, a more intensive meristematic activity is observed at levels where new leaves are initiated than at the tip; and during the elongation of the stem, cell division extends several internodes below the apical meristem.[15]

In meristematic activity, cell division is combined with an enlargement of the products of division. Generally, from a younger to an older meristematic tissue the amount of cell enlargement increases (fig. 2.3) and eventually becomes the main factor in the increase in width and length of the particular region of root or shoot. The no longer dividing cells—they may be still enlarging, however—gradually differentiate into the specific cells characteristic of the region of shoot or root where these cells occur. Thus, the various phenomena of growth and differentiation overlap in the same cell; furthermore, at the same level of shoot or root, different regions may be in different stages of growth and differentiation.

The relationship between cell division and cell enlargement in meristematic activity has been analyzed in terms of genetic control of plant form. Stebbins[17] distinguishes between mitosis-determined and elongation-determined (more appropriately, enlargement-determined) conditions in the meristem and proposes the following hypothesis. The genes controlling the shape of leaves and other parts of higher plants, as well as the genes governing the number and arrangement of plant parts, exert their action by affecting, first, the tempo and distribution of mitoses and, second, the timing of the transition from mitosis-determined to enlargement-determined state.

In view of the gradual change from apical meristems to the adult primary tissues and the intergrading of the phenomena of cell division, cell enlargement, and cell differentiation, one cannot restrict the term meristem to the apex of shoot or root. The shoot and root parts where the future tissues and organs are already partly determined but where cell division and cell enlargement are still in progress are also meristematic. If a distinction between the apical meristem and the subjacent levels is desired, one may speak of the apical meristems and the primary meristematic tissues below them, and one may use the terms root tip and shoot tip in a broad sense to include the apical meristem and the subjacent primary meristematic tissues.

DIFFERENTIATION, SPECIALIZATION, AND MORPHOGENESIS

The progressive change from the structurally simple meristematic tissue to the complex and variable tissues and combinations of tissues in the adult plant body is referred to as *differentiation.* The change from the undifferentiated meristematic state to the differentiated adult state involves the chemical constitution of cells as well as their morphologic characteristics, and it may be analyzed in terms of single cells, a tissue, a tissue system, an organ, or the plant as a whole. Differentiation may be looked upon as a double process, first, of becoming different from the meristematic precursors and, second, of becoming different from the neighboring cells or tissues.

When we compare cells that have completed their differentiation we recognize that some become much more distinct from the meristematic cells than others and that the higher degree of change is associated with a more pronounced *specialization* with reference to the functions that the cells perform in the plant body. A high degree of specialization is attained, for example, by the water-conducting cells in the xylem, cells that have relatively thick walls and no living contents at maturity, and by the food-conducting sieve elements in the phloem, cells that lack nuclei at maturity. A less profound change occurs during the differentiation of a photosynthetic parenchyma cell in the mesophyll of a leaf. Such a cell may assume a noticeably different shape than its meristematic precursor, but its wall thickens only moderately and its protoplast remains complete. The most distinguishing feature in the development of such a cell is the acquisition of numerous chloroplasts.

If a differentiated cell retains a protoplast, it may be stimulated to resume meristematic activity. Wounding of a leaf, for example, may induce division of mesophyll cells and subsequent formation of suberized protective tissue. Mesophyll cells may be also cultured in vitro.[12] Formation of callus tissue along cut surfaces is another example of a resumption of cell division by parenchyma, and such parenchyma may be many years old.[2] Differentiated living cells may become meristematically active spontaneously, as is demonstrated by the formation of periderm in stems. The development of adventitious shoots or roots by a reactivation of cell division in differentiated parenchyma is often a spontaneous phenomenon as well.

During the development, or ontogeny, of a plant, growth and cell differentiation are coordinated in such a way that the resultant plant assumes a specific form and thus exhibits the phenomenon of *morphogenesis*. The term morphogenesis can be used with reference to both the development of the external form and that of the internal organization. In this context, differentiation and specialization are elements of morphogenesis.

To understand the causal relations and control mechanisms in cell differentiation and morphogenesis, investigators conduct various developmental and experimental studies.[3] The research may deal with normal development of cells, tissues, embryos, parts of plants, or entire plants; with abnormal growth; with responses of plants to various external stimuli or surgical treatments;[23] or with the particularly fruitful studies of cells, tissues, or organs excised from plants and cultured in vitro.[7,18,20] Smaller or larger parts of plants, from organs to single cells, or even free protoplasts,[8] may be induced to grow and to produce complete plants. Formation of a plant in culture is essentially a matter of association of the newly emerging root and shoot into a whole organism.[9]

The ability of single cells to produce complete plants when cultured in vitro indicates their developmental totipotency. According to a favored view, cells are totipotent because they ultimately originate from the same single cell, the zygote, and thus may be genetically identical. The problem is then to elucidate how the progenies of similar, totipotent cells become so diverse in structure and function during plant development. The model devised to explain the diversification of genetically identical cells postulates that not all genes of the same complement are active in a cell at the same time. As Heslop-Harrison[10] has expressed it, the developmental changes in plants are based on shifting patterns of gene repression and derepression. Appropriate external stimuli can cause activation of certain genes and repression of others. As a result, cells become competent to react to the particular stimuli and to follow specific paths of development. At the molecular level, an activation of certain genes by some agent, a hormone for example,[1,19] evokes a synthesis of specific enzymes which determine the path of differentiation of the cell. Since the concentration of particular enzymes may precede the morphologic changes in cells and tissues, one can say that the competence of a cell to react to stimuli in a certain way is established in advance of the open manifestation of differentiation.

In addition to explaining how an individual cell may be induced to pursue a certain path of development, one must account for the phenomenon that specific cell types differentiate in specific positions in a plant body; that, in other words, cell differentiation in an organism follows a pattern. The development of tissue systems and organs characteristic of a plant clearly indicates that the individual cells composing the growing plant do not display their totipotency when they are structural components of an organism. The cells are subject to positional restraint and manifest only some of their potentialities. The type of restraint (kind of genes repressed) varies from position to position so that the cell in one position follows a different pathway of development than a cell in another position. Considerations of this kind have led to the concept of interaction between neighboring cells, although the thoughts about the agency of the interaction are somewhat vague.

It has been postulated that cells contribute to each other some substances vital for growth and differentiation.[24] This interaction may be expressed in inductive effects resulting in the appearance of specific cells (similar or different) side by side, as well as in mutual incompatibility between similar structures or conditions. Inductive effects are well illustrated by the sequence of events in regeneration of severed vascular bundles (chapter 8) and in differentiation of guard cell-subsidiary cell combinations in leaf epidermis (chapter 7). Mutual incompatibility is seen by some investigators in the occurrence of circumscribed regions of active cell division, as for example in the emergence of a leaf primordium, that seemingly prevent the neighboring cells from becoming engaged in meristematic activity.[4]

A study, in which growth of cells in suspension was compared with growth of the same kind of cells in an aggregate, has provided an experimental evidence of cell interaction.[24] Known numbers of cells growing in suspension were forced to reaggregate by transfer to a confined space in the form of a nylon cylinder. In cell suspensions, the exponential rate of growth (measured by numbers of cells and weights of cultures) was independent of cell number, whereas in the reaggregates the growth rate was influenced by cell number: the rate was higher in presence of larger numbers of cells. The same study further produced evidence that cell interaction affects differentiation of cells, in this instance, that of sclereids.[24] As the authors point out, the development of an embryo, either from a fertilized egg or from a cell in suspension culture, also indicates that cell interaction is a limiting factor in organized development and that the sequential character of differentiation in the embryo reflects the number of cells present.

Considerations of growth and differentiation at levels of cells and tissues leads to questions regarding determination of form of plant organs and, accordingly, of the plant as a whole. In Heslop-Harrison's[10] words, at the level of organogenesis the control extends beyond the effects of neighboring cells upon each other. There may be interactions between different tissues and organs, possibly through transfer of hormones and nutrients. Occurrence of such long-distance interaction would be one of the evidences of integrated development in a plant body through interdependence of its parts. Another phenomenon indicating presence of an overall regulatory mechanism in plant morphogenesis is polarity, that is, occurrence of physiological and structural differences between two ends of a plant or its individual organs; it is a form of asymmetry in organization. Although there is evidence that environment is important in establishing polarity,[4,6] the bipolar type of development may be detected as soon as an embryo begins to develop, and it continues to be one of the dominant factors in differentiation.[16] Asymmetry and polarity in development appear in a cell culture as soon as the products of cell division become cell aggregates.[18] The introduction of polarity leads to divergent patterns of development among cells of the same aggregate.

The effects of polar growth in a plant may be disclosed by changes, from level to level, in the character of structures or their arrangement or both, as well as in differential physiological activities of cells and tissues. In a developing seedling, for example, the root pole and the shoot pole appear to dominate the subjacent parts of the plant body but, in the transition between the two parts, a gradual shift from one structure to the other is present and secures a harmonious accommodation between root and shoot organization (chapter 24). The tendency toward gradual transition between dissimilar structures is interpreted as evidence of graded influences in differentiation, that is, of gradients of differentiation.[16] The gradients may be those of diffusion of metabolic products, of temperature, of pH, and others.

The question has been asked whether polarity is a factor controlling differentiation or a manifestation of differentiation.[6] This question is especially pertinent in view of the evidence that polarization may be detected in individual cells. The egg and the zygote show a distinctly polarized arrangement of contents, the nucleus occupying the future apical end, the vacuole the future basal end (chapter 21). There are other examples of differential distribution of cell contents and, when the dissimilar parts of the cell are separated by a cell wall in mitosis, unequal daughter cells result. Unequal divisions may be observed, for example, in the epidermis of roots in which the future hair cell may be quite distinct from its sister cell in morphology and biochemistry (chapter 7). Unequal divisions are common in leaf epidermis as well, especially in the development of stomata (chapter 7). Clearly, the development of a higher plant indicates that phenomena of differentiation may be observed at various levels of organization. Specialization in plant structure results from complex interactions between the genetic complement and biochemistry, and is to a high degree controlled by external factors.[14,16]

REFERENCES

1. Abeles, F. B. *Ethylene in plant biology.* New York, Academic Press. 1973.
2. Barker, W. G. Proliferative capacity of the medullary sheath region in the stem of *Tilia americana. Amer. J. Bot.* 40: 773–778. 1953.
3. Brookhaven Symposium: Basic mechanisms in plant morphogenesis. Proc. Symp. Upton, N.Y., 1973. *Brookhaven Symposia in Biology* No. 25. Upton, N.Y., Brookhaven National Laboratory. 1974.
4. Bünning, E. Morphogenesis in plants. *Surv. Biol. Prog.* 2:105–140. 1952.
5. Buvat, R. Structure, évolution et fonctionnement du méristème apical de quelques Dicotylédones. *Ann. Sci. Nat., Bot. Sér.* 11. 13:202–300. 1952.
6. Cutter, E. G. *Plant anatomy: experiment and interpretation.* Part 1. *Cells and tissues.* London, Edward Arnold. 1969.
7. Gautheret, R. J. Factors affecting differentiation of plant tissues grown in vitro. In: *Cell differentiation and morphogenesis. Internatl. Lecture Course.* Wageningen, North Holland Publishing Company. 1966.
8. Grambow, H. J., K. N. Kao, R. A. Miller, and O. L. Gamborg. Cell division and plant development from protoplasts of carrot cell suspension cultures. *Planta* 103:348–355. 1972.
9. Halperin, W. Morphogenesis in cell cultures. *Ann. Rev. Plant Physiol.* 20:395–418. 1969.
10. Heslop-Harrison, J. Differentiation. *Ann. Rev. Plant Physiol.* 18:325–348. 1967.
11. Hillman, W. S. *The physiology of flowering.* New York, Holt, Rinehart and Winston. 1962.
12. Joshi, P. C., and E. Ball. Growth of isolated palisade cells of *Arachis hypogaea* in vitro. *Devel. Biol.* 17:308–325. 1968.
13. Meyer, C. F. Cell patterns in early embryogeny of the McIntosh apple. *Amer. J. Bot.* 45:341–349. 1958.
14. Mohr, H. *Lectures on photomorphogenesis.* New York, Springer Verlag. 1972.
15. Sachs, R. M. Stem elongation. *Ann. Rev. Plant Physiol.* 16:73–96. 1965.

16. Sinnott, E. W. *Plant morphogenesis.* New York, McGraw-Hill. 1960.
17. Stebbins, G. L. Some relationships between mitotic rhythm, nucleic acid synthesis, and morphogenesis. In: *Brookhaven Symposia in Biology* No. 18. 1965.
18. Steward, F. C. From cultured cells to whole plants: the induction and control of their growth and morphogenesis. *Proc. Roy. Soc. London B.* 175:1–30. 1970.
19. Steward, F. C., ed. *Plant physiology. A treatise.* Volume IVB. *Physiology of development: the hormones.* New York, Academic Press. 1972.
20. Street, H. E., ed. *Plant tissue and cell culture.* Berkeley, University of California Press. 1973.
21. Sweeney, B. M. *Rhythmic phenomena in plants.* New York, Academic Press. 1969.
22. Torrey, J. G. *Development in flowering plants.* New York, Macmillan. 1967.
23. Wardlaw, C. W. *Morphogenesis in plants.* 2nd ed. London, Methuen. 1968.
24. Wilbur, F. H., and J. L. Riopel. The role of cell interaction in the growth and differentiation of *Pelargonium hortorum* cells in vitro.
    I. Cell interaction and growth. *Bot. Gaz.* 132:183–193. 1971.
    II. Cell interaction and differentiation. *Bot. Gaz.* 132:193–202. 1971.

3

The Cell

Living organisms consist of single cells or of complexes of cells. In a multicellular organism the cells are not merely aggregated, but are connected and coordinated into a harmonious whole. Cells vary greatly in size, form, structure, and function. Some are measured in microns, others in millimeters, or even centimeters (fibers in certain plants). Some cells are relatively simple in internal organization, others are complex. Some cells perform various functions, others are specialized in their activities. As Sitte[55] has stated, it was a great achievement that, despite the extraordinary diversity among cells, the early microscopists recognized the homology of these structural units.

Cells of plants and animals are variations of one basic type of unit of structure. This generally accepted postulate is based on the *cell theory* formulated in the first half of the nineteenth century by Mathias Schleiden and Theodor Schwann. Many other workers, however, contributed to the recognition of cells as structural units of living organisms after Robert Hooke first referred to *cells* in 1665 by describing them as compartments in the cork tissue.

Originally, Robert Hooke saw only a delimitation of spaces by cell walls, but later he detected "juices" in the cavities. Eventually, the contents of cells were interpreted as living matter and received the name of *protoplasm.* A derivative of this word, *protoplast,* refers to the contents of an individual cell. An important step toward recognition of the complexity of a protoplast was the discovery of the nucleus by Robert Brown in 1831. This discovery was soon followed by reports on cell division. In 1846 Hugo von Mohl called attention to the distinction between protoplasmic material and cell sap, and in 1862 Kölliker used the term *cytoplasm* for the material surrounding the nucleus. The most conspicuous inclusion in this cytoplasm, the *plastids,* were long considered to be merely condensations of protoplasm. The concept of independent identity and continuity of these organelles was established late in the nineteenth century.[29]

The interpretation of cells as structural

units has stood the test of time. It is well supported by modern research in tissue culture, which shows that individual cells, isolated from a complex organism, behave as independent entities. Isolated cells grow and divide and can produce replicas of parent organisms. Ultramicroscopic and biochemical studies, moreover, underscore the basic similarity among cells, be they those of unicellular or multicellular organisms, or those of plants or animals.

As an organic unit, the cell has a means of isolating its contents from the external environment. A membrane called *plasma membrane,* or *plasmalemma,* brings about this isolation. Plant cells (and some animal cells) have, in addition, a rigid coat, the *cell wall,* deposited outside the plasmalemma. A cell can release and transfer the energy necessary for growth and for the maintenance of metabolic processes. A cell is organized to retain and transfer information so that its development and that of its progeny can occur in an orderly manner. In this way, the integrity of the individual of which the cells are part is maintained.

The degree of internal organization of cells permits the recognition of two basic types of cell. The first type, the *prokaryotic cell,* is simple in morphology; it has no separate cell units to perform specific functions. The hereditary material, deoxyribonucleic acid (DNA), is distributed through a large portion of the cell and is not enclosed in a membranous envelope. Organisms characterized by prokaryotic cells are bacteria and blue-green algae.

The second type of cell, the *eukaryotic cell,* which is characteristic of all organisms other than bacteria and blue-green algae, is compartmentalized into distinct parts (fig. 3.1) that perform different functions. The DNA responsible for storage and transfer of information is located in the *chromosomes,* which are enclosed in a membrane-bound organelle, the *nucleus.* The nucleus also contains the *nucleolus* (often more than one), a body involved in information transfer. Photosynthesis is carried on in the *chloroplasts,* a form of *plastids* containing chlorophyll. Aerobic respiration is performed by another organelle, the *mitochondrion.* The *dictysome* is concerned with secretion of cell wall materials and other products. Synthesis of protein is the function of *ribosomes* and a membrane system called *endoplasmic reticulum* (ER). Certain organelles, the *microbodies,* contain enzymes essential in a number of metabolic pathways. Still other structures are present the role of which has not been definitely determined. Clowes and Juniper[11] give the following figures for the possible number of organelles per cell: 1 nucleus, 20 plastids, 700 mitochondria, 400 dictysomes, 500,000 ribosomes, 500,000,000 or more enzyme molecules, which occur as 10,000 different kinds. All the discrete parts of the protoplast are embedded in a fluid matrix, the ground cytoplasm, in which electron microscopy has not yet revealed a definite structure.

In contrast to most animal cells, plant cells develop an internal aqueous phase, the *vacuole,* which is bound by a membrane called *tonoplast.* Cellular activities not only release energy for growth and differentiation but also produce reserve and waste materials. These materials are called *ergastic substances.* Their representatives are starch, fatty materials, protein inclusions, tannins, and crystals of various compositions.

The compartmentalization in eukaryotic cells is accomplished by means of membranes. The appearance of membranes under the electron microscope is remarkably similar in various living organisms. With suitable fixation, a membrane is revealed as two dark layers, each about 25 Å (Ångström) thick, enclosing between them a lightly stained

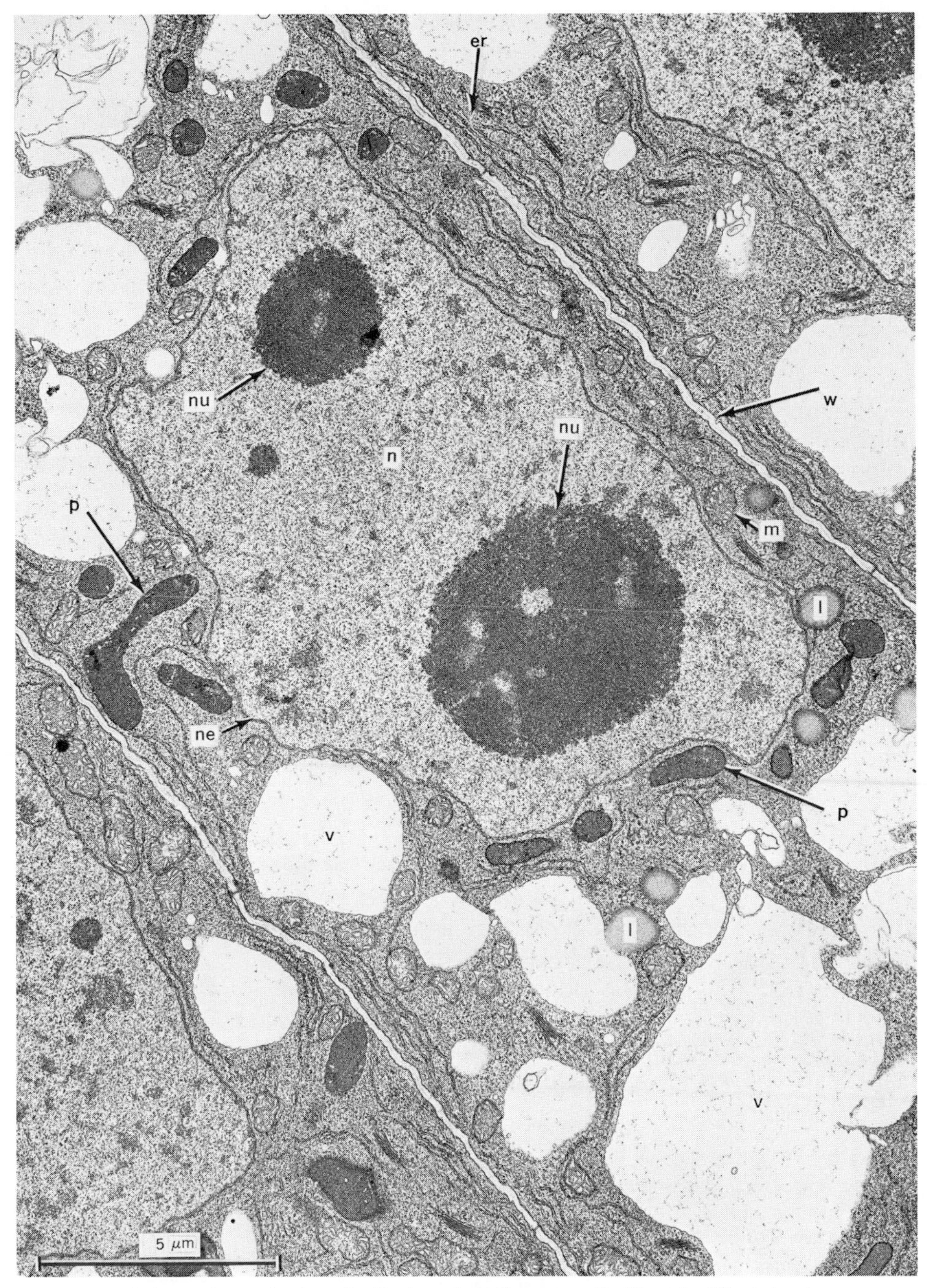

Figure 3.1 *Nicotiana tabacum* (tobacco) root tip. Longitudinal section of young cells. The identified structures are: *er*, endoplasmic reticulum; *l*, lipid globule; *m*, mitochondrion; *n*, nucleus; *ne*, nuclear envelope; *nu*, nucleolus; *p*, plastid; *v*, vacuole; *w*, cell wall.

layer 35 Å thick. (See the plasmalemma in fig. 3.13,*A*.) This type of membrane was named *unit membrane* by Robertson.[49] Its general structure was early predicted by Danielli and Davson.[12] The unit membrane is interpreted as a bimolecular lipid layer covered on each side with a layer of protein. This model is no longer sufficient to interpret the relation between structure and function for all known cellular membranes. (The model is shown in one of its numerous modifications in fig. 3.2.) Nevertheless, the notion of a bimolecular lipid leaflet continues to be the best explanation of the unspecialized properties of biological membranes[7] and "unit membrane" remains a useful designation for a visually definable structural component of membranous formations in the cell. In the freeze-etching technique of preparation of material for electron microscopy the unit membranes are cleaved internally through the lipid layer and reveal particles on the inner surfaces exposed by fracturing of the membrane (figs. 3.3 and 3.4). The number, size, and distribution of the particles differ in different membranes and vary also in response to functional states of the cells.[40,61]

Membranes separate protoplasts from the environment and delimit the various organ-

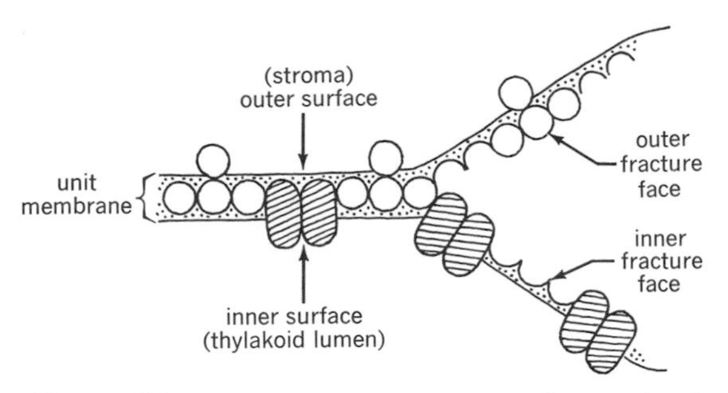

Figure 3.3 Diagram of a partly freeze-fractured unit membrane from a chloroplast thylakoid. The fracture occurred through the lipid layer and exposed the interior particle-bearing surfaces of the membrane. The surfaces are designated inner and outer fracture faces in relation to the orientation of the membrane in the thylakoid. (Adapted from K. Mühlethaler. Studies on freeze-etching of cell membranes. *Internatl. Rev. Cytol.* 31: 1–19, 1971.)

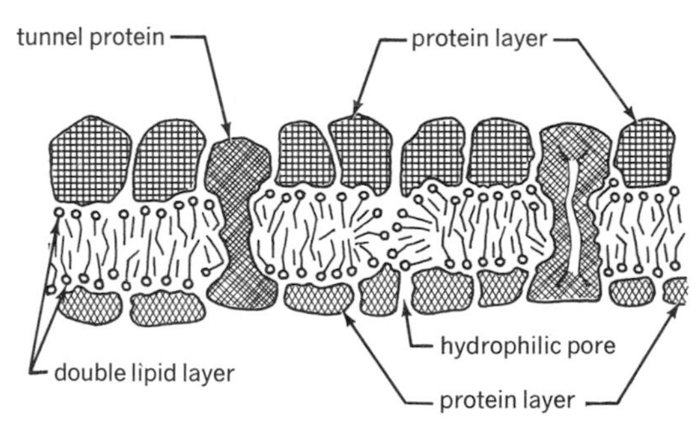

Figure 3.2 Diagram of a model of a three-layered biological membrane (protein-lipid-protein). In the bimolecular lipid layer, the lipid molecules are oriented with their hydrophilic, or polar, ends (circles) toward the protein layers. In the hydrophilic pore region, the polar ends of lipid molecules face each other. The protein layers are composed of discrete units. Some units extend through the entire membrane (tunnel proteins). (Adapted from P. Sitte, Biomembranen: Structur und Funktion. *Ber. Deut, Bot. Ges.* 82: 329–383, 1969.)

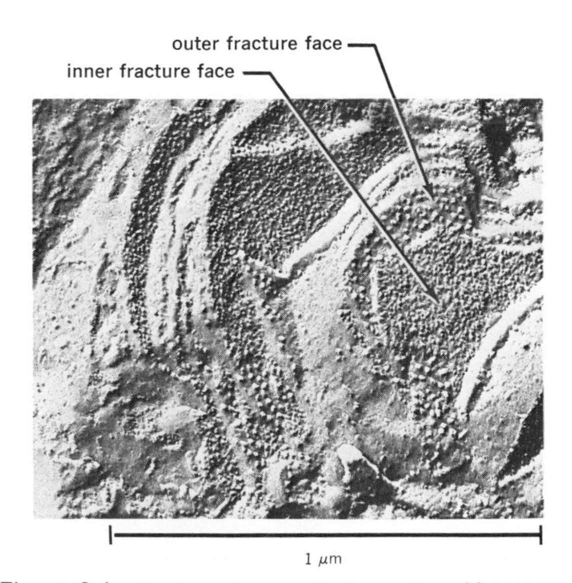

Figure 3.4 Electron micrograph of a replica of freeze-fractured thylakoid membranes from an isolated chloroplast of *Portulaca devacea.* (Courtesy of W. M. Laetsch.)

elles in the cell. Membranes are not merely physical boundaries but are fluid, dynamic, functional surfaces.[54] They are sites of many biochemical processes such as active uptake of organic and inorganic substances, oxidative and photosynthetic phosphorylations, quantum conversion of photosynthesis, and others. Membranes permit a spatial separation of biochemical events and thus ensure their sequential operation in one and the same cell.[56]

Compartmentalization of cellular contents means division of labor at the subcellular level. In a multicellular organism, a division of labor occurs also at the cellular level as the cells differentiate and become more or less specialized with reference to certain funcions. Functional specialization finds its expression in morphologic differences between cells, a feature that accounts for the complexity of structure in a multicellular organism.

CYTOPLASM

As mentioned, the term cytoplasm was introduced to designate the protoplasmic matrix surrounding the nucleus. In time, discrete structures were discovered in this matrix, first only those that were within the resolving power of the light microscope; later smaller entities were observed with the electron microscope. Discoveries of further ultramicroscopic structures may be anticipated. Thus, the concept of cytoplasm is undergoing an evolution but continues to be imprecise. Most commonly it is treated together with certain discrete entities (ribosomes, microtubules) and membranes that are not parts of clearly circumscribed organelles (ER, plasmalemma, tonoplast). At the same time, one visualizes the existence of an unstructured ground substance apart from those entities and membranes. To differentiate this substance from the cytoplasm in its wide sense, the words groundplasm or hyaloplasm are sometimes used.[11] Ground cytoplasm or cytoplasmic matrix are equally satisfactory terms for specifying the apparently unstructured portion of the cytoplasm.

The ground cytoplasm contains proteins, lipids, nucleic acids, and other substances soluble in water, that is, not bound to discrete structures. The cytoplasm is viscous and capable of forming a gel. Cytoplasmic movement, or streaming, as revealed by displacement of organelles and other particles, is frequently observed in living cells. It is not certain whether the streaming depends on some specific component of the protoplast, but many reports in the literature associate cytoplasmic movement with the presence of bundles of fine microfilaments in the cells.[42] The cytoplasm is delimited from the cell wall by the plasmalemma, a unit membrane, and from the vacuole by the tonoplast, another unit membrane (fig. 3.6). The outermost region of the cytoplasm is sometimes called cortex, an adaptation from animal cytology. The various constituents embedded in the cytoplasmic matrix are described individually in the following paragraphs.

NUCLEUS

Most cells of higher plants are uninucleate (fig. 3.1). Certain specialized cells may be multinucleate (coenocytic) either only during their development or for life. A multiplication of DNA occurs in some cells that remain uninucleate. The nucleus then becomes polyploid (endopolyploidy), a phenomenon frequently accompanying somatic differentiation.

As a depository of genetic information, the

nucleus plays a major role in cell division. It undergoes cyclical changes in relation to stages of division. Between divisions, that is, at interphase, the nucleus is a discrete organelle surrounded by an envelope and containing one or more nucleoli (fig. 3.1). The chromosomes are in an uncoiled state and are not easily distinguished from the nuclear matrix, or *nucleoplasm.* In many species, however, some chromatin remains in condensed form and is conspicuous as dense masses at interphase. This chromatin is called *heterochromatin* in contrast to the *euchromatin,* which forms the bulk of the chromosomes. Nuclei may contain proteinaceous inclusions of unknown function in crystalline, fibrous, or amorphous form.[63]

The nuclear envelope consists of a pair of unit membranes with a *perinuclear space* between them (fig. 3.5,*A*). Apparently, no specific information on the contents of this space is available. The envelope resembles an endoplasmic reticulum cisterna in structure. The two kinds of membrane complexes may be connected in a manner that makes the perinuclear space and the lumen of the ER cisterna continuous with one another. The nuclear envelope has *pores* (fig. 3.5), which are distributed in a rather regular pattern in some plants. The membranes of the envelope are

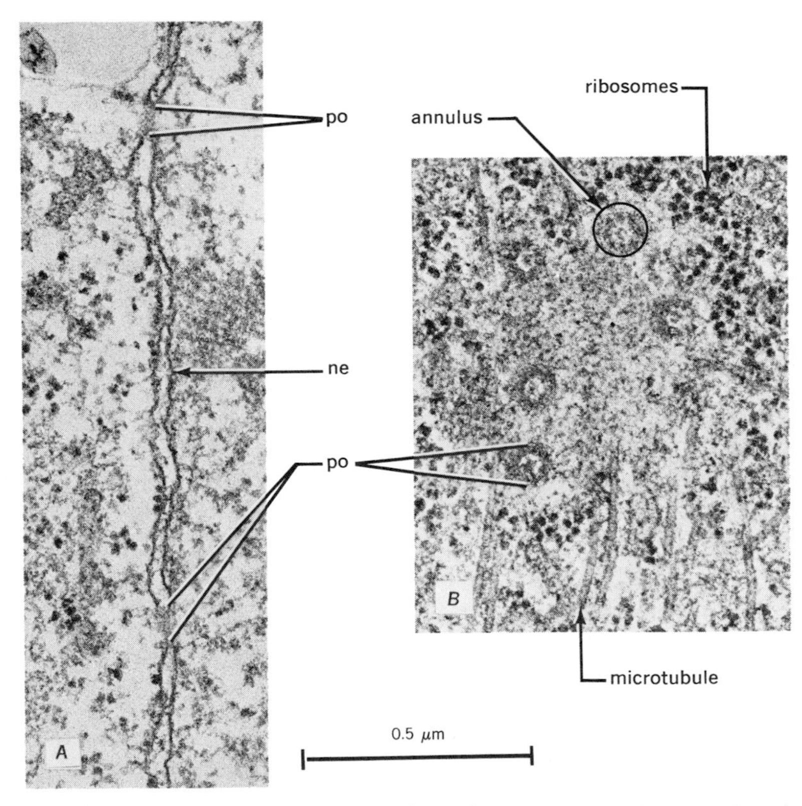

Figure 3.5 Nuclear envelope (*ne*) in profile (*A*) and from the surface (*B*, central part) showing pores (*po*). The electron-dense material in the pores in *A* is shown, in *B*, to have a form of an annulus with a central granule. From a parenchyma cell in *Mimosa pudica* petiole.

joined around each pore and thus form the pore margin. A nuclear pore is not an unobstructed opening but is occupied by a complex structure in which several parts may be distinguished:[16] an *annulus,* circular structure surrounding the pore, a central granule, and a system of fibrils between the granule and the annulus (fig. 3.5,*B*). Despite the apparent structural restriction of the pores, experiments indicate that the pores are open to molecules of certain size.[14]

In many diploid species, the nucleus contains one nucleolus to each haploid set of chromosomes and thus has two nucleoli. But diploid species in which more than one pair of chromosomes function in nucleolar formation are known.[65] Frequently the nucleoli are coalesced into one body during the interphase. The nucleolus shows a dense structure in which two kinds of elements, one granular the other fibrillar, may be visible. Some of the granules contain ribonucleic acid (RNA) and are comparable to cytoplasmic ribosomes in size. Smaller granules may be protein. The fibrillar component is known to contain DNA.[11] Nucleoli show lightly stained regions commonly referred to as vacuoles (fig. 3.1). In living cultured cells these vacuoles are found to be undergoing a repeated contraction, a phenomenon that may be related to synthesis of ribosomal RNA in the nucleolus.[24] The nucleolus has no bounding membrane and is often seen associated with some chromatin. This material may be the *nucleolus organizing region,* a part of the chromosome concerned with the formation of the nucleolus after nuclear division.[27]

Nuclear divisions are of two kinds: *mitosis,* during which the chromosomes are duplicated and the daughter cells have the same number of chromosomes as the original cell; *meiosis,* during which the daughter cells receive half of the original number of chromosomes. Mitosis gives rise to somatic cells (chapter 4), meiosis to reproductive cells (chapter 21). In both kinds of division (with some exceptions), the nuclear envelope breaks into fragments, which become indistinguishable from ER cisternae. When new nuclei are assembled during telophase, ER cisternae join into two new envelopes. The nucleoli disperse during division (with some exceptions) and are newly organized during telophase.

The series of events occurring between the division of one cell and that of a cell of the next generation is the *cell cycle* or *mitotic cycle.*[11] The interphase of the cycle is not a resting stage but one during which DNA synthesis takes place in preparation for the replication of chromosomes. The period of DNA synthesis is referred to as the S (synthesis) period. It is interpolated between G_1 and G_2, the two gap periods during which the cell undergoes a biochemical preparation for DNA synthesis (G_1) and mitosis (G_2).

In addition to the replication of genetic material, the nucleus has another major function, control of protein synthesis in the cell. A special kind of RNA, the messenger RNA, is synthesized by transcription of DNA. The coded message, incorporated in the m-RNA, is carried to the ribosomal RNA in the cytoplasm where synthesis of protein, mainly enzymes, occurs. The enzymes control the pattern of metabolism and thus determine the course of development of the cell. Since every cell inherits the same DNA from the fertilized egg its ribosomes may receive potentially the same information from the nucleus. Yet, cells in the same organism develop along divergent paths and assume diverse forms and functions. According to a widely accepted view, such differentiation is possible because different cells utilize different genetic information depending on which genes (linear segments of chromosomal DNA) are active in producing the message.

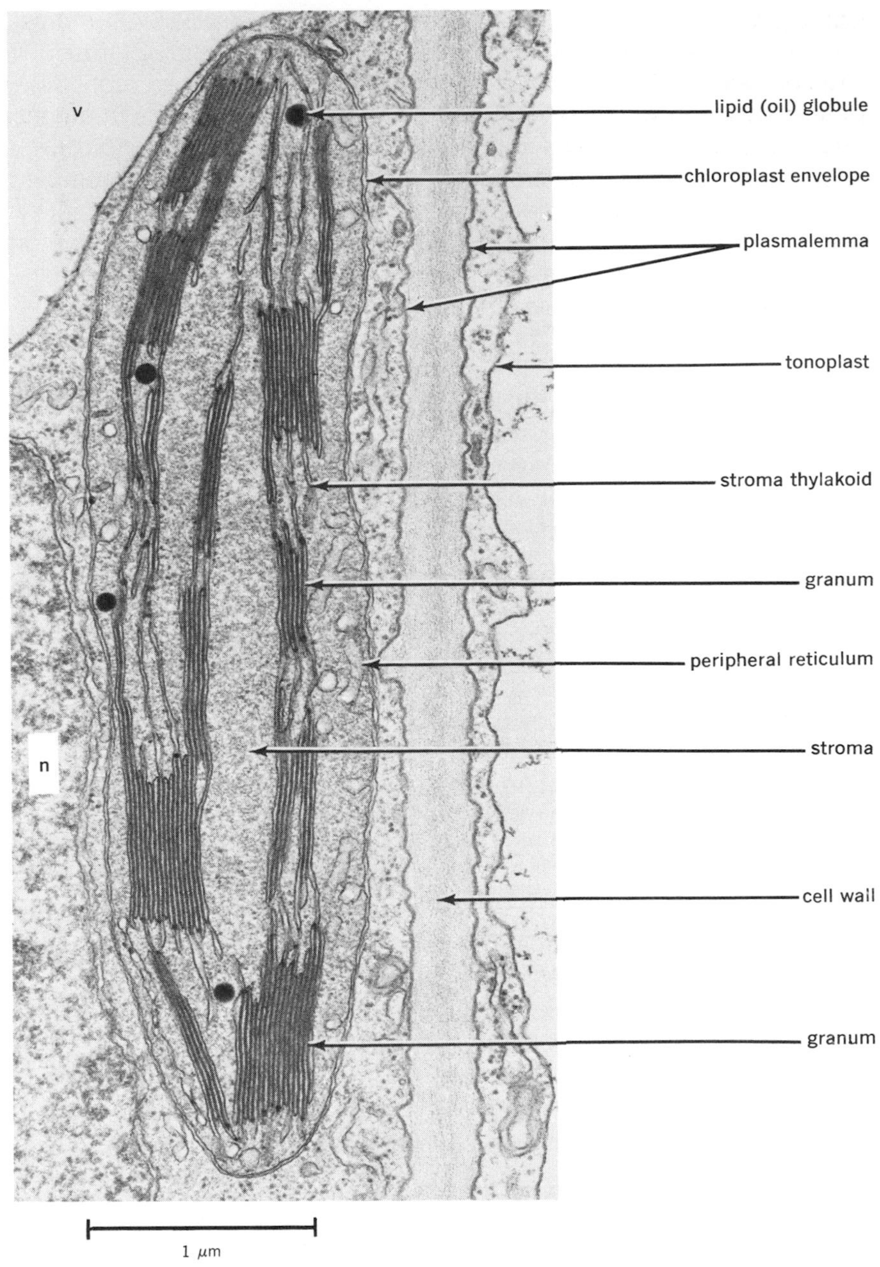

Figure 3.6 Chloroplast with grana seen in profile. From a leaf of *Nicotiana tabacum* (tobacco).

PLASTIDS

Plastids are characteristic organelles of eukaryotic plant cells. They appear in many forms and sizes and are categorized chiefly on the basis of presence or absence and type of pigmentation. Plastids frequently combine characteristics of more than one type, and one type of plastid may change into another. The principal categories of plastids are *chloroplasts, chromoplasts,* and *leucoplasts.* Chloroplasts (fig. 3.6) contain chlorophyll and are concerned with photosynthesis. They occur in green plant parts and are particularly numerous and well differentiated in leaves. Chromoplasts commonly contain yellow and orange carotenoid pigments. They are found in petals and other colored flower parts, in fruits, and some roots. Leucoplasts, the nonpigmented plastids, belong to a poorly circumscribed category. Sometimes they are identified with young, relatively undifferentiated plastids, that is, the *proplastids,* which occur in meristematic cells. The term is also used to designate nonpigmented plastids located in tissues removed from light and storing certain plant products: starch (*amyloplasts*), proteins (*proteinoplasts*), fats (*elaioplasts*), or combinations of these products.[38] Starch, phytoferritin (an iron compound), and lipid in the form of globules (plastoglobuli[30]) may be present in various plastids, including chloroplasts.

The origin of plastids is discussed from two aspects, ontogeny and phylogeny. The concept that plastids are continuous from generation to generation, at least through the egg cell, is widely accepted.[55] The plastids thus inherited divide and differentiate into the different categories of this organelle depending on the kind of cell in which they become located. The ontogenetic relation between plastids in younger and older cells gives a theoretical meaning to the term proplastid. The proplastid stage, however, may be extremely short. The development of chloroplasts from proplastids in a young shoot, for example, occurs so close to the apical meristem that plastids continue to divide after they become chloroplasts in structure and function.[43]

Plastids contain DNA and ribosomes and thus could be genetically autonomous. The indication of such autonomy has led to the concept that, phylogenetically, plastids originated as free-living prokaryotes that were later enclosed in primitive eukarytoic cells and became stabilized as permanent symbiotic elements within the host cells.[46] This view is not generally accepted, however.[5]

A plastid is surrounded by an envelope consisting of two unit membranes[44] (fig. 3.6). Internally, the plastid is differentiated into two main components, the membrane system and the embedding matrix or stroma. The membrane system consists of flattened sacks called *thylakoids.*[35] The degree of development of the thylakoid system is variable in relation to the type of plastid. The least differentiated proplastid has few or no thylakoids. As the proplastid differentiates, thylakoids proliferate. Flattened vesicles develop from the inner membrane of the envelope and eventually align themselves into the characteristic system.

In the chloroplasts of higher plants, the thylakoid system consists of *grana,* stacks of disc-shaped thylakoids, and the *stroma thylakoids,* or *frets,* traversing the stroma between the grana and interconnecting them (figs. 3.7 and 3.8). The thylakoids in the grana are closely appressed so that the membranes from two adjacent thylakoids form a double layer, the *partition,* separating adjacent compartments, or lumina, of thylakoids[62] (figs 3.7 and 3.8,*A*). Continued research on the organization of the thylakoid system in the chloroplast has led to the view that thylakoids are

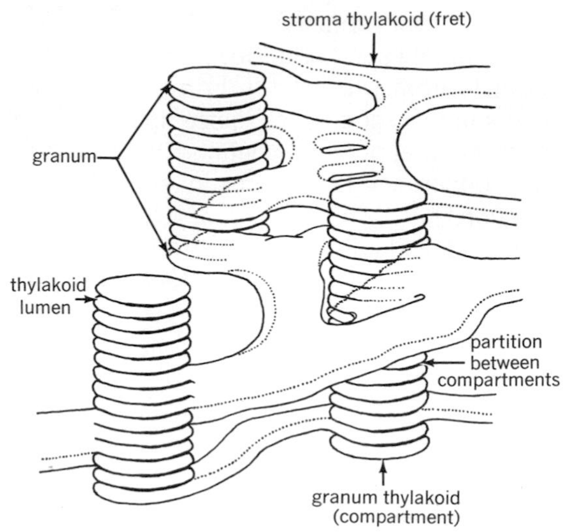

Figure 3.7 Sketch of three grana and interconnecting frets (stroma thylakoids). (Adapted from Weier, Stocking, and Shumway.[62]

not separate entities but are interconnected in such a way that the spaces within the thylakoids may be continuous.[41] Chloroplast membranes are composed of about equal amounts of lipid and protein. The chlorophyll is localized in the thylakoid membranes. The interior faces of chloroplast membranes exposed by freeze fracturing (fig. 3.3) show more particles per unit area than any other membrane system thus far examined[7] (fig. 3.4). Certain kinds of granules have been interpreted as subunits of the membrane, the *quantasomes,* representing the morphological units for light reactions of photosynthesis. According to subsequent research, the quantasome granule is not necessarily a functional unit.[41] Chloroplasts contain small ribosomes and frequently show stroma-free regions enclosing the fine network of DNA. The stroma contains enzymes responsible for fixation of carbon dioxide into sugar. Under certain metabolic conditions chloroplasts form and accumulate starch (fig. 3.11,*A*).

When a leaf is grown in the dark, the vesicles derived from the inner membrane develop into tubes which fuse and form a paracrystalline lattice, the *prolamellar body*[20] (fig. 3.9). When the dark-grown leaf is placed in light the prolamellar body gives rise to the thylakoid system. For convenience, chloro-

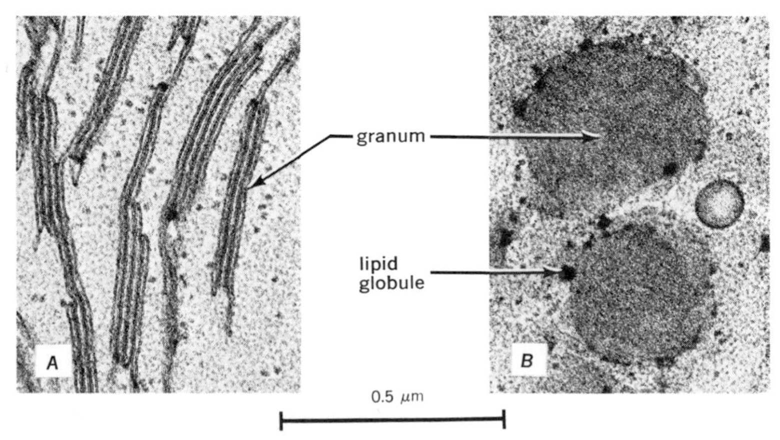

Figure 3.8 Parts of mesophyll chloroplasts of *Beta vulgaris* (sugar beet) showing grana in profile (*A*) and from the surface (*B*).In *A,* the doubleness of the partitions between grana compartments is discernible.

plasts with prolamellar bodies in plants grown in the dark are called *etioplasts.*[28]

Chromoplasts are often derived from chloroplasts but may arise from less differentiated plastids. The essence in the differentiation of chromoplasts is synthesis and localization of carotenoid pigments such as carotene (fig. 3.16,*D*; carrot, *Daucus*) or lycopene (fig. 3.16,*C*; tomato, *Lycopersicon*). Pigment development is associated with a modification and even a complete breakdown of the thylakoids, and, in the process, lipid globules become more abundant. In some chromoplasts the globules store the pigment (petals of *Ranunculus repens* and yellow fruits of *Capsicum,*[11] perianth of *Tulipa,*[31] *Citrus* fruit[57]). In others, the pigment accumulates in numerous protein fibrils (red fruits of *Capsicum,*[11] rose hypanthium[55]). The third form of pigment deposit is a crystalloid. In the red tomato, crystalline lycopene develops in association with thylakoid membranes. Some crystals become very long, and the thylakoids elongate while the lycopene is formed.[21] The crystalloids of carotene in the carrot root are formed during disorganization of the internal structure of the plastid and remain associated with the lipoprotein envelope.[18] Chromoplast development is not an irreversible phenomenon. The chromoplasts of citrus fruit and of carrot root have been found capable of reverse differentiation into chloroplasts; they

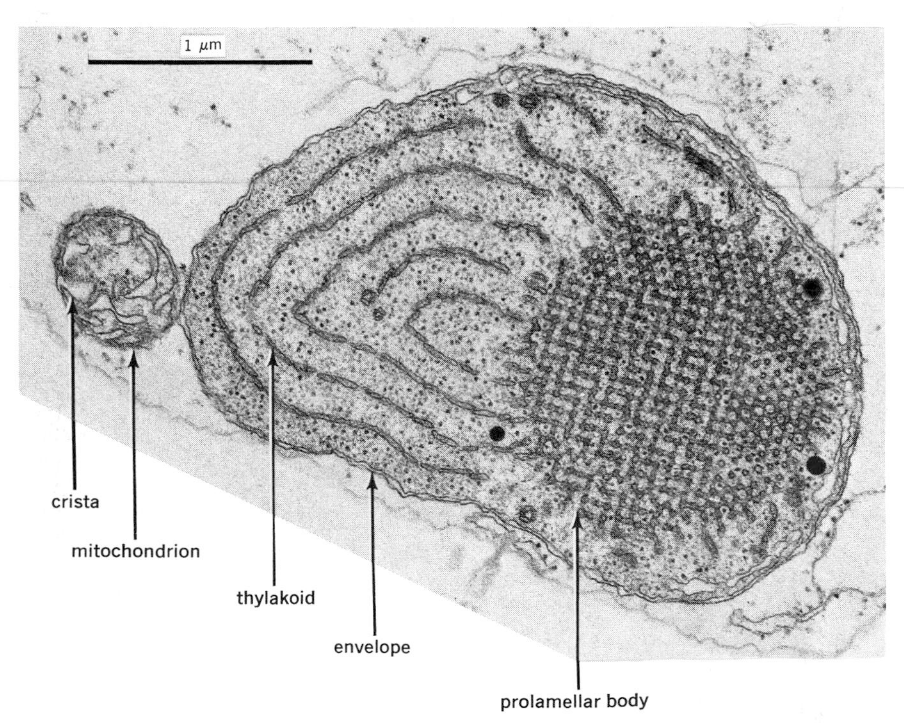

Figure 3.9 Etiolated chloroplast with a prolamellar body in a leaf cell of *Saccharum officinarum* (sugar cane). The connection between the labeled crista and the envelope is discernible in the mitochondrion. Ribosomes are conspicuous in the plastid. (Courtesy of W. M. Laetsch.)

lose the carotene pigment and develop a thylakoid system and chlorophyll[19,58] (chapter 22).

MITOCHONDRIA

Mitochondria are discernible at the highest magnifications with a light microscope as spheres and rods. An electron microscope reveals spherical, elongated, bowllike, and sometimes lobed bodies with a double-membraned envelope and internal membranous structures called *cristae*[44] (fig. 3.10). The cristae are derived from the inner membrane of the envelope (fig. 3.9) and have the form of folds or tubules. Enzymes, including those of the Krebs cycle, are integrated into the membranes of the cristae. Mitochondria are concerned with energy-releasing respiration and with conservation of energy in a form usable for energy-requiring functions.

Growth of mitochondria, followed by division, is accepted as the most likely method of replication of these organelles.[4] Mitochondria

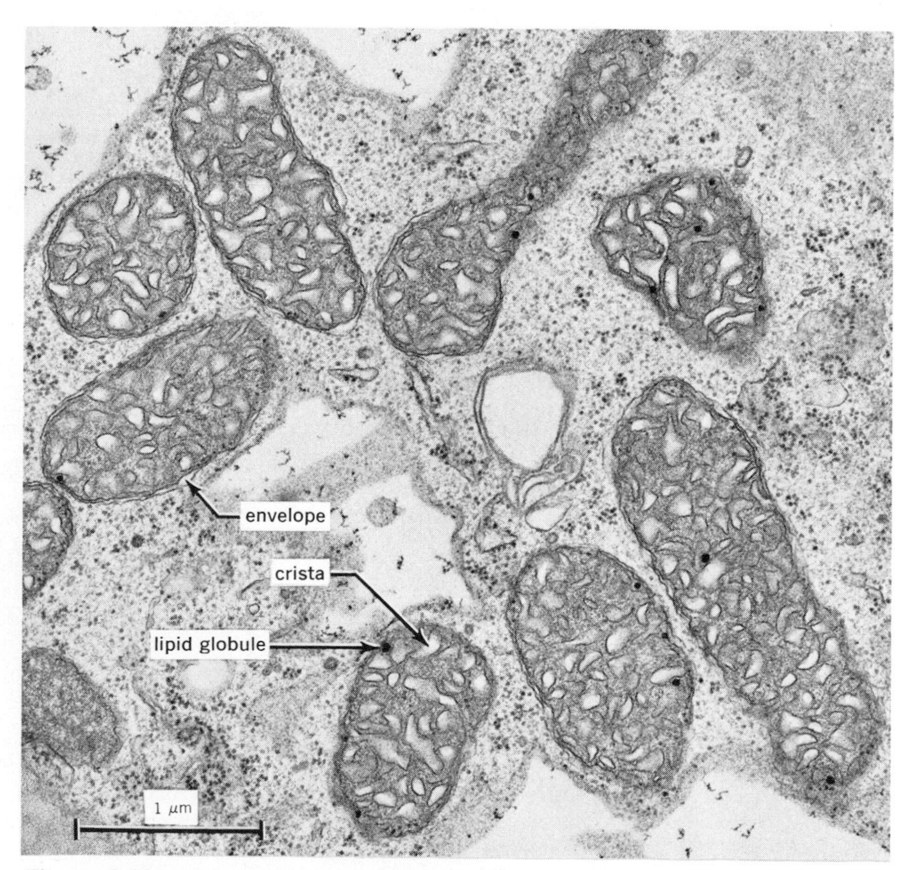

Figure 3.10 Mitochondria in a leaf cell of *Nicotiana tabacum* (tobacco). The envelope consists of two membranes, and the cristae are embedded in a dense stroma. Ribosomes are in the cytoplasm.

possess DNA and ribosomes but their genetic capability is limited. Nevertheless, the idea that mitochondria, just like plastids, are autonomous in phylogenetic origin and became associated with the eukaryotic cell as an "invading symbiont" is strongly advocated by some investigators.[46] Others think that eukaryotic cells evolved along pathways not involving symbiosis.[5]

MICROBODIES

Microbodies occur in a variety of plant species and tissues. They are common in chlorenchyma of dicotyledons and monocotyledons where they are often closely associated with chloroplasts. Microbodies have a single bounding membrane and their matrix is granular or fibrillar (fig. 3.11,*A*). A single

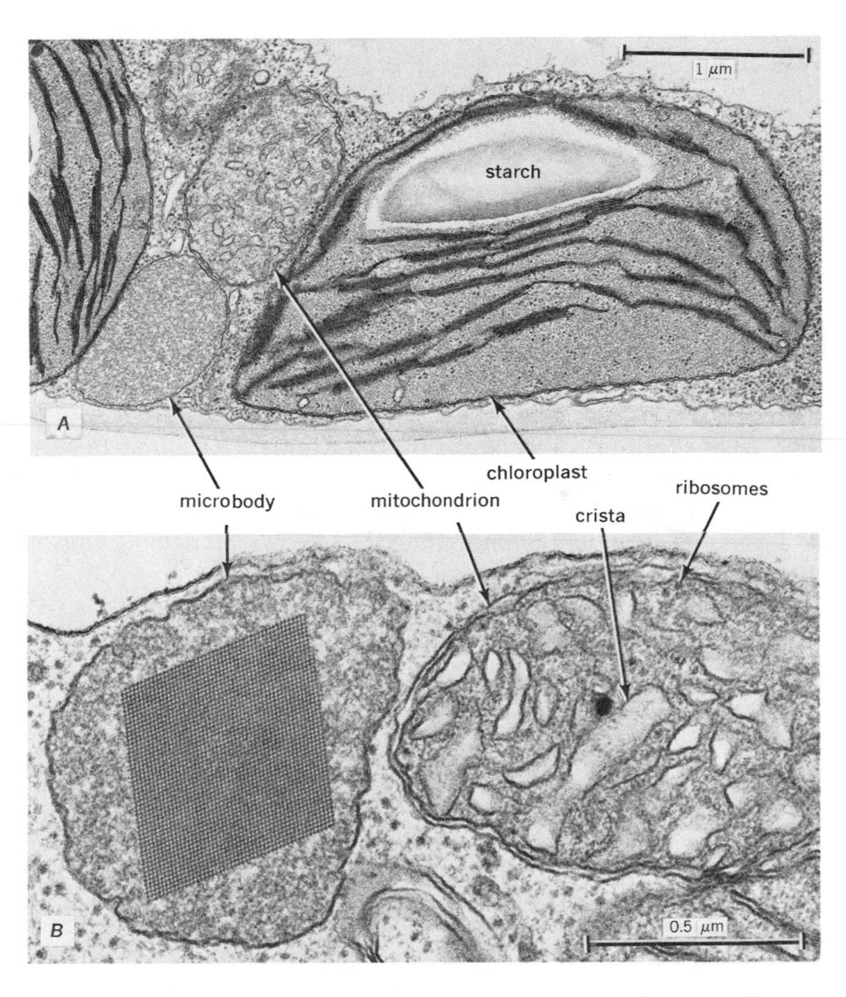

Figure 3.11 Organelles in leaf cells of *Beta vulgaris* (sugar beet, *A*) and *Nicotiana tabacum* (tobacco, *B*). The unit membranes enclosing the microbodies may be contrasted with the double-membraned envelopes of the other organelles. The microbody in *B* contains a crystal. Some ribosomes are perceptible in the chloroplast in *A* and in the mitochondrion in *B*.

crystal may be present in a microbody (fig. 3.11,*B*). Microbodies contain enzymes that vary according to type of cell and its state of differentiation.[17,47,60] Microbodies are *peroxisomes* if they function in the glycolate metabolism and *glyoxysomes* if they contain enzymes of the glyoxylate cycle involved in lipid degradation. Microbodies are also called *cytosomes*.[55]

VACUOLES

The vacuole is an important component of a plant protoplast (fig. 3.1). It contains water and a variety of organic and inorganic substances, many of which are present in dissolved state. The substances may be reserve compounds such as sugars, organic acids, proteins, phosphatids; or they may be excretory products such as calcium oxalate, tannin compounds, anthocyanin pigments. Some substances in the vacuole become crystalline, or the entire content congeals into a firm body (tannin inclusions, protein bodies in dry fruits and seeds). The tonoplast enclosing the vacuole is differentially permeable and, therefore, involved in the regulation of osmotic phenomena associated with the vacuoles, notably the maintenance of turgor in the cell.

Electron microscopy combined with enzyme localization studies has revealed that vacuoles do not merely passively accumulate metabolic products but participate in the biochemical recycling of materials in the cell.[10,15,34] They bring about the breakdown of cell constituents or reserves. Substances are thus mobilized for synthetic activities of the cell or are transformed in connection with differentiation of some specialized cells (e.g., secretory cells). Thus, the vacuole can act as an organelle having a vital function in such metabolic processes as senescence, differentiation, and mobilization of reserves.

Storage cells offer a good example of hydrolytic activities in vacuoles.[9,34] In the cotyledons of leguminous seeds the stored protein occurs in the form of grains, each bounded by the tonoplast of the vacuole in which the grain was formed. During germination, acid hydrolases are detectable in these vacuoles. The protein is digested and the vacuoles fuse into one large central vacuole.

Digestion of cellular constituents may occur in well-developed central vacuoles of parenchyma cells.[15] In various places around a vacuole, the tonoplast invaginates into the vacuole and becomes detached. The resulting vesicle, suspended in the vacuole, may contain various cell components: mitochondria, plastids, ribosomes, and others. After the contents are lysed, the limiting membrane disappears.

The ontogenetic origin of vacuoles is a much debated topic. Meristematic cells (other than vascular cambium cells, which are highly vacuolated) have numerous small vacuoles in an otherwise dense cytoplasm (fig. 3.1). As the cell enlarges and differentiates, the vacuoles enlarge also and fuse into a single large vacuole occupying the center of the cell. According to some investigators, the initial small vacuoles arise anew through attraction of water to localized regions; according to others, the vacuoles have precursors in the form of small organelles which swell into vacuoles by absorbing water. Some electron microscopists think that the vacuoles arise by dilation of ER cisternae or of vesicles derived from the ER. Pinocytotic vesicles formed by an invagination of plasmalemma also have been associated with vacuole formation.[10] Probably vacuoles originate by more than one method.

In their hydrolytic action vacuoles resemble *lysosomes*. Organelles of that name were first reported for animal cells and described as polymorphic organelles bound by a single membrane and containing hydrolytic enzymes.[11] The term lysosome does not designate a specific morphologic entity in plant cytology. Plant cells contain a variety of hydrolases capable of digesting cytoplasmic constituents and metabolites but these enzymes occur in several different kinds of membrane-bound structures, among which vacuoles occupy a prominent place. Hydrolases even occur in the cell wall, that is, in the extracellular space. The terms *lysosomal cell compartment*[34] and *lysosomal system*[64] have been proposed to designate the totality of structures in which hydrolysis occurs in plant cells. In this connection, Matile[34] suggests that the term lysosome be used only in a biochemical sense.

PARAMURAL BODIES

Invaginations of plasmalemma are frequently observed in material prepared for electron microscopy. These invaginations assume various aspects. Some form pockets between the cell wall and the cytoplasm. The frequent presence of tubules and vesicles in such pockets has led to the assumption that the pockets are functional organelles concerned with cell wall growth and possibly with other relations between the cytoplasm and the cell wall. A plasmalemma invagination may intrude into the vacuole, pushing the tonoplast forward. The invaginations are often detached from the plasmalemma and embedded in the cytoplasm as the so-called multivesiculate bodies or are suspended in the vacuole.

Similar formations were first observed in fungi and named *lomasomes*.[11] Subsequently, the term *paramural bodies*[32] was proposed for all membranous structures associated with the plasmalemma. They are called *plasmalemmasomes* if their formation involves only the plasmalemma, lomasomes if some other membranes also participate in their formation.[32] It is not certain whether all such structures are normal cell components. Possibly some of them at least are induced to form by the manipulations preparatory for electron microscopy.

RIBOSOMES

Ribosomes are particles approximately 170–230 Å in diameter (figs. 3.5,*B* and 3.13,*B*) that are the sites of protein synthesis from amino acids.[6] Ribosomes consist of subunits of different sedimentation coefficients held together by magnesium ions. Ribosomes contain about equal amounts of protein and RNA. In protein synthesis, the ribosomes are united into *polyribosomes* (or *polysomes;* fig. 3.13,*B*,*C*) by the messenger RNA carrying the genetic message from the nucleus. The amino acids from which the proteins are synthesized are brought to the polyribosomes by the soluble RNA (transfer RNA) located in the cytoplasm. The source of energy for the synthesis is guanosine triphosphate.

The polyribosomes are commonly associated with the ER (fig. 3.14,*A*), unattached ribosomes are distributed in the cytoplasm singly or in groups. The nuclear envelope also may bear polyribosomes. Ribosomes occur in nuclei, plastids (figs. 3.8 and 3.9), and mitochondria (fig. 3.11,*B*). The organelle ribosomes appear to be different from those located in the cytoplasm.

DICTYOSOMES

Dictyosomes are organelles composed of stacks of flat circular cisternae, each bound by a unit membrane (figs. 3.12 and 3.13,*A*). The margin of a cisterna is often more or less deeply fenestrated. When the fenestration is extensive the part of the cisterna involved appears as a tubular network, a feature that is

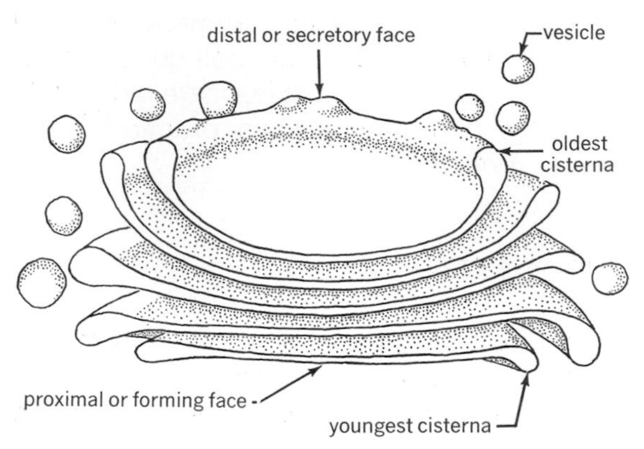

Figure 3.12 Sketch of a dictyosome with five cisternae and some secretory vesicles. (Redrawn from K. Esau and V. I. Cheadle, *Univ. Calif. Pubs. Bot.* 36: 253–344, 1965.

reflected in the name dictyosome (dictyo, net; fig. 3.13,*C*). From time to time vesicles are pinched off the network or off the margin of the not fenestrated cisterna. The totality of dictyosomes and vesicles derived from them in a given cell is the *Golgi apparatus*[36] and, accordingly, the dictyosomes are frequently called *Golgi bodies*.

Plant dictyosomes consist of two to seven (sometimes many more) cisternae. There is no physical continuity between the individual cisternae of a stack but they remain at a certain minimal distance from one another, even when they are isolated from the cells. Functionally, dictyosome cisternae act in unison, so that informational continuity may be assumed. Fibrous or tubular elements in parallel arrangement have been seen in intercisternal position. Their function is unknown.

Dictyosomes are concerned with secretion. The product to be secreted accumulates in the vesicles which carry the material to its destination. In actively secreting dictyosomes the production of vesicles is vigorous and eventually the secreting cisternae completely break up into vesicles. A disappearing cisterna is replaced by a new one. In these activities, the dictyosomes show polarity: production of vesicles and final vesiculation of a cisterna occurs at one face of the stack (distal or secretory face), the addition of a new cisterna at the other face (proximal or forming face) (fig. 3.12). Accordingly, the fenestration increases from the proximal to the distal face. The new cisternae appear to be produced from membrane material derived from the ER[37] (fig. 3.13,*A*). Dictyosomes also increase in number in a given cell, but the mode of multiplication is not certain. Fragmentation of existing dictyosomes is one of the methods visualized.

The secreted product is not synthesized exclusively in the dictyosome, but may be derived from an outside source, such as the endoplasmic reticulum, and condensed and transformed in the dictyosome. The products secreted are mainly polysaccharides or polysaccharide-protein complexes of high viscosity. The materials may be incorporated into cell walls or excreted to the outside (e.g., mucilages in root tips). When the vesicle transporting the material to the wall reaches the plasmalemma the vesicle membrane and the plasmalemma fuse and the contents are released toward the cell wall. Dictyosome-derived vesicles are involved also in the initiation of a new cell wall after mitosis (chapter 4).

ENDOPLASMIC RETICULUM

The endoplasmic reticulum (ER) is a complex membrane system that has not been completely revealed in its three-dimensional aspect. High tension electron microscopy combined with stereoscopic micrography shows promise in this respect.[33] The extended parts of the reticulum are flat cisternae that appear, in sectional view (profile),

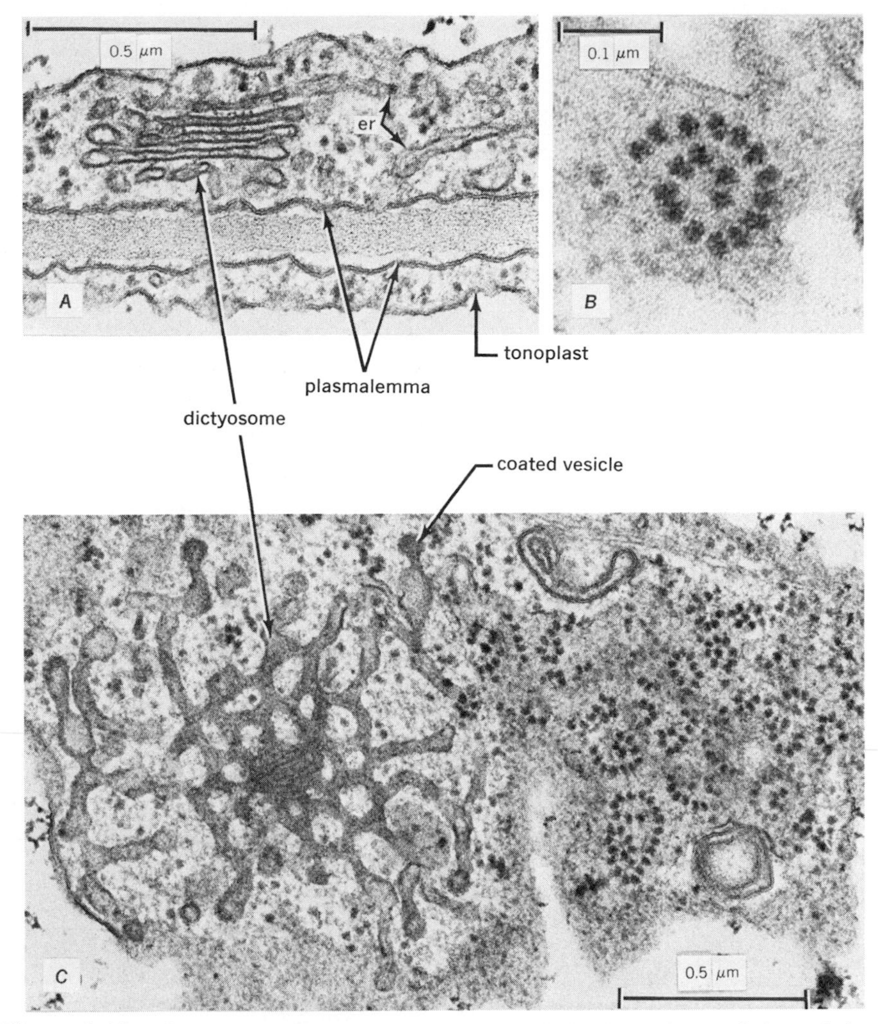

Figure 3.13 Dictyosomes and ribosomes from a leaf of *Nicotiana tabacum* (tobacco). *A,* dictyosome in profile with the fenestrated distal face turned toward the cell wall. *B,* polyribosome attached to the surface of ER. *C,* dictyosome is seen from its fenestrated distal face. Some of the vesicles to be pinched off are coated. To the right is a surface view of an ER cisterna covered with polyribosomes.

as two-unit membranes with a narrow space between them (fig. 3.14). Freeze-etch preparations of onion root tip cells show the ER as extensive fenestrated sheets.[8] It also may form tubules instead of cisternae. The ER appears to be an anastomosing system of an indefinite extent and, therefore, the term organelle is seldom applied to this cell component. The ER may be connected with the nuclear envelope which is interpreted as part of the ER system. The various interconnections within the ER system are probably changeable in relation to conditions in the cell. The ER has some relation to the cyto-

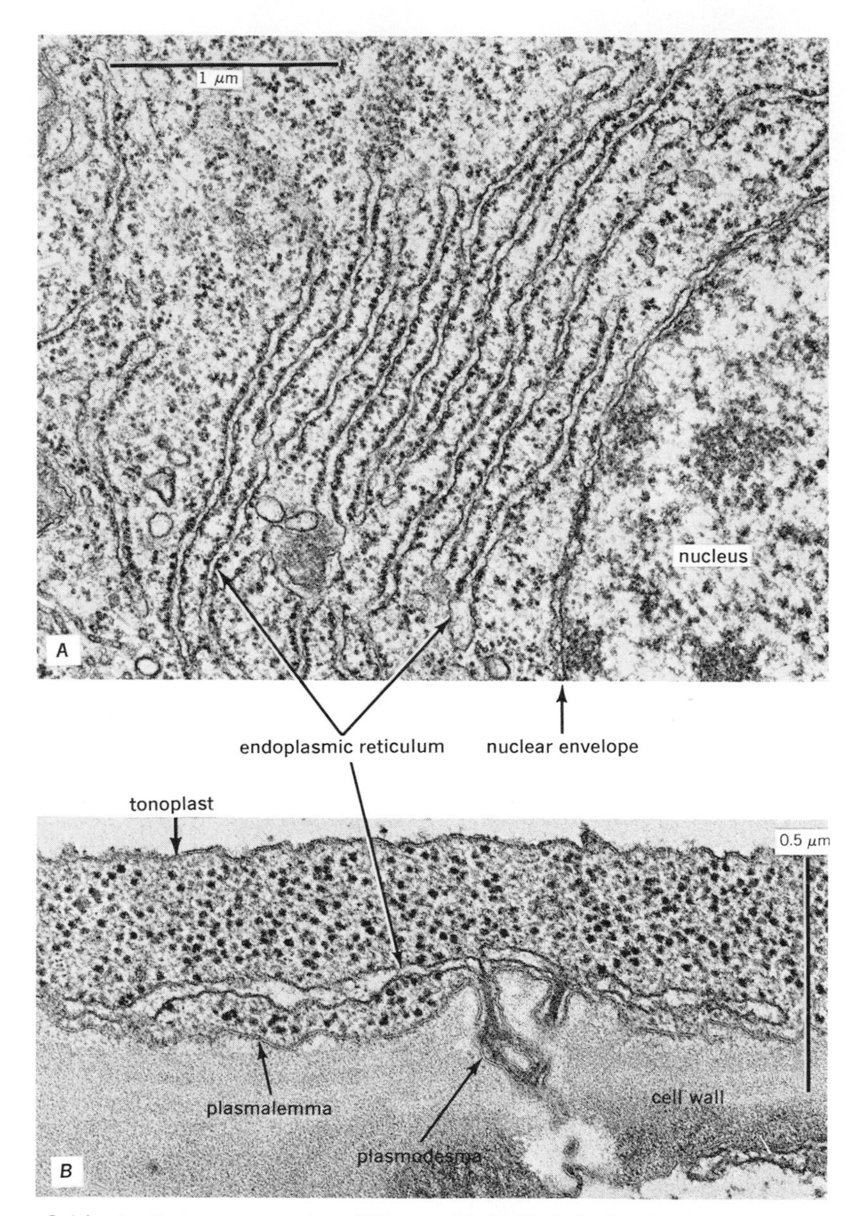

Figure 3.14 Endoplasmic reticulum (ER) seen in profile in leaf cells of *Nicotiana tabacum* (tobacco, *A*) and Beta vulgaris (sugar beet, *B*). The ER is associated with numerous ribosomes (rough ER) in *A*, with fewer in *B*. The ER in *B* is connected to the electron dense cores (desmotubules) of plasmodesmata (seen only in part). Plasmalemma lines the plasmodesmatal canals. Note triple nature of tonoplast and plasmalemma in *B*.

plasmic strands traversing the cell walls (*plasmodesmata;* fig. 3.14,*B*). The common interpretation is that ER cisternae on the two sides of a cell wall are interconnected by an ER tubule (desmotubule[48]) forming the core of a plasmodesma, but the tubular structure is rarely discernible. The ER is termed rough-surfaced, rough, or granular when ribosomes adhere to its surface (fig. 3.14,*A*). In the absence of ribosomes, the ER is called smooth-surfaced, smooth, or agranular.

The association of ribosomes with the ER is interpreted as evidence that the ER is involved in protein synthesis. The morphology of the ER system also suggests that the ER may serve as an intracellular circulatory system transporting sugars, amino acids, and ATP (adenosine triphosphate) to sites of usage or storage. Connections through plasmodesmata would make intercellular transport possible too. The ER may serve as a compartment for the condensation of certain products and may become dilated into protein-containing cisternae. The large surface of the ER may allow for a differential distribution of enzymes. The ER is capable of considerable increase in volume without any evidence that membrane material is transferred into it. This observation suggests a capacity for membrane assembly.

LIPID GLOBULES

Globules that become more or less electron dense after fixation with osmium tetroxide are common in cells of various tissues (fig. 3.1). The reaction to osmium indicates presence of lipids (triglycerides [34]) in the globules and the difference in density is determined by the degree of saturation of the lipid. Intense osmiophily indicates a high degree of unsaturation.[13] The globules appear to be identical to the highly refractive granules visible in living cells under high magnification with a light microscope and also to the osmiophilic globules in plastids[30] (fig. 3.6).

Electron microscopy has not led to a unanimous interpretation of the nature and structure of the osmiophilic globules. The globules are described either as organelles (*spherosomes*) enclosed in a unit membrane or as lipid droplets having no bounding membrane. A concept combining the two views suggests that the globules originate as oil-containing vesicles cut off from the ER and are later converted into oil droplets. Lipid globules have also been identified as membrane-bound organelles in which hydrolases are associated with the triglycerides.[34] In most instances, the globules appear to be droplets surrounded by cytoplasm, with lipid as the dominant component. An intense reduction of osmium at the interface between the lipid and the cytoplasm frequently results in the appearance of a thin, dense line around the periphery of the droplet.[13] This line is often taken to be a membrane. Some microscopists interpret the thin line as one-half of a unit membrane.[52]

MICROTUBULES

The microtubule is recognized as a common constituent of eukaryotic cells. It is a hollow, straight tube averaging 240 Å in diameter (fig. 3.15) composed of globular protein subunits.[39] Microtubules are components of mitotic and meiotic spindles and of the phragmoplast that is concerned with the formation of the new cell wall in cytokinesis (chapter 4). They also occur in the peripheral cytoplasm near the growing cell wall. The occurrence and disposition of microtubules in cells has led to the concept that they act in development of form in the cell and in intracellular movements.[59] In plant cells, the mi-

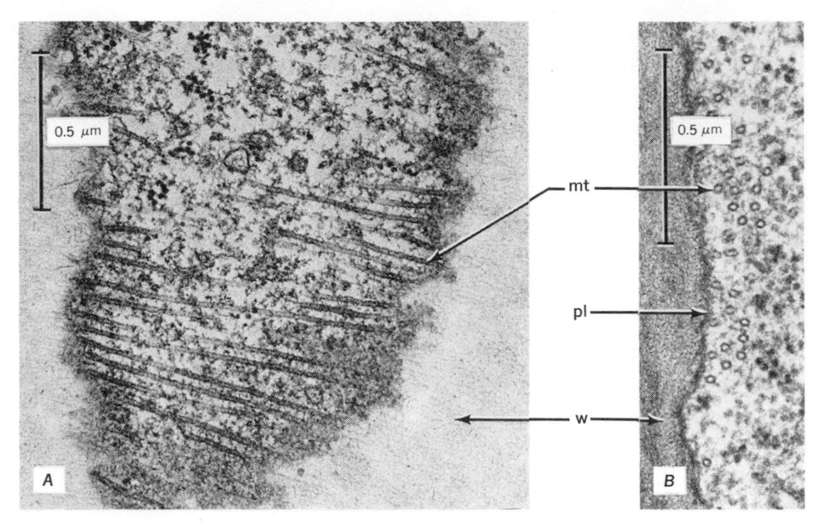

Figure 3.15 Microtubules (*mt*) seen in longitudinal (*A*) and transectional (*B*) views. From leaf cells of *Mimosa pudica* (*A*) and *Nicotiana tabacum* (*B*). Other details: *pl*, plasmalemma; *w*, cell wall.

crotubules appear to direct the positioning of the new wall in dividing cells and the subsequent orderly growth in thickness of the wall. This concept is supported by experimental work; destruction of microtubules in differentiating cells with colchicine results in conspicuous abnormalities in cell wall development.[22,23]

ERGASTIC SUBSTANCES

Starch

Starch, a common ergastic substance, develops in the form of grains in plastids. Next to cellulose, starch is the most abundant carbohydrate in the plant world.[45] During photosynthesis, starch is formed in chloroplasts. Later, it is broken down and resynthesized as storage starch in amyloplasts. An amyloplast (fig. 3.16,*F*) may contain one or more starch grains. If several starch grains originate together, they may form a compound starch grain (fig. 3.16,*G*).

Starch grains are varied in shape (fig. 3.16,*E–G,M,N*) and commonly show layering centered around a point, the *hilum* (fig. 3.16,*E*), which may be in the center of the grain or to one side. Splits, often radiating from the hilum, appear to result from dehydration of the grains. Layering is ascribed to an alternation of two carbohydrates, an amylose (linear molecules) and an amylopectin (branched molecules). The amylose is more highly soluble in water than the amylopectin, and when the grain is placed in water the differential swelling of the two substances emphasizes the layering. In cereal starch, the layering depends on the daily rhythm, in potato starch the periodicity causing the layering is of endogenous origin.[18] Radial arrangement of molecules in the layers and parallel bonding by hydrogen result in appearance of crystalline regions.[2] The starch grain is a spherocrystal which shows a figure of a Maltese cross under polarized light. Starch

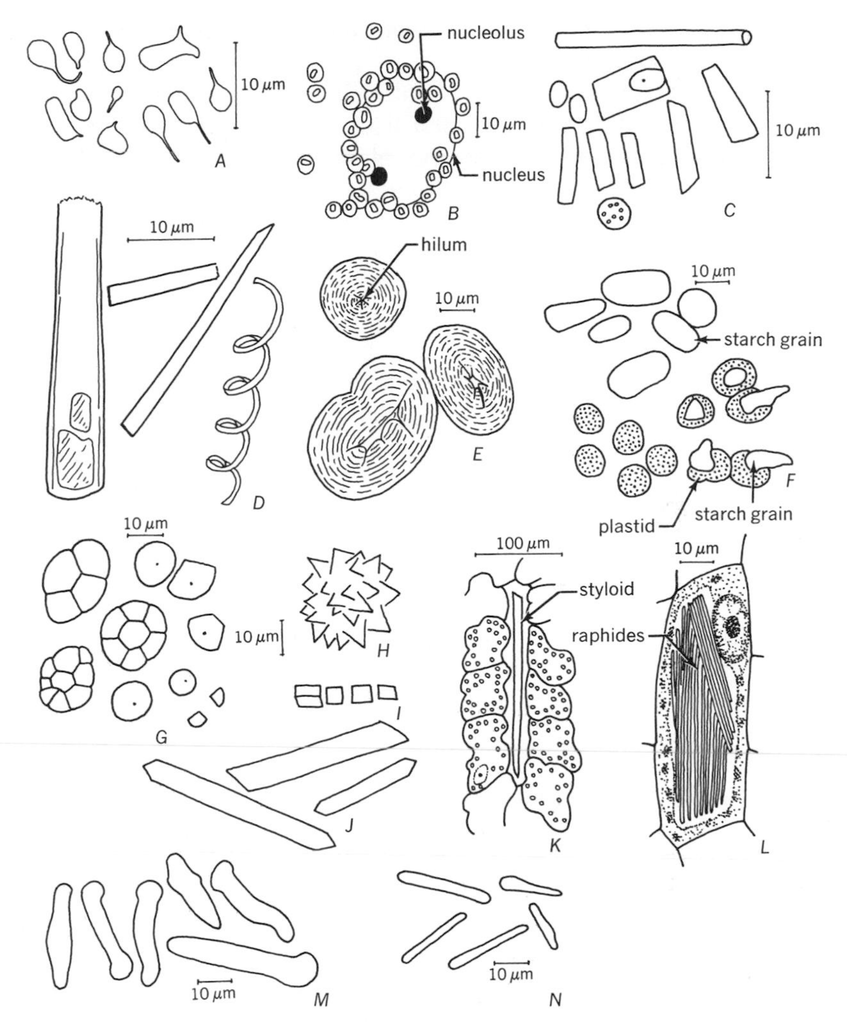

Figure 3.16 Plastids and ergastic inclusions of protoplasts. *A*, chromoplasts from disc flowers of *Gaillardia*. *B*, nucleus (with two nucleoli) and chromoplasts from pericarp of red pepper (*Capsicum*). The bodies in the plastids are deep orange-red in fresh material. *C*, pigment bodies from pericarp of tomato (*Lycopersicon*). *D*, pigment bodies from the root of carrot (*Daucus*). *E*, starch grains from seed of bean (*Phaseolus*). *F*, starch grains and plastids from rhizome of *Iris*. The plastids are leucoplasts (or elaioplasts) that form starch and oil. *G*, *H*, simple and multiple starch grains (*G*) and a druse crystal (*H*) from root of sweet potato (*Ipomoea*). *I*, *J*, crystals from secondary phloem of *Juglans* (*I*) and *Pinus* (*J*). *K*, styloid crystal in an elongated cell among mesophyll cells of *Iris*. *L*, cell with raphide crystals from root tip of *Vitis*. *M*, *N*, starch grains from laticifers of two different species of *Euphorbia*.

commonly stains bluish-black with iodine in potassium iodide.

Storage starch is found in parenchyma cells of cortex, pith, and vascular tissues of stems and roots; in parenchyma of fleshy leaves (bulb scales), rhizomes, tubers, fruits, cotyledons; and in the endosperm of seeds. Commercial starches are obtained from various sources[51] as, for example, endosperm of cereals, fleshy roots of the tropical *Manihot esculenta* (tapioca starch), tuber of potato, tuberous rhizomes of scitaminaceous plants (arrowroot starch), stem of *Metroxylon sagu* (sago starch).

Tannins

Tannins are a heterogeneous group of phenol derivatives that are widely distributed in the plant body. In some of their forms the tannins are highly conspicuous in sectioned material. They appear as coarsely or finely granular masses or as bodies of various sizes colored yellow, red, or brown. No tissues lack tannins entirely. They are abundant in leaves of many plants, in vascular tissues, in the periderm, in unripe fruits, in seed coats, and in pathologic growths. They occur in the cytoplasm and the vacuole and may impregnate walls. They may be present in many cells of a given tissue or in isolated cells scattered through the tissue (tannin idioblasts). They may be located in much enlarged cells called tannin sacs. Phenolic compounds are useful as supplementary indicators of taxonomic relationships.[3] Tannins are used commerically, especially in the centuries-old industry of tanning of animal hides to obtain stable leather.

Proteins

Proteins are stored as solid protein bodies, or *aleurone grains,* in fruits and seeds of many species. Examples of such storage are found in the cotyledons of Fabaceae seeds and in the outermost endosperm layer, the aleurone layer, of the caryopsis in Poaceae. Protein bodies are enclosed in a unit membrane and may have an amorphous matrix or a matrix including a protein crystalloid and amorphous nonproteinaceous (phytin) globoid.[25,50] Some protein bodies contain calcium oxalate crystals (Apiaceae). Storage proteins accumulate in vacuoles. In this process, large vacuoles break up into smaller ones and, at ripening of the storage tissue, each small vacuole is converted into a protein body with the tonoplast remaining as the bounding membrane.[9,29] Upon germination, the protein is digested and the numerous vaculoes again become a single large one.[9,26] Protein may occur in the form of crystalloids in the cytoplasm, for example, in peripheral parenchyma of the potato tuber, among starch grains in the seed of *Musa,* and in the fruit parenchyma of *Capsicum.*

Fats, oils, and waxes

Fats, oils, and waxes are a particularly important class of ergastic substances of plants used commercially.[51] Waxes are fatty acid esters of monohydric alcohols; oils and fats are glycerides of certain organic acids. The distinction between oils and fats is chiefly physical, fats being solid at ordinary temperatures, oils liquid. Fats and oils are storage forms of lipids. In this section the word fat refers to both oils and fats.

Fats occur in all plant taxa and probably are present in every cell, at least in small amounts.[29] They are found in solid form, but more frequently in liquid form as oil droplets. Crystalline fat is rare. An example was reported for the endosperm of the palm *Elais* in which the cells are filled with short needle-shaped crystals of fat.[29] Fats apparently originate in the cytoplasm or in plastids (elaioplasts; fig. 3.16,*F*). Waxes occur as protective coating on the epidermis but may become deposited within cells (pericarp of *Rhus vernicifera*). Most plants contain too little wax to be valuable for commercial use. Exceptions are the palm *Copernicia cerifera* which yields the carnauba wax of commerce, and *Simmondsia chinensis* (jojoba) the cotyledons of which contain a liquid wax similar in quality to the oil of the sperm whale.[53] Oils having small molecules are the essential oils that are secreted in special cells or excreted into intercellular cavities (chapter 13). Oils and fats may be identified by a reddish color when they are treated with Sudan III or IV.

Crystals

Calcium oxalate is most prominently represented among crystals in plants. The formation of calcium oxalate crystals is one of many aspects of calcium accumulation in plant tissues.[1] Calcium oxalate crystals assume several different forms. They appear as *raphides,* bundles of needles (fig. 3.16,*L*; leaves of grapevine, *Impatiens, Arum;* stem of *Tradescantia*); *styloids,* elongated columnar crystals (fig. 3.16,*K*; leaf of *Eichhornia crassipes*); *prisms,* rectangular or pyramidal (fig. 3.16,*I*,*J*; phloem of *Acer, Tilia, Ficus, Quercus;* leaves of *Begonia, Hyoscyamus niger, Vicia sativa*); *druses,* spheroidal aggregates of prismatic crystals (fig. 3.16,*H*; rhizome of rhubarb; root of *Ipomoea batatas;* leaves of *Datura stramonium, Ruta graveolens*); *crystal sand,* very small crystals, usually in masses (stem of *Aucuba japonica, Sambucus nigra;* leaf of *Atropa balladonna*). Appearance and location of crystals may be specific and useful in taxonomic classification.[29] Calcium carbonate crystals are not common in higher plants. This compound is sometimes associated with the cell wall in the form of cystoliths (chapter 13). Crystals usually develop in vacuoles. In raphide cells, a chambered envelope appears in the vacuole and subsequently crystals are formed in the chambers, so that each crystal is included in an individual sheath.[1] Crystals are often classified as excretory products, but possibly some of the calcium is recycled.

REFERENCES

1. Arnott, H. J., and F. C. E. Pautard. Calcification in plants. In: *Biological calcification: cellular and molecular aspects.* H. Schraer, ed. pp. 375–446. New York, Appleton-Century-Crofts. 1970.
2. Badenhuizen, N. P. Occurrence and development of starch in plants. In: *Starch: chemistry and technology.* R. L. Whistler and E. F. Paschall, eds. Vol. 1. pp. 65–103. New York, Academic Press. 1965.
3. Bate-Smith, E. C. The phenolic constituents of plants and their taxonomic significance. *J. Linn. Soc. London, Bot.* 58:95–173. 1962.
4. Baxter, R. Origin and continuity of mitochondria. In: *Origin and continuity of cell organelles.* J. Reinert and H. Ursprung, eds. pp. 46–64. New York, Springer. 1971.
5. Bogorad, L. Evolution of organelles and eukaryotic genomes. *Science* 188:891–898. 1975.
6. Brachet, J. Quelques aspects moleculaires de la cytologie et de embryologie. *Biol. Rev.* 43:1–16. 1968.
7. Branton, D. Membrane structure. *Ann. Rev. Plant Physiol.* 20:209–238. 1969.
8. Branton, D., and H. Moor. Fine structure in freeze-etched *Allium cepa* L. root tips. *J. Ultrastruct. Res.* 11:401–411. 1964.
9. Briarty, L. G., D. A. Coult, and D. Boulter.

Protein bodies of developing seeds of *Vicia faba. J. Exp. Bot.* 20:358–372. 1970. Protein bodies of germinating seeds of *Vicia faba. J. Exp. Bot.* 21: 513–524. 1970.

10. Buvat, R. Origin and continuity of cell vacuoles. In: *Origin and continuity of cell organelles*. J. Reinert and H. Ursprung, eds. pp. 127–157. New York, Springer. 1971.
11. Clowes, F. A. L., and B. E. Juniper. *Plant cells.* Oxford, Blackwell Scientific Publications. 1968.
12. Danielli, J. F., and H. Davson. A contribution to the theory of permeability of thin films. *J. Cell. Compar. Physiol.* 5:495–508. 1935.
13. Fawcett, D. W. *The cell. Its organelles and inclusions. An atlas of fine structure.* Philadelphia and London, W. B. Saunders Company. 1966.
14. Feldherr, C. M., and C. V. Harding. The permeability characteristic of the nuclear envelope at interphase. *Protoplasmatologia* 5 (part 2):35–50. 1964.
15. Fineran, B. A. Ultrastructure of vacuolar inclusions in root tips. *Protoplasma* 72:1–18. 1971.
16. Franke, W. W. On the universality of nuclear pore complex structure. *Z. Zellforsch. u. Mikroskop. Anat.* 105:405–429. 1970.
17. Frederick, S. E., and E. H. Newcomb. Cytochemical localization of catalase in leaf microbodies (peroxisomes). *J. Cell Biol.* 43:343–353. 1969.
18. Frey-Wyssling, A., and K. Mühlethaler. *Ultrastructural plant cytology, with an introduction to molecular biology.* Amsterdam, Elsevier. 1965.
19. Grönegress, P. The greening of chromoplasts in *Daucus carota* L. *Planta* 98: 274–278. 1971.
20. Gunning, B. E. S., and M. P. Jagoe. The prolamellar body. In: *Biochemistry of chloroplasts.* T. W. Goodwin, ed. pp. 655–676. New York, Academic Press. 1967.
21. Harris, W. M., and A. R. Spurr. Chromoplasts of tomato fruits. II. The red tomato. *Amer. J. Bot.* 56:380–389. 1969.
22. Hepler, P. K., and D. E. Fosket. The role of microtubules in vessel member differentiation. *Protoplasma* 72:213–236. 1971.
23. Hepler, P. K., and E. H. Newcomb. Fine structure of cell plate formation in the apical meristem of *Phaseolus* roots. *J. Ultrastruct. Res.* 19:498–513. 1967.
24. Johnson, J. M. A study of nucleolar vacuoles in cultured tobacco cells using radio-autography, actinomycin D, and electron microscopy. *J. Cell Biol.* 43: 197–206. 1969.
25. Jones, R. L. The fine structure of barley aleurone cells. *Planta* 85:359–375. 1969.
26. Jones, R. L., and J. M. Price. Gibberellic acid and the fine structure of barley aleurone cells. III. Vacuolation of the aleurone cells during the phase of ribonuclease release. *Planta* 94:191–203. 1970.
27. Jordan, E. G., and J. M. Chapman. Ultrastructural changes in the nucleoli of Jerusalem artichoke (*Helianthus tuberosus*) tuber discs. *J. Exp. Bot.* 22:627–634. 1971.
28. Kirk, J. T. O., and R. A. E. Tilney-Bassett. *The plastids: their chemistry, structure, growth and inheritance.* San Francisco, Freeman. 1967.
29. Küster, E. *Die Pflanzenzelle.* 3rd ed. Jena, Gustav Fischer. 1956.
30. Lichtenthaler, H. K. Plastoglobuli und Plastidenstrukturen. *Ber. Deut. Bot. Ges.* 79:(82)–(88). 1966.
31. Lichtenthaler, H. K. Die Feinstruktur

der Chromoplasten in plasmochromen Perigon-Blättern von *Tulpia. Planta* 93:143–151. 1970.

32. Marchant, R., and A. W. Robards. Membrane systems associated with the plasmalemma of plant cells. *Ann. Bot.* 32:457–471. 1968.
33. Marty. F. O. Observation au microscope électronique au haute tension (3 MeV) de cellules végétales au coupes épaisses de 1 à 5 μ. *C. R. Acad. Sci. Paris D.* 277:2681–2684. 1973.
34. Matile, P. Lysosomes. In: *Dynamic aspects of plant ultrastructure.* A. W. Robards, ed. Chapter 5, pp. 178–218. London, McGraw-Hill Book Company (UK) Limited. 1974.
35. Menke, W. Structure and chemistry of plastids. *Ann. Rev. Plant Physiol.* 13: 27–44. 1962.
36. Mollenhauer, H. H., and D. J. Morré. Golgi apparatus and plant secretion. *Ann. Rev. Plant Physiol.* 17:27–46. 1966.
37. Morré, D. J., H. H. Mollenhauer, and C. E. Bracker. Origin and continuity of Golgi apparatus. In: *Origin and continuity of cell organelles.* J. Reinert and H. Ursprung, eds. pp. 82–126. New York, Springer. 1971.
38. Newcomb, E. H. Fine structure of protein-storing plastids in bean root tips. *J. Cell Biol.* 33:143–163. 1967.
39. Newcomb, E. H. Plant microtubules. *Ann. Rev. Plant Physiol.* 20:253–288. 1969.
40. Parish, G. R. Seasonal variations in the membrane structure of differentiating shoot cambial zone cells demonstrated by freeze etching. *Cytobiologie* 9: 131–143. 1974.
41. Park, R. B., and P. V. Sane. Distribution of function and structure in chloroplast lamellae. *Ann Rev. Plant Physiol.* 22:395–430. 1971.
42. Parthasarathy, M. V., and K. Mühlethaler. Cytoplasmic microfilaments in plant cells. *J. Ultrastruct. Res.* 38:46–62. 1972.
43. Possingham, J. V., and W. Saurer. Changes in chloroplast number per cell during leaf development in spinach. *Planta* 86:186–194. 1969.
44. Racker, E., ed. *Membranes of mitochondria and chloroplasts.* New York, Reinhold. 1970.
45. Radley, J. A. *Starch and its derivatives.* Vol. 1. 3rd ed. New York, John Wiley & Sons. 1954.
46. Raven, P. H. A multiple origin of plastids and mitochondria. *Science* 169:641–646. 1970.
47. Richardson, M. Microbodies (glyoxysomes and peroxisomes) in plants. *Sci. Progr. Oxford* 61:41–61. 1974.
48. Robards, A. W. Plasmodesmata. *Ann. Rev. Plant Physiol.* 26:13–29. 1975.
49. Robertson, J. D. The membrane of the living cell. *Sci. Amer.* 206:64–72. 1962.
50. Rost, T. L. The ultrastructure and physiology of protein bodies and lipids from hydrated dormant and nondormant embryos of *Setaria lutescens* (Gramineae). *Amer. J. Bot.* 59:607–616. 1972.
51. Schery, R. W. *Plants for man.* New York, Prentice-Hall. 1952.
52. Schwarzenbach, A. M. Observations on spherosomal membranes. *Cytobiologie* 4:145–147. 1971.
53. Sherbroke, W. C., and E. F. Haase. Jojoba: a wax-producing shrub of the Sonoran desert. *Arid Lands Resource Information Paper* No. 5. Tucson, Arizona. 1974.
54. Singer, S. J., and G. L. Nicolson. The fluid mosaic model of the structure of cell membranes. *Science* 175:720–731. 1972.
55. Sitte, P. *Bau und Funktion der Pflanzen-*

zelle. Stuttgart, Gustav Fischer. 1965.

56. Steward, F. C., and R. L. Mott. Cells, solutes, and growth: salt accumulation in plants reexamined. *Internatl. Rev. Cytol.* 28:275–370. 1970.
57. Thomson, W. W. Ultrastructural development of chromoplasts in Valencia oranges. *Bot. Gaz.* 127:133–139. 1966.
58. Thomson, W. W., L. N. Lewis, and C. W. Coggins. The reversion of chromoplasts to chloroplasts in Valencia oranges. *Cytologia* 32:117–124. 1967.
59. Tilney, L. G. Origin and continuity of microtubules. In: *Origin and continuity of cell organelles.* J. Reinert and H. Ursprung, eds. pp. 222–260. New York, Springer. 1971.
60. Tolbert, N. E. Microbodies—peroxisomes and glyoxysomes. *Ann. Rev. Plant Physiol.* 22:45–74. 1971.
61. Wang, A. Y.-I., and L. Packer. Mobility of membrane particles in chloroplasts. *Biochem. Biophys. Acta* 305:488–492. 1973.
62. Weier, T. E., C. R. Stocking, and L. K. Shumway. The photosynthetic apparatus in chloroplasts of higher plants. *Brookhaven Symposia in Biology* No. 19:353–374. 1966.
63. Wergin, W. P., P. J. Gruber, and E. H. Newcomb. Fine structural investigation of nuclear inclusions in plants. *J. Ultrastruct. Res.* 30:533–557. 1970.
64. Wilson, C. L. A lysosomal concept for plant pathology. *Ann. Rev. Plant Pathol.* 11:247–272. 1973.
65. Wilson, G. B., and J. H. Morrison. *Cytology.* New York, Reinhold. 1961.

4

Cell Wall

The cell wall is a typical component of a plant cell. Because of the presence of walls, the distention of the protoplast by the osmotically active vacuole is restricted, and the size and shape of the cell becomes fixed at maturity. The kind of cell walls present determines the texture of a tissue. In peripheral tissues, cell walls contain materials that protect the underlying cells from desiccation. Walls serve as mechanical support of plant organs, especially the thick, rigid walls. Cell walls have an effect on such important activities of plant tissues as absorption, transpiration, translocation, and secretion.

MACROMOLECULAR COMPONENTS AND THEIR ORGANIZATION IN THE WALL

The principal compound in plant cell walls is *cellulose,* a polysaccharide with the empirical formula $(C_6H_{10}O_5)n$. Its molecules are linear chains of glucose, which may reach four microns in length. In the cell wall, cellulose is associated with other polysaccharides, the *hemicelluloses* and *pectic substances* (polyuronide compounds). *Lignin,* a polymer of phenylpropanoid units, incrusts the walls of many kinds of cells. Lignin is a complex, heterogeneous substance, which imparts rigidity to the cell wall.[20] Many other substances, organic and inorganic, as well as water, are present in cell walls in amounts varying in relation to the nature of the cell. Among the organic substances, *cutin, suberin,* and *waxes* are fatty substances that are found most commonly in the protective surface tissues of the plant, cutin in the epidermis, suberin in the secondary protective tissue, the cork (phellem). Waxes occur in combination with cutin and suberin and also on the surface of the *cuticle,* that is, the layer of cutin covering the outer wall of the epidermis.

The architecture of the cell wall is largely determined by cellulose. This carbohydrate forms the framework interpenetrated by the *matrix* represented by the noncellulosic carbohydrates. Some hemicelluloses appear to serve as an important cross-link between the noncellulosic polymers and the cellulose.[12]

The incrusting substances, such as lignin or suberin, are deposited in the matrix. According to an electron microscope study, in which lignin was identified by its selective reaction to the potassium permanaganate fixative, the compound appeared to fill the spaces within the fibrillar cellulose framework.[11] During deposition, lignin becomes chemically linked to the polysaccharides.[7]

The cellulose framework is a system of fibrils composed of cellulose molecules (fig. 4.1). The fibrils are of different classes of magnitude. The largest are visible with a light microscope and are called *macrofibrils.* With an electron microscope, these fibrils are resolvable into about 100 Å wide *microfibrils*[22] (fig. 4.2). With the increase in resolving power of electron microscopes, smaller and smaller fibrils are visualized and reported as subunits of the microfibril.[10]

Microfibrils display a dense, textile-like pattern in electron microscope preparations. This appearance is caused, in part, by dehydration of walls during preparation for microscopy (fig. 4.2). In walls of fresh tissues, the microfibrils are farther apart from each other.

Cellulose has crystalline properties as a result of orderly arrangement of cellulose molecules in the microfibrils. Such arrangement is restricted to parts of microfibrils that are referred to as *micelles* (fig. 4.1). Less regularly arranged glucose chains occur between and around the micelles and constitute the paracrystalline regions of the microfibril. The crystalline structure of cellulose makes the cell wall anisotropic and consequently doubly refractive (birefringent) when viewed in polarized light (fig. 4.3).

Although the cell wall may be regarded as an ergastic product of the protoplast, it maintains a close relation to the cytoplasm. Proteins, commonly containing the rare amino acid hydroxyproline, are known to be constituents of primary walls.[13] The protein and the polysaccharides of the wall are visualized as forming a glycoprotein by covalent attachment. Increase in amount of this component in the wall has been found to have a causal relation to the cessation of extension growth of walls.[27]

Cell walls contain enzymes, which could be concerned with synthesis, transfer, and hydrolysis of cell wall macromolecules, as well as with a modification of extracellular metabolites that facilitates the transport of these metabolites into the cell. Cytochemical localization of enzymes in cell walls was successful in tests for acid phosphatase[9] and peroxidase.[8]

CELL WALL LAYERS

The method of growth of cell wall, the amount of this growth, and the arrangement of microfibrils in the successive increments of the wall produce a more or less pronounced layering. Each protoplast forms its wall from the outside inward so that the oldest layer of a given wall is in the outermost position in the cell, the most recent in the innermost position, next to the protoplast. The layers formed first constitute the *primary wall.* In many cell types, additional wall layers are deposited; these form the *secondary wall* (fig. 4.7).

The region of union of the primary walls of two contiguous cells is called *middle lamella* (also *intercellular lamella* and *intercellular substance*). Electron microscopy rarely reveals the middle lamella as a well-delimited layer except along the corners of cells where the intercellular material is most abundant (fig. 4.4). Recognition of the middle lamella is based chiefly on microchemical tests and maceration techniques. The middle lamella is largely pectic in nature but often becomes lignified in older cells.

In thick-walled wood cells, the secondary

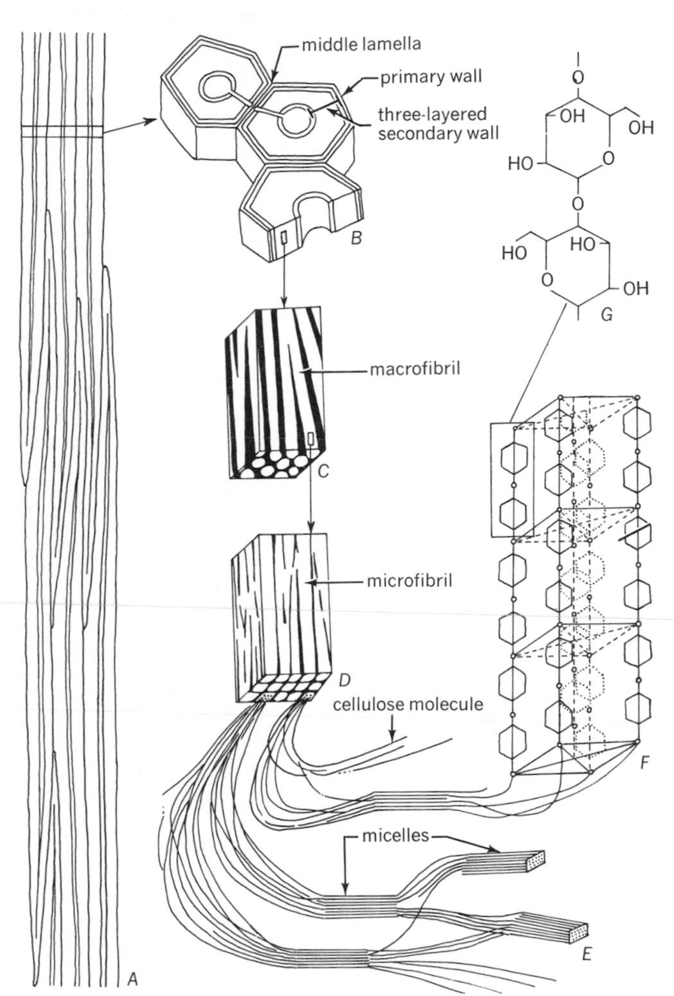

Figure 4.1 Detailed structure of cell walls. *A*, strand of fiber cells. *B*, cross section of fiber cells showing gross layering: a layer of primary wall and three layers of secondary wall. *C*, fragment from middle layer of secondary wall showing macrofibrils (white) of cellulose and interfibrillar spaces (black), which are filled with noncellulosic materials. *D*, fragment of a macrofibril showing microfibrils (white), which may be seen in electron micrographs (fig. 4.2). The spaces among microfibrils (black) are filled with noncellulosic materials. *E*, structure of microfibrils: chainlike molecules of cellulose, which in some parts of microfibrils are orderly arranged. These parts are the micelles. *F*, fragment of a micelle showing parts of chainlike cellulose molecules arranged in a space lattice. *G*, two glucose residues connected by an oxygen atom—a fragment of a cellulose molecule.

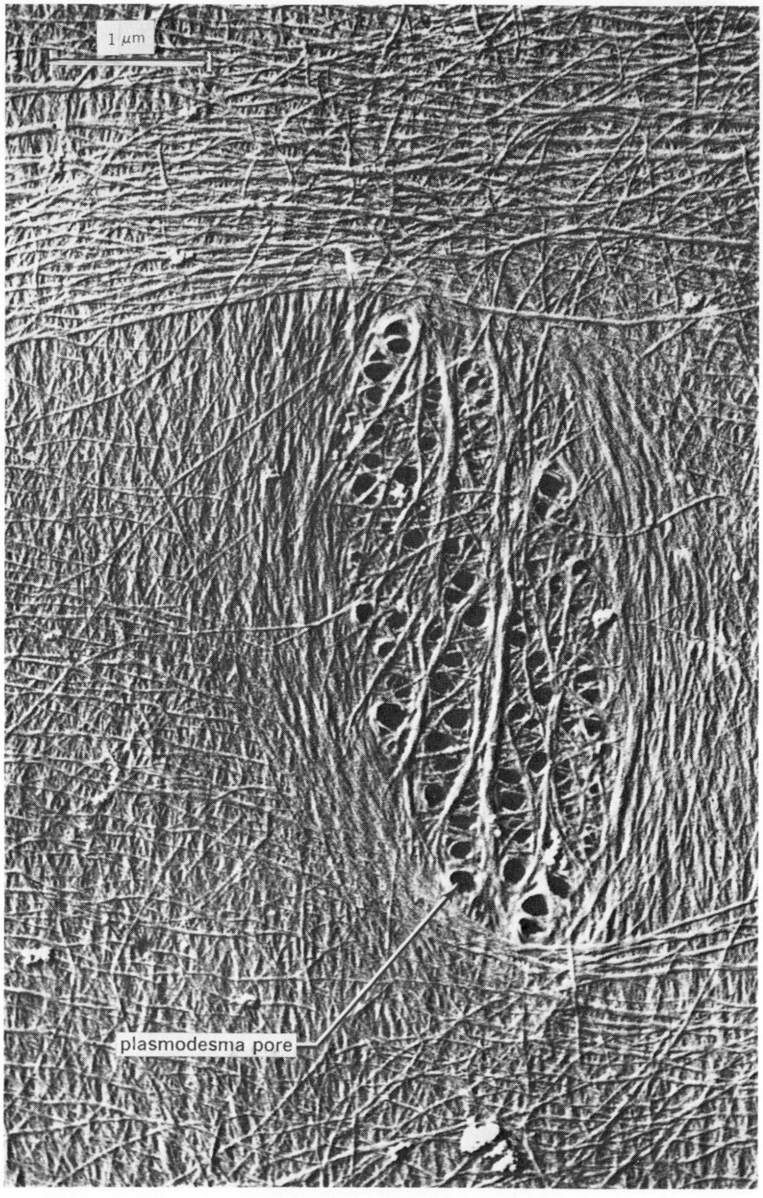

Figure 4.2 Electron micrograph of primary wall of a parenchyma cell from *Avena* coleoptile. The longitudinal axis of the cell was in the direction of the 1μm scale. The parallel-oriented microfibrils occurred in one of the angles of the cell. Elsewhere the microfibrils are less definitely oriented. The plasmodesmata pores are grouped in an oval area—a primordial pit. (From H. Böhmer. Untersuchungen über das Wachstum und den Feinbau der Zellwände in der *Avena*-Koleoptile. *Planta* 50:461–497, 1958.)

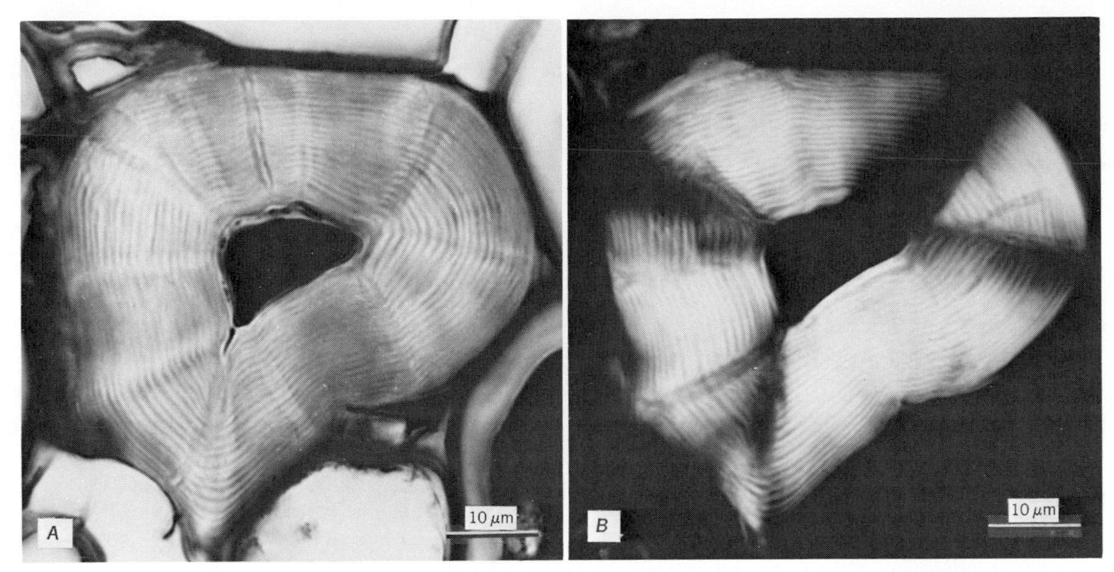

Figure 4.3 Sclereid from root cortex of fir (*Abies*), as seen with nonpolarized (*A*) and polarized (*B*) light. Because of crystalline nature of cellulose, cell wall shows double refraction and appears bright in polarized light (*B*). Wall has concentric lamellation.

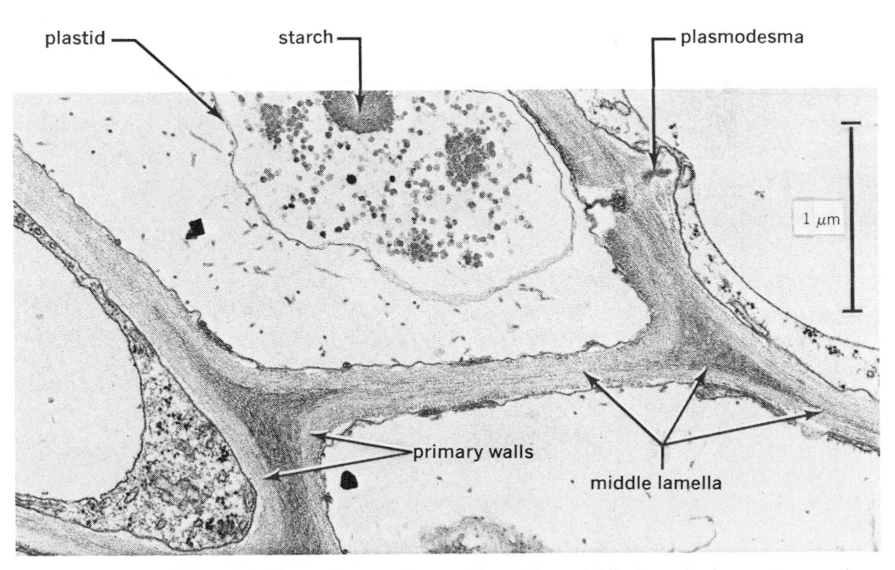

Figure 4.4 Parts of cells from the phloem tissue in a petiole of *Nelumbo nucifera*. The middle lamella is most conspicuous in cell corners; is faintly discernible elsewhere. Parts of a plasmodesma occur in a thickened wall region. The plastid is in a sieve element.

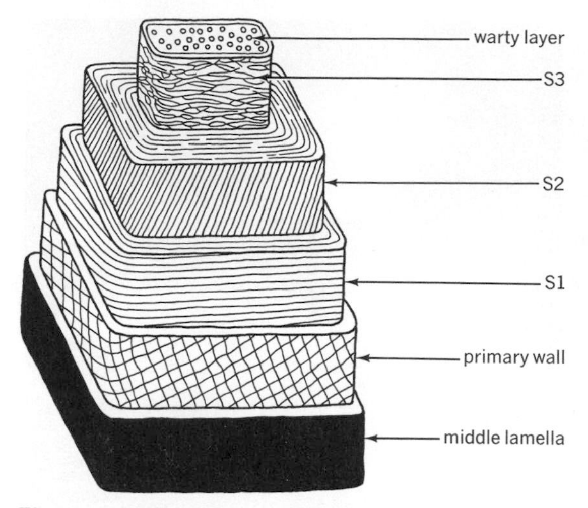

Figure 4.5 Diagram of a piece of tracheid wall illustrating the layers and their microfibrillar organization. *S* refers to secondary wall. *S*3 is sometimes interpreted as a tertiary wall layer. (Adapted from Liese.[14])

wall frequently consists of three major layers (figs. 4.1 and 4.5). Beginning with the outermost, these layers are designated *S*1, *S*2, and *S*3. The *S*2 layer is the thickest. The *S*3 layer may be very thin or lacking entirely. Some wood anatomists consider the *S*3 layer sufficiently distinct from *S*1 and *S*2 layers to be called *tertiary wall.*[14] The distinction between the various layers is often obscured, especially that between the primary wall and the middle lamella. In cells with secondary walls, the two adjacent primary walls and the middle lamella may appear as one layer, the *compound middle lamella.*

The separation of the secondary wall into the three *S* layers results mainly from different orientations of microfibrils in the three layers. Typically, microfibrils are helically oriented in the cell wall (fig. 4.5). In *S*1, the slope of the helix makes a large angle with the long axis of the cell so that the disposition of the microfibrils is nearly horizontal. In *S*2, the angle is small and the slope of the helix steep. In *S*3, the microfibrils are deposited as in *S*1, at a large angle to the long axis of the cell. The primary wall differs from the secondary in having a rather random arrangement of microfibrils. In fibers and tracheids of most woody species, the inner surface of the *S*3 layer is covered with a noncellulosic film often bearing lumps called warts. This *warty layer* is thought to be composed of remnants of a disorganized protoplast.[2]

The primary wall is usually thin in cells having secondary walls. It is also relatively thin in various kinds of metabolically active parenchyma cells, for example, mesophyll cells in the leaf and storage parenchyma cells in roots and tubers. But it may attain considerable thickness as in collenchyma of stems and leaves and endosperm of some seeds. Thick primary walls may show layering caused by variations in relative amounts of cellulose, noncellulosic components, and water in the different increments of the wall.

INTERCELLULAR SPACES

A large volume of the plant body is occupied by an intercellular space system. Although the intercellular spaces are most characteristic of mature tissues (fig. 4.6), they also extend into meristematic tissues where the dividing cells are intensively respiring. Examples of tissues having large and well-interconnected intercellular spaces are found in foliage leaves (chapter 18) and submerged organs of water plants.

The most common intercellular spaces develop by separation of contiguous primary walls through the middle lamella. The process starts in the corner, where more than two cells are joined, and spreads to other wall parts. This type of intercellular space is called *schizogenous,* that is, arising by splitting, although most likely an enzymic removal

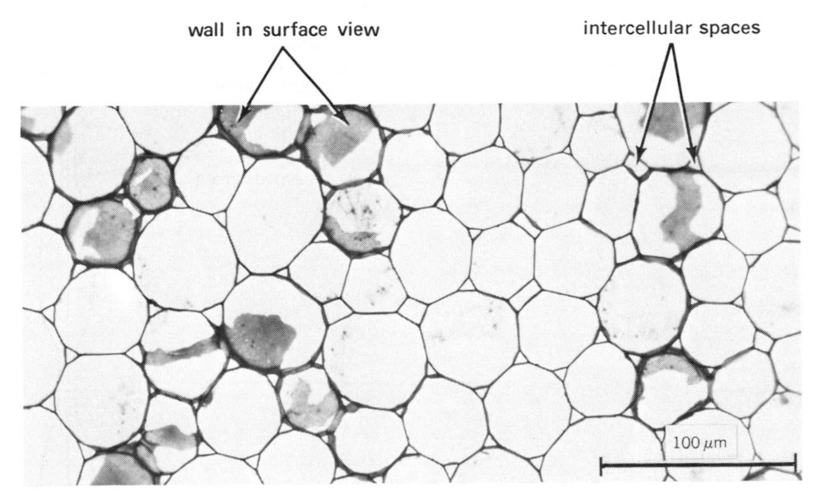

Figure 4.6 A thin-walled type of parenchyma, with regularly shaped cells and schizogenous intercellular spaces; from petiole of celery (*Apium*).

of pectins is involved. Electron microscopy indicates that the formation of intercellular spaces is a complex phenomenon, for an accumulation of membranous structures may precede the development of cavities between cells.[4] Some intercellular spaces result from a breakdown of entire cells and are called *lysigenous* (arising by dissolution). Some roots develop extensive lysigenous intercellular spaces. Intercellular spaces of both kinds may also serve as containers for various secreted materials (chapter 13). Schizogeny and lysigeny may be combined in the formation of spaces.

PITS, PRIMARY PIT-FIELDS, AND PLASMODESMATA

Definitions and structure

Walls with secondary layers show characteristic depressions called *pits* (fig. 4.7,*B*). The pits of two contiguous cells usually oppose one another. The two opposing pits together are called *pit-pair.* Each pit of a pair has a *pit cavity,* and the two cavities are separated from each other by a thin wall part, the *pit membrane.* Pits arise during ontogeny of the cell wall and result from differential deposition of secondary wall material; none is deposited over the pit membrane so that the pits are actual discontinuities in the secondary wall. In a pit-pair, the pit membrane consists of two primary walls and the middle lamella (fig. 4.7,*B*).

Primary walls also may have depressions, the *primary pit-fields* or *primordial pits.*[3] In this book, these structures are also called *primary pits.* The primordial pit is a thin place in the wall penetrated by plasmodesmata (fig. 4.7,*A*). No interruptions occur in the primary wall in the pit-field region except where plasmodesmatal canals are present; that is, the primary wall is continuous over the pit-field membrane. During the deposition of secondary wall the pits in that wall are formed over the primary pit-fields. The use of the word field in the term pit-field takes into account the fact that in the secondary wall several pits may arise over one primordial pit.

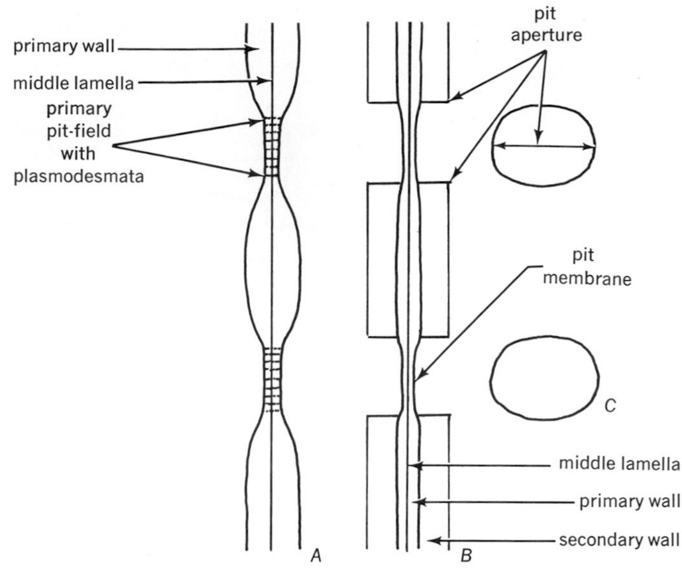

Figure 4.7 Primary pit-fields (*A*) and pits (*B*). *A*, cell wall composed of middle lamella and two primary wall layers. Plasmodesmata traverse the pit membranes of the pit-fields. *B*, wall composed of middle lamella, two primary wall layers, and two secondary wall layers. *C*, outlines of pits of the type shown in *B* as they would appear in a surface view of the wall.

A physiologically significant feature of primary pit-pairs are the plasmodesmata traversing the pit membrane (figs. 4.7,*A*, and 4.8). As was mentioned in chapter 3, plasmodesmata are thought to provide cytoplasmic continuity between adjacent cells. The plasmodesmatal canal in the wall is lined with plasmalemma and its center is occupied by the desmotubule continuous with ER cisternae lodged against the plasmodesmatal openings. Cytoplasmic matrix fills the rest of the canal. The relation of the structural features to the function of plasmodesmata as connections between protoplasts is not yet fully understood.[24]

When a secondary wall develops, the plasmodesmata remain in the pit membrane as connections between the cytoplasmic masses filling the pit cavities in the secondary wall. As the latter continues to thicken, the cavities become canals. A decrease of the inner circumference of the growing wall may cause a fusion of adjacent pit canals. Thus, the so called *branched* or *ramiform pit* develops (chapter 6). If the cell becomes devoid of a living protoplast at maturity, the plasmodesmata and the cytoplasm in the pit-pair cavities disappear.

Plasmodesmata are not restricted to pits. A scattering of plasmodesmata through a wall of uniform thickness is of common occurrence. Moreover, electron microscopy reveals that in many instances the primary wall is thickened specifically where plasmodesmata occur (fig. 4.4). Plasmodesmata may be branched on one or both sides of the middle lamella. A cavity is frequently formed in the middle lamella region of the plasmodesmatal canal (fig. 4.8,*A*,*C*). The *median cavity* is particularly large when plasmodesmata are branched because the branches are often interconnected through the cavity. For a time, plasmodesmata were thought to occur in outer epidermal walls. They appeared as

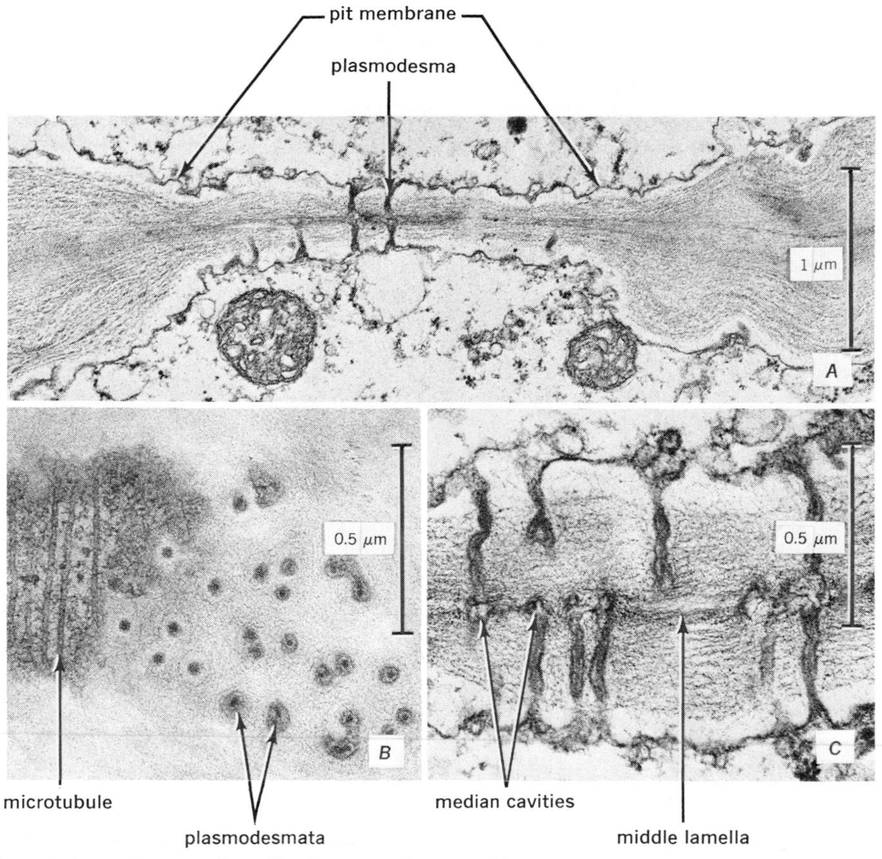

Figure 4.8 Primary pit-fields and plasmodesmata in walls of a parenchyma cell from a petiole of *Mimosa pudica*. Plasmodesmata appear in longitudinal sections in *A* and *C*, in transverse section in *B*. To the left in *B*, the section barely touched the wall and passed through a group of microtubules. Desmotubules discernible in *B* and *C*.

lines demonstrable with certain dyes and were named *ectodesmata*. Subsequent research has shown that cytoplasmic strands do not occur in outer epidermal walls but that channels filled with a coarse reticulum of cellulose fibrils may extend from plasmalemma to the cuticle and serve as polar pathways in foliar absorption and excretion.[5,16] Franke[6] has proposed a substitution of the term *teichode* (from the Greek words teichos, wall, and hodos, path) for ectodesma in order to remove the implication of cytoplasmic nature of the pathway.

Types of pits

Pits vary in size and detailed structure (chapters 6 and 8). The secondary wall may end abruptly at the pit cavity, which thus retains approximately the same diameter through the depth of the secondary wall. This kind of pit is called *simple pit* and the combination of two simple pits, *simple pit-pair* (fig. 4.7,*B,C*). The secondary wall may overarch the pit cavity, forming a *border*. A *bordered pit* or *bordered pit-pair* is the result (fig. 4.9,*A*). The pit cavity enclosed by the border opens into the cell lumen through a discontinuity in the border, the *pit aperture* (figs. 4.9,*A–D*, and 4.10,*A*). Combinations of bordered and simple pits, called *half-bordered pit-pairs,* are found in the xylem (fig. 4.9,*D,E*).

In gymnosperm tracheids, especially in those of Pinaceae, the membrane of a bordered pit-pair shows a highly specialized structure. A thickening in the middle of the membrane forms the *torus* (fig. 4.9,*A*), whereas the surrounding part of the membrane, the *margo,* consists of bundles of microfibrils, most of them radiating from the torus (fig. 4.10,*B*). The open structure of the

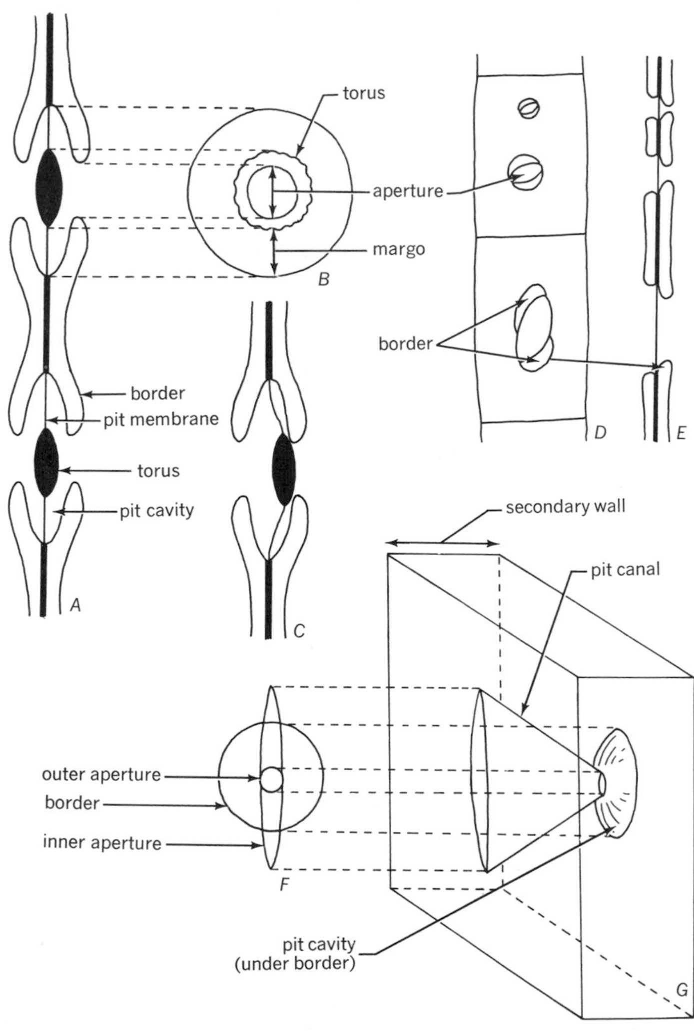

Figure 4.9 Diagrams of bordered and half-bordered pit-pairs. *A*, two bordered pit-pairs, each with a torus, in side view. *B*, bordered pit in surface view. *C*, aspirated bordered pit-pair. *D, E,* half-bordered pit-pairs in surface (*D*) and side (*E*) views. *F, G,* bordered pit with extended inner aperture and reduced border. (*F, G,* after S. J. Record, *Timbers of North America,* John Wiley & Sons, 1934.)

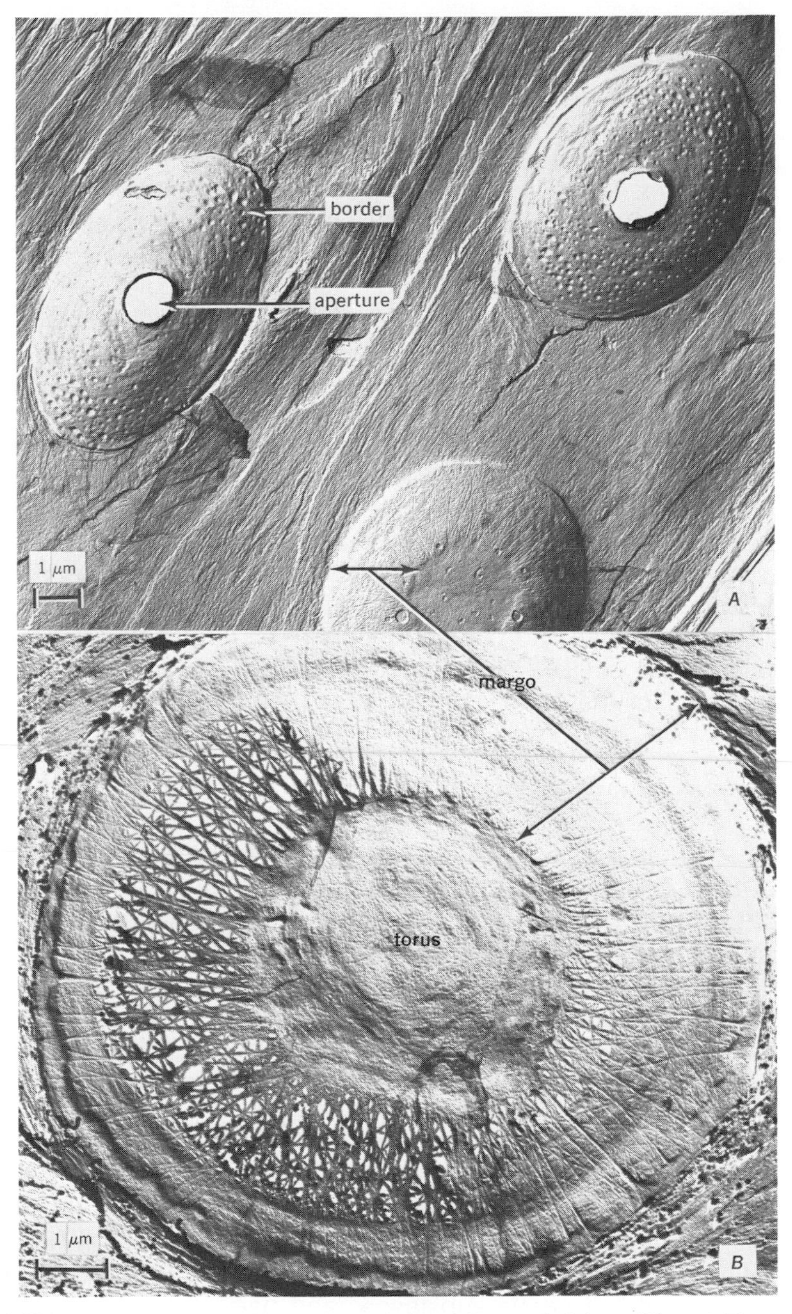

Figure 4.10 Bordered pits in early wood tracheid of *Pinus virginiana* (*A*) and *Pinus pungens* (*B*). The border with its aperture is present in the upper two pits in A. In the lower pit in *A* and in the pit in *B*, the border was cut away and the pit membrane is exposed. The solid torus contrasts with the fibrillar network of the margo. (Courtesy of W. A. Côté, Jr.)

margo results from removal of the noncellulosic matrix of the primary wall and middle lamella.[14] The margo is flexible and under certain conditions of stress (chapter 8) it moves toward one or the other side of the border, closing the aperture by means of the torus (fig. 4.9,*C*). In this condition the pit is not functioning in conduction and is called *aspirated.*

If the secondary wall is very thick, the pit border is correspondingly thick (fig. 4.9,*F,G*). The cavity of such a pit is rather small and is connected with the cell lumen through a narrow passage in the border, the *pit canal.* The canal has an *outer aperture* facing the pit cavity and an *inner aperture* facing the lumen of the cell. In certain pits, the pit canal resembles a compressed funnel, and its two apertures differ in size and form. The outer aperture is small and circular, the inner extended and slitlike. In a pit-pair, the inner apertures of the two pits are crossed with respect to each other (chapter 8). This arrangement is related to the helical deposition of microfibrils in the secondary wall.

ORIGIN OF CELL WALL DURING CELL DIVISION

During vegetative growth, cell division (*cytokinesis*) usually follows nuclear division (*karyokinesis;* fig. 4.11). The mother cell divides in two daughter cells. Because the nature of the initial partition between the daughter cells is not perfectly known, the partition is called *cell plate* (fig. 4.12–4.14). Since, imperceptibly, the cell plate becomes a cell wall, it may be regarded as the first layer of cell wall. The assumption that the cell plate is composed of pectic substances and becomes the middle lamella between the primary walls of the two daughter cells has not been fully investigated.

The cell plate arises through fusion of vesicles which are deposited in the equatorial plane of the *phragmoplast,* an assemblage of microtubules extending between the two daughter nuclei (figs. 4.11,*A,* and 4.13). In somatic mitosis, the formations of the mitotic spindle and of the phragmoplast are closely integrated so that the spindle and the phragmoplast appear to share the same microtubules, although new microtubules are added to the phragmoplast before cell plate formation is completed. The cell plate is initiated as a disc suspended in the phragmoplast (figs. 4.11,*A*, and 4.12). At this stage, the phragmoplast does not extend to the walls of the mother cell and, consequently, the cell plate is isolated from these walls. Phragmoplast microtubules disappear where the cell plate has been formed but are successively regenerated at the free margins of the cell plate (figs. 4.11,*B,C,* and 4.14,*A*). The expanding phragmoplast enables the plate to extend laterally until it joins the mother cell walls (fig. 4.14,*B*).

According to a prevalent view,[19] the vesicles forming the cell plate are derived from dictyosomes in the vicinity of the phragmoplast (fig. 4.14,*A*), but ER vesicles also may participate in cell plate growth. The microtubules of the phragmoplast appear to be involved in directing the vesicles toward the equatorial region. The dictyosome-derived vesicles are carrying polysaccharides, including pectic substances, that become the building materials of the cell plate. When the vesicles fuse, their membranes become the plasmalemma. Fusion of vesicles into the cell plate leaves small gaps, the plasmodesmatal canals (fig. 4.14,*B*). These canals are lined with plasmalemma from their inception. ER tubules are sometimes seen entrapped in these gaps.

Before the cell divides, the nucleus assumes a proper position for the event. If the cell is conspicuously vacuolated, a layer of cytoplasm, the *phragmosome,* spreads

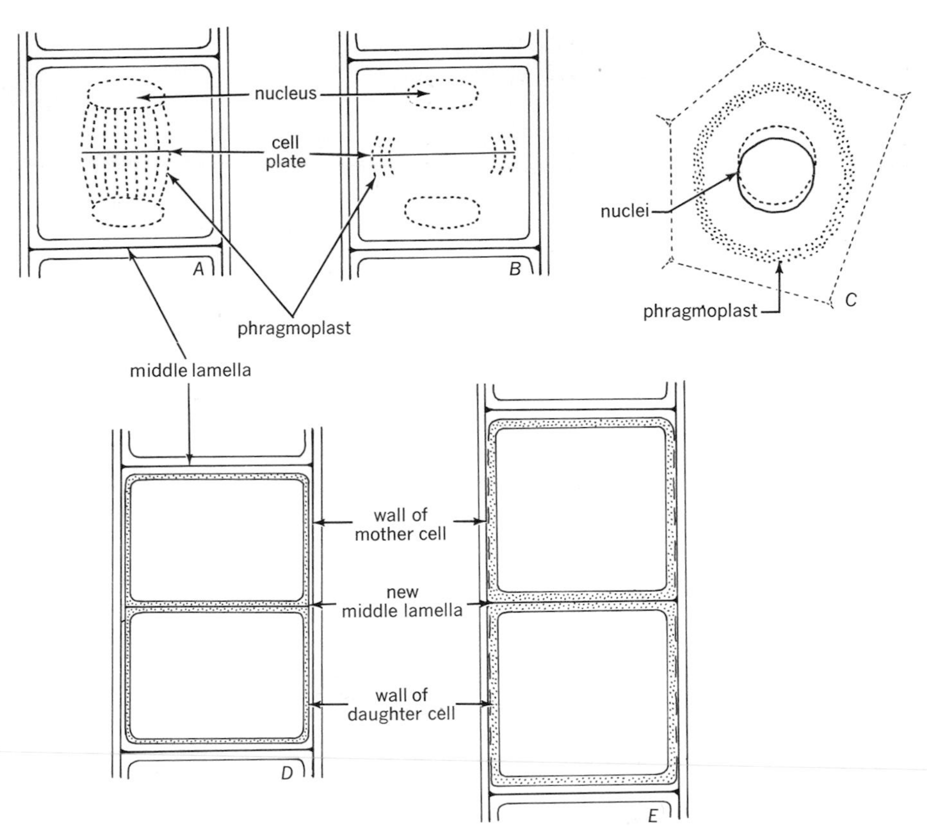

Figure 4.11 Formation of wall during cell division. *A*, formation of cell plate in the equatorial plane of phragmoplast at telophase. *B, C*, phragmoplast now appears along the margin of the circular cell plate (in side view in *B;* in surface view in *C*). *D*, cell division is completed and each sister cell has formed its own primary wall (stippled). *E*, sister cells have enlarged, their primary walls have thickened, and the mother cell wall has been torn along the vertical sides of cells.

across the plane of future division and the nucleus becomes located in this layer. Microtubules may be involved in the positioning of the nucleus, for they form a ringlike band, *preprophase band,*[21] several layers in depth, outlining the equatorial plane of the future mitotic spindle and of the phragmoplast.

GROWTH OF CELL WALL

The fusion of vesicles into the cell plate is followed by deposition of additional wall material on either side of the original plate as is indicated by the increase in thickness of the new partition. New primary wall material is also deposited over the old mother cell wall so that each daughter cell forms a complete primary wall (fig. 4.11,*D,E*). Dictyosome-derived vesicles continue to be involved in wall growth in both primary and secondary stages of wall development. Autoradiographic studies have led to the concept that dictyosomes are the source of polysaccharide wall material serving as the matrix, whereas cellulose is synthesized in close association with the plasmalemma.[18,26,29] Experimental work and electron microscopy have given in-

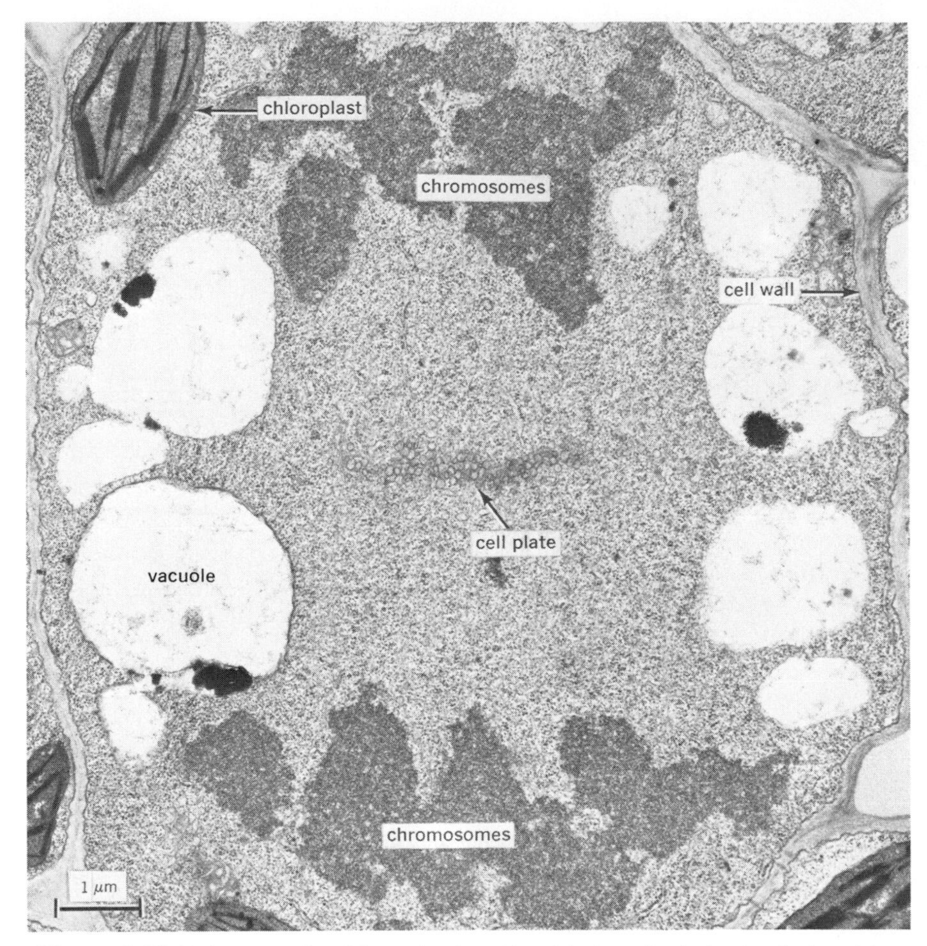

Figure 4.12 Early stage of wall formation (cytokinesis) after mitosis (early telophase) in a *Nicotiana tabacum* mesophyll cell. A layer of vesicles in the midplane between the two sets of daughter chromosomes forms an incomplete cell plate. In its further growth, the cell plate would have traversed the cytoplasm between the vacuoles and reached the mother cell wall.

dications of the importance of microtubules for the orderly growth of the cell wall (chapter 3), including direction of dictyosome-derived vesicles toward the wall[15] and control of the alignment of microfibrils in the wall.[17]

According to the classical concept, growth of wall in thickness occurs by two methods of deposition of wall material, *apposition* and *intussusception.* In apposition, the building units are placed one on top of another, in intussusception the units of new material are inserted into the existing structure. Intussusception is probably the rule when lignin or cutin are incorporated into the wall. With regard to microfibrils of cellulose, intussusception would result in an interweaving of the fibrils. In some walls, microfibrils appear to be interwoven, in others their arrangement suggests appositional growth.[2]

Cell walls grow not only in thickness but

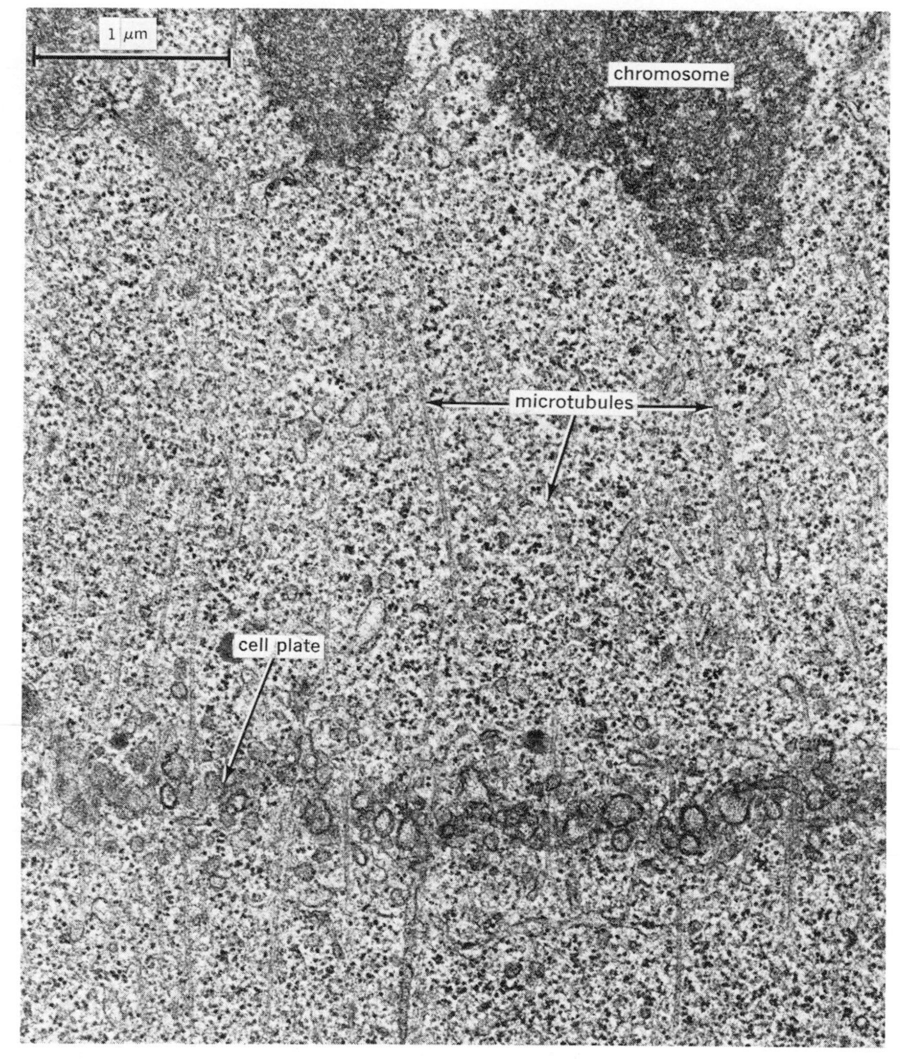

Figure 4.13 Details of early cytokinesis in a *Nicotiana tabacum* mesophyll cell. Cell plate is still composed of individual vesicles. Phragmoplast microtubules occur on both sides of the cell plate, some traversing the plate. Some chromosome material of one of the two future daughter nuclei is shown.

also in surface. The extension growth of the wall is a complex process that has not been fully explained.[1,23] It requires a loosening of the wall structure, a phenomenon that can be regulated by supply of auxin, turgor pressure, protein synthesis, and respiration. Thus, studies on extension growth of walls stress the dependence of the process on activities of a living protoplast. Accordingly, investigators concerned with cell wall growth emphasize the discoveries of proteins and enzymes within the walls.

Surface or extension growth of walls occurs in cells that are still increasing in size. These

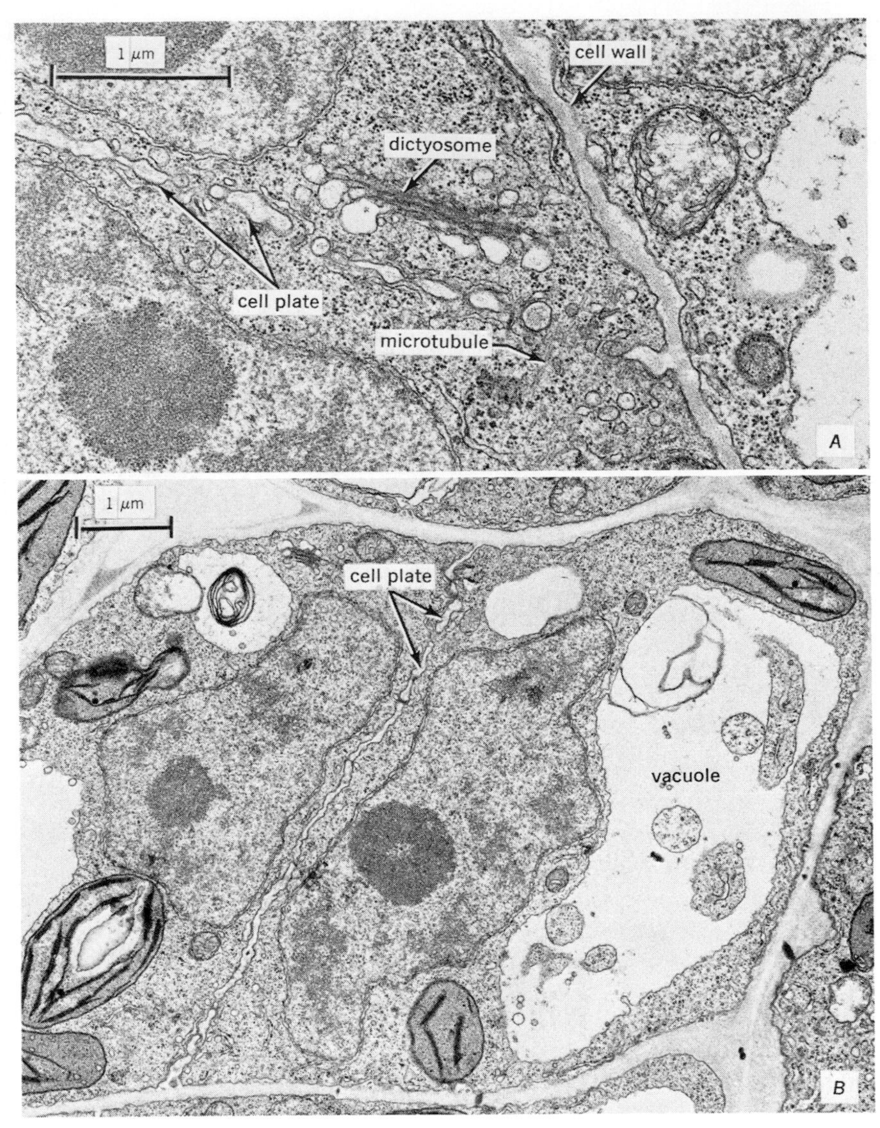

Figure 4.14 Completion of cytokinesis in mesophyll cells of sugar beet (*Beta vulgaris*). In *A*, vesicles of the cell plate have fused in groups, leaving large interruptions in the plate. At right, the cell plate has not yet reached the mother cell wall, which has formed a protrusion toward the advancing cell plate. Some phragmoplast microtubules are discernible in the gap region. A dictyosome with recently produced vesicles is near the plate. In *B*, the cell plate is complete. The remaining discontinuities are potential plasmodesmatal canals. Vacuoles contain cytoplasmic material, an indication of autophagy.

cells have nonlignified primary walls with a relatively small amount of cellulose. The microfibrils respond to wall extension by change in their orientation. From an initially almost horizontal position they are reoriented to a more upright one (according to Veen,[28] from a flat helix to a steep one). The subsequent layers deposited over the older layers are stretched to a successively smaller degree. The primary wall thus appears as a stack of nets with different orientations and different densities of microfibrils in the successive nets in the stack. This interpretation of wall growth is referred to as the multinet theory of wall growth.[25] The hypothesis has found wide acceptance but there are some primary walls that appear to require a different interpretation regarding their growth.[22]

REFERENCES

1. Cleland, R. Cell wall extension. *Ann. Rev. Plant Physiol.* 22:197–222. 1971.
2. Clowes, F. A. L., and B. E. Juniper. *Plant cells.* Oxford, Blackwell Scientific Publications. 1968.
3. Committee on Nomenclature. International Society of Wood Anatomists. International glossary of terms used in wood anatomy. *Trop. Woods* 107:1–36. 1957.
4. Fagerlind, F., and A. Massalski. The development of cell walls and intercellularies in the root of *Lemna minor* L. *Svensk. Bot. Tidskr.* 68:64–93. 1974.
5. Franke, W. Mechanisms of foliar penetration of solutions. *Ann. Rev. Plant Physiol.* 18:281–300. 1967.
6. Franke, W. Über die Natur der Ektodesmen und einen Vorschlag zur Terminologie. *Ber. Deut. Bot. Ges.* 84:533–537. 1971.
7. Freudenberg, K. Lignin: Its constitution and formation from p-hydroxycinnamyl alcohols. *Science* 148:595–600. 1965.
8. Goff, C. W. A light and electron microscopic study of peroxidase localization in the onion root tip. *Amer. J. Bot.* 62:280–291. 1975.
9. Halperin, W. Ultrastructural localization of acid phosphatase in cultured cells of *Daucus carota. Planta* 88:91–102. 1969.
10. Hanna, R. B., and W. A. Côté, Jr. The sub-elementary fibril of plant cell wall cellulose. *Cytobiologie* 10:102–116. 1974.
11. Hepler, P. K., D. E. Fosket, and E. H. Newcomb. Lignification during secondary wall formation in *Coleus:* an electron microscopic study. *Amer. J. Bot.* 57:85–96. 1970.
12. Keegstra, K., K. W. Talmadge, W. D. Bauer, and P. Albersheim. The structure of plant cell walls. III. A model of the walls of suspension-cultured sycamore cells based on the interconnections of the macromolecular components. *Plant Physiol.* 51:188–196. 1973.
13. Lamport, D. T. A. Cell wall metabolism. *Ann. Rev. Plant Physiol.* 21:235–270. 1970.
14. Liese, W. Elektronenmikroskopie des Holzes. *Handbuch der Mikroskopie in der Technik.* Band V. Teil 1. pp. 109–170. Frankfurt am Main, Umschau Verlag. 1970.
15. Maitra, S. C., and D. N. De. Role of microtubules in secondary thickening of differentiating xylem element. *J. Ultrastruct. Res.* 34:15–22. 1971.
16. Merkens, W. S. W., G. A. de Zoeten, and G. Gaard. Observations on ectodesmata and the virus infection process. *J. Ultrastruct. Res.* 41:397–405. 1972.
17. Newcomb, E. H. Plant microtubules.

Ann. Rev. Plant Physiol. 20:253–288. 1969.

18. Northcote, D. H. The Golgi apparatus. *Endeavour* 30:26–33. 1971.
19. O'Brien, T. P. The cytology of cell-wall formation in some eukaryotic cells. *Bot. Rev.* 38:87–118. 1972.
20. Pearl, I. A. *The chemistry of lignin.* New York, Dekker. 1967.
21. Pickett-Heaps, J. D., and D. H. Northcote. Organization of microtubules and endoplasmic reticulum during mitosis and cytokinesis in wheat meristems. *J. Cell Sci.* 1:109–120. 1966.
22. Preston, R. D. Plant cell walls. In: *Dynamic aspects of plant ultrastructure.* A. W. Robards, ed. Chapter 7, pp. 256–309. London, McGraw-Hill Book Company (UK) Limited. 1974.
23. Ray, P. M. The action of auxin on cell enlargement in plants. *Devel. Biol.* Suppl. 3:172–205. 1969.
24. Robards, A. W. Plasmodesmata. *Ann. Rev. Plant Physiol.* 26:13–29. 1975.
25. Roelofsen, P. The plant cell wall. *Handbuch der Pflanzenanatomie.* Band 3. Teil 4. 1959.
26. Roland, J.-C. The relationship between the plasmalemma and plant cell wall. *Internatl. Rev. Cytol.* 36:45–92. 1973.
27. Sadava, D., and M. J. Chrispeels. Hydroxyproline-rich cell wall protein (extensin): role in the cessation of elongation in excised pea epicotyls. *Devel. Biol.* 30:49–55. 1973.
28. Veen, B. W. Orientation of microfibrils in parenchyma cells of pea stem before and after longitudinal growth. *Proc. Koninkl. Nederl. Akad. Wetenschap. Amsterdam* Ser. C. 73:114–117. 1970.
29. Wilson, J. H. M., and E. C. Cocking. Microfibril synthesis at the surfaces of isolated tobacco mesophyll protoplasts, a freeze-etch study. *Protoplasma* 84:147–159. 1975.

5

Parenchyma and Collenchyma

PARENCHYMA

Parenchyma is the main representative of the ground tissue system and is found in plant organs forming a continuous tissue, as in the cortex and pith of stems, cortex of roots, ground tissue of petioles, mesophyll of leaves, or individual or groups of cells in the complex tissue systems, xylem and phloem.

Parenchyma cells carry on a variety of functions in relation to their position in the plant body and to their participation in the activities of other cells. Although parenchyma cells may be more or less highly specialized in their functions they are capable of changing their activities. This physiologic plasticity is determined by the presence of a complete protoplast in the cell. The protoplast is, in fact, able to perform numerous functions at the same time because its various membrane systems produce extensive compartmentalization (chapter 3). A parenchyma cell with a normal nucleate protoplast can also resume meristematic activity. The phenomena of wound healing, regeneration, formation of adventitious roots and shoots, and union of grafts are made possible through a resumption of meristematic activity by parenchyma cells. Furthermore, isolated groups of parenchymatic cells, or even single cells, are capable of producing entire plants (chapter 2).

Contents of cells

The variations in activities of parenchyma cells are usually reflected in variations in the balance of protoplasmic components. Parenchyma intensively engaged in photosynthesis contains numerous chloroplasts and is called *chlorenchyma*. This tissue finds its highest expression in the leaf mesophyll (chapter 18), but chloroplasts may be abundant also in the cortex of a stem and may occur in deeper tissues, even in the pith.[22] Photosynthesizing cells are usually conspicuously vacuolated and form a highly lacunate tissue. Secretory types of parenchyma cells often have dense protoplasts, especially rich in ribosomes, and have either numerous dictyosomes or a massively developed endoplasmic reticulum in relation to the type of secretory product formed (chapter 13).

Parenchyma cells may assume distinctive characteristics by accumulating specific ergastic substances. In starch storing cells the plastids have a rather simple internal organization and may be defined as amyloplasts, as in many seeds and underground storage organs. Some seeds store solid protein or fats in the parenchyma of the storage tissue or the embryo. Parenchyma cells in flowers and fruits often contain chromoplasts. In various organs of the plant, parenchyma cells may become conspicuous by accumulating anthocyanins or tannins in their vacuoles or depositing crystals of one or another form.

Shape and arrangement of cells

Parenchyma cells vary in form (fig. 5.1), but typically the ground tissue parenchyma consists of cells that are not much longer than wide and may be nearly isodiametric. However, parenchyma cells may also be considerably elongated or variously lobed (fig. 5.1,*C–E*). Even if the parenchyma cells are approximately isodiametric, they are not spherical but have many facets along which

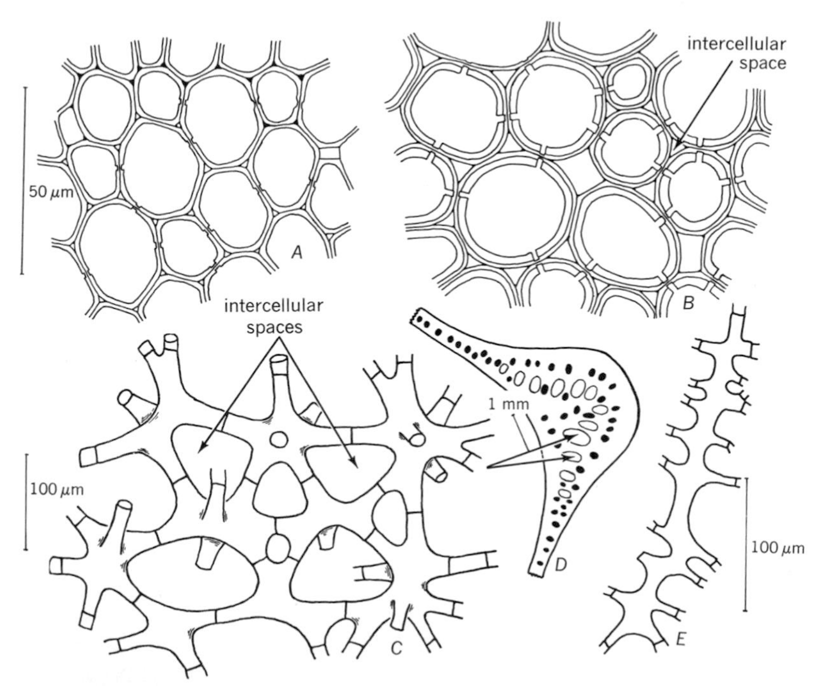

Figure 5.1 Shape and wall structure of parenchyma cells. (Cell contents are omitted.) *A, B,* parenchyma from the stem pith of birch (*Betula*). In younger stem (*A*) the cells have only primary walls; in older (*B*), secondary walls occur also. *C, D,* parenchyma of the aerenchyma type (*C*), which occurs in lacunae of petioles and midribs (*D*) of *Canna* leaves. The cells have many "arms." *E,* long "armed" cell from the mesophyll of a disc flower of *Gaillardia*.

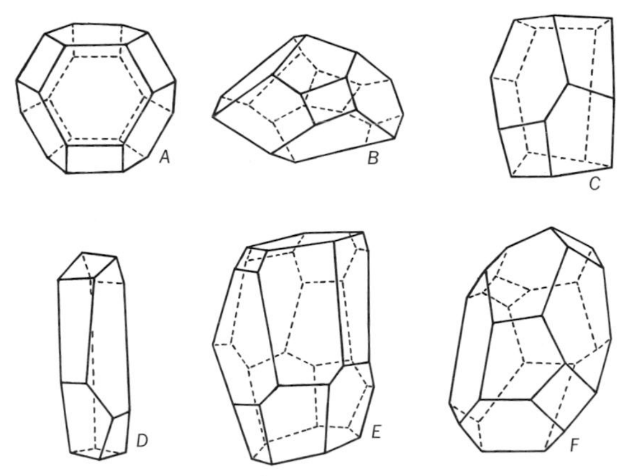

Figure 5.2 The shape of parenchyma cells. *A,* diagram of the orthic tetrakaidecahedron, a 14-sided polyhedron. *B,* diagram of a cell from the pith of *Ailanthus.* It has 1 heptagonal, 4 hexagonal, 5 pentagonal, and 4 quadrilateral faces, a total of 14 faces. An example of a cell approximating an ortic tetrakaihedron. *C–F,* diagrams of pith cells of *Eupatorium.* The numbers of facets are 10 (*C*), 9 (*D*), 16 (*E*), and 20 (*F*). (*A, B,* after Matzke;[11] *C–F,* after Marvin.[10])

they are in contact with neighboring cells (fig. 5.2). In relatively homogeneous parenchyma, the number of facets approaches fourteen. A geometrically perfect 14-sided figure is a polyhedron with 8 hexagonal and 6 quadrilateral faces (fig. 5.2,*A*; orthic tetrakaidecahedron). Plant cells rarely approach this ideal form[11] (fig. 5.2,*B*) and show variable numbers of facets even in such homogeneous parenchyma as is often found in the pith of stems[10] (fig. 5.2,*C–F*). The occurrence of smaller and larger cells in the same tissue, the development of intercellular spaces, and the change of cells from nearly isodiametric to some other shape are factors associated with changes in the average number of facets per cell.[12]

The arrangement of cells varies in different kinds of parenchyma. Storage parenchyma of fleshy stems and roots has abundant intercellular spaces but the endosperm of seeds is usually a compact tissue, with small intercellular spaces if any are present. The extensive development of intercellular spaces in the mesophyll, and chlorenchyma in general, is associated with the requirement for gaseous exchange in a photosynthetic tissue. Air spaces are particularly well developed in angiosperms growing in waterlogged soils or aquatic habitats.[8,15,16] Because of the prominence of intercellular spaces the tissue is called *aerenchyma.* The tissue is considered to be important in aerating the plant (chapter 14). Moreover, it appears to be structurally efficient, for it provides the necessary strength with the smallest amount of tissue.[21]

The intercellular spaces in the various tissues just described are commonly of schizogenous origin (chapter 4). Such spaces can become very large if the cells separate along considerable area of their contact with other cells. The separation is combined with an expansion of the tissue as a whole. In the growing tissue, the cells maintain their limited connection with one another by differential growth and assume a lobed or "armed" form[8] (fig. 5.1, *C*,*E*). In some species, the cells not only grow but also divide next to the intercellular spaces. In these divisions, the new walls are formed perpendicular to the walls outlining the spaces.[7]

Cell wall

The cell walls in active vegetative ground parenchyma, including the mesophyll of leaves, are relatively thin and are classified as primary (chapter 4). The middle lamella may or may not be distinguishable (fig. 5.1,*A*). Plasmodesmata are common in such walls, sometimes concentrated in primary pits or in thickened wall portions, sometimes distributed in walls of even thickness. The submicroscopic structure of primary parenchyma walls appears to be rather variable

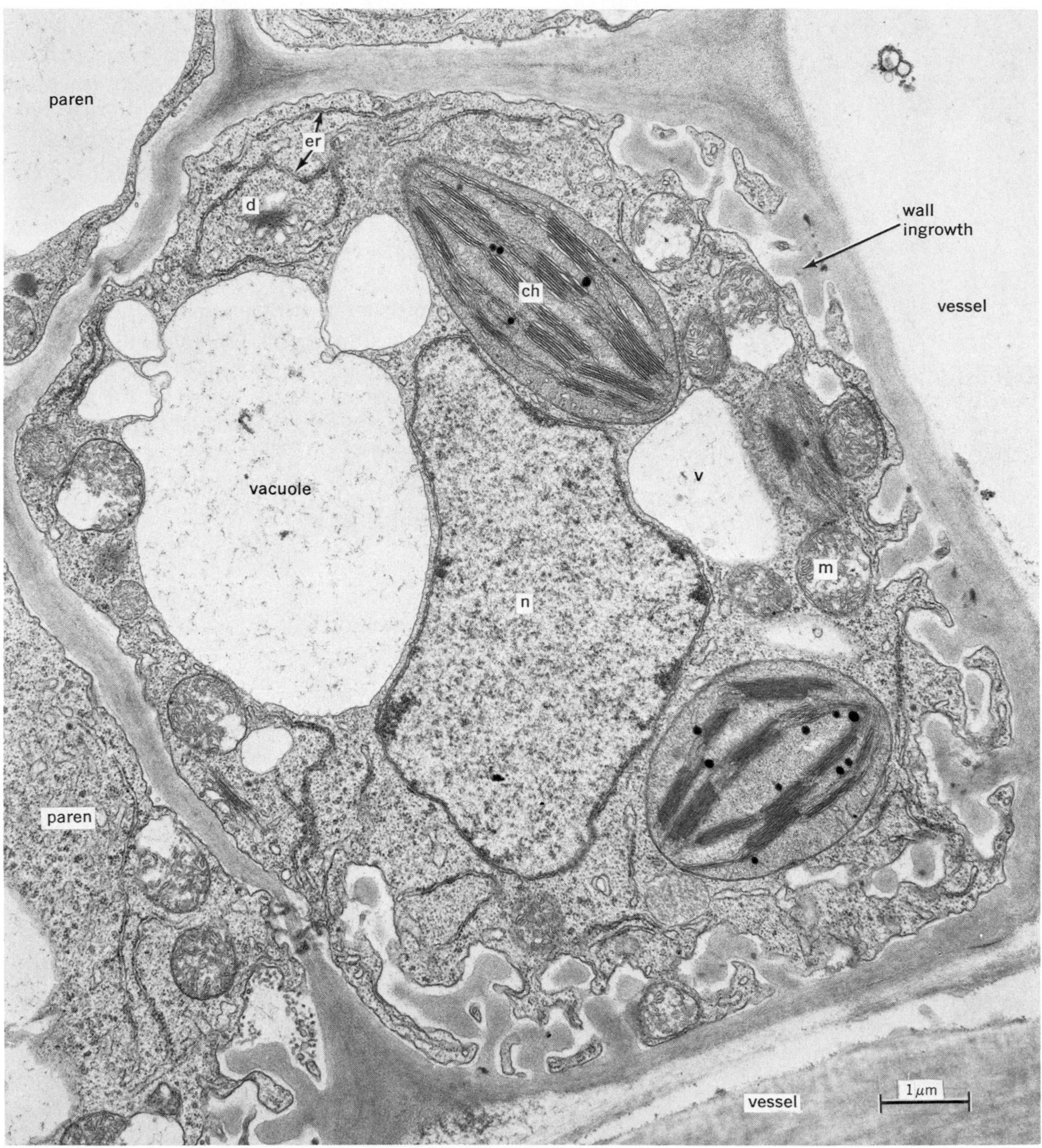

Figure 5.3 Transverse section of parenchyma cell with wall ingrowths (*transfer cell*) from primary xylem of stem of *Phaseolus vulgaris*. Wall ingrowths are limited to walls adjoining vessels. Details: *ch*, chloroplast; *d*, dictyosome; *er*, endoplasmic reticulum; *m*, mitochondrion; *n*, nucleus; *paren*, parenchyma cell; *v*, vacuole. (Negative, courtesy of J. Cronshaw.)

even within a given cell.[4] In walls disposed parallel to the vertical axis of the plant, the cellulose microfibrils are nearly horizontally oriented (adapted to bending) in the stem, helically oriented (adapted to pull) in the root. In transverse walls, the microfibrils are either crossed at approximately right angles to one another or have a spheroidal orientation giving a Maltese cross pattern in polarized light.

Sometimes the primary wall becomes very thick,[1] a feature prominently displayed by the storage parenchyma of some seeds, as those of *Asparagus, Coffea arabica, Diospyros* (persimmon), and *Phoenix dactylifera* (date palm). The carbohydrates of such walls are regarded as reserve material utilized by the embryo during germination. In many taxa, a special type of primary wall thickening characterizes parenchyma cells concerned with short-distance transport of solutes between cells (chapters 8, 13, and 18). The wall thickening occurs in the form of localized ingrowths of various lengths (fig. 5.3). Since the plasmalemma is applied to the cell wall, it is greatly increased in surface area by the ingrowths and is brought in close relation to much of the protoplast. The cells with wall ingrowths are referred to as *transfer cells,*[6] although cells without such wall modification may be similarly concerned with transfer of materials between cells. Parenchyma cells develop secondary walls in certain positions (fig. 5.1,*B*). In the wood and the pith, parenchyma cells often have secondary lignified walls which make it difficult to distinguish between these sclerified parenchyma cells and the typical representatives of sclerenchyma called sclereids (chapter 6).

COLLENCHYMA

Collenchyma consists of thick-walled cells and is regarded as a supporting tissue. It is closely related to parenchyma. Both tissues have complete protoplasts capable of resuming meristematic activity. In both tissues, the cell walls are typically primary and not lignified. The difference between the two tissues lies chiefly in the thicker walls of collenchyma; moreover, the more highly specialized collenchyma cells are longer than are most kinds of parenchyma cells. Where the two tissues are in contact, however, they intergrade in wall thickness and cell form.

Collenchyma differs from the other representative of supporting tissues, sclerenchyma, in wall structure and condition of the protoplast. Collenchyma has relatively soft, pliable, nonlignified primary walls, whereas sclerenchyma has hard, more or less rigid, secondary walls, which are commonly lignified. Collenchyma retains active protoplasts capable of removing the wall thickenings when the cells are induced to resume meristematic activity, as in formation of cork cambium (chapter 12) or in response to wounding (fig. 5.6,*C*). Sclerenchyma walls are more permanent than those of the collenchyma. They do not appear to be readily removed even if the protoplast is retained in the cell. Many sclerenchyma cells lack protoplasts at maturity.

Cell wall

The structure of cell walls in collenchyma is the most distinctive characteristic of this tissue. The walls are thick and glistening in fresh sections, and often the thickening is unevenly distributed (fig. 5.4). They contain, in addition to cellulose, large amounts of pectin and hemicelluloses[14] but no lignin. Since the pectic substances are hydrophilic, collenchyma walls are rich in water. This feature can be demonstrated by treating fresh sections of collenchyma with alcohol. The dehydrating action of alcohol causes a noticeable

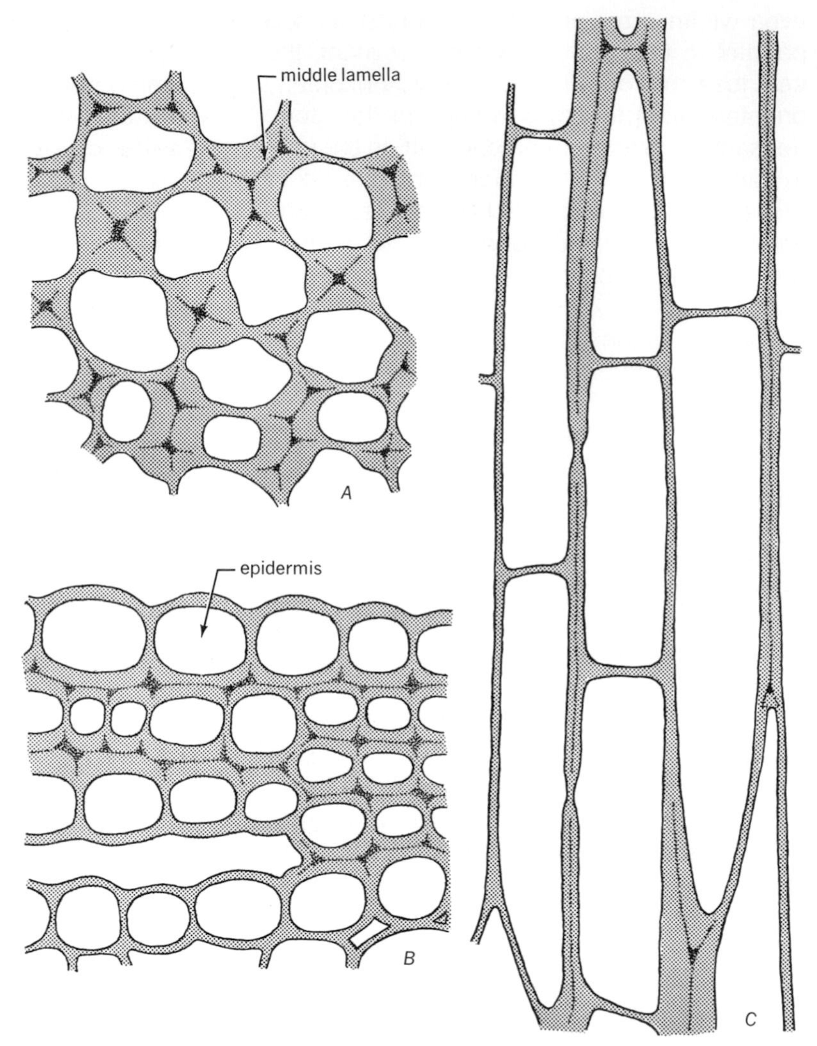

Figure 5.4 Walls of collenchyma cells. *A, C,* angular collenchyma of *Salvia* in transverse (*A*) and longitudinal (*C*) sections. *B,* lamellar collenchyma of *Astrantia.* (Adapted from G. Haberlandt, *Physiologische Pflanzenanatomie,* 1904, p. 146.)

contraction of collenchyma walls. Ultrastructurally, collenchyma walls of various types show layering—lamellae with transverse orientation of microfibrils alternate with those having longitudinal orientation of microfibrils.[3,20] Primary pits are often present in collenchyma walls, especially in those that are rather uniform in thickness.[5]

The distribution of wall thickening in collenchyma shows several patterns. If the wall is unevenly thickened, it attains its greatest thickness either in the corners of the cell or on two opposite walls, the inner and the outer tangential walls. Collenchyma with wall thickenings localized in the corners is commonly called *angular collenchyma* (figs. 5.4,*A*,*C*, and 5.5), the one with the thickenings on the tangential walls, *lamellar* or *plate collenchyma* (fig. 5.4,*B*). As the wall ages, its pattern may change through deposition of additional wall layers. Thus the initially angular pattern, for example, may be obscured (fig. 5.6,*A*) as the outline of the cell lumen becomes smaller in transverse sections.[5]

Collenchyma may or may not contain intercellular spaces. If spaces are present in the angular type of collenchyma, the thickened walls occur next to the intercellular spaces. Collenchyma with such distribution of wall thickening is sometimes classified as a special type, the *lacunar collenchyma* (fig. 5.6,*B*). When collenchyma develops no intercellular spaces, the corners where several cells meet show a thickened middle lamella (figs. 5.4 and 5.5,*B*). Such thickening is sometimes exaggerated by an accumulation of intercellular material in the potential intercellular spaces. The rate of this accumulation apparently varies, for intercellular spaces may arise in early stages of development, only to

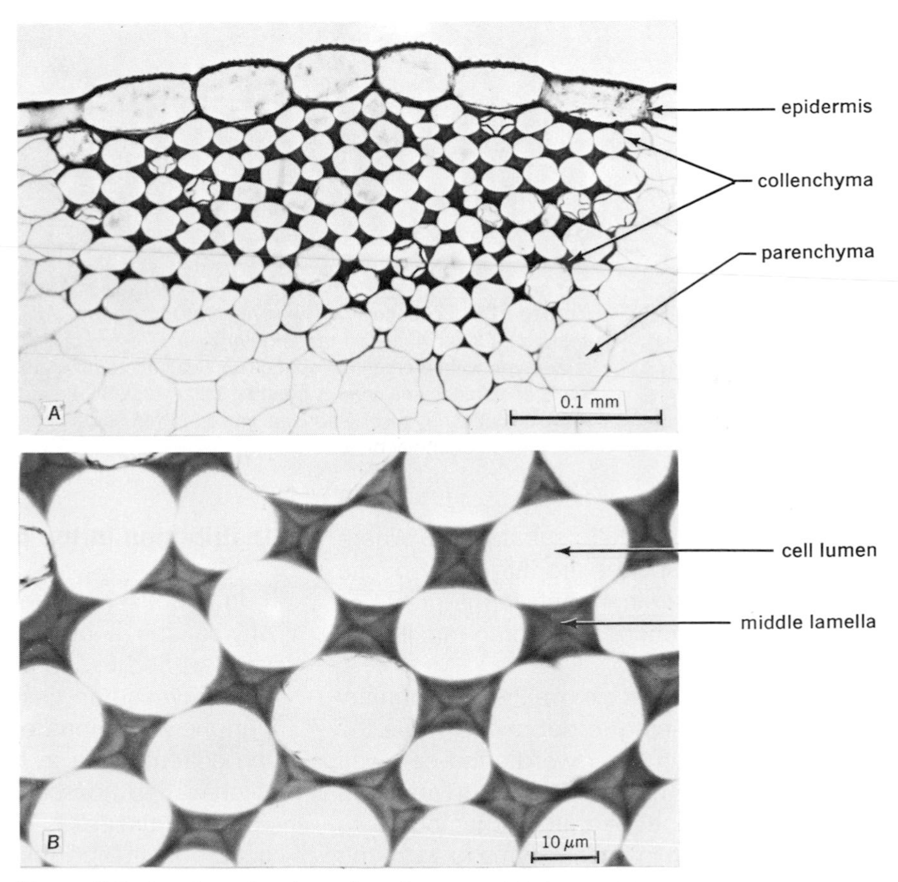

Figure 5.5 Angular collenchyma from *Rumex* petiole in transverse section. (Slide, courtesy of E. M. Engleman.)

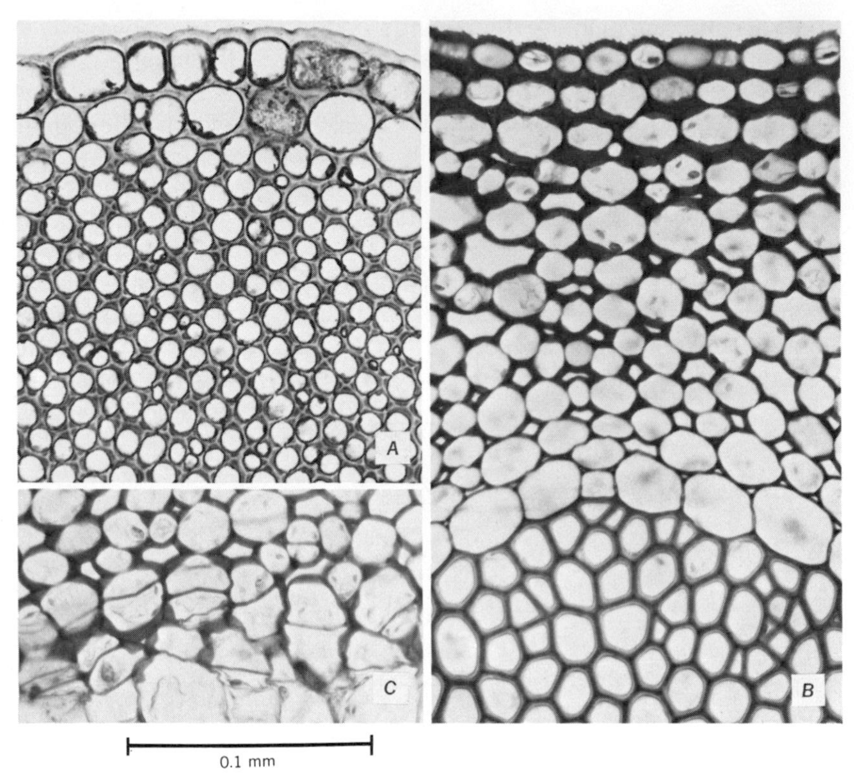

Figure 5.6 Collenchyma tissue in cross sections. *A*, petiole of celery (*Apium graveolens*). Collenchyma of the angular type. *B*, stem of *Ambrosia*. Collenchyma of the lacunar type with intercellular spaces and wall thickening next to these. Unevenly thickened walls of collenchyma sharply contrast with evenly thickened walls of sclerenchyma (below in *B*). *C*, collenchyma of *Ambrosia* with cells that underwent divisions near a surface injury. (*B, C,* slide courtesy of N. H. Boke.)

be closed later by pectic substances. Where the intercellular spaces are large, the pectic material fails to fill them and forms crests or wartlike accumulations protruding into the intercellular spaces.[2,5]

Collenchyma walls exemplify thick primary walls. The thickening is deposited while the cell is growing. In other words, the cell wall increases simultaneously in surface area and in thickness. Because this thickening is so great, wall growth in collenchyma is a striking and complicated phenomenon that has not been completely clarified with regard to development.[9,19]

Distribution in the plant

In discussing the distribution of collenchyma one should distinguish between the thick-walled tissue that arises independently of the vascular tissues and occurs in the peripheral regions of stem and leaf, that is, the collenchyma in the strict sense, and the thick-walled parenchyma associated with vascular tissues.[5] This parenchyma, which occurs in the peripheral part of the phloem (outer part of a vascular bundle), or the peripheral part of the xylem (inner part of a vascular bundle), or completely surrounds a vascular bundle, consists of long cells with thick primary walls. The wall thickening may resemble that of collenchyma, especially the type with uniformly thickened walls. It is often called collenchyma, but because of its association with the vascular tissues it has a history of development somewhat different from that of the independent collenchyma. The fine structure of walls in the two kinds of tissue may also be different. The elongated cells with thick primary walls associated with the vascular bundles may be referred to as collenchymatous parenchyma cells if their resemblance to collenchyma cells must be stressed. This designation may be applied to parenchyma resembling collenchyma in any location in the plant. The present discussion deals only with the independent peripheral collenchyma.

The peripheral position of collenchyma is highly characteristic. The tissue is located either directly beneath the epidermis (fig. 5.6,*B*) or one or a few layers removed from it (fig. 5.6,*A*). In stems, collenchyma frequently forms a continuous layer around the circumference of the axis (chapter 17). Sometimes it occurs in strands, often within externally visible ridges (ribs) found in many herbaceous stems and in those woody stems that have not yet undergone secondary growth. In the petioles, the distribution of collenchyma shows patterns similar to those encountered in the stems. In the leaf blade it occurs in the ribs accompanying the larger vascular bundles (veins), sometimes on both sides of the rib, sometimes on one side only, usually the lower (chapter 18). Roots rarely have collenchyma.

Structure in relation to function

Collenchyma appears to be particularly adapted for support of growing leaves and stems. Its walls begin to thicken early during the development of the shoot, but this thickening is plastic and capable of extension. It therefore does not hinder the elongation of stem and leaf. In a more advanced state of development collenchyma continues to be a supporting tissue in plant parts (many leaves, some herbaceous stems) that do not develop much sclerenchyma. In connection with the discussion of the supporting role of collenchyma it is of interest that in developing plant parts subjected to mechanical stresses (by exposure to wind, attachment of weights to inclined shoots) the wall thickening in collenchyma begins earlier and becomes more massive than in plants not subjected to such stresses.[13,17,18]

Mature collenchyma is a strong flexible tissue consisting of long overlapping cells (they may reach 2 millimeters in length[5]) with thick nonlignified walls. In their tensile strength collenchyma cells favorably compare with fibers. In old plant parts collenchyma may harden, or it may change into sclerenchyma by a deposition of secondary lignified walls.[5,20] If it does not undergo these changes, its role as a supporting tissue may become less important because of the development of sclerenchyma in the deeper parts of the stem or petiole. Moreover, in stems with secondary growth, the xylem becomes the chief supporting tissue because of the predominance of cells with lignified secondary walls and the abundance of long overlapping cells in that tissue.

REFERENCES

1. Bailey, I. W. Cell wall structure of higher plants. *Indus. Eng. Chem.* 30:40–47. 1938.
2. Carlquist, S. On the occurrence of intercellular pectic warts in Compositae. *Amer. J. Bot.* 43:425–429. 1956.
3. Chafe, S. C. The fine structure of the col-

lenchyma cell wall. *Planta* 90:12–21. 1970.
4. Czaja, A. T. Untersuchungen über die submikroskopische Struktur der Zellwände von Parenchymzellen in Stengelorganen und Wurzeln. *Planta* 51:329–377. 1958.
5. Duchaigne, A. Les divers types de collenchymes chez les Dicotylédones: leur ontogénie et leur lignification. *Ann. Sci. Nat., Bot. Sér.* 11. 16:455–479. 1955.
6. Gunning, B. E. S., and J. S. Pate. "Transfer cells"—plant cells with wall ingrowths, specialized in relation to short distance transport of solutes: their occurrence, structure and development. *Protoplasma* 68:107–133. 1969.
7. Hulbary, R. L. The influence of air spaces on the three-dimensional shapes of cells in *Elodea* stems, and a comparison with pith cells of *Ailanthus*. *Amer. J. Bot.* 31:561–580. 1944.
8. Kaul, R. B. Diaphragms and aerenchyma in *Scirpus validus*. *Amer. J. Bot.* 58:808–816. 1971.
9. Magin, T. L'ontogénie du collenchyme chez *Lamium album* L. *Rev. Cytol. Biol Vég.* 17:219–258. 1956.
10. Marvin, J. W. Cell shape and cell volume relations in the pith of *Eupatorium perfoliatum* L. *Amer. J. Bot.* 31:208–219. 1944.
11. Matzke, E. B. What shape is a cell? *Teach. Biol.* 10:34–40. 1940.
12. Matzke, E. B., and R. M. Duffy. Progressive three-dimensional shape changes of dividing cells within the apical meristem of *Anacharis densa*. *Amer. J. Bot.* 43:205–225. 1956.
13. Razdorskij, V. F. *Arkhitektonika rastenij.* [*Architectonics of plants.*] Moskva, Sovetskaya Nauka. 1955.
14. Roelofsen. P. A. The plant cell wall. *Handbuch der Pflanzenanatomie.* Band 3. Teil 4. 1959.
15. Stant, M. Y. Anatomy of the Alismataceae. *Bot. J. Linn. Soc.* 59:1–42. 1964.
16. Stant, M. Y. Anatomy of the Butomaceae. *Bot. J. Linn. Soc.* 60:31–60. 1967.
17. Venning, F. D. Stimulation by wind motion of collenchyma formation in celery petioles. *Bot. Gaz.* 110:511–514. 1949.
18. Walker, W. S. The effect of mechanical stimulation and etiolation on the collenchyma of *Datura stramonium*. *Amer. J. Bot.* 47:717–724. 1960.
19. Wardrop, A. B. The mechanism of surface growth in parenchyma of *Avena* coleoptiles. *Aust. J. Bot.* 3:137–148. 1955.
20. Wardrop, A. B. The structure of the cell wall in lignified collenchyma of *Eryngium* sp. (Umbelliferae). *Aust. J. Bot.* 17:229–240. 1969.
21. Williams, W. T., and D. A. Barber. The functional significance of aerenchyma in plants. *Soc. Exp. Biol. Symp.* 15:132–144. 1961.
22. Zavalishina, S. F. Khloroplasty v tkanyakh steli u pokrytosemennykh rastenij. [Chloroplasts in stelar tissues in angiosperms.] *Dok. Akad. Nauk SSSR* 78:137–139. 1951.

6

Sclerenchyma

The basic structural features of sclerenchyma were introduced in chapter 5 by comparing sclerenchyma and collenchyma, the two principal supporting tissues. Sclerenchyma cells have secondary walls that are deposited over the primary after the latter complete their extension growth. Secondary walls are present also in water-conducting cells of the xylem, and frequently in the parenchyma cells of that tissue as well. Moreover, parenchyma cells in various other tissue regions may become sclerified. Thus, secondary walls are not unique in sclerenchyma cells, and, therefore, the delimitation between typical sclerenchyma cells and sclerified parenchyma on the one hand and water-conducting cells on the other is not sharp. Sclerenchyma also poses a problem of delimitation as a tissue because sclerenchyma cells are often scattered among cells of other types, individually or in small groups.

In this chapter sclerenchyma is discussed with reference to those mechanical, or supporting, cells that chiefly lend hardness or rigidity to tissues. Sclerenchyma cells are usually divided into two categories, sclereids and fibers. These two classes of cells are not sharply separated from each other, but in general the fiber is a long slender cell, many times longer than wide, whereas sclereids vary in form from an approximately isodiametric to a considerably elongated one, and some kinds of sclereids are much branched. Sclerenchyma cells may or may not retain their protoplasts at maturity. This variability adds to the difficulty of distinguishing between sclerenchyma cells and sclerified parenchyma cells.

SCLEREIDS

Sclereids are widely distributed in the plant body and vary greatly in shape. These cells usually have thick secondary walls, strongly lignified and provided with numerous, commonly simple, pits. Sclereids have been categorized on the basis of form, but the classification is of limited utility because the various forms frequently intergrade. Commonly listed categories are: *brachysclereids,* stone cells, roughly isodiametric or somewhat elongated (fig. 6.1,*A–D*); *macrosclereids,* rod cells, elongated and columnar (fig. 6.1,*Q*); *osteosclereids,* bone cells, also columnar but with enlarged ends; *astrosclereids,* star cells, with lobes, or arms, diverging from a central body (fig. 6.1,*J*); *trichosclereids,* internal hairlike cells, with branches projecting into intercellular spaces; *filiform sclereids,* long, fiberlike, sometimes branched (fig. 6.1,*L,M*). Astrosclereids and trichosclereids are structurally similar, and trichosclereids intergrade with the filiform sclereids. Osteosclereids may be branched at their ends (as in fig. 6.1,*G*) and consequently similar to trichosclereids.

Distribution of sclereids

The distribution of sclereids among other cells is of special interest with regard to problems of cell differentiation in plants. Sclereids may occur in more or less extensive layers or clusters, but frequently they appear isolated among other types of cells from which they may differ sharply by their thick walls and often bizarre shapes. As isolated cells they are classified as idioblasts. The differentiation of idioblasts poses many still unresolved questions regarding causal relations in the development of tissue patterns in plants.[7]

Sclereids occur in the epidermis, the ground tissue, and the vascular tissues. In the following paragraphs, sclereids are described by examples from different parts of the plant body excluding those sclereids that occur in the vascular tissues.

SCLEREIDS IN STEMS.

A continuous cylinder of sclereids occurs on the periphery of the vascular region in the stem of *Hoya carnosa* and groups of sclereids in the pith of stems of *Hoya* and *Podocarpus*. These sclereids have moderately thick walls and numerous pits (fig. 6.1,*C,D*). In shape and size they resemble the adjacent parenchyma cells. This resemblance is often taken as an indication that such sclereids are by origin sclerified parenchyma cells. Their sclerification, however, has advanced so far that they may be grouped with the sclereids rather than parenchyma cells. This simple type of sclereid exemplifies a stone cell, or brachysclereid. A much branched astrosclereid is found in the cortex of *Trochodendron* stem (fig. 6.1,*J*). Somewhat less profusely branched sclereids occur in the cortex of the douglas fir, *Pseudotsuga taxifolia.*

SCLEREIDS IN LEAVES.

Leaves are an especially rich source of sclereids with regard to variety of form. In the mesophyll, two main distributional patterns of sclereids are recognized: the diffuse, with *sclereids dispersed in the leaf tissue (Trochodendron, Osmanthus, Olea, Pseudotsuga*), and the terminal, with the sclereids confined to the ends of the small veins.[6,7,8] In some protective foliar structures like the clove scales of garlic (*Allium sativum*) the sclereids form part of or the entire epidermis (fig. 6.1,*K*).

Sclereids with definite branches or only with spicules occur in the ground tissue of *Camellia* petiole (fig. 6.1,*H,I*) and in the mesophyll of *Trochondendron* leaf. The mesophyll of *Osmanthus* and *Hakea* contains columnar sclereids, ramified at each end (fig. 6.1,*G*). *Monstera deliciosa, Nymphaea* (water lily), and *Nuphar* (yellow pond lily) have typical trichosclereids with branches extending into large intercellular spaces, or air chambers, characteristic of the leaves of these species (chapter 19). Branched

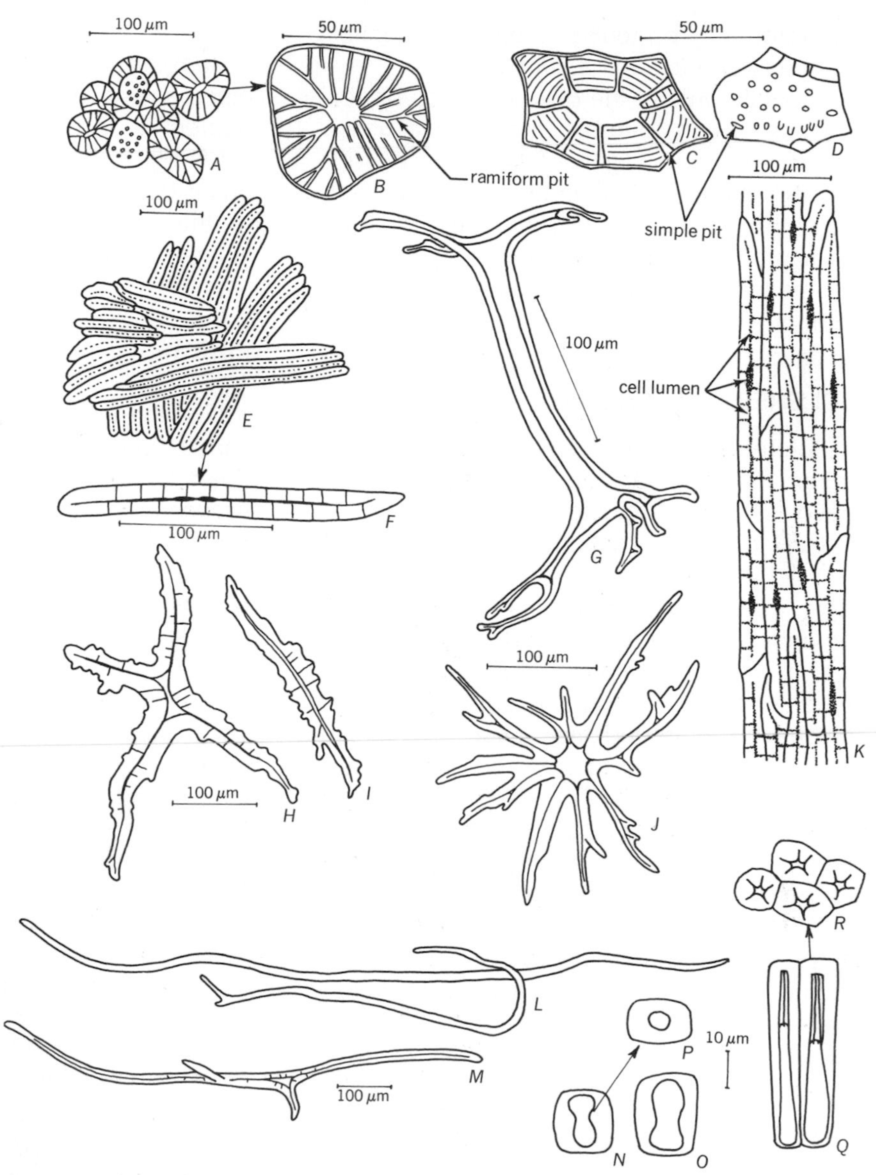

Figure 6.1 Sclereids. *A, B,* stone cells from fruit flesh of pear (*Pyrus*). *C, D,* sclereids from stem cortex of wax plant (*Hoya*), in sectional (*C*) and surface (*D*) views. *E, F,* sclereids from endocarp of fruit of apple (*Malus*). *G,* columnar sclereid with ramified ends; from palisade mesophyll of *Hakea*. *H, I,* sclereids from petiole of *Camellia, J,* astrosclereid from stem cortex of *Trochodendron*. *K,* layer of sclereids from epidermis of clove scale of garlic (*Allium sativum*). *L, M,* filiform sclereids from leaf mesophyll of olive (*Olea*). *N–P,* sclereids from subepidermal layer of seed coat of bean (*Phaseolus*), "hourglass cells"; seen from the side (*N, O*) and from above (*P*). *Q, R,* epidermal sclereids, macrosclereids, with fluted wall thickenings; from seed coat of bean (*Phaseolus*) seen from the side (*Q*) and from above (*R*).

sclereids may be found in leaves of conifers such as *Pseudotsuga taxifolia.*

The sclereids of the olive (*Olea europaea*) leaf are of considerable interest because of their great length (fig. 6.1,*L,M*). They average 1 millimeter in length[2] and may appropriately be called fiberlike or filiform sclereids. They originate in both palisade and spongy parenchyma and permeate the mesophyll in the form of a dense mat.

SCLEREIDS IN FRUITS.

Sclereids occupy various positions in fruits. *Pyrus* (pear) and *Cydonia* (quince) have single or clustered stone cells, or brachysclereids, scattered in the fruit flesh (fig. 6.1,*A,B*). In the formation of clusters in the pear, cell divisions occur concentrically around some earlier formed sclereids, and the new cells also become sclereids[19] (chapter 22). The radiating pattern of parenchyma cells around the mature clusters of sclereids is related to this mode of development. The sclereids of pear and quince often show ramiform pits resulting from a fusion of one or more cavities during the increase in thickness of the wall (fig. 6.1,*B*).

The apple (*Malus*) furnishes another example of sclereids in the fruit. The cartilaginous endocarp enclosing the seeds consists of obliquely oriented layers of elongated sclereids (fig. 6.1,*E,F*). Sclereids also compose the hard shells of nutlike fruits and the stony endocarp of stone fruits.

SCLEREIDS IN SEEDS.

The hardening of seed coats during ripening of the seeds often results from a development of secondary walls in the epidermis and in the layer or layers beneath the epidermis. The leguminous seeds furnish a good example of such sclerification (chapter 23). In seeds of bean (*Phaseolus*), pea (*Pisum*), and soybean (*Glycine*), columnar macrosclereids (fig. 6.1,*R,Q*) compose the epidermis and prismatic sclereids (fig. 6.1,*N–P*) or bone-shaped osteosclereids occur beneath the epidermis. The seed coat of the coconut (*Cocos nucifera*) contains sclereids with numerous ramiform pits.

FIBERS

Like the sclereids, fibers may be found in various parts of the plant. In dicotyledons, fibers are particularly common in the vascular tissues. They are the phloem fibers and the xylem, or wood, fibers. In monocotyledons, fibers may completely enclose each vascular bundle like a sheath, form a strand on one or both sides of a vascular bundle ("bundle caps"), or form strands or layers that appear to be independent of the vascular tissues. The fibers of the xylem are considered in chapter 8. The present chapter is concerned mainly with the fibers located outside the xylem, the *extraxylary fibers,* which include the phloem fibers of the dicotyledons and the fibers of the monocotyledons, whether associated with the vascular bundles or not.

The fibers are long cells, with more or less thick secondary walls, and usually occur in strands (fig. 6.2). These strands constitute the "fibers" of commerce. The process of retting used in the extraction of fibers from the plant results in a separation of the fiber bundles from the associated nonfibrous cells. Within a strand, the fibers overlap, a feature that imparts strength to the fiber bundles. In contrast to collenchyma walls, the fiber walls are not highly hydrated. They are, therefore, harder than the collenchyma walls and are elastic rather than plastic. The fibers serve as supporting elements in plant parts that are no longer elongating. The degree of lignification

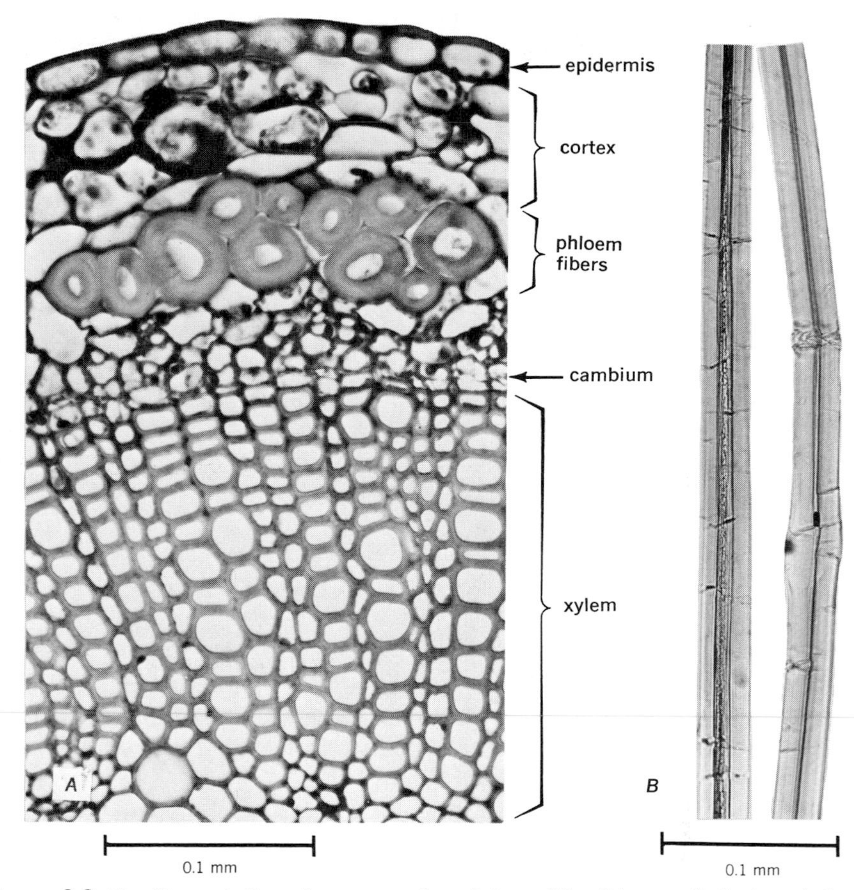

Figure 6.2 Flax fibers. *A,* fibers in cross section of stem of flax (*Linum usitatissimum*). *B,* fragments of isolated fibers. (*B,* from C. H. Carpenter and L. Leney, *91 Papermaking Fibers,* Tech. Publ. 74, College of Forestry at Syracuse, 1952.)

varies, and pits are relatively scarce and commonly slitlike.

Phloem fibers occur in many stems. The flax (*Linum usitatissimum*) stem has only one band of fibers, several cell layers in depth, located on the outer periphery of the vascular cylinder (fig. 6.2). These fibers originate in the earliest part of the primary phloem but mature as fibers after this part of the phloem ceases to function in conduction. Flax fibers are, therefore, primary phloem fibers. In stems of *Sambucus* (elderberry), *Tilia* (basswood), *Liriodendron* (tulip tree), *Vitis* (grapevine), *Robinia pseudoacacia* (black locust) and many others, fibers occur on the periphery of the phloem (primary phloem fibers) and also within the secondary phloem (secondary phloem fibers; fig. 6.5). Conifers may have secondary phloem fibers (*Sequoia, Thuja*).

Stems of some dicotyledons have primary fibers on the periphery of the vascular cylinder that do not originate as part of the phloem tissue but outside of it. These fibers are commonly referred to as pericyclic fibers. However, the designation pericyclic is often used with reference to the primary phloem fibers as well. (See chapter 16 for an evaluation of the term pericycle.)

Economic fibers

The phloem fibers of dicotyledons represent the bast fibers of commerce.[10] These fibers are classified as soft fibers because, whether lignified or not, they are relatively soft and flexible. Some of the well-known sources and usages of bast fibers are hemp (*Cannabis sativa*), cordage; jute (*Corchorus capsularis*), cordage, coarse textiles; kenaf (*Hibiscus cannabinus*), coarse textiles; flax (*Linum usitatissimum*), textiles (e.g., linen), thread; and ramie (*Boehmeria nivea*), textiles. Phloem fibers of some dicotyledons are used for making paper.[4]

The fibers of monocotyledons—usually called leaf fibers[10] because they are obtained from leaves—are classified as hard fibers. They have strongly lignified walls and are hard and stiff. Examples of sources and uses of leaf fibers are abaca or Manila hemp (*Musa textilis*), cordage; bowstring hemp (*Sansevieria,* entire genus), cordage; henequen (*Agave,* various species), cordage, coarse textiles; New Zealand hemp (*Phormium tenax*), cordage; pineapple fiber (*Ananas comosus*), textiles; and sisal (*Agave sisalana*), cordage. Leaf fibers of monocotyledons (together with the xylem) serve as raw material for making paper:[4] corn (*Zea mays*), sugar cane (*Saccharum officinarum*), esparto grass (*Stipa tenacissima*), and others.

The length of individual fiber cells varies considerably in different species. Examples of ranges of lengths in millimeters may be cited from Harris'[10] handbook. Bast fibers: jute, 0.8–6.0; hemp, 5–55; flax, 9–70; ramie, 50–250. Leaf fibers: sisal, 0.8–8.0; bowstring hemp, 1–7; abaca, 2–12; New Zealand hemp, 2–15.

In commerce, the term fiber is often applied to materials that include, in botanical sense, other types of cells beside fibers and also to structures that are not fibers at all. In fact, the leaf fibers of the monocotyledons commonly include vascular elements. Cotton fibers are epidermal hairs of seeds of *Gossypium* (chapter 7); raffia is composed of leaf segments of *Raphia* palm; rattan is made from stems of *Calamus* palm.

DEVELOPMENT OF SCLEREIDS AND FIBERS

The development of branched and long sclereids and of the usually long fibers involves remarkable intercellular adjustments that suggest a degree of independence of these cells from the positional influences stressed in chapter 2. The primordium of a branched sclereid may not differ in appearance from neighboring parenchyma cells. Later, however, instead of enlarging uniformly it develops processes that elongate into branches. During this elongation, the branches not only invade the intercellular spaces but also force their way between walls of other cells[9] (fig. 6.3). The sclereid thus establishes new contacts during its growth and is able to attain a much larger size than its neighbors. If the tissue is lacunate, branches of the sclereid protrude freely into intercellular spaces.

The growth of cells by intrusion between walls of other cells is called *intrusive,* or *interpositional, growth,* and is contrasted with the *coordinated growth* that involves no separation of walls. A group of similar cells in a homogeneous parenchyma tissue undergo coordinated growth, with the pairs of conjoined primary walls presumably expanding at the same rate without a breakage of connection along the middle lamella. Coordinated growth does not prevent some cells

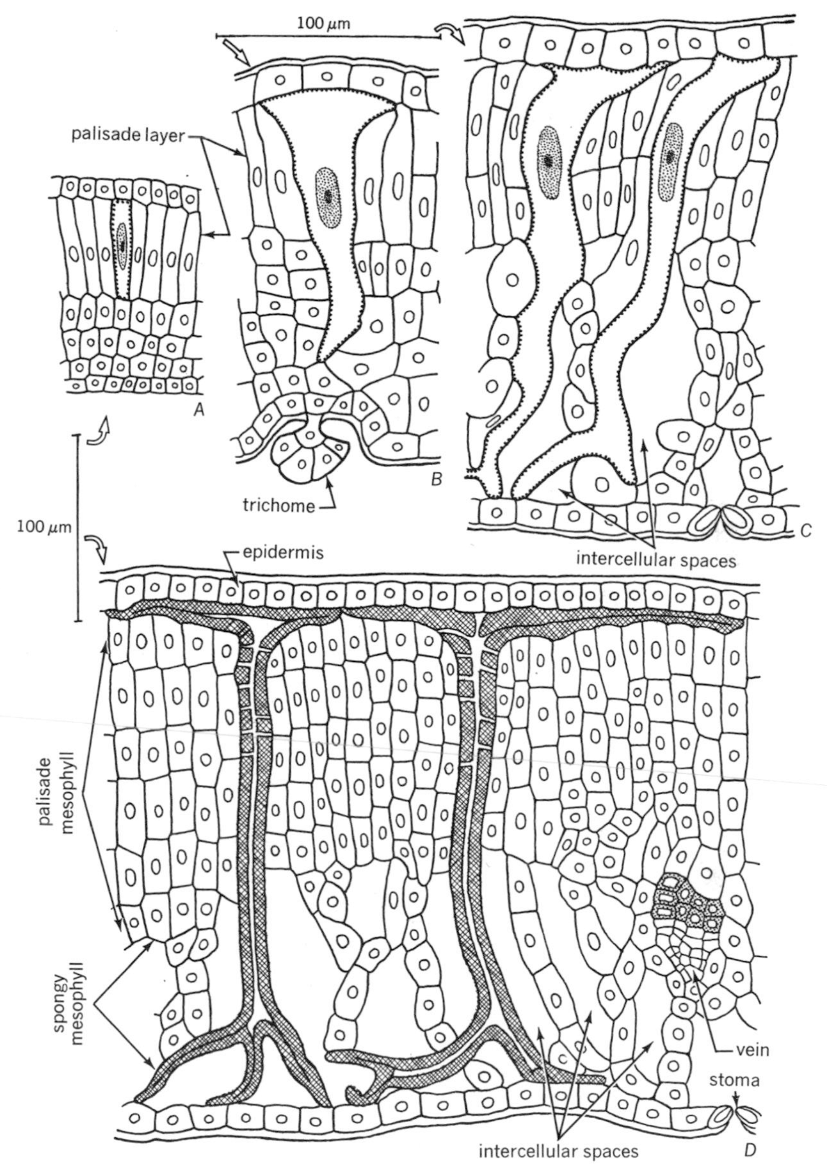

Figure 6.3 Development of sclereids in the leaf of *Osmanthus* (Oleaceae). *A–C*, differentiating sclereids, indicated by large nuclei and dots along the walls; *D*, mature sclereids, indicated by cross-hatched secondary walls. In all drawings, the mesophyll and epidermis cells are marked with circles or ovals. The narrow intercellular spaces characteristic of palisade parenchyma have been omitted. *A*, future sclereid is indicated symbolically; it was not yet differentiated from other palisade cells. *B*, young sclereid has extended beyond the palisade layer. *C*, two young sclereids have reached the lower epidermis by growing through the spongy mesophyll. *D*, mature sclereids have some branches extended parallel with the epidermis and others projected into intercellular spaces. Pits in the secondary wall are located in the parts of sclereids that do not sever connections with adjacent cells during growth. (Adapted from Griffith.[9])

from becoming longer than others. If a cell ceases to divide while its neighbors continue to do so, the nondividing cell will become longer than its neighbors without affecting the cellular relation of walls. In the growth of a sclereid, coordinated growth of the main body of cells is combined with intrusive growth in the elongating part of its branches. In figure 6.3,*D*, pits indicate the parts of the sclereids that maintained their initial contact with adjacent cells by coordinated growth. The unpitted parts grew intrusively.

A fiber also shows a combination of coordinated and intrusive growth. A young fiber primordium increases in length without changing cellular contacts while the adjacent parenchyma cells are actively dividing. Somewhat later, the fiber primordium attains

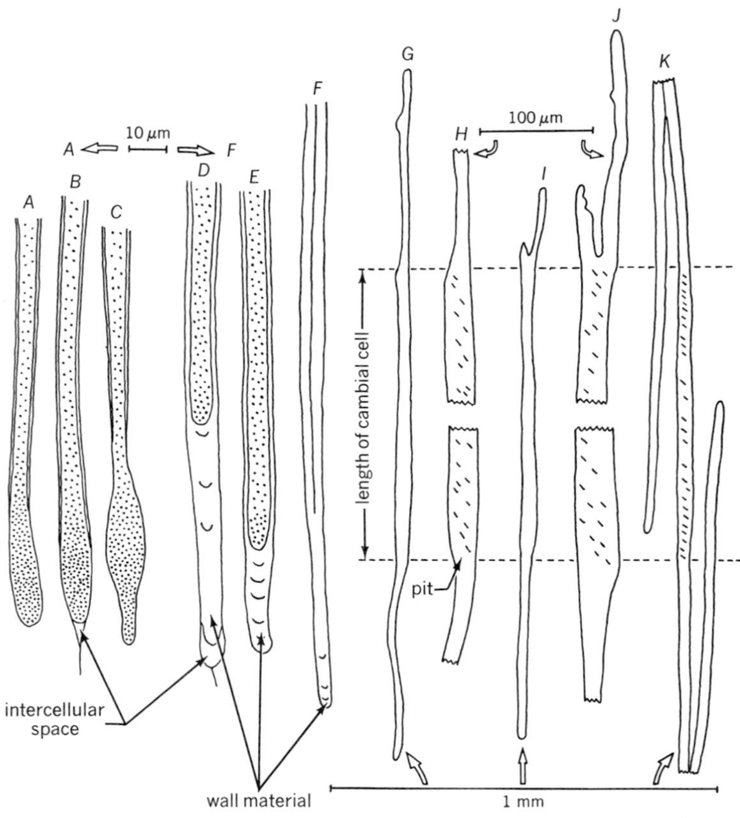

Figure 6.4 Apical intrusive growth in fibers from stems. *A–F*, from phloem of flax (*Linum perenne*) *G–J*, from xylem and, *K*, from phloem of *Sparmannia* (Tiliaceae). *H* and *J*, enlarged views of parts of *G* and *I*, respectively. *A–C*, the intrusively growing tips of fibers (below) have thin walls and dense cytoplasm. *D–F*, the tips of fibers have become filled with wall material after completion of growth. *G–K*, fibers have extended in both directions from the original position in the cambium (between broken lines). Pits occur only in the original cambial parts. The phloem fiber (*K*) is considerably longer than the xylem fibers (*G*, *I*). (*A–F*, adapted from Schoch-Bodmer and Huber;[17] *G–K*, adapted from Schoch-Bodmer.[15])

additional length by intrusive growth carried out at both its apices. During elongation, the fiber may become multinucleate as a result of repeated nuclear divisions not followed by formation of new walls. While the fiber is still a living cell its cytoplasm shows rotational streaming.[11] By means of tests with dyes, this phenomenon has been related to intercellular transport of materials.[20]

Apical intrusive growth has been studied in detail in flax fibers.[17] By measuring young and old internodes and the fibers contained in these internodes the authors calculated that by coordinated growth alone the fibers could become 1 to 1.8 cm long. Actually they found fibers ranging in length between 0.8 and 7.5 cm. Thus, lengths over 1.8 cm. must have been attained by apical intrusive growth. The growing tips of young fibers dissected out of living stems showed thin walls, contained dense cytoplasm (fig. 6.4,*A–C*) with chloroplasts, and were not plasmolyzable. When the tips ceased to grow they became filled with secondary wall material (fig. 6.4,*D–F*).

Intrusive growth may be identified in transverse sections of stems and roots by the appearance of small cells—transections of growing tips—among the wider, not elongating parts of fiber primordia. The secondary vascular tissues of *Sparmannia* (Tiliaceae) offer a graphic illustration of this phenomenon[16] (fig. 6.5,*A*). The orderly radial alignment of cells seen in the cambium is replaced by a mosaic pattern in the axial system of phloem. In a given transection, three to five growing fiber tips are added to each wider median portion of a fiber primordium (indicated by diagonal hatching in fig. 6.5,*A*) by intrusive elongation. The radial alignment in the axial system of xylem is less strongly affected because xylary fibers elongate less than do phloem fibers (fig. 6.4,*G–K*). As seen in radial longitudinal sections, the bipolar apical growth of fibers makes these cells extend above and below the horizontal levels of cambial cells among which they are initiated (fig. 6.5,*B*).

When during intrusive growth a fiber tip is obstructed by other cells, the tip curves or forks (fig. 6.4,*I,J*). Thus, bent and forked ends in fibers (and sclereids) is another evidence of intrusive growth. The intrusively growing parts usually fail to develop pits in their secondary walls and thus serve as a measure of the amount of apical elongation[15] (fig. 6.4,*G–K*).

Prolonged apical intrusive growth of fibers and some sclereids makes the secondary thickening of the walls in these cells a rather complex phenomenon. As was mentioned previously, the secondary wall commonly develops over the primary after the latter ceases to expand. In intrusively growing fibers and sclereids, the older part of the cell stops growing while the apices continue to elongate. The older (that is, median) part of the cell begins to form secondary wall layers before the growth of the tips is completed. From the median part of the cell, the secondary thickening progresses toward the tips and is completed after the tips cease to grow.

Some fibers and sclereids undergo regular mitotic divisions after the secondary wall is deposited and are partitioned into two or more compartments by septae of primary walls.[12] The septae are in contact but not fused with the secondary wall and are separated by the latter from the original primary wall of the cell as a whole (fig. 6.6,*A*). There is apparently no evidence that this type of cytokinesis is associated with a deposition of primary wall layers over the secondary wall in each compartment but additional secondary wall may develop after the division and cover the septae also[3] (fig. 6.6,*B*). Fibers and sclereids partitioned by septae after secondary wall development are called *septate*.

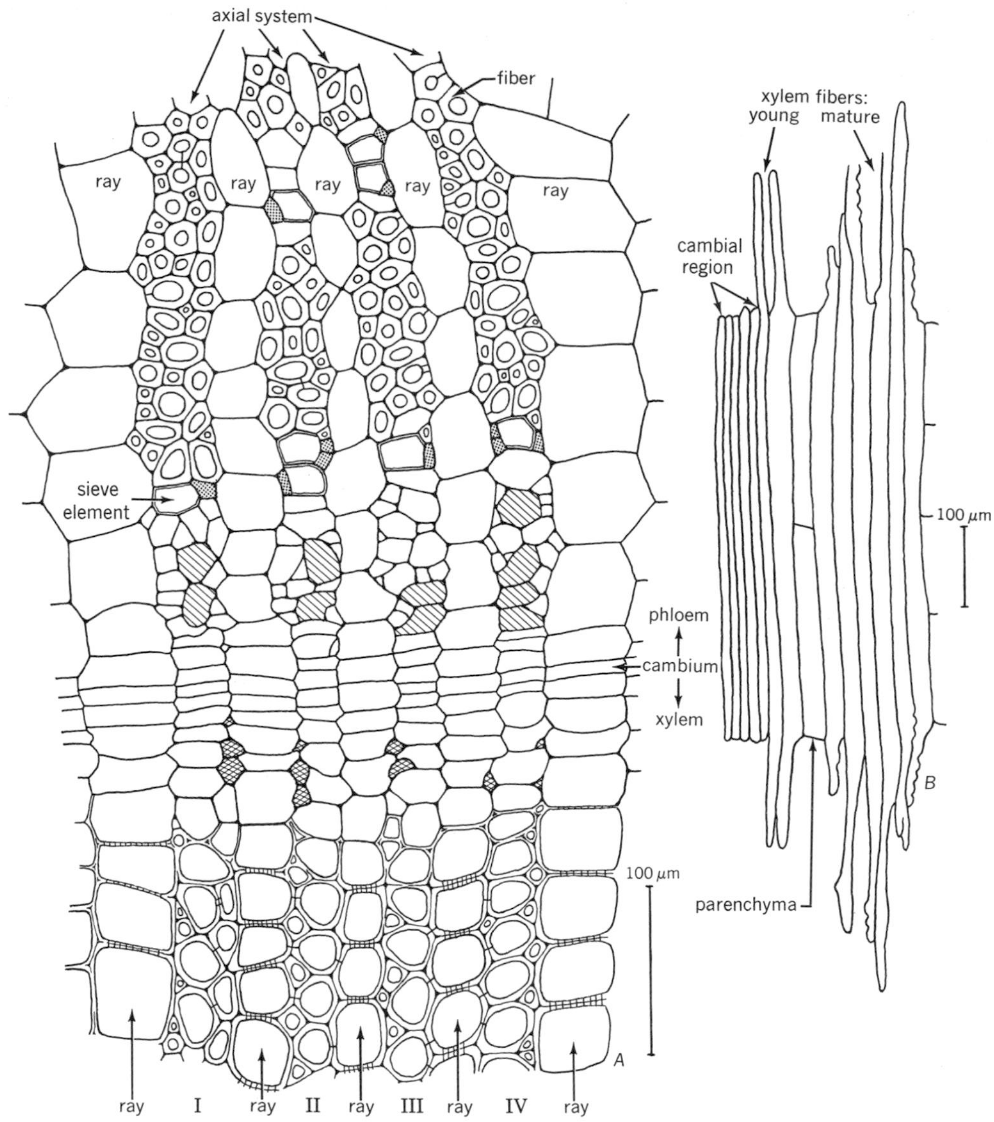

Figure 6.5 Development of fibers in secondary phloem and xylem of *Sparmannia* (Tiliaceae) as seen in transverse (*A*) and radial longitudinal (*B*) sections of stem. In *A*, I–IV are the files of cells in the axial (longitudinal) system. The files alternate with rays. Phloem and xylem are immature next to the cambium. Mature xylem has secondary walls. In mature phloem, dotted companion cells serve to identify the sieve elements, secondary walls mark the fibers. Cells with diagonal lines are median parts of young fiber cells. They are accompanied by small cells most of which are tips of intrusively growing fibers. The cross-hatched cells on the xylem side are intrusively growing tips of xylem fibers. *B*, xylem fibers extend beyond the cambial region in both directions. (Adapted from Schoch-Bodmer and Huber.[16])

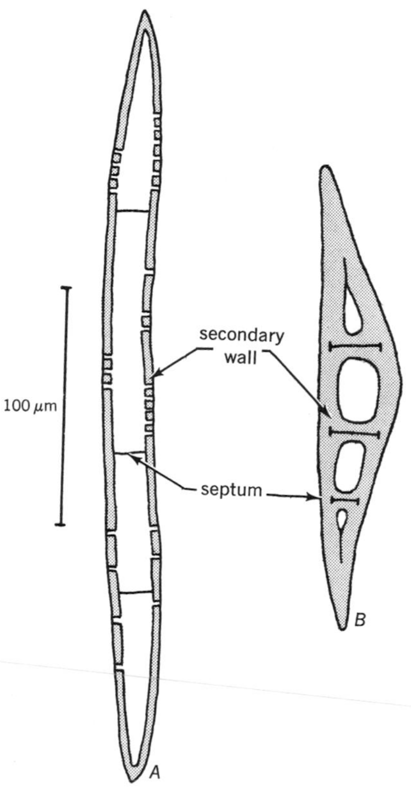

Figure 6.6 *A,* septate fiber from phloem of grapevine (*Vitis*) stem. The septae are in contact with the pitted secondary wall. *B,* septate sclereid from the phloem of *Pereskia* (Cactaceae) in which the septae are covered with secondary wall material. (*B,* adapted from Bailey.[3])

The factors that control development of sclerenchyma have been considered by some investigators. Cutting into leaves (*Camellia,*[5] *Fagraea*[14]) induced differentiation of sclereids near the new surface, a result interpreted as evidence that position plays a large role in sclerenchyma development. Investigations of hormonal factors indicated that auxin levels in the leaf influence sclereid development.[1,13] When auxin concentration was high, the development was suppressed, whereas at low concentrations of auxin the cell walls remained thin and did not become lignified. Fiber development also was object of experimental studies.[18] Treatment with gibberellic acid induced a significant increase in cell length and diameter in primary phloem fibers of hemp, jute, and kenaf. The slitlike pits became more steeply inclined, their frequency and width reduced, their length increased. The increase in fiber length was a concomitant of greater internodal elongation in treated plants. As was stated previously, lengths of fibers and internodes are also correlated in normal development. The effect on pits was traceable to a change in cellulose microfibril orientation evidently determined by the intensive stretching of the cell wall. There was, in addition, an increase in wall thickness, an indication of accelerated carbohydrate synthesis.

REFERENCES

1. Al-Talib, K. H., and J. G. Torrey. Sclereid distribution in the leaves of *Pseudotsuga* under natural and experimental conditions. *Amer. J. Bot.* 48:71–79. 1961.
2. Arzee, T. Morphology and ontogeny of foliar sclereids of *Olea europaea.* I. Distribution and structure. *Amer. J. Bot.* 40:680–687. 1953. II. Ontogeny. *Amer. J. Bot.* 40:745–752. 1953.
3. Bailey, I. W. Comparative anatomy of the leaf-bearing Cactaceae, II. Structure and distribution of sclerenchyma in the phloem of *Pereskia, Pereskiopsis* and *Quiabentia. J. Arnold Arb.* 42:144–150. 1961.
4. Carpenter, C. H., L. Leney, H. A. Core, W. A. Côté, Jr., and A. C. Day. *Papermaking fibers.* Tech. Publ. No. 74 of State University College of Forestry at Syracuse University. 1963.

5. Foard, D. E. Pattern and control of sclereid formation in the leaf of *Camellia japonica* L. *Nature* 184:1663–1664. 1959.
6. Foster, A. S. Structure and ontogeny of terminal sclereids in *Boronia serrulata. Amer. J. Bot.* 42:551–560. 1955.
7. Foster, A. S. Plant idioblasts: remarkable examples of cell specialization. *Protoplasma* 46:184–193. 1956.
8. Govindarajalu, E., and N. Parameswaran. On the morphology of the foliar sclereids in the Salvadoraceae. *Beitr. Biol. Pflanz.* 43:41–57. 1967.
9. Griffith, M. M. Development of sclereids in *Osmanthus fragrans* Lour. *Phytomorphology* 18:75–79. 1968.
10. Harris, M., ed. *Handbook of textile fibers.* Washington, Harris Research Laboratories. 1954.
11. Mitchell, J. W., and J. F. Worley. Intracellular transport apparatus of phloem fibers. *Science* 145:409–410. 1964.
12. Parameswaran, N., and W. Liese. On the formation and fine structure of septate wood fibres of *Ribes sanguineum. Wood Sci. Techn.* 3:272–286. 1969.
13. Rao, A. N., and M. Singarayar. Controlled differentiation of foliar sclereids in *Fagraea fragrans. Experientia* 24:298–299. 1968.
14. Rao, A. N., and S. J. Vaz. Morphogenesis of foliar sclereids. II—Effects of experimental wounds on leaf sclereid development and distribution in *Fagraea fragrans. J. Singapore Acad. Sci.* 1:1–7. 1970.
15. Schoch-Bodmer, H. Spitzenwachstum und Tüpfelverteilung bei sekundären Fasern von *Sparmannia Beih. Z. Schweiz. Forstver.* 30:107–112. 1960.
16. Schoch-Bodmer, H., and P. Huber. Wachstumstypen plastischer Pflanzenmembranen. *Mitt. Naturforsch. Ges. Schaffhausen* 21:29–43. 1946.
17. Schoch-Bodmer, H., and P. Huber. Das Spitzenwachstum der Bastfasern bei *Linum usitatissimum* und *Linum perenne. Ber. Schweiz. Bot. Ges.* 61:377–404. 1951.
18. Stant, M. Y. The effect of gibberellic acid on fibre-cell length. *Ann. Bot.* 25:453–462. 1961. The effect of gibberellic acid on cell width and the cell wall of some phloem fibres. *Ann. Bot.* 27:185–196. 1963.
19. Staritsky, G. The morphogenesis of the inflorescence, flower and fruit of *Pyrus nivalis* Jacquin var. *orientalis* Terpó. *Meded. Landbouwhogesch. Wageningen* 70:1–91. 1970.
20. Worley, J. F. Rotational streaming in fiber cells and its role in translocation. *Plant Physiol.* 43:1648–1655. 1968.

7

Epidermis

The epidermis is a system of cells, variable in structure and function, that constitutes the covering of the primary plant body. Many of the structural features of the epidermis can be related to the roles this tissue plays as the layer of cells in contact with the external environment of the plant. The presence of the fatty material, cutin, within the outer wall and on its surface (cuticle) restricts transpiration. Stomata are concerned with gaseous exchange. Because of the compact arrangement of cells and the presence of the relatively tough cuticle, the epidermis offers mechanical support. Its thin walls and root hairs indicate that the epidermis of young roots is specialized with regard to absorption. There is good evidence that the epidermis is the site of light perception involved in circadian leaf movements and photoperiodic induction.[20] The epidermis may last through the life of a given plant part, or it may be later replaced by another protective tissue, the periderm (chapter 12).

The epidermis is usually one layer of cells in thickness (fig. 7.1). In some plants, however, protoderm cells in leaves divide parallel with the surface (periclinally), and their derivatives may divide again so that a tissue of several ontogenetically related layers is produced. Such a tissue is referred to as *multiple epidermis* (fig. 7.2). The velamen of roots (chapter 14) is also an example of a multiple epidermis. The outermost layer of a multiple epidermis in a leaf assumes epidermal characteristics, whereas those beneath commonly develop into a tissue with few or no chloroplasts. One of the functions ascribed to such a tissue is storage of water. In some plants subepidermal layers resemble those of a multiple epidermis but are derived from the ground tissue. Multiple epidermis, therefore, can be identified as such only by developmental studies.

COMPOSITION

The epidermis consists of relatively unspecialized cells composing the groundmass of the tissue and the more specialized cells dis-

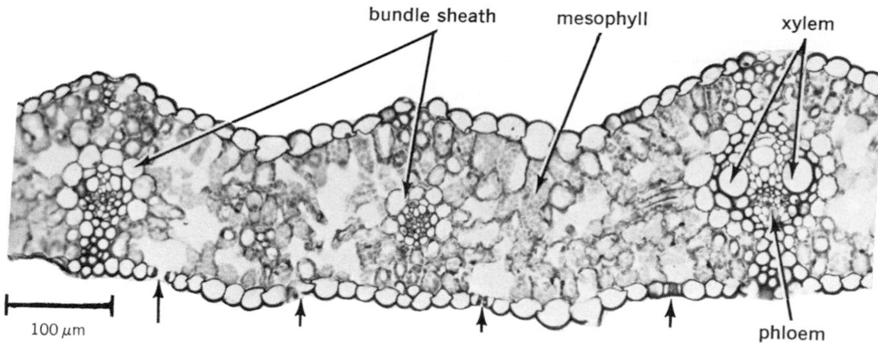

Figure 7.1 Cross section of leaf of barley (*Hordeum vulgare*) showing a single-layered epidermis on both sides of the blade. Stomata are indicated by arrows. The vascular bundles of various sizes are delimited from the mesophyll by bundle sheaths poor in contents. (From K. Esau, *Hilgardia* 27: 15–69, 1957.)

persed through this mass. The ground cells of the epidermis vary in depth but are often tabular in shape (fig. 7.3). In elongated plant parts such as stems, petioles, vein ribs of leaves, and leaves of most monocotyledons, the epidermal cells are elongated parallel with the long axis of the plant part (fig. 7.4). In leaves, petals, ovaries, and ovules the epidermal cells may have wavy vertical (anticlinal) walls.

Epidermal cells have living protoplasts and may store various products of metabolism. The cells contain plastids which usually develop only few grana and are, therefore, deficient in chlorophyll. Photosynthetically active chloroplasts, however, occur in the

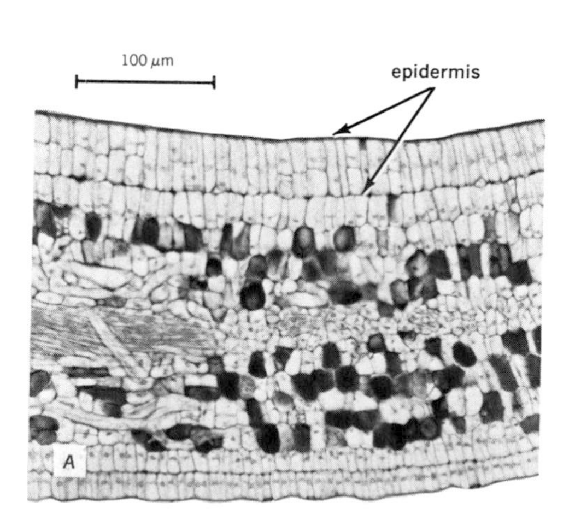

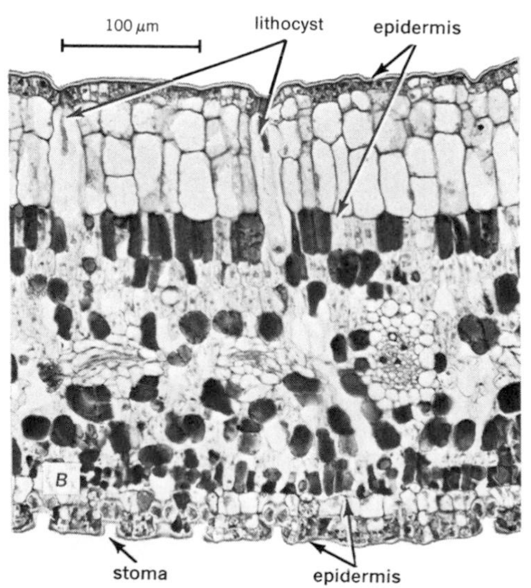

Figure 7.2 Cross sections of leaf of *Ficus elastica* showing development of multiple epidermis. In upper epidermis of *A*, many cells have divided periclinally; in *B*, this epidermis is three to four cells in depth. Lithocysts do not divide. Lower epidermis also becomes multiple.

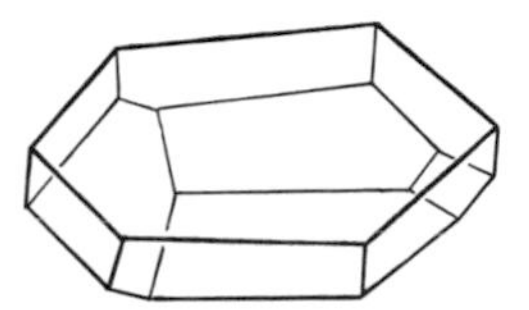

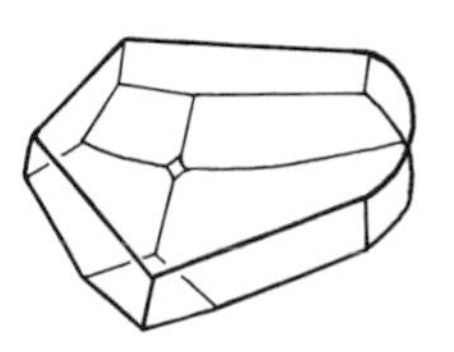

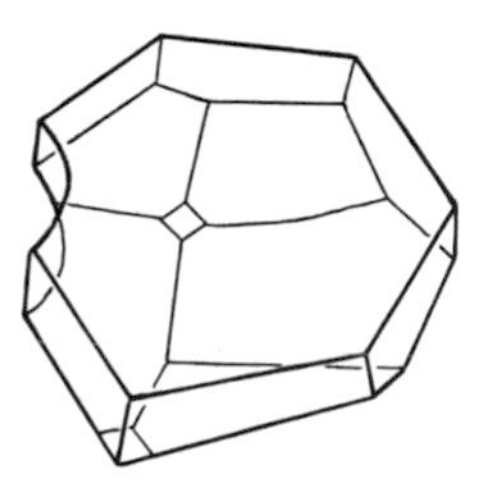

Figure 7.3 Three-dimensional aspect of epidermal cells of *Aloe aristata* (Liliaceae) leaf. The upper face in each drawing is the outer face of the cell. On the opposite side are the faces of contact with the subjacent mesophyll cells. (Redrawn from E. B. Matzke, *Amer. J. Bot.* 34: 182–195, 1947.)

epidermis of plants of shady habitats. Starch and protein crystals may be present in epidermal plastids, anthocyanins in vacuoles.

Specialized epidermal cells are represented, first of all, by the guard cells of the stomata, which may or may not be associated with subsidiary cells. Many plants bear epidermal appendages called trichomes, diverse in form, structure, and function. Plant hairs are common representatives of trichomes. Cells containing tannins, oils, crystals, and other materials are often scattered as idioblasts or arranged in specific patterns. In Poaceae (grasses), for example, small cells (silica cells) filled with silica solidified into bodies of various shapes and cells with suberized walls (cork cells) occur in pairs which alternate with elongated epidermal cells (fig. 7.4,*B*). Poaceae and other monocotyledons often have rows of enlarged bubblelike epidermal cells called bulliform cells (chapter 19). Plants of various taxa contain sclerenchyma cells in the epidermis. Poaceae may have epidermal fibers over 300 microns long. Sclereids occupy considerable expanses of, or the entire epidermis, in some

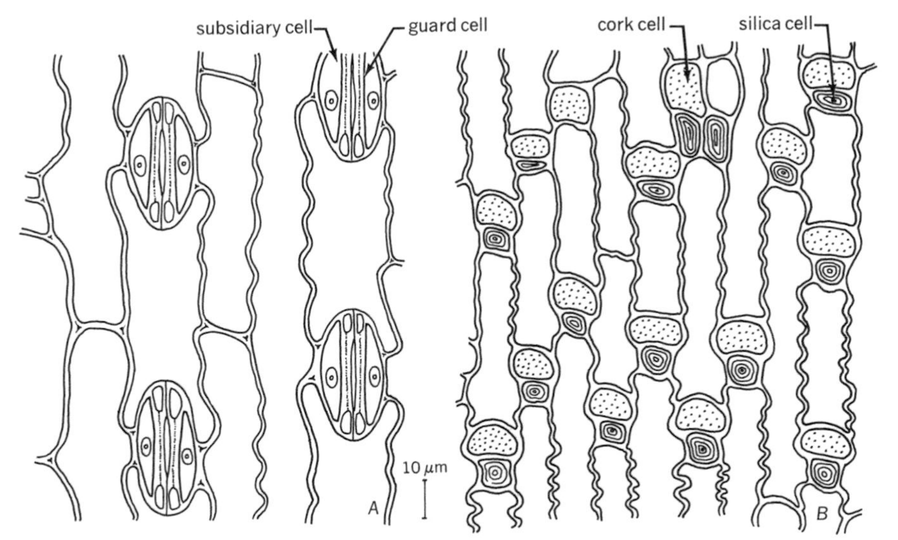

Figure 7.4 Epidermis of a grass—sugar cane (*Saccharum*)—in surface views. *A*, lower epidermis of leaf with stomata. *B*, epidermis of stem with cork cells and silica cells.

seeds and scales (chapter 6). Secretory cells may form the epidermis in some plant parts (chapter 13).

DEVELOPMENTAL ASPECTS

The epidermis differentiates from the meristematic surface layer, the protoderm, or dermatogen, which originates by periclinal divisions in the developing embryo (chapter 24). The various phenomena that cause the diversification of cells within an originally homogeneous meristematic tissue (chapter 2) may be recognized in the differentiation of the epidermis. In grass epidermis, for example, asymmetric divisions result in the formation of short and long cells and only the short cells produce the various specialized cells or cell complexes: trichomes, guard cells of the stomata, and cork-silica cell pairs.[13] The dense appearance of the short cell as contrasted with the highly vacuolated state of the long cell is interpreted as a result of cytoplasmic polarization preceding cell division, the cytoplasm accumulating at the end of the cell that will become the smaller of the two daughter cells.[3] A more accurate description of the course of events is that the nucleus and associated cytoplasm migrate to the part of the cell where mitosis is to occur. Similar nuclear migrations occur in cells contiguous with the precursor of guard cells when subsidiary cells are formed (fig. 7.12).

In some taxa, unequal divisions occur when root hair-forming cells, the *trichoblasts,* originate[5] (fig. 7.5). The small dense trichoblastic daughter cells show considerable cytologic and biochemical differentiation. In *Hydrocharis,* for example, the trichoblasts differ from their long sister cells in having larger nuclei and nucleoli, simpler plastids, more intense enzymic activity, and larger amounts of nucleohistone, total protein, RNA, and nuclear DNA.[7] All these features suggest a de-

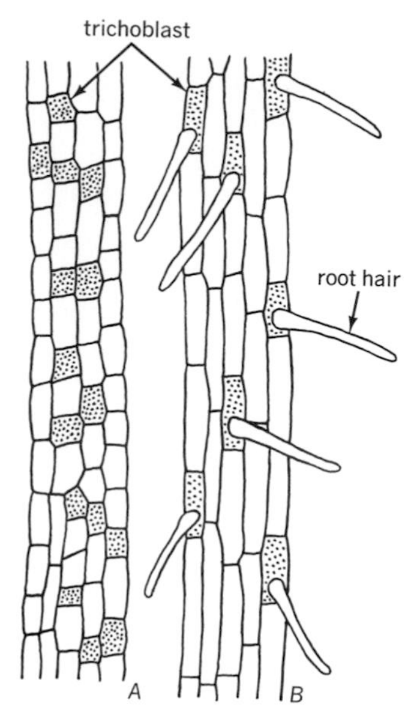

Figure 7.5 Strips of epidermis of *Elodea canadensis* roots showing young, root-hair forming, small trichoblasts (stippled cells in *A*) and similar cells after root hairs had developed from them (*B*). (Adapted from Cormack.[5])

layed maturation of the cell that is preparing to undergo the intensive unilateral growth giving rise to a root hair.

The cork and silica cells of the grass epidermis (fig. 7.4), which arise from a common precursor, show contrasting developments. The cork cell expands and is suberized while the silica cell becomes devoid of its protoplast and silicifies.[13] The deposition of silica apparently occurs by a passive nonmetabolic mechanism.[30]

The patterned distribution of specialized cells in the epidermis is usually explained by the concept that localized cells or groups of cells engaged in meristematic activity (*meristemoids*) within a differentiating tissue prevent other meristemoids from appearing in the immediate vicinity. Thus, developing stomata are thought to inhibit the initiation of other stomata nearby.[3,16] Exceptions to this rule are not uncommon, however.

Although the mature epidermis is generally passive with regard to meristematic activity, it often retains the potentiality for growth for a long time. In perennial stems in which the periderm arises late in life, or not at all, the epidermis continues to divide in response to the circumferential expansion of the axis. If a periderm is formed, the source of its meristem, the phellogen (chapter 12), may be the epidermis. Adventitious buds can arise in the epidermis.[10] With improvements in tissue culture techniques, production of embryo-like structures from epidermal cells has become possible.[15]

CELL WALL

Epidermal cell walls vary in thickness in different plants and in different parts of the same plant. In the thin-walled epidermis, the outer wall is frequently the thickest. An epidermis with exceedingly thick walls is found in the leaves of conifers (chapter 19); the wall thickening, which is probably secondary, almost obliterates the cell lumina and is lignified. The cell wall may become silicified as in grasses. Primary pits and plasmodesmata generally occur in the anticlinal and the inner periclinal (next to the mesophyll) walls of the epidermis. As was mentioned in chapter 4, regions with wide interfibrillar spaces, the teichodes (formerly called ectodesmata), are sometimes demonstrable in the outer walls of the epidermis.

An outstanding feature of the epidermis is the presence of the fatty wall substance, cutin. This material occurs as an incrustation of the outer walls and as a separate layer, the cuticle, on the outer surface of the epidermis.[19] (The process of impregnation with cutin is called *cutinization,* the formation of cuticle, *cuticularization.*) The cuticle varies in thickness and its development is affected by environmental conditions.[5] Cuticle is characteristic of all plant surfaces exposed to the air and is sometimes reported to be present in the absorbing region of the root, even the root hairs. The cuticle is commonly covered with wax either in smooth flat-lying forms or as rods or filaments growing outward from the surface.[11] The latter type of deposit appears as the "bloom" that makes the leaf glaucous. The epicuticular waxes form a variety of patterns which are effectively revealed by scanning electron microscopy.[8] The cuticle, the cutinized cell wall beneath it, and the surface wax serve to reduce loss of water. The commercial practice of dipping grapes in chemicals that accelerate the drying of fruit causes a close adpression of wax platelets and their parallel orientation. This change probably facilitates the movement of water from the fruit to the atmosphere.[25] The epicuticular wax is also responsible for enhancing the ability of the epidermal surface to shed water.[9,27]

The combination of cuticle and cutinized wall is highly complex. The structural components in this combination are cellulose, pectic compounds, cutin, waxes, and several other incrusting substances. The outer wall of the upper epidermis of pear leaf depicted in figure 7.6 exemplifies the relation between the various wall components.[23] Beginning at bottom of the diagram, the layers are: cellulose cell wall; layer of pectin that partly penetrates the cellulose wall below and the cuticle above; cuticle with birefringent waxes embedded near its surface; platelets of epicuticular wax. The cuticle is penetrable where the birefringent waxes are interrupted by isotropic material. Birefringence is often discontinuous and the wax layers are thinner over the anticlinal walls. The surface over the anticlinal walls is known to lose water and to take up materials applied to the leaf more readily than other regions of the leaf surface. The diagram indicates that the cuticle is depressed above the anticlinal wall, a common feature in leaf epidermis.

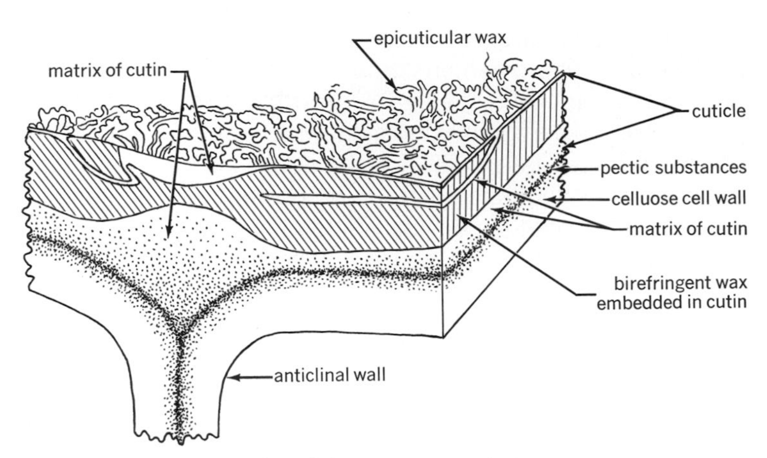

Figure 7.6 Diagrammatic representation of the outer cell wall of the upper epidermis in pear (*Pyrus communis*) leaf with details of the cuticular and waxy layers. (Adapted from Norris and Bukovac.[23])

Cutin and waxes (or their precursors) are synthesized in living protoplasts and migrate to the surface through the cell wall. There is no agreement as to whether these materials pass by moving throughout the wall or through special channels (teichodes, for example). Similarly, epicuticular waxes are visualized as either having no special pathways for passing to the surface or having access to pores in the cuticle. As cutin and waxes migrate through the wall to the surface they also impregnate this wall to a smaller or a greater extent. In exceptional instances, cuticular layers are also formed in cortical cells and give rise to a protective tissue called *cuticular epithelium*.[4]

STOMATA

Occurrence and function

The stomata are openings (the stomatal pores, or apertures) in the epidermis bound by two specialized epidermal cells, the *guard cells* (fig. 7.7), which by changes in shape bring about the opening and closing of the aperture. It is convenient to apply the term stoma to the entire unit, the pore and the two guard cells. The stoma may be surrounded by cells that do not differ from other ground cells of the epidermis (fig. 7.13,*A*). Contrariwise, in many plants the stomata are flanked or surrounded by cells that differ in shape, and sometimes in content, from the ordinary epidermal cells. These distinct cells are called *subsidiary cells* of the stoma (fig. 7.13,*B,C*). The subsidiary cells participate in the osmotic changes involved in movements of the guard cells.

Stomata occur on all aerial parts of the plant, but they are most abundant on leaves. Roots usually lack stomata. Stomatal frequency varies greatly. It varies on different parts of the same leaf and on different leaves on the same plant, and it is influenced by environmental conditions. In leaves, stomata may occur on both sides or only on one side, usually the lower. Stomata also vary in the level of their position in the epidermis. Some are even with the other epidermal cells, others are raised above or sunken below the surface.

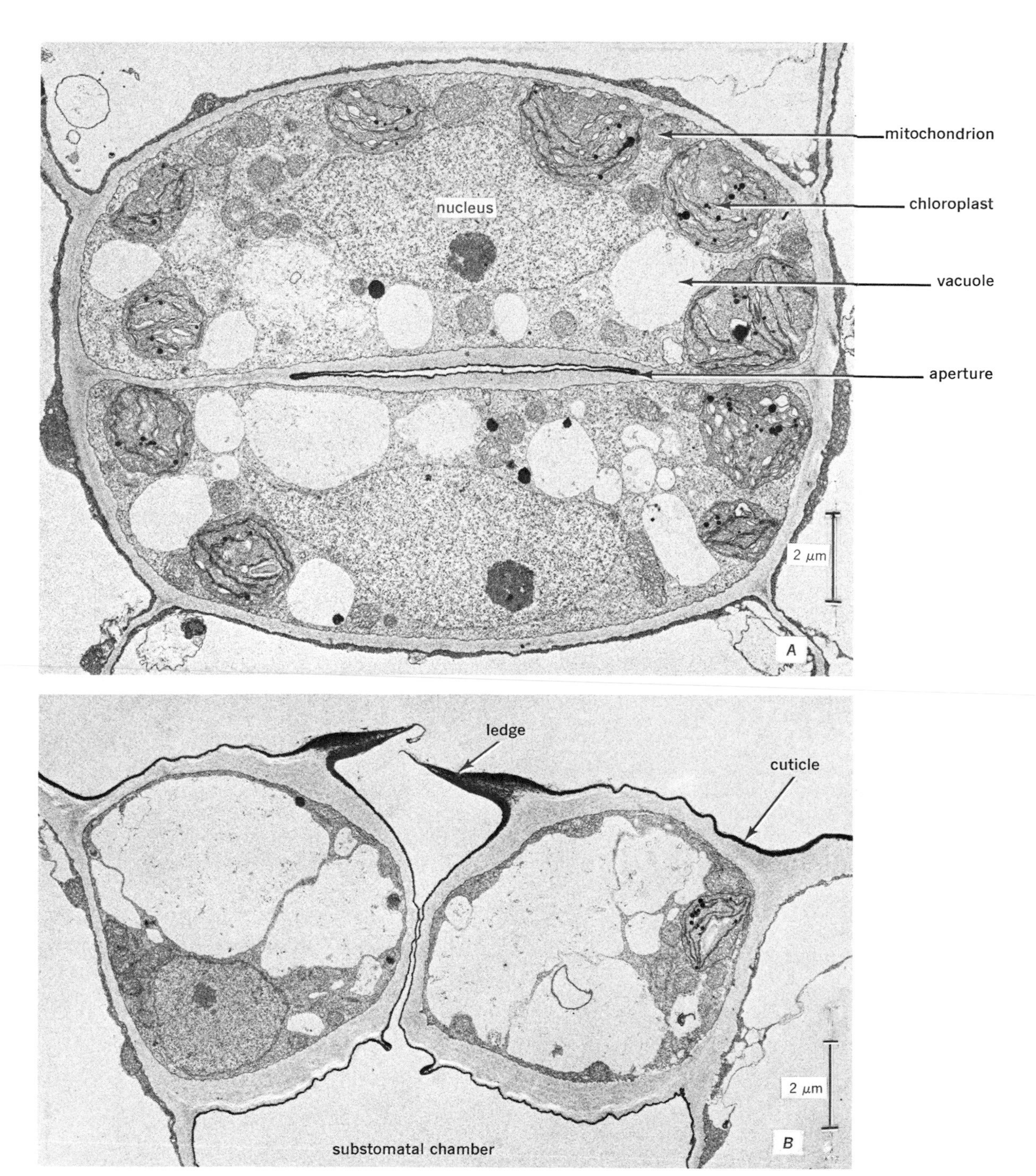

Figure 7.7 Electron micrographs of stomata from a sugar beet (*Beta vulgaris*) leaf seen from the surface (*A*) and in transection (*B*).

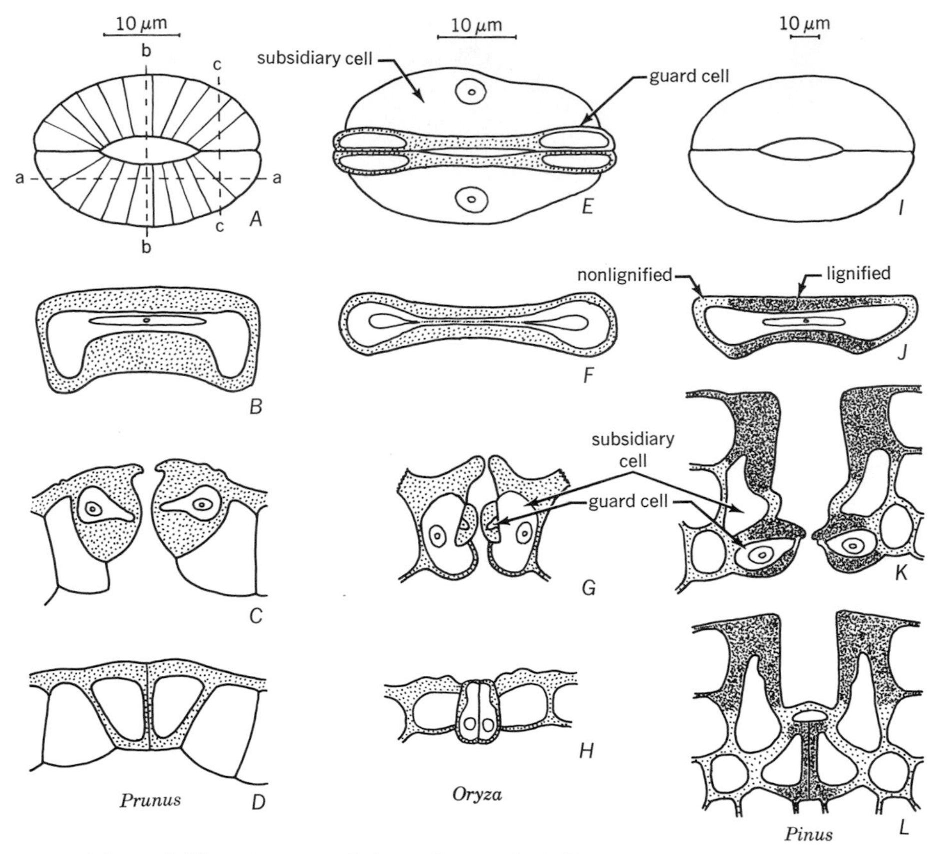

Figure 7.8 Stomata in representatives of different groups of plants. Stomata in *A, E, I* are shown from the surface. Other drawings show sections of stomata made in the planes indicated in *A: B, F, J,* plane *aa; C, G, K,* plane *bb; D, H, L,* plane *cc. E,* guard cells are shown in a high focal plane so that the lumen is not visible in the narrow part of the cell. The radial lines in *A* indicate radial micellation in the cell wall. (*Prunus* and *Pinus* after Esau, *Plant Anatomy,* 2nd ed. John Wiley & Sons, 1965.)

The guard cells of dicotyledons (fig. 7.8,*A–D*) are commonly crescent shaped, with rounded ends (kidney shaped) as seen from the surface, and have ledges of wall material on the upper or both the upper and the lower sides. The cells are covered with a cuticle that extends over the surfaces facing the stomatal pore and the substomatal chamber (fig. 7.7). Stomata may be completely covered with wax.[28] Each guard cell has a prominent nucleus and chloroplasts which periodically accumulate starch. The vacuolar system is dissected to variable degrees.

Among the guard cells of monocotyledons, those of the Poaceae have a rather uniform, specific structure (fig. 7.8,*E–H*). As seen from the surface, the cells are narrow in the middle and enlarged at both ends. Electron microscopy has revealed that the two guard cell protoplasts are interconnected through pores in the common wall between the enlarged bulbous ends. Because of this protoplasmic continuity, the guard cells must be regarded as a single physiological unit in which changes in turgor are immediately equalized. The pores appear to result from incomplete development of the wall.[14,31] The nucleus extends through the entire cell and is ovoid at either end and almost threadlike in the middle. There are two subsidiary cells, one on each side of the stoma.

The stomata of the gymnosperms (fig. 7.8,*I–L*) are commonly deeply sunken and sometimes appear as though suspended from the subsidiary cells which overarch them. The outstanding feature of these stomata is that the walls of the guard cells and of the subsidiary cells are partly lignified and that a strip of thin wall facing the pore is not lignified (fig. 7.8,*K*).

Although the guard cells of the major taxa have their distinguishing characteristics, they share the feature that the anticlinal wall away from the pore (dorsal wall) is thinner, and therefore more flexible, than the other walls, especially the tangential walls. This feature is commonly cited as having a causal relation to the ability of the guard cells to change their shape in response to turgor changes and thus to control the size of the stomatal opening. According to a different hypothesis, the radial arrangement of cellulose microfibrils (radial micellation[1,26]) in the guard cell wall (indicated by radially arranged lines in fig. 7.8,*A*) plays a more important role in stomatal movements than the distribution of wall thickening. The radial orientation of microfibrils in guard cell walls was recognized by polarization optics and electron microscopy.[26]

The probable effect of the radial micellation on the movement of guard cells was explored by mathematical analysis and study of appropriately constructed models. The models consisted of inflatable latex cylinders connected in pairs at their ends and made rigid with masking tape. The tape was placed so as to simulate the radial micellation and to prevent the cylinders from increasing in length during inflation. Figure 7.9 depicts the results of some of the experiments on the basis of which the authors singled out the radial orientation of microfibrils as the crucial feature in the transmission of the movement of the dorsal guard cell wall to the stomatal slit.

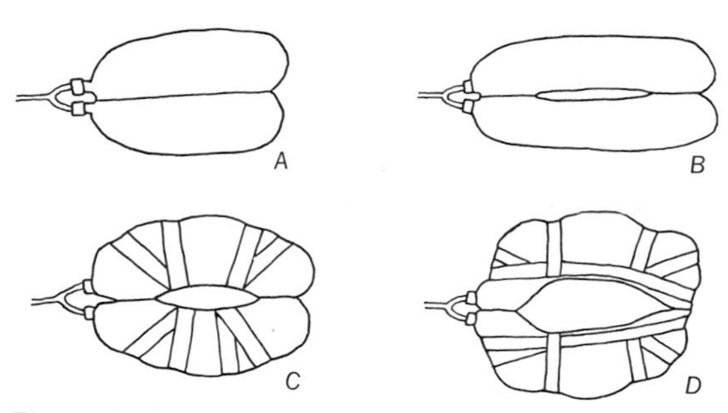

Figure 7.9 Models for studying the effect of radial arrangement of microfibrils in guard cell walls on the opening of stomata. *A,* two latex cylinders connected at their ends and partially inflated. *B,* the same model at higher pressure. A narrow slit is visible. *C,* bands of tape simulate radial micellation on the cylinders, which are inflated. Slit wider than in *B*. *D,* radial micellation extends farther to the ends of the cylinders, and some tape is present along the "ventral wall." Inflation has induced the formation of a slit wider than in *C*. (From photographs in Aylor, Parlange, and Krikorian.[1])

The mechanism of stomatal movement is a subject of intensive studies and discussions.[18,26] Transport of potassium between guard cells and contiguous cells is one of the factors in guard cell movement, the stoma being open in the presence of increased amounts of potassium ion. In stomatal responses to light, changes in CO_2 are involved. During stomatal opening, starch disappears from chloroplasts at the same time as K^+ ions enter the guard cells and, during stomatal closure, the reappearance of starch parallels the loss of K^+ ion. The early theory that the breakdown of starch contributes to the increase of osmotic pressure in the guard cells, because of formation of sugar, has been replaced by the concept that starch hydrolysis may provide the organic anions with which potassium uptake is associated.

The question whether plasmodesmata are involved in the exchange of materials between the stoma and the adjacent cells has not been answered unequivocally. In some species, plasmodesmata have not been detected in guard cell walls or were seen only in immature stomata. In other species, plasmodesmata have been recognized in mature stomata.[29,31]

Development and mature configurations of stomatal complexes

Stomata begin to develop in a leaf shortly before the main period of meristematic activity in the epidermis is completed and continue to arise through a considerable part of the later extension of the leaf by cell enlargement. In leaves with parallel venation, as in most monocotyledons, and with the stomata arranged in longitudinal rows, the formation of stomata begins at the apices of the leaves and progresses in the downward direction. In the netted-veined leaves, as in most dicotyledons, different developmental stages are mixed in a mosaic fashion.

In the development of stomata in angiosperms, the mother cell, or precursor, of the guard cells commonly originates by an unequal division of a protodermal cell and is the smaller of two cells resulting from such a division (fig. 7.10,*A,D*). The small cell divides into two guard cells (fig. 7.10,*A,B,E*) which, through differential expansion, acquire their characteristic shape. The intercellular substance between the guard cells swells (fig. 7.10,*B*), and the connection between the cells is weakened. They separate in their median parts, and thus the stomatal opening is formed (fig. 7.10,*C,F*). Various spatial readjustments occur between the guard cells and the adjacent cells so that the guard cells may be elevated above or lowered below the surface of the epidermis. The adjacent cells may overarch the guard cells or grow under them into the substomatal chamber.

Subsidiary or other kinds of neighboring

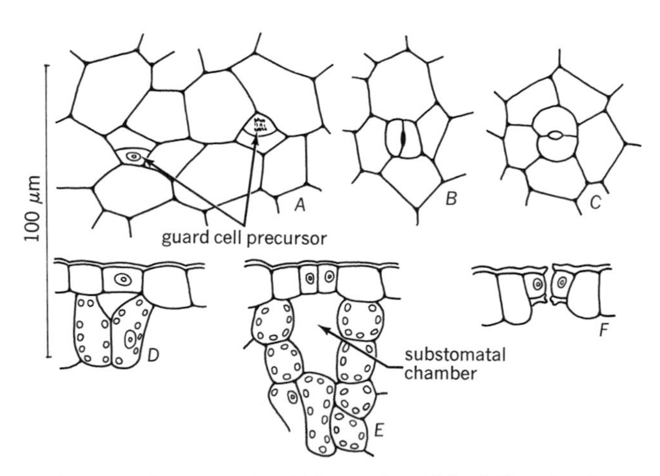

Figure 7.10 Development of stomata in a sugar beet (*Beta vulgaris*) leaf. Guard cell precursors (mother cells) have been formed by a division of a protodermal cell (*A, D*). The precursor has divided into two guard cells (*B, E*). Stomatal opening has been formed (*C, D*). Surface (*A–C*) and sectional (*D–F*) views.

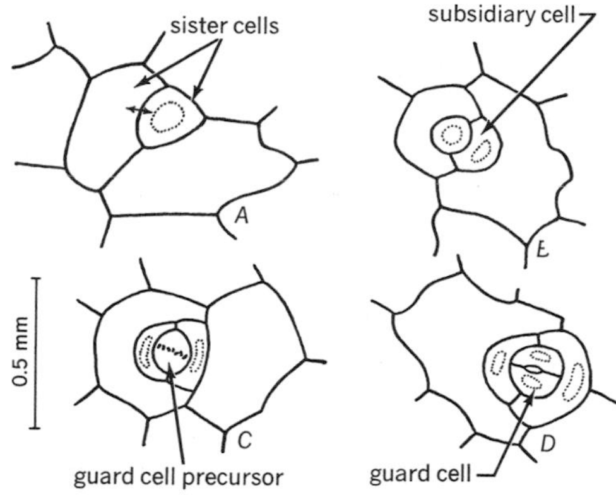

Figure 7.11 Development of stoma with mesogenous subsidiary cells in a leaf of *Thunbergia erecta*. *A*, epidermal cell has divided and given rise to a small precursor of the stomatal complex. *B*, the precursor has divided setting apart one subsidiary cell. *C*, the second subsidiary cell and the guard cell precursor have been formed. *D*, the stomatal complex has been completed by division of the guard cell precursor. (Adapted from G. S. Paliwal, *Phytomorphology* 16: 527–539, 1966.)

cells may arise from the same precursor as the stoma or from cells that are not directly related ontogenetically with the mother cell of the guard cells. On this basis, three major stomatal categories have been formulated:[24] *mesogenous*—guard cells and neighboring cells, which may or may not differentiate as subsidiary cells, have common origin (fig. 7.11); *perigenous*—neighboring or subsidiary cells do not have common origin with the guard cells (fig. 7.12); *mesoperigenous*—at least one neighboring or subsidiary cell is directly related to the stoma, the others are not.

In the development of a stoma with mesogenous subsidiary cells (fig. 7.11), the precursor of the stomatal complex is formed by an asymmetric division of a protodermal cell, and two subsequent divisions result in the partitioning of the precursor into a mother cell of the guard cells and two subsidiary cells. One more division leads to the formation of two guard cells.

The origin of perigenous subsidiary cells is graphically illustrated in the differentiation of a grass stoma (fig. 7.12). The guard cell precursor is the short daughter cell formed through an unequal division of a protodermal cell. The subsidiary cells are formed along the sides of this short cell by unequal divisions of two contiguous cells. Growth adjustments after the formation of guard cells make the subsidiary cells appear as integral parts of the stomatal complex.

Different developmental sequences result in different configurations of stomatal complexes. The patterns formed by fully differentiated guard cells and neighboring cells, as seen from the surface, are used for taxonomic purposes. It must be remembered, however, that morphologically similar configurations may be dissimilar developmentally and that the developmental sequence cannot be judged from the final pattern.[32] Metcalfe and Chalk,[21] therefore, use their terminology concerning stomatal complexes in dicotyledons solely in a descriptive manner without reference to the development. They distinguish the following principal types of stomatal configurations: *anomocytic* (irregular-celled)—no subsidiary cells are present (fig. 7.13,*A*); *anisocytic* (unequal-celled)—three subsidiary cells, one distinctly smaller than the other two, surround the stoma (fig. 7.13,*B*); *paracytic* (parallel-celled)—one or more subsidiary cells border the stoma parallel with the long axes of the guard cells (fig. 7.13,*C*); *diacytic* (cross-celled)—two subsidiary (or neighboring) cells, their common walls at right angles to the long axes of the guard cells, surround the stoma (fig. 7.13,*D*); *actinocytic* (radiate-celled)—several subsidiary cells, their long axes perpendicular to the outline of the guard cells, surround the stoma. The same species may show more

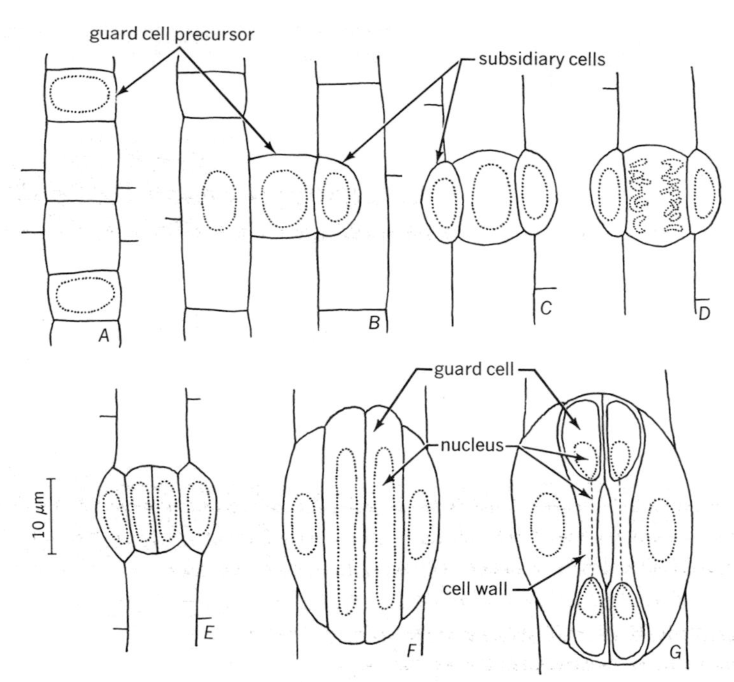

Figure 7.12 Development of stomatal complex in oat (*Avena sativa*) internode. The subsidiary cells are perigenous. *A*, the two short cells are guard cell precursors. *B*, left, the nucleus of a long cell is in position to divide to form a subsidiary cell; right, subsidiary cell has been formed. *C*, guard cell precursor before mitosis. *D*, guard cell precursor in anaphase. *E*, stomatal complex of two guard cells and two subsidiary cells is still immature. *F*, cells of the stomatal complex have elongated. *G*, stomatal complex is mature. (From photomicrographs in Kaufman et al.[14])

than one type of stoma and the pattern may change during leaf development.

In most monocotyledons, the configuration of the stomatal complex is rather precisely related to the developmental sequence. Having examined about 100 species representing most families of the monocotyledons Tomlinson[32] recognized the following main configurations of stomatal complexes resulting from specific developmental sequences (fig. 7.14). The guard cell precursor is usually in contact with four neighboring cells (*B*). These may not divide and remain as *contact cells* in the mature complex (*F*). The neighboring cells may divide and produce *derivatives* (stippled cells). The stomatal complex now becomes definable as guard cells and a combination of neighboring cells and their derivatives (*G*) or as guard cells and derivatives of neighboring cells (*E,H*). Thus the contact cells of the stoma are either all derivatives (*E,H*) or derivatives and undivided neighboring cells (*G*). In some families, the early derivatives of the neighboring cells are formed by divisions in planes oblique to the precursor of the guard cells (*C–E*), in others no oblique divisions occur (*F–H*).

TRICHOMES

Trichomes (fig. 7.15 and chapter 13) are highly variable appendages of the epidermis, including glandular (or secretory) and non-

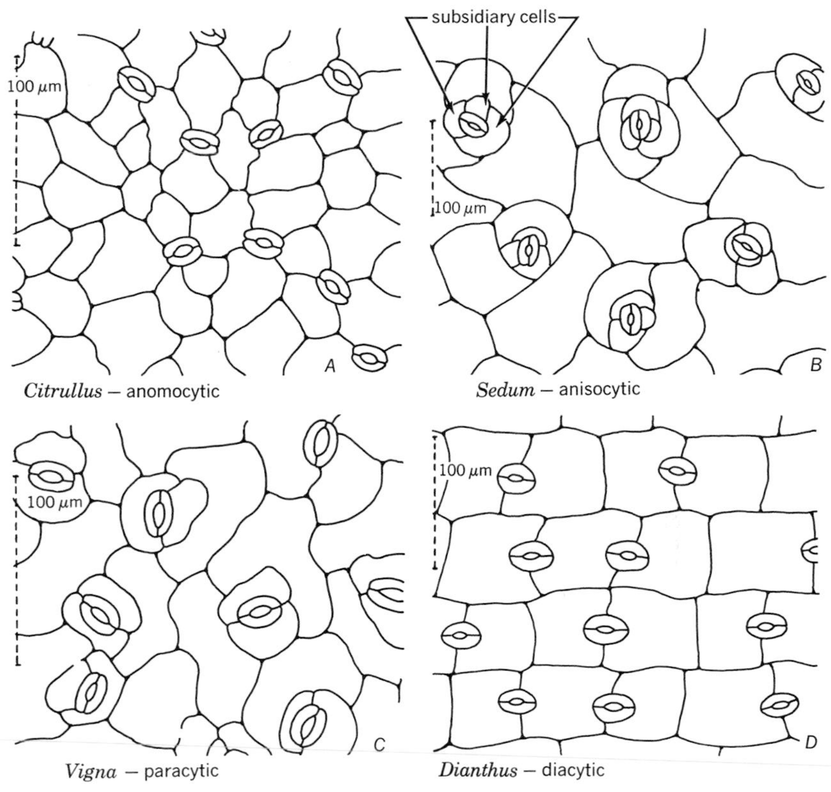

Figure 7.13 Epidermis in surface views illustrating patterns formed by guard cells and neighboring cells.

glandular hairs, scales, papillae, and the absorbing hairs of roots.[33] They occur in all parts of the plant and may persist through the life of a plant part or may fall off early. Some of the persisting hairs remain alive; others die and become dry. Although trichomes vary widely in structure within larger and smaller groups of plants, they are sometimes remarkably uniform in a given taxon and may be used for taxonomic purposes.[6,21]

The function of the various trichomes is obscure. Studies of the effect of plant hairs upon loss of water have not produced uniform results, but it is likely that trichomes can insulate the mesophyll from excessive heat.[2] A study of vesiculated hairs on leaves of *Atriplex halimus* (chapter 13) suggests that this type of hair may serve to remove salts from the leaf tissue and thus prevent an accumulation of toxic salts in the plant.[22] Trichomes may provide a defense against insects.[17] In numerous species, trichome density is negatively correlated with insect responses in feeding and oviposition and with the nutrition of larvae. Hooked trichomes impale insects and larvae. Secretory trichomes participate in chemical defense.

Trichomes are classified into morphological categories. Some of these categories are: (1) hairs, which may be unicellular or multicellular, glandular (chapter 13) or nonglandular (fig. 7.15,*A,B,J,Q*); (2) scales, or peltate hairs (fig. 7.15,*H,I*); (3) water vesicles, which are enlarged epidermal cells (fig. 7.15,*N*); (4) root hairs (chapter 14). The hairs may be tufted (fig. 7.15,*C,D*), star shaped

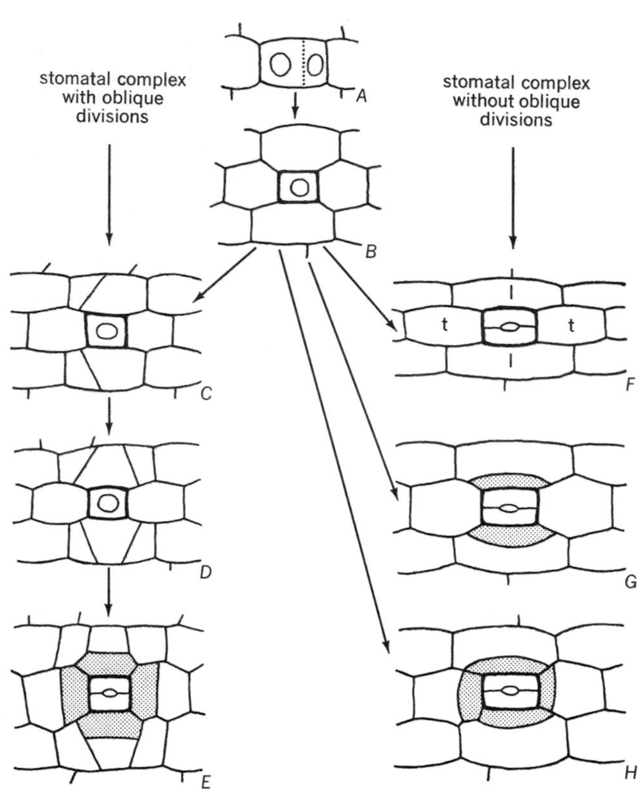

Figure 7.14 Examples of types of stomatal development in monocotyledons. Diagrammatic. *A,* nonequational division results in formation of, *B,* small guard-cell precursor surrounded by four neighboring cells in cruciate arrangement. *C–E,* oblique and other divisions in neighboring cells result in formation of four derivatives (stippled) in contact with the guard cells. *F–H,* no oblique divisions in formation of stomatal complexes: *F,* original neighboring cells, two lateral (*l*) and two terminal (*t*), become contact cells; *G,* derivatives (stippled) of the two lateral neighboring cells and the two undivided terminal neighboring cells become contact cells; *H,* derivatives (stippled) of four neighboring cells become contact cells. *E,* palm type; *G,* grass type. (Adapted from Tomlinson.[32])

(stellate, fig. 7.15,*E*), or branched (dendroid, fig. 7.15,*F*). The hairs of cotton seeds (source of commercial cotton[12]) are unicellular and develop secondary walls at maturity (fig. 7.15,*K–M*).

The development of trichomes varies in complexity in relation to their final form and structure. The unicellular cotton hair, for example, is formed through elongation of a protodermal cell (fig. 7.15,*K,L*). The secondary wall develops after the elongation is completed (fig. 7.15,*M*). Multicellular trichomes show characteristic patterns of cell division and cell growth, some simple (fig. 7.15,*O–Q*) others complex.[12]

REFERENCES

1. Aylor, D. E., J.-Y. Parlange, and A. D. Krikorian. Stomatal mechanics. *Amer. J. Bot.* 60:163–171. 1973.

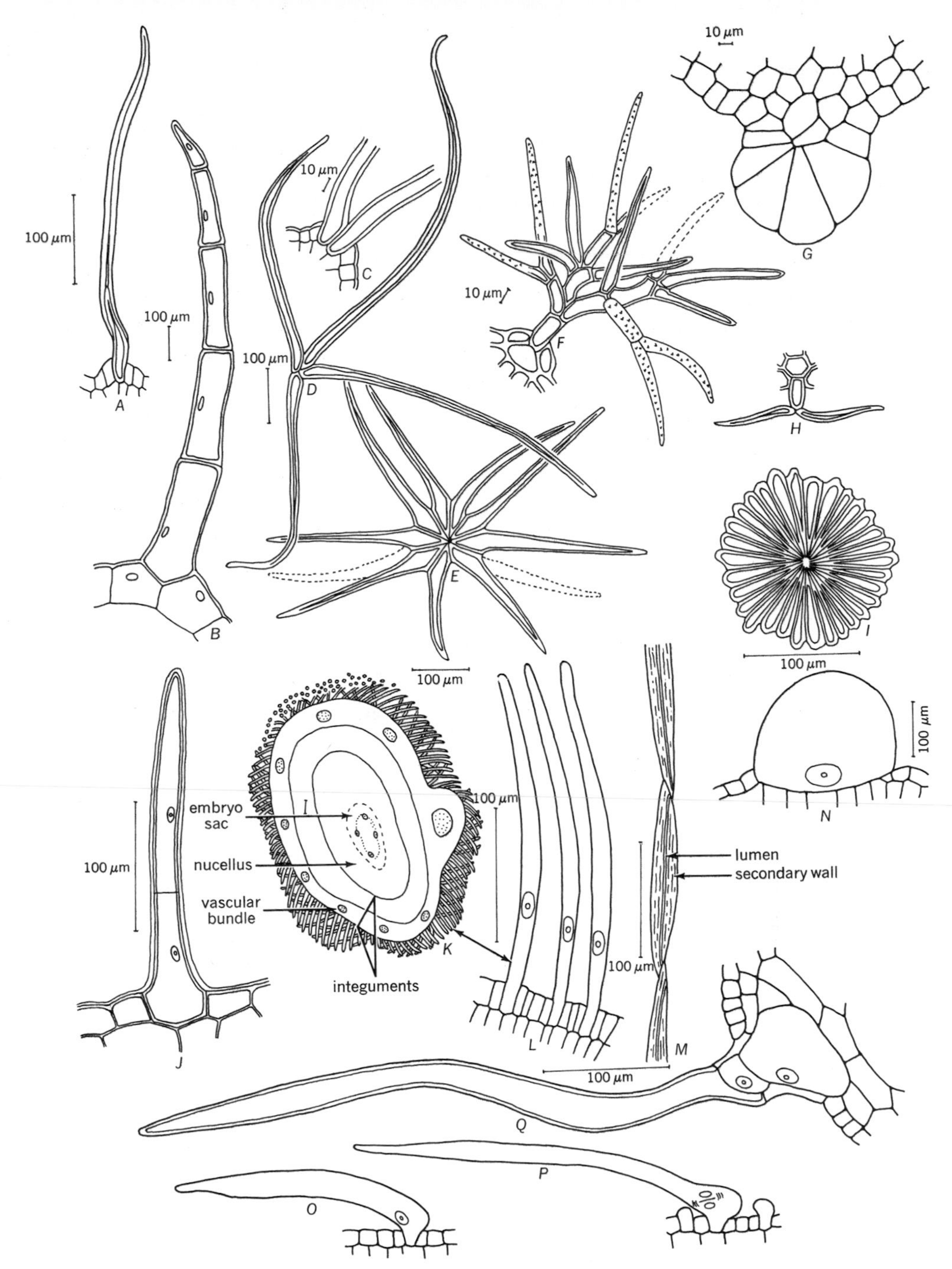

Figure 7.15 Trichomes. *A*, simple hair from *Cistus* leaf. At its base is a compartment formed by deposition of a siliceous wall. *B*, uniseriate hair from *Saintpaulia* leaf. *C, D,* tufted hair from leaf of cotton (*Gossypium*). E, stellate hair from leaf of alkali mallow (*Sida*). *F*, dendroid hair from lavender leaf (*Lavandula*). *G*, short multicellular hair from leaf of potato (*Solanum*). *H, I,* peltate scale from leaf of olive (*Olea*). *J*, bicellular hair from stem of *Pelargonium*. *K–M,* cotton (*Gossypium*). Epidermal hairs from seed (*K*) in young stage (*L*) and mature, with secondary walls (*M*). *N*, water vesicle of *Mesembryanthemum*. *O–Q,* hairs in three stages of development from leaf of soybean (*Glycine*).

2. Black, R. F. The leaf anatomy of Australian members of the genus *Atriplex*. I. *Atriplex vesicaria* Heward and *A. nummularia* Lindl. *Aust. J. Bot.* 2:269–286. 1954.
3. Bünning, E. Morphogenesis in plants. *Survey of Biological Progress* 2:105–140. New York, Academic Press. 1952.
4. Calvin, C. L. Anatomy of the aerial epidermis of the mistletoe, *Phoradendron flavescens*. *Bot. Gaz.* 131:62–74. 1970.
5. Cormack, R. G. H. The development of root hairs by *Elodea canadensis*. *New Phytol.* 36:19–25. 1937.
6. Cowan, J. M. *The Rhododendron leaf; a study of epidermal appendages*. Edinburgh, Oliver and Boyd. 1950.
7. Cutter, E. G., and L. J. Feldman. Trichoblasts in *Hydrocharis*. I. Origin, differentiation, dimensions, and growth. *Amer. J. Bot.* 57:190–201. 1970. II. Nucleic acids, proteins and a consideration of cell growth in relation to polyploidy. *Amer. J. Bot.* 57:202–211. 1970.
8. Davis, D. G. Scanning electron microscopic studies of wax formation on leaves of higher plants. *Can. J. Bot.* 49:543–546. 1971.
9. Eglington, G., and R. J. Hamilton. Leaf epicuticular waxes. *Science* 156:1322–1335. 1967.
10. Gulline, H. F. Experimental morphogenesis in adventitious buds in flax. *Aust. J. Bot.* 8:1–10. 1960.
11. Hall, D. M., A. I. Matus, J. A. Lamberton, and H. N. Barber. Infraspecific variation in wax on leaf surfaces. *Aust. J. Biol. Sci.* 18:323–332. 1965.
12. Joshi, P. C., A. M. Wadhwani, and B. M. Johri. Morphological and embryological studies of *Gossypium* L. *Proc. Natl. Inst. Sci. India* 33:37–93. 1967.
13. Kaufman, P. B., L. B. Petering, and S. L. Soni. Ultrastructural studies on cellular differentiation in internodal epidermis of *Avena sativa*. *Phytomorphology* 20:281–309. 1970.
14. Kaufman, P. B., L. B. Petering, C. S. Yocum, and D. Baic. Ultrastructural studies on stomata development in internodes of *Avena sativa*. *Amer. J. Bot.* 57:33–49. 1970.
15. Konar, R. N., and K. Nataraja. Experimental studies in *Ranunculus sceleratus* L. Development of embryos from the stem epidermis. *Phytomorphology* 15:132–137. 1965.
16. Korn, R. W. Arrangement of stomata on the leaves of *Pelargonium zonale* and *Sedum stahlii*. *Ann. Bot.* 36:325–333. 1972.
17. Levin, D. A. The role of trichomes in plant defense. *Quart. Rev. Biol.* 48 (Pt. 1):3–15. 1973.
18. Levitt, J. The mechanism of stomatal movement—once more. *Protoplasma* 82:1–17. 1974.
19. Martin, J. T., and B. E. Juniper. *The cuticles of plants*. London, Edward Arnold. 1970.
20. Mayer, W., I. Moser, and E. Bünning. Die Epidermis als Ort der Lichtperzeption für circadiane Laubblattbewegungen und photoperiodische Induktionen. *Z. Pflanzenphysiol.* 70:66–73. 1973.
21. Metcalfe, C. R., and L. Chalk. *Anatomy of the dicotyledons*. 2 vols. Oxford, Clarendon Press. 1950.
22. Mozafar, A., and J. R. Goodin. Vesiculated hairs: a mechanism for salt tolerance in *Atriplex halimus* L. *Plant Physiol.* 45:62–65. 1970.
23. Norris, R. F., and M. J. Bukovac. Structure of the pear leaf cuticle with special

reference to cuticular penetration. *Amer. J. Bot.* 55:975–983. 1968.

24. Pant, D. D. On the ontogeny of stomata and other homologous structures. *Plant Sci. Ser. Allahabad* 1:1–24. 1965.
25. Possingham, J. V. Surface wax structure in fresh and dried Sultana grapes. *Ann. Bot.* 36:993–996. 1972.
26. Raschke, K. Stomatal action. *Ann. Rev. Plant Physiol.* 26:309–340. 1975.
27. Rentschler, I. Die Wasserbenetzbarkeit von Blattoberflächen und ihre submikroskopische Wachsstruktur. *Planta* 96:119–135. 1971.
28. Rentschler, I. Electron-microscopial investigations on wax-covered stomata. *Planta* 117:153–161. 1974.
29. Robards, A. W. Plasmodesmata. *Ann. Rev. Plant Physiol.* 26:13–29. 1975.
30. Sangster, A. G., and D. W. Parry. Silica deposition in the grass leaf in relation to transpiration and the effect of dinitrophenol. *Ann. Bot.* 35:667–677. 1971.
31. Srivastava, L. M., and A. P. Singh. Stomatal structure in corn leaves. *J. Ultrastruct. Res.* 39:345–363. 1972.
32. Tomlinson, P. B. Development of the stomatal complex as a taxonomic character in the monocotyledons. *Taxon* 23:109–128. 1974.
33. Uphof, J. C. T. Plant hairs. *Handbuch der Pflanzenanatomie.* Band 4. Teil 5. 1962.

8

Xylem: General Structure and Cell Types

The xylem is the principal water-conducting tissue in a vascular plant. It is usually spatially associated with the phloem (fig. 8.1), the principal food-conducting tissue. The two tissues together are called the vascular tissue or tissues. The combination of xylem and phloem forms a continuous vascular system throughout all parts of the plant, including all branches of stem and root.

Developmentally, it is convenient to distinguish between primary and secondary vascular tissues. The primary tissues differentiate during the formation of the primary plant body, and the meristem directly concerned with the formation of the primary vascular tissues is the procambium. The secondary vascular tissues are produced during the second major stage of plant development, in which an increase in thickness results from lateral additions of new tissues to the axial parts of the plant (that is, stem and root) and their larger branches. It results from the activity of the vascular cambium (fig. 8.1). As was mentioned in chapter 2, secondary growth is absent in small annuals of the dicotyledons and in most monocotyledons.

The primary and the secondary xylem have histologic differences, but both are complex tissues containing at least water-conducting elements and parenchyma cells and usually also other types of cells, especially supporting cells. The characteristics of these various types of cells and their interrelations in the tissue may be best introduced by a consideration of the secondary xylem, or wood.

GROSS STRUCTURE OF SECONDARY XYLEM

Axial and radial systems

With the aid of low magnification, a study of a block of wood reveals the presence of two distinct systems of cells (fig. 8.1): the *axial* (longitudinal or vertical) and the *radial* (transverse or horizontal) or *ray* system. The axial system contains cells or files of cells with their long axes oriented vertically in the stem or the root, that is, parallel to the main, or

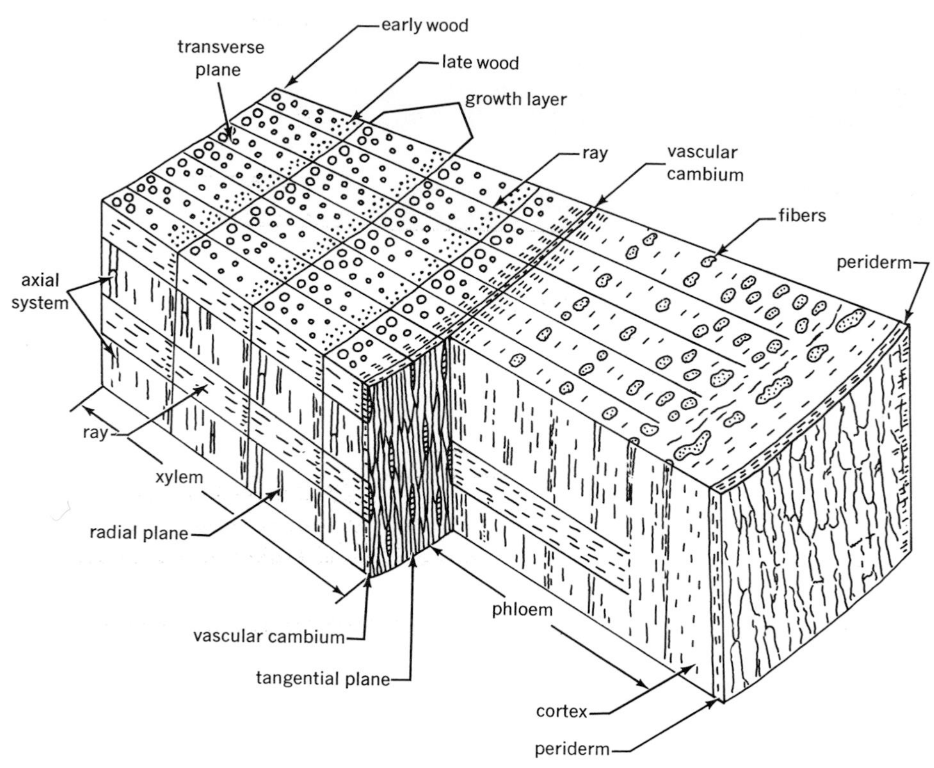

Figure 8.1 Block diagram illustrating basic features of secondary vascular tissues and their spatial relation to one another and to vascular cambium and periderm.

longitudinal, axis of these organs (or their branches); and the radial is composed of files of cells oriented horizontally with regard to the axis of stem or root.

Each of the two systems has its characteristic appearance in the three kinds of sections employed in the study of wood (chapter 9). In the transverse section, that is, the section cut at right angles to the main axis of stem or root, the cells of the axial system are cut transversely and reveal their smallest dimensions. The rays, in contrast, are exposed in their longitudinal extent in a cross section. When stems or roots are cut lengthwise, two kinds of longitudinal sections are obtained: the radial (parallel to a radius) and the tangential (perpendicular to a radius). Both show the longitudinal extent of cells of the axial system, but they give strikingly different views of the rays. Radial sections expose the rays as horizontal bands lying across the axial system (fig. 8.1). When a radial section cuts a ray through its median plane it reveals the height of the ray. A tangential section cuts a ray approximately perpendicularly to its horizontal extent and reveals its height and width. In tangential sections, therefore, it is easy to measure the height of the ray—this is usually done in terms of number of cells—and to determine whether the ray is uniseriate (one cell wide, fig. 8.6,*A*) or multiseriate (two to many cells wide, fig. 8.6,*C*).

Growth layers

With little or no magnification the wood discloses the layering resulting from the presence of more or less sharp boundaries between successive growth layers—growth rings in transections (fig. 8.1). Each growth layer may be a product of one season's growth, but various environmental conditions may induce the formation of more than one growth layer in one season. When conspicuous layering is present each growth layer is divisible into early and late wood. The early wood is less dense than the late wood because wider cells with thinner walls predominate in the early wood, narrower cells with thicker walls in the late. The late wood forms a distinct boundary in a growth ring because of its sharp contrast to the early wood of the following season, but the change from early wood to late wood of the same growth layer is more or less gradual. The relative amounts of early and late wood are affected by environmental conditions and specific differences. Adverse growth conditions, for example, increase the relative amount of late wood in a pine but decrease the relative amount of late wood in an oak. In general, the ability to develop growth layers is determined by the genetic constitution of the individual species and is found in trees of both temperate and tropical zones.

Sapwood and heartwood

The earlier increments of secondary xylem gradually become nonfunctional in conduction and storage. The relative amounts of the nonfunctioning wood, the heartwood, vary in different species and are also affected by environmental conditions. In a few species no heartwood is formed.[33] The heartwood is generally darker in color than the active wood, or sapwood. Formation of heartwood involves removal of reserve materials or their conversion into heartwood substances and eventual death of protoplasts of the parenchymatic and other living cell types of the wood. Heartwood formation is explained[33] as one of the processes enabling the plant to remove from regions of growth those metabolic byproducts that may be inhibitory or even toxic to living cells. The translocation of these substances, possibly after their detoxication, occurs through the rays, in part toward the outer bark, in part toward the center of the tree where the accumulation of excreta results in the death of cells. As the process continues, the sapwood-heartwood boundary moves outward. The number of growth increments retaining sapwood characteristics varies greatly among species.[29]

CELL TYPES IN THE SECONDARY XYLEM

The cellular structure of the xylem is studied in preparations of the three kinds of sections mentioned earlier and in macerated wood, that is, wood dissociated into groups of cells or individual cells by treatments that dissolve the middle lamella.

The principal cell types of the secondary xylem (figs. 8.2 and 8.3) may be thus tabulated (see Table 8.1).

Tracheary elements

The tracheary elements are the most highly specialized cells of the xylem and are concerned with the conduction of water and substances dissolved in water. They are more or less elongated cells, nonliving at maturity.

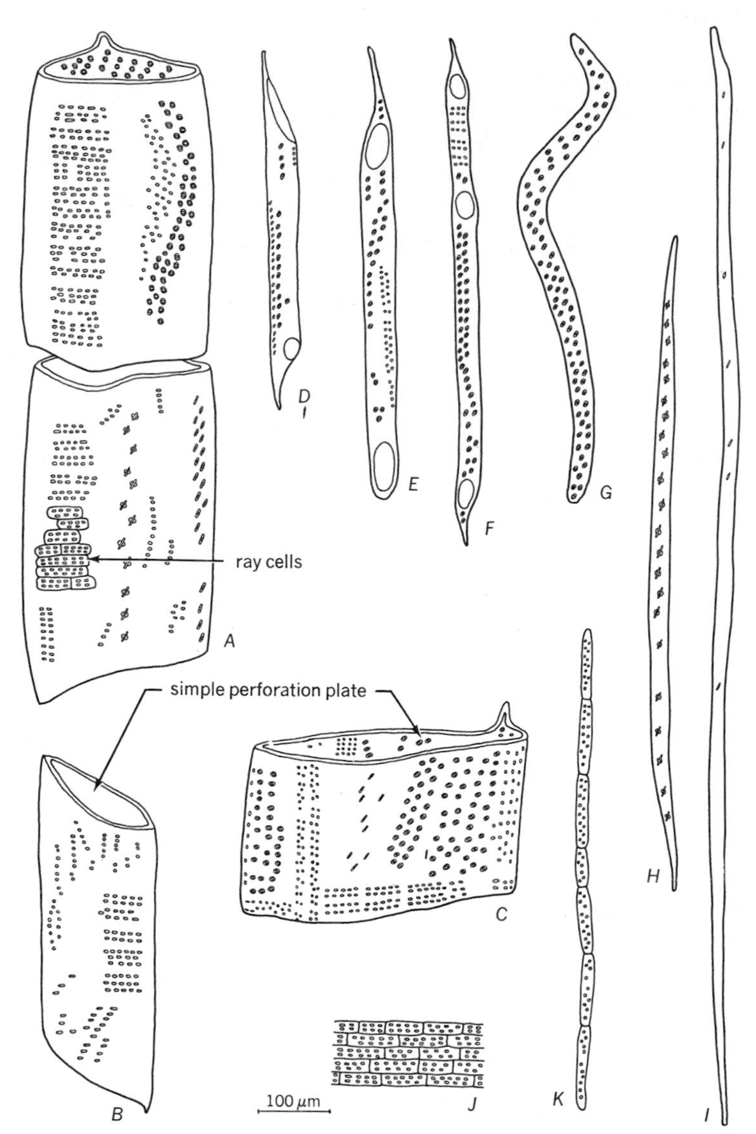

Figure 8.2 Cell types in secondary xylem as illustrated by dissociated wood elements of *Quercus*, oak. Various pits appear on cell walls. *A–C*, wide vessel members. *D–F*, narrow vessel members. *G*, tracheid. *H*, fiber-tracheid. *I*, libriform fiber. *J*, ray parenchyma cells. *K*, axial parenchyma strand. (*A–I*, from photographs in C. H. Carpenter and L. Leney, *91 Papermaking Fibers*, Tech. Publ. 74, College of Forestry at Syracuse, 1952.)

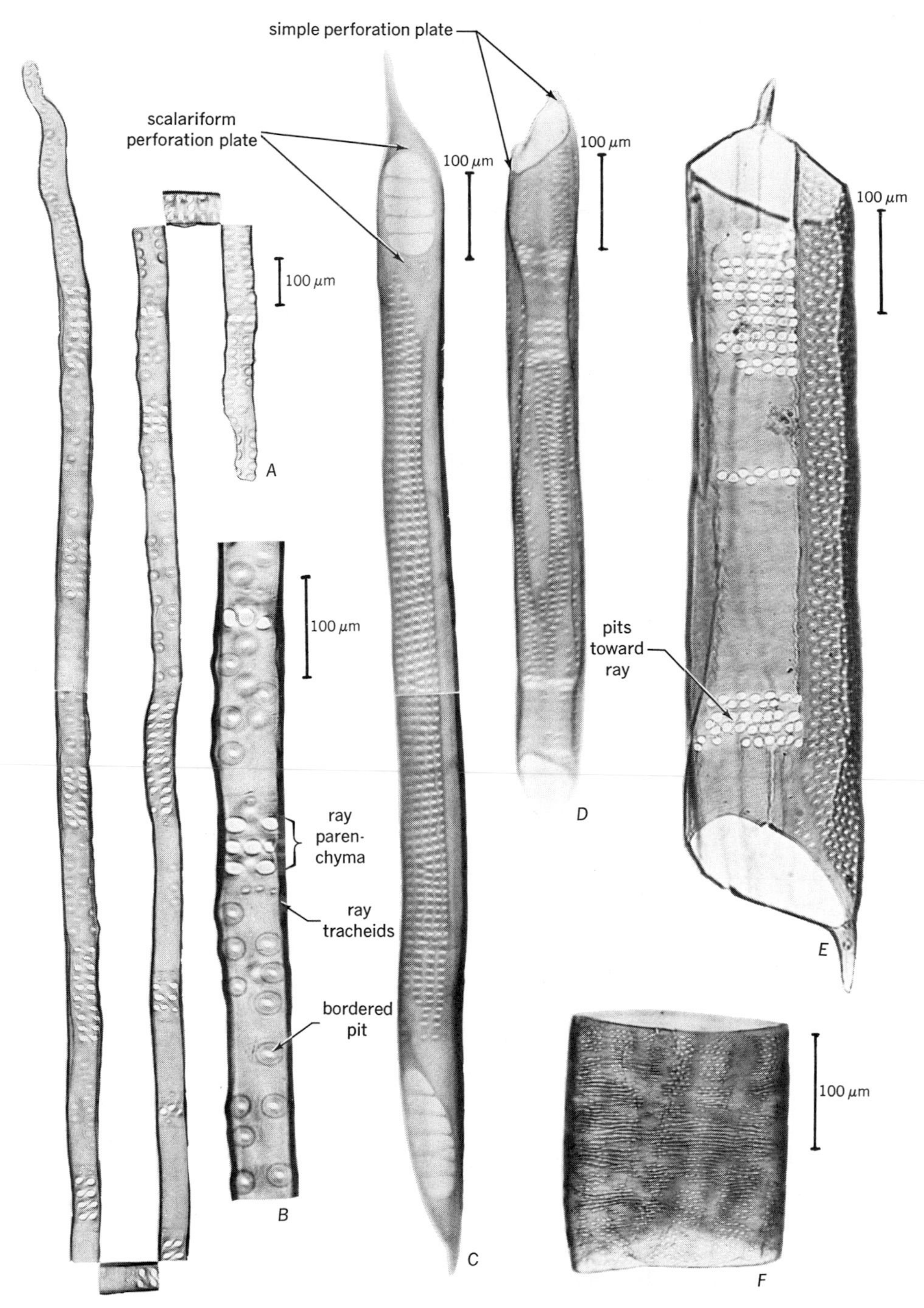

Figure 8.3 Tracheary elements. *A*, early wood tracheid of sugar pine (*Pinus lambertiana*). *B*, enlarged part of *A*. *C–F*, vessel members of tulip tree, *Liriodendron tulipifera (C)*, beech, *Fagus grandifolia* (*D*), black cottonwood, *Populus trichocarpa* (*E*), tree-of-heaven, *Ailanthus altissima* (*F*). (From C. H. Carpenter and L. Leney, *91 Papermaking Fibers,* Tech. Publ. 74, College of Forestry at Syracuse, 1952.)

Table 8.1

Cell Types	Principal Function
Axial system	
Tracheary elements	
Tracheids	Conduction of water
Vessel members	Conduction of water
Fibers	
Fiber-tracheids	Support; sometimes storage
Libriform fibers	Support; sometimes storage
Parenchyma cells	Storage and translocation of ergastic substances
Ray system	
Parenchyma cells	Storage and translocation of ergastic substances
(Tracheids in some conifers)	

They have lignified walls with secondary thickenings and a variety of pits.

The two kinds of tracheary cells, the tracheids and the vessel members, differ from each other in that the tracheid is an imperforate cell whereas the vessel member has perforations, one or more at each end (fig. 8.2), sometimes also on a side wall (fig. 8.2,*F*). Cells resembling vessel members in form and arrangement but lacking perforations are called vascular tracheids and are regarded as incompletely developed vessel members.[23] In tracheids, the passage of water from cell to cell occurs mainly through pit-pairs, in which the pit membranes may be assumed to be highly penetrable to water and dissolved substances. Ultrastructural studies indicate that the noncellulosic components of pit membranes are enzymically removed toward the end of cell ontogeny.[28] In vessel members, the water moves freely through perforations in the wall.

Longitudinal series of vessel members interconnected through their perforations (in a manner indicated in fig. 8.2,*A*) are called vessels. Vessels are not of indefinite lengths, although in some species with especially wide vessels in the early wood (ring-porous wood) they have been reported to extend through almost the entire height of a tree.[19] As measured in numbers of vessel members, primary xylem vessels were calculated to consist of two to fifty cells in the stem of *Scleria* (Cyperaceae).[6] The question of length of vessels needs further study.[26] Movement of water from vessel to vessel occurs through pits that are not penetrable to particulate matter. By injecting India ink into the xylem, that feature can be used to determine the presence of vessels and their lengths.

The perforated part of a wall of a vessel member is called the perforation plate. A plate may be simple, with only one perforation, or multiperforate, with more than one perforation (fig. 8.3). The multiperforate plates are scalariform if the perforations are elongated and arranged parallel to each other (fig. 8.3,*C*) and reticulate if the perforations form a net-like pattern.

Simple and bordered pits are encountered in the secondary walls of tracheids and vessel members (figs. 8.2 and 8.3). The number and arrangement of these pits are highly variable, even on different wall facets of the same cell, because they depend on the type of cell contiguous with the particular wall facet. Usually numerous bordered pit-pairs occur between contiguous tracheary ele-

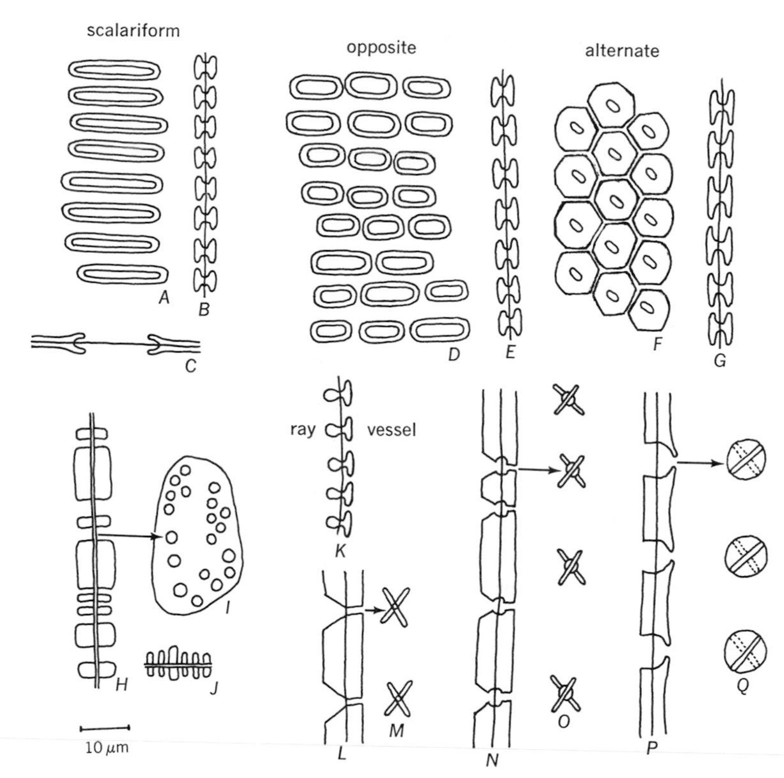

Figure 8.4 Pits and patterns of pitting. *A–C*, scalariform pitting in surface (*A*) and side (*B*, *C*) views (*Magnolia*). *D–E*, opposite pitting in surface (*D*) and side (*E*) views (*Liriodendron*). *F–G*, alternate pitting in surface (*F*) and side (*G*) views (*Acer*). *A–G*, all bordered pit-pairs in vessel members. *H–J*, simple pit-pairs in parenchyma cells in surface (*I*) and side (*H*, *J*) views; *H*, in side wall; *J*, in end wall (*Fraxinus*). *K*, half-bordered pit-pairs between a vessel and a ray cell in side view (*Liriodendron*). *L*, *M*, simple pit-pairs with slitlike apertures in side (*L*) and surface (*M*) views (libriform fiber). *N*, *O*, bordered pit-pairs with slitlike inner apertures extended beyond the outline of the pit border; *N*, side view, *O*, surface view (fiber-tracheid). *P*, *Q*, bordered pit-pairs with slitlike inner apertures included within the outline of the pit border; *P*, side view, *Q*, surface view (tracheid). *L–Q*, *Quercus*.

ments (fig. 8.4, *A–G*,*P*,*Q*); few or no pit-pairs occur between tracheary elements and fibers: half-bordered or simple pit-pairs are found between tracheary elements and parenchyma cells. In half-bordered pit-pairs the border is on the side of the tracheary cell (fig. 8.4,*K*).

The bordered pit-pairs of conifers are large, particularly in early wood. One of the common forms of such pit-pairs is circular in face view (fig. 8.5,*A*,*B*) and the borders enclose a conspicuous cavity (fig. 8.5,*C*). The pit membrane has a torus in the center. It is surrounded by the thin part of the membrane, the margo (chapter 4). The bundles of microfibrils forming the margo are sometimes aggregated in fixed material and therefore visible with the light microscope (fig. 8.5,*A*). A thickening of middle lamella and primary wall covered by the secondary wall, the *crassula*, may be present between pits (fig. 8.5,*A*). The

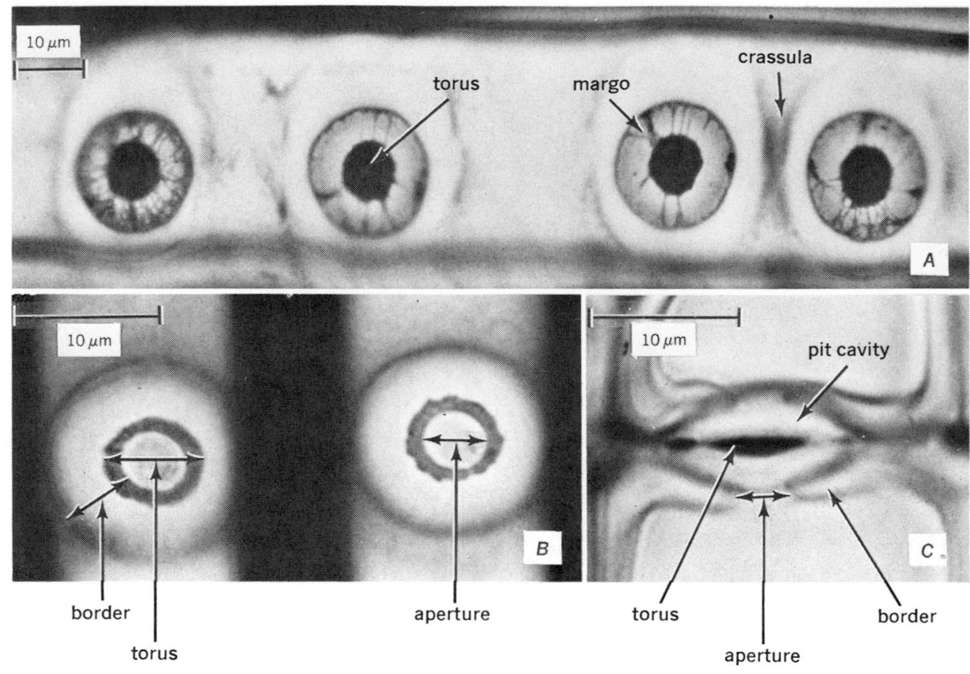

Figure 8.5 Bordered pits in conifer tracheids in radial (*A*, *B*) and transverse sections (*C*) of wood. *A*, *Tsuga canadensis*. *B*, *Pinus strobus*. *C*, *Abies nobilis*.

torus is wider than the aperture in the border (fig. 8.5,*B*,*C*) and consequently blocks the aperture when the pit-pair is aspirated (chapter 4). Aspirated bordered pit-pairs are common in the heartwood. The aspiration of pits begins gradually in the sapwood and is thought to be related to the drying out of the central core of the wood; it seems that the displacement of pit membranes occurs where a tracheid wall lies between a tracheid containing gases and one containing water.[21] The bordered pits thus act as valves preventing the entry of air into conducting tracheids, which would disrupt the water column in the latter.[20] No torus develops in the membrane of the half-bordered pit-pairs that occur in walls between conifer tracheids and parenchyma cells.

Fibers

The fibers are long cells with secondary, commonly lignified, walls. The walls vary in thickness but are usually thicker than the walls of tracheids in the same wood. Two principal types of xylem fibers are recognized, the fiber-tracheids (fig. 8.2,*H*) and the libriform fibers (fig. 8.2,*I*). If both occur in the same wood, the libriform fiber is longer and has thicker walls than the fiber-tracheid. The fiber-tracheids have bordered pits with cavities smaller than the pit cavities of tracheids or vessels in the same wood (fig. 8.4,*N*,*O*). These pits have a pit canal with a circular outer aperture and an elongated or slit-like inner aperture (chapter 4).

The pit in a libriform fiber has a slit-like

aperture toward the cell lumen and a canal resembling a much flattened funnel, but no pit cavity (fig. 8.4,*L*,*M*). In other words, the pit has no border; it is simple. The reference to the pits of libriform fibers as simple implies a sharper distinction between fibers and fiber-tracheids than actually exists. The fibrous xylem cells show a graduated series of pits between those with pronounced borders and those with vestigial borders or no borders. The intergrading forms with recognizable pit borders are placed, for convenience, in the fiber-tracheid category.[29]

Fibers of both categories may be septate (chapter 6). Septate fibers are widely distributed in dicotyledons and usually retain their protoplasts in mature sapwood, being concerned with storage of reserve materials.[17,18] Thus, the living fibers approach xylem parenchyma cells in structure and function. The distinction between the two is particularly tenuous when the parenchyma cells develop secondary walls and septa. The retention of protoplasts by fibers is an indication of evolutionary advance,[1,5] and where living fibers are present, the axial parenchyma is small in amount or absent.[27]

Another modification of fiber-tracheids and libriform fibers are the so-called gelatinous fibers. These fibers have a nonlignified wall layer of gelatinous aspect (*G*-layer) which is deposited over the S_3, S_2, or even S_1 secondary layer (chapter 4) of the secondary wall. Gelatinous fibers are common components of the reaction wood (chapter 9) in dicotyledons.

Phylogenetic specialization of tracheary and fiber cells

The lines of specialization of cells and tissues are better understood for the xylem than for any other tissue of the vascular plant. Among the individual lines, those pertaining to the evolution of the tracheary elements have been studied with particular thoroughness.

The specialization of tracheary elements was a concomitant of the separation of functions of conduction and strengthening in the vascular plant that occurred during the evolution of land plants.[1] In the less specialized state, conduction and support are combined in tracheids. With increasing specialization woods evolved with conducting elements—the vessel members—more efficient in conduction than in support. In contrast, fibers evolved as principally strengthening elements. Thus, from primitive tracheids two lines of specialization diverged, one toward the vessels, the other toward the fibers.

Vessels evolved independently in several taxa of vascular plants. A large body of evidence suggests that in dicotyledons the vessels originated and underwent specialization first in the secondary xylem, then in the late primary xylem (metaxylem), and last in the early primary xylem (protoxylem). In the primary xylem of the monocotyledons, origin and specialization of vessels also occurred first in the metaxylem, then in the protoxylem; furthermore, in this taxon, vessels appeared first in the roots, then at progressively higher levels in the shoot.[12,16] The relation between the first appearance of vessels and type of organ in dicotyledons has not been fully explored but some data indicate an evolutionary lag in leaves, floral appendages, and seedlings.[2]

Vessels may undergo an evolutionary loss. Absence of vessels in some aquatic plants, saprophytes, parasites, and succulents, for example, is interpreted as a result of a reduction in the xylem tissue. Reduction in this sense implies failure of potential xylem elements, including vessel members, to undergo typical ontogenetic differentiation. These vesselless plants are highly specialized as contrasted with the some ten known genera of primitively vesselless dicotyledons (*Trochodendron, Tetracentron, Drimys, Pseudowintera,* and others) belonging to the lowest taxa of dicotyledons.[1,13,24] The vascular tracheids found in some advanced families of dicotyledons (Cactaceae, Asteraceae) also originated by an evolutionary degeneration of vessel members.[4,8]

The evolutionary sequence of vessel member types in the secondary xylem of dicotyledons began with long scalariformly pitted tracheids similar to those still found in some lower dicotyledons. These tracheids were succeeded by vessel members of long narrow shape with tapering ends (fig. 8.3,*C*). The cells shortened progressively, became wider, and their end walls became less inclined and finally transverse (fig. 8.3,*D–F*). In the more primitive state the perforation plate was scalariform, with numerous bars, resembling a wall with scalariformly arranged pits devoid of pit membranes. Increase in specialization resulted in a decrease in the number of bars (fig. 8.3,*C*) and finally their total elimination and the appearance of a simple perforation (fig. 8.3,*D–F*). Retrogressive modifications recorded in this line of evolution are perforation plates with reticulate or other anomalous patterns found in some specialized families.[8]

The pitting of vessel walls also changed during the evolution. In intervessel pitting, bordered pit-pairs in scalariform arrangement (fig. 8.4,*A*) were replaced by smaller bordered pit-pairs, first in opposite (fig. 8.4,*D*), later in alternate arrangement (fig. 8.4,*F*). The pit-pairs between vessels and parenchyma cells changed from bordered, through half-bordered, to simple.

The tracheids were not eliminated when vessels evolved, but underwent phylogenetic changes. They became shorter—not as short as the vessel members, however—and the pitting of their walls evolved parallel to that of the associated vessel members. They generally did not increase in width.

In the specialization of xylem fibers the emphasis on mechanical function became apparent in the increase in wall thickness and decrease in cell width. Concomitantly the pits changed from elongated to circular, the borders became reduced (fig. 8.4,*N,O*) and eventually disappeared (fig. 8.4,*L,M*). The inner apertures of the pit became elongated and then slitlike. Thus, the evolutionary sequence was from tracheids, through fiber-tracheids, to libriform fibers.

The matter of evolutionary change in length of fibers is rather complex. The shortening of vessel members is correlated with a shortening of the fusiform cambial initials (chapter 10) from which the axial cells of the xylem are derived. Thus, in woods with shorter vessel members, the fibers are derived ontogenetically from shorter initials than in more primitive woods with longer vessel members. In other words, with increase in xylem specialization the fibers become shorter. Because, however, during ontogeny fibers undergo intrusive growth whereas vessel members do so only slightly or not at all, the fibers are longer than the vessel members in the mature wood, and, of the two categories of fibers, the libriform fibers are the longer ones. Nevertheless, the fibers of specialized woods are shorter than their ultimate precursors, the primitive tracheids.

The evolutionary lines in the xylem have been reconstructed from comparative studies of existing plants.[3] The vascular plants now living show a wide range in the degree of specialization of their cells, tissues, and organs. These variations are useful in the identification of woods (chapter 9).

Parenchyma cells

The parenchyma of the secondary xylem is represented by the axial parenchyma and the ray parenchyma. These two kinds of cell complexes are fundamentally alike regarding wall structure and contents, and in both the cells may vary considerably in structure and contents.[10,36] The parenchyma cells store starch, oils, and many other ergastic substances of unknown function. Tanniniferous compounds and crystals are common inclusions. The types of crystals and their arrangements may be sufficiently characteristic to serve in identification of woods.[11,32]

The walls of radial and axial parenchyma cells may have secondary thickenings and be lignified.[36] If secondary walls are present, the pit-pairs between the parenchyma cells may be bordered, half-bordered, or simple. Some parenchyma cells become sclerified by deposition of thick walls. These are sclerotic cells, or sclereids. Crystalliferous parenchyma cells frequently have lignified walls with secondary thickenings and may be chambered by septa, each chamber containing one crystal.

The axial parenchyma cells are derived from elongated fusiform cambial cells. If the derivative of such a cambial cell differentiates into a parenchyma cell without transverse (or oblique) divisions, a fusiform *parenchyma cell* results. If such divisions occur, a *parenchyma strand* is formed (fig. 8.2,*K*). Neither type undergoes intrusive growth.

The radial parenchyma cells are divided into categories according to their form. The two most common types are the *procumbent* and the *upright* ray cells (fig. 8.6). A ray cell the longest diameter of which is oriented radially is procumbent; the cell elongated axially is upright. The two kinds of ray cells are often combined in the same ray, the upright cells typically appearing at the upper and lower margins of the ray (fig. 8.6,*A*). Rays composed of one kind of cell are called *homocellular* (fig. 8.6,*C*,*D*), those containing procumbent and upright cells, *heterocellular* (fig. 8.6,*A*,*B*).

PRIMARY XYLEM

The primary xylem contains the same basic cell types as the secondary xylem: tracheary elements—both kinds, tracheids and vessel members—fibers, and parenchyma cells. It is, however, not organized into the combina-

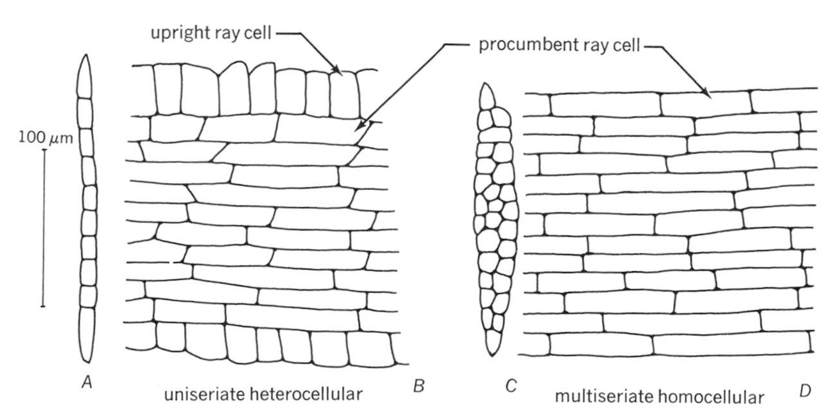

Figure 8.6 Two types of rays as seen in tangential (*A*, *C*) and radial (*B*, *D*) sections. *A*, *B*, *Fagus grandifolia*. *C*, *D*, *Acer saccharum*.

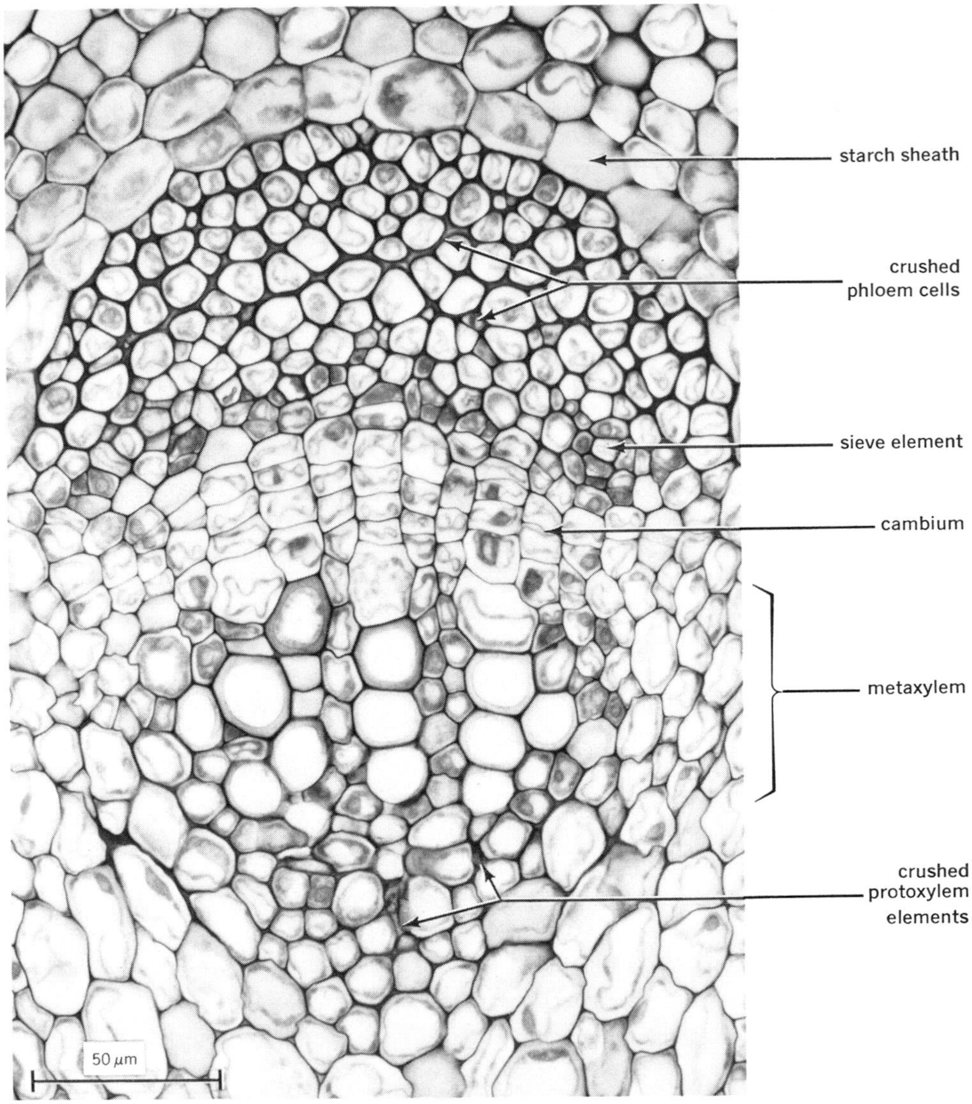

Figure 8.7 Vascular bundle from stem of *Medicago sativa* (alfalfa) in cross section. Illustrates primary xylem and phloem. The cambium has not yet produced secondary tissues. The earliest xylem (protoxylem) and phloem (protophloem) are no longer functioning in conduction. Their conducting cells have been obliterated. The functional tissues are metaxylem and metaphloem.

tion of axial and radial systems, for it contains no rays. In stems and leaves, and floral parts, the primary xylem and the associated primary phloem commonly occur in strands, the vascular bundles (fig. 8.7). Panels of parenchyma, the interfascicular regions, occur between the vascular bundles in stems (chapter 16). These panels are often called medullary rays and are considered to be part of the ground tissue. In the root the primary xylem forms a core with or without parenchyma in the center or is arranged in strands (chapter 14).

Protoxylem and metaxylem

Developmentally, the primary xylem usually consists of an earlier part, the *protoxylem,* and a later part, the *metaxylem* (figs. 8.7 and 8.8,*B*). Although the two parts have some distinguishing characteristics, they intergrade so that the delimitation of the two can be made only approximately.

The protoxylem differentiates in the parts of the primary plant body that have not completed their growth and differentiation. In fact, in the shoot, the protoxylem matures among actively elongating tissues and is, therefore, subjected to stresses. Its mature nonliving tracheary elements are stretched and eventually destroyed. In the root, the protoxylem elements persist longer because here they mature beyond the region of maximum growth.

The metaxylem is commonly initiated in the still growing primary plant body, but it matures largely after the elongation is completed. It is, therefore, less affected by the primary extension of the surrounding tissues than the protoxylem.

The protoxylem usually contains only tracheary elements embedded in parenchyma that is considered to be part of the protoxylem. When the tracheary elements are destroyed they may become completely obliterated by the surrounding parenchyma cells (fig. 8.7). In the shoot xylem of many monocotyledons, the stretched nonfunctioning elements are partly collapsed but not obliterated; instead, open canals, the so-called protoxylem lacunae, surrounded by parenchyma cells, appear in their place (chapter 11). If preserved in sectioning, the secondary walls of the nonfunctioning tracheary cells may be seen along the margin of the lacuna.

The metaxylem is somewhat more complex than the protoxylem and may contain fibers in addition to tracheary elements and parenchyma cells. The parenchyma cells may be dispersed among the tracheary elements or may occur in radial rows simulating rays. Longitudinal sections reveal them as axial parenchyma cells. The radial seriation often encountered in the metaxylem, and also in the protoxylem, has given rise to a misleading tendency in the literature to interpret the primary xylem of many plants as secondary, for radial seriation is characteristic of the secondary vascular tissues.

The tracheary elements of the metaxylem are retained after the primary growth is completed but become nonfunctioning after some secondary xylem is produced. In plants lacking secondary growth the metaxylem remains functional in mature plant organs.

Cell wall in primary tracheary elements

The primary tracheary cells have a variety of secondary wall thickenings. The different forms of wall appear in a specific ontogenetic series that indicates a progressive increase in the extent of the primary wall area covered by secondary wall material (fig. 8.8). In the earliest tracheary elements the secondary walls may occur as rings (*annular* thickenings) not connected with one another. The elements differentiating next have *helical* (*spiral*) thickenings. Then follow cells with thickenings that may be characterized as helices with coils interconnected (*scalariform* thickenings). These are succeeded by cells with netlike, or *reticulate,* thickenings, and finally by *pitted* elements.

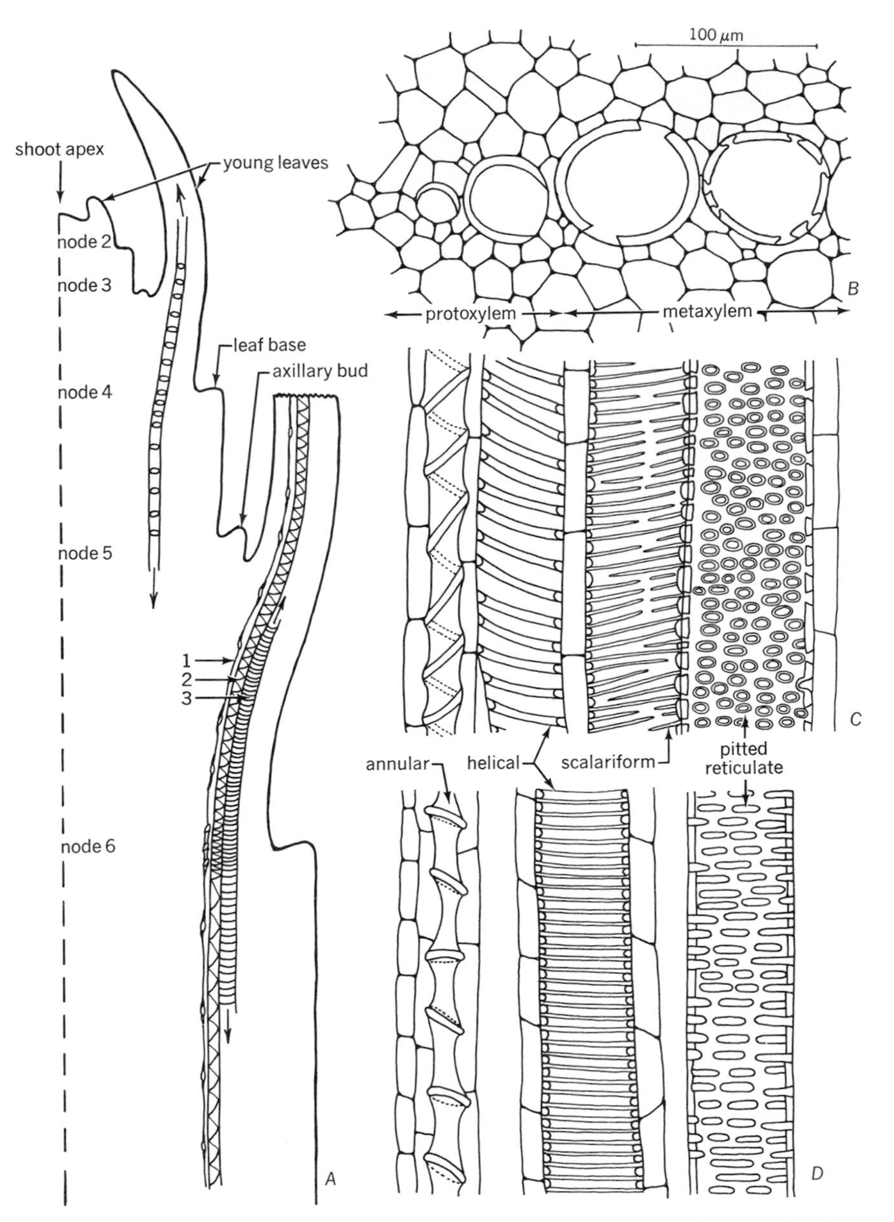

Figure 8.8 Details of structure and development of primary xylem. *A*, diagram of a shoot tip showing stages in xylem development at the different levels. *B–D*, primary xylem of castor bean, *Ricinus*, in cross (*B*) and longitudinal (*C*,*D*) sections.

Not all types of secondary thickenings are necessarily represented in the primary xylem of a given plant or plant part, and the different types of wall structure intergrade. The annular thickenings may be interconnected here and there; annular and helical or helical and scalariform thickenings may be combined in the same cell; and the difference between scalariform and reticulate is sometimes so tenuous that the thickening may best be called scalariform-reticulate. The pitted elements also intergrade with the earlier ontogenetic type. The openings in a scalariform reticulum of the secondary wall may be comparable to pits, especially if a slight border is present. A border-like overarching of the secondary wall is common in the various types of secondary wall in the primary xylem. Rings, helices, and the bands of the scalariform-reticulate thickenings may be connected to the primary wall by narrow bases, so that the secondary wall layers widen out toward the lumen of the cell and overarch the exposed primary wall parts (fig. 8.12,*A*).

The intergrading nature of the secondary wall thickenings in the primary xylem makes it impossible to assign distinct types of wall thickenings to the protoxylem and the metaxylem with any degree of consistency. Most commonly the first tracheary elements to mature, that is, protoxylem elements, have the minimal amounts of secondary wall material. Annular and helical thickenings predominate. These types of thickening do not hinder materially the stretching of the mature protoxylem elements during the extension growth of the primary plant body. The evidence that such stretching occurs is easily perceived in the increase in distance between rings in older xylem elements, the tilting of rings, and the uncoiling of the helices (fig. 8.8,*A*).

The metaxylem, in the sense of xylem tissue maturing after the extension growth, may have helical, scalariform, reticulate, and pitted elements; one or more types of thickening may be omitted. If many elements with helical thickenings are present, the helices of the succeeding elements are less and less steep, a condition suggesting that some stretching occurs during the development of the earlier metaxylem elements.

Convincing evidence exists that the type of wall thickening in primary xylem is strongly influenced by the internal environment in which these cells differentiate. Annular thickenings develop when the xylem begins to mature before the maximum extension of the plant part occurs, as, for example, in the shoots of normally elongating plants (fig. 8.8,*A*, nodes 3–5); they may be omitted if the first elements mature after this growth is largely completed, as is common in the roots. If the elongation of a plant part is suppressed before the first xylem elements mature, one or more of the early ontogenetic types of thickenings are omitted. On the contrary, if elongation is stimulated, as for example by etiolation, more than the usual number of elements with annular and helical thickenings will be present.

According to a comprehensive study of mature and developing protoxylem and metaxylem of angiosperms,[6] the elements with more extensive secondary thickenings than that represented by a helix deposit the secondary wall in two stages. First, a helical framework is built (first-order secondary wall). Then, additional secondary wall material is laid down as sheets or strands or both between the gyres of the helix (second-order secondary wall). This concept may be used to explain the effect of environment on the wall pattern in terms of inhibition or induction of second-order secondary wall deposition, depending on circumstances.

The intergrading of the different types of thickening of tracheary elements is not limited to the primary xylem. The delimitation between the primary and the secondary xylem may also be vague. To recognize the limits of the two tissues it is necessary to consider many features, among these the length of tracheary cells—the last primary elements are typically longer than the first secondary—and the organization of the tissue, particularly the appearance of the combination of ray and axial systems characteristic of secondary xylem. Sometimes the appearance of one or more identifying features of the secondary xylem is delayed, a phenomenon referred to as paedomorphosis.[9]

In the primary xylem, the protoxylem elements may be the narrowest, but not necessarily so. Successively differentiating metaxylem elements often are increasingly wider, whereas the first secondary xylem cells may be rather narrow and thus be distinct from those of the latest wide-celled metaxylem. On the whole, however, it is difficult to make precise distinctions between successive developmental categories of tissues.

DIFFERENTIATION OF TRACHEARY ELEMENTS

During the stage of cell growth and secondary wall deposition the living protoplast is concerned with the phenomena of differentiation (fig. 8.9,*A*,*B*). The protoplast contains the full complement of organelles including a nucleus and tonoplast-bound vacuoles. The nucleus is known to become polyploid and to increase in size.[22] The endoplasmic reticulum (ER) is seen in extensive profiles along, and especially between, the secondary wall thickenings (fig. 8.10). This membrane system may be concerned with channeling material to the active sites of wall deposition and excluding it from other sites. Dictyosomes are

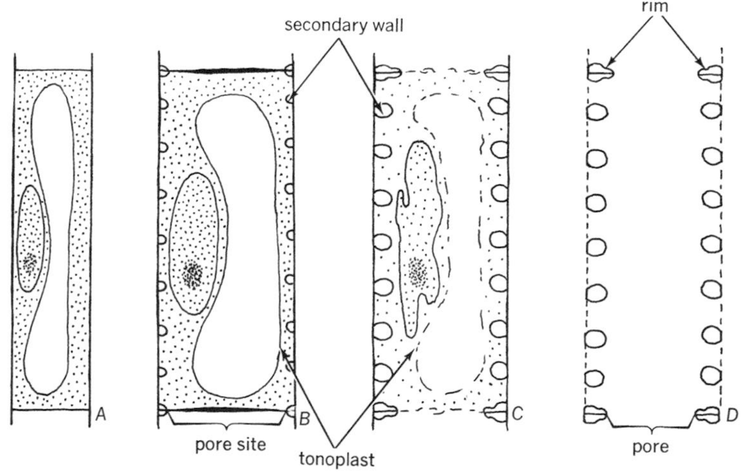

Figure 8.9 Diagrams illustrating development of a vessel member with a helical secondary thickening. *A*, cell without secondary wall. *B*, cell has attained full width, nucleus has enlarged, secondary wall has begun to be deposited, primary wall at the pore site has increased in thickness. *C*, cell at stage of lysis: secondary thickening completed, tonoplast ruptured, nucleus deformed, wall at pore site partly disintegrated. *D*, mature cell without protoplast, open pores at both ends, primary wall partly hydrolyzed between secondary thickenings.

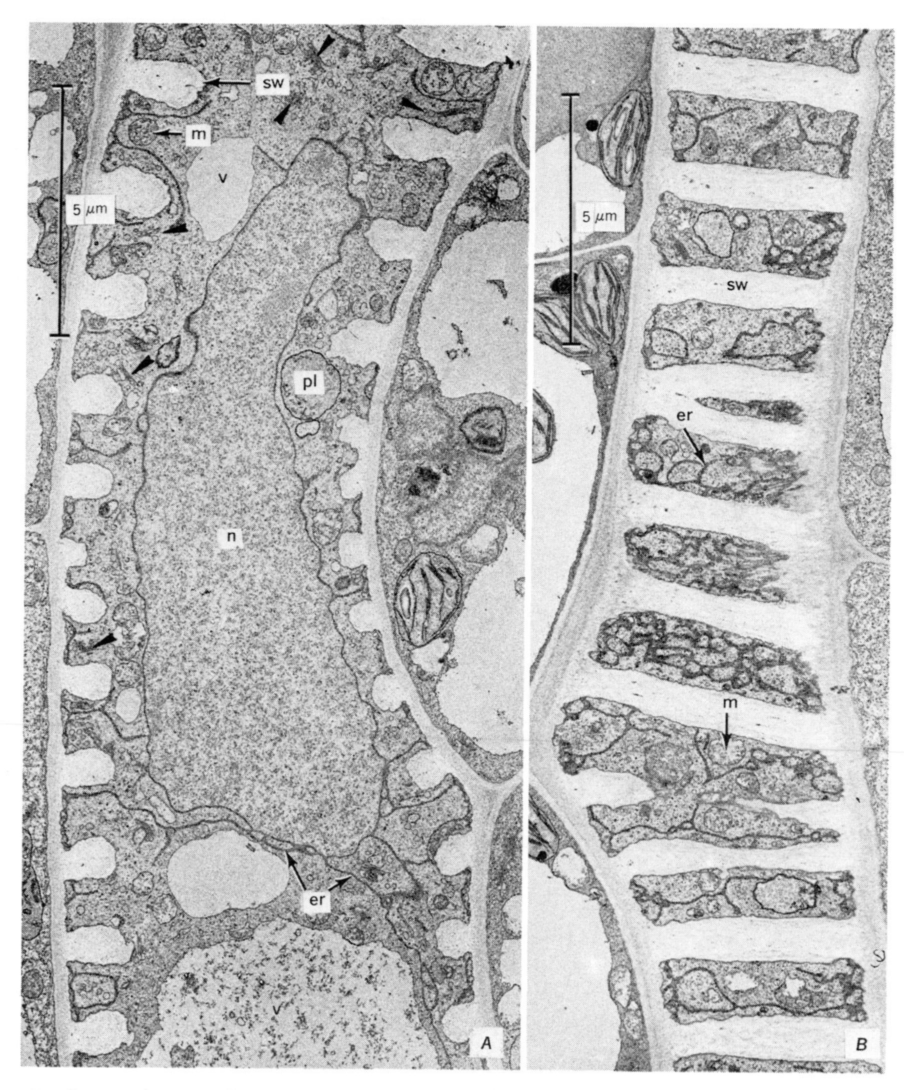

Figure 8.10 Differentiating tracheary elements in a leaf blade of sugar beet (*Beta vulgaris*). The secondary thickening is helical with transition to scalariform. *A*, section through cell lumen. *B*, section through the secondary thickening. Details: *er*, endoplasmic reticulum; *m*, mitochondrion; *n*, nucleus; *pl*, plastid; *sw*, secondary wall; *v*, vacuole; arrowheads, dictyosomes.

also conspicuous (figs. 8.10,*A*, arrowheads; 8.11,*B*). Radioactive labeling of cytoplasm indicates that the ER and dictyosomes are associated with incorporation of material into the wall,[31] possibly by means of vesicles that move toward the periphery of the cell, unite with the plasmalemma, and release their contents toward the wall (chapter 4). These deductions, however, were made on the basis of static images and need confirmation.[28]

Microtubules are prominently displayed during the growth of cell wall. At first, they are

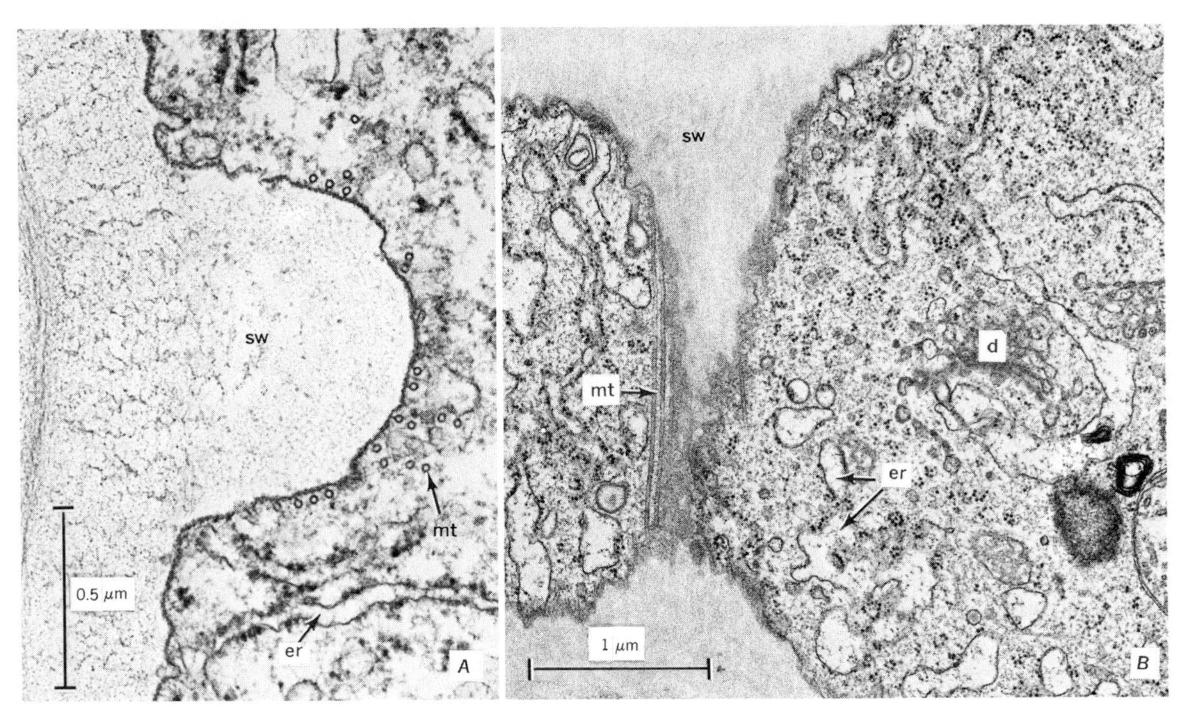

Figure 8.11 Parts of differentiating tracheary elements from leaves of *A*, bean (*Phaseolus vulgaris*) and, *B*, sugar beet (*Beta vulgaris*). Microtubules associated with the secondary thickening are seen in cross section in *A*, in longitudinal section in *B*. Details: *d*, dictyosome; *er*, endoplasmic reticulum; *mt*, microtubule; *sw*, secondary wall.

spread evenly along the entire wall, but later are concentrated in bands at the sites of secondary thickenings[30] (fig. 8.11). The importance of the association of microtubules with the growing wall has been tested experimentally. Treatment with colchicine, which causes the disappearance of microtubules, results in conspicuous irregularities in the form and distribution of wall thickenings (chapter 4).

After the secondary wall is deposited, the cell enters the stage characterized by phenomena of lysis affecting the protoplast and certain parts of the cell wall. It appears that the vacuoles are acting as do lysosomes (chapter 3) in providing the hydrolytic enzymes for autodigestion. The cytoplasm is exposed to the hydrolases possibly through rupture of the tonoplast[37] (fig. 8.9,*C*). The hydrolases also reach the cell walls and attack the primary wall parts that are not covered by lignified secondary wall layers. The lateral walls are partly digested. Removal of noncellulosic components leaves a fine network of cellulose microfibrils[28] (fig. 8.12,*A*). In the sites of the perforations, the entire primary wall part disappears (figs. 8.9,*D*; 8.12,*B*).

The wall occupying the site of the future perforation is clearly set off from the secondary wall (figs. 8.12,*C*; 8.13). It is thicker than the primary wall elsewhere and, in unstained fine sections, is much whiter than the other wall parts of the same cell (fig. 8.12,*C*). The removal of this wall (fig. 8.14) appears to be a gradual dissolution process indicated by the appearance of small holes before the final breakdown.[25] In a lysed scalariform perforation plate, fine networks of fibrils have been

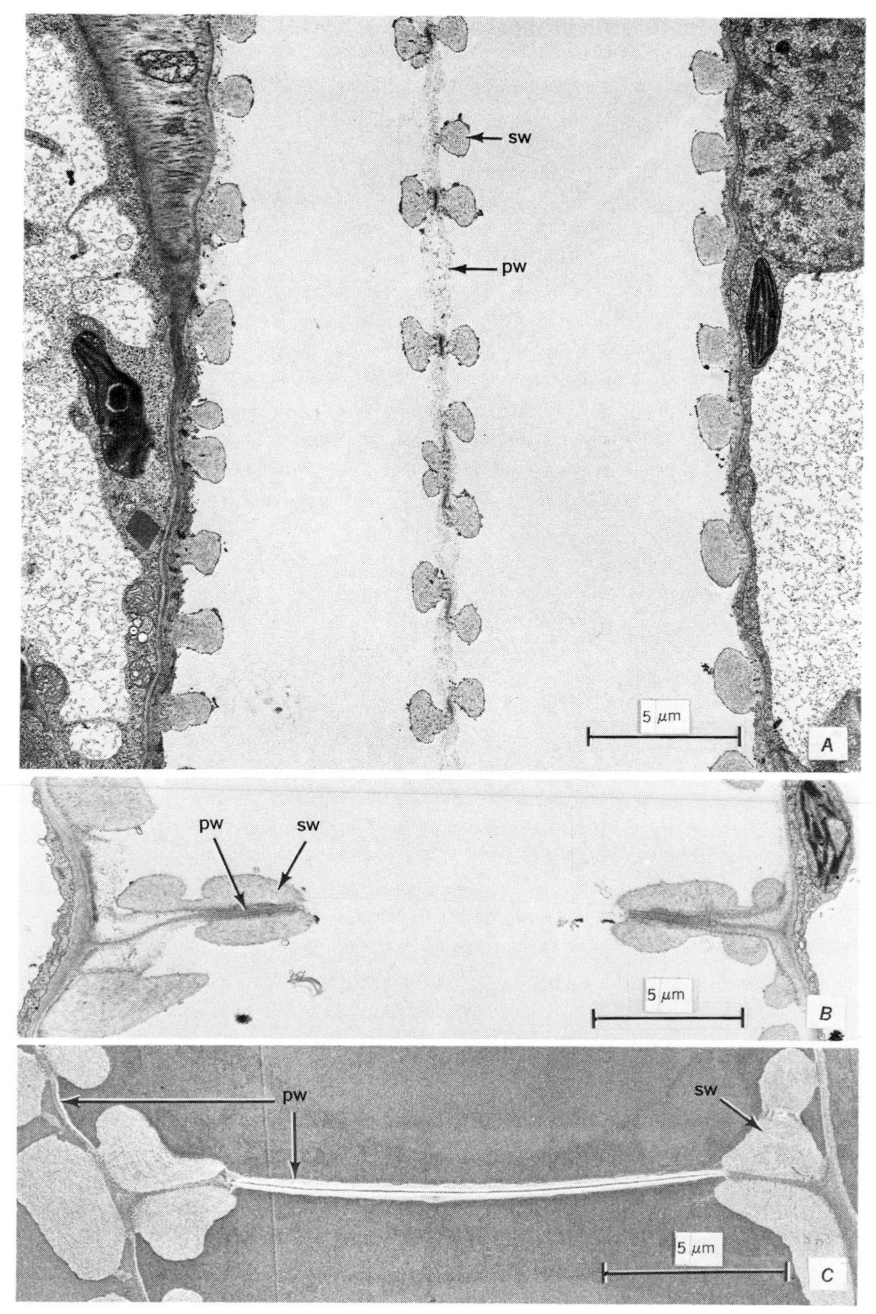

Figure 8.12 Parts of tracheary elements in longitudinal sections from leaves of, *A*, *B*, tobacco (*Nicotiana tabacum*) and, *C*, bean (*Phaseolus vulgaris*) showing details of walls. In *A*, the wall between two tracheary elements (center) illustrates the effect of hydrolysis upon the primary wall between the secondary thickenings: the primary wall is reduced to fibrils. In *B*, the end-wall perforation is delimited by a rim in which secondary thickenings are present. In *C*, the primary wall at the pore site has not yet disappeared. It is considerably thicker than the primary wall elsewhere and is supported by a secondarily thickened rim. Details: *pw*, primary wall; *sw*, secondary wall.

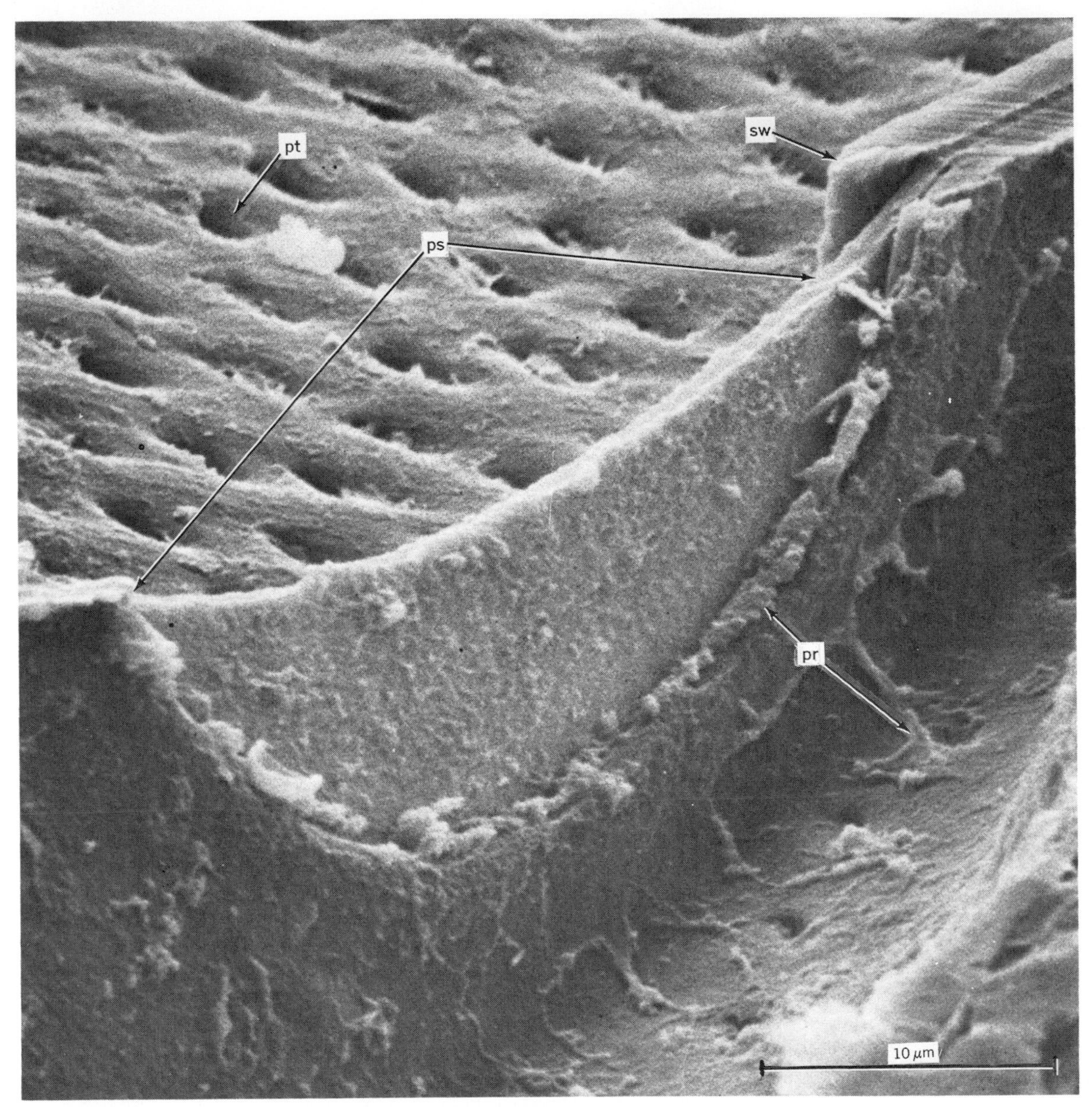

Figure 8.13 Scanning electron microscope view of half of a potential perforation plate in a vessel from wood of *Knightia excelsa*. The primary wall is still present at the pore site. Details: *pr*, dried protoplast; *ps*, pore site; *pt*, pit in side wall; *sw*, secondary wall of rim of perforation plate. (Courtesy of Dr. B. A. Meylan. From B. A. Meylan and B. G. Butterfield.[25])

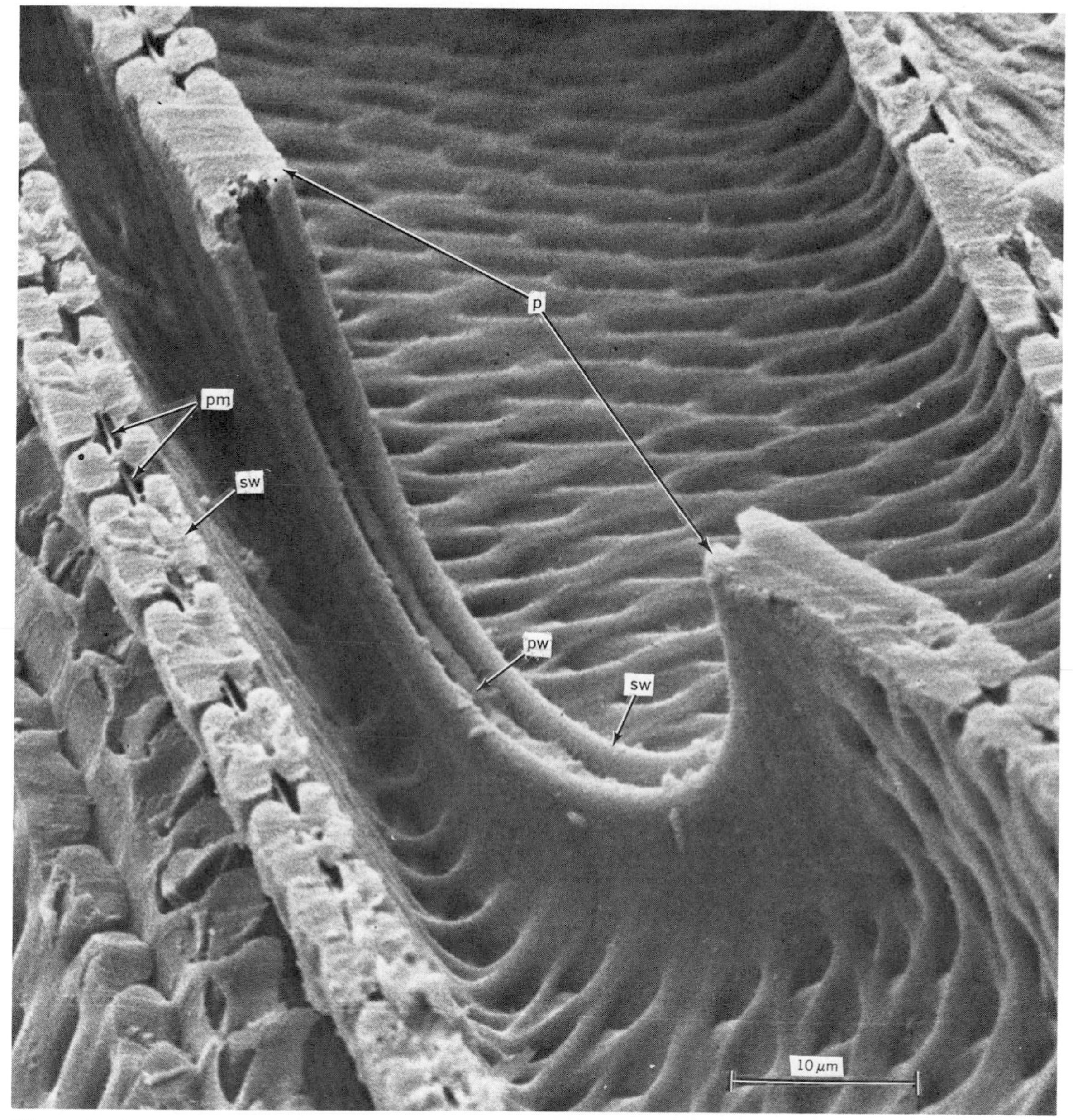

Figure 8.14 Scanning electron microscope view of half of a mature perforation plate in a vessel from wood of *Knightia excelsa*. Details: *p*, perforation; *pm*, pit membranes in pit-pairs in a side wall; *pw*, primary wall in rim embedded in, *sw*, secondary wall. (Courtesy of Dr. B. A. Meylan. From B. A. Meylan and B. G. Butterfield.[25])

seen stretching across the narrow perforations.[7] Since the networks are not present in the conducting tissue their removal by the transpiration stream is assumed.

The differentiation of tracheary elements is frequently studied to elucidate causal relations in morphogenesis.[14,15] Many morphogenetic studies have provided the evidence of hormonal control in xylem differentiation. Auxin from developing leaves is the normal limiting factor for the differentiation of xylem during regeneration of connections in severed vascular bundles. No regeneration occurs if the leaf and the bud located above the wound are removed, but it does occur if the leaf stump is covered with lanolin containing auxin. The regeneration is polar in relation to the polarity of auxin movement. Xylem differentiation can be induced in tissue cultures by grafting buds into callus tissue or treating the latter with auxins. Normal xylem differentiation also is limited by auxin, and it may be considerably enhanced by the addition of growth substances to young leaves or the culture solution.

Using the results of experimental work and observations on normal xylem development investigators discuss the probable cardinal steps in the differentiation of tracheary elements. Commonly the cell that is to become a tracheary element originates by cell division. Some authors suggest that mitosis and the associated DNA synthesis are necessary for the initiation of xylem differentiation.[34] This concept is being questioned, however, for in cultured explants of pith tissue of *Lactuca sativa* wound vessels were found differentiating without preceding DNA synthesis or cell division.[35] The future treacheary cell formed by cell division enlarges. During this step DNA may be synthesized in connection with an endomitotic reduplication of chromosomes.[22] Synthesis of protein and primary wall material may be expected to occur at this time also. In the next step, secondary wall material is synthesized, deposited against the primary wall and is lignified. During the steps described thus far, in which the living protoplast is actively participating in the metabolic changes, specific events are regulated by specific hormones.[34] After the secondary wall is deposited, phenomena of lysis become dominant in the cell. The protoplast is broken down, perforations are formed in the wall of a vessel member, and the pit membranes are reduced to networks of fibrils. The cell becomes functional in conduction.

REFERENCES

1. Bailey, I. W. Evolution of the tracheary tissue in land plants. *Amer. J. Bot.* 60:4–8. 1953.
2. Bailey, I. W. *Contributions to plant anatomy.* Waltham, Massachusetts, The Chronica Botanica Co. 1954.
3. Bailey, I. W. The potentialities and limitations of wood anatomy in the study of the phylogeny and classification of angiosperms. *J. Arnold Arb.* 38:243–254. 1957.
4. Bailey, I. W. Additional notes on the vesselless dicotyledon, *Amborella trichopoda* Baill. *J. Arnold Arb.* 38:374–378. 1957.
5. Bailey, I. W., and L. M. Srivastava. Comparative anatomy of the leaf-bearing Cactaceae. IV. The fusiform initials of the cambium and the form and structure of their derivatives. *J. Arnold Arb.* 43:187–202. 1962.
6. Bierhorst, D. W., and P. M. Zamora. Primary xylem elements and element associations of angiosperms. *Amer. J. Bot.* 52:657–710. 1965.
7. Butterfield, B. G., and B. A. Meylan. Scalariform perforation plate development in *Laurelia novae-zealandiae* A. Cunn.:

a scanning electron microscope study. *Aust. J. Bot.* 20:253–259. 1972.

8. Carlquist, S. *Comparative plant anatomy.* New York, Holt, Rinehart and Winston. 1961.
9. Carlquist, S. A theory of paedomorphosis in plants. *Phytomorphology* 12:30–45. 1962.
10. Chattaway, M. M. Morphological and functional variations in the rays of pored timbers. *Aust. J. Sci. Res. Ser. B., Biol. Sci.* 4:12–29. 1951.
11. Chattaway, M. M. Crystals in woody tissues. Part I. *Trop. Woods* (102):55–74. 1955.
12. Cheadle, V. I. Independent origin of vessels in the monocotyledons and dicotyledons. *Phytomorphology* 3:23–44. 1953.
13. Cheadle, V. I. Research on xylem and phloem—progress in fifty years. *Amer. J. Bot.* 43:719–731. 1956.
14. Cutter, E. G. *Plant anatomy: experiment and interpretation.* Part I. *Cells and tissues.* London, Edward Arnold Ltd. 1969.
15. Esau, K. *Vascular differentiation in plants.* New York, Holt, Rinehart and Winston. 1965.
16. Fahn, A. Metaxylem elements in some families of the Monocotyledoneae. *New Phytol.* 53:530–540. 1954.
17. Fahn, A., and B. Leshem. Wood fibres with living protoplasts. *New Phytol.* 62:91–98. 1963.
18. Frison, E. De la présence d'amidon dans le lumen des fibres du bois. *Bull. Agr. Congo Belge Brussels.* 39:869–874. 1948.
19. Greenidge, K. N. H. An approach to the study of vessel length in hardwood species. *Amer. J. Bot.* 39:570–574. 1952.
20. Gregory, S. G., and J. A. Petty. Valve action of bordered pits in conifers. *J. Exp. Bot.* 24:763–767. 1973.
21. Harris, J. M. Heartwood formation in *Pinus radiata* (D. Don.). *New Phytol.* 53:517–524. 1954.
22. Innocenti, A. M., and S. Avanzi. Some cytological aspects of the differentiation of metaxylem in the root of *Allium cepa. Caryologia* 24:283–292. 1971.
23. Jane, F. W. *The structure of wood.* 2nd ed. Revised by K. Wilson and D. J. B. White. London, Adam & Charles Black. 1970.
24. Lemesle, R. Les éléments du xylème dans les Angiospermes à charactères primitifs. *Bull. Soc. Bot. France* 103:629–677. 1956.
25. Meylan, B. A., and B. G. Butterfield. Perforation plate development in *Knightia excelsa* R. Br.: a scanning electron microscope study. *Aust. J. Bot.* 20:79–86. 1972.
26. Milburn, J. A., and P. A. K. Covey-Crump. A simple method for determination of conduit length and distribution in stems. *New Phytol.* 70:427–434. 1971.
27. Money, L. L., I. W. Bailey, and B. G. L. Swamy. The morphology and relationships of the Monimiaceae. *J. Arnold Arb.* 31:372–404. 1950.
28. O'Brien, T. P. Primary vascular tissues. In: *Dynamic aspects of plant ultrastructure.* A. W. Robards, ed. Chapter 12, pp. 414–440. London, McGraw-Hill Book Company (UK) Limited. 1974.
29. Panshin, A. J., and C. de Zeeuw. *Textbook of wood technology.* Vol. I. *Structure, identification, uses, and properties of the commerical woods of the United States and Canada.* New York, McGraw-Hill Book Company. 1970.
30. Pickett-Heaps, J. D. Incorporation of radioactivity into wheat xylem walls. *Planta* 71:1–14. 1966.

31. Pickett-Heaps, J. D. Xylem wall deposition. Radioautographic investigations using lignin precursors. *Protoplasma* 65:181–205. 1968.
32. Scurfield, G., A. J. Michel, and S. R. Silva. Crystals in woody stems. *Bot. J. Linn. Soc.* 66:277–289. 1973.
33. Stewart, C. M. Excretion and heartwood formation in living trees. *Science* 153:1068–1074. 1966.
34. Torrey, J. G., D. E. Fosket, and P. K. Hepler. Xylem formation: a paradigm of cytodifferentiation in higher plants. *Amer. Sci.* 59:338–352. 1971.
35. Turgeon, R. Differentiation of wound vessel members without DNA synthesis, mitosis or cell division. *Nature* 257:806–808. 1975.
36. Wardrop, A. B., and H. E. Dadswell. The cell wall structure of xylem parenchyma. *Aust. J. Sci. Res. Ser. B., Biol. Sci.* 5:223–236. 1952.
37. Wodzicki, T. J., and C. L. Brown. Organization and breakdown of the protoplast during maturation of pine tracheids. *Amer. J. Bot.* 60:631–640. 1973.

9

Xylem: Variations in Wood Structure

Woods are usually classified in two main groups, the softwoods and the hardwoods. The term softwood is applied to gymnosperm wood, that of hardwood to the dicotyledon wood. The two kinds of wood show basic structural differences, but they are not necessarily distinct in degree of density and hardness. The gymnosperm wood is homogeneous in structure—with long straight elements predominating—and is, therefore, easily workable. It is highly suitable for papermaking. Many commercially used dicotyledon woods are especially strong, dense, and heavy because of high proportion of fiber tracheids and libriform fibers (*Quercus, Carya, Eucalyptus, Acacia*), but some are light and soft (the lightest and softest is balsa, *Ochroma*). The main sources of commercial timbers are the conifers among the gymnosperms and the dicotyledons among the angiosperms. The monocotyledons having secondary growth do not produce a commercially important homogeneous body of secondary xylem.

CONIFER WOOD

The secondary xylem of conifers is relatively simple in structure[17] (figs. 9.1 and 9.2), simpler than that of most of the dicotyledons. One of its outstanding features is the lack of vessels. The tracheary elements are imperforate and are mainly tracheids. Fiber-tracheids may occur in the late wood, but libriform fibers are absent. The tracheids are narrow elongated cells averaging 2 to 5 millimeters in length (fig. 8.3,*A*). Their overlapping ends may be curved and branched because of intrusive growth. Basically, the ends are wedge shaped, with the approximately truncated end of the wedge exposed in the radial section (fig. 9.1).

The early wood tracheids have circular bordered pits with circular inner apertures (fig. 8.5). The late wood tracheids (or fiber-tracheids) have somewhat reduced borders with oval inner apertures. This difference in pit structure is a concomitant of the increase in wall thickness in the late wood cells. The

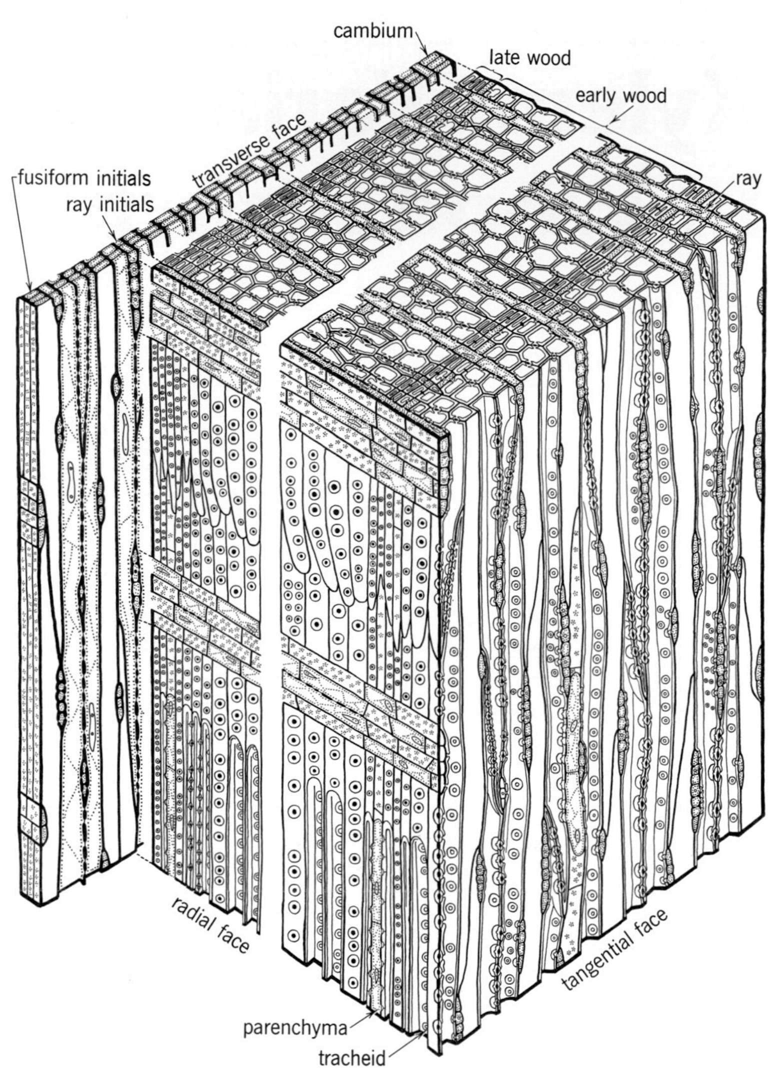

Figure 9.1 Block diagram of vascular cambium and wood of *Thuja occidentalis* (white cedar), a conifer. The axial system consists of tracheids and some parenchyma cells. The rays contain only parenchyma cells. (From Esau, *Plant Anatomy,* 2nd ed. John Wiley & Sons, 1965.)

pit-pairs between tracheids usually have tori. Throughout most of a growth layer the pits are restricted to the radial walls (fig. 9.1); the tangential walls may bear pits in the late wood. The pit-pairs are abundant on the overlapping ends of tracheids. The pits are typically in one row. In Taxodiaceae and Pinaceae some wide early wood tracheids may have two or more rows of pits in opposite arrangement. Conifer tracheids may have

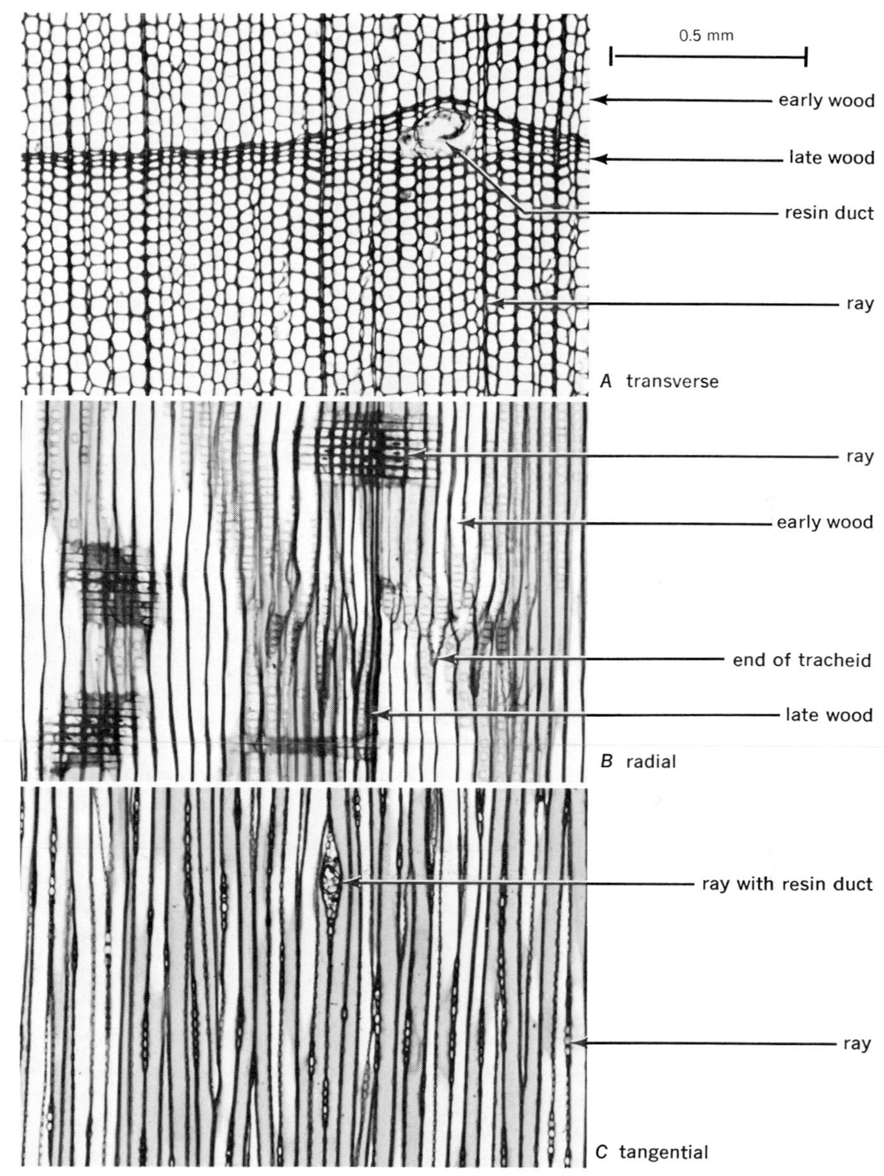

Figure 9.2 Wood of pine (*Pinus strobus*), a conifer in three kinds of sections.

helical thickenings in addition to the pitted secondary wall.

Axial parenchyma may or may not be present in conifer wood. In Podocarpaceae, Taxodiaceae, and Cupressaceae parenchyma is prominent in the wood (fig 9.1). It is scantily developed or absent in Araucariaceae, Pinaceae, and Taxaceae. In some genera, axial parenchyma is restricted to that associated with resin ducts (*Pinus, Picea,*

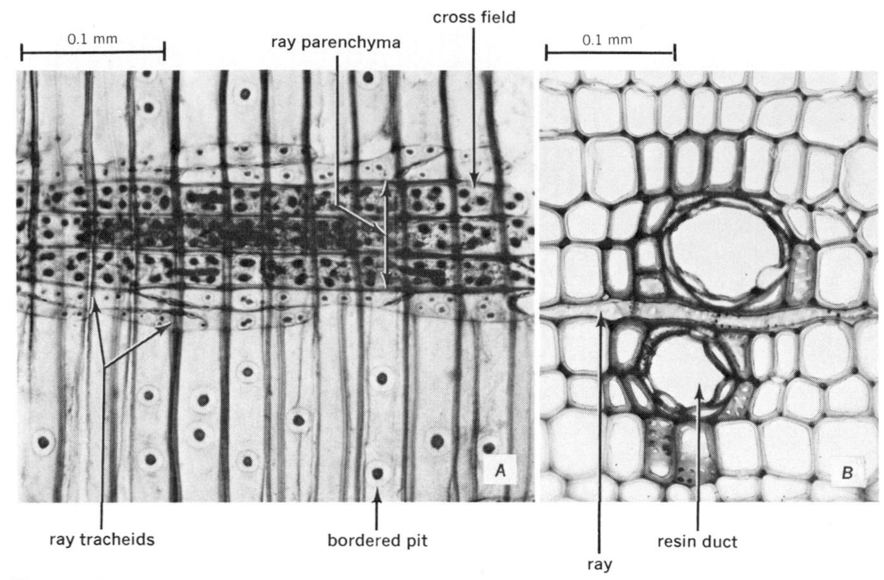

Figure 9.3 Details of conifer wood. *A*, radial section of *Larix laricina* wood showing parts of axial tracheids with bordered pits and of a ray consisting of parenchyma cells in the middle and of ray tracheids at the margins. The ray parenchyma cells show accumulations of stained cytoplasm in the pits in the cross fields. The ray tracheids have bordered pits and no cytoplasm. *B*, transverse section of wood of *Pseudotsuga taxifolia* showing two resin ducts with thick-walled epithelial cells.

Larix, Pseudotsuga). Resin ducts (figs. 9.2 and 9.3,*B*) appear as a constant feature of some woods (Pinaceae), but they also develop as a result of injury (traumatic resin ducts.[9]) Resin ducts occur in the axial and in the radial systems.

The rays of conifers are mostly one cell wide (figs. 9.1 and 9.2), occasionally biseriate and from one to twenty or even to fifty cells high. Presence of resin ducts makes the normally uniseriate rays appear multiseriate (fig. 9.2,*C*). The rays consist of parenchyma cells or may also contain ray tracheids. These tracheids resemble parenchyma cells in shape but are devoid of protoplasts at maturity and have secondary walls with bordered pits (fig 9.3,*A*). Ray tracheids are normally present in most Pinaceae, occasionally in *Sequoia* and the Cupressaceae. The ray tracheids commonly occur along the margins of rays, one or more cells in depth.

Each axial tracheid is in contact with one or more rays (fig. 9.1). The pit-pairs between the axial tracheids and ray parenchyma cells are half-bordered, with the border on the side of the tracheid (chapter 4); those between the axial and the ray tracheids are fully bordered. The pitting between ray parenchyma cells and axial tracheids form such characteristic patterns in radial sections that the cross-field, that is, the rectangle formed by the radial wall of a ray cell against an axial tracheid (fig. 9.3,*A*), is utilized in classification and in taxonomic studies of conifer woods.

DICOTYLEDON WOOD

The wood of dicotyledons is more varied than that of gymnosperms. The wood of the primitively vesselless dicotyledons is relatively simple, but that of the vessel-containing

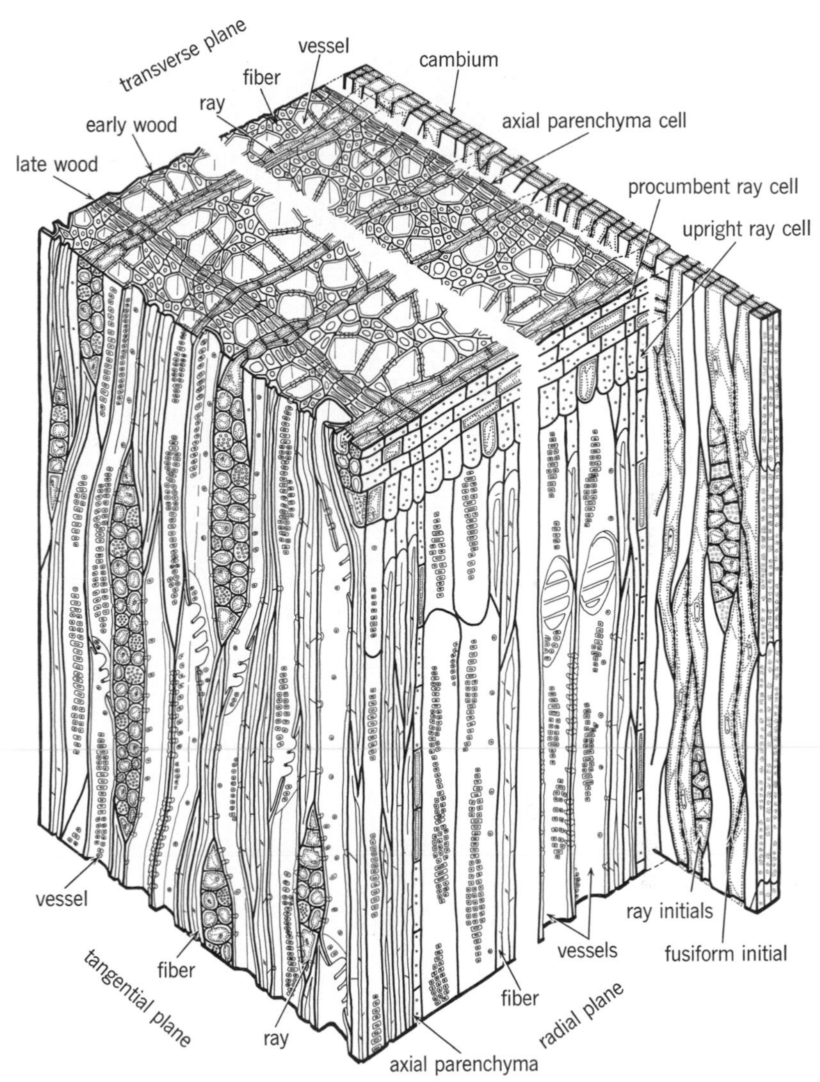

Figure 9.4 Block diagram of vascular cambium and wood of *Liriodendron tulipifera* (tulip tree), a dicotyledon. The axial system consists of vessel members with scalariform perforation plates, fiber-tracheids, and axial xylem parenchyma strands in terminal position. (From Esau, *Plant Anatomy,* 2nd ed. John Wiley & Sons, 1965.)

species is usually complex. Wood of the latter species may have both vessels and tracheids, one or more categories of fibers (chapter 8), axial parenchyma, and rays of one or more kinds (figs. 9.4–9.6).

Storied and nonstoried wood

In transverse sections the secondary xylem shows more or less orderly radial seriation of cells—a result of the origin of cells from

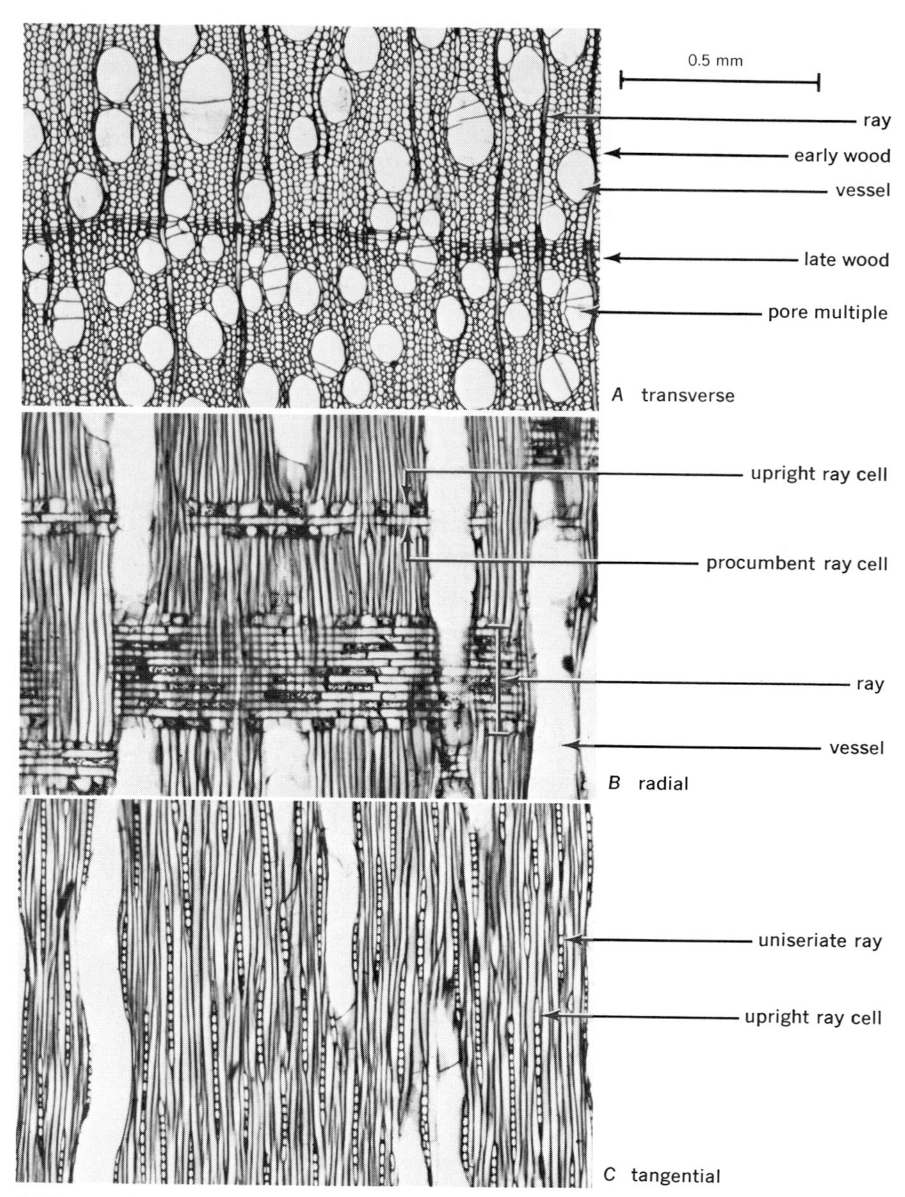

Figure 9.5 Wood of willow (*Salix nigra*), a dicotyledon, in three kinds of sections. Diffuse-porous nonstoried wood with uniseriate heterocellular rays.

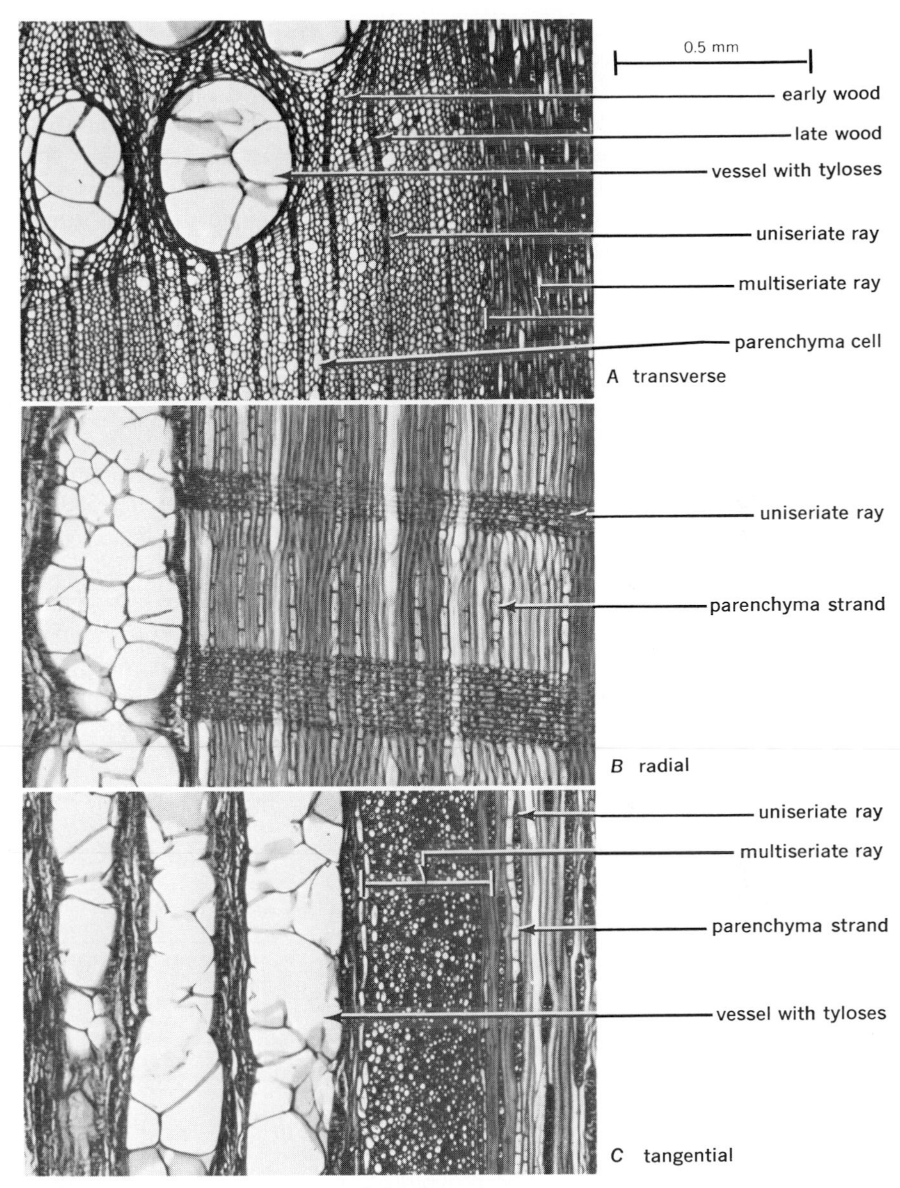

Figure 9.6 Wood of oak (*Quercus alba*), a dicotyledon, in three kinds of sections. Ring-porous nonstoried wood with high multiseriate and low uniseriate rays. The large vessels are occluded by tyloses.

tangentially dividing cambial cells. In the homogeneous conifer wood this seriation is pronounced (fig. 9.2); in vessel-containing dicotyledons it may be somewhat obscured by the ontogenetic enlargement of the vessel members and the consequent displacement of adjacent cells (figs. 9.5 and 9.6). Radial sections also reveal the radial seriation, and they indicate that the radial series of the axial system are superimposed one upon the other in horizontal layers, or tiers (fig 9.6). The tangential sections are varied in their appearance in different woods. In some, the cells of one tier unevenly overlap those of another; in others the horizontal layers are clearly displayed in tangential sections. Thus, some woods are nonstratified, or nonstoried, in tangential sections (fig. 9.7,*A; Castanea, Fraxinus, Juglans, Quercus*), others stratified, or storied (fig. 9.7,*B; Aesculus, Cryptocarya, Ficus, Tilia,* numerous Fabales). The storied condition is especially pronounced when the height of the ray matches that of a horizontal layer of the axial system. From the evolutionary aspect the storied woods are more highly specialized than the nonstoried. They are derived from vascular cambia with short fusiform initials. Many intermediate patterns are found between the strictly storied woods and the strictly nonstoried woods derived from cambia with long fusiform initials.

Distribution of vessels

The wood anatomist refers to a vessel in cross section as a pore. Two principal types of woods are recognized on the basis of distribution of pores in a growth layer: diffuse-porous wood with pores rather uniform in size and distribution throughout a growth ring (figs. 9.5 and 9.8,*A*,*B*; species in *Acer, Betula, Carpinus, Fagus, Juglans, Liriodendron, Platanus, Populus, Pyrus*); ring-porous wood with pores distinctly larger in the early wood than in the late wood (figs. 8.1, 9.6, and 9.8,*D*; species in *Castanea, Catalpa, Celtis, Fraxinus, Gleditsia, Morus, Quercus, Robinia, Ulmus*). Intergrading patterns occur between the two types of pattern. The ring-porous condition appears to be an indication of evolutionary specialization and occurs in comparatively few species, nearly all characteristic of the north temperate zone. A ring-

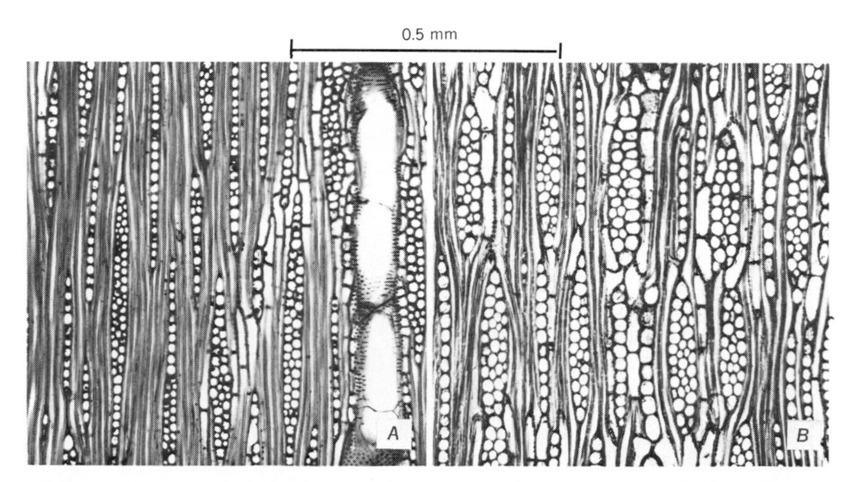

Figure 9.7 *A*, nonstoried wood of pecan (*Carya pecan*). *B*, storied wood of persimmon (*Diospyros virginiana*). Both tangential sections.

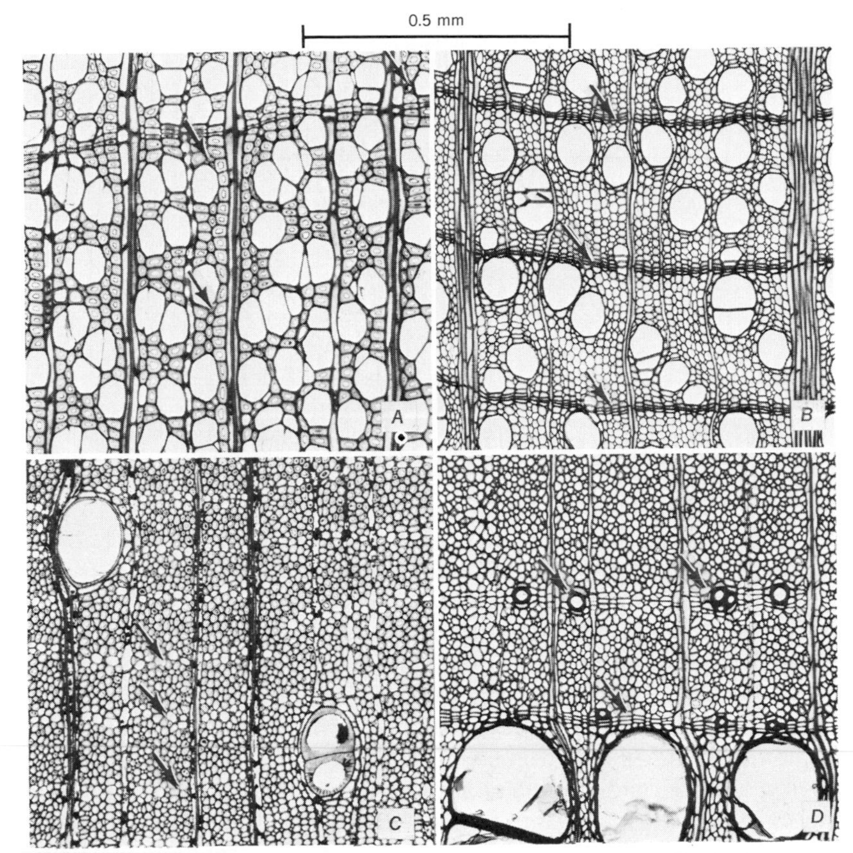

Figure 9.8 Distribution of axial parenchyma (arrows) in wood. *A, Liquidambar styraciflua,* parenchyma very sparse. *B, Acer saccharum,* boundary parenchyma. *C, Carya pecan,* apotracheal banded parenchyma. *D*, *Fraxinus* sp., paratracheal and boundary parenchyma. All cross sections.

porous wood conducts water almost entirely in the outermost growth increment, at a speed that is about ten times greater than is recorded for diffuse-porous woods.

Within the main distributional patterns of vessels, minor variations occur in the spatial relation of the pores to each other. A pore is called solitary when the vessel is completely surrounded by other types of cells (figs. 9.6 and 9.8,*C*). A group of two or more pores appearing together form a pore multiple (fig. 9.5,*A*). This may be a radial pore multiple, with pores in a radial file, or a pore cluster, with an irregular grouping of pores. Although vessels or vessel groups may appear isolated in wood transections, in the three-dimensional space the vessels are interconnected in various planes. In some species the vessels are interconnected only within individual growth increments, in others connections occur across the boundaries of growth increments.[1]

Distribution of axial parenchyma

The distribution of the axial xylem parenchyma shows many intergrading patterns. The spatial relation to vessels, as seen in transec-

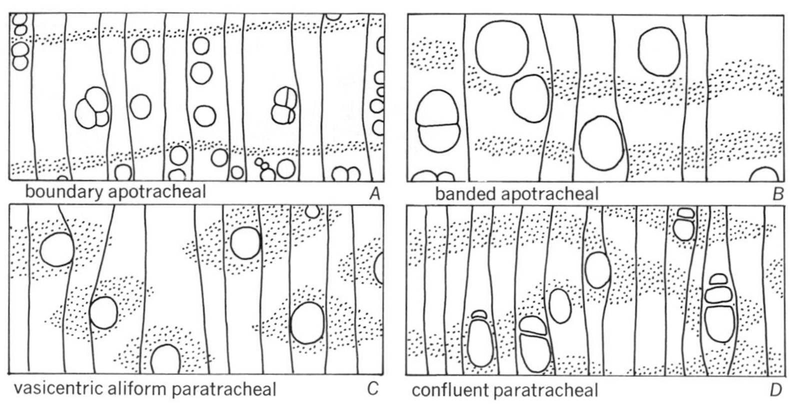

Figure 9.9 Distribution of axial parenchyma (stipples) in wood of *A*, *Michelia; B*, *Saccopetalum; C*, a leguminous species; *D*, *Terminalia.* (Drawn from photographs in S. J. Record, *Timbers of North America,* John Wiley & Sons, 1934.)

tions, serves for the division in two main patterns: *apotracheal,* parenchyma not definitely associated with the vessels (fig. 9.9,*A*,*B*); *paratracheal,* parenchyma consistently associated with the vessels (fig. 9.9,*C*,*D*). The apotracheal parenchyma is further subdivided into: *diffuse,* single parenchyma cells or parenchyma strands scattered among fibers (fig. 9.6); apotracheal *banded* (figs. 9.8,*C* and 9.9,*B*); boundary[12] or *marginal*[2] parenchyma, with single cells or a band at the end (terminal) or at the beginning (initial) of a growth layer (figs. 9.8,*B*,*D*, and 9.9, *A*). Diffuse apotracheal parenchyma may be sparse (fig. 9.8,*A*). The paratracheal parenchyma appears in the following forms: *scanty vasicentric,* forming complete sheaths around vessels; *aliform* vasicentric with winglike tangential extensions (fig. 9.9,*C*); and *confluent,* coalesced aliform forming irregular tangential or diagonal bands (fig. 9.9,*D*). If septate fibers instead of axial parenchyma occur in the xylem, they show distributional patterns similar to those assumed by the axial xylem parenchyma. From the evolutionary aspect the apotracheal and diffuse patterns are primitive.[14]

The paratracheal parenchyma shows physiologic differences from parenchyma scattered among fibers.[1,5] During the mobilization of stored carbohydrates in the spring, starch dissolves earlier in paratracheal cells than in the scattered ones. Paratracheal cells also show a high phosphatase activity. They release sugar into the vessels for rapid transport to the buds and appear to participate in refilling with water those vessels that have accumulated gases during dormancy. Parenchyma cells having the distinct physiologic relation to the vessels have been named *contact cells.*[18] They are analogous to the companion cells that serve in the sugar exchange with the sieve elements in the phloem (chapter 11).

Structure of rays

In contrast to the predominantly uniseriate rays of conifers, those of the dicotyledons may be one to many cells wide, that is, they may be uniseriate (fig. 9.5) or multiseriate (figs. 9.4, 9.6, and 9.7), and range in height from one to many cells (from a few mm to 3 cm or more).

The multiseriate rays frequently have uniseriate margins (fig. 9.7,*A*). Small rays may be grouped so as to appear to be one large ray. Such groups are called aggregate rays (*Carpinus*).

The appearance of rays in radial and tangential sections can be used as a basis for their classification. Individual rays may be homocellular, that is, composed of cells of one form only (figs. 8.6,*C*,*D*, and 9.6), either procumbent or upright, or heterocellular, that is, having two morphological cell types, procumbent and upright (figs. 8.6,*A*,*B*,9.4, and 9.5). The entire ray system of a wood may consist of either homocellular or heterocellular rays or of combinations of the two types of rays. On this basis the ray tissue system is classified into homogeneous, rays all homocellular (procumbent cells only), or heterogeneous, rays all heterocellular or combinations of homocellular and heterocellular.[12] Further variations between homogeneous and heterogeneous ray tissues result from combinations of uniseriate and multiseriate rays or absence of multiseriate rays.

The different ray combinations have a phylogenetic significance. The primitive ray tissue may be exemplified by that of the Winteraceae (*Drimys*). The rays are of two kinds: one homocellular—uniseriate composed of upright cells; the other heterocellular—multiseriate composed of radially elongated or nearly isodiametric cells in the multiseriate part and upright cells in the uniseriate marginal parts. Both kinds of ray are many cells in height. From such primitive ray structure other ray systems, more specialized, have been derived. For example, multiseriate rays may be eliminated (*Aesculus hippocastanum*) or increased in size (*Quercus*), or both multiseriate and uniseriate rays may be decreased in size (*Fraxinus*).

The evolution of rays strikingly illustrates the maxim that phylogenetic changes depend on successively modified ontogenies. In a given wood the specialized ray structure may appear gradually. The earlier growth layers may have a more primitive ray structure than the later because the vascular cambium commonly undergoes successive changes before it begins to produce a ray pattern of a more specialized type.

The ray cells share some functions with the axial parenchyma cells and are also concerned with radial transport of assimilates.[1] Ray cells that are connected through pits with tracheary elements (illustrated for a conifer wood in fig. 9.3,*A*) function as contact cells in delivering carbohydrates to the vessels.[18] They mobilize their stored starch precociously in the spring, show a periodic high phosphatase activity and a periodic increase in size of nucleoli, and have a high fat content. Contact ray cells may be upright or procumbent but they always show prominent pit connections with vessels. Ray cells that have no contacts with vessels—such cells are particularly numerous in multiseriate rays—deposit starch in the early summer and mobilize it in the early spring. The cells are mainly procumbent and show polarized phosphatase activity in the spring; in a given cell, the activity is concentrated next to the periclinal wall that is facing the cambium. The cells appear to be concerned with periodic radial transport of mobilized carbohydrate toward the reactivated cambium.

Tyloses

In many species, axial and ray parenchyma cells located next to the vessels form outgrowths through the pit cavities into the lumina of the vessels when the latter become inactive (fig. 9.6). These outgrowths are called tyloses. The membranes of the pits from which tyloses issue are modified by deposition of a so-

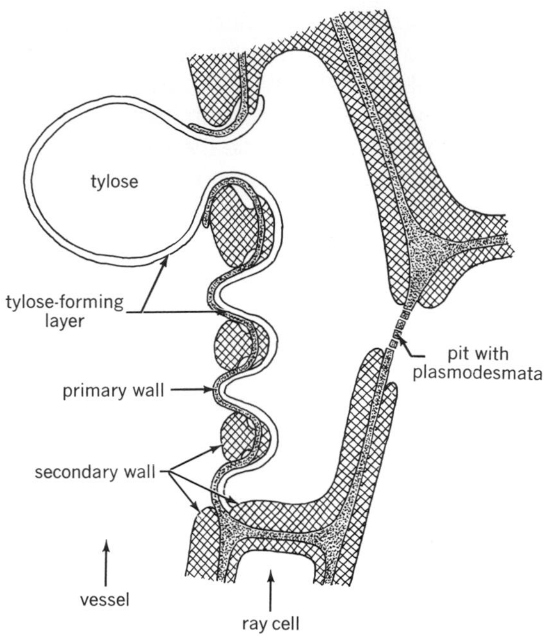

Figure 9.10 Diagram of ray cell that has formed a tylose protruding through a pit into the lumen of a vessel. The tylose-forming layer is also called protective layer. (Constructed from data in Foster[10] and Meyer and Côté.[13])

called protective layer on the side of the parenchyma cell.[13] This layer has a loose fibrillar structure and consists of polysaccharides and pectins.[6] The deposition occurs toward the end of tissue differentiation and seals off the parenchyma cell from the mature vessel element. The additional wall is laid down not only on the pit membranes but on the entire wall facing the vessel member[10] (fig. 9.10), and may occur as a thin layer on walls between contiguous parenchyma cells. The pit membrane becomes degraded enzymically whereas the unlignified protective layer undergoes surface growth and baloons out as a tylose into the vessel lumen. The nucleus and part of the cytoplasm of the parenchyma cell commonly migrate into the tylose. Tyloses store ergastic substances and may develop secondary walls, or even differentiate into sclereids. It seems that tylose development is possible only if the pit aperture on the vessel side is no less than 10 microns wide.[4] Examples of woods with abundant development of tyloses are those of *Quercus* (white oak species), *Robinia, Vitis, Morus, Catalpa, Juglans nigra, Maclura.* Tyloses occur in primary xylem also and, as observed in oat coleoptiles and bean leaves, originate from a wall enriched with acidic polyuronides.[16]

Tyloses block the lumina of vessels and reduce the permeability of the wood. Technically this phenomenon is important in the treatment of wood with preservatives and in its selection for tight coopering. With regard to conduction in the xylem, the significance of tyloses is not clearly understood. They are known to block vessels during the formation of heartwood and, in the sapwood, in response to wounding and infection with diseases.

Intercellular canals and cavities

Intercellular canals similar to the resin ducts of gymnosperms occur in the dicotyledon woods. They are often called gum ducts although they may contain resins. They occur in both the axial and the radial systems and may be normal or traumatic. Intercellular canals vary in extent, and some are more appropriately called intercellular cavities. Intercellular canals and cavities may be schizogenous, but gummosis of surrounding cells may also occur. Canals associated with gummosis are well known in such genera as *Amygdalus* and *Prunus.*

SOME FACTORS IN DEVELOPMENT OF SECONDARY XYLEM

The secondary xylem is produced by the vascular cambium (chapter 10) and consequently its development is greatly affected by factors controlling cambial activity. The intermittent functioning of the cambium, which is mainly seasonal in the temperate regions, is reflected in the production of growth increments in the wood. The increments, or rings in transections, are delimited from one another by anatomic differences between early wood and late wood (chapter 8). The causes of these differences have not been fully revealed, and most of the pertinent studies were conducted on coniferous seedlings.[25]

According to a common concept, auxin attains high levels under conditions promoting shoot growth and continued leaf development. The cells produced in the xylem at this time are wide and thus of the early wood type. Conversely, conditions adversely affecting shoot growth lower the level of diffusible auxin and induce the formation of narrow flattened cells of the late wood type. It should be stressed, however, that the auxin involved in secondary growth is only partly derived from the growing shoots. The differentiating vascular tissues, and specifically xylem, appear to be important sources of auxin that maintains cambial activity after its initial reactivation under the influence of expanding buds.[20] The increase in cell wall thickness in late wood cells, which is causally unrelated to cell diameter, is usually explained in terms of promotion of synthesis of cell wall material by seasonal assimilation.

This brief summary does not do justice to the complexity of the phenomenon of annual growth, for, undoubtedly, several types of growth promoting substances and some natural inhibitors are involved in secondary development, and the activity of the substances is modified by nutritional conditions and availability of water. Many variations in width of wood rings within a tree may be attributed to changing supplies of food. The nutritional factors in turn are affected by the climate. Recognition of these relations has led to the development of dendrochronology, that is, study of yearly growth patterns in trees and use of the information for evaluating past fluctuations in climate and dating past events.[11]

Sometimes individual factors can be related to specific aspects of xylem differentiation. When *Xanthium* seedlings were decapitated cambial activity continued but fibers failed to differentiate in the secondary xylem.[21] Fiber differentiation could be induced by applying naphthaleneacetic acid to the decapitated plants, a result suggesting that auxin directly affects differentiation of xylem cells[22] rather than indirectly through induction of cell division (chapter 8). Detached *Fraxinus* stems, grown in a culture medium, in which the water potential (controlled by addition of polyethylene glycol) and the concentration of growth substances were varied, showed that the width of new tissue produced was influenced more by the water potential than by the concentrations of indoleacetic and gibberellic acids.[8] Growth relations in the wound callus formed on the detached *Fraxinus* stems provided the additional evidence that, besides a certain concentration of auxin, xylem differentiation requires physical pressure.[8] The importance of mechanical pressure in secondary tissue development was demonstrated also by manipulating longitudinal strips of bark partly separated along the differentiating xylem from stems of *Populus* and *Pinus*.[25] A strip of bark attached to the bole above but left suspended below formed callus on its inner side. Later, a cambium developed in the callus in continuity with the existing cambium in the strip converting the latter into a stemlike structure. In about three weeks after the strip was isolated orderly formation of xylem and phloem was in progress. In a strip of bark that was not left suspended but placed back against the wood of the bole (with a plastic film separating the two) and subjected to some pressure, callus formation was very restricted and normal production of xylem and phloem was shortly restored.

Differentiation of specific cell types of xylem and their distribution in the tissue might also lend itself to analyses in terms of indentifiable factors. The previously mentioned experiments with *Fraxinus*[8] indicate that vessel development is particularly sensitive to indoleacetic acid supply. Increase of this supply results in the formation of wider vessels but the effect diminishes when the water potential is reduced. The authors suggest that the nonuniform distribution of vessels may be the result of differential distribution of the necessary growth factors among xylem mother cells. Possibly individual cells or cell groups form metabolic sinks and, by sequestering growth factors, increase solute concentration and take up water preferentially in competition with neighboring cells. The successfully competing cells expand rapidly and differentiate into vessel members.

The developmental effects of vessels upon contiguous cells at the histological level clearly indicate the presence of a competitive factor in cell adjustments during vessel differentiation. When a future vessel element begins to expand in the cambial mother cell zone, production of cells ceases in one or more rows adjacent to the row containing the expanding cell. Divisions are resumed in these rows after the vessel has expanded and the cambium has been displaced outward. If the space thus provided does not suffice to accommodate the expanding vessel, the latter forces adjacent cells apart. The separating cells often assume irregular shapes and, at maturity, are called *disjunctive parenchyma cells* or *disjunctive tracheids* depending on their final differentiation.

The recognition of specific relations between availability of auxins and water and the differentiation of vessels is just a beginning in analyses of factors that bring about the profound differences among cells derived from apparently identical cambial precursors. The intrusive elongation of fibers (chapter 6), the idioblastic behavior of cells accumulating specific ergastic substances or differentiating into sclereids, the physiologic specializations of the parenchymatic members of the xylem are some examples of programmed developments that await interpretation with reference to controlling factors.

Reaction wood

The reaction type of wood is formed on the lower sides of branches and leaning or crooked stems of conifer trees, and on the upper sides of similar structures in dicotyledon trees. This wood is called reaction wood (compression wood in conifers and tension wood in dicotyledons) because its development is assumed to result from the tendency of the branch or stem to counteract the force inducing the inclined position.[23] Reaction wood occurs in roots also.[24]

Research involving experimental modifications in position of plant axes has provided evidence that the stimulus of gravity and the distribution of endogenous growth substances are important factors in evoking the development of reaction wood.[3,24] Experiments with auxins and anti-auxins indicate that the tension wood of dicotyledons is formed where auxin concentration is low.[15] In contrast, the compression wood of gymnosperms is formed in regions of high auxin concentration.[24] As to the force counteracting the inclined position, it is thought to reside in the region where the differentiating reaction wood cells are undergoing lignification and to arise from the swelling of cell walls as a result of the lignification.[19]

The reaction wood differs from the normal in both anatomy and chemistry.[7] The compression wood of conifers is typically denser and darker than the surrounding tissue, and its tracheids are shorter than those in normal wood. The cell walls appear rounded in transections and contain more or less strongly lignified layers. The inner layer of the usually three-layered secondary wall is missing. In the tension wood of dicotyledons the vessels are reduced in width and number and the fibers have a thick highly refractive inner layer—the so-called gelatinous layer—consisting largely of cellulose. The walls of these fibers may be two to four layered; the gelatinous layer is usually the innermost (chapter 8).

IDENTIFICATION OF WOOD

The use of wood for purposes of identification requires a very sound knowledge of wood structure and of factors modifying that structure. The search for diagnostic features is best based on an examination of collections from more than one tree of the same species, made with proper attention to the location of the sample on the tree. The wood acquires its mature character not at the beginning of cambial activity but in the later growth increments. Thus, the wood of a twig would be of a different ontogenetic age than that of a trunk of the same tree. Furthermore, in certain sites, the wood has reaction wood properties that deviate more or less strongly from features considered to be typical of the taxon in question. Adverse or unusual environmental conditions and improper methods of preparation of sample for microscopy also may obscure the diagnostic features.

A further complicating aspect of wood identification is that the anatomical characteristics of woods are often less differentiated than the external features of the taxa involved. Although woods of large taxa differ considerably from one another, within groups of closely related taxa, such as species, or even genera, the wood may be so uniform that no consistent differences are detectable. Under such circumstances, it is imperative to use aggregates of macroscopic and microscopic characters of woods, as well as odor and taste.

Characters used in identification of woods

The two large taxa that yield commercial timber and the wood anatomy of which is well known are the conifers (softwoods) and the dicotyledons (hardwoods). In keys for wood identification these two taxa are invariably separated on the basis of absence or presence of vessels: nonporous wood (vessels absent), conifers; porous wood (vessels present), dicotyledons. The exceptional vesselless lower dicotyledons (*Drimys*, *Zygogynum*, etc.) can be identified by their ray structure, which is more varied than in the conifers. In the following, the important diagnostic features of softwoods and hardwoods are summarized with reference to examples of genera and species showing these features.

CONIFERS.

Several genera (*Pinus*, *Picea*, *Larix*, and *Pseudotsuga*) have normal resin canals in axial and ray systems, although in *Picea* the canals are often less abundant than in the other three genera. Other conifers (*Abies*, *Sequoia*, *Taxodium*) may have traumatically induced resin canals, but these are recognizable in the axial system by their alignment in tangential groups or rows. The epithelial cells of the resin canals have thin walls in *Pinus*, thick walls in *Picea*, *Larix*, and *Pseudotsuga*.

The thin-walled cells may be collapsed in sectioned wood, particularly in the axial system; their identification is more certain in rays.

Ray tracheids are present in *Pinus, Picea, Larix, Pseudotsuga, Cedrus, Tsuga, Chamaecyparis nootkatensis,* and sporadically in *Abies balsamea.* The ray tracheids have dentate wall ornamentations in some pines (*Pinus strobus,* white pine, *P. lambertiana,* sugar pine), smooth walls in others (*P. ponderosa,* western yellow pine).

Ray parenchyma cells also have diagnostic characteristics. Their tangential walls (end walls), as seen in radial sections, may be smooth (*Thuja plicata, Chamaecyparis lawsoniana, Araucaria, Podocarpus chilinus, Sequoia sempervirens, Taxodium distichum, Taxus baccata*) or they may appear beaded because of deep pitting (species of *Cedrus, Tsuga, Abies pectinata, Pseudotsuga, Picea, Larix*). The pits on the radial walls of ray parenchyma cells visible within the confines of a cross-field vary in size, number, arrangement, and degree of development of the border in the contiguous axial tracheid. The pines have either one or two large pits to a cross-field (*Pinus lambertiana*) or several smaller ones (*P. ponderosa*), all with none or barely perceptible borders in the adjacent tracheids. Most other conifers have one to several pits with borders in the adjoining axial tracheids (half-bordered pit-pairs). The pits in a cross field may be in one or two rows or rather irregularly arranged.

Helical thickenings deposited over the secondary walls are normally present in axial tracheids of *Pseudotsuga,* being especially well developed in the early wood. Helical thickenings are found occasionally in late wood tracheids of *Larix* and some species of *Picea,* but not in *Pinus. Taxus* has helices in all axial tracheids but differs from *Pseudotsuga* in having no normal resin canals and no ray tracheids. *Agathis* and *Araucaria* differ from other conifers in having alternate arrangement of pits on the axial tracheids if the pits are in two or more rows. A striking feature of the axial tracheids of *Cedrus* is the crenulated edge in the torus of the bordered pit.

The size and shape of rays, as seen in tangential sections, may be useful in conifer wood identification. In *Pinus, Picea,* and *Larix* the resin canal region in the fusiform ray (resin canal-containing ray in tangential section) bulges abruptly and often has uniseriate extensions above and below. In *Pseudotsuga,* the ray outline is commonly more regularly fusiform. The rays in *Sequoia sempervirens* have rather large cells and are partly biseriate. In the closely related *Taxodium distichum* the rays have smaller cells and rarely contain biseriate portions. Rays in one taxon may be markedly higher than in another. *Cedrus,* for example, has higher rays (sometimes exceeding 40 cells) than *Tsuga* and *Chamaecyparis nootkatensis,* both of which otherwise have wood rather similar to that of *Cedrus.* Presence of axial parenchyma, as in *Sequoia* and *Taxodium,* may be a reliable diagnostic feature.

DICOTYLEDONS.

The hardwoods are usually subdivided into two major groups on the basis of presence or absence of rings of pores (early wood vessels) in cross sections: ring porous, early wood vessels appreciably larger than those of the late wood; diffuse porous, early wood vessels not larger or only slightly larger than those of the late wood. Ring porousness is visible with a hand lens on a clean surface of a piece of timber. The choice between the two alternatives in identifying a wood may be made difficult by the following conditions. (1) Early wood vessels are much larger than others but grade into those of the late wood (*Carya*). (2) Early wood vessels are only moderately larger than the late wood vessels and intergrade with them; this condition is sometimes classified as semi-ring porous or semi-diffuse porous, and the same species may fit into the ring-porous or one of the intermediate categories (*Carya, Catalpa, Robinia*). (3) Vessels predominate in the early wood but are not appreciably larger than those of the late wood; this condition is classified as diffuse porous (*Juglans, Tilia*).

Ring-porous woods are distinguished from one another by the degree of expression of ring porousness and other characters. In *Fraxinus, Quercus,* and *Ulmus* the change from early wood to late wood is abrupt, in *Castanea* and *Paulownia* it is gradual. In *Ulmus* and *Celtis,* the late wood vessels, vascular tracheids, and paratracheal parenchyma are aggregated into undulating tangential bands as seen in cross sections. The deciduous oaks may be separated from other ring-porous woods by the combination of very broad high rays visible with the unaided eye and narrow, mostly uniseriate, low rays. These oaks are divisible in two groups: red oaks (*Quercus borealis, Q. palustris, Q. velutina*) in which the early wood vessels are usually free of tyloses in the heartwood, the late wood vessels are rounded in transections and have thick walls, and the large rays average 6 to 12 mm in height; white oaks (*Q. alba, Q. bicolor, Q. macrocarpa*) in which the early wood vessels become occluded with tyloses, the late wood vessels are not rounded in transections and have thin walls, and the large rays average 12 to 32 mm in height. Tyloses are common also in *Carya, Maclura, Morus,* and *Robinia.*

A number of ring-porous woods have helical thickenings in the narrow late wood vessels, for example *Catalpa, Celtis,* and *Gymnocladus. Sassafras* has occasional scalariform perforation plates in late wood vessels, thin-walled fibers, and oil cells in the 1–4 seriate rays. The representatives of the

Table 9.1 EXAMPLES OF WOODS WITH DIFFERENT DISTRIBUTIONS OF VESSELS

Ring Porous
Carya pecan (pecan)
Castanea dentata (American chestnut)
Catalpa speciosa
Celtis occidentalis (hackberry)
Fraxinus americana (white ash)
Gleditsia triacanthos (honey locust)
Gymnocladus dioicus (Kentucky coffee tree)
Maclura pomifera (Osage orange)
Morus rubra (red mulberry)
Paulownia tomentosa
Quercus spp. (deciduous oaks)
Robinia pseudoacacia (black locust)
Sassafras albidum
Ulmus americana (American elm)

Semi-Ring Porous or Semi-Diffuse Porous
Diospyros virginiana (persimmon)
Juglans cinerea (butternut)
Juglans nigra (black walnut)
Lithocarpus densiflora (tanbark oak)
Populus deltoides (cottonwood)
Prunus serotina (black cherry)
Quercus virginiana (live oak)
Salix nigra (black willow)

Diffuse Porous
Acer saccharinum (silver maple)
Acer saccharum (sugar maple)
Aesculus glabra (buckeye)
Aesculus hippocastanum (horse chestnut)
Alnus rubra (red alder)
Betula nigra (red birch)
Carpinus caroliniana (blue beech)
Cornus florida (dogwood)
Fagus grandifolia (American beech)
Ilex opaca (holly)
Liquidambar styraciflua (American sweet gum)
Liriodendron tulipifera (tulip tree)
Magnolia grandiflora (evergreen magnolia)
Nyssa sylvatica (black gum)
Platanus occidentalis (American plane tree)
Tilia americana (basswood)
Umbellularia californica (California laurel)

Fabales, *Cercis, Gleditsia, Gymnocladus,* and *Robinia* have vestured bordered pits (see Glossary).

The distribution of axial parenchyma and character of rays can be usefully combined with other characters. The rays are seldom uniformly homocellular or heterocellular. Uniseriate homocellular rays occur in *Castanea,* and multicellular, essentially homocellular rays are found in *Fraxinus, Gymnocladus, Paulownia,* and *Ulmus. Fraxinus* has conspicuous vasicentric parenchyma. The apotracheal parenchyma forms 1–4 seriate tangential bands in cross sections of *Carya* wood.

Among the taxa with semi-ring porous wood, *Diospyros* is distinguished by an essentially storied wood. It also has minute intervessel pitting and thick vessel walls. *Prunus* shows helices in vessels and has very sparse axial parenchyma. *Salix* has uniseriate, heterocellular rays, *Populus* uniseriate, homocellular rays. *Quercus virginiana* and *Lithocarpus* resemble the deciduous oaks in having rays of two sizes numerous low uniseriate rays and high, broad multiseriate rays. The latter may be simple or aggregate. The aggregate condition is more distinct in *Lithocarpus.* In both genera, vasicentric tracheids and sparse paratracheal parenchyma form sheaths about the vessels.

In the taxa with diffuse-porous woods, a number of genera typically have scalariform perforation plates: *Alnus, Betula, Cornus, Ilex, Liriodendron, Liquidambar, Nyssa;* mostly scalariform, some simple: *Magnolia;* simple, occasionally scalariform: *Carpinus, Fagus, Platanus.* Intervessel pitting may be a helpful diagnostic feature. It is commonly scalariform in *Cornus* and *Magnolia,* opposite in *Liriodendron* and *Nyssa,* opposite or linear in *Liquidambar.* The pitting is minute in *Betula,* crowded in *Carpinus* and *Fagus,* not crowded in *Alnus* and *Platanus.* Helical thickenings are typical in *Ilex, Magnolia,* and *Tilia;* may be present, sometimes only in the tapering ends of vessel members, in *Aesculus, Liriodendron,* and *Nyssa.*

Some characteristic ray structures are: homocellular 1-3 seriate rays in *Betula;* narrow simple rays and wide aggregate rays in *Alnus* and *Carpinus;* 3-14 seriate homocellular rays, up to 3 mm in height, in *Platanus.* Rays of two distinct sizes occur in *Fagus,* narrow, mostly uniseriate low rays and broad multiseriate high rays, a combination similar to that in the oaks. Two sizes of rays are found also in *Acer* (intergrading in some species) and *Tilia.* In *Cornus* and *Ilex,* the uniseriate rays are composed entirely or largely of upright cells, the multiseriate rays contain procumbent cells in the middle portions and upright cells in the uniseriate margins. *Liquidambar* has a similar ray system except that the proportions of upright cells are relatively small. *Liquidambar* may have traumatic gum canals. *Umbellularia* has oil cells in the rays.

REFERENCES

1. Braun, H. J. Funktionelle Histologie der sekundären Sprossachse. I. Das Holz. *Handbuch der Pflanzenanatomie.* Band 9. Teil 1. 1970.
2. Carlquist, S. *Comparative plant anatomy.* New York, Holt, Rinehart and Winston. 1961.
3. Casperson, G. Zur Kambiumphysiologie von *Aesculus hippocastanum* L. *Flora* 155:515–543. 1965.
4. Chattaway, M. M. The development of tyloses and secretion of gum in heartwood formation. *Aust. J. Sci. Res. B, Biol. Sci.* 2:227–240. 1949.
5. Czaninski, Y. Étude du parenchyme ligneux du Robinier (parenchyme à réserves et cellules associées aux vaisseaux) au cours du cycle annuel. *J. Microscopie* 7:145–164. 1968.

6. Czaninski, Y. Observations sur une nouvelle couche pariétale dans les cellules associées aux vaisseaux du Robinier et du Sycomore. *Protoplasma* 77:211–219. 1973.
7. Dadswell, H. E., A. B. Wardrop, and A. J. Watson. The morphology, chemistry and pulp characteristics of reaction wood. In: *Fundamentals of papermaking fibres,* pp. 187–219. British Paper and Board Makers' Association. 1958.
8. Doley, D., and L. Leyton. Effects of growth regulating substances and water potential on the development of secondary xylem in *Fraxinus. New Phytol.* 67:579–594. 1968. Effects of growth regulating substances and water potential on the development of callus in *Fraxinus. New Phytol.* 69:87–102. 1970.
9. Fahn, A., and E. Zamski. The influence of pressure, wind, wounding and growth substances on the rate of resin duct formation in *Pinus halepensis* wood. *Israel J. Bot.* 19:429–446. 1970.
10. Foster, R. C. Fine structure of tyloses in three species of the Myrtaceae. *Aust. J. Bot.* 15:25–34. 1967.
11. Fritts, H. C. Growth rings of trees: their correlation with climate. *Science* 154: 973–979. 1966.
12. Jane, F. W. *The structure of wood.* 2nd ed. Revised by K. Wilson and D. J. B. White. London, Adam & Charles Black. 1970.
13. Meyer, R. W., and W. A. Côté, Jr. Formation of the protective layer and its role in tylosis development. *Wood Sci. Tech.* 2:84–94. 1968.
14. Money, L. L., I. W. Bailey, and B. G. L. Swamy. The morphology and relationships of the Monimiaceae. *J. Arnold Arb.* 31:372–404. 1950.
15. Morey, P. R., and J. Cronshaw. Developmental changes in the secondary xylem of *Acer rubrum* induced by gibberellic acid, various auxins and 2,3,5-tri-iodobenzoic acid. *Protoplasma* 65:315–326. 1968.
16 O'Brien, T. P. Primary vascular tissues. In: *Dynamic aspects of plant ultrstructure.* A. W. Robards, ed. Chapter 12, pp. 414–440. London, McGraw-Hill Book Company (UK) Limited. 1974.
17. Panshin A. J., and C. de Zeeuw. *Textbook of wood technology.* Vol 1. *Structure, identification, uses, and properties of the commercial woods of the United States and Canada.* New York, McGraw-Hill Book Company. 1970.
18. Sauter, J. J., W. Iten, and M. H. Zimmermann. Studies on the release of sugar into the vessels of sugar maple (*Acer saccharum*). *Can. J. Bot.* 51:1–8. 1973.
19. Scurfield, G. Reaction wood: its structure and function. *Science* 179:647–655. 1973.
20. Sheldrake, A. R. Auxin in the cambium and its differentiating derivatives. *J. Exp. Bot.* 22:735–740. 1971.
21. Shininger, T. L. The production and differentiation of secondary xylem in *Xanthium pennsylvanicum. Amer. J. Bot.* 57:769–781. 1970.
22. Shininger, T. L. The regulation of cambial division and secondary xylem differentiation in *Xanthium* by auxins and gibberellin. *Plant Physiol.* 47:417–422. 1971.
23. Sinnott, E. W. Reaction wood and the regulation of tree form. *Amer J. Bot.* 39:69–78. 1952.
24. Westing, A. H. Formation and function of compression wood in gymnosperms. *Bot. Rev.* 31:381–480. 1965; 34:51–78. 1968.
25 Zimmermann, M. H., and C. L. Brown. *Trees. Structure and function.* New York, Springer Verlag. 1971.

10

Vascular Cambium

The vascular cambium is the meristem that produces the secondary vascular tissues. It is a lateral meristem, for in contrast to the apical meristem it occupies a lateral position in stem and root. In the three-dimensional aspect, the cambium is a continuous sheath about the xylem of stem and root and their branches, and it extends in the form of strips into the leaves if the latter have secondary growth.

ORGANIZATION OF CAMBIUM

The cells of the vascular cambium do not fit the usual descriptions of meristematic cells, as those that have dense cytoplasm, large nuclei, and are approximately isodiametric in shape. Although the resting cambial cells have relatively few small vacuoles, the active cambial cells are highly vacuolated.[18] Morphologically, cambial cells occur in two forms. One type of cell, the *fusiform initial* (fig. 10.1,*A*), is several to many times longer than wide; the other, the *ray initial* (fig. 10.1,*B*), is slightly elongated to nearly isodiametric. The term fusiform implies that the cell is shaped like a spindle. A fusiform initial, however, is an approximately prismatic cell in its middle part and wedge shaped at the ends. The pointed end of the wedge is seen in tangential sections, the truncated end in radial sections (fig. 10.1,*A*). The tangential sides of the cell are wider than the radial.

In the cambial zone, the fusiform initials and their derivatives constitute the axial system, the ray initials the radial system. The cambium may be storied or nonstoried depending on whether or not, as seen in tangential sections, the cells are arranged in horizontal tiers (fig. 10.2). In a storied cambium the fusiform initials are shorter and less strongly overlapping than in a nonstoried cambium. The arrangement of the cambial initials determines the organization of the secondary vascular tissues. The cells of the axial systems in these tissues are derived from the similarly arranged fusiform initials

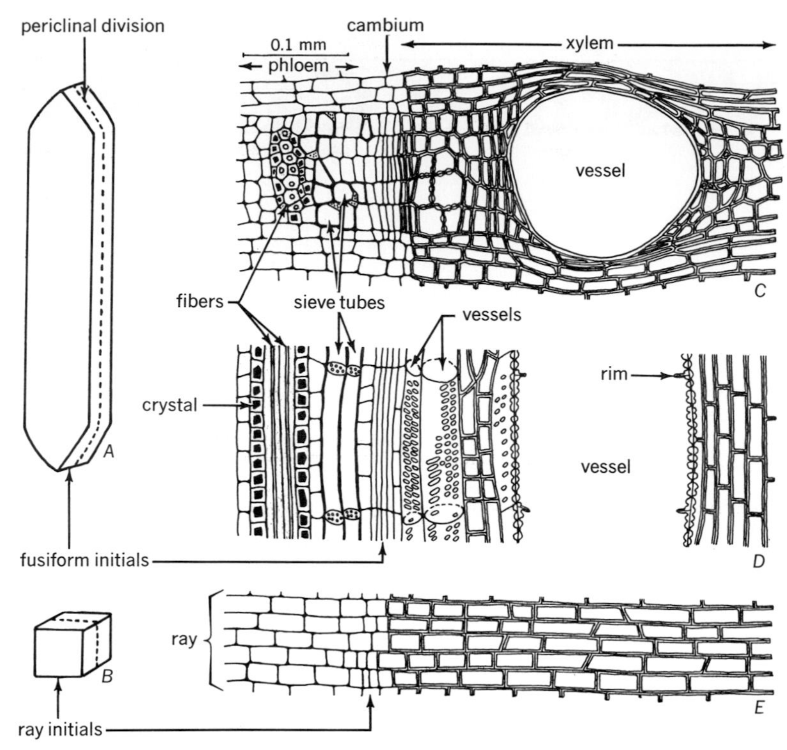

Figure 10.1 Vascular cambium in relation to derivative tissues. *A*, diagram of fusiform initial; *B*, of ray initial. In both, orientation of division concerned with formation of phloem and xylem cells (periclinal division) is indicated by broken lines. *C*, *D*, *E*, *Robinia pseudoacacia;* sections of stem include phloem, cambium, and xylem. *C*, transverse; *D*, radial (axial system only); *E,* radial (ray only).

and the ray systems from the ray initials; the storied and nonstoried cambia give rise to storied and nonstoried woods, respectively.

When the cambial initials produce secondary xylem and phloem cells they divide periclinally (fig. 10.1,*A*,*B*). At one time a derivative cell is produced toward the xylem, at another time toward the phloem, although not necessarily in alternation. Thus, each cambial initial produces radial files of cells, one toward the outside, the other toward the inside, and the two files meet at the cambial initial (figs. 10.1,*C*, and 10.3). Cambial divisions that add cells to the secondary vascular tissues are called *additive divisions.*[2]

During the height of cambial activity, cell addition occurs so rapidly that older cells are still meristematic when new cells are produced by the initials. Thus, a wide zone of more or less undifferentiated cells accumulates. Within this zone, the *cambial zone,* only one cell in a given radial file is considered to be an initial in the sense that after it divides periclinally, one of the two resulting cells remains as the initial and the other is given off toward the differentiating phloem or xylem. The initials are difficult to distinguish from their recent derivatives because these derivatives divide periclinally one or more times before they begin to differentiate into xylem or

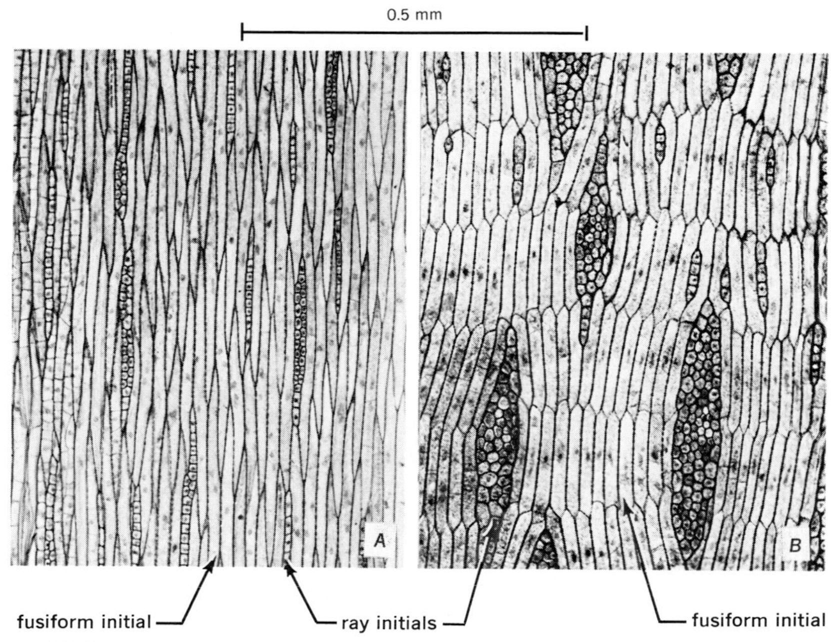

Figure 10.2 Arrangement of cells in vascular cambium as seen in tangential sections. *A*, nonstoried cambium of *Rhus typhina*. *B*, storied cambium of Wisteria sp.

phloem cells. Some workers, therefore, prefer to use the word cambium to designate the entire cambial zone.[7]

The cambial zone thus constitutes a more or less wide stratum of periclinally dividing cells organized into axial and ray systems. In the approximately median plane (it is more often an off-median plane) of this stratum one commonly visualizes a single layer of cambial initials flanked along their two tangential walls by initials of the vascular tissues, the phloem initials (or phloem mother cells) toward the periphery, the xylem initials (or xylem mother cells) toward the inside (fig. 10.4). The initial of a given radial file of cells in the cambial zone does not necessarily have an accurate tangential alignment with the initials in neighboring radial files.[3] In one radial file, the initial may be located closer to the xylem or the phloem than in another file. Moreover, a given initial may cease to participate in additive divisions and be displaced by its derivative which then assumes the role of a cambial initial.

DEVELOPMENTAL CHANGES IN THE INITIAL LAYER

As the core of secondary xylem increases in thickness, the cambium is displaced outward and its circumference increases. This increase is accomplished by division of cells, but in arborescent species it also involves complex phenomena of intrusive growth, elimination of initials, and formation of ray initials from fusiform initials. The changes in the cambium are reflected in cell relationships of the derivative tissues so that serial transverse and tangential sections, particularly of the xylem, may be used to analyze the behavior of the cambium in the past.

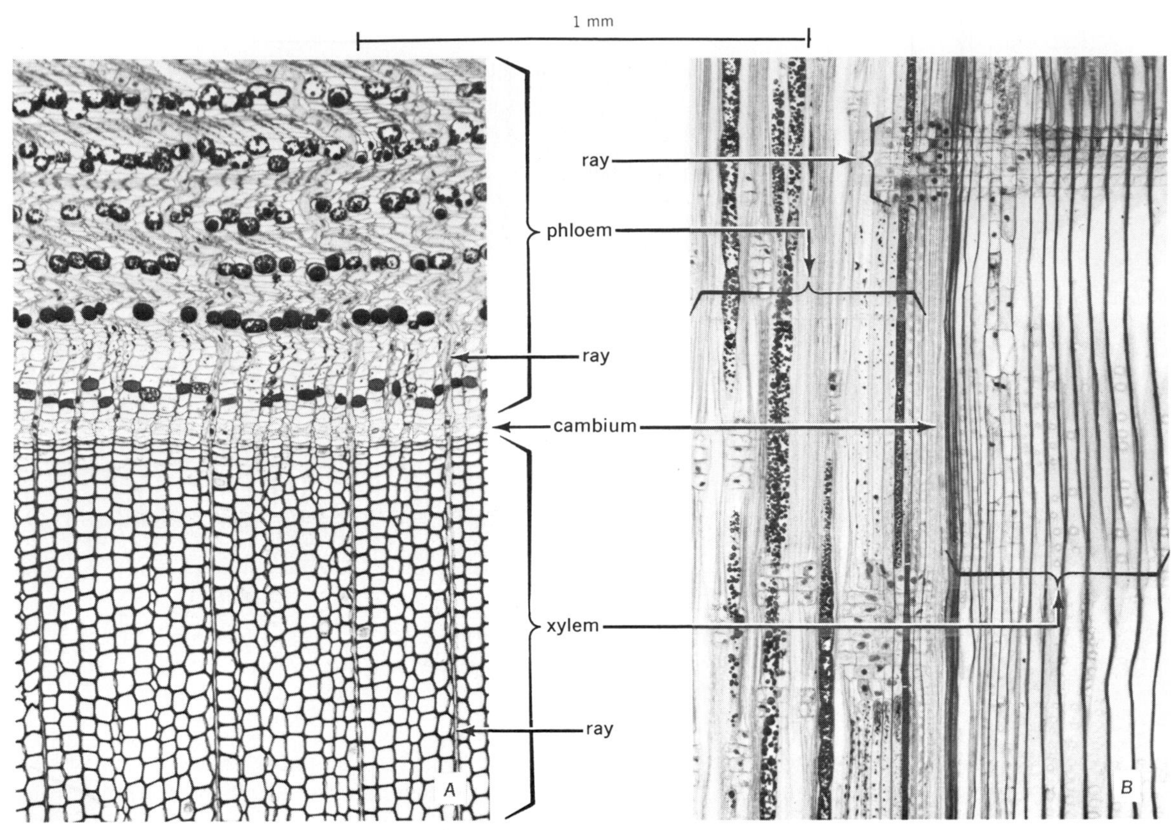

Figure 10.3 Vascular tissues and cambium in stem of pine (*Pinus* sp., a conifer) in cross (*A*) and radial (*B*) sections.

The divisions increasing the number of initials are called *multiplicative divisions.*[2] In cambia with short fusiform initials, the multiplicative divisions are mostly *radial anticlinal*[6] (fig. 10.5,*A*). Thus, two cells appear side by side where one was present formerly, and each enlarges tangentially. In herbaceous and shrubby dicotyledons the anticlinal divisions are frequently *lateral;* that is, they intersect twice the same mother cell wall[10] (fig. 10.5,*B*). Long fusiform initials divide by more or less inclined anticlinal walls[3] (*pseudotransverse divisions;* figs. 10.5,*C–E*, and 10.6,*A*), and each new cell elongates by apical intrusive growth (fig. 10.5,*F*,*G*). As a result of this growth the new sister cells come to lie side by side in the tangential plane (fig. 10.5,*G*), and they thus increase the circumference of the cambium. During the intrusive growth the ends of the cells may fork (fig. 10.5,*H*,*I*). The ray initials also divide radially anticlinally if the plant has biseriate or multiseriate rays.

The formation of ray initials from fusiform initials, or their segments, is a common phenomenon. If one compares growth layers in the xylem near the pith with those farther outward, a relative constancy in the ratio between the rays and the axial components may be observed.[5] This constancy results from the addition of new rays as the column of xylem increases in girth; that is, new ray ini-

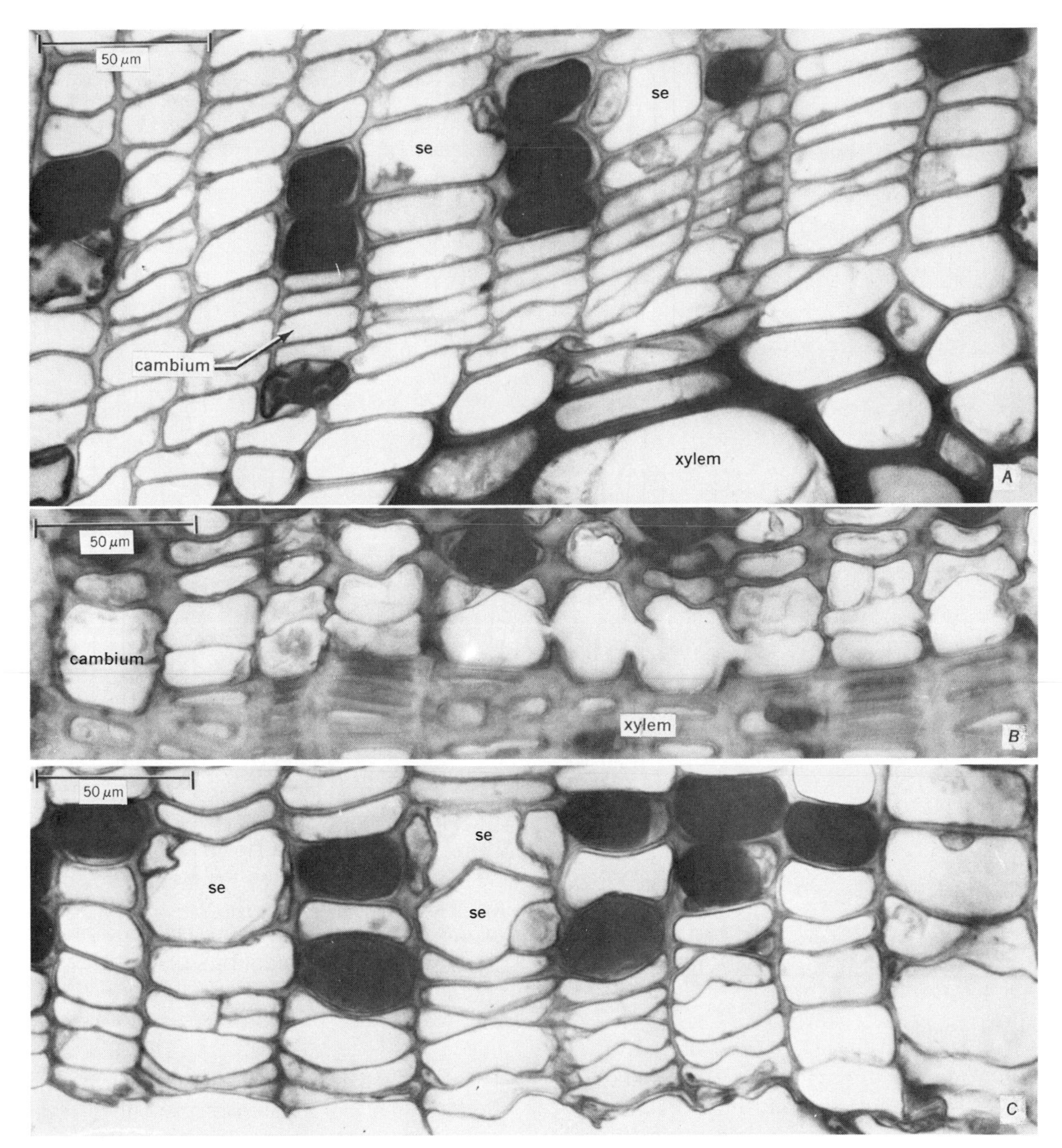

Figure 10.4 Vascular cambium from stem of grapevine (*Vitis vinifera*) in cross sections. *A*, active cambium late in the season. *B*, cambium at beginning of reactivation with first tangential walls of the season and some ruptures in anticlinal walls. *C*, active cambium at later stage than in *B*; breakage occurred through the youngest cambial cells and caused the bark to slip, that is, to separate from the xylem. Detail: *se*, sieve elements, each with one or two companion cells. (From K. Esau, *Hilgardia* 18:217–296, 1948.)

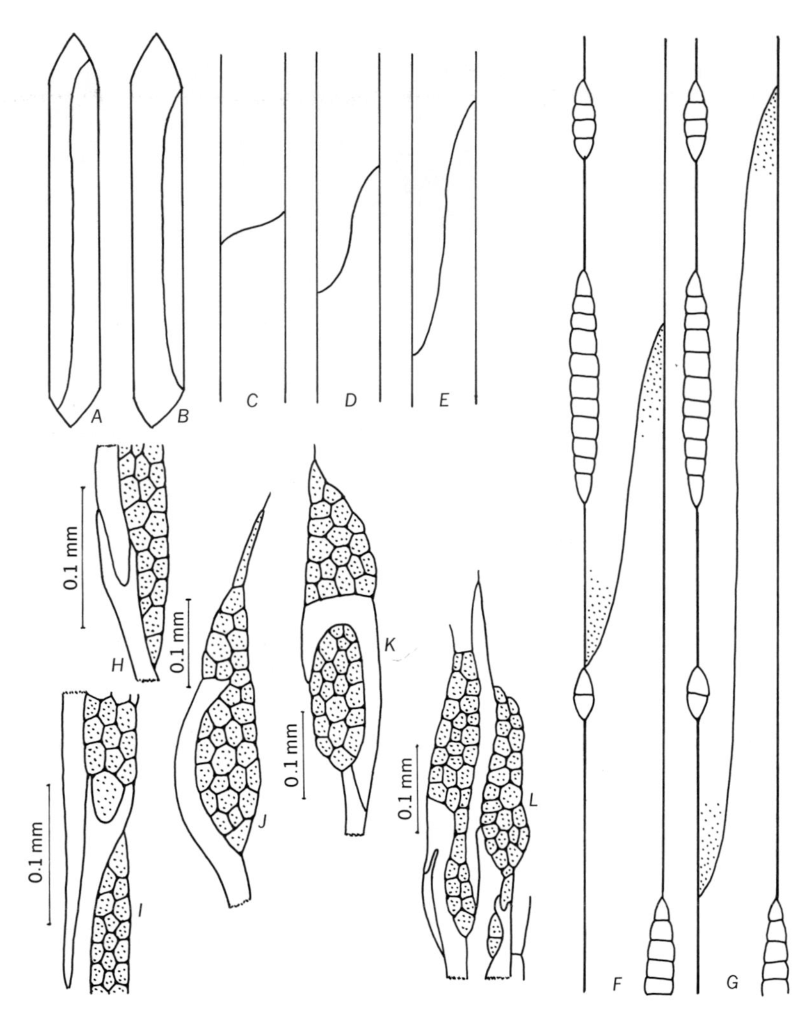

Figure 10.5 Division and growth of fusiform initials. Initial divided: *A*, by radial anticlinal wall; *B*, by lateral anticlinal wall; *C–E*, by various oblique anticlinal walls. *F*, *G*, oblique anticlinal division is followed by apical intrusive growth (growing apices are stippled). *H*, *I*, forking of fusiform initials during intrusive growth (*Juglans*). *J–L*, intrusion of fusiform initials into rays (*Liriodendron*). (All tangential views.)

tials appear in the cambium. These new ray initials are derived from fusiform initials.

The initials of new uniseriate rays may arise as unicellular segments cut out from fusiform initials at their apices or in the middle parts (conifers[5]) or by transverse divisions of such initials (herbaceous and shrubby dicotyledons[8–10]). The origin of rays, however, may be a highly complicated process involving a transverse subdivision of fusiform initials into several cells, loss of some of the products of these divisions, and the transformation of others into ray initials.[2] The loss, or elimination, of an initial is a displacement of this cell

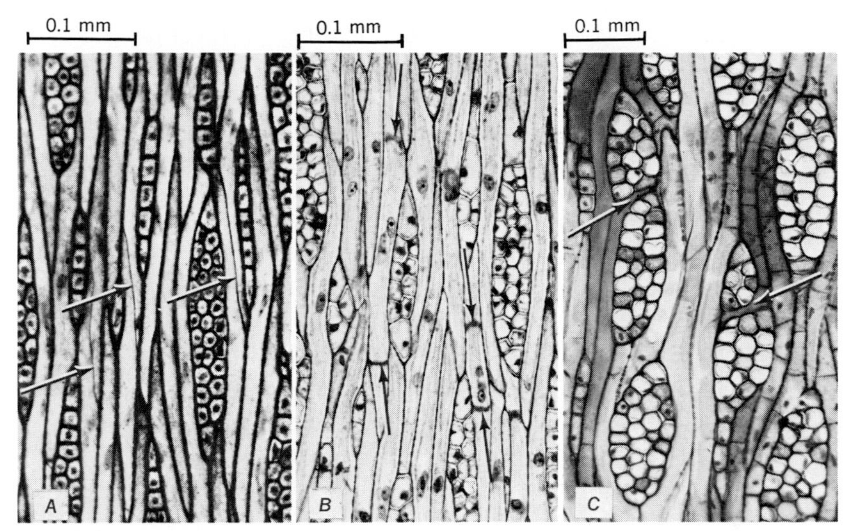

Figure 10.6 Division and growth of fusiform initials. *A*, cambium of *Juglans* with three fusiform initials recently divided by oblique anticlinal walls (arrows). *B*, cambium of *Cryptocarya;* two periclinally dividing fusiform initials with phragmoplasts (arrows), which indicate extent of cell plates. *C*, phloem of *Liriodendron* with two rays penetrated by axial cells, as a result of intrusive growth, while the tissue was still in cambial state. All, tangential sections.

toward the xylem or the phloem and eventual maturation into a xylem or a phloem cell, often after a gradual reduction in size while the cell is still in the initial layer.[1]

In conifers and dicotyledons new uniseriate rays begin as rays one or two cells high and only gradually attain the height typical for the species.[5] The increase in height occurs through transverse divisions of the established ray initials and through fusion of rays located one above the other. In the formation of multiseriate rays radial anticlinal divisions and fusions of laterally approximated rays are involved. Indications are that in the process of fusion some fusiform initials intervening between rays are converted into ray initials by transverse divisions; others are displaced toward the xylem or the phloem and are thus lost from the initial zone. The reverse process, a splitting of rays, also occurs. A common method of such splitting is a breaking up of a panel of ray initials by a fusiform initial that intrudes among the ray initials (figs. 10.5,*I–L*, and 10.6,*C*).

The multiplicative and additive divisions commonly occur toward the end of the maximal growth concerned with the seasonal production of xylem and phloem.[3,5] In plants with nonstoried cambia this timing in divisions means that the cambium contains, on the average, shorter fusiform initials at the end of the season than earlier. Subsequently the new cells elongate so that the average length of the initials increases until a new period of divisions ensues at the end of the growth season.

The periodic changes in length of the fusiform initials are reflected in the variation in length in the resulting xylem cells. In both gymnosperms and angiosperms the length of the elongated types of cells (tracheids, fibers) rises from the first-formed early wood to the last formed late wood.[4] There is also an overall increase in the length of the fusiform initials from the beginning of secondary growth and through the successive years until the length is more or less stabilized or perhaps reduced.[12]

In some conifers, the anticlinal divisions of fusiform initials occur according to a precise pattern.[14] The inclined walls formed during the multiplicative divisions are tilted in one direction through panels ("domains") of cambium of considerable size. The panels vary in size, and the orientation of new walls changes periodically throughout a given panel. The unidirectional orientation of anticlinal walls and of the intrusively growing tips of the new cells, combined with a frequent loss of initials, appears to be causally related to the development of spiral grain in the wood.

The foregoing discussion of developmental transformations in the initial region of the cambium clearly indicates that this meristem is in continuous state of change. The concept of cambial initials must take into account this lack of stability. Elimination of initials is a particularly significant feature in this regard. The initials have no continuing individuality, but the function of initiation of new cells is sustained; it is "inherited" by one cell after another.[16] Recognition of the impermanence of cellular composition of the cambium also affects the concept of the uniseriate cambial initial layer. The assumption that only one specific layer in the cambium merits the name of initial layer has frequently been questioned.[17] Studies on conifer cambium have demonstrated, however, that the multiplicative divisions, which establish new patterns of cell alignment in xylem and phloem, occur mainly in one specific layer.[3] Thus, the several layers of a cambial zone, which are similar cytologically and are undergoing divisions, are not equivalent in the degree of their impact upon the architecture of the secondary vascular tissues. At a given time, a single layer functions as the initial layer by perpetuating bidirectionally the pattern of its cellular arrangement.

Cytokinesis, or formation of new cells, in the cambium is of special interest when the cells divide longitudinally and the new wall is formed along the long diameter of the cell. In such a division, the diameter of the initial phragmoplast originating during telophase (fig. 10.7,*A*) is very much shorter than the long diameter of the cell. The phragmoplast and the cell plate reach the longitudinal walls of the cambial cell soon after nuclear division (fig. 10.7,*E*), but the progress of the cell plate toward the ends of the cell is an extended process (fig. 10.7,*A–C*). Before the side walls are reached the phragmoplast appears as a circular halo in front view (fig. 10.7,*D*). After those walls are intersected by the cell plate—but before the ends of cells are

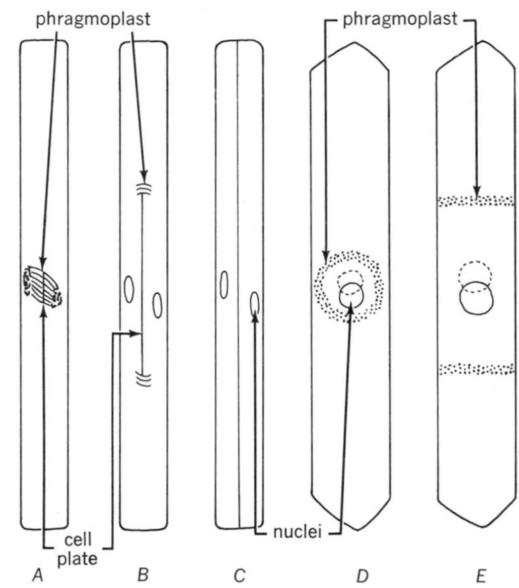

Figure 10.7 Cell division in fusiform initials. *A–C*, three stages in formation of cell plate as seen in radial sections. *D*, *E*, two stages of cell-plate formation as seen in tangential sections. The cell plate has extended through about one-third of the cell in *B* and *E*. All views illustrate tangential divisions.

reached—the phragmoplast in the front view forms two bars intersecting the side walls (figs. 10.6,*B*, and 10.7,*E*).

The ultrastructural features of cytokinesis of the long cambial cells dividing longitudinally are similar to those observed during the division of shorter cells (chapter 4). Since the dividing cambial cells are considerably vacuolated, the formation of a cytoplasmic layer, the phragmosome, bridging the vacuole would seem to be a particularly important prerequisite for the progress of the phragmoplast and the cell plate from the site of the nucleus to the ends of the cell. A phragmosome has been recognized in transections of dividing fusiform initials,[13] but remains to be demonstrated in longitudinal views. It may not appear as a continuous layer at once but be formed successively in advance of the progressing phragmoplast in continuity with the parietal cytoplasm.

PATTERNS AND CAUSAL RELATIONS IN CAMBIAL ACTIVITY

The seasonal changes in the activity of vascular cambium is a much explored topic but continues to reveal new aspects as the causal relations in growth and differentiation are elucidated.[15] In the temperate regions, winter rest is succeeded by reactivation of the vascular cambium. Cambial cells take up water, enlarge radially, and divide periclinally (fig. 10.4,*B*). While the cells enlarge, their radial walls become thinner and, as a result, the bark (phloem and tissues outside of it; chapter 12) may be easily peeled off, that is, "the bark slips" (fig. 10.4,*C*). Cell division does not necessarily start in the initial layer. The first divisions may appear in the overwintering mother cells of the xylem or the phloem, followed later by additive divisions in the initial layer. The slippage of bark occurs not only through this layer but often also through the differentiating xylem. In dicotyledons, the expanding vessels constitute a particularly weak connection between bark and wood during cambial growth. The maximum of additive divisions is reached a few weeks after the cambium is reactivated. Periodicity in cambial activity occurs in both deciduous and evergreen species and is not confined to the temperate regions. In the tropics, however, the periodicity is less clearly related to seasonal changes in environmental conditions and may be weakly expressed or absent.

The first additions from the initial layer may be made toward either the xylem or the phloem, the variation depending on plant species. Generally, however, more cells are added to the xylem than to the phloem. This well-known pattern of cell addition has been confirmed by the use of radioactively labeled CO_2 and the resulting ^{14}C marking of the walls in new secondary tissue in a *Eucalyptus*.[23] Cell production toward the xylem was about four times that toward the phloem. A greater difference was observed in a conifer.[1] In a vigorously growing *Thuja occidentalis*, 12 to 16 new cells were formed on the phloem side and 100 or more on the xylem side.

Initiation of cambial activity in the spring is often clearly related to resumption of bud growth. The relation is somewhat variable but is well expressed in dicotyledons with diffuse-porous wood. Cambial activity, as determined by bark slippage, begins beneath the emerging new shoots and proceeds basipetally toward the trunk and root. Many weeks may pass between the time of cambial reactivation under the buds and that in the roots. The relationship is less clear in dicotyledons with ring-porous woods and in conifers, but experimental work involving removal of buds and leaves indicates that primary growth in the shoot generally provides the stimulus for the inception of secondary growth in the axis below. Correspondingly, the cessation of cambial activity at the end of the growth season may be correlated with the completion of shoot extension.[11]

The hormonal nature of the stimulus inducing cambial activity was postulated by some of the earliest students of secondary growth. Subsequent intensive research has repeatedly associated the initial stimulation of cambial activity with the downward movement of growth substances from the expanding buds.[19] This movement establishes gradients of hormones to which the spread of cambial activity is clearly related.[11] Although the growing buds provide the initial hormonal stimulus for the resumption of cambial activity, the maintenance of that activity appears to be independent of the auxin from the shoots. The continued cambial growth has a local source of auxin as is evidenced by studies on secondary growth in excised internodes.[21] Analysis of auxin content in the three layers of the cambial region, the differentiating xylem, the differentiating phloem, and the cambium, indicates that the differentiating xylem may be the main source of this locally supplied auxin. It is proposed[20] that this auxin is released by the autolysing tracheary cells as they mature into nonliving conducting cells.

Auxin appears to be one of the most important growth substances associated with cambial growth but other substances, such as cytokinins and gibberellic acid, may interact with auxins in activating the cambium and affecting the pattern of differentiation of derivatives. Moreover, the growth-regulating substances act in conjunction with other growth factors, namely, availability of food (especially sugar[22]) and water, appropriate temperature and photoperiod, and the endogenously determined rhythm of growth characteristic of a given species.

REFERENCES

1. Bannan, M. W. The vascular cambium and radial growth in *Thuja occidentalis* L. *Can. J. Bot.* 33:113–138. 1955.
2. Bannan, M. W. Some aspects of the elongation of fusiform cambial cells in *Thuja occidentalis* L. *Can. J. Bot.* 34:175–196. 1956.
3. Bannan, M. W. Anticlinal divisions and the organization of conifer cambium. *Bot. Gaz.* 129:107–113. 1968.
4. Bisset, I. J. W., and H. E. Dadswell. The variation in cell length within one growth ring of certain angiosperms and gymnosperms. *Aust. Forestry* 14:17–29. 1950.
5. Braun, H. J. Beiträge zur Entwicklungsgeschichte der Markstrahlen. *Bot. Stud.* No. 4:73–131. 1955.
6. Butterfield, B. G. Developmental changes in the vascular cambium of *Aeschynomene hispida* Willd. *New Zeal. J. Bot.* 10:373–386. 1972.
7. Catesson, A. M. Origine, fonctionnement et variations cytologiques saisonnières du cambium de l'*Acer pseudoplatanus* L. (Aceracées). *Ann. Sci. Nat., Bot. Sér.* 12. 5:229–498. 1964.
8. Cumbie, B. G. Development and structure of the xylem in *Canavalia* (Leguminosae). *Bull. Torrey Bot. Club.* 94:162–175. 1967.
9. Cumbie, B. G. Developmental changes in the vascular cambium in *Leitneria floridana*. *Amer. J. Bot.* 54:414–424. 1967.
10. Cumbie, B. G. Developmental changes in the vascular cambium of *Polygonum lapathifolium*. *Amer. J. Bot.* 56:139–146. 1969.
11. Digby, J., and P. F. Wareing. The relationship between endogenous hormone

levels in the plant and seasonal aspects of cambial activity. *Ann. Bot.* 30:608–622. 1966.

12. Dinwoodie, J. M. Tracheid and fibre length in timber. A review of literature. *Forestry* (London) 34:125–144. 1961.
13. Evert, R. F., and B. P. Deshpande. An ultrastructural study of cell division in the cambium. *Amer. J. Bot.* 57:942–961. 1970.
14. Hejnowicz, Z. Orientation of the partition in pseudotransverse division in cambia of some conifers. *Can. J. Bot.* 42:1685–1691. 1964.
15. Kozlowski, T. T. *Growth and development of trees.* Vol. 2. *Cambial growth, root growth, and reproductive growth.* New York, Academic Press. 1971.
16. Newman, I. V. Pattern in meristems of vascular plants. I. Cell partition in living apices and in the cambial zone in relation to the concept of initial cells and apical cells. *Phytomorphology* 6:1–19. 1956.
17. Philipson, W. R., J. M. Ward, and B. G. Butterfield. *The vascular cambium.* London, Chapman & Hall. 1971.
18. Robards, A. W., and P. Kidwai. A comparative study of ultrastructure of resting and active cambium of *Salix fragilis* L. *Planta* 84:239–249. 1969.
19. Samish, R. M. Dormancy in woody plants. *Ann. Rev. Plant Physiol.* 5:183–204. 1954.
20. Sheldrake, A. R. Auxin in the cambium and its differentiating derivatives. *J. Exp. Bot.* 22:735–740. 1971.
21. Sheldrake, A. R., and D. H. Northcote. The production of auxin by tobacco internode tissues. *New Phytol.* 67:1–13. 1968.
22. Siebers, A. M., and C. A. Ladage. Factors controlling cambial development in the hypocotyl of *Ricinus communis* L. *Acta Bot. Neerl.* 22:416–432. 1973.
23. Waisel, Y., I. Noah, and A. Fahn. Cambial activity in *Eucalyptus camaldulensis* Dehn. II. The production of phloem and xylem elements. *New Phytol.* 65:319–324. 1966.

11 Phloem

The food-conducting tissue of a seed plant, the phloem, is associated with the xylem in the vascular system. Like the xylem, the phloem consists of several types of cells and may be classified, developmentally, into a primary and a secondary tissue. The primary phloem is derived from procambium. The secondary phloem originates in the vascular cambium and reflects the organization of this meristem in that it has an axial and a ray system. The rays are continuous through the cambium with those of the xylem.

Thus, the overall development and structure of the phloem tissue parallel those of the xylem, but the distinct function of the phloem is associated with structural characteristics peculiar to this tissue. The phloem tissue is less sclerified and less persisting than the xylem tissue. Because of its usual position near the periphery of stem and root the phloem becomes much modified in relation to the increase in circumference of the axis and is eventually cut off by the periderm. The old xylem, in contrast, remains relatively unchanged in its basic structure.

CELL TYPES

Primary and secondary phloem tissues contain the same categories of cells. The primary phloem, however, is not organized into two systems, the axial and the radial; it has no rays. The summary illustration (fig. 11.1) and the list of phloem cells in Table 11.1 are based on the characteristic composition of the secondary phloem.

Sieve elements

The sieve elements are the most highly specialized cells in the phloem. Their principal characteristics are the ontogenetically modified protoplasts with restricted metabolic activity and the close connection with contiguous sieve elements through wall areas (*sieve areas*) penetrated by pores.

CELL WALL AND SIEVE AREAS

Sieve element walls vary in thickness but are usually distinctly thicker than those of the

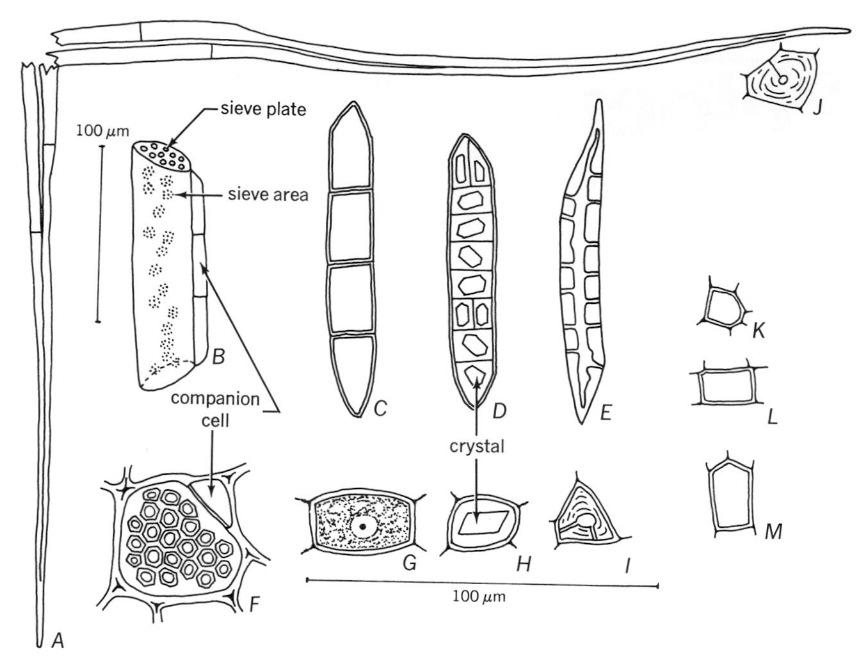

Figure 11.1 Cell types in secondary phloem of a dicotyledon, *Robinia pseudoacacia*. *A–E*, longitudinal views; *F–J*, cross sections. *A*, *J*, fiber *B*, sieve-tube member and, *F*, sieve plate. *C*, *G*, phloem parenchyma cells (parenchyma strand in *C*). *D*, *H*, crystal-containing parenchyma cells. *E*, *I*, sclereids. *K*, *L*, *M*, ray cells in tangential (*K*), radial (*L*), and cross (*M*) sections of phloem.

surrounding parenchymatic cells (fig. 11.11,*A*). Because of their shiny appearance in fresh sections the sieve element walls received the name of *nacreous* (having a pearly luster) walls. In some taxa, the walls are remarkably thick (fig. 11.19,*D*), sometimes almost occluding the cell lumen.[14] When the wall is considerably thickened the inner

Table 11.1

Cell Types	Principal Function
Axial system	
Sieve elements	Long-distance conduction of food materials
Sieve cells	
Sieve-tube members (with companion cells)	
Sclerenchyma cells	
Fibers	Support; sometimes also storage of food materials
Sclereids	
Parenchyma cells	
Ray system	Storage and radial translocation of food substances
Parenchyma cells	

layers are more or less clearly set off from the outer and are sometimes interpreted as secondary wall.[11] Ultrastructural studies of the thick nacreous wall of certain dicotyledons have revealed interconnected microfibrils in parallel orientation and arranged concentrically with respect to the wall.[1] In *Pinus strobus,* the thickened wall has a distinct polylamellate structure, with the microfibrils inclined to the longitudinal cell axis at an angle larger than 45° and arranged parallel to the plane of the wall.[3] The nacreous wall has a relatively loose structure and consists of cellulose and pectic compounds, but even the greatly thickened walls are not lignified. The innermost layer frequently appears electron dense (fig. 11.11,*A*) and shows radial striation (fig. 11.10,*C*).

The sieve areas are wall areas with pores through which the protoplasts of vertically or laterally adjoining sieve elements are interconnected (figs. 11.2,*A*,*B*,*D*,*E*; 11.3,*A*,*D*, and 11.10,*C*). The sieve area takes its name from its resemblance to a sieve. The pores range in width from those that are comparable to plasmodesmatal pores and those several microns in diameter. In the lower vascular plants and up to and including the gymnosperms the pores are commonly narrow and rather uniform in size in the sieve areas on different walls of the same cell. In the angiosperms the size of pores varies considerably, even on different walls of the same cell[15] (fig. 11.3,*A*–*C*). Sieve areas with the larger pores usually occur on the end walls (fig. 11.2,*A*,*B*), occasionally on the side walls. The wall parts bearing the more highly differentiated sieve areas, that is, areas with comparatively large pores, are called *sieve plates* (figs. 11.1,*B*, 11.2, and 11.3,*A*). This term parallels the designation perforation plate used in describing walls of vessel members having perforations.

In most preparations of conducting phloem, each pore is lined with callose (fig. 11.2,*D*,*E*),

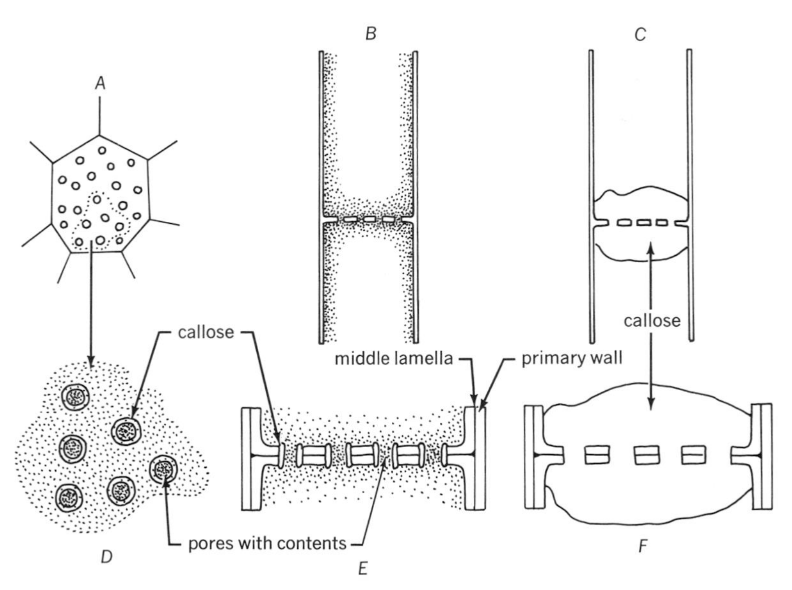

Figure 11.2 Details of a sieve plate. *A*, *D*, surface views. *B*, *C*, *E*, *F*, side views. *A*, *B*, *D*, *E*, sieve plate at functioning stage; *C*, *F*, at cessation of function or during dormancy.

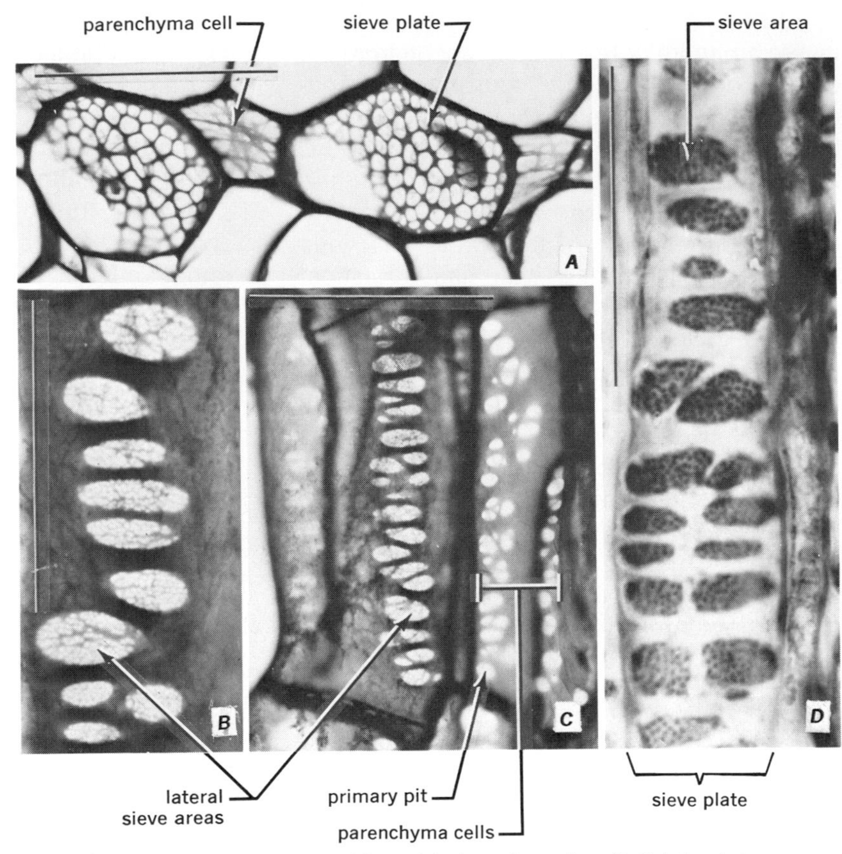

Figure 11.3 *A*, simple sieve plates of *Cucurbita* in surface view. *B*, *C*, lateral sieve areas in sieve elements and primary pits in parenchyma cells of *Cucurbita* in surface view. *D*, compound sieve plate of *Pyrus*, with numerous sieve areas. (Lines are 0.1 mm in *A* and *C*, 50 μm in *B* and *D*; from Esau, Cheadle, and Gifford, *Amer. J. Bot.* 40:9–19, 1953.)

a carbohydrate that stains blue with aniline blue and resorcin blue and yields glucose upon hydrolysis. As the sieve element ages, more callose accumulates. The layer within a pore thickens, and some callose also appears on the surface of the sieve area. The pore is increasingly more constricted and later completely obliterated as the sieve element becomes dormant or dies. At this stage, the callose forms a pad on the sieve area (fig. 11.2,*C*,*F*). In old, completely inactive sieve elements callose is absent, and open perforations are exposed in the sieve areas. If the phloem is only dormant, callose decreases in amount and continuity of protoplasts through the pores is reestablished during the reactivation of the tissue in the spring.

Although callose is a characteristic wall component in sieve elements, considerable research indicates (1) that when the cells are in their functional state callose is small in amount or absent; (2) that callose is synthesized remarkably fast in response to injury and, therefore, appears in conducting cells when the material for study is not prepared with minimal disturbance of living cells; (3) that callose deposition in early ontogeny of the sieve area and at the end of functional period is a response to internal stimuli and not to outside influences.

SIEVE CELLS AND SIEVE-TUBE MEMBERS

The degree of specialization of the sieve areas and differences in their distribution on the walls of a given cell serve to classify the sieve elements into sieve cells and sieve tube members. In *sieve cells,* the sieve areas are not highly specialized and are not markedly aggregated on restricted wall parts into sieve plates (fig. 11.4,*A*). In *sieve tube members,* the more highly differentiated sieve areas

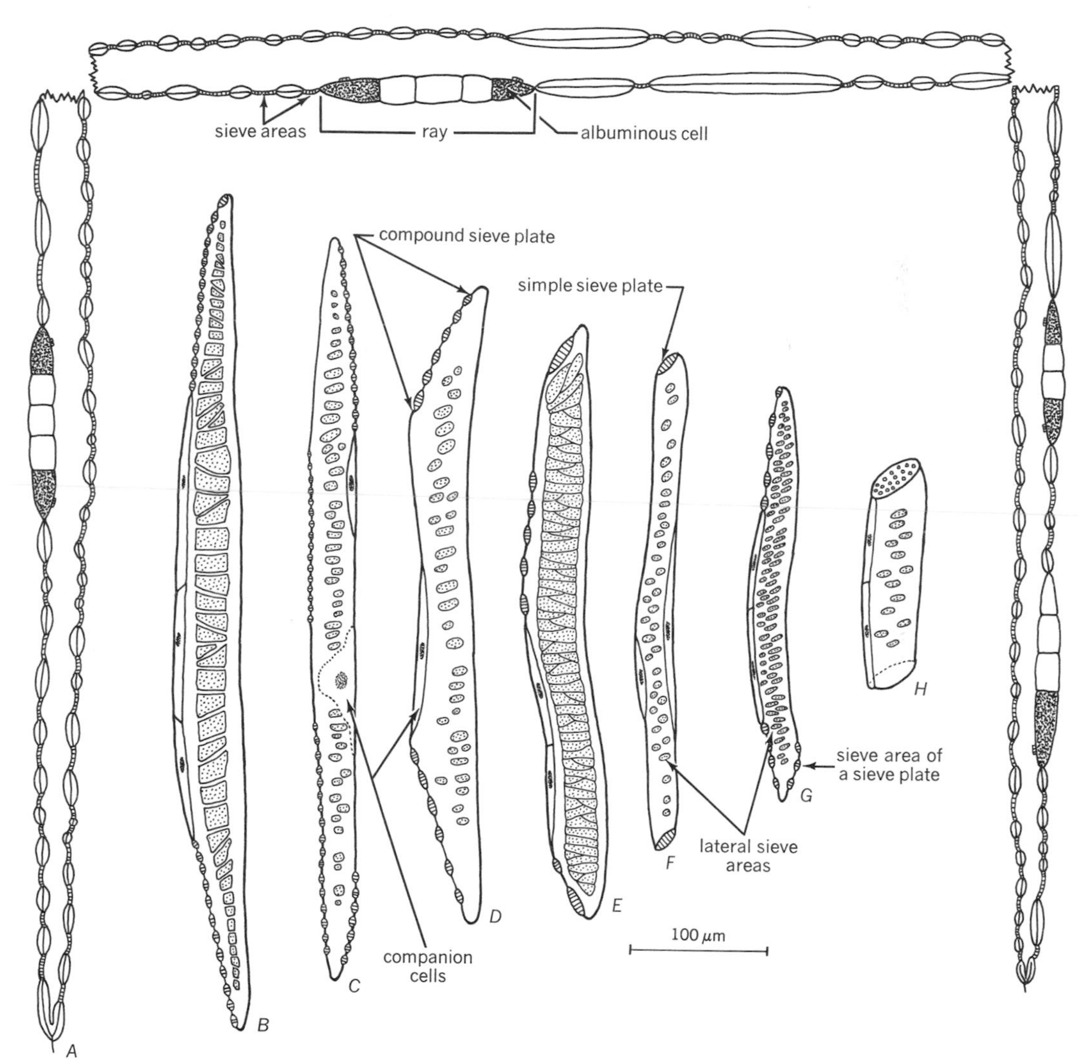

Figure 11.4 Variations in structure of sieve elements. *A*, sieve cell of *Pinus pinea,* with associated rays, as seen in tangential section. Others are sieve-tube members with companion cells from tangential sections of phloem of the following species: *B*, *Juglans hindsii; C*, *Pyrus malus; D*, *Liriodendron tulipifera; E*, *Acer pseudoplatanus; F*, *Cryptocarya rubra; G*, *Fraxinus americana; H*, *Wisteria* sp. In *B–G*, the sieve plates appear in side views and their sieve areas are thicker than the intervening wall regions because of deposition of callose.

occur on restricted wall parts—the sieve plates—usually at ends of the cells (fig. 11.4,*B–H*). Sieve plates provide a more complete interconnection between sieve tube members than do the less differentiated lateral sieve areas or the sieve areas in sieve cells. The evolutionary significance of the difference appears to be that cells with sieve plates have developed a higher degree of specialization for longitudinal conduction than the sieve cells. This relation parallels the comparative efficiency in conduction of vessel members interconnected by perforation plates as contrasted with tracheids interconnected by pits. The presence of perforation plates in vessel members and of sieve plates in sieve elements makes the longitudinal series of these two kinds of cell appear as functional units. In the xylem, the conducting units are called vessels (chapter 8), in the phloem, *sieve tubes.* A sieve tube is thus a series of sieve tube members connected end to end by means of sieve plates (fig. 11.13,*A*). As in vessels, the ends of the component cells may be transverse or more or less inclined and overlapping. Comparative studies carried out thus far suggest that, with some exceptions, gymnosperms and lower vascular plants have sieve cells; angiosperms, both dicotyledons and monocotyledons, have sieve tube members and sieve tubes.[11]

Sieve areas and sieve plates of angiosperms vary widely in differentiation and arrangement. To a degree, these differences are related to the length and form of cells.[15] Long sieve tube members with much inclined end walls commonly have compound sieve plates (fig. 11.4,*B–D*), that is, sieve plates composed of several sieve areas. The ends of such elements are wedge shaped, and the sieve areas are borne on the oblique face of the wedge which is part of the radial facet of the cell (fig. 11.3,*D*, and 11.4,*D*). The arrangement of the multiple sieve areas may be scalariform or less orderly.

The much compounded sieve plates—a concomitant of much inclined end walls—often have relatively narrow pores (fig. 11.3,*D*) and their sieve areas are not strikingly different from those on the lateral walls of the same cells. Such sieve tube members are interpreted as being rather primitive for angiosperms. Increasing specialization is characterized by a decrease in the inclination of end walls, reduction in number of sieve areas in the sieve plates, increase in size of pores in the sieve plates, and the concomitant increase in the difference in specialization between lateral sieve areas and those of the sieve plates (fig. 11.4,*B–H*). Considerable evidence suggests that the most highly specialized sieve element has simple sieve plates with large pores on transverse end walls and sieve areas of low degree of specialization on lateral walls (fig.11.4,*H*).

The foregoing discussion suggests an evolutionary sequence of changes in the sieve elements similar in many respects to that in the vessel members of the xylem. The length of cells, which is such a useful criterion in evaluation of the evolutionary level of vessel members, cannot be used, however, with the same consistency in comparative studies of sieve tube members. Although short fusiform cambial cells give rise to short sieve tube members, the derivatives of long fusiform initials on the phloem side may become divided so that the potential length of sieve tube members is reduced.[13]

PROTOPLAST

The sieve element protoplast undergoes a profound change during ontogeny (fig. 11.5). The individual cell components are differentially affected by this change, and variations in detail are encountered in different taxa and

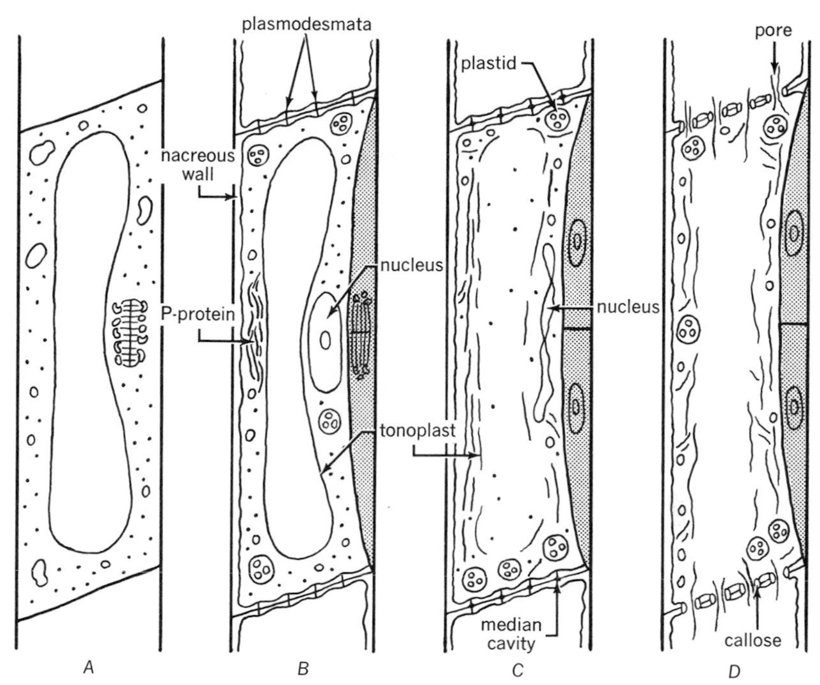

Figure 11.5 Diagrams illustrating differentiation of a sieve element. *A*, precursor of sieve element in division. *B*, after division: sieve element with nacreous wall and P-protein body; dividing companion cell precursor (stippled). *C*, nucleus degenerating, tonoplast partly broken down, P-protein dispersed; median cavities in future sieve plates; two companion cells (stippled). *D*, mature sieve element; pores in sieve plates open; they are lined with callose and contain some P-protein. No endoplasmic reticulum is shown.

in different sites in the plant body. In seed plants, the nucleus usually degenerates. It either remains in the cell as a collapsed body with no internal organization or, more commonly, disappears as a definable entity. The endoplasmic reticulum becomes smooth and most of it is aggregated into stacks (stacked ER), with proteinaceous material, possibly enzymes, accumulating between the the cisternae (fig. 11.6). Unstacked smooth endoplasmic reticulum occurs next to the plasmalemma where it forms a network recognizable as such in glancing sections through the cell wall.[12] The endoplasmic reticulum diminishes in amount as the cell advances in maturation, the unstacked kind next to the plasmalemma persisting longer than the stacked. Dictyosomes, which are actively producing vesicles during formation of the thickened wall, are not discernible in the mature cell. Ribosomes also disappear. Two kinds of organelles are retained, plastids and mitochondria (figs. 11.10,*C*, and 11.11,*A*). The plastids occur in two basic types, one with little internal organization but accumulating starch (figs. 11.6 and 11.10,*C*); the other containing protein as fibrils (fig. 11.10,*A*,*B*) or crystals. (Starch formation may occur in protein-containing plastids.) Plastid differences in sieve elements are taxonomically useful.[2]

The two limiting cytoplasmic membranes, plasmalemma and tonoplast, show con-

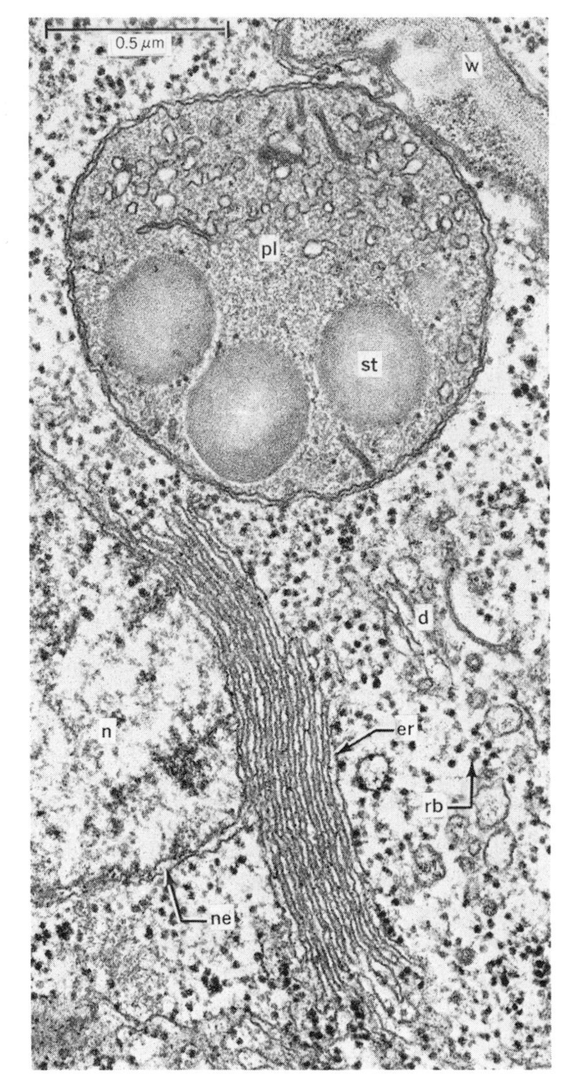

Figure 11.6 Part of immature sieve element protoplast from bean (*Phaseolus vulgaris*) root tip showing a plastid (*pl*) with starch (*st*), aggregated (stacked) endoplasmic reticulum (*er*) next to the nucleus (*n*), small part of a dictyosome (*d*) with vesicles, and numerous ribosomes (*rb*). The *er* stack is applied to the nuclear envelope (*ne*), which resembles an *er* cisterna. Cell wall at *w*.

trasting behavior. The plasmalemma persists and is probably responsible for the retention of differential permeability by the protoplast as indicated by its plasmolysability in mature cells.[8] The tonoplast breaks down so that the delimitation between vacuole and cytoplasm disappears (figs. 11.5,*C*,*D*, and 11.10,*C*).

Sieve elements commonly form a proteinaceous inclusion, the P-protein, formerly called slime.[11] This protein often appears in filaments of tubular form, with subunits arranged helically (figs. 11.7 and 11.8,*A*). The tubules may become stretched, especially in mature sieve elements. In that condition the helical structure imparts a striated appearance to the filaments (fig. 11.8,*B*). Initially, the P-protein is aggregated into one or more bodies (fig. 11.5,*B*) but subsequently it spreads out in the cytoplasm (fig. 11.5,*C*) forming strands or networks. In some taxa, the bodies disperse only partially or not at all. After the tonoplast disappears, the dispersed P-protein is found in parietal position in the cell lumen and sieve plate pores (fig. 11.5,*D*), provided care is taken to disturb the phloem as little as possibe during sampling.[17] Otherwise it is found throughout the cell lumen and plugging the pores in the sieve areas.

DIFFERENTIATION OF SIEVE PLATES

In a young sieve element, the sieve area (or areas) of the incipient sieve plate is penetrated by typical plasmodesmata (fig. 11.5,*B*,*C*), with which endoplasmic reticulum cisternae are associated (figs. 11.9 and 11.10,*D*). These plasmodesmata mark the sites of the future pores. The endoplasmic reticulum cisternae located on the two sides of a pore site are interconnected by a desmotubule extending through the plasmodesmatal canal (chapters 3 and 4). While the protoplast is still intact, callose appears around each plasmodesma except in the middle lamella region. The paired callose masses assume the form of platelets or cones interrupted in the center where the plasmodesma is located. When the nucleus is undergoing degeneration plasmodesmatal canals are en-

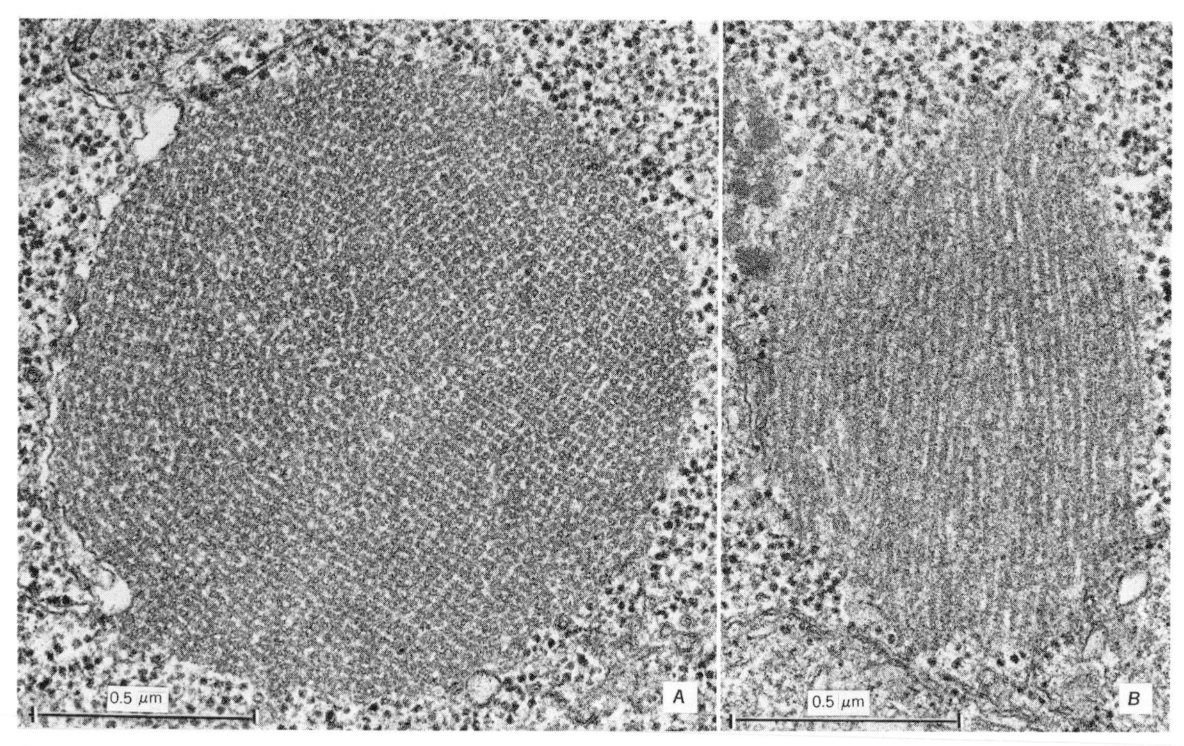

Figure 11.7 P-protein bodies from protophloem of bean (*Phaseolus vulgaris*) root tip in transverse (*A*) and longitudinal (*B*) sections with reference to the component tubular filaments. Paracrystalline arrangement of tubules in *A*.

larging, often developing a cavity in the middle lamella region (figs. 11.5,*C*, and 11.10,*D*). The endoplasmic reticulum cisternae are removed from the pore sites as the pores attain their final size. The plasmalemma lining the original plasmodesmatal pore remains (fig. 11.8,*B*) but the endoplasmic reticulum tubule disappears. The pore thus becomes open in the sense that its contents are a continuation of the contents of the contiguous sieve elements (fig. 11.5,*D*). The maturation of the sieve plate is one of the end results of autolysis that occurs in the sieve element as it is converted into a specialized conducting cell. The sequence parallels that in differentiating xylem vessels in which the perforation of the end wall occurs in connection with the degradation of the protoplast (chapter 8).

CONDITION OF PORES IN MATURE SIEVE ELEMENTS

Since the past century, physiological research on translocation has been providing evidence that the sieve element sap is under positive pressure reaching some 30 atmospheres.[24] Modern studies on translocation include use of radioactively labeled translocates and honeydew released by aphids feeding on the phloem. It has been well established that such aphids penetrate the sieve tubes,[18] ingest nitrogenous substances, and release the sap containing sugar (honeydew). Various manipulations of these "research tools" consistently reveal presence of positive pressure in the conduit and a movement of assimilates with a velocity of 10 to 100 cm an hour.

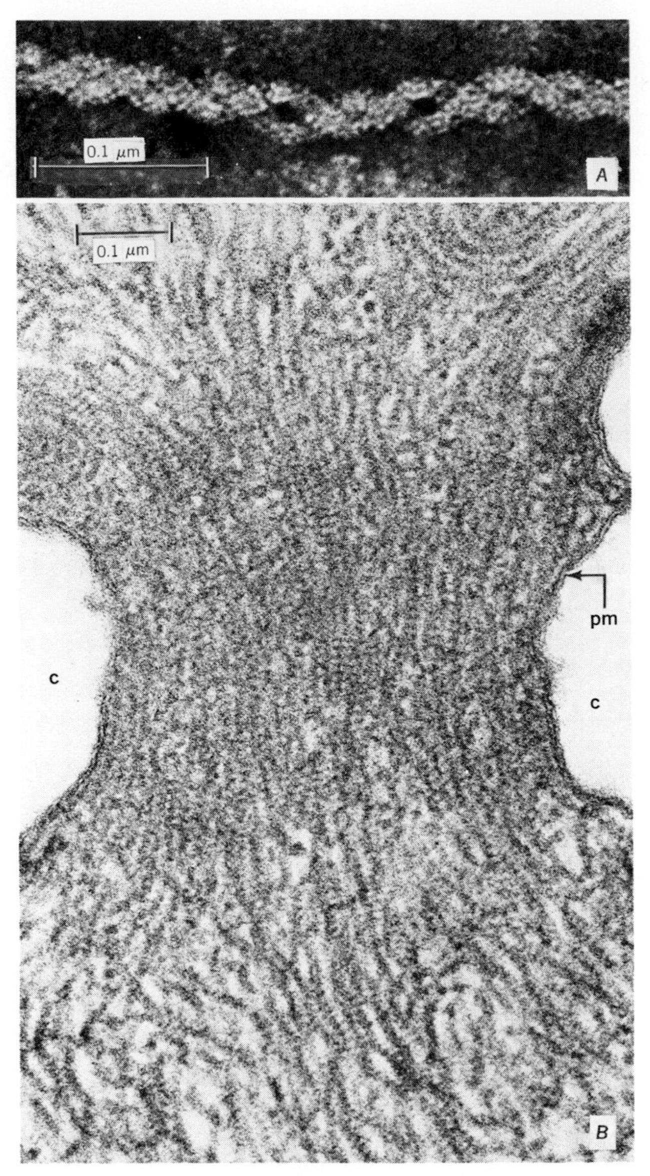

Figure 11.8 P-protein from sieve elements of, *A*, tobacco (*Nicotiana tabacum*), and *B*, *Nelumbo nucifera*. *A*, high magnification of negatively stained phloem exudate reveals the double-stranded structure of the P-protein filament. *B*, P-protein accumulated in a sieve plate pore shows horizontal striations in the extended filaments. At *c*, callose lining the pore beneath the plasmalemma (*pm*). (*A*, from Cronshaw, Gilder, and Stone.[7])

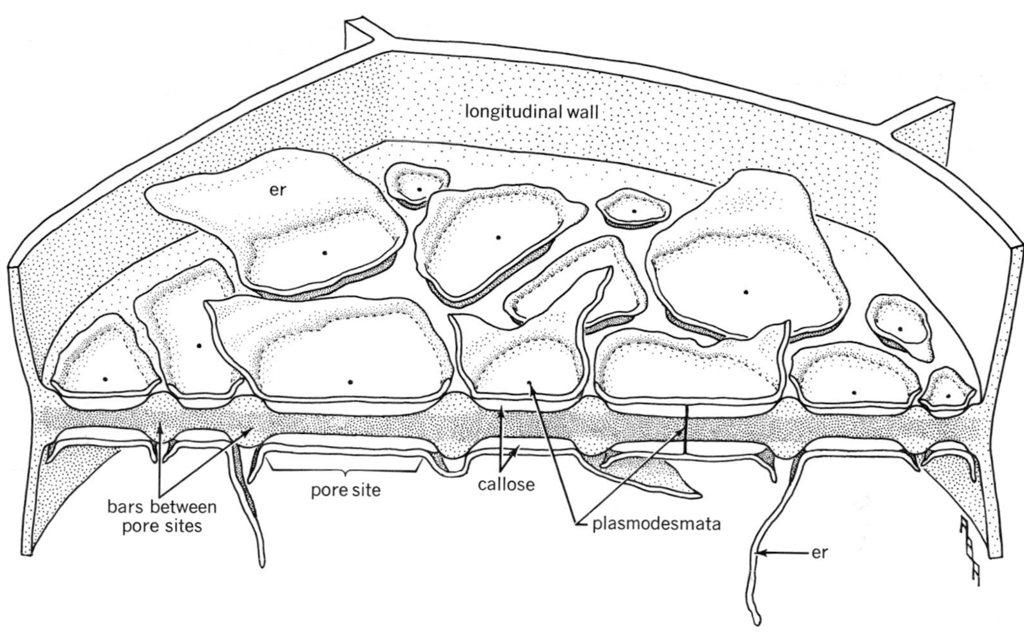

Figure 11.9 Three-dimensional drawing of half of a future sieve plate interpreting the relation between the plasmodesmata at the pore sites and endoplasmic reticulum (*er*) cisternae. Plasmalemma occurs between the *er* and the callose at the pore sites but is not indicated separately in the drawing. (From Esau and Cheadle, *Univ. Calif. Pubs. Bot.* 36: 253–344, 1965.)

As a consequence of the physical conditions in the sieve tube, its contents readily flow out when the plant is cut. The exudate from cut sieve tubes ("phloem exudate") represents the translocated material, but it also contains some protoplast components, notably P-protein.[7] Not all of the P-protein, however, flows out. Some remains lodged within the pores (figs. 11.8,*B*, and 11.10,*C*) and often forms accumulations ("slime plugs") on the sieve plate. The plugging prevents continued exudation from the sieve tubes. Wound callose rapidly appearing after the phloem is severed[10] enhances the blockage of pores.

The preceding description agrees with the concept that plugging of pores in conducting sieve elements with P-protein and callose is a response to a release of pressure and other disturbances resulting from cutting of the phloem and that in intact plants the pores are open to the passage of translocated materials. For the well-known mass-flow theory of translocation[5] open pores are a prerequisite. Some investigators, however, think that sieve area pores are obstructed in undisturbed tissue and that this circumstance must be taken into account when theories of translocation are formulated.

The interpretation just mentioned is not borne out by experiments designed to minimize the effect of injuries to the phloem when the plant is cut. To achieve this result, a plant may be rapidly frozen before the phloem is sampled[6] or the solute concentration in the sieve tubes may be reduced by removal of leaves (conveniently done in a seedling[17]) a couple of days before the plant is cut. Treatments of this kind indicate that the pores are not obstructed with callose and P-protein in intact conducting sieve elements.

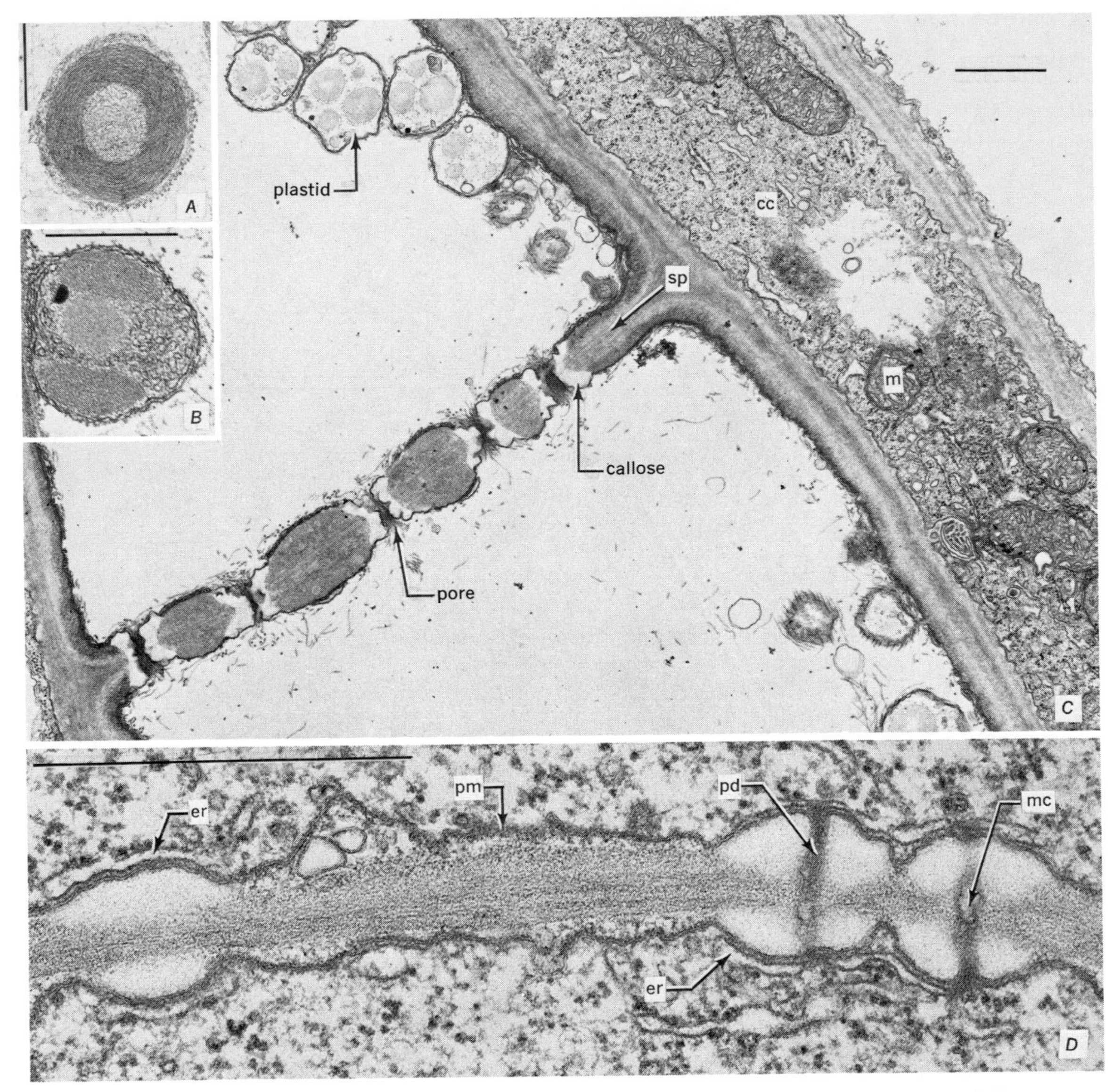

Figure 11.10 Plastids from sieve elements of *Tetragonia expansa* (*A*) and *Beta vulgaris*, sugar beet (*B*). Fibrous protein ring in face view in *A*, in section in *B*. In *C*, parts of two sieve elements with a mature sieve plate (*sp*) between them. Some P-protein occurs in the pores. A companion cell at *cc* shows a dense cytoplasm and several mitochondria (*m*). In *D*, a differentiating sieve plate with three pore sites, each with a pair of callose pads (white). A plasmalemma (*pm*) covers the wall including the callose at the pore sites. Endoplasmic reticulum (*er*) cisternae cover the pore sites. To the right, *er* appears discontinuous over the plasmodesmata (*pd*); it is merged with their contents (desmotubules, which are not visible). Median cavities (*mc*) at the middle lamella. *C* and *D*, from phloem in leaf of *Nicotiana tabacum*. (Lines are 0.5 μm in *A* and *B*, 1 μm in *C* and *D*.) (In *C*, the sieve element to which the *cc* is directly related is not in view.)

Companion cells

Research on translocation has provided much evidence that movement of organic materials in the phloem depends on the physiologic interaction between sieve elements and the contiguous parenchymatic cells. In tissues where sugars become available for transport, such as, photosynthesizing leaf mesophyll or reactivated storage parenchyma, sugars are transmitted to the conduits (loading of sieve elements) by the contiguous parenchymatic cells. At sites of utilization of sugars, that is, wherever growth occurs or storage materials are sequestered, parenchymatic cells remove sugars from the conduit (unloading of sieve elements). Thus, the phloem is an integrated system of conduits and contiguous cells concerned with loading and unloading of the conduits along the path of translocation at sites of sources of sugars and of sinks for the same. The constant exchange of sugars between the conduits and the associated nucleate cells establishes sucrose gradients which are involved in determining the rate and direction of movement.[5] It is possible that substances secreted into sieve elements by the neighboring nucleate cells are also the source of energy for the translocation mechanism in the conduits themselves.[21]

Companion cells are parenchyma cells specialized with regard to the functional association with sieve elements regulating the translocation. They are closely connected with sieve elements by plasmodesmata (usually branched on the side of the companion cell, fig. 11.11,*B*) and their longevity depends on that of the contiguous sieve elements. The protoplasts of companion cells are those characteristic of metabolically active cells (fig. 11.11,*A*). Their nuclei and nucleoli are relatively large. The cells contain plastids, often differentiated as chloroplasts, numerous large mitochondria, and some endoplasmic reticulum. Most outstanding is the abundance of ribosomes, frequently combined with a high density of background cytoplasm. The cells are vacuolated to various degrees, often containing many small vacuoles.

In general, companion cells resemble secretory cells in their ultrastructure, and their ability to deliver sugar into the conduit against a positive gradient suggests a secretory function. In certain taxa of dicotyledons, companion cells develop wall ingrowths in parts of the phloem system where the exchange between sieve elements and contiguous cells is particularly intense (chapter 18). The ingrowths result in an increase of the internal surface of the wall and of the contiguous plasmalemma. The latter is thus brought into close contact with much of the protoplast. Cells with wall ingrowths, usually referred to as transfer cells,[20] are found in many of those positions in plants where active short-distance transport of solutes is expected to occur. The development of such ingrowths in companion cells emphasizes the involvement of these cells in retrieval of solutes and their transfer to the sieve elements. Absence of ingrowths does not mean, however, that companion cells without wall ingrowths are not functioning as transfer cells. Presence or absence of wall ingrowths is a matter of degree of structural expression of the characteristics associated with the function of transfer of materials between cells.

Companion cells are formed by divisions of sieve element precursors so that sieve elements and their companion cells are ontogenetically related (fig. 11.5). One or more companion cells may be associated with one sieve element and they may occur on one or more sides of the sieve element. In some taxa, the companion cells appear in vertical files (*companion cell strands*) as a result of

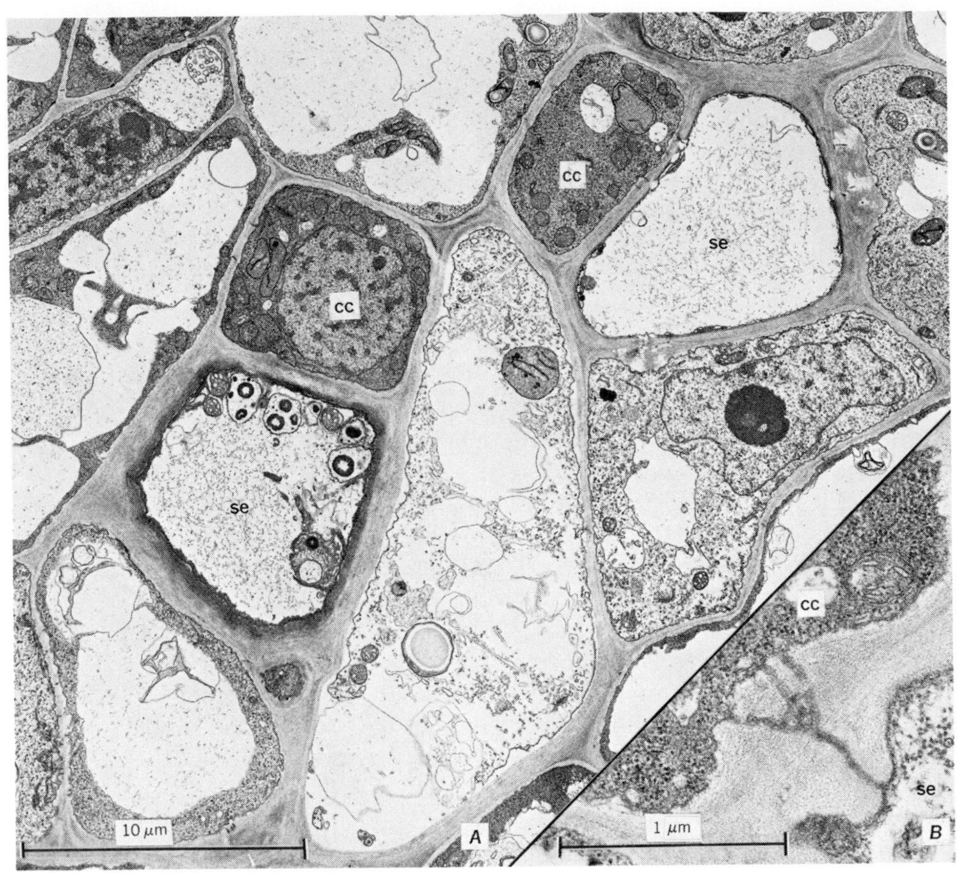

Figure 11.11 *A*, two sieve elements (*se*), each with a companion cell (*cc*), surrounded by phloem parenchyma cells. From a cross section of a vein in a *Nicotiana tabacum* leaf. *B*, wall between a sieve element (*se*) and companion cell (*cc*) with a branched plasmodesma. From a cross section of a vein in a *Mimosa pudica* leaf.

divisions in their immediate precursor. The ontogenetic relation of companion cells to sieve elements is usually regarded as a specific characteristic of these cells. The relation is typical in angiosperms, and the presence of companion cells is included in the definition of the sieve tube member as contrasted with the sieve cell. The functional counterparts of companion cells in gymnosperms, the *albuminous cells* (Strasburger cells) (fig. 11.4,*A*), are rarely derived from the same precursor as the sieve cell but their cytologic and functional characteristics parallel those of companion cells. Cytochemical studies on respiration in the phloem of *Larix* (larch) have shown increased respiratory and acid phosphatase activities in albuminous cells that were associated with fully differentiated sieve cells. Moreover, the increased activities were restricted to those periods when loading (source action: mobilization of starch in the spring) and unloading (sink action: storage starch accumulation in early summer) were taking place.[21]

Parenchyma cells

Parenchyma cells containing various ergastic substances, such as starch, tannins, and crystals, are regular components of the phloem. In the secondary phloem they are classified into axial parenchyma cells (fig. 11.1,*C*,*D*) and ray parenchyma cells (fig. 11.1,*K–M*). The axial cells may occur in parenchyma strands or as single fusiform parenchyma cells. A strand results from division of a precursor cell in two or more cells.

Crystal-forming parenchyma cells may become subdivided into small cells, each containing a single crystal (fig. 11.1,*D*). Such cells are commonly associated with fibers or sclereids and have lignified walls with secondary thickenings.

Parenchyma cells located next to sieve elements (fig. 11.11,*A*) function in loading and unloading the conduits, and in some taxa and in some positions in the plant body they become differentiated as transfer cells of the type developing wall ingrowths.[20] There may be also an ontogenetic association of parenchyma cells and sieve elements. In many dictoyledons, sieve tube members and some parenchyma cells are derived from the same phloem initial, and these parenchyma cells are apt to die at the same time as the associated sieve element.[13] Phloem parenchyma thus ranges from mainly storage cells to those approaching companion cells in their relation to the sieve elements.

Sclerenchyma cells

Fibers (fig. 11.1,*A*) are common components of both primary and secondary phloem. In the primary phloem, fibers occur in the outermost part of the tissue; in the secondary, in various distributional patterns among the other phloem cells of the axial system. The fibers may be nonseptate or septate and may be living or nonliving at maturity. Living fibers serve as storage cells as they do in the xylem. In many species, primary and secondary fibers are long cells with thick walls and are used as a commercial source of fiber (*Linum, Cannabis, Hibiscus*).

Sclereids (fig. 11.1,*E*,*I*) are also frequently found in the phloem. They may occur in combination with fibers or alone, and they may be present in both axial and radial systems of the secondary phloem. Sclereids typically differentiate in older parts of the phloem as a result of sclerification of parenchyma cells. This sclerification may or may not be preceded by intrusive growth of the cells. During such growth the sclereids often become branched or much elongated. The distinction between fibers and sclereids is not always sharp, especially if the sclereids are long and slender. The intermediate cell types are fiber-sclereids.

PRIMARY PHLOEM

The primary phloem is classified into *protophloem* and *metaphloem* on the same basis that the primary xylem is classified into protoxylem and metaxylem. The protophloem matures in plant parts that are still undergoing extension growth, and its sieve elements are stretched and soon become nonfunctional. Eventually they are completely obliterated (fig. 11.12). The metaphloem differentiates later and, in plants without secondary growth, constitutes the only conducting phloem in adult plant parts.

The protophloem sieve elements of angiosperms are usually narrow and inconspicuous, but they are enucleate and have sieve areas with callose. They may or may not have companion cells. They appear either in groups or singly among parenchyma cells or,

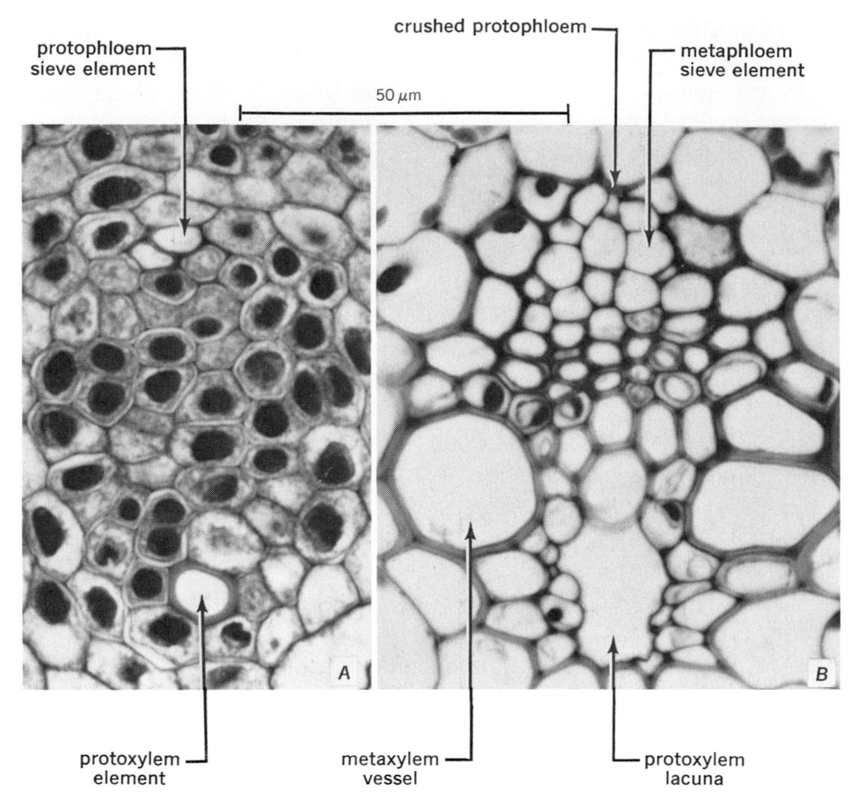

Figure 11.12 Cross sections of vascular bundles of oat (*Avena sativa*) in two stages of differentiation. *A*, first elements of protophloem and protoxylem have matured. *B*, metaphloem and metaxylem are mature; protophloem is crushed; protoxylem is replaced by a lacuna. (From Esau: *A*, *Amer. J. Bot.* 44:245–251, 1957; *B*, *Hilgardia* 27:15–69, 1957.)

in many dicotyledons, among conspicuously elongated living cells. In numerous species these elongated cells are fiber primordia. While the sieve elements cease to function and are obliterated the fiber primordia increase in length, develop secondary walls, and mature as fibers (fig. 11.13,*B*). Such fibers are found on the periphery of the phloem region in numerous dicotyledon stems and are often called pericyclic fibers (chapters 6 and 16). Protophloem fibers occur in roots also.

The metaphloem has more and commonly wider sieve elements than the protophloem. Companion cells are regularly present in the metaphloem of angiosperms (fig. 11.13,*B*), but fibers are usually absent. Parenchyma cells may become sclerified after the phloem ceases to conduct.

SECONDARY PHLOEM

The secondary phloem constitutes a much less prominent part of a branch, a tree trunk, or a root than the secondary xylem. The amount of phloem produced by the vascular cambium is usually smaller than that of xylem,

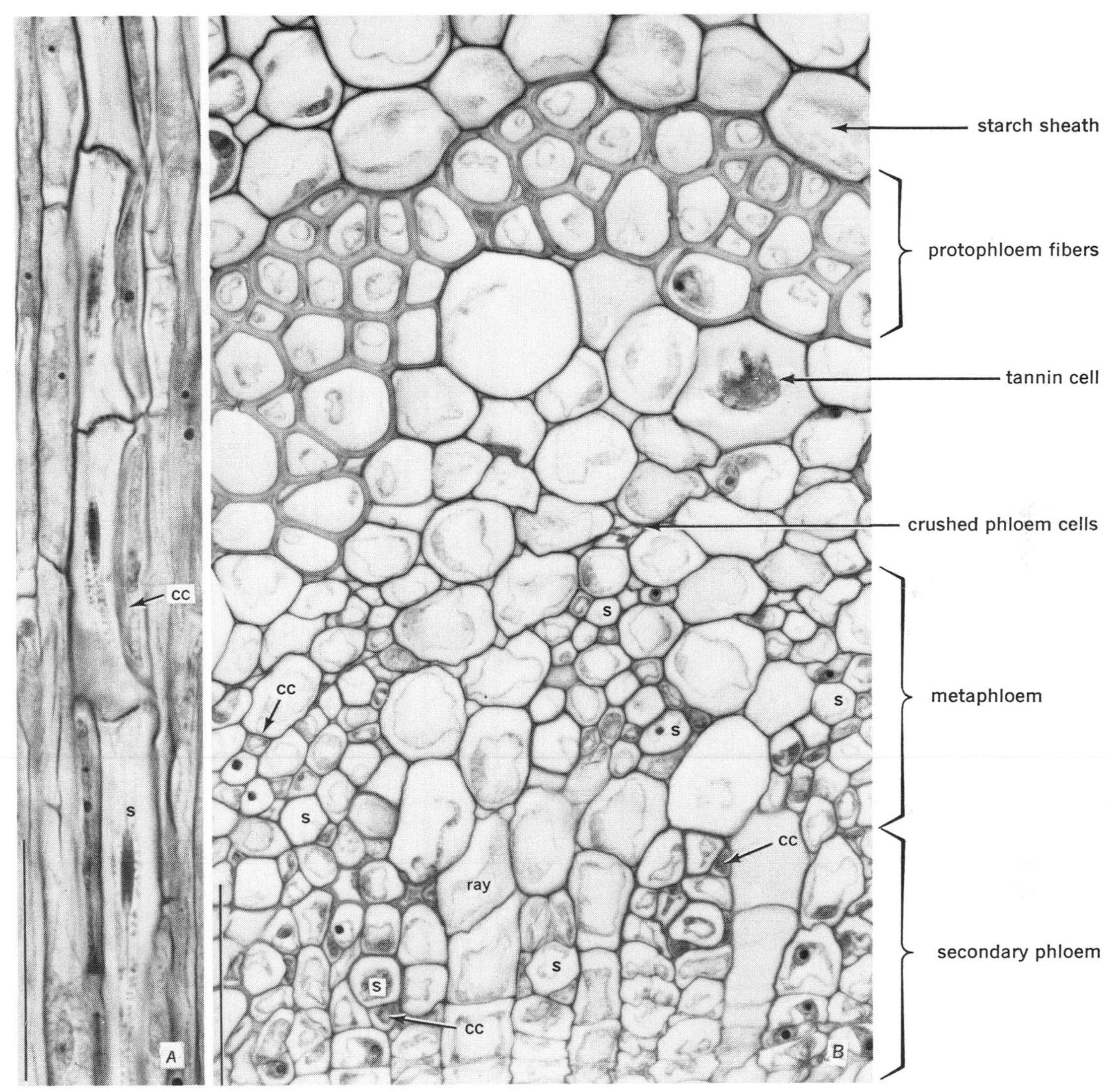

Figure 11.13 Phloem from bean (*Phaseolus vulgaris*) leaf. *A*, longitudinal section showing part of a sieve tube with two complete and two incomplete sieve tube members (*s*). Companion cell at *cc*. Spindle-shaped P-protein bodies in three of the elements. *B*, cross section including no longer conducting protophloem with fibers, metaphloem, and some secondary phloem. Sieve element at *s*, companion cell at *cc*. (Lines are 50 μm.)

the old phloem becomes more or less conspicuously crushed, and eventually the nonfunctional phloem is separated from the axis by the periderm. Thus, whereas the successive increments of xylem accumulate in the branch, trunk, or root, the phloem remains restricted in amount.

Under the term bark the phloem is included with the tissues located outside the vascular cambium (chapter 12). The functional (conducting) phloem constitutes the innermost part of the bark of woody stems and roots.

Conifer phloem

The secondary phloem in conifers is generally simpler and less variable among species than it is in dicotyledons.[9,11] The axial system contains sieve cells and parenchyma cells (figs. 11.14 and 11.15), some of which may be differentiated as albuminous cells. Fibers and sclereids also may be present. The rays are uniseriate and contain parenchyma cells and albuminous cells, if these cells are present in the species. Most commonly, albuminous cells are located at the margins of the rays (fig. 11.4,*A*). Resin ducts may be present in both systems.

The sieve elements are long cells that have numerous sieve areas, commonly restricted to the radial faces (fig. 11.4,*A*). The parenchyma cells form strands (fig. 11.15,*B*,*C*) or are single cells. Fibers are consistently absent in *Pinus* (fig. 11.16) and consistently present in Taxaceae, Taxodiaceae, and Cupressaceae. When they are present the fibers occur, as a rule, in uniseriate tangential bands (fig. 11.15,*A*) alternating with similar bands of parenchyma cells and sieve cells.

In a given section of conifer phloem only a narrow band, approximately one growth layer, may be in active state; the rest is no longer conducting. If fibers are absent, collapse of the sieve cells gives the tissue a distorted appearance, especially because the rays assume a wavy course (fig. 11.16). The parenchyma cells are enlarged in the nonfunctional (nonconducting) phloem and remain alive until they are cut off by the periderm (figs. 11.15 and 11.16). Ray parenchyma cells also remain active but not the albuminous cells, which collapse in the nonfunctional phloem.

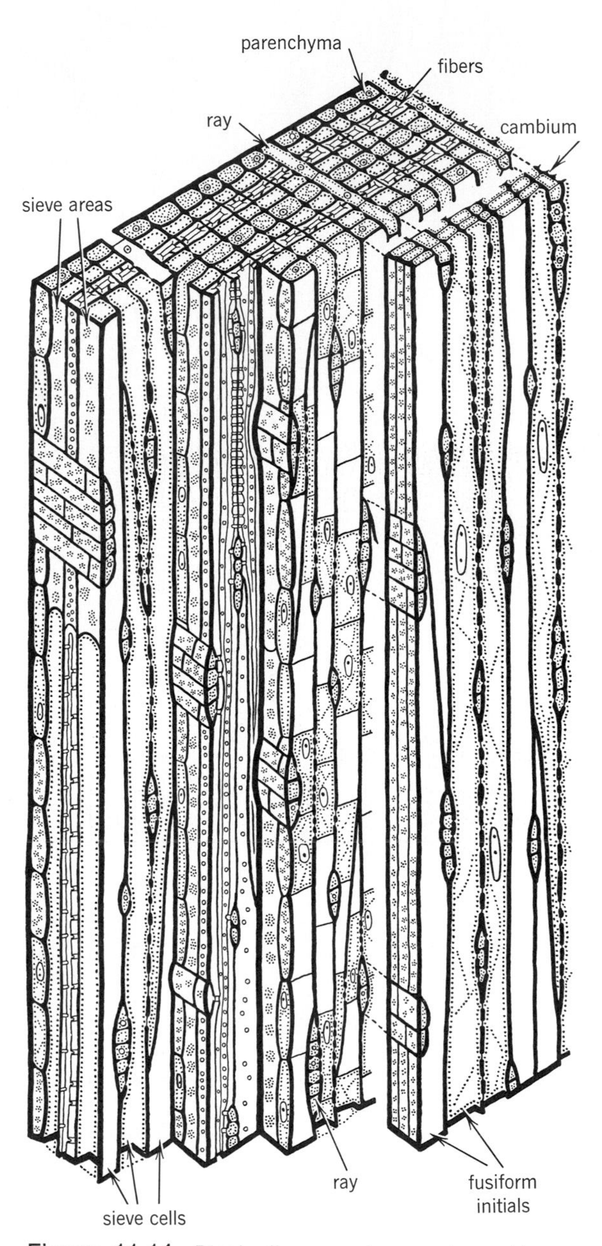

Figure 11.14 Block diagram of secondary phloem and vascular cambium of *Thuja occidentalis* (white cedar), a conifer. (From Esau, *Plant Anatomy,* 2nd ed. John Wiley & Sons, 1965.)

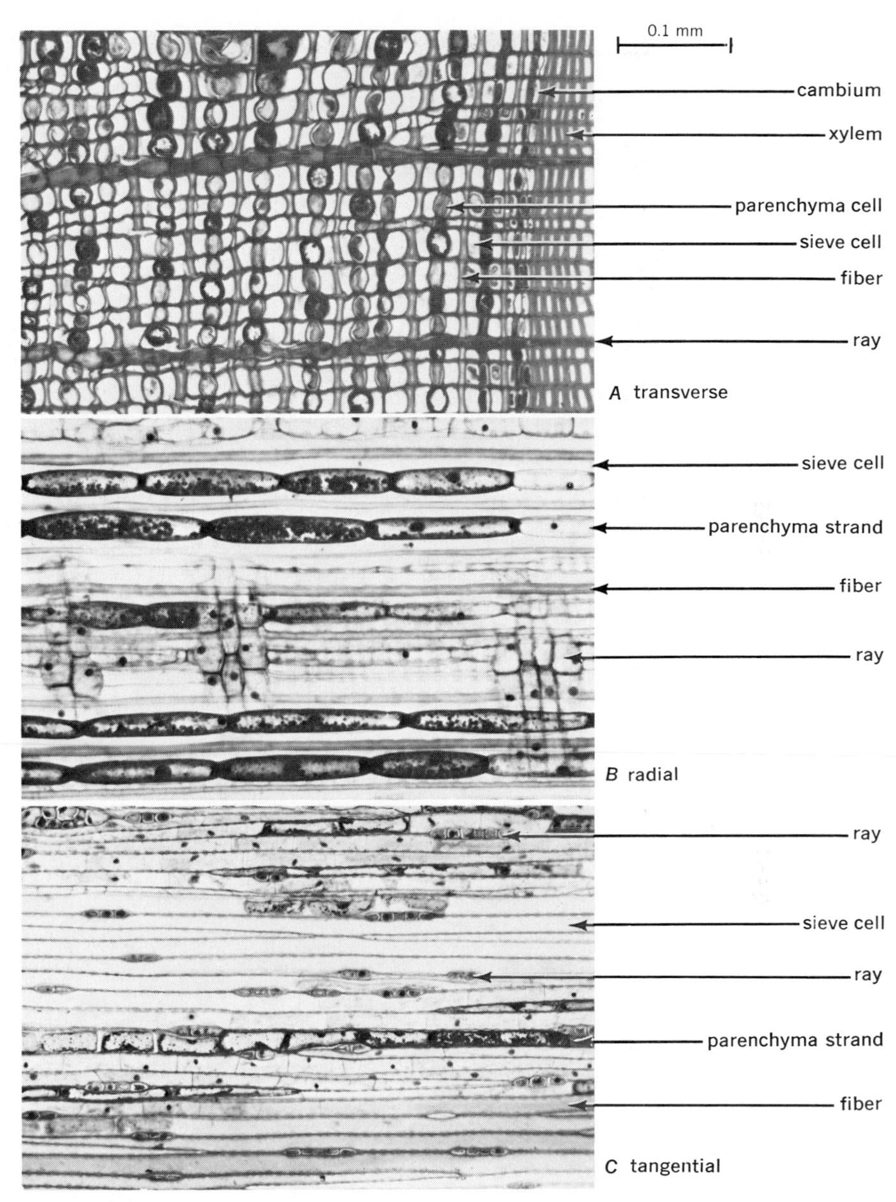

Figure 11.15 Secondary phloem of *Thuja occidentalis* (white cedar), a conifer, in three kinds of sections. Sieve cells, phloem parenchyma cells, and fibers alternate in tangential bands.

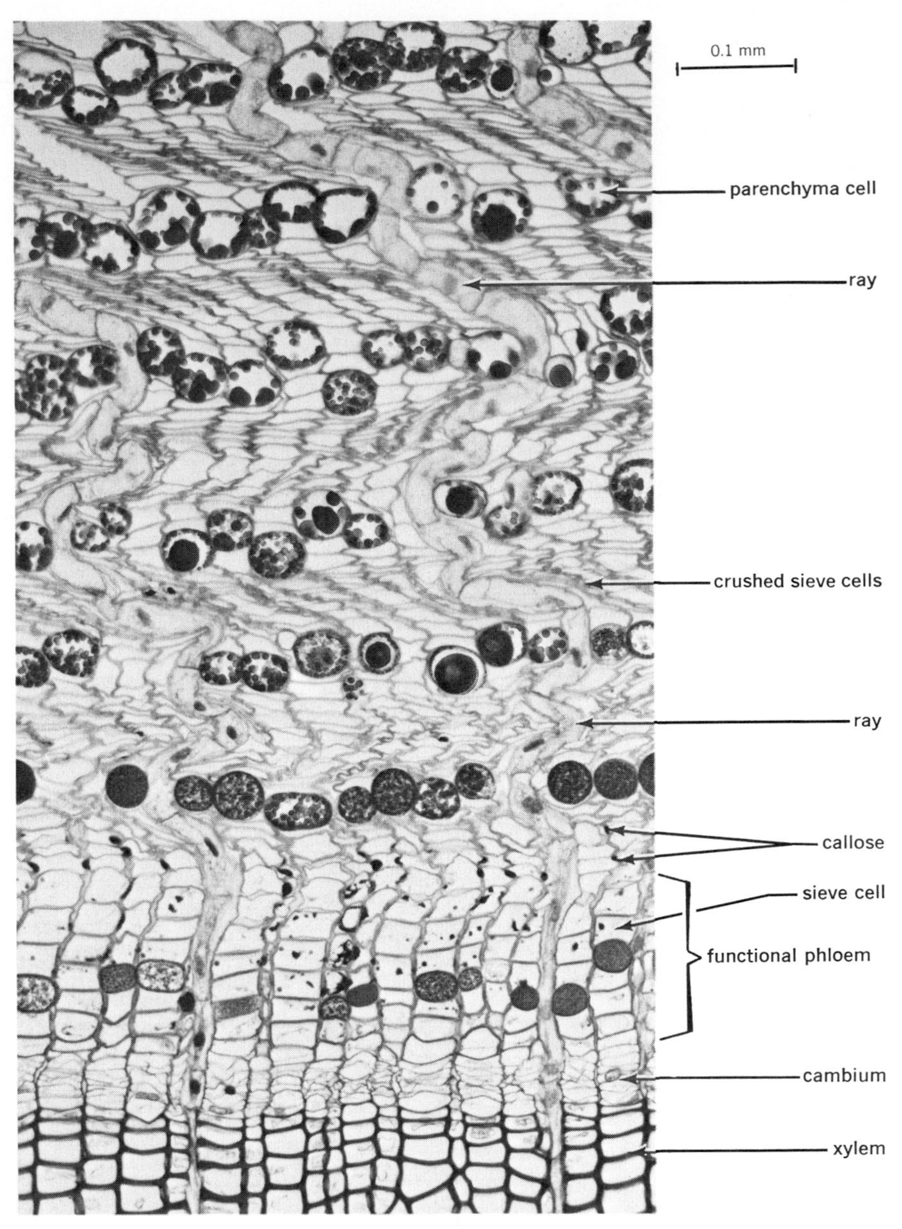

Figure 11.16 Cross section of secondary phloem of *Pinus.* Functional phloem much smaller in amount than the nonfunctional (only partly shown in the figure), in which all sieve cells are crushed and only axial (with dark contents) and ray parenchyma cells are intact.

Dicotyledon phloem

The secondary phloem of dicotyledons varies with regard to composition, arrangement, and size of cells as well as the characteristics of the nonfunctional phloem (fig. 11.17 and 11.18). Sieve tubes, companion cells, and parenchyma cells are constant elements of the axial system, but fibers may be absent (*Aristolochia*). When they are present, the fibers may be scattered (*Campsis,* fig. 11.19,*C; Cephalanthus, Laurus*) or may appear in tangential bands in parallel arrangement (*Fraxinus,* fig. 11.18,*A*; *Liriodendron, Magnolia, Robinia, Tilia,* fig. 11.19,*A*) or somewhat scattered (*Ostrya,* fig. 11.19,*B*). Fibers may be so abundant that the sieve tubes and parenchyma cells occur as small groups surrounded by fibers (*Carya*). In some species sclerenchyma cells, usually sclereids or fiber-sclereids, differentiate only in the nonfunctional part of the phloem (*Prunus*). The septate fibers of *Vitis* are living cells concerned with storage of starch.

Depending on the characteristic of the cambium, the secondary phloem may be storied (*Robinia*) or nonstoried (*Betula, Quercus, Populus, Liriodendron, Juglans*). The long sieve-tube members in nonstoried phloem typically have inclined end walls with compound sieve plates. In more or less definitely storied phloem the sieve elements have slightly inclined or almost transverse end walls, and their sieve plates have few sieve areas or only one.

The rays resemble the xylem rays of the same plant and may be uniseriate or multiseriate, high or low; and different kinds of rays may be represented in the same tissue. The rays are composed of parenchyma cells (fig. 11.18) but may contain sclereids or sclerified parenchyma cells with crystals. In older parts of the phloem the rays may become dilated in response to the increase in

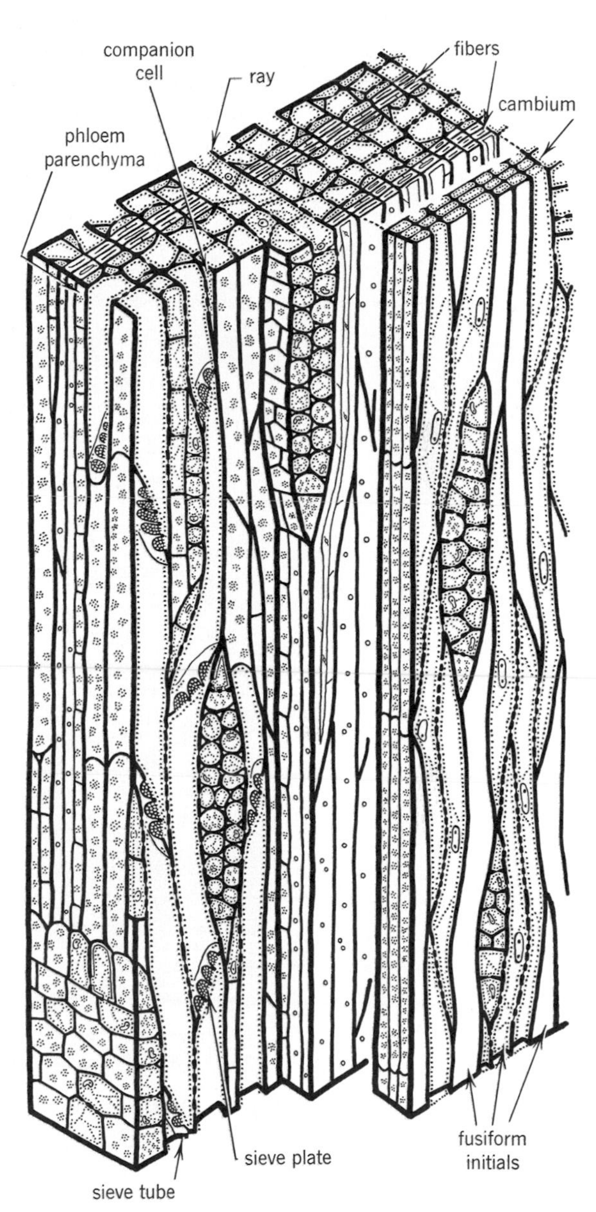

Figure 11.17 Block diagram of secondary phloem and vascular cambium of *Liriodendron tulipifera* (tulip tree), a dicotyledon. (From Esau, *Plant Anatomy,* 2nd ed. John Wiley & Sons, 1965.)

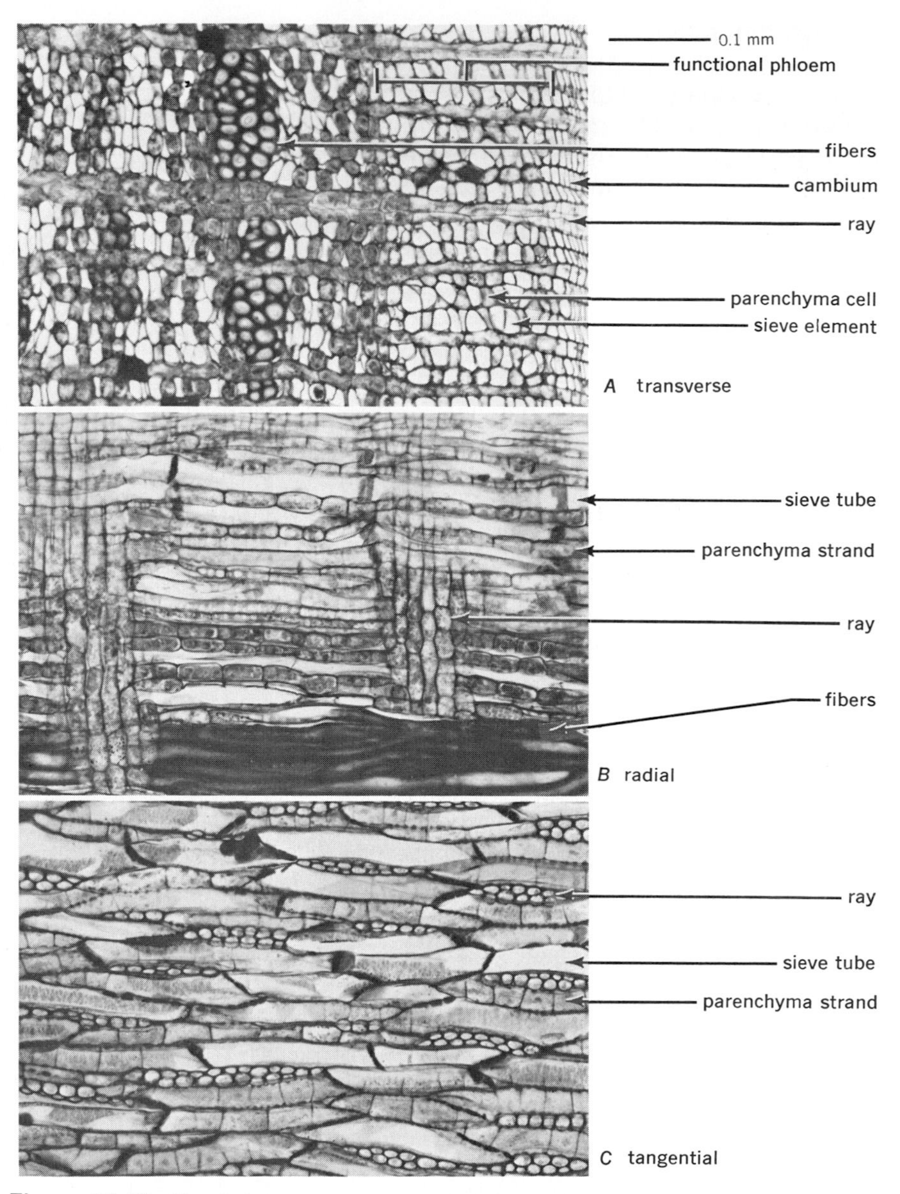

Figure 11.18 Secondary phloem of *Fraxinus americana* (white ash) in three kinds of sections.

circumference of the axis (fig. 11.19,*A*). Radial anticlinal cell division and tangential cell enlargement bring about the dilatation. Sometimes the dilatational divisions are restricted to the median position in a ray, and the region of dividing cells resembles a meristem.[19,22] Usually only some rays become dilated; others remain as wide as they were at

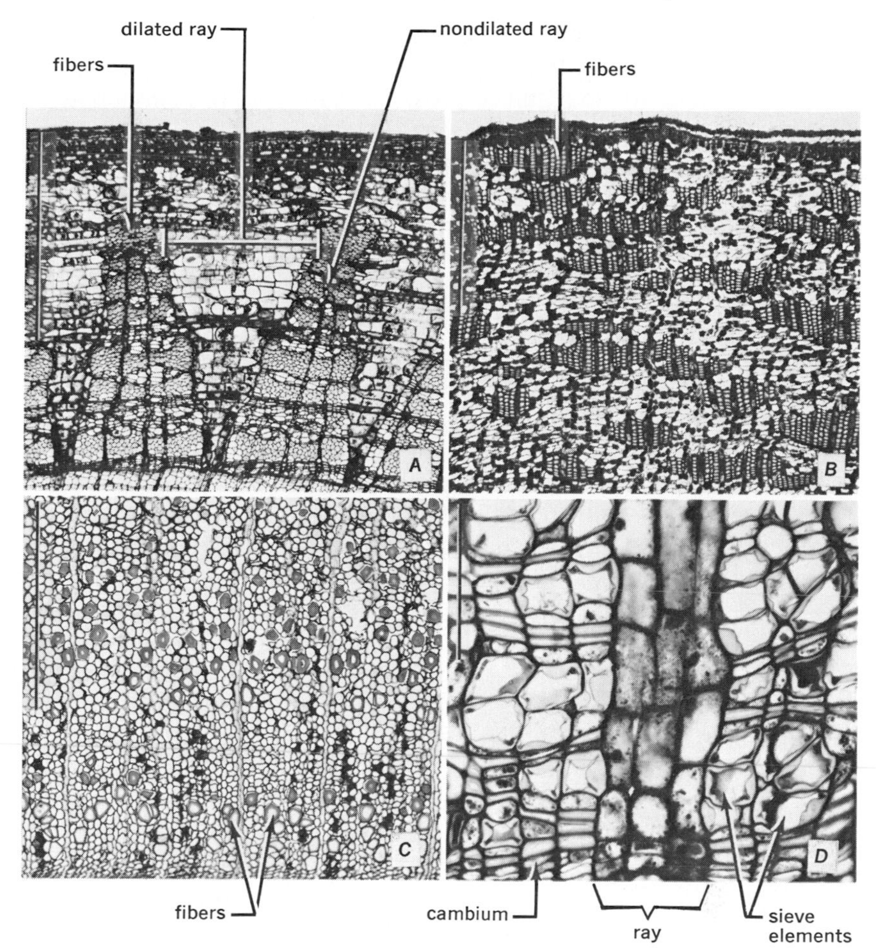

Figure 11.19 Secondary phloem of dictotyledons in cross sections. *A*, *Tilia,* fibers in parallel tangential bands. *B*, *Ostrya,* fibers in groups and bands. *C*, *Campsis,* fibers scattered singly. *D*, *Liriodendron,* sieve elements with nacreous walls. (Lines are 0.5 mm in *A*–*B*, 100 μm in *D*.)

the time of origin in the cambium (fig. 11.19,*A*). In some species of *Eucalyptus* wide wedges of tissue develop in the dilatating phloem by division of phloem parenchyma cells.[4]

The nonfunctional phloem assumes different aspects depending on the types of cells present and on their behavior. The dilatation of rays is one of the characteristics of the old phloem. Sieve tubes are either completely crushed, or they remain open and become filled with gases. Parenchyma cells often enlarge and thus compress the sieve tubes or invade their lumina (*thylosoids*). If the tissue shrinks because of collapse of cells, the rays may be bent. The parenchyma cells in the nonfunctional phloem continue to store starch until they are cut off by the periderm.

The amount of functional phloem is commonly limited to one growth increment because the sieve elements originating from the cambium in the spring usually cease to conduct and die in the fall (*Pyrus*,[16] *Acer negundo*[23]). There are exceptions to this sequence, however. In *Vitis*, for example, the sieve elements of one year become dormant in the winter but resume activity in the following spring. The amount of nonfunctional phloem is highly variable. If periderm is formed repeatedly at short intervals, the old phloem does not accumulate. In some species growth layers are detectable in the phloem—the early sieve elements are wider than those formed in the late phloem or there is a band of sclerenchyma formed in the late phloem—but generally the delimitation of the layers is obscured by the changes affecting the nonfunctioning phloem.

REFERENCES

1. Behnke, H.-D. Über den Feinbau verdickter (nacré) Wände und der Plastiden in den Siebröhren von *Annona* und *Myristica*. *Protoplasma* 72:69–78. 1971.
2. Behnke, H.-D. Sieve-tube plastids in relation to angiosperm systematics—an attempt towards a classification by ultrastructural analysis. *Bot. Rev.* 38: 155–197. 1972.
3. Chafe, S. C., and M. E. Doohan. Observations on the ultrastructure of the thickened sieve cell wall in *Pinus strobus* L. *Protoplasma* 75:67–78. 1972.
4. Chattaway, M. M. The anatomy of bark. VI. Peppermints, boxes, ironbarks and other eucalypts with cracked and furrowed barks. *Aust. Jour. Bot.* 3:170–176. 1955.
5. Crafts, A. S., and C. E. Crisp. *Phloem transport in plants.* San Francisco, California, Freeman and Company. 1971.
6. Cronshaw, J., and R. Anderson. Sieve plate pores of *Nicotiana*. *J. Ultrastruct. Res.* 27:134–148. 1969.
7. Cronshaw, J., J. Gilder, and D. Stone. Fine structural studies of P-proteins in *Cucurbita, Cucumis,* and *Nicotiana*. *J. Ultrastruct. Res.* 45:192–205. 1973.
8. Currier, H. B., K. Esau, and V. I. Cheadle. Plasmolytic studies of phloem. *Amer. Jour. Bot.* 42:68–81. 1955.
9. Den Outer, R. W. Histological investigations of the secondary phloem of gymnosperms. *Med. Landbouwhogesch. Wageningen,* Nederland, 67(7), 119 p. 1967.
10. Engleman, E. M. Sieve element of *Impatiens sultani*. I. Wound reaction. *Ann. Bot.* 29:83–101. 1965.
11. Esau, K. The phloem. *Handbuch der Pflanzenanatomie.* Band 5. Teil 2. 1969.
12. Esau, K. Changes in the nucleus and the endoplasmic reticulum during differentiation of a sieve element in *Mimosa pudica* L. *Ann. Bot.* 36:83–101. 1972.
13. Esau, K., and V. I. Cheadle. Significance of cell divisions in differentiating secondary phloem. *Acta Bot. Neerl.* 4: 348–357. 1955.
14. Esau, K., and V. I. Cheadle. Wall thickening in sieve elements. *Proc. Natl. Acad. Sci.* 44:546–553. 1958.
15. Esau, K., and V. I. Cheadle. Size of pores and their contents in sieve elements of dicotyledons. *Proc. Natl. Acad. Sci.* 45:156–162. 1959.
16. Evert, R. F. Phloem structure in *Pyrus communis* L. and its seasonal changes. *Univ. Calif. Pubs. Bot.* 32:127–194. 1960.
17. Evert, R. F., W. Eschrich, and S. E. Eich-

horn. P-protein distribution in mature sieve elements of *Cucurbita maxima*. *Planta* 109:193–210. 1973.

18. Evert, R. F., W. Eschrich, S. E. Eichhorn, and S. T. Limbach. Observations on penetration of barley leaves by the aphid *Rhopalosiphum maidis* (Fitch). *Protoplasma* 77:95–110. 1973.
19. Holdheide, W. Anatomie mitteleuropäischer Gehölzrinden. *Handbuch der Mikroskopie in der Technik.* Band 5. Teil 1:195–367. 1951.
20. Pate, J. S., and B. E. S. Gunning. Transfer cells. *Ann. Rev. Plant Physiol.* 23:173–196. 1972.
21. Sauter, J. J., and H. J. Braun. Cytochemische Untersuchung der Atmungsaktivität in den Strasburger-Zellen von *Larix* und ihre Bedeutung für den Assimilattransport. *Z. Pflanzenphysiol.* 66:440–458. 1972.
22. Schneider, H. Ontogeny of lemon tree bark. *Amer. Jour. Bot.* 42:893–905. 1955.
23. Tucker, C. M., and R. F. Evert. Seasonal development of the secondary phloem in *Acer negundo*. *Amer. J. Bot.* 56:275–284. 1969.
24. Weatherley, P. E. The mechanism of sieve-tube translocation: observation, experiment and theory. *Adv. Sci.* 18:571–577. 1962.

12

Periderm

The periderm is a protective tissue of secondary origin replacing the epidermis in stems and roots that increase in thickness by secondary growth. Woody dicotyledons and gymnosperms furnish the best examples of periderm development. Leaves usually produce no periderm although scales of winter buds may do so. Periderm occurs in herbaceous dicotyledons, especially in the oldest parts of stem and root. Some monocotyledons have periderm; others a different kind of secondary protective tissue.

Periderm develops along surfaces that are exposed after abscission of plant parts, such as leaves and branches. Periderm formation is also an important stage in the development of protective layers near injured or dead (necrosed) tissues (wound periderm or wound cork; fig. 12.1,*C*), whether resulting from mechanical wounding[15] or invasion of parasites.[21] In several families of dicotyledons, periderm is formed in the xylem—interxylary cork—in relation to a normal dying back of annual shoots or a splitting of perennating roots and stems.[16]

The nontechnical term bark must be distinguished from the term periderm. Although the word bark is employed loosely, and often inconsistently, it is a useful term if properly defined. Bark may be used most appropriately to designate all tissues outside the vascular cambium. In secondary state, bark includes the secondary phloem, the primary tissues that may still be present outside the secondary phloem, the periderm, and the dead tissues outside the periderm. Death of cells isolated outside the periderm brings about a distinction between the outer nonliving bark and the inner living bark. The functional phloem is the innermost part of the living bark. The term bark is sometimes used for stems in primary state of growth. It then includes the primary phloem, the cortex, and the epidermis. Because of the radially alternate arrangement of xylem and phloem in roots in primary state, the primary phloem of a root cannot be conveniently included with the cortex under the term bark.

STRUCTURE OF PERIDERM AND RELATED TISSUES

The periderm includes the *phellogen* (*cork cambium*), the meristem that produces the periderm; *phellem* (commonly called *cork*), the protective tissue formed outward by the phellogen; and *phelloderm,* a living parenchyma tissue formed inward by the meristem (fig. 12.1,*A*). Death of tissues lying outside the periderm results from an insertion of nonliving cork between these tissues and the living inner tissues of the axis.

The phellogen is relatively simple in structure. In contrast to the vascular cambium it

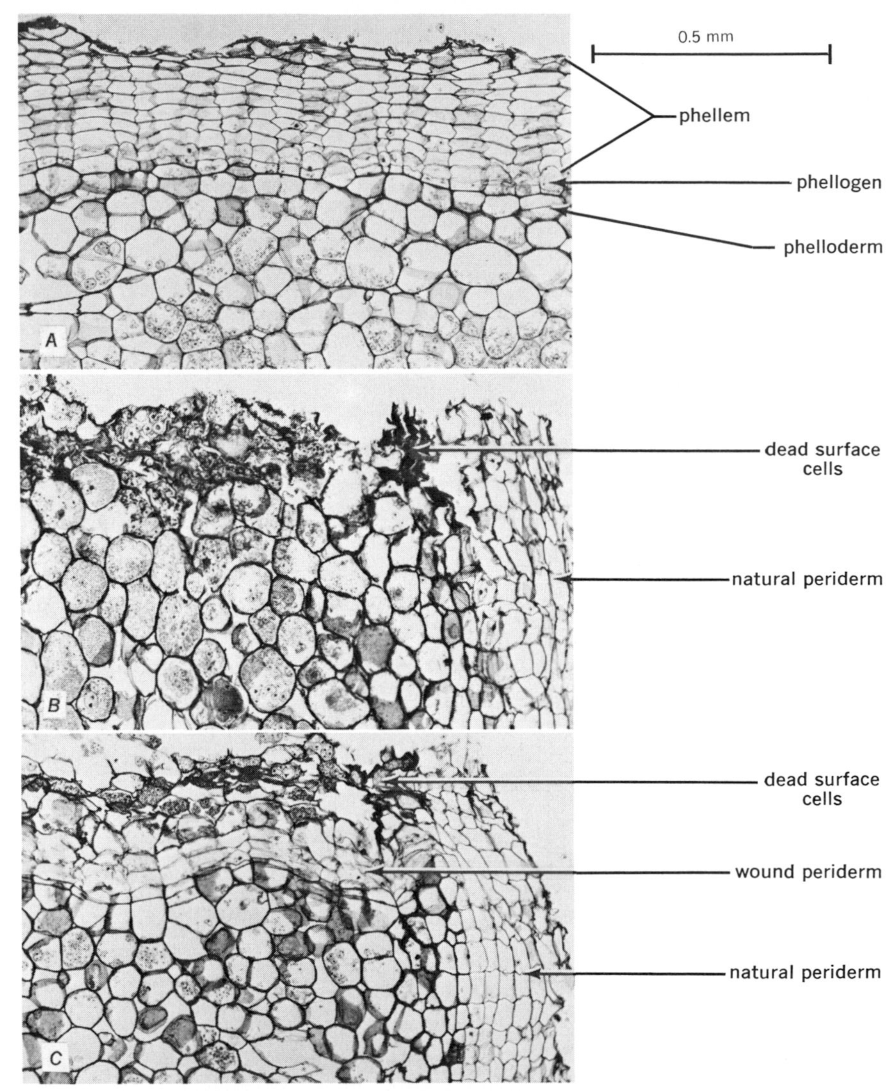

Figure 12.1 Periderm of root of sweet potato (*Ipomoea batatas*). *A*, natural periderm. *B*, *C*, wound healing on broken ends at end of curing period: *B*, after cure in a pile in a field; *C*, after cure in a warm house. In *B*, broken end covered with dead cells. In *C*, a wound periderm has developed beneath the dead surface cells and has become connected (right) with the natural periderm. (From Morris and Mann.[15])

has only one form of cell. In cross section the phellogen commonly appears as a continuous tangential layer (lateral meristem) of rectangular, radially flattened cells, each with the derivatives in a radial file extending outward through the cork cells and inward through the phelloderm cells (fig. 12.1,*A*). In longitudinal sections the phellogen cells are rectangular or polygonal in outline, sometimes somewhat irregular.

The cork cells are often approximately prismatic in shape (fig. 12.2,*A*,*B*) although they may be rather irregular in the tangential plane (fig. 12.2,*F*). They may be elongated vertically (fig. 12.2,*E*,*F*), radially (fig. 12.2,*B*–*E*), or tangentially (figs. 12.2,*A*, narrower cells; 12.7,*A*). They are usually arranged compactly, that is, the tissue lacks intercellular spaces. The cells are nonliving at maturity, but may have fluid or solid contents, some colorless, others pigmented.

The cork cells are characterized by suberization of their walls. The suberin, a fatty substance, usually occurs as a distinct lamella that covers the original primary cellulose wall, which may be lignified. The suberin lamella appears layered under the electron microscope probably because of alternation of suberin and waxes.[9] The walls of cork cells vary in thickness. In thick-walled cells a lignified cellulose layer occurs on the inside of the suberin lamella, which thus may be embedded between two cellulose layers. The walls of cork cells may be colored brown or yellow.

Cork used commercially as bottle cork has thin walls and lumina filled with air. It is highly impervious to water and resistant to oil. It is light in weight and has thermal insulating qualities. Mature cork of this type is also a compressible, resilient tissue. The commercially valuable properties, imperviousness to water and insulating qualities, also make the cork effective as a protective layer on the plant surface. The dead tissue that becomes isolated by the periderm adds to the insulating effect of the cork.

In many species the phellem consists of cork cells and of nonsuberized cells called *phelloid cells.* Like the cork cells, the nonsuberized cells may have thick or thin walls, and they may differentiate as sclereids (fig. 12.2,*D*). In a species of *Abies,* layers of cork cells alternate with those of cells having thick sclerified outer tangential walls.[14] This layering indicates seasonal increments.

The phelloderm cells resemble cortical parenchyma cells and may be distinguished from the latter by their position in the same radial files as the phellem cells (fig. 12.1,*A*).

POLYDERM

A special type of protective tissue called polyderm occurs in roots and underground stems of Hypericaceae, Myrtaceae, Onagraceae, and Rosaceae.[13,18] It consists of alternating tissue layers: layers, one cell deep, of partly suberized cells and layers, several cells deep, of nonsuberized cells (fig. 12.3). The polyderm may become twenty or more layers in total thickness, but only the outermost layers are dead. In the living part, the nonsuberized cells function as storage cells.

RHYTIDOME

As a tree grows older, periderms often arise at successively greater depths and thus cause an accumulation of dead tissues on the surface of stem and root (figs. 12.4 and 12.8). This dead part of the bark, composed of layers of tissues isolated by the periderms and of layers of no longer growing periderms, is called rhytidome. Rhytidome thus constitutes the outer bark and is especially well developed in older stems and roots of trees. In shrubs, early exfoliation of older bark is

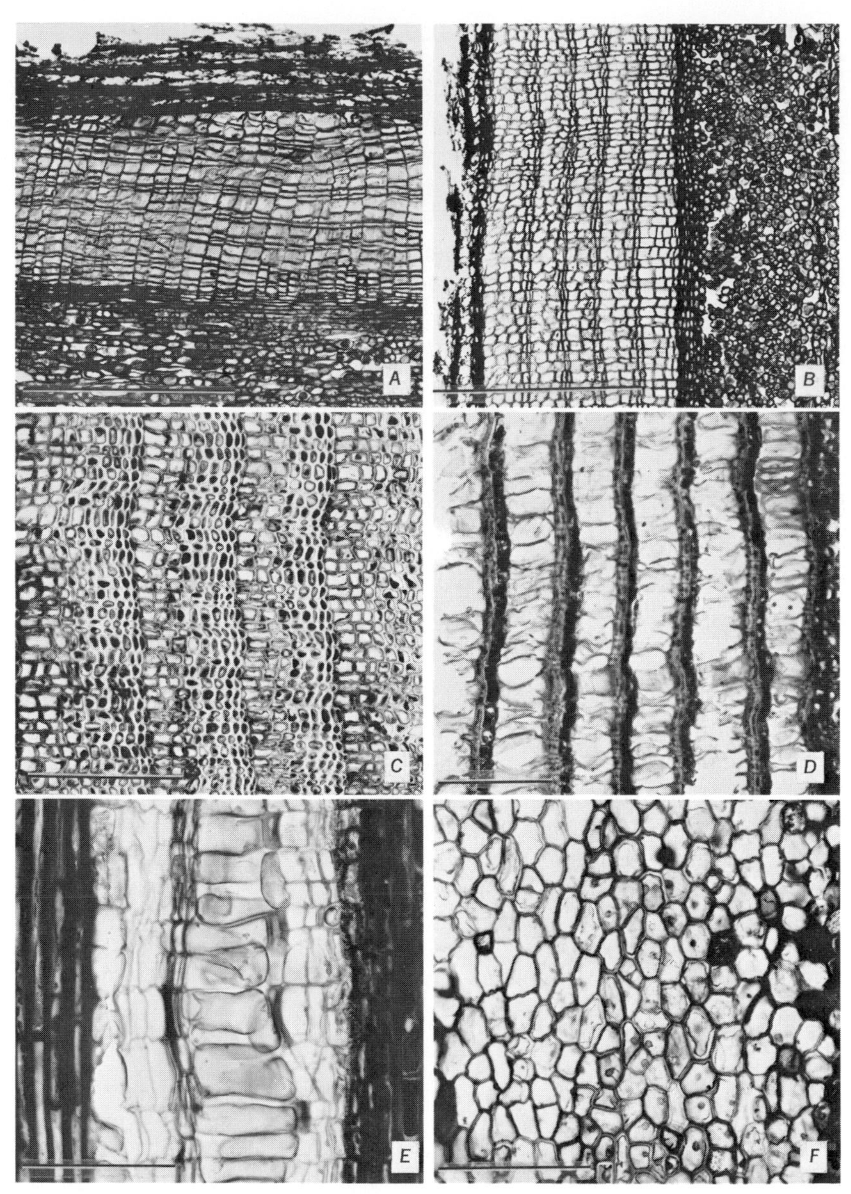

Figure 12.2 Variation in structure of phellem in stems. *A*, *B*, *Rhus typhina.* Phellem in transverse (*A*) and radial (*B*) sections of stem shows growth layers revealed in the alternation of narrower and wider cells. *C*, birch (*Betula populifolia*). Phellem with thick cell walls and conspicuous growth layers; radial section. *D*, *Rhododendron maximum.* Heterogeneous phellem consisting of cells of different sizes; sclereids compose some of the layers of small cells; radial section. *E*, *F*, *Vaccinium corymbosum.* Phellem in radial (*E*, light-colored cells in the middle) and tangential (*F*) sections. Phellem cells vary in form in *E*. (Lines are: 0.5 mm in *A*, *B*; 0.1 mm in *C–F*.)

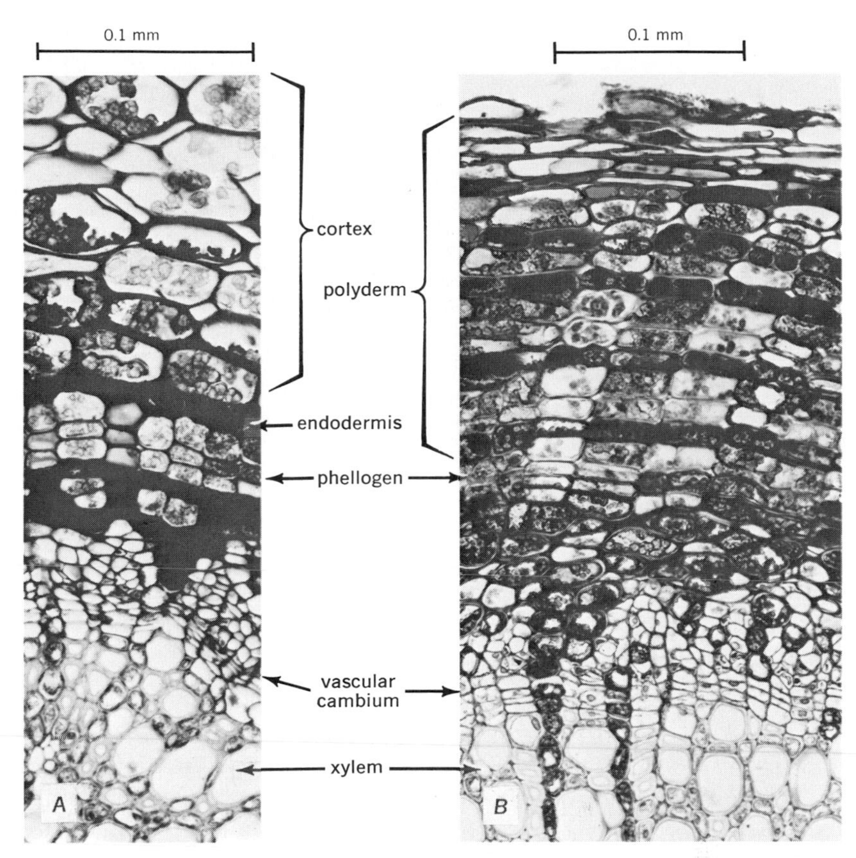

Figure 12.3 Polyderm of root of strawberry (*Fragaria*) in cross sections. *A*, root in early stage of secondary growth. Phellogen has been initiated, but the cortex is still intact. *B*, older root. Wide layer of polyderm has been formed by the phellogen. The cells composing the darkly stained bands in the polyderm are suberized. These cells alternate with nonsuberized cells. Both kinds of cells are living. Nonliving suberized cells form the outer covering. No cortex is present. (From Nelson and Wilhelm.[18])

common and precludes an accumulation of a thick rhytidome.

Studies of bark based on the use of cryofixation have revealed that two kinds of periderm are involved in rhytidome formation in three species of conifers.[17] The initial periderm and some sequent periderms are brown, other sequent periderms are reddish-purple. In addition to the color, the two kinds of periderm have other specific physical and chemical characters, and also differ in their position in the rhytidome. The reddish sequent periderms occur next to the dead phloem embedded in the rhytidome and appear to serve in protecting living tissues from effects associated with cell death. The brown sequent periderms appear sporadically and are separated from the dead phloem by the reddish-purple periderms. The brown initial periderm and the brown sequent periderms are similar in all their characteristics; and both act in protecting living tissues against the

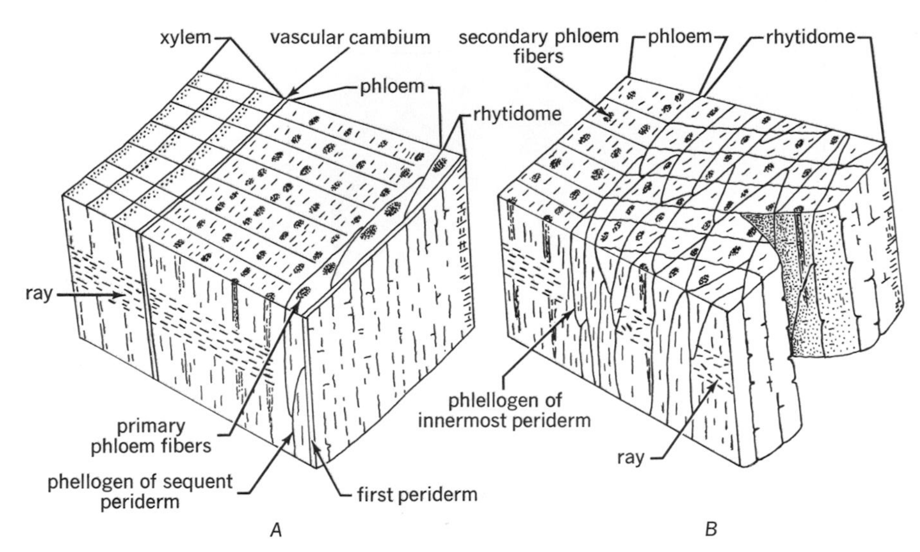

Figure 12.4 Diagrams showing an earlier (*A*) and a later (*B*) stage in the development of rhytidome. In *A* cortex and primary phloem are included in the rhytidome; in *B*, many layers of secondary phloem. The earlier layers of rhytidome have been shed in *B*.

external environment, the first periderm before the formation of rhytidome, the brown sequent periderms after the shedding of rhytidome layers. In view of their presumptive specific functions identifying terms have been proposed for the two kinds of periderm: *necrophylactic* for the reddish-purple sequent periderms, *exophylactic* for the brown periderms, the first and the sequent.[17]

DEVELOPMENT OF PERIDERM

The first periderm commonly appears during the first year of growth of stem and root. The subsequent, deeper periderms may be initiated later the same year or many years later (species of *Abies, Carpinus, Fagus, Quercus*[14]) or may never appear. In addition to specific differences, environmental conditions influence the appearance of both the initial and the sequent periderms. Availability of water, temperature, and intensity of light all affect the timing of the development of periderms.[3,6,14]

The first periderm of a stem originates most commonly in the subepidermal layer (fig. 12.5,*A*), occasionally in the epidermis. In some species, however, the first periderm arises rather deeply in the stem (*Berberis, Ribes, Vitis;* fig. 12.6), usually in the primary phloem. In most roots the first periderm originates in the pericycle (chapter 14), but it may appear near the surface as, for example, in some trees and perennial herbaceous plants in which the root cortex serves for food storage. The sequent periderms arise in successively deeper layers beneath the first (fig. 12.4) and thus, eventually, originate from parenchyma of the secondary phloem, including ray cells.

The first phellogen is initiated either uniformly around the circumference of the axis or in localized areas and becomes continuous by a lateral spread of meristematic activity. The sequent periderms appear most com-

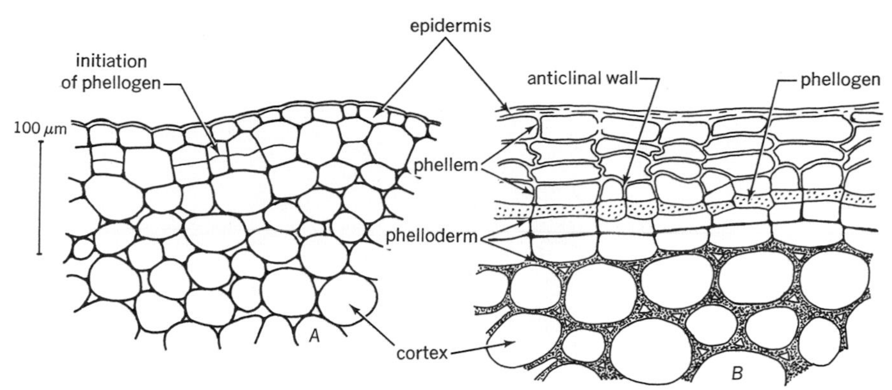

Figure 12.5 Origin of periderm in *Pelargonium* stem as seen in cross sections. *A*, periclinal divisions in subepidermal layer have produced phellogen cells toward the outside and phelloderm cells toward the inside, one of each to a divided cell. *B*, periderm is established.

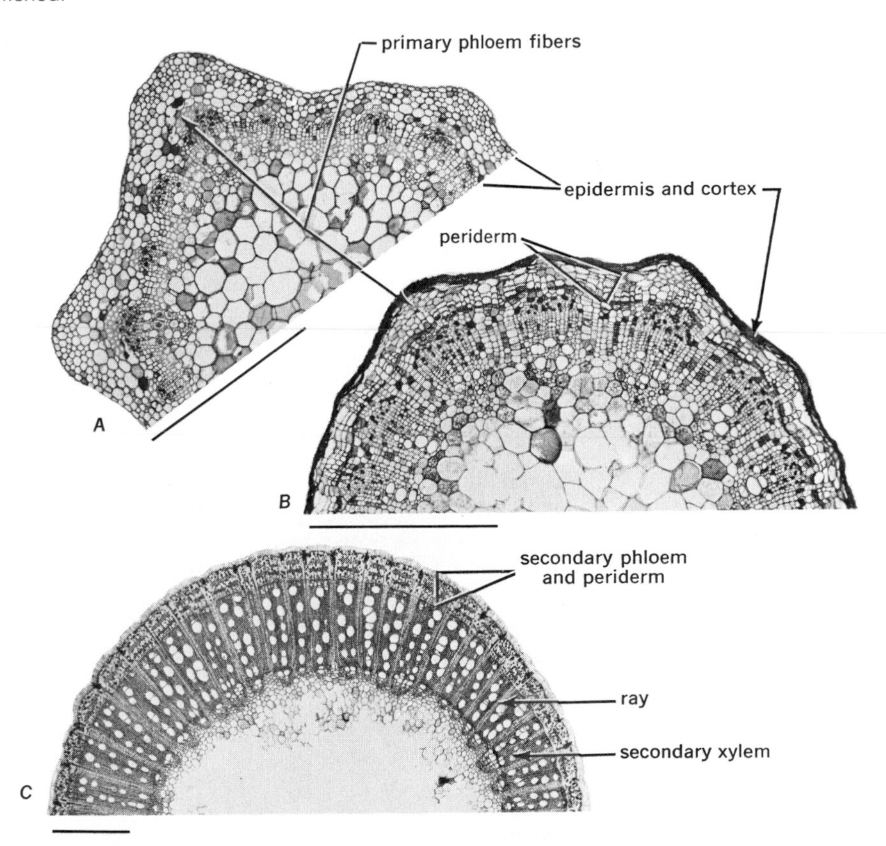

Figure 12.6 Origin of periderm in grapevine (*Vitis vinifera*) as seen in cross sections. *A*, seedling stem without periderm. *B*, older seedling stem with periderm that originated in the primary phloem and caused death and collapse of cortex. Primary phloem fibers also appear outside the periderm. *C*, one-year-old cane with periderm outside the secondary phloem. (Lines are: 0.5 mm in *A*, *B*; 1 mm in *C*. From Esau, *Hilgardia* 18:217–296, 1948.)

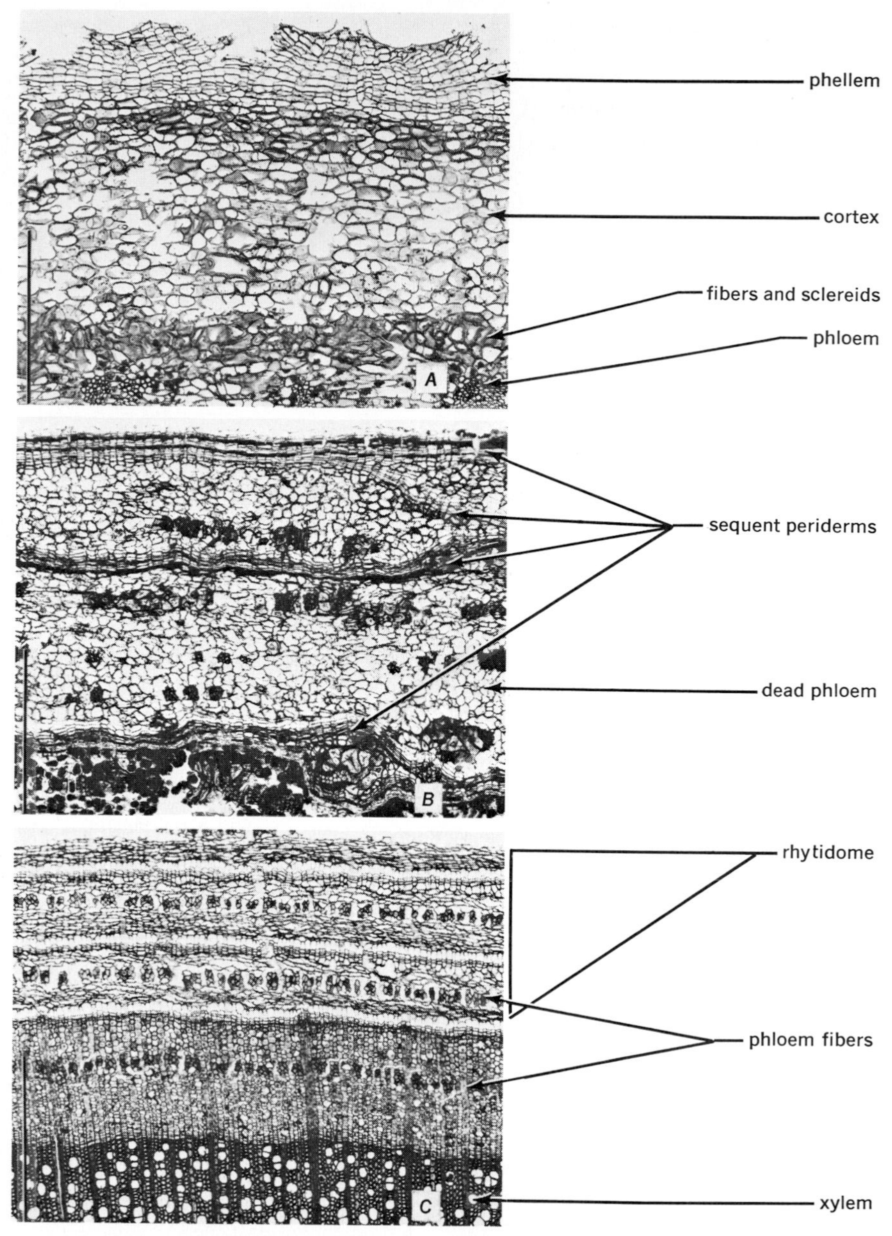

Figure 12.7 Periderm and rhytidome in cross sections of stems. *A*, *Talauma*. Phellem with deep cracks. *B*, *Quercus alba* (oak). Rhytidome with narrow layers of periderm and wide layers of dead phloem tissue. *C*, *Lonicera tartarica*. Rhytidome in which periderm layers alternate with layers derived from secondary phloem containing phloem fibers. (Lines are 0.5 mm.)

monly as discontinuous but overlapping layers (figs. 12.4, 12.7,*B*, and 12.8). These approximately shell-shaped layers originate beneath cracks of overlying periderms. The sequent periderms may also be continuous around the circumference or at least for considerable parts of the circumference (fig. 12.7,*C*).

The phellogen, the first or the sequent, is initiated by divisions of cells of various kinds. Depending on the position of the phellogen these may be cells of the epidermis, subepidermal parenchyma or collenchyma, parenchyma of the pericycle or phloem including that of the phloem rays. Usually these cells are indistinguishable from other cells of the same categories; all are living cells and, therefore, potentially meristematic. The initiating divisions may begin in the presence of chloroplasts and ergastic substances, such as starch and tannins, and while the cells still have thick primary walls, as in collenchyma. Eventually the chloroplasts change into leucoplsts and the ergastic substances and wall thickenings disappear. Sometimes the subepidermal cells, in which phellogen is to arise, have no collenchymatic thickenings and show an orderly and compact arrangement.

The phellogen is initiated by periclinal divisions and produces the phellem and the phelloderm by periclinal divisions (fig. 12.5). The phellogen keeps pace with the increase in circumference of the axis by periodic division of its cells in the radial anticlinal plane (fig. 12.5,*B*).

The sequence of divisions initiating the periderm is somewhat variable, even in plants of the same species growing under different environmental conditions. The phellogen either becomes immediately restricted to a single layer of cells, usually the outer of the two layers formed by the initial periclinal divisions; or there are some preparatory divi-

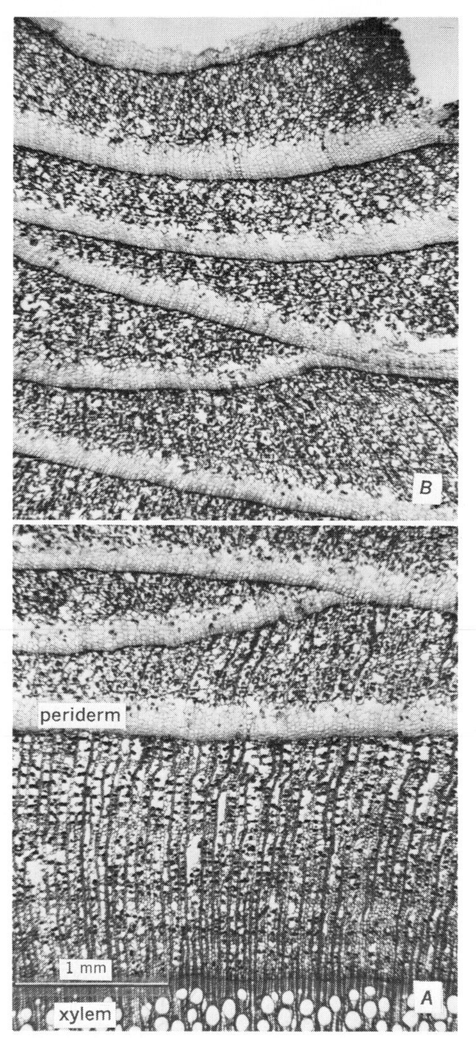

Figure 12.8 Rhytidome in cross section of 11-year-old stem of *Cephalanthus occidentalis* (Rubiaceae). *A*, inner part of bark with functional phloem and three layers of periderm. *B*, outermost part of rhytidome with six layers of periderm alternating with nonconducting secondary phloem layers. Total width of rhytidome was 4.5 mm. (From Esau p. 44 in M. H. Zimmerman, ed. *The formation of wood in forest trees*. Academic Press, New York, 1964.)

sions before the phellogen is defined. The latter sequence is common in roots.

The phellogen of the initial periderm produces most of the cells outward. Consequently the phelloderm is commonly small in amount, sometimes restricted to the single layer of cells left on the inner side of the phellogen after the first periclinal divisions (fig. 12.1,*A*). A relatively wide phelloderm was observed in stems and roots of certain Cucurbitaceae.[7] The deeper seated, sequent periderms also have phelloderm. In general a given phellogen cell produces only a few cork cells yearly. In many species the yearly increments are not discernible; in some, however, the early formed cork cells are wider and have thinner walls than those formed later (*Betula,* fig. 12.2,*C*; *Prunus, Robinia*[24]), and the later cork may have dark contents.[25]

PROTECTIVE TISSUE IN MONOCOTYLEDONS

Monocotyledons rarely produce a periderm similar to that in dicotyledons.[20] In many monocotyledons, the epidermis is permanent and, therefore, the surface layers are not replaced. These layers may become suberized and sclerified, with little or no cell division preceding the change (see *metacutisation* in glossary). In woody monocotyledons, including the palms[23] a special method of development of a protective tissue has been observed. Parenchyma cells in successively deeper positions divide several times periclinally, and the products of these divisions become suberized. Because of its storied appearance in transections, the tissue is called storied cork (fig. 12.9).

WOUND PERIDERM

Wounding induces a series of metabolic events and related cytologic responses that lead, under favorable conditions, to a complete closure of the wound.[12] Wound healing is a developmental process requiring synthesis of DNA and protein.[2]

Natural and wound periderms are basically alike in method of origin and growth and may

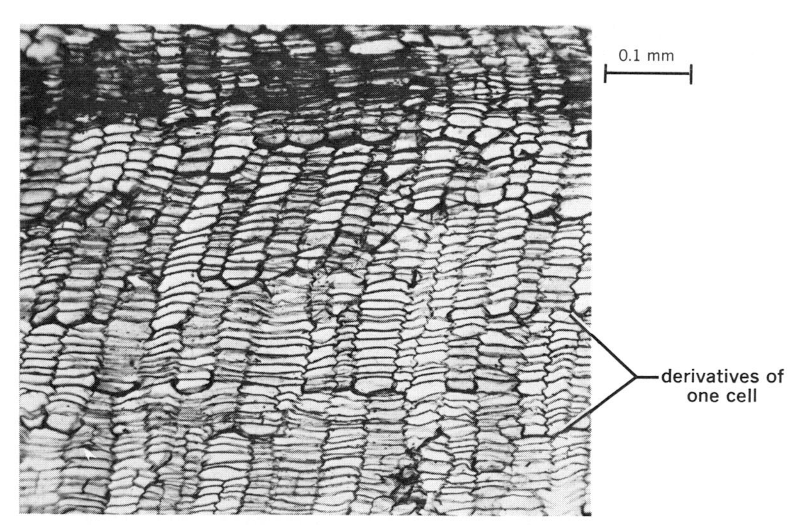

Figure 12.9 Storied cork of *Cordyline terminalis* in cross section. (Slide, courtesy of V. I. Cheadle.)

have the same cellular components.[15] The natural periderm develops beneath a surface which is effectively sealed by the cutinized epidermis. Correspondingly, the formation of wound periderm is preceded by a sealing of the newly exposed surface by scar (*cicatrice*) tissue. This tissue includes dead (*necrosed*) cells on the surface (fig. 12.1,*B*) and living cells beneath, which become suberized and lignified and form the so-called *closing layer.*[8] The wound phellogen arises beneath the closing layer and, when cork develops, the dead scar tissue is carried outward (fig. 12.1,*C*). Successful development of wound periderm is important in horticultural practice when plant parts used for propagation must be cut (e.g., potato tubers, sweet potato roots). Experiments in which wound-healing phenomena in sliced potato tubers were retarded or inhibited by chemical treatment showed the importance of wound periderm in protection from infection by decay organisms.[1] Environmental conditions markedly influence the development of wound periderm.[15] The ability to develop wound periderm in response to invasion by parasites may distinguish resistant from susceptible plants.[21]

Plant taxa vary with regard to the histological aspects of wound healing, as they do in details of natural development of the protective tissue.[8, 22] In general, monocotyledons are less responsive to wounding than the dicotyledons. In dicotyledons and certain monocotyledons (Liliales, Araceae, Pandanaceae) healing includes formation of both, closing layer and wound periderm. In other monocotyledons no wound periderm is detectable. Among these, the Zingiberales produce a slightly suberized closing layer, whereas the Arecaceae and the Poaceae form a lignified closing layer.

A wound reaction occurs when the periderm is peeled off to the living cells underneath. The newly exposed cells die, and a new periderm arises below them. This reaction is utilized in the production of commercial cork from the cork oak. The first cork, which is of inferior quality, is removed to the phellogen and the new phellogen developing beneath the scar tissue produces massive cork of superior quality.

OUTER ASPECT OF BARK IN RELATION TO STRUCTURE

The external features of periderm and rhytidome vary in relation to structure and development of periderm and the kinds of tissues isolated by the periderms. If there is only a superficial periderm with thin cork, the surface is smooth. Massive cork is usually cracked and fissured (fig. 12.7,*A*). When the yearly production of cork occurs in isolated positions, the outer cork layers are sloughed off in these positions so that the surface resembles that produced by some scaly rhytidomes. The stems of some species produce a so-called winged cork, a form that results from a symmetrical longitudinal splitting of the cork in relation to the uneven expansion of different sectors of the stem (*Ulmus* sp.[19]). Winged cork may also result from an initially localized activity of the phellogen (*Euonymus alatus*[4]).

The rhytidome presents a variety of aspects. On the basis of manner of formation, two forms are distinguished, scale bark and ring bark. Scale bark occurs when the sequent periderms develop in restricted overlapping strata, each cutting out a "scale" of tissue (figs. 12.4, 12.7,*B*, and 12.8; *Pinus, Pyrus*). Ring bark is less common and results from the formation of successive periderms approximately concentrically around the axis (*Vitis, Clematis, Lonicera;* fig. 12.7,*C*).

With regard to the nonperidermal tissue enclosed in the rhytidome, the fibrous tissue lends a characteristic aspect to the bark.[11] If fibers are absent, the bark breaks into individual scales or shells (*Pinus, Acer pseudoplatanus*). In fibrous bark, a netlike pattern of splitting occurs (*Tilia, Fraxinus*).

The scaling off of the bark may have different structural bases. If thin-walled cells of cork or of phelloids are present in the periderms of the rhytidome, the scales may exfoliate along these cells (fig. 12.2,*D*). Breaks in the rhytidome can occur also through cells of the nonperidermal tissues. In *Eucalyptus*, breaks occur through phloem parenchyma cells,[5] in *Lonicera tartarica*, between fibers and parenchyma of the phloem (fig. 12.7,*C*). Cork is frequently a strong tissue and renders the bark persistent, even if deep cracks develop (species of *Betula*, fig. 12.2,*C*; *Pinus, Quercus, Robinia, Salix, Sequoia*). Such barks wear off without forming scales.

LENTICELS

A lenticel may be defined as a limited part of the periderm in which the phellogen is more active than elsewhere and produces a tissue that, in contrast to the phellem, has numerous intercellular spaces. The lenticel phellogen itself also has intercellular spaces. Because of this relatively open arrangement of cells, the lenticels are regarded as structures permitting the entry of air through the periderm.

Lenticels are usual components of periderm of stems and roots. Outwardly, a lenticel often appears as a vertically or horizontally elongated mass of loose cells that protrudes above the surface through a fissure in the periderm (fig. 12.10, *B*). Lenticels vary in size from structures barely visible without magnification to those 1 centimeter and more in length. They occur singly or in rows. Vertical rows of lenticels frequently occur opposite the wide vascular rays, but in general there is no constant positional relation between lenticels and rays.

The phellogen of a lenticel is continuous with that of the corky periderm but usually bends inward so that it appears more deeply situated (fig. 12.10). The loose tissue formed by the lenticel phellogen toward the outside is the *complementary* or *filling tissue;*[25] the tissue formed toward the inside is the phelloderm.

The degree of difference between the filling tissue and the neighboring phellem varies in different species. In the gymnosperms the filling tissue is composed of the same types of cells as the phellem. The main difference between the two is that the tissue of the lenticel has intercellular spaces. Lenticel cells may also have thinner walls and be radially elongated instead of radially flattened like the phellem cells of so many species. In lenticels of the potato tuber, scanning microscopy has revealed waxy outgrowths on the cell walls facing the intercellular spaces.[10] This wax may serve in the regulation of water loss from the tuber and in deterring the entry of water, and possibly pathogens, through the lenticels.

In dicotyledons, three structural types of lenticels are recognized.[25] The first and simplest, exemplified by species of *Liriodendron, Magnolia, Malus, Persea* (fig. 12.10,*B*), *Populus, Pyrus,* and *Salix,* has a filling tissue composed of suberized cells. This tissue, though having intercellular spaces, may be more or less compact and may show annual growth layers, with thinner-walled looser tissue appearing earlier, and thicker-walled more compact tissue later.

Lenticels of the second type, as found in species of *Fraxinus, Quercus, Sambucus,* and *Tilia,* consist mainly of a mass of more or less loosely structured nonsuberized filling

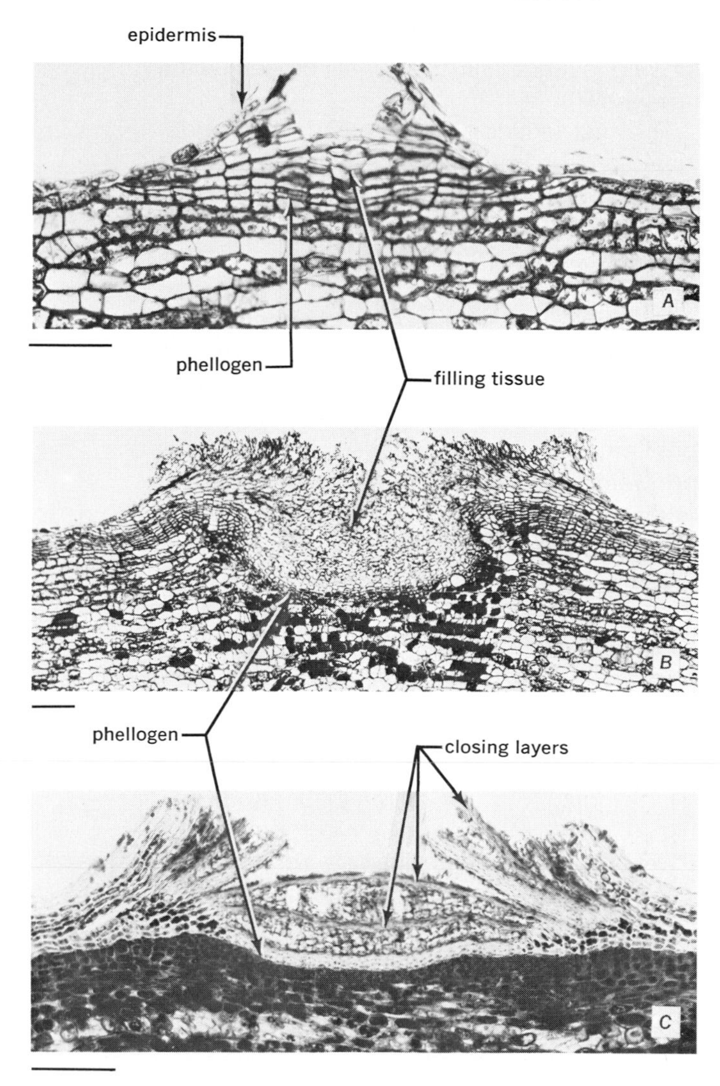

Figure 12.10 Lenticels in cross sections of stems. *A*, *B*, avocado (*Persea americana*). Young lenticel in *A*, older in *B*. No closing layers are present. *C*, beech (*Fagus grandifolia*). Lenticel with closing layers. (Lines are 0.1 mm.)

tissue, succeeded at the end of the season by a more compact layer of suberized cells.

The third type, illustrated by lenticels of species of *Betula*, *Fagus* (fig. 12.10,*C*), *Prunus*, and *Robinia*, shows the highest degree of specialization. The filling tissue is layered because loose nonsuberized tissue regularly alternates with compact suberized tissue. The compact tissue forms the closing layers, each one to several cells in depth, that hold together the loose tissue, usually in layers several cells deep. Several strata of each kind of tissue are produced yearly. The closing layers are successively broken by the new growth.

The first lenticels frequently appear under stomata. The parenchyma cells beneath a stoma undergo some preparatory divisions; then a phellogen pushes the overlying cells outward and ruptures the epidermis (fig. 12.10,*A*).

Lenticels are maintained in the periderm as long as the periderm continues to grow, and new ones arise from time to time by change in the activity of the phellogen from formation of phellem to that of lenticel tissue. The deeper periderms also have lenticels, which usually appear at bottom of cracks in the rhytidome. The lenticels of the rhytidome are basically similar to those of the initial periderm, but their phellogen is less active, and, therefore, they are not as well differentiated.

REFERENCES

1. Audia, W. V., W. L. Smith, Jr., and C. C. Craft. Effects of isopropyl N-(3-chlorophenyl) carbamate on suberin, periderm, and decay development by Katahdin potato slices. *Bot. Gaz.* 123:255–258. 1962.
2. Borchert, R., and J. D. McChesney. Time course and localization of DNA synthesis during wound healing of potato tuber tissue. *Devel. Biol.* 35:293–301. 1973.
3. Borger, G. A., and T. T. Kozlowski. Effect of water deficits on first periderm and xylem development in *Fraxinus pennsylvanica. Can. J. For. Res.* 2:144–151. 1972. Effects of light intensity on early periderm and xylem development in *Pinus resinosa, Fraxinus pennsylvanica,* and *Robinia pseudoacacia. Can. J. For. Res.* 2:190–197. 1972. Effects of temperature on first periderm and xylem development in *Fraxinus pennsylvanica, Robinia pseudoacacia,* and *Ailanthus altissima. Can. J. For. Res.* 2:198–205. 1972.
4. Bowen, W. R. Origin and development of winged cork in *Euonymus alatus. Bot. Gaz.* 124:256–261. 1963.
5. Chattaway, M. M. The anatomy of bark. I. The genus *Eucalyptus. Aust. Jour. Bot.* 1:402–433. 1953.
6. De Zeeuw, C. Influence of exposure on the time of deep cork formation in three northeastern trees. *New York State Col. Forestry, Syracuse Univ. Bul.* 56. 1941.
7. Dittmer, H. J., and M. L. Roser. The periderm of certain members of the Cucurbitaceae. *Southwest Nat.* 8:1–9. 1963.
8. El Hadidi, M. N. Observations on the wound-healing process in some flowering plants. *Mikroskopie* 25:54–69. 1969.
9. Falk, H., and M. N. El Hadidi. Der Feinbau der Suberinschichten verkorkter Zellwände. *Z. Naturforsch.* 16b:134–137. 1961.
10. Hayward. P. Waxy structures in the lenticels of potato tubers and their possible effects on gas exchange. *Planta* 120:273–277. 1973.
11. Holdheide, W. Anatomie mitteleuropäischer Gehölzrinden. *Handbuch der Mikroskopie in der Technik.* Band 5. Teil 1:195–367. 1951.
12. Lipetz, J. Wound healing in higher plants. *Internatl. Rev. Cytol.* 27:1–28. 1970.
13. Luhan, M. Das Abschlussgewebe der Wurzeln unserer Alpenpflanzen. *Ber. Deut. Bot. Ges.* 68:87–92. 1955.

14. Mogensen, H. L. Studies on the bark of the cork bark fir: *Abies lasiocarpa* var. *arizonica* (Merriam) Lemmon. I. Periderm ontogeny. *Ariz. Acad. Sci. J.* 5:36–40. 1968. II. The effect of exposure on the time of initial rhytidome formation. *Ariz. Acad. Sci. J.* 5:108–109. 1968.

15. Morris, L. L., and L. K. Mann. Wound healing, keeping quality, and compositional changes during curing and storage of sweet potatoes. *Hilgardia* 24:143–183. 1955.

16. Moss, E. H., and A. L. Gorham. Interxylary cork and fission of stems and roots. *Phytomorphology* 3:285–294. 1953.

17. Mullick, D. B., and G. D. Jensen. New concepts and terminology of coniferous periderms: necrophylactic and exophylactic periderms. *Can. J. Bot.* 51: 1459–1470. 1973.

18. Nelson, P. E., and S. Wilhelm. Some aspects of the strawberry root. *Hilgardia* 26:631–642. 1957.

19. Smithson, E. Development of winged cork in *Ulmus* x *hollandica* Mill. *Proc. Leeds Philos. Lit. Soc., Sci. Sec.* 6:211–220. 1954.

20. Solereder, H., and F. J. Meyer. *Systematische Anatomie der Monkotyledonen.* Heft III. Berlin, Gebrüder Borntraeger. 1928.

21. Struckmeyer, B. E., and A. J. Riker. Wound periderm formation in white-pine trees resistant to blister rust. *Phytopathology* 41:276–281. 1951.

22. Swamy, B. G. L., and D. Sivaramakrishna. Wound healing responses in monocotyledons. I. Responses in vivo. *Phytomorphology* 22:305–324. 1972.

23. Tomlinson, P. B. *Anatomy of the monocotyledons.* II. *Palmae.* Oxford, Clarenden Press. 1961.

24. Waisel, Y., N. Liphschitz, and T. Arzee. Phellogen activity in *Robinia pseudoacacia* L. *New Phytol.* 66:331–335. 1967.

25. Wutz, A. Anatomische Untersuchungen über System and periodische Veränderungen der Lenticellen. *Bot. Stud.* No. 4:43–72. 1955.

13

Secretory Structures

Secretion refers to the complex phenomena of separation of substances from the protoplast or their isolation in parts of the protoplast. The secreted substances may be surplus ions, which are removed in the form of salts; surplus assimilates eliminated as sugars or as cell wall substances; compounds that may or may not be end products of metabolism but are not utilizable or only partially utilizable physiologically (alkaloids, tannins, terpenes, resins, various crystals); substances that have a special physiologic function after they are secreted (enzymes, hormones). The removal of substances that no longer participate in the metabolism of a cell is sometimes referred to as excretion. In the plant, however, no sharp distinction can be made between excretion and secretion.[35] The same cells accumulate various substances, some being waste products, others materials that are again utilized. Furthermore, the exact role of many of the secreted substances, perhaps of most, is not known. In this book the term secretion is used broadly to include synthesis, segregation, and release of materials that are either functionally specialized or destined for storage or for excretion. Secretion covers both removal of materials from the cell (either to the surface of the plant or into internal spaces) and accumulation of secreted materials in some compartment of the cell.

Discussions of secretion phenomena in plants usually emphasize activities of specialized secretory structures such as glandular hairs, nectaries, resin canals, laticifers, and others. In reality, secretory activities occur in all living cells as part of the normal metabolism. Secretion characterizes various steps in the accumulation of temporary deposits in organelles and vacuoles; in mobilization of enzymes involved in synthesis and breakdown of cellular components; in interchange of materials between organelles; and in phenomena of transport between cells. The ubiquity of secretory processes in the living plant must not be lost sight of when the specialized secretory structures are studied.

The visibly differentiated secretory structures occur in many forms and vary with

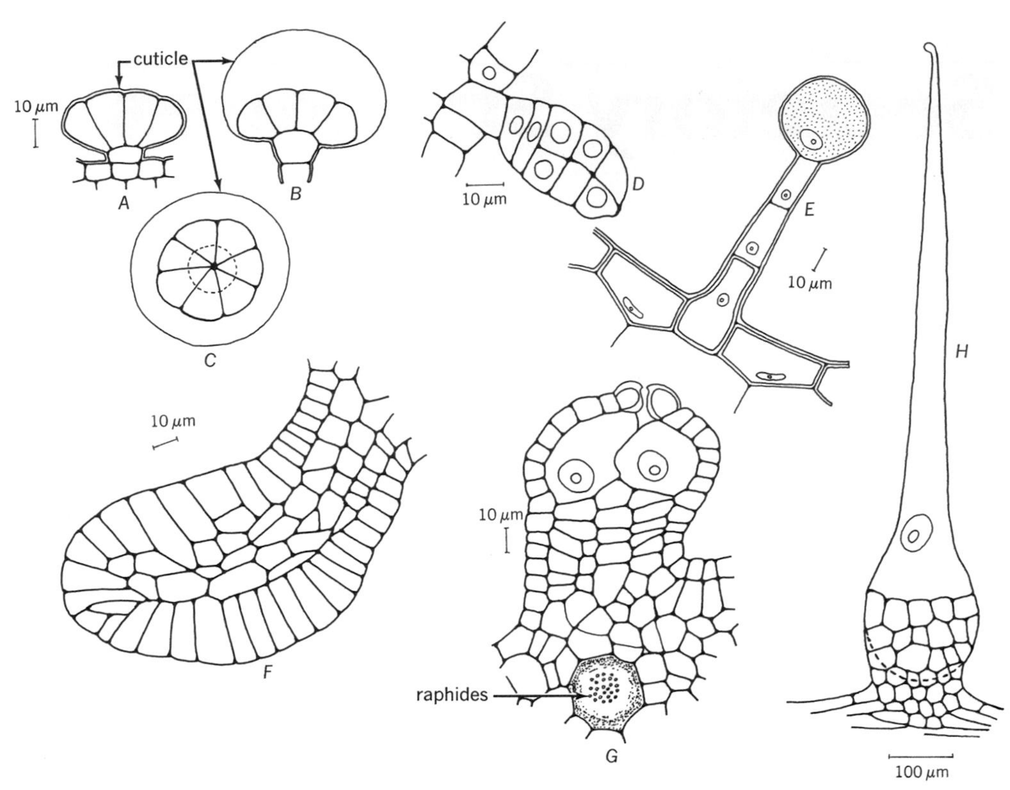

Figure 13.1 Secretory trichomes. *A–C*, glandular hairs from leaf of lavender (*Lavandula vera*) with cuticle undistended (*A*) and distended (*B, C*) by accumulation of secretion. *D*, glandular hair from leaf of cotton (*Gossypium*). *E*, glandular hair with unicellular head from stem of *Pelargonium*. *F*, colleter from young leaf of *Pyrus*. *G*, pearl gland from leaf of grapevine (*Vitis vinifera*). *H*, stinging hair of nettle (*Urtica urens*).

regard to the kind of substances they secrete. Highly differentiated secretory structures consisting of many cells are referred to as glands (fig. 13.1,*G*), the simpler ones are qualified as glandular, such as glandular hairs (fig. 13.1,*A*,*D*,*E*), glandular epidermis, or glandular cells. The distinction is vague, however, and a variety of secretory structures, large and small, the hairlike and the more elaborate ones, are often called glands.

Glands may be highly specific in their activities as is indicated by the predominance of one compound or one group of compounds in the material exported by a given gland.[21] Some glandular structures secrete hydrophilic substances (hydathodes, mucilage glands, nectaries, salt glands), others release lipophilic substances (oil glands, epithelial cells of resin ducts). Ultrastructural studies have revealed some cytologic differences between the two kinds of glands. Secretion of hydrophilic substances occurs in the presence of abundant mitochondria and voluminous complements of endoplasmic reticulum or dictyosomes. In oil glands, a progressive degeneration of the initially dense protoplasts appears to be associated with the secretion.[20] Dictyosomes, which function so prominently in the secretion of polysaccharides, pectic materials, and mucilages,[28] are apparently not involved in the secretion of terpenoids.[39]

The methods of elimination of secretions from cells have been only partially determined. If dictyosome vesicles are the carriers of secretions, the latter are released toward the outside of the protoplasts through fusion of the vesicles with the plasmalemma (*granulocrinous secretion*). Passage through the wall presumably follows. In many glands the outward movement through the wall (apoplastic transfer) is assured by cutinization of radial walls in an endodermoid layer of cells located beneath the secretory cells.[34] A direct passage through plasmalemma and wall is visualized when the secreted molecules are small (*eccrinous secretion*). This passage is passive if it is controlled by concentration gradients, active if it requires metabolic energy. Cells secreting hydrophilic substances, for example those in glands secreting salts or carbohydrates, may be differentiated as transfer cells characterized by wall protuberances.[29] Ethereal oils frequently accumulate between the cell wall and the cuticle in secretory trichomes (fig. 13.1,*A–C*). Later, the cuticle may be ruptured thus completing the release of the oil. Specific examples of secretory structures are, in the following sections, grouped into those that are found on the surface of the plant and those that are embedded in various tissues.

EXTERNAL SECRETORY STRUCTURES

Trichomes and glands

The external secretory structures vary in complexity. Sometimes simply a part of the epidermis is secretory, sometimes the secretory cells are components of appendages—glandular trichomes and glands—derived from the epidermis or from both epidermis and subepidermal layers. Trichomes often have a unicellular or multicellular head composed of cells producing the secretion and borne on a stalk of nonglandular cells (fig. 13.1,*A–E*).

In the oil secreting trichomes of *Mentha*,[1] oil appears as droplets of osmiophilic material in the cytoplasm. In the multicellular trichomes of *Dictamnus* oil has been detected in plastids.[2] These plastids and, later, the entire cells in the center of the trichome disintegrate and leave the oil in the resulting cavity. The oil is released to the surface when the cutinized beaklike extension at the upper end of the trichome is broken.

A somewhat similar mechanism of release of contents (muscle-stimulating substances[40]) is found in the stinging hair of nettle (*Urtica urens;* fig. 13.1,*H*). The bladderlike end of the hair is embedded in epidermal cells, which are raised above the surface. The upper part of the hair resembles a fine capillary tube closed at the top by means of a spherical tip. When the hair comes in contact with the skin, the spherical tip breaks off along a predetermined line and leaves a sharp edge. This edge readily penetrates the skin, and pressure upon the bulbous part forces the liquid into the wound. In the pearl gland of Vitaceae, the epidermis covering the large secretory cells has an open stoma permitting the release of the secretion to the outside (fig. 13.1,*G*).

The trichomes on the leaves of insectivorous plants secrete mucopolysaccharides that trap insects and proteolytic enzymes that digest the insects. In *Pinguicula,* for example, stalked glands produce mucous material, and sessile glands yield proteolytic enzymes. Fluid containing the enzymes is secreted upon the leaf surface on stimulation by suitable nitrogenous material trapped on this surface. Autoradiographic studies—use of ^{14}C-

labeled protein—indicate that the sessile glands also serve to resorb the digested food.[16] A variety of other adaptations for trapping and digesting insects by leaves of certain taxa are characterized by specialized secretory trichomes.[5]

Salt-excreting formations vary in structure and in methods of salt release. In species of saltbush (*Atriplex*), part of the ions carried in the transpiration stream are eventually delivered through the cytoplasm and plasmodesmata to living bladderlike cells attached to the epidermis (fig. 13.2 and chapter 19). In the bladder cell the ions are secreted into the large central vacuole (short arrows in fig. 13.2). After the trichomes collapse salt is deposited on the surface of the leaf. Since there is a considerable positive gradient of salt concentration from the mesophyll to the bladder cells, the delivery of ions into the bladder vacuoles is an energy consuming process.[21]

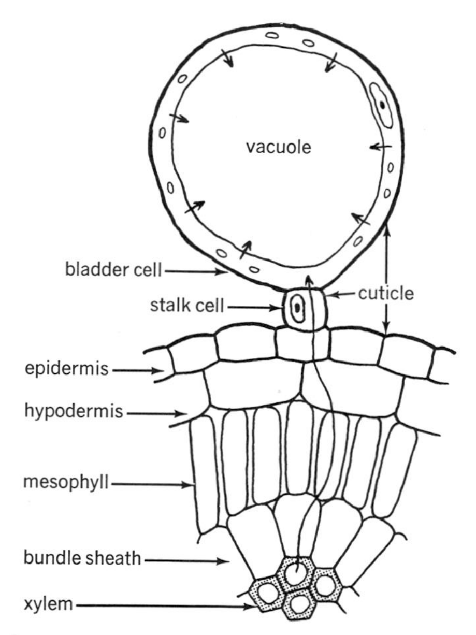

Figure 13.2 Diagram of salt-secreting trichome shown attached to part of leaf of *Atriplex* (saltbush). The long arrow indicates the path of movement of ions from the xylem of a vein to the bladder cell of the trichome. Short arrows indicate release of ions into the vacuole. (For relation of trichomes to the leaf as a whole see chapter 19.)

An example of a gland secreting salt directly to the outside is found in the tamarisk (*Tamarix aphylla*). The gland is a complex of eight cells, six of which are secretory and two are basal collecting cells (fig. 13.3). The group of secretory cells is enclosed in a cuticular layer except where the lowermost secretory cells are connected by plasmodesmata with the collecting cells. The process of salt secretion is visualized as follows.[37] Salt enters the gland through plasmodesmata between the collecting cells and the mesophyll and moves from cell to cell in the gland also through plasmodesmata. It accumulates in microvacuoles which eventually fuse with the plasmalemma lining the wall and wall protuberances. The salt is released into the pectin-rich wall and moves toward the top of the gland where it exits through pores in the surface layer. A backflow of salt through the walls into the mesophyll is prevented by the cuticular layer.

Salt glands are considered to be desalination devices that appear to maintain the salt balance in leaves by secreting excess salt. Accordingly, the composition of secreted salt depends on the salt composition of the root environment.

In many woody plants (*Aesculus, Betula, Carya, Malus*) glandular hairs and more complex appendages called colleters (fig. 13.1,*F*) develop on young leaf primordia and produce a sticky secretion that both permeates and covers the entire bud. When the bud opens and the leaves expand, the glandular appendages commonly dry up and fall off. They seem to provide a protective coating for the dormant buds.

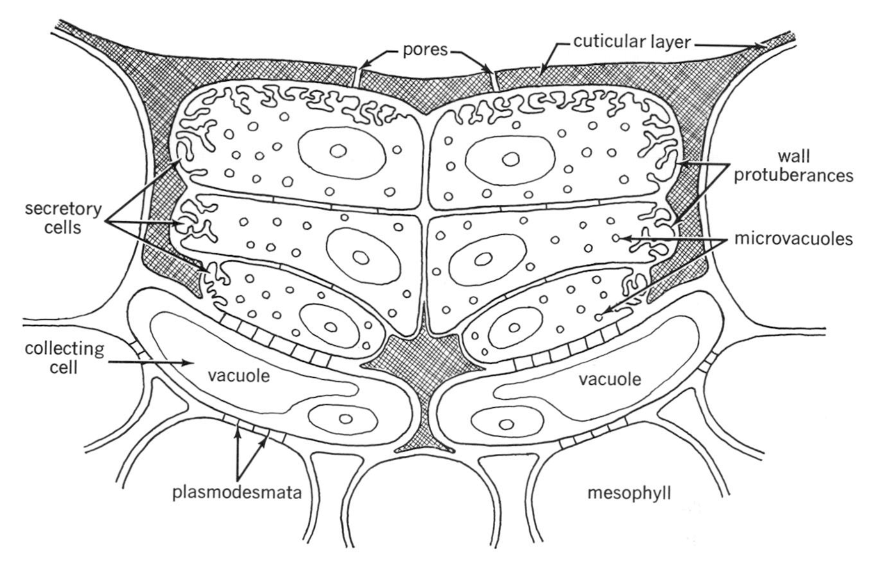

Figure 13.3 Diagram of a salt-secreting gland of *Tamarix aphylla* (tamarisk). The complex of eight cells, six of which are secretory and two are the so-called collecting cells, is embedded in the epidermis and is in contact with the mesophyll (below). Cuticle and cutinized wall are indicated jointly (cuticular layer) by cross hatching. (Constructed from data in Thomson, Berry, and Liu[37])

Nectaries

In the salt glands, the principal source of secreted solutes is the transpiration stream; in nectaries, both phloem and xylem contribute to the secreted fluid. The nectaries secrete a sugar-containing liquid. They occur on flowers (floral nectaries) and on vegetative parts of the plant (extrafloral nectaries). The nectaries may have the form of glandular surfaces (fig. 13.4,*D*,*E*), or they may be differentiated into specialized structures (fig. 13.4,*A–C*,*F–H*). The floral nectaries occupy various positions on the flower; they are found on sepals, petals, stamens, ovaries, or the receptacle.[8,9,18] The extrafloral nectaries occur on stems, leaves, stipules, and pedicels of flowers.

The secretory tissue of a nectary may be restricted to the epidermis, or it may be several layers of cells deep (fig. 13.4). The secretory tissue is covered on the outside with a cuticle. Vascular tissue occurs more or less close to the secretory tissue. Sometimes this vascular tissue is merely a trace to some other part of the flower (fig. 13.4,*B*,*D*), but some nectaries have their own vascular bundles (fig. 13.4,*G*), often consisting of phloem only.[12,18] A close relation exists between the relative amount of phloem in the vascular tissue supplying the nectary and the concentration of sugar in the nectar. If phloem predominates, the nectar may have up to 50 percent sugar; at the opposite extreme—xylem predominating in the vascular supply—the content of sugar may fall to as low as 8 percent.[13] Nectaries do not simply release the sugar as it is derived from the phloem but variously transform it by means of enzymes.[35] Nectaries with vascular supply in which

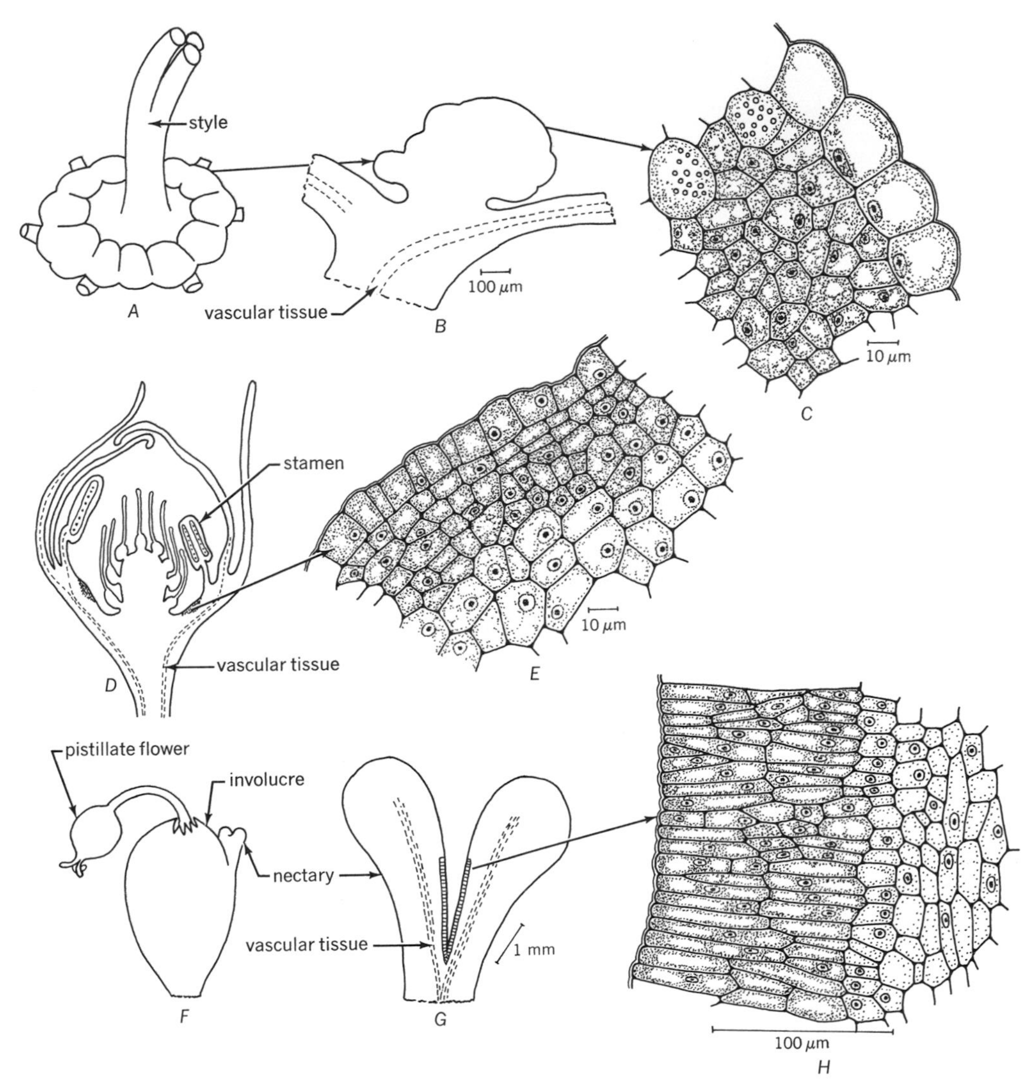

Figure 13.4 Floral nectaries. *A–C*, *Ceanothus*. Nectary is a lobed disc inserted at base of gynoecium (*A*). *D,E*, strawberry (*Fragaria*). Nectary tissue lines floral tube beneath stamens (*D*). *F–H*, poinsettia (*Euphorbia pulcherrima*). Lobed nectary (*G*) is attached to involucre investing inflorescence.

xylem predominates intergrade physiologically with the hydathodes.

The secretory cells in nectaries have dense cytoplasm and small vacuoles, which often contain tannins. Numerous mitochondria with well developed cristae indicate that the cells respire intensively. The endoplasmic reticulum is abundant and may be stacked or convoluted.[7] This ER reaches its maximum volume at the stage of nectar secretion. In some nectaries (*Lonicera japonica*[10]), vesicles derived from ER, rather than those re-

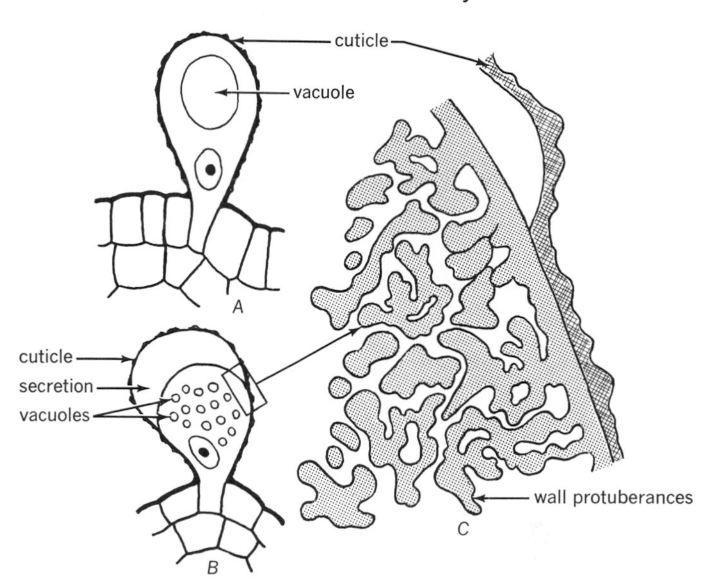

Figure 13.5 Details of nectary of *Lonicera japonica*. *A*, *B*, nectar-secreting hairs from inner epidermis of corolla tube shown before secretion (*A*) and during secretion (*B*). *C*, part of cell wall with protuberances that characterize an actively secreting hair. The black line outlining the protuberances indicates the plasmalemma. The apparently detached protuberances are to be thought of as connected to the wall at levels other than that of the drawing. (Constructed from data in Fahn and Rachmilevitz.[10])

leased by dictyosomes, are thought to be concerned with the secretion of sugar.

In the nectary of *Lonicera,* the nectar-secreting cells are short hairs located in a limited area of the inner epidermis of the corolla tube. Young hairs have a single large vacuole and a tightly fitting cuticle (fig. 13.5,*A*). In actively secreting hairs, the vacuolar volume is reduced, small vacuoles replace the single large one, and the cuticle is detached (fig. 13.5,*B*). At this stage, the wall of the hairs bears numerous protuberances covered with the plasmalemma (fig. 13.5,*C*).

Hydathodes

Hydathodes discharge water from the interior of the leaf to its surface. This process is called guttation. Structurally, hydathodes are modified parts of leaves, usually located at leaf margins or tips, in which water released from the xylem is enabled to reach the surface of the leaf. It does so by passing through a more or less modified mesophyll, the *epithem,* and leaving the leaf through openings in the epidermis. These openings are stomata (fig. 13.6) that are incapable of closing and opening movements. The epithem is frequently a chlorophyll-free parenchyma, which may be compact[33] or may have prominent intercellular spaces. Epithem cells are also known to differentiate as transfer cells provided with wall ingrowths.[30] The xylem of the hydathodes consists of tracheids (fig. 13.6) and may be enlarged into nodulelike complexes.

Some hydathodes appear to release the water passively under the control of root pressure. Others are true glands, with the epithem

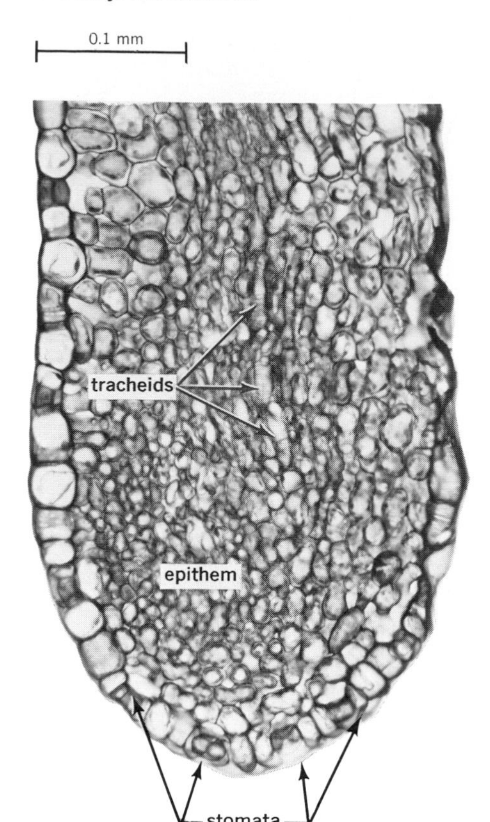

Figure 13.6 Hydathode from cabbage leaf. Section shows a strand of tracheids ending at a small-celled tissue, the epithem. Guttation water passes through intercellular spaces of epithem and is released through stomata. (From Esau, *Plant Anatomy,* 2nd ed. John Wiley & Sons, 1965.)

actively secreting the water. Hydathodes may be differentiated also as glandular trichomes.[15]

Certain experiments suggest that guttation serves for mineral nutrition when transpiration is suppressed, solutes being absorbed from the guttation water as it moves upward in the plant,[19] possibly with the aid of transfer cells in the epithem.[30] Guttation products may also cause injury to plants through accumulation and concentration or through interaction with pesticides.[17]

INTERNAL SECRETORY STRUCTURES

Secretory cells

Internal secretory cells have a wide variety of contents. The secretory cells often appear as specialized cells dispersed among other, less specialized cells. They are then called idioblasts, more specifically excretory idioblasts[11] if their contents seem to be waste products. The secretory cells may be much enlarged, especially in length, and are then called sacs or tubes. The secretory cells are usually classified on the basis of their contents, but many secretory cells contain mixtures of substances, and in many the contents have not been identified. Nevertheless, the secretory cells, as well as the secretory cavities and canals, are useful for diagnostic purposes in taxonomic work.[26]

Some plant families, as, for example, Calycanthaceae, Lauraceae, Magnoliaceae, Simaroubaceae, and Winteraceae, have secretory cells with oily contents. These cells appear like enlarged parenchyma cells (fig. 13.7,*A*) and are known to occur in vascular and ground tissues of stem and leaf. Cells similar in appearance to the oil cells but with unspecified contents occur in many other families and are often referred to as oil cells (Clusiaceae, Hypericaceae, Rutaceae, Tetracentraceae, Trochodendraceae). Some dicotyledon families contain resiniferous (Meliaceae), others mucilaginous cells (Cactaceae, Lauraceae, Magnoliaceae, Malvaceae, Tiliaceae). Mucilage cells often include raphide crystals (fig. 13.7,*B*). Cells containing the enzyme myrosinase have been identified in such families as Capparidaceae, Brassicaceae, and Resedaceae. The myrosin cells may be elongated and even branched.

In oil cells the formation, or at least the accumulation, of lipoidal secretions occurs in

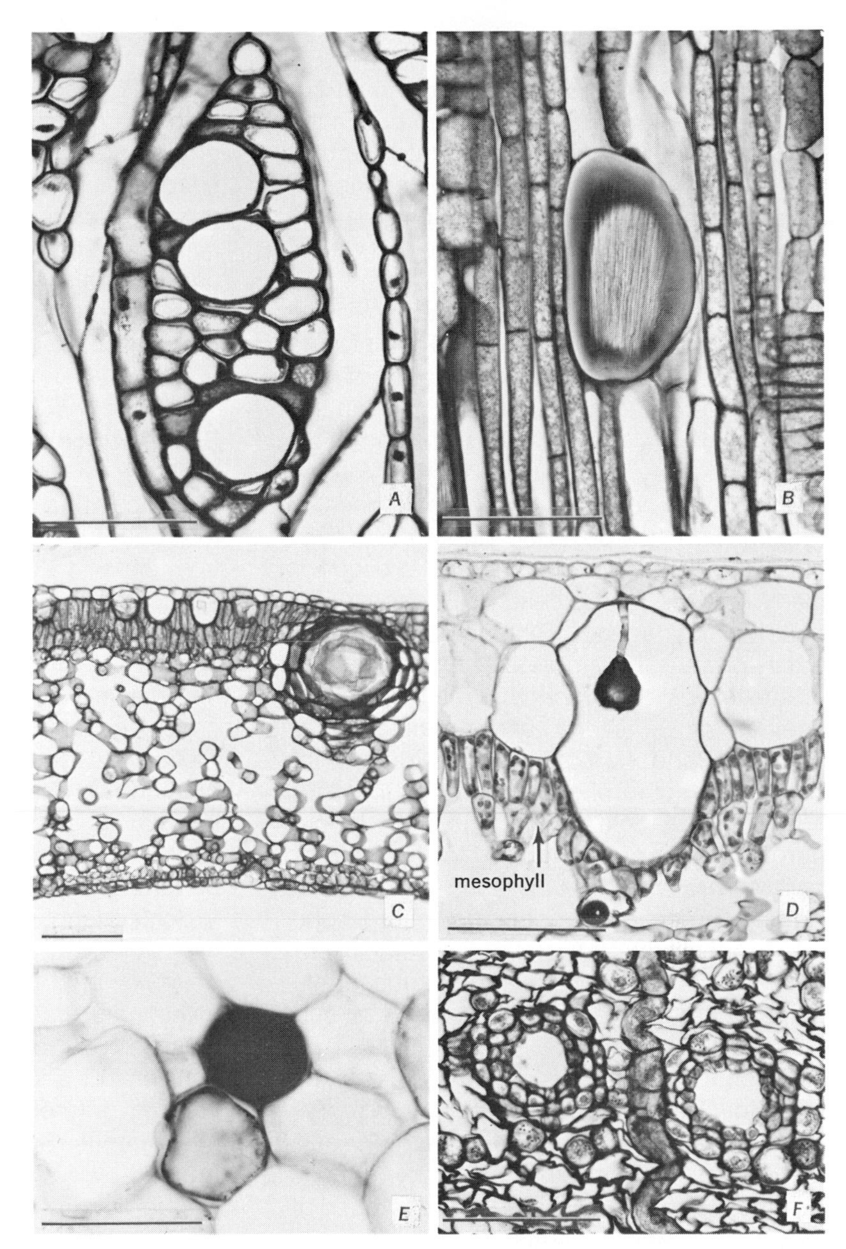

Figure 13.7 Various internal secretory structures. *A*, tulip tree (*Liriodendron*). Oil cells in phloem ray cut tangentially. *B*, *Hydrangea paniculata.* Idioblast containing mucilage and raphides in a radial section of phloem. *C*, lemon (*Citrus*). Lysigenous oil cavity in upper part of leaf to the right. *D*, rubber plant (*Ficus elastica*). Enlarged epidermal cell contains a cystolith—calcium carbonate precipitation on a stalk of cellulose. The cell is part of multiple epidermis of leaf (all cells above mesophyll in *D*). Cystolith is cut off median and does not show base of stalk. *E*, elderberry (*Sambucus*). Tannin sacs in pith of stem in cross section. *F*, *Rhus typhina.* Schizogenous secretory canals in cross section of nonfunctioning phloem. (Lines are 0.1 mm.)

the thylakoids of plastids.[35] Later, the oil appears as droplets in the cytoplasm but, in the end, all cell components degenerate. In many oil cells, the secretion becomes confined to oil sacs having their own cellulose wall attached by a stalklike extension to the wall of the cell (*Persea,* avocado[36]).

In numerous secretory cells tannin is the most conspicuous inclusion. Tannin is a common ergastic substance in parenchyma cells (chapter 3), but some cells contain this material in great abundance, and, in addition, such cells may be conspicuously enlarged. The tannin cells often form connected systems and may be associated with vascular bundles. Tanniniferous idioblasts occur in many families (Crassulaceae, Ericaceae, Fabaceae, Myrtaceae, Rosaceae, Vitaceae). Easily procured examples are the tannin cells in the leaves of *Sempervivum tectorum* and species of *Echeveria* and tubelike tannin cells, one and more centimeters long, in the pith and phloem of stems of *Sambucus* (fig. 13.7,*E*). The tannin compounds in the tanniniferous cells are oxidized to brown and reddish-brown phlobaphenes which are readily perceived under the microscope. Cells in the ground tissue of the fruit of *Ceratonia siliqua* contain solid tannoids, inclusions of tannins combined with other substances.

Certain investigators include among the excretory idioblasts cells containing crystals[11] (chapter 3). Crystal-containing cells may not differ from other parenchyma cells, but they also may be more or less specialized in form and contents. Striking examples of such specializations are the cystolith-containing cells in *Ficus elastica* leaves and the raphide cells. The cystoliths are structures combining wall material, including cellulose and callose, with calcium carbonate. In *Ficus elastica* the cystoliths occur singly in epidermal cells, and each is attached by means of a cellulose stalk to the outer epidermal wall (fig. 13.7,*D*). The raphides are often found in long saclike cells filled with mucilage. In the secondary vascular tissues a cell forming crystals becomes subdivided into small cells, each depositing one crystal. In another modification the crystal is walled off by cellulose from the living part of the protoplast.

Secretory cavities and canals

The cavities and canals differ from secretory cells in that they are spaces resulting from either a dissolution of cells (lysigenous spaces) or a separation of cells from one another (schizogenous spaces). Lysigeny and schizogeny may be combined in the formation of secretory spaces. In the lysigenous secretory cavities (*Citrus, Gossypium*) the secretion is formed in cells that eventually break down and release the substances into the cavity resulting from the breakdown. Partly disintegrated cells occur along the periphery of the cavity (fig. 13.7,*C*).

Schizogenous secretory cavities are usually lined with intact cells (fig. 13.7,*F*). The development of a schizogenous oil gland, as investigated in the embryo of *Eucalyptus,*[4] is illustrated in figure 13.8. The gland arises by divisions of a single epidermal cell and becomes differentiated into epithelial and casing cells. Some of the latter may be contributed by a subepidermal cell (fig. 13.8,*B*). During germination of the seed, the epithelial cells separate from one another in the center of the gland and secrete the oil into the intercellular space.

The contents of secretory cavities and canals are often characterized as oily, although their chemistry is not known exactly.[26] An example of lysigenous mucilaginous canals is found in the bud scales of

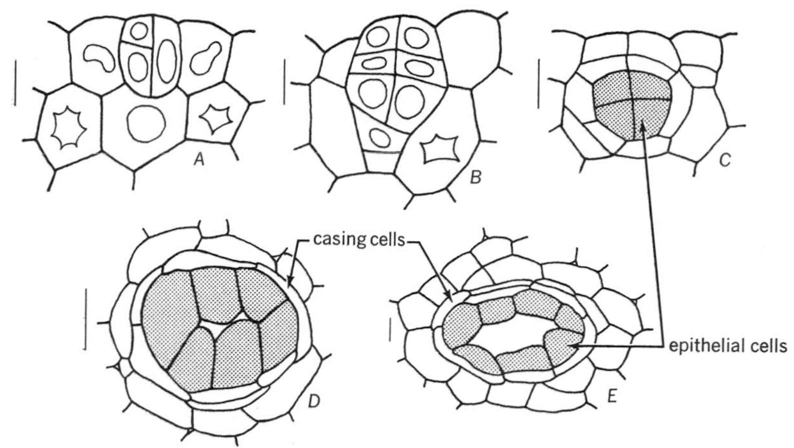

Figure 13.8 Development of epidermal oil glands in the embryo of *Eucalyptus* shown in longitudinal (*A–C*) and transverse (*D*, *E*) sections. *A*, *B*, two stages in division of a gland initial and its derivatives. *C*, after completion of divisions: secretory cells (stippled) are surrounded by casing cells. *D*, schizogenous formation of cavity between secretory cells. *E*, mature gland with secretory cells forming the epithelium around the oil cavity. (From photomicrographs in Carr and Carr.[4] Lines are 10 μm.)

Tilia cordata. Schizogenous canals with resiniferous contents occur in the Asteraceae and with unknown contents in the Apiaceae. The copal-yielding secretory canals of certain tropical Fabales also arise as schizogenous spaces.[27]

The best-known schizogenous canals are the resin ducts of the conifers. Similar ducts in the dicotyledons are arbitrarily called gum ducts. The conifer resin ducts occur in vascular tissues (chapter 9) and ground tissues of all plant organs and are, structurally, long intercellular spaces lined with resin-producing epithelial cells.[42]

As seen with the electron microscope,[43] epithelial cells in the pine contain plastids with scanty thylakoids and sheathed by endoplasmic reticulum cisternae. Osmiophilic droplets have been observed in the stroma of plastids, in the plastid envelope, within the endoplasmic reticulum cisternae near the plastids, and on both sides of the plasmalemma. The material in the droplets resembles the resin in the duct itself. The distribution of the droplets in the cell possibly indicates the site of formation and the manner of migration of the resin from the cell into the duct.

Secretory cavities and canals resulting from normal development may be difficult to distinguish from canals and cavities arising under the stimulus of injury. Resin and gum ducts and pockets are frequently traumatic formations in the wood of gymnosperms and dicotyledons, but their development and contents may parallel those constituting normal features of these woods.

Laticifers

Laticifers are cells or series of connected cells that contain latex, a fluid of complex composition. In common with other secretory structures, the laticifers are depositories of substances some of which may be classified

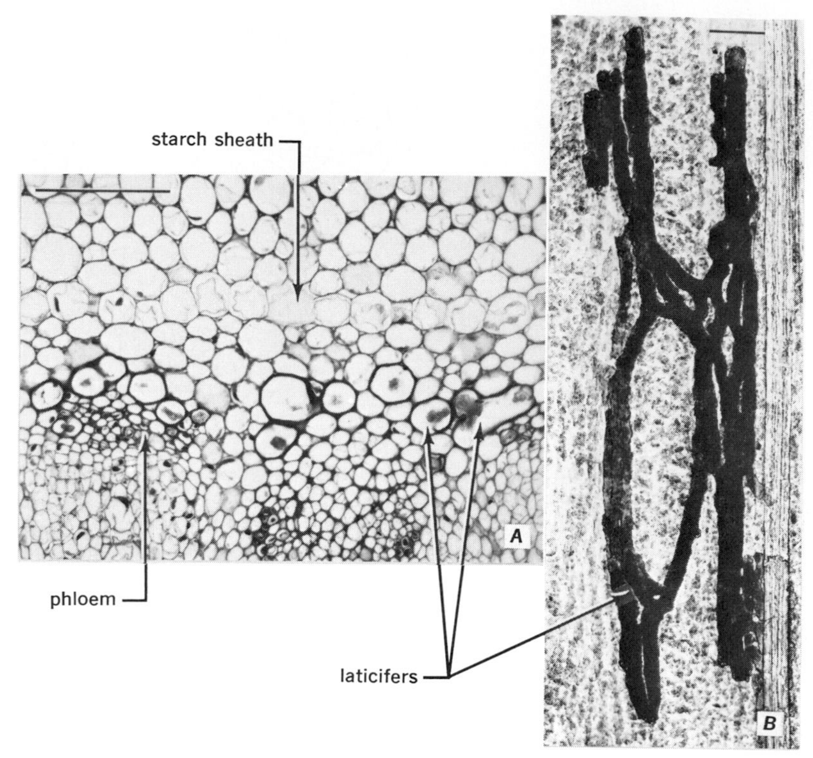

Figure 13.9 Articulated anastomosing laticifers of *Lactuca scariola*. *A*, cross section of stem. Laticifers are outside the phloem. *B*, longitudinal view of laticifers in partly macerated tissue from stem. (Lines are 0.1 mm.)

as excretions (terpenes, resins) and others as secretions in the narrow sense (enzymes).

The laticifers may be simple or compound by origin. The simple laticifers are single cells; the compound laticifers are derived from series of cells. In a more highly specialized state the series of cells in a compound laticifer become united by dissolution of intervening walls. Because of this junction of cells the compound laticifers are commonly called *articulated* laticifers. In contrast, the simple laticifers are called *nonarticulated.* Both kinds of laticifers may be branched or unbranched.

Laticifers occur in various tissues and organs of the plant but may be restricted to the phloem.[32,38] The wide distribution of laticifers in the plant results from their mode of development. The articulated laticifers (fig. 13.9) extend into new tissues by addition of cells from these tissues; that is, certain cells in newly formed tissues differentiate as laticiferous cells when they are in contact with the older laticiferous cells. The nonarticulated laticifers originate as single cells in the embryo[22] (fig. 13.10,*A*) and follow the growth of the plant by penetrating tissues newly formed by the apical meristems. A combination of coordinated and apical intrusive growth characterizes the development of the single-cell laticifers. Their growth among the cells resembles that of a haustorium or a fungal

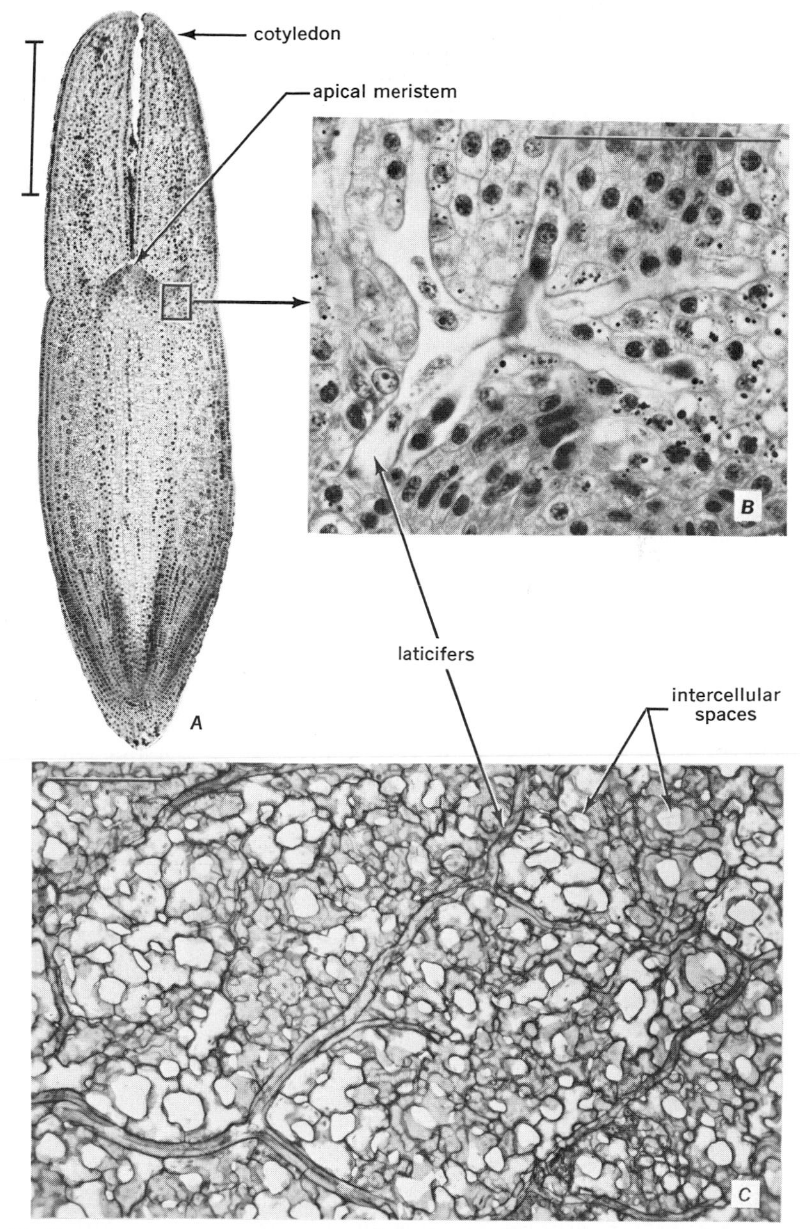

Figure 13.10 Nonarticulated branching laticifers of *Euphorbia* sp. *A*, embryo. Rectangle indicates a locus of origin of laticifers. *B*, section through laticifers showing multinucleate condition. *C*, laticifers branching within spongy parenchyma as seen in a paradermal section of a leaf. (Slide for *A* and *B*, courtesy of K. C. Baker. Lines are: *A*, 0.5 mm; *B*, *C*, 0.1 mm.)

hypha (fig. 13.10,*B*,*C*). In plants with secondary growth, laticifers develop in the secondary tissues also. A nonarticulated laticifer is able to penetrate the vascular cambium and to maintain its continuity during subsequent growth of tissues from the cambium.[41]

Laticifers have primary nonlignified cell walls variable in thickness. The articulated laticifers become multinucleate when series of cells fuse by dissolution of common walls. Nonarticulated laticifers also become multinucleate as they elongate and their nuclei divide repeatedly.[22] The protoplasts of the laticifers are living when the latex is formed, but some cell constituents undergo autolysis in the process.[6]

The relation of the latex to the protoplast is not entirely clear. The term latex refers to the fluid that can be extracted from a laticifer. The fluid varies in appearance and in composition. It is frequently milky but may be completely clear and colorless or brown or orange. Among the common components of latices are the terpenoids, rubber being one of their well-known representatives. Plants having large amounts of this polyterpene (*Hevea brasiliensis, Ficus elastica*) are the source of the natural commercial rubber. The rubber and other polyterpenes occur as particles in the cytoplasm or in small vesicles, which are also referred to as vacuoles.[3,14,23] In the more mature laticifers, the particles are released into a large central vacuole.[6] Many other substances are found in latices, such as, alkaloids (morphine, codeine, and papaverine in the opium poppy, *Papaver somniferum*), sugar (in representatives of Asteraceae), waxes, proteins, enzymes (proteolytic enzyme in *Carica papaya*), crystals, tannins, and starch, which often assumes the form of large grains of unusual shape (chapter 3).

The small vacuoles mentioned above are a conspicuous structural component of latices. These vacuoles vary in size and may be as small as mitochondria. They have been reported arising as dilations in endoplasmic reticulum cisternae[38] or as dictyosome vesicles.[24] The vacuoles are bound by a unit membrane and are surrounded by cytoplasm but may fuse into larger units. In rubber-yielding plants, the small vacuoles in laticifers are usually referred to as lutoids.[31] An increasing body of information indicates that laticifer vesicles, or vacuoles, contain enzymes of hydrolysis typical of lysosomes and that they act as autophagic vacuoles causing an intracellular digestion of a considerable fraction of the cytoplasm.[24,25] This activity leads to a selective accumulation of cytoplasmic constituents in the vacuoles, such as citric acid, mineral phosphorus, and several anions,[31] and possibly also to a breakdown of polyterpene particles.[24]

Laticifers occur in several families and in certain genera of some families of dicotyledons and monocotyledons. The familiar plants with latex are representatives of the spurge family, Euphorbiaceae, which includes genera of commercial source of rubber (*Hevea* and *Manihot*); several genera in the Asteraceae (dandelion, *Taraxacum;* sow thistle, *Sonchus;* lettuce, *Lactuca*); and the Moraceae, which include the Indian rubber plant, *Ficus elastica.*

REFERENCES

1. Amelunxen, F. Elektronenmikroskopische Untersuchungen an den Drüsenhaaren von *Mentha piperita* L. *Planta Med.* 12:121–139. 1964.
2. Amelunxen, F., and H. Arbeiter. Untersuchungen an den Spritzdrüsen von *Dictamnus albus* L. *Z. Pflanzenphysiol.* 58:49–69. 1967.

3. Arreguín, B. Rubber and latex. *Handb. Pflanzenphysiol.* 10:223–248. 1958.
4. Carr, D. J., and S. G. M. Carr. Oil glands and ducts in *Eucalyptus* l'Hérit. II. Development and structure of oil glands in the embryo. *Aust. J. Bot.* 18:191–212. 1970.
5. Cutter, E. G. *Plant anatomy: experiment and interpretation.* Part I. *Cells and tissues.* London, Edward Arnold. 1969.
6. Esau, K. Laticifers in *Nelumbo nucifera* Gaertn.: distribution and structure. *Ann. Bot.* 39:713–719. 1975.
7. Eymé, J. Nouvelles observations sur l'infrastructure de tissus nectarigènes floraux. *Botaniste* 1/6:169–183. 1967.
8. Fahn, A. On the structure of floral nectaries. *Bot. Gaz.* 113:464–470. 1952.
9. Fahn, A. The topography of the nectary in the flower and its phylogenetic trend. *Phytomorphology* 3:424–426. 1953.
10. Fahn, A., and T. Rachmilevitz. Ultrastructure and nectar secretion in *Lonicera japonica. Bot. J. Linn. Soc.* Suppl. 1. 63:51–56. 1970.
11. Foster, A. S. Plant idioblasts: remarkable examples of cell specialization. *Protoplasma* 46:184–193. 1956.
12. Frei, E. Die Innervierung der floralen Nektarian dikotyler Pflanzenfamilien. *Ber. Schweiz. Bot. Ges.* 65:60–114. 1955.
13. Frey-Wyssling, A. The phloem supply to the nectaries. *Acta Bot. Neerl.* 4:358–369. 1955.
14. Heinrich, G. Elektronenmikroskopische Untersuchung der Milchröhren von *Ficus elastica. Protoplasma* 70:317–323. 1970.
15. Heinrich, G. Die Feinstruktur der Trichom-Hydathoden von *Monarda fistulosa. Protoplasma* 77:271–278. 1973.
16. Heslop-Harrison, Y., and R. B. Knox. A cytochemical study of the leaf-gland enzymes of insectivorous plants of the genus *Pinguicula. Planta* 96:183–211. 1971.
17. Ivanoff, S. S. Guttation injuries in plants. *Bot. Rev.* 29:202–229. 1962.
18. Kartashova, N. N. *Stroenie i funktsiya nektarnikov tsvetka dvudol'nykh rastenij.* [*Structure and function of nectaries of flowers of dicotyledonous plants.*] Tomsk, USSR, Izdatel'stvo Tomskogo Universiteta. 1965.
19. Klepper, B., and M. R. Kaufmann. Removal of salt from xylem sap by leaves and stems of guttating plants. *Plant Physiol.* 41:1743–1747. 1966.
20. Loomis, W. D., and R. Croteau. Biochemistry and physiology of lower terpenoids. *Recent Adv. Phytochem.* 6:147–185. 1973.
21. Lüttge, U. Structure and function of plant glands. *Ann. Rev. Plant Physiol.* 22:23–44. 1971.
22. Mahlberg, P. G., and P. S. Sabharwal. Mitosis in the non-articulated laticifer of *Euphorbia marginata. Amer. J. Bot.* 54:465–472. 1967. Origin and early development of nonarticulated laticifers in embryos of *Euphorbia marginata. Amer. J. Bot.* 55:375–381. 1968.
23. Marty, F. Infrastructure des laticifères différenciés d'*Euphorbia characias. C. R. Acad. Sci. Paris D.* 267:299–302. 1968.
24. Marty, F. Vésicules autophagiques des laticifères différenciés d'*Euphorbia characias* L. *C. R. Acad. Sci. Paris D.* 272:399–402. 1971.
25. Matile, Ph., B. Jans, and R. Rickenbacher. Vacuoles of *Chelidonium* latex: lysosomal property and accumulation of alcaloids. *Biochem. Physiol. Pflanz.* 161:447–458. 1970.

26. Metcalfe, C. R., and L. Chalk. *Anatomy of the dicotyledons.* 2 Vols. Oxford, Clarendon Press. 1950.
27. Moens, P. Les formations sécrétrices des copaliers congolais. Étude anatomique, histologique et histogénétique. *Cellule* 57:33–64. 1955.
28. Mollenhauer, H. H., and D. J. Morré. Golgi apparatus and plant secretion. *Ann. Rev. Plant Physiol.* 17:27–46. 1966.
29. Pate, J. S., and B. E. S. Gunning. Transfer cells. *Ann. Rev. Plant Physiol.* 23: 173–196. 1972.
30. Perrin, A. Présence de "cellules de transfer" au sein de l'épithème de quelques hydathodes. *Z. Pflanzenphysiol.* 65:39–51. 1971.
31. Ribaillier, D., J.-L. Jacob, and J. d'Auzac. Sur certains caractères vaculolaires des lutoides du latex d'*Hevea brasiliensis* Mull. Arg. *Physiol. Vég.* 9:423–437. 1971.
32. Rosowski, J. R. Laticifer morphology in the mature stem and leaf of *Euphorbia supina. Bot. Gaz.* 129:113–120. 1968.
33. Rost, T. L. Vascular pattern and hydathodes in leaves of *Crassula argentea* (Crassulaceae). *Bot. Gaz.* 130:267–270. 1959.
34. Schnepf, E. Zur Feinstruktur der schleimsezernierenden Drüsenhaare auf der Ochrea von *Rumex* und *Rheum. Planta* 79:22–34. 1968.
35. Schnepf, E. Gland cells. In: *Dynamic aspects of plant ultrastructure.* A. W. Robards, ed. Chapter 9, pp. 331–357. London, McGraw-Hill Book Company (UK) Limited. 1974.
36. Scott, F. M., B. G. Bystrom, and E. Bowler. *Persea americana,* mesocarp, cell structure, light and electron microscope study. *Bot. Gaz.* 124:423–428. 1963.
37. Thomson, W. W., W. L. Berry, and L. L. Liu. Localization and secretion of salt by the salt glands of *Tamarix aphylla. Proc. Natl. Acad. Sci.* 63:310–317. 1969.
38. Thureson-Klein, Å. Observations on the development and fine structure of the articulated laticifers of *Papaver somniferum. Ann. Bot.* 34:751–759. 1970.
39. Vasil'ev, A. E. O lokalizatsii sinteza terpenoidov v rastitel'noi kletke. [On localization of terpenoid synthesis in a plant cell.] In: *Rastitel'nye resursy.* 5:29–45. Akad. Nauk SSSR. 1970.
40. Vialli, M., F. Barbetta, L. Zanotti, and K. Mihalyi. Estendibilità del concetto di sistema cellulare enterocromaffine ai vegetali. I. Inquadramento della questione e contributo alla conoscenza istochimica dei peli di *Urtica dioica* L. *Acta Histochem.* 45:270–282. 1973.
41. Vreede, M. C. Topography of the laticiferous system in the genus *Ficus. Ann. Jard. Bot. Buitenzorg.* 51:125–149. 1949.
42. Werker, E., and A. Fahn. Resin ducts of *Pinus halepensis* Mill.—Their structure, development and pattern of arrangement. *Bot. J. Linn. Soc.* 62:379–411. 1969.
43. Wooding, F. B. P., and D. H. Northcote. The fine structure of mature resin canal cells in *Pinus pinea. J. Ultrastruct. Res.* 13:233–244. 1965.

14

The Root: Primary State of Growth

TYPES OF ROOTS

The first root of a seed plant develops from the apical meristem at the root end of the embryo. This root is called *taproot,* or primary root. In the gymnosperms and dicotyledons, the taproot and its branched lateral roots comprise the root system (chapter 2). In the monocotyledons, the first root commonly lives a relatively short time and the root system is formed by *adventitious roots* arising on the shoot, often in connection with axillary buds. Although these roots become branched they form a rather homogeneous system referred to as the *fibrous root system.* A taproot system generally penetrates the soil more deeply than the fibrous roots do, but the latter bind more firmly the superficial layers of the soil. The taproot and its larger branches undergo secondary growth but the small absorbing roots remain in the primary state and are often ephemeral. The adventitious roots of the monocotyledons may or may not have secondary growth.

The taproot and the fibrous root systems are the most common in seed plants, and both are concerned with anchorage, absorption, storage, and conduction. Some roots or root parts are specialized with reference to a particular function. The fleshy part of the root of the carrot (*Daucus*), radish (*Raphanus*), beet (*Beta*), sweet potato (*Ipomoea*), yam (*Dioscorea*), and others is specialized as storage organ. A fleshy storage root is often associated with an equally fleshy hypocotyl and may have an anomalous form of secondary growth (chapter 15).

Tropical swamp plants (mangroves) have large aerial prop roots, some resembling flying buttresses. These plants also have aerating roots (pneumatophores) that grow upward and rise above the surface of the mud. Some vines and epiphytes develop aerial roots capable of attaching themselves to the surface upon which the shoot may be growing.

In the parasitic associations between higher plants, the roots of the parasite develop haustoria establishing a connection along which nutrients pass from the host to the parasite. In a nutritional study of the hemiparasite *Odonites,* parasitizing roots of barley and *Stellaria media,* xylem was recognized in the haustorial connection, and use of tracers showed that solutes moved across this pathway directly to the parasite.[25]

Contractile roots

Anchorage of a plant by means of roots may involve root contraction, a process that pulls the shoot closer to the ground or, in bulbous plants (fig. 14.1), deeper into the soil. Root contraction is widely distributed among monocotyledons and herbaceous perennial dicotyledons. The contraction usually occurs in certain individual roots and is limited to parts of these roots (fig. 14.1). The shortening of the root depends mainly on the change in shape of the inner cortical cells which, at a certain stage of development, expand radially and contract longitudinally.[53] In a species of *Hyacinthus,* the change in cell shape was found to be a growth phenomenon, for the radial longitudinal wall increased in surface area and in thickness. It also showed a change in the angle of helical striations and an obliteration of pit fields.[77] The root tissues not involved in this growth, that is, the central vascular tissues and the peripheral cortical tissues, became contorted and wrinkled.

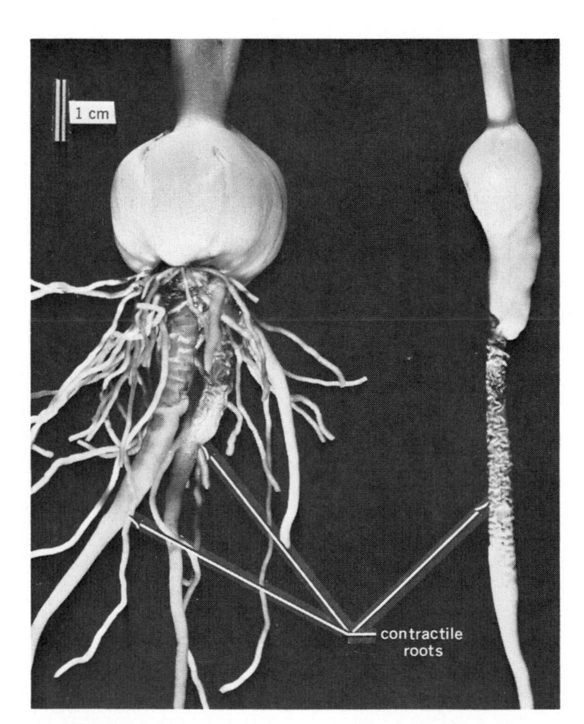

Figure 14.1 *Scilla* (squill, Liliaceae) bulbs with contractile and noncontractile roots. The contractile roots show the result of contraction in the wrinkling of the surface tissue.

Mycorrhiza

The physiological activity of the root concerned with providing the plant with water and nutrients may be enhanced by a symbiotic association with a specific fungus. Such association is called *mycorrhiza* (plural mycorrhizae). The fungus invades the cortex but the root cells develop no pathologic symptoms and retain their vital characteristics.

Mycorrhizae are classified according to the relation of the fungus to the cortical cells of the host. The two principal forms of mycorrhizae are (1) *ectomycorrhizae,* in which the fungus envelops the entire root tip with a dense sheath called hyphal mantle and enters the intercellular spaces; (2) *endomycorrhizae,* in which the fungus forms an inconspicuous mantle but invades the interior of the cells. Specific characteristics of the host determine the type of mycorrhizal fungus with which it becomes associated.

Ectomycorrhizal roots are short, branched, and appear swollen. In such roots the development of root hairs is depressed and the volumes of apical meristem and rootcap may be reduced.[14] The endomycorrhizal roots are similar to uninfected roots in form but are darker in color.

In natural habitats, mycorrhizal associations are the rule rather than an exception.[27] The chief role of the mycorrhizal fungi appears to be conversion of minerals of the soil and of the decaying organic material into forms accessible to the host. The host is presumably secreting sugars, amino acids, and other organic materials, making them available to the fungus. In an experimental study with soybean, mycorrhizal infection was found to enhance the growth of the host plant and to cause a reduction in the resistance to water transport in the root.[60]

Root nodules

Roots may have an association with bacteria (nitrogen-fixing bacteria of species of *Rhizobium*) that is beneficial to the plant. The association leads to a development of *root nodules,* a phenomenon particularly characteristic of Fabaceae.[6] The bacteria invade the root, chiefly through the root hairs, and, in multiplying, form an infection thread by becoming enclosed in a sheath of gumlike material. The thread penetrates deeply into the root and induces a proliferation of the inner cortical cells. This proliferation, superficially resembling a branch root primordium, becomes the nodule. Some authors propose that nodules are modified lateral roots but developmental studies do not support this concept.[39] The nodule long retains a meristematic zone in its outermost abaxial part and harbors the bacteria in its inner part. Branching vascular bundles, which are connected with the vascular cylinder of the root, surround the bacterial tissue. Each bundle has a parenchymatic sheath and an endodermis. In some species, the sheath cells develop wall protuberances characteristic of cells concerned with short-distance transport between cells (transfer cells). This feature indicates the existence of a transport system for an exchange of nutrients between the bacteria and the host.[45]

PRIMARY STRUCTURE

The internal organization of the root is variable but generally simpler and phylogenetically more primitive than that of the stem. The root is an axial structure with no leaflike organs and with no division into nodes and internodes. Correspondingly, the arrangement of tissues in the root shows relatively little difference from level to level, whereas in the stem the connection of the axis with the leaves results in differences in structure between nodes and internodes and even between different levels of a given internode.

A transverse section through a root in the primary state of growth shows a clear separation between the usual three tissue systems, the epidermis (dermal tissue system), the cortex (ground tissue system), and the vascular tissue system (fig. 14.2). The vascular tissues form a solid cylinder or, if pith is present, a hollow cylinder (fig. 14.3,*A*). Each system has some structural features that are characteristic of roots. The rootcap, which covers the apical meristem of the root, is also a part of the primary body.

Epidermis

In young roots the epidermis is specialized as an absorbing tissue and usually bears root hairs, which are tubular extensions of epidermal cells (fig. 14.4). Root hairs markedly

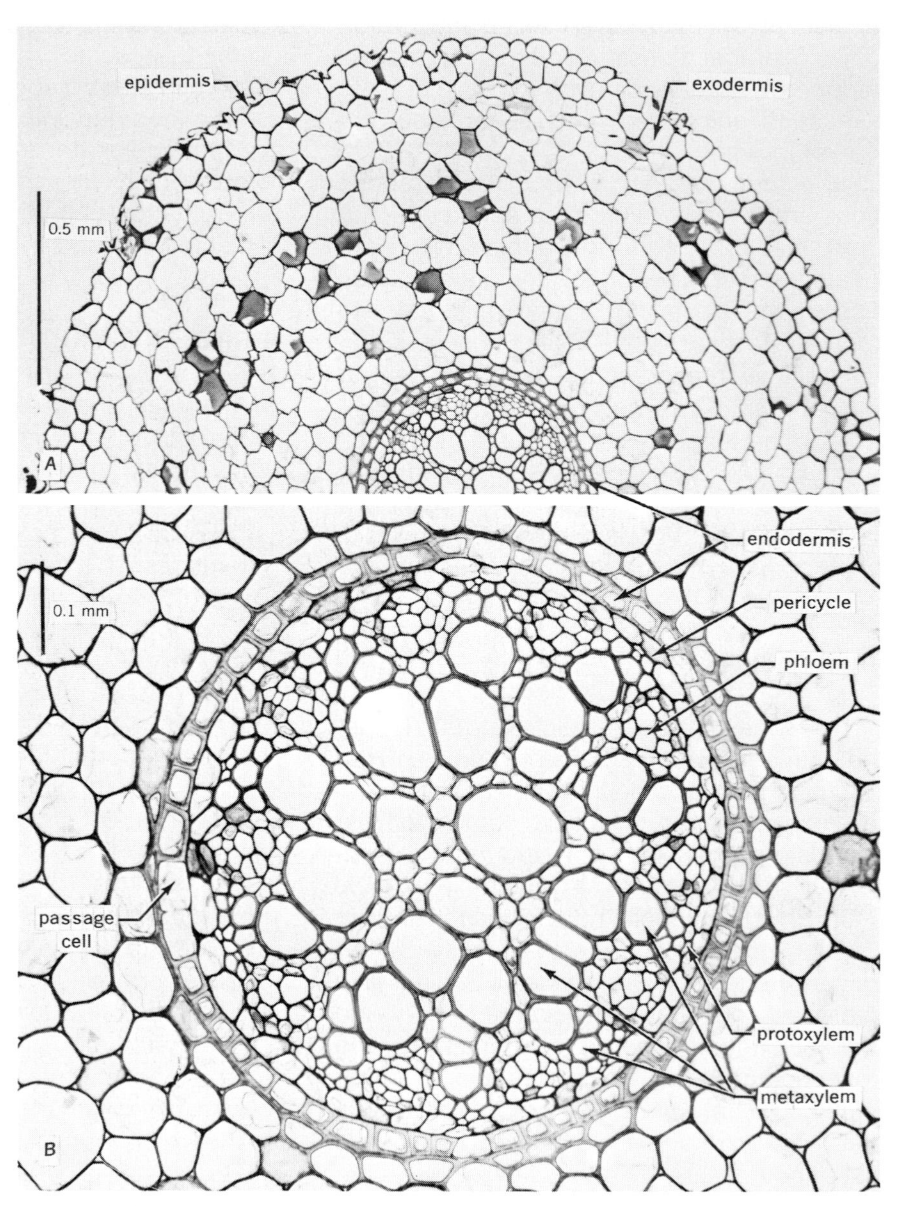

Figure 14.2 Transverse sections of mature *Lilium* root. *A*, the wide cortex consists of parenchyma. The epidermis is partly broken down so that the exodermis is exposed in some regions. The vascular cylinder has no pith. *B*, enlarged view from *A* showing the thick-walled endodermis with passage cells, a mostly single-layered pericycle, ten groups of phloem cells alternating with the same number of groups of protoxylem cells.

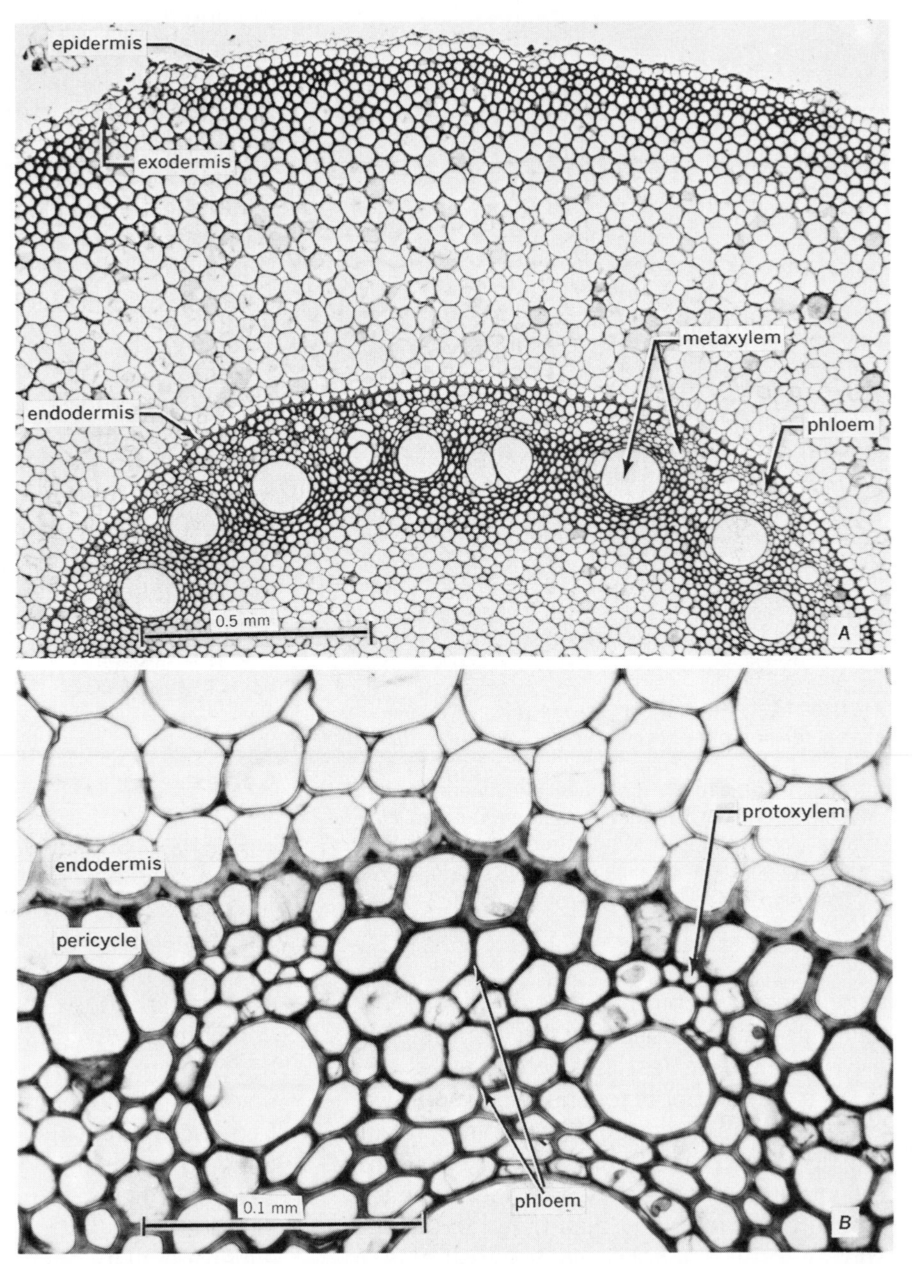

Figure 14.3 Transverse sections of mature *Zea mays* roots. *A*, the wide cortex is sclerified beneath the single-layered exodermis; the epidermis is partly broken down. The vascular tissue surrounds a pith. *B*, enlarged view from *A* showing the thick-walled endodermis and a group of phloem cells between two radial files of xylem cells.

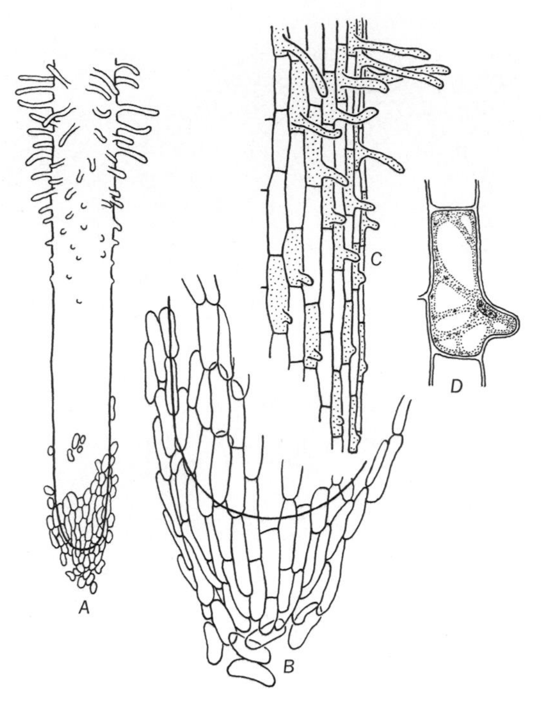

Figure 14.4 Root tip of *Tradescantia*. *A*, illustrates spatial relation between rootcap and root hair region. *B*, enlarged view of rootcap. Some cells are partly separated from the cap. *C*, root hairs in different stages of development. Root hair-forming cells are stippled. *D*, enlarged view of a cell with a root hair primordium. (Adapted from Braune, Leman, and Taubert. *Pflanzenanatomisches Praktikum.* 2nd ed. Jena, Gustav Fischer, 1971.)

extend the absorbing surface of the root. Certain calculations based on the root system of a rye plant[58] suggest that comparatively few of the total number of root hairs can supply all the water necessary for transpiration and growth of the plant. Such efficiency would be particularly important when the available moisture is irregularly distributed in the soil. The absorbing function is not restricted to root hairs, however, for other epidermal cells also absorb.

In the absorbing part of the root, a thin cuticle has been identified on the epidermis, including the root hairs. If the epidermis is persistent it may show eventually a conspicuous cutinization. In fact, in some herbaceous perennials the epidermis remains for a long time, or permanently, as a protective tissue.[40] Its walls increase in thickness, and the cell lumina sometimes become filled with deeply colored substances.

In the aerial roots of tropical Orchidaceae and epiphytic Araceae, as well as in some terrestrial monocotyledons, the epidermis develops into a multiseriate tissue (multiple epidermis) called *velamen*. The velamen consists of nonliving cells, compactly arranged and often bearing secondary wall thickenings in the form of numerous narrow strips. The velamen is commonly interpreted as an absorptive tissue, but some physiological studies on orchid velamen indicate that the principal roles of this tissue are mechanical protection and reduction in loss of water from the cortex.[19]

Cortex

The root cortex is often composed of parenchyma cells only, but if it is persistent, it may develop sclerenchyma, or it may become collenchymatous. Intercellular spaces are characteristic of the root cortex (fig. 14.2 and 14.3). The cortex may differentiate as an aerenchyma, with the intercellular spaces developing into large lacunae. Such cortex is common in plants growing in moist habitats (e.g., rice) but is also encountered in species growing in drier regions. A comparison of several species of monocotyledons showed that the large air spaces in roots may have either a schizogenous or a lysigenous origin or may arise by a combination of the two processes.[50,56] Aerenchyma of roots is regarded as a tissue serving in gas transport and as a reservoir of oxygen required in the respiration of tissues having no access to the oxygen of the air. Some of the oxygen also diffuses from the root and apparently serves to ameliorate unfavorable soil conditions, in part by the oxidation of toxic products.[1] Cortical cells are highly vacuolated. Their plastids are usually devoid of chlorophyll but are accumulating various amounts of starch. The innermost layer of the cortex is differentiated as an endodermis, and one or more layers at the periphery may develop into an exodermis (figs. 14.2 and 14.3).

ENDODERMIS

In the absorbing region of the root the endodermal cell wall contains suberin in a bandlike region extending completely around the cell within the radial and transverse walls[46] (fig. 14.5). This band, called *casparian strip,* is not merely a wall thickening but an integral part of the primary wall, for the deposition of suberin is continuous across the middle lamella. The electron microscope reveals the casparian strip as a slightly thickened part of the wall showing an intense, homogeneous staining reaction.[7,57] The plasmalemma next to the strip is thicker and more regular in profile than it is elswhere in the cell (fig. 14.6,*A*), and it tightly adheres to the strip region of the wall. In plasmolysed cells, this membrane remains in contact with the strip while it separates from the rest of the wall (fig. 14.6,*B*). The tenaceous adherence of the plasmalemma to the strip does not depend on plasmodesmatal connections, for such connections have not been detected in the strip region. A possible cause of the phenomenon may be an interaction between the membrane lipids or the hydrophobic portions of membrane proteins and the hydrophobic suberized wall in the strip.[7]

The presence of the suberized casparian strip tightly connected with the plasmalemma has an important effect on the transport of soil solution between the cortex and the vascular cylinder. In a pertinent study,[68] several species of plants grown in a culture solution were treated with lead chelate in water, then

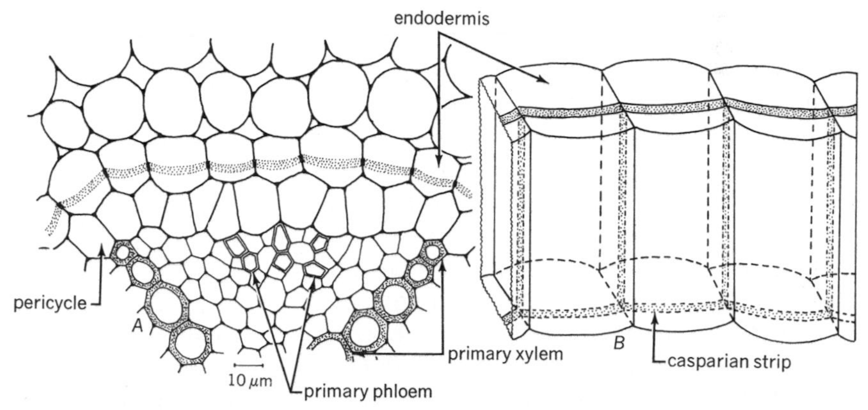

Figure 14.5 Structure of endodermis. *A*, cross section of part of root of morning glory (*Convolvulus arvensis*) showing position of endodermis with regard to xylem and phloem. The endodermis is shown with transverse walls bearing casparian strips in focus. *B*, diagram of three connected endodermal cells oriented as they are in *A*; casparian strip occurs in transverse and radial walls (that is, in all anticlinal walls) but is absent in tangential walls.

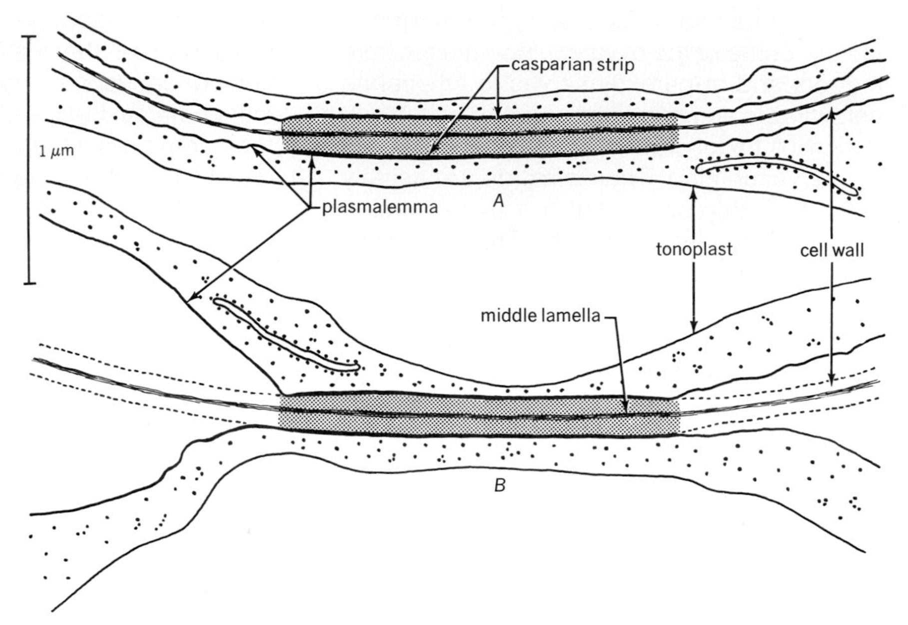

Figure 14.6 Structure of endodermis as revealed by electron microscopy. *A*, plasmalemma is smooth along the casparian strip region, wavy elsewhere. *B*, in plasmolyzed cells plasmalemma adheres to the casparian strip region but separates from the wall elsewhere. (Constructed from electron micrographs in Bonnett.[7])

exposed to hydrogen sulphide that caused a precipitation of lead sulphide along the transpiration stream. The distribution of the precipitate indicated that in transpiring plants the solute moves through the cortex in the cell walls (free space or apoplastic pathway), is blocked in its apoplastic progress by the casparian strip, is forced to pass through the protoplasts of the endodermal cells, and resumes the apoplastic course in the vascular cylinder until the tracheary cells are entered. Thus, the movement of solutes is subjected to the control of the differentially permeable protoplast membranes within the endodermis. The effectiveness of the casparian strip as a barrier to apoplastic movement has been further demonstrated at the electron microscope level by the use of maize roots that were made to absorb lanthanum.[43] Since the lanthanum cation does not penetrate cell membranes it was found only in the cell walls of the cortex, having been completely stopped from further progress by the casparian strip.

In roots with considerable amount of secondary growth the endodermis is usually cast off together with the cortex, but in those remaining in primary state (commonly roots of monocotyledons) it often develops thick secondary walls (figs. 14.2 and 14.3). These walls typically consist of a suberin lamella covered by layers of lignified cellulose (fig. 14.7). The thickening is often most voluminous on the inner tangential wall (fig. 14.3), but may be even (fig. 14.2). Formation of the secondary wall may be delayed in endodermal cells opposite the xylem. Such thin-walled cells (with casparian strips) in an otherwise thick-walled endodermis are called *passage cells* (fig. 14.2). The endodermis is usually uniseriate, but in many Asteraceae

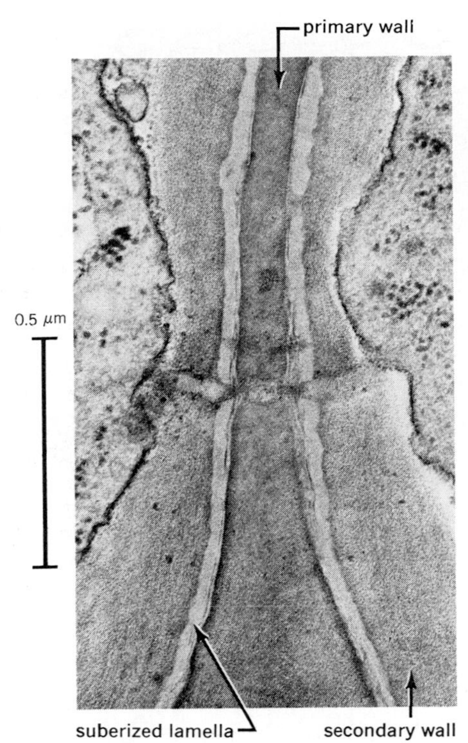

Figure 14.7 Electron micrograph of a section through an anticlinal wall between two endodermal cells from *Zea mays* root. Late stage of differentiation of endodermis: suberized lamellae covered by cellulose wall layers appear on both sides of the primary wall. (From Karas and McCully.[37])

certain young endodermal cells divide tangentially, and schizogenous secretory canals develop in the two-layered zones.[26] In some species—pear and apple are good examples—the cortical layers outside the endodermis develop prominent wall thickenings, often restricted to the radial walls.[55]

EXODERMIS

The exodermis occurs beneath the epidermis. Its cells may have casparian strips, but more commonly they appear to have a suberin lamella covered by a cellulose wall. The exodermis contains either one kind of cell or short and long cells. The exodermis may be several layers in thickness. The cortex is sclerified beneath the exodermis in Poaceae (fig. 14.3,*A*), Arecaceae, and Cyperaceae. If the epidermis collapses, the exodermis becomes the outermost protective layer in roots that do not shed their cortex.[41]

Vascular cylinder

The vascular cylinder (stele) comprises the vascular tissues and one or more layers of nonvascular cells—the pericycle. The inclusion of the pericycle in the vascular cylinder is done in part on historical ground: the concept of the stele (chapter 16) defines the pericylce as the limiting layer of the stele. In the seed plants, moreover, the pericycle of the root arises from the same part of the apical meristem as the vascular tissues. The pericycle may be composed entirely of parenchyma (figs. 14.2, 14.5,*A*, and 14.8,*A*), or it may contain sclerenchyma (fig. 14.3). It may be interrupted by elements of protoxylem (fig. 14.18). The pericycle is commonly one layer of cells in thickness but may be multiseriate. Schizogenous secretory canals occur in the pericycle of the Apiaceae.[12] Lateral roots, part of the vascular cambium, and, in many roots the phellogen, arise in the pericycle.

The xylem frequently forms a solid core with ridgelike projections, seen as radial files of cells in transections, extending toward the pericycle (fig. 14.9). Strands of phloem alternate with the xylem ridges. If the xylem does not differentiate in the center of the root, a pith consisting of parenchyma or sclerenchyma is present (figs. 14.3, 14.8,*A*, and 14.9,*D*).

The number of xylem ridges varies in different species and among roots of the same plant, and in relation to this variation the roots are called *diarch, triarch, tetrarch,* etc., or

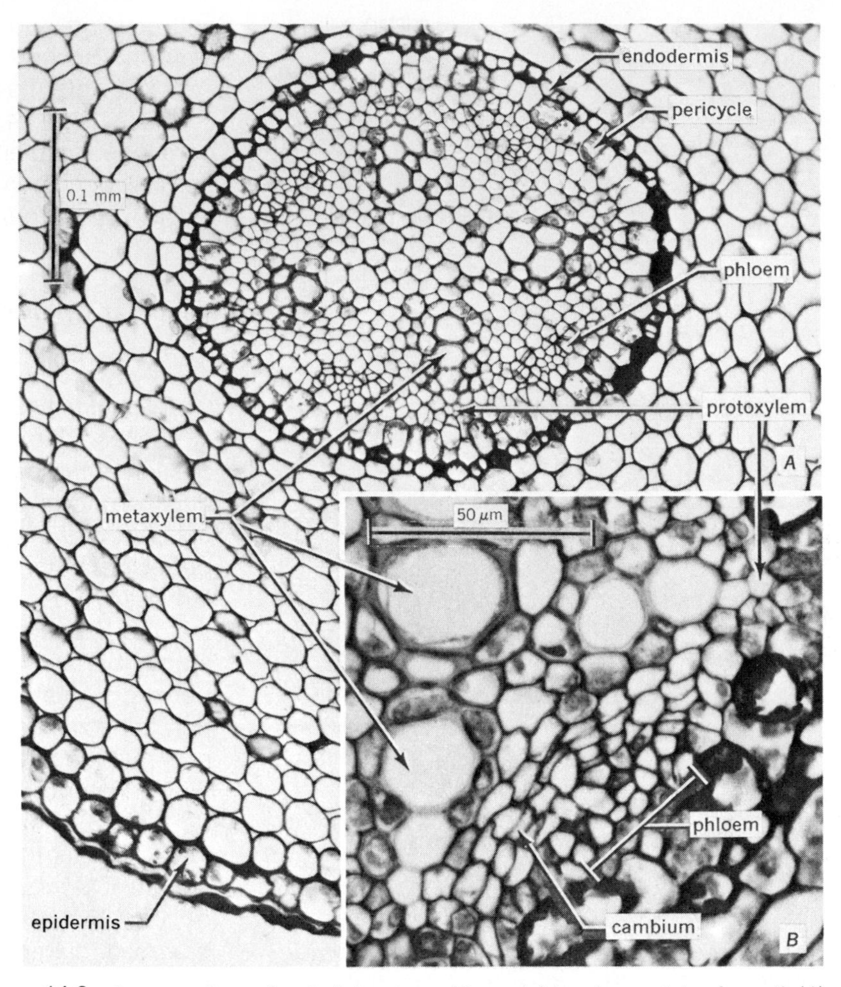

Figure 14.8 Cross sections of root of strawberry (*Fragaria*) in primary state of growth (*A*) and at initiation of vascular cambium (*B*). (From Nelson and Wilhelm, *Hilgaria* 26: 631–642, 1957.)

polyarch (fig. 14.9). The tracheary cells in the outermost position in each xylem file are the narrowest and are the earliest to mature. They constitute the protoxylem and have helical or scalariform-reticulate or sometimes annular secondary thickenings. Closer to the center are the increasingly wider metaxylem elements, most of which, especially the latest, commonly have secondary walls with bordered pits. The xylem with the centripetal direction of maturation as seen in roots is termed *exarch*.

The primary vascular structure of monocotyledon roots—these rarely have secondary growth—is highly variable and often complex. In some the center is occupied by a single metaxylem vessel (shown in immature state in fig. 14.15,*B*); in others a circle of such vessels

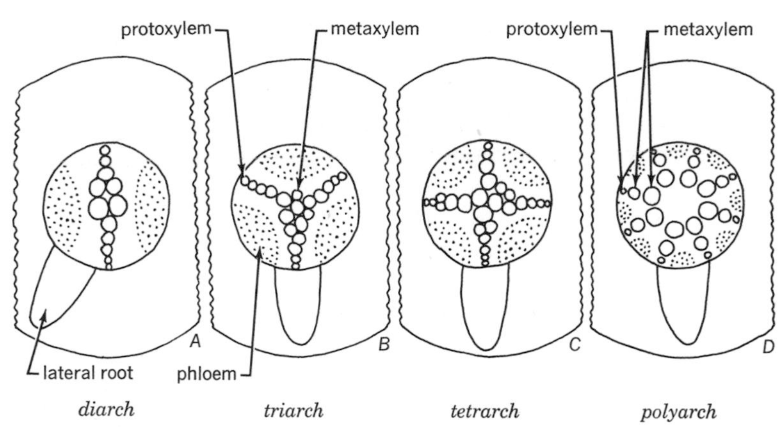

Figure 14.9 Different patterns formed by primary xylem in cross sections of roots and position of lateral root with regard to xylem and phloem of main root. The patterns *A–C* are characteristic of dicotyledons; *D* is found in many monocotyledons.

surrounds a pith (figs. 14.3 and 14.18). In the lily root, the radial files of metaxylem are not precisely arranged and reach the center of the root (fig. 14.2). The number of protoxylem points may exceed a hundred in large palm roots, and in some monocotyledons, phloem strands are scattered throughout the xylem core.[70]

The first mature sieve elements, those of the protophloem, occupy a peripheral position in the vascular cylinder; the metaphloem occurs farther inward. Thus, the primary phloem shows a centripetal order of differentiation (figs. 14.3,*B*, and 14.15) as does the xylem. Companion cells are characteristic of the metaphloem but may be absent in the protophloem. In grasses, each protophloem sieve element is associated with two companion cells, the three cells commonly forming a consistent, symmetrical pattern in transections (figs. 14.3,*B*, and 14.15,*B*). In roots having secondary growth the cells located between the xylem and the phloem eventually function as a vascular cambium (fig. 14.8,*B*). In roots without secondary growth the cells in the same location mature as parenchyma or sclerenchyma cells (figs. 14.2 and 14.3).

Rootcap

The rootcap (fig. 14.4) is interpreted as a structure protecting the apical meristem and assisting the growing root in penetrating the soil. It consists of living parenchyma cells derived from the apical meristem by divisions contributing cells away from the apex (fig. 14.12). This part of the apical meristem often appears as a distinct meristem called *calyptrogen.* As new cells are produced, cells on the periphery of the rootcap are sloughed off (fig. 14.4,*A*,*B*). Determinations of duration of mitotic cycles in rootcap initials and of the number of cells in the longitudinal cell files in the rootcap make it possible to estimate the time interval between the origin of a rootcap cell and its shedding from the cap. In the oat (*Avena sativa*), for example, this time interval was calculated to be 5 to 6 days at the most.[30] Environmental conditions affect the development of the rootcap. When plants normally growing in soil are raised in a water culture their roots may fail to develop rootcaps.[54] Aquatic plants, however, often have particularly large rootcaps.

Root tips growing in the soil are coated with

more or less large amounts of mucilage.[34,38] The rootcap appears to be the main source of this mucilage, although the material occurs also on the surface of the young root at levels free of the rootcap and extending to the root hair zone. The well-known effect of this coating of mucilage is the adherence of soil particles to the root tips and the root hairs.

Chemical analyses, electron microscopy, and authoradiographic studies of root tips incubated with radioactive glucose[36,42,44] have elucidated the nature of the mucilage and the method of its secretion. The secreted product is a highly hydrated polysaccharide, probably a pectic substance. In the rootcap, it is secreted by the outer cells. The secretion process is a function of dictyosomes, which show hypertrophied cisternae and produce large vesicles containing the secretory product. The vesicles fuse with the plasmalemma along the outer walls of the cells and release the mucilage into the space between the plasmalemma and the wall. From here, the secretion passes through the wall to the outside where it appears as a droplet. The functions attributed to the coat of mucilage are protection from harmful products of the soil, prevention of desiccation of the root tip, serving as an absorbing surface, effecting an exchange of ions, dissolving and possibly chelating certain nutrients.[38]

The rootcap has long been regarded as an organ controlling the georeaction of the root.[62,75] When roots normally growing downward are placed horizontally, the subsequent growth in the elongation zone occurs along a curve until the normal downward orientation is attained (fig. 14.10). When the tip is excised or the rootcap removed, growth of the root is not stopped but the geotropic reaction is lost until the rootcap is regenerated.[49]

The site of perception of gravity in the rootcap is the central column of cells in which starch-containing amyloplasts act as statoliths, that is, gravity sensors. In vertically growing roots the amyloplasts are sedimented on the distal horizontal walls of the cells in the central column (fig. 14.11,*A*). A few minutes after the root is placed horizontally the amyloplasts slide toward the longitudinal walls directed downward (fig.14.11,*B*). Within some 24 hours the root assumes its normal vertical position by growing along a curve (fig. 14.10) and the amyloplasts return to their previous location next to the distal horizontal walls.

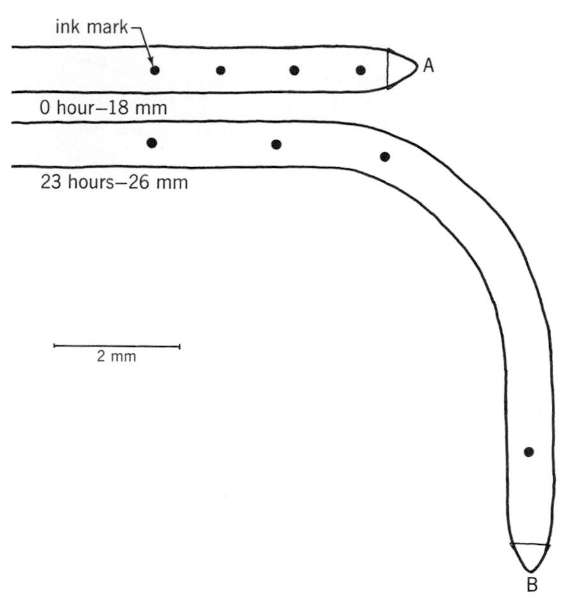

Figure 14.10 Diagrams illustrating response of root tip to gravity. *A*, root tip of root normally growing downward is placed horizontally. *B*, 23 hours later the root tip grows along a curve and returns to a vertical orientation with regard to the soil. (Adapted from Pilet.[49])

According to some studies involving removal of rootcaps, amyloplasts start developing in the exposed apical meristem of the root and the root becomes responsive to gravity just before rootcap regeneration begins.[3] This observation does not negate the concept that the rootcap is the normal seat of gravity perception but it indicates that the perception is not limited to the rootcap, and it seems to support the interpretation of amylo-

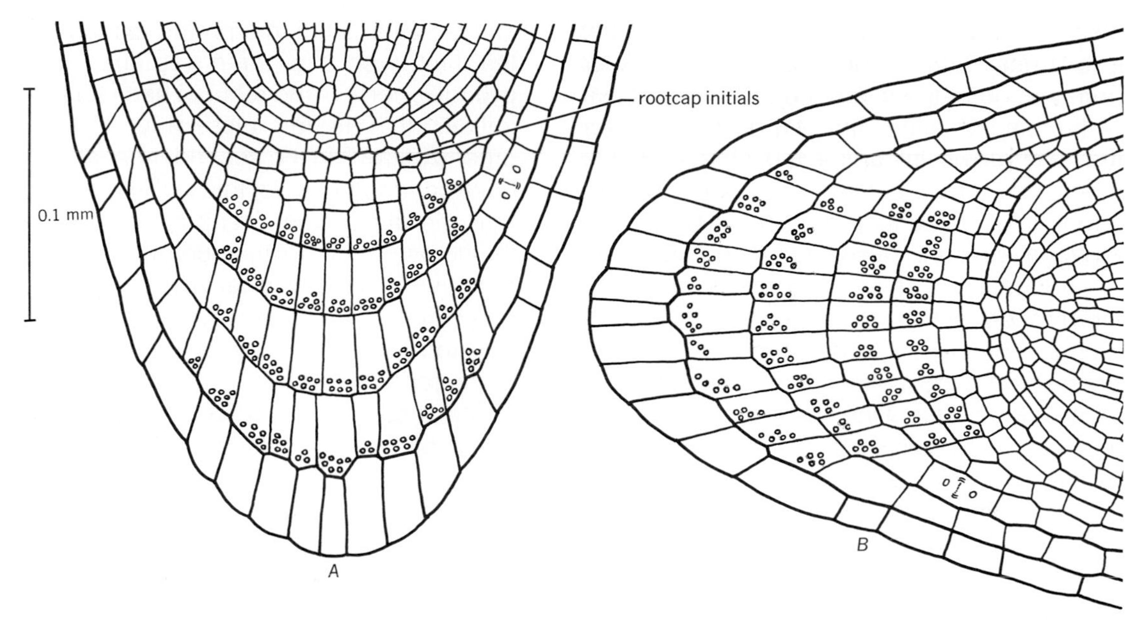

Figure 14.11 Diagrams illustrating response of amyloplasts with starch (small circles), the statoliths, to gravity. *A*, amyloplasts are sedimented on distal horizontal walls in rootcap of a root growing vertically downward. *B*, in the same root, placed horizontally, the amyloplasts slide toward the normally vertical walls that are now parallel with the soil. This change in position is involved in the georeaction shown in fig. 14.10.

plasts as gravity sensors. In contrast, studies in which the curvature of the root and various quantitative data regarding the amyloplasts and their movements were statistically analyzed failed to reveal a direct link between the behavior of the amyloplasts and the georeaction.[47] Thus, the role of the amyloplasts in the response of roots to gravity is obscure. Moreover, some studies suggest that cell components other than amyloplasts such as endoplasmic reticulum[71] and dictyosomes[64] also participate in the geotropic response.

The site of perception of gravity, the root tip, is widely separated from the site of gravity reaction, the growing region of the root[33,49] (fig. 14.10). The nature of the stimulus transmitted from the root tip to the zone of cell elongation is not known. Some workers postulate that an inhibitor mediates the geotropic response.[3,47] At the site of georeaction, endogenous growth substances are redistributed so that auxins accumulate on the lower side of the horizontally oriented root and are depleted on its upper side, the upper cells elongate, and the root curves.

The geotropic reaction of the root is influenced by light, as well as gravity. The response to light is mediated by phytochrome, and experiments with microbeam irradiation suggest that the pigment is located in the root apex.[69]

DEVELOPMENT

Apical meristem

The main phenomenon in the origin of the root in the embryo is the organization of the apical meristem of the root[26] at the lower end of the hypocotyl; sometimes not only a meristem but also an embryonic root, the radicle, is present in the embryo (chapter 24). After germination of the seed, the apical meristem of the root forms the taproot (primary root). As the latter grows, the apical meristem acquires a definite cellular pattern, which varies in different taxa. Branch roots and adventitious roots, if present, also show characteristic arrangements of cells in the apical meristem, more or less similar to that in the taproot. The architecture of apical meristems of roots has been studied most often for the purpose of revealing the origin of the tissue systems. When this architecture was found to have characteristic differences, investigators began relating the differences to the taxonomic groupings of plants, and attempts were made to discover the trends in the evolution of apical organization.[72]

By analyzing the pattern of cells in an apical meristem it is possible to trace out planes of cell division and the direction of growth. In one type of analysis, the differentiating tissues are followed to the apex of the root in order to determine whether there are specific cells that appear to be the source of one or more of the discrete tissues. Thus, the implication is made that a spatial correlation of tissues with certain cells or cell groups at the apex indicates an ontogenetic relation between the two, that, in other words, the apical cells function as initials (chapter 2) of those tissues. In seed plants, the roots show two main patterns of spatial relation between tissue regions and the cells at the apex. In one, the vascular cylinder, the cortex, and the rootcap are traceable to independent layers of cells in the apical meristem, with the epidermis differentiating from the outermost layer of the cortex (fig. 14.12,*A*,*B*) or from cells having common origin with the rootcap cells (fig. 14.12,*C*,*D*). In the other type, all regions, or at least the cortex and the rootcap, converge in one group of cells oriented transversely (fig. 14.12,*E*,*F*). These patterns are interpreted to mean that in the first type the three regions, vascular cylinder, cortex, and rootcap each has its own initials; this is the closed type of apical organization.[26] In the second type, all regions have common initials; this is the open type of apical organization. An apical meristem with the common initials may be phylogenetically primitive.[72] In many lower vascular plants only one cell, the *apical cell,* appears in the focal position and is the common initial for all parts of the root.

The analysis of origin of root tissues in terms of distinct initials at the apex corresponds to the approach used by Hanstein[28,29] when he formulated the *histogen theory*. According to this theory, the body of the plant arises from a massive meristem of considerable depth comprising three precursors of tissue regions, the *histogens,* each beginning with one to several initials at the apex arranged in superposed stories. The histogens are the *dermatogen* (precursor of the epidermis), the *plerome* (precursor of the central vascular cylinder), and the *periblem* (precursor of the cortex). The subdivision into the three histogens does not have a universal application because it is seldom discernible in shoots, but it is often convenient for descriptions of the differentiation of tissue regions in the root.

Another approach to an analysis of the relation between cell patterns and growth in root tips is that represented in the *body-cap* (Körper-Kappe) concept,[61] which emphasizes the planes of those divisions that are responsible for the increase in the number of vertical cell files in the meristematic region of the root. In this growth, longitudinal and transverse divisions are combined in such a way that, at the point of doubling of files, transverse and longitudinal walls in a given file form a pattern resembling a T (or a Y). If the rootcap has its own initials, the direction of

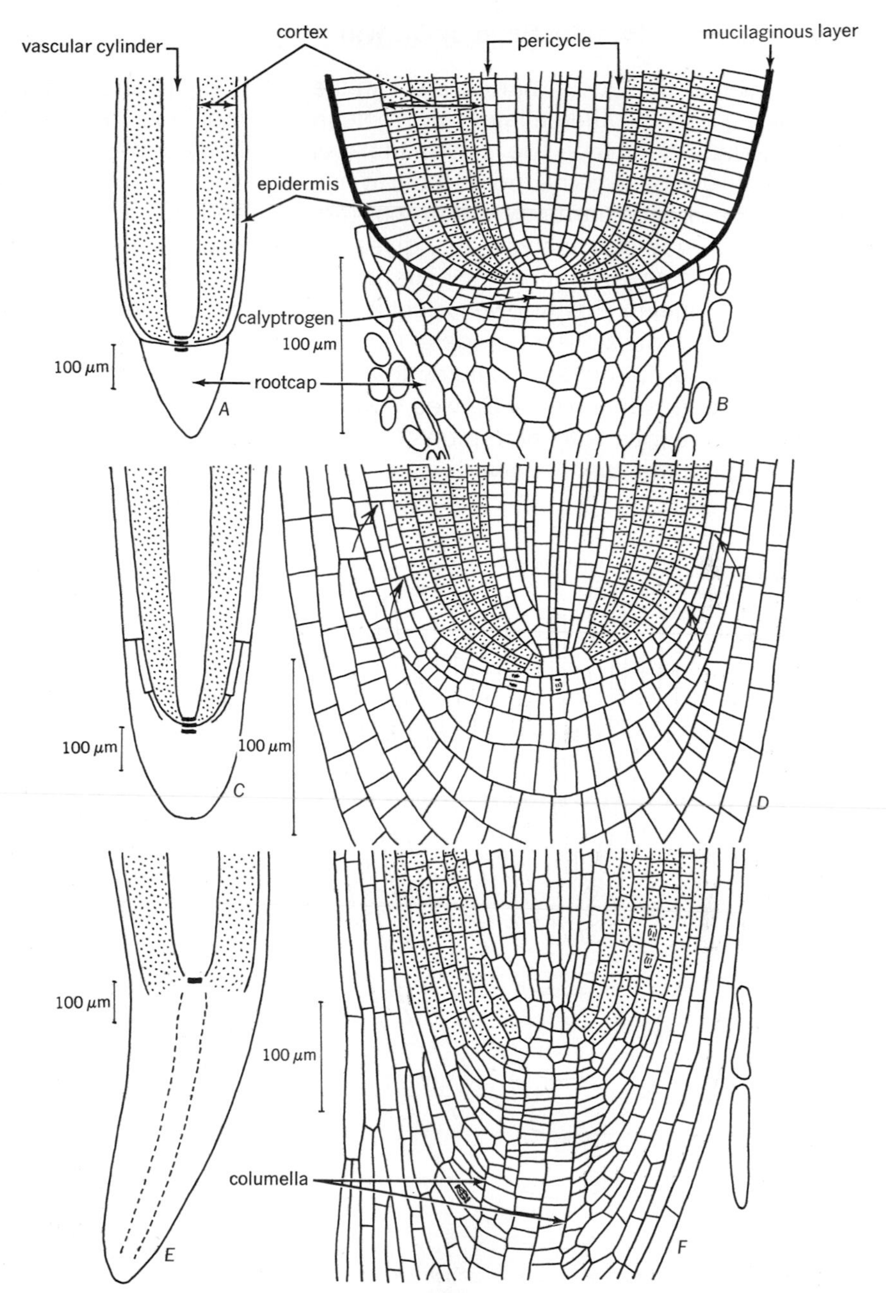

Figure 14.12 Apical meristem and derivative regions in roots. *A*, *B*, grass (*Stipa*). Three stories of initials, those of rootcap forming a calyptrogen. Epidermis has common origin with cortex. *C*, *D*, radish (*Raphanus*). Three stories of initials. Epidermis has common origin with rootcap and becomes delimited on the sides of root by periclinal walls (arrows in *D*). *E*, *F*, spruce (*Picea*). All regions of root arise from one group of initials. Rootcap has a central columella of transversely dividing cells. The columella also gives off derivatives laterally.

the top stroke (horizontal bar) of the T in the cell files sharply differentiates the rootcap from the body of the root. The horizontal bar of the T faces the base of the root in the cap, the apex of the root in the body of the root (fig. 14.13). The body-cap analysis also gives the idea of the minimal number of cells at the apex (only one in many lower plants) that could by their divisions propagate the pattern. This group of cells constitutes the minimal constructional center of Clowes,[16] which probably corresponds to the region of initials in the original histogen theory.

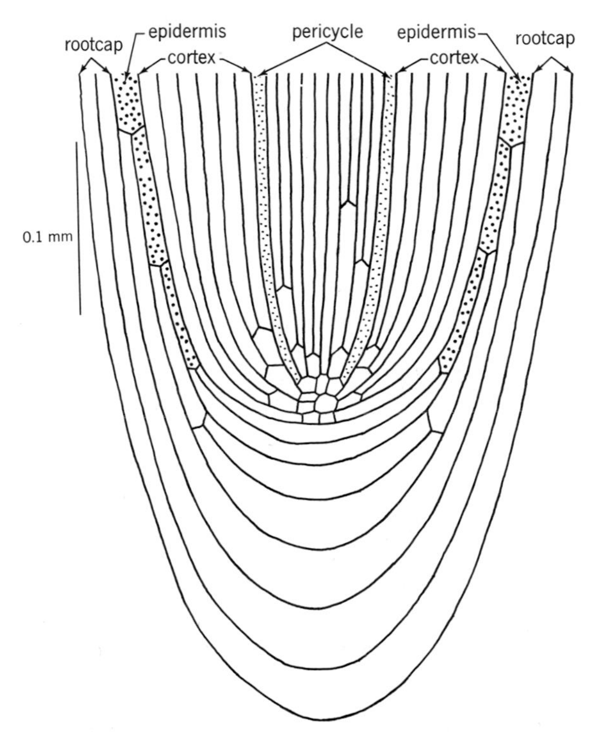

Figure 14.13 Diagram of *Linum usitatissimum* root tip (fig. 2.3) analyzed according to the body-cap concept of apical meristem activity. The T pattern formed by a combination of longitudinal and transverse walls distinguishes between the cap type of growth (the upper stroke of the T faces the base of the root) and the body type of growth (the upper stroke of the T faces the apex of the root).

By means of analyses of the organization of root tips in terms of the histogen and the body-cap concepts information is obtained about growth that has already taken place and has produced the pattern now discernible. Physiological and biochemical techniques applied to root growth studies, however, have revealed that addition of cells at the tip may not depend on a continued activity of the most distal minimal group of apical cells. Extensive research on normally developing roots and those treated surgically, also on irradiated roots, and on roots that were fed labeled compounds involved in DNA synthesis, has shown that, as a general phenomenon, the initials which are responsible for the original cell pattern largely cease to be mitotically active during the later growth of the root[15–17]. They are supplanted in this activity by cells located somewhat deeper in the body of the root. These observations have led to the concept of the *quiescent center* in the apical meristem. The concept states that the most distal cells in the body of the root (the former initials of plerome and periblem) divide infrequently, show little change in size, and have low rates of synthesis of nucleic acids and protein. The quiescent center excludes the initials of the rootcap, is hemispherical or discoid in shape (fig. 14.14), and in some species studied contains 500 to 1000 cells in apices with between 125,000 to 250,000 meristematic cells. The quiescent center is variable in volume apparently in relation to root size, for it is smaller or entirely absent in thin roots. In roots with a quiescent center the mitotically active initials (the promeristem according to Clowes[15]) occur just outside the proximal surface of the quiescent center.

The quiescent state of the distal apical cells does not mean that these cells have become permanently nonfunctional.[15,17] The quiescent center is not completely devoid of

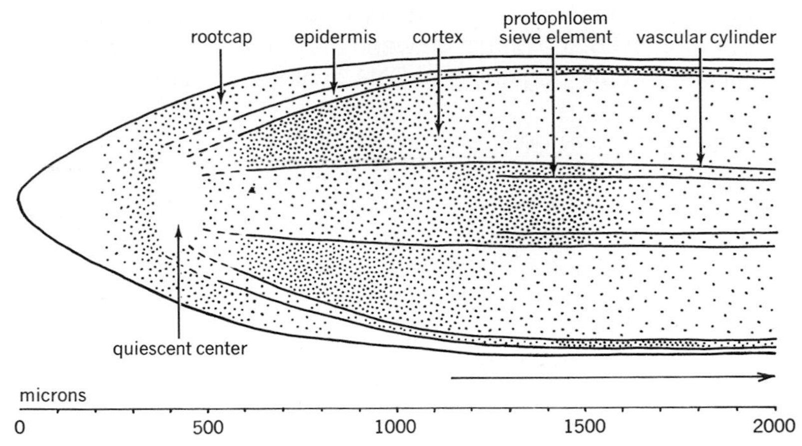

Figure 14.14 Diagram of root tip of onion in longitudinal view illustrating distribution of meristematic activity. Frequency of mitosis is indicated by density of stippling. (After Jensen and Kavaljian.[35])

divisions under normal conditions. In roots injured experimentally by radiation or surgical treatments the quiescent center is able to repopulate the meristem. It also resumes division during recovery from a period of dormancy induced by cold. When the rootcap is removed the cells of the quiescent center begin to grow and undergo a controlled sequence of divisions that regenerate the rootcap.[2]

By labeling nuclei with thymidine and by blocking the cell cycle at metaphase by means of inhibitors one can obtain quantitative data on the duration of the mitotic cycle in the different regions of the root meristem.[17] These data show that the cells of the quiescent center divide approximately ten times slower than the adjacent cells (table 1). Pulse labeling with thymidine has shown moreover that the differences in the duration of mitotic

Table 14.1 AVERAGE DURATION OF MITOTIC CYCLE IN HOURS CALCULATED FROM METAPHASE ACCUMULATIONS IN DIVIDING NUCLEI IN ROOT MERISTEMS TREATED WITH INHIBITORS BLOCKING MITOSIS. (ADAPTED FROM CLOWES,[17] TABLE 1, P. 7.)

			Central Cylinder	
Species	Quiescent Center	Rootcap Initials	Just above[a] QC[b]	200–250 μm Above[a] QC[b]
Zea mays	174	12	28	29
Vicia faba	292	44	37	26
Sinapis alba	520	35	32	25
Allium sativum	173	33	35	26

[a] Toward the base of root.
[b] Quiescent center.

cycles are largely caused by differences in the duration of G_1, the phase between the end of mitosis and the beginning of DNA synthesis.

Many views have been expressed on possible causes of the appearance of a quiescent center in a growing root. According to a proposal based on analyses of growth patterns in root tips, quiescence in the particular location of the root meristem results from antagonistic directions of cell growth in the various parts of the meristem, the rootcap being particularly important in the suppression of growth.[2] Another view singles out the presence of rapidly dividing cells around the quiescent center as the cause of the inactivity in that center.[73] This view is based on the following test. When excised roots, in which DNA synthesis and mitosis were arrested by starvation, were supplied with sucrose both processes were resumed in the root, including the quiescent center. But the latter soon ceased to participate in cell proliferation. One can visualize the mitotic activity of cells surrounding the quiescent center and the "antagonistic" directions of growth of these cells as complementary phenomena in their effect on the quiescent center.

Growth of the root tip

The region of actively dividing cells in a root tip extends for a considerable distance basipetally from the apex, that is, toward the older part of the root. These divisions are combined with cell enlargement and also overlap with cell differentiation in certain regions of the root. The maximum mitotic activity occurs not in the apical region but some distance from it, and this distance varies in different tissue regions (fig. 14.14). In *Zea* root, the rate of cell formation rises to a maximum 1.25 mm from the tip of the rootcap and declines to zero at about 2.5 mm.[21] Basipetally from this level, cell elongation alone brings about the further elongation of the root.

The overall elongation of roots was studied by making microscopic photographic records from live growing roots and calculating elemental growth rates (growth of infinitely small parts). In *Phleum* root, maximum elemental growth rate occurred 600 to 650 microns from the apex.[24] In *Zea,* the elemental growth rate was small near the apex, rose to a maximum 4 mm from the rootcap tip, where cells were elongating but no longer dividing, and fell to a minimum at 10 mm.[20]

Auxins regulate root growth in a complex way.[62] Auxin is necessary for the elongation of root cells but it also inhibits cell elongation and thus regulates its duration. The mechanism of auxin action in cell elongation has not been fully revealed but it appears most likely that auxin has its first effect upon the cell wall. Auxin reaches the growing region predominantly by an active polar transport directed from the base of the root toward the tip.

The tip of the root does not grow continuously at the same rate, expecially in perennial plants. In noble fir (*Abies procera*), for example, the roots show periodic deceleration of growth and have periods of dormancy.[74] Dormancy is preceded by a deposition of fatty materials, probably suberin, in the cortex and rootcap throughout a layer of cells that is continuous with the endodermis and covers the apical meristem. The latter thus becomes sealed off by a protective layer on all sides except toward the base of the root. Externally, such root tips appear brown. When growth is resumed, the brown covering is broken and the root tip pushed beyond it. Studies on excised roots indicate that roots may have a growth rhythm not dependent on seasonal changes but determined by internal factors.[66]

Primary differentiation

At various distances from the apical meristem cells enlarge and develop their specific characteristics in relation to their position in the root; the cells become differentiated. The meristems of the three regions of the root, the epidermis, the cortex, and the vascular cylinder, are delimited close to the apical meristem (figs. 14.12 and 14.14). Differences in distribution of mitoses and in the degree of the early cell enlargement contribute to the initial differentiation of tissue regions[35] (figs. 2.3 and 14.14).

EPIDERMIS

The protoderm of monocotyledon roots is set apart as a derivative of the initials that also give rise to the cortex (fig. 14.12,*A*,*B*). In the dicotyledons, the protoderm is separated by stepwise periclinal divisions (T divisions) from the derivatives of the layer of initials that forms the rootcap (figs. 14.12,*C*,*D*, and 14.13). In roots with no distinct stories of initials, the protoderm has a common origin with the rootcap and cortex (fig. 14.12,*E*,*F*) and is not clearly defined in early stages of development.

In the root region where vascular tissues are differentiating, the young epidermis proceeds with the formation of root hairs (fig. 14.4). Root hair-forming cells may not differ from those remaining devoid of hairs. In some plants, however, the protodermal cells that give rise to root hairs are smaller than their sister cells and show other features indicating specialization as hair-forming cells, or trichoblasts (chapter 7). The occurrence of cytologically distinct trichoblasts in the root epidermis has been found to be a taxonomically significant feature.[59]

Root hairs arise as small papillae in the region where cell division is subsiding and reach their full development at a root level where at least the first xylem is mature. The extension of the hair occurs at its tip where consequently the cell wall is soft. The proximal part of the hair shows a gradual, acropetally advancing hardening of the cell wall. Availability of calcium is often discussed as one of the factors in the control of elongation of the root hair.[67] Calcification of pectic substances is cited as a cause of hardening of the cell wall and the consequent termination of wall extension.[18] The extension of root hair wall proceeds rapidly (0.1 mm per hour in radish root[10]). Dictyosome vesicles appear to be actively involved in this wall growth.

CORTEX

The cortical ground meristem consists of longitudinal files of cells attaining their maximum number by periclinal divisions (T divisions) close to the apex (figs. 14.12 and 14.13). The increase in circumference of the cortical layers occurs by anticlinal divisions. In many roots, the periclinal divisions progress in a centripetal order, that is, they are repeated several times in the innermost layer of the nascent cortex (fig. 14.13). The repetition of divisions in the same relative position makes the resulting tissues appear orderly, arranged in more or less distinct radial files (fig. 14.15,*A*). In some roots the outer part of the cortex has a different pattern of divisions and its cell arrangement does not coincide with that of the inner cortex. After the periclinal divisions in the inner cortex are completed the innermost layer differentiates as the endodermis.

The intercellular spaces, so prominent in the root cortex, arise close to the apical meristem. In living roots observed in water mounts with the light microscope, the inter-

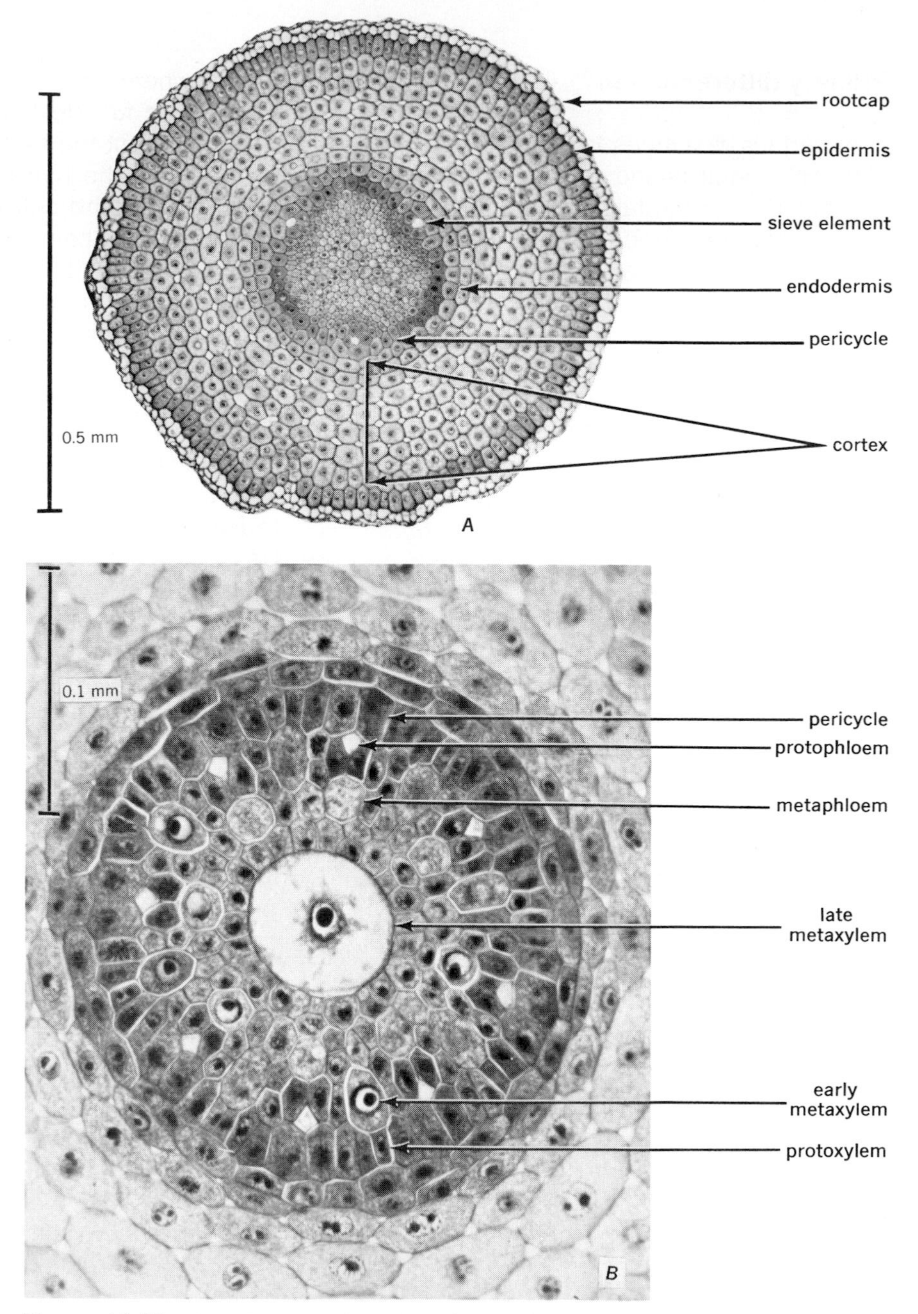

Figure 14.15 Transverse sections of differentiating roots. *A*, tomato (*Lycopersicon esculentum*). Three mature protophloem sieve elements. The vacuolated central part of the vascular cylinder indicates the triarch form of the future xylem. *B*, barley (*Hordeum vulgare*). Vascular cylinder with eight mature protophloem sieve elements, each accompanied by two companion cells. The metaphloem sieve elements are immature. In the xylem, which is all immature, the metaxylem elements are more highly vacuolated than the protoxylem elements. (*A*, from Rasa and Esau, *Hilgardia* 30:469–515, 1961; *B*, from Esau, *Hilgardia* 27:15–69, 1957.)

cellular spaces appear as black streaks because of light refraction from the gases contained in these spaces.

VASCULAR CYLINDER

The vascular cylinder is derived from procambium, or provascular tissue, appearing as a column in the center of the root. If the root has a pith, the latter is often interpreted as potential vascular tissue that, in the course of evolution, ceased to differentiate as such. In this context, the pith is regarded as a part of the vascular cylinder originating from the procambium. The contrary view is that the pith in the root is derived from the ground meristem as it is in the stem (chapter 16).

The pericycle is usually the first cell layer in the vascular cylinder that can be idenified as such (fig. 14.13). It delimits the vascular core close to the apical meristem. Within the core, enlargement and increasing vacuolation of certain cells block out the xylem pattern (diarch, triarch, etc.; figs. 14.15 and 14.16). Because the metaxylem cells exceed the protoxylem cells in width, they are the first to become recognizable at the onset of enlargement and vacuolation. However, the deposition of secondary walls and functional maturation occur first in the protoxylem and later in the metaxylem. Thus, the primary xylem differentiates centripetally (as seen in transections of the root) and is exarch. Phloem differentiation occurs in the same direction, that is, centripetally, for the protophloem appears next to the pericycle, the metaphloem deeper in the vascular core (fig. 14.15).

The sequence of xylem differentiation described above is common in monocotyledon and dicotyledon roots[23,51,56] but exceptions occur. In *Cucurbita pepo,* for example, while the protoxylem and the early metaxylem are maturing the central region of the root, as seen in transections, resembles a pith. Then, some 13 cm from the tip, a cell in the center enlarges to about four times of its original diameter and differentiates as a metaxylem vessel member, causing in the process a considerable rearrangement of the surrounding cells.[31,32]

The longitudinal differentiation of the primary vascular tissues in the root is acropetal, the first phloem maturing closer to the apical meristem than the first xylem (fig. 14.16). The distances between the apex and the first mature vascular elements, especially those of the xylem, vary[23] and are affected by age of root, rate of growth, plant species, presence of a disease, type of root (short or long type, terminal or lateral), and other factors.[48,56,63] In general, slowly growing roots have mature vascular elements closer to the apical meristem than rapidly growing roots. The proximity of mature vascular tissues to the apex may change in the same root under the influence of environmental conditions. During dormancy of perennial plants, for example, root growth decelerates and the maturation of vascular cells advances toward the apex.

Lateral roots

Lateral roots arise at the periphery of the vascular cylinder at variable distances from the apical meristem. Because of their deep-seated origin lateral roots are spoken of as being endogenous. Depending on the rank of the root that gives rise to laterals, the latter are frequently designated as secondary roots (laterals on a taproot), tertiary roots (laterals on a secondary root), and so on.

The lateral roots of gymnosperms and angiosperms, whether they arise on taproots or their branches or on adventitious roots, originate most commonly in the pericycle (fig. 14.17). The endodermis also may contribute some cell layers to the root primordium.[4,26] Not uncommonly, the cells of endodermal ori-

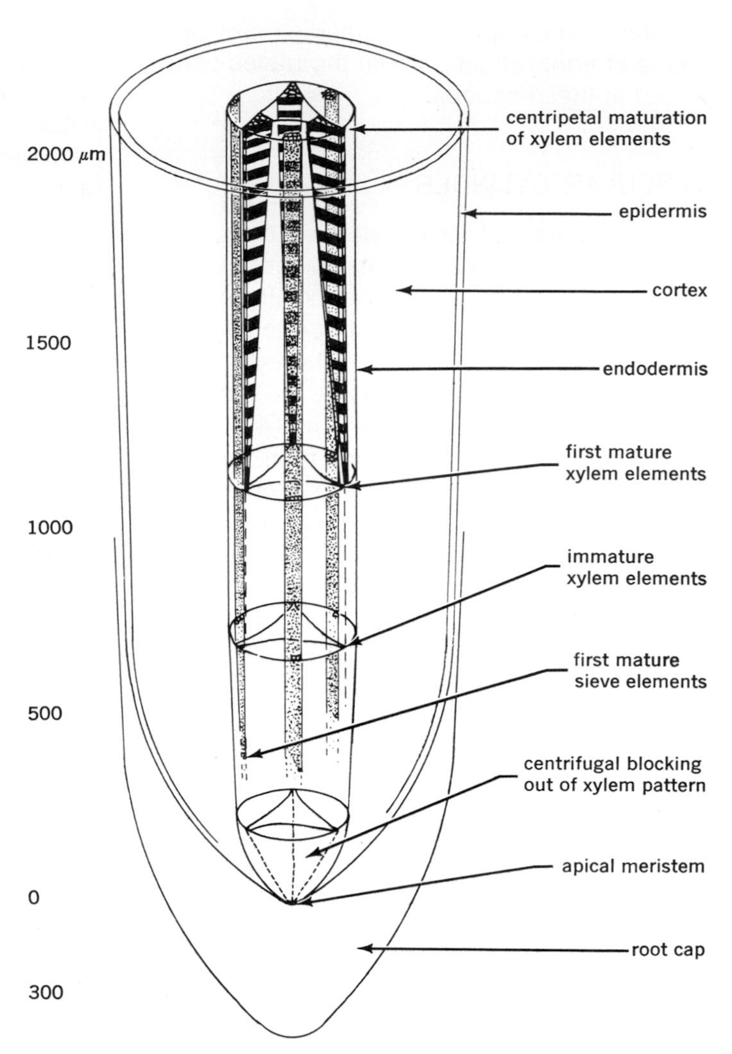

Figure 14.16 Diagram of primary vascular differentiation in a root of pea (*Pisum sativum*). Root was grown isolated from the plant in a nutrient medium, but sequence and pattern of differentiation correspond with those observed in roots attached to plants. (From Torrey, *Amer. J. Bot.* 40:525–533, 1953.)

gin are sloughed off after the lateral emerges from the parent root.[13,63] Lateral root formation is stimulated by auxins and other growth regulators,[5] but its suppression by endogenous inhibitors may be responsible for the frequency and distribution of the laterals on the parent root.[66]

The pericyclic origin of the lateral root places it in close juxtaposition with the vascular tissues of the parent root with which the vascular tissues of the lateral root eventually become connected. The position of the lateral root with regard to xylem ridges of the parent root varies in relation to the vascular pattern of

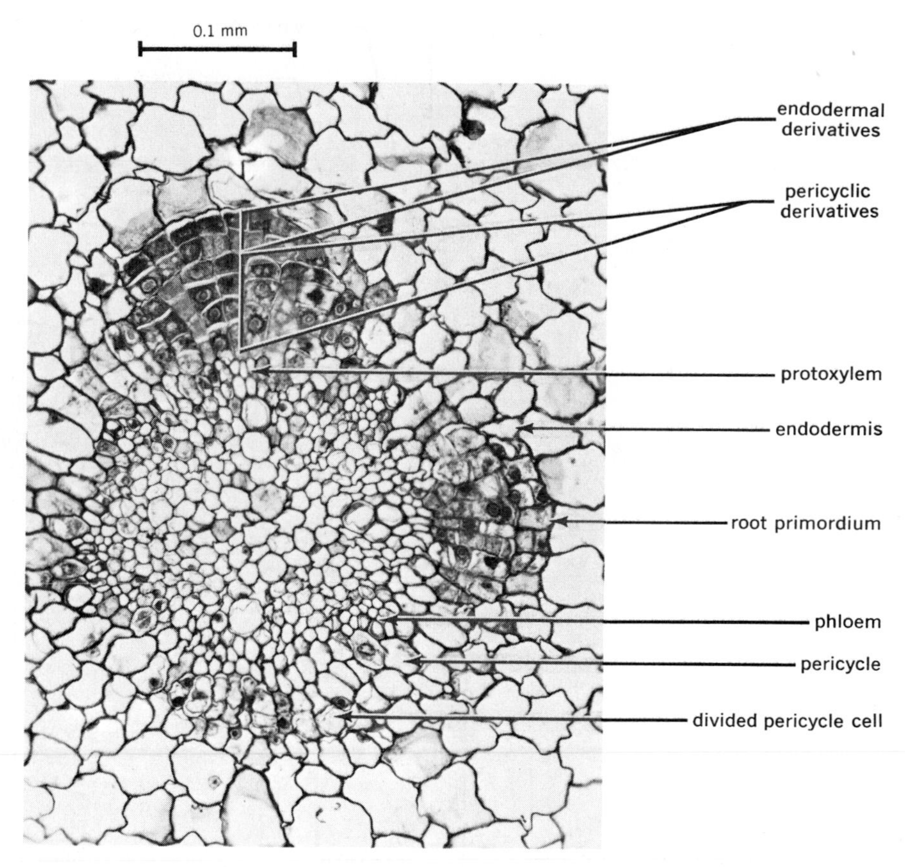

Figure 14.17 Origin of lateral roots in sunflower (*Helianthus annuus*). Two root primordia are present. Metaxylem of parent root is still immature. Position of lateral roots is similar to that in fig. 14.9, *C*. Parent root is in transection.

the parent root but is stable in a root with a given pattern (fig. 14.9). In a diarch root, the lateral root arises between phloem and xylem, in a triarch, tetrarch, and so forth, opposite the xylem, in a polyarch monocotyledon root, opposite the phloem.

When a lateral root is initiated, several contiguous pericyclic cells acquire dense cytoplasm and divide periclinally. The products of these divisions divide again, periclinally and anticlinally. The accumulating cells form a protrusion, the root primordium (fig. 14.17). As the primordium increases in length, it penetrates the cortex and emerges on the surface. The endodermis often divides only anticlinally and thus keeps pace with the growth of the primordium, but the other cortical cells are deformed, crushed, pushed aside, and probably partly degraded by enzymic activity.[4,8]. The lack of connection between the emerging lateral root and the cortex of the parent root leaves an opening into the interior of the parent root that may serve as an entry for pathogens. During growth of the young primordium through the cortex, the apical meristem and rootcap are initiated and the

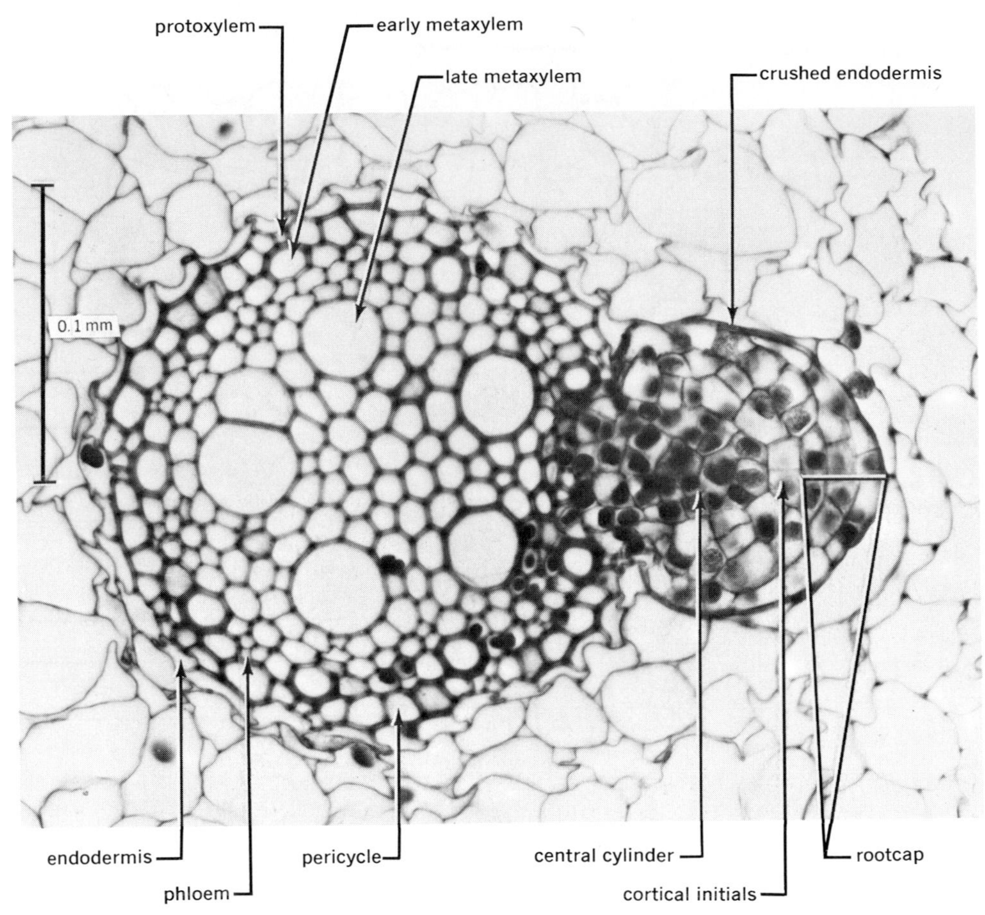

Figure 14.18 Origin of lateral root in brome-grass (*Bromus mollis*) root. One root primordium is present. Its position is similar to that in fig. 14.9, *D*. Parent root is in transection. (From Esau, *Hilgardia* 27:15–69, 1957.)

vascular cylinder and cortex blocked out behind the apical meristem (fig. 14.18).

At the site of origin of a lateral root the cells of the parent root are variously affected by the new growth. If the lateral root arises close to the apical meristem, the parent root cells are not yet fully differentiated and initiate the meristematic activity with no profound histologic changes. If the endodermis has casparian strips where the laterals arise it usually divides in front of the proliferating pericycle and may form casparian strips in the new cells, at least for a time. Lateral roots, however, may arise also at root levels where the pericycle and the endodermis have lignified secondary walls[4,37] (*Zea mays*). In such circumstances, major changes involving delignification and removal of secondary walls precede the meristematic activity concerned with the initiation of the lateral root.

When phloem and xylem begin to differentiate in the lateral root these tissues become connected with the equivalent tissues in the parent root by a differentiation of the intervening parenchyma cells into vascular elements. These parenchyma cells are first of all those derived from the divided pericycle. Depending on the extent of the vascular connection to be established, cells in the parent vascular cylinder beneath the pericycle may also participate by division and vascular differentiation in forming the connection.[4,23]

The development of new increments of growth by continued elongation of existing roots and by initiation of new lateral roots is considered to be an important feature with regard to absorption by roots. This growth creates new absorbing surfaces, and it brings these surfaces in contact with new areas of soil.

Buds on roots

Formation of buds on roots is a familiar phenomenon,[22,52] which is utilized in propagation of plants by root cuttings and serves as a means of spread of certain noxious weeds[9] and other harmful plants.[76] Root buds frequently arise endogenously, as do lateral roots. In comparative studies on the initiation of the two kinds of structures in *Convolvulus arvensis* roots cultured in vitro,[11] the earliest stages of root and bud formation appeared to be identical, but later stages differed not only in rates of development but also in the histologic organization and kind of participation of parent tissues in the growth of the primordium. Studies on phytochrome regulation of root bud development in the same species of *Convolvulus* have led to the understanding of the remarkable interrelation between (*1*) root growth, as regulated by the gravity perception mechanism (statoliths) and a light perception system (phytochrome), and (*2*) bud growth, as regulated by light received by the root during its occasional approach to the surface of the soil.[9]

REFERENCES

1. Armstrong, W. A re-examination of the functional significance of aerenchyma. *Physiol. Plant.* 27:173–177. 1972.
2. Barlow, P. W. Mitotic cycles in root meristems. In: *The cell cycle in development and differentiation.* M. Balls and F. S. Billet, eds. Cambridge University Press. 1973.
3. Barlow, P. W. Recovery of geotropism after removal of the root cap. *J. Exp. Bot.* 25:1137–1146. 1974.
4. Bell, J. K., and M. E. McCully. A histological study of lateral root initiation and development in *Zea mays. Protoplasma* 70:179–205. 1970.
5. Blakely, L. M., S. J. Rodaway, L. B. Hollen, and S. G. Croker. Control and kinetics of branch root formation in cultured root segments of *Haplopappus ravenii. Plant Physiol.* 50:35–42. 1972.
6. Bond, L. Origin and developmental morphology of root nodules of *Pisum sativum. Bot. Gaz.* 109:411–434. 1948.
7. Bonnett, H. T., Jr. The root endodermis: fine structure and function. *J. Cell Biol.* 37:109–205. 1968.
8. Bonnett, H. T., Jr. Cortical cell death during lateral root formation. *J. Cell Biol.* 40:144–159. 1969.
9. Bonnett, H. T. Phytochrome regulation of endogenous bud development in root cultures of *Convolvulus arvensis. Planta* 106:325–330. 1972.
10. Bonnett, H. T., Jr., and E. H. Newcomb. Coated vesicles and other cytoplasmic components of growing root hairs of radish. *Protoplasma* 62:59–75. 1966.
11. Bonnett, H. T., Jr., and J. G. Torrey. Comparative anatomy of endogenous bud and lateral root formation in *Convolvulus arvensis* roots cultured in vitro. *Amer. J. Bot.* 53:496–507. 1966.

12. Bruch, H. Beiträge zur Morphologie und Entwicklungsgeschichte der Fenchelwurzel (*Foeniculum vulgare* Mill.). *Beitr. Biol. Pflanz.* 32:1–26. 1955.
13. Byrne, J. M. The root apex of *Malva sylvestris*. III. Lateral root development and the quiescent center. *Amer. J. Bot.* 60:657–662. 1973.
14. Chilvers, G. A., and L. D. Pryor. The structure of eucalypt mycorrhizas. *Aust. J. Bot.* 13:245–259. 1965.
15. Clowes, F. A. L. *Apical meristems*. Botanical Monographs Vol. 2. Oxford, Blackwell. 1961.
16. Clowes, F. A. L. The functioning of meristems. *Sci. Prog. Oxford* 55:529–542. 1967.
17. Clowes, F. A. L. Anatomical aspects of structure and development. In: *Root growth.* Proc. Fifteenth Easter School in Agric. Sci. Univ. Nottingham 1968. London, Butterworth. 1969.
18. Cormack, R. G. H. The development of the root hairs in angiosperms. *Bot. Rev.* 15:583–612. 1949. II. *Bot. Rev.* 28:446–464. 1962.
19. Dycus, A. M., and L. Knudson. The role of the velamen of the aerial roots of orchids. *Bot. Gaz.* 119:78–87. 1957.
20. Erickson, R. O., and K. B. Sax. Elemental growth rate of the primary root of *Zea mays. Proc. Amer. Phil. Soc.* 100:487–498. 1956.
21. Erickson, R. O., and K. B. Sax. Rates of cell division and cell elongation in the growth of the primary root of *Zea mays. Proc. Amer. Phil. Soc.* 100:499–514. 1956.
22. Esau, K. *Plant anatomy*. 2nd ed. New York, John Wiley & Sons. 1965.
23. Esau, K. *Vascular differentiation in plants.* New York, Holt, Rinehart and Winston. 1965.
24. Goodwin, R. H., and C. J. Avers. Studies on roots. III. An analysis of root growth in *Phleum pratense* using photomicrographic records. *Amer. J. Bot.* 43:479–487. 1956.
25. Govier, R. N., J. G. S. Brown, and J. S. Pate. Hemiparasitic nutrition in angiosperms. II. Root haustoria and leaf glands of *Odonites varna (Bell) Dum.* and their relevance to the abstraction of solutes from the host. *New Phytol.* 67:963–972. 1968.
26. Guttenberg, H. von. Der primäre Bau der Angiospermenwurzel. *Handbuch der Pflanzenanatomie,* Band VIII, Teil 5. 1968.
27. Hacskaylo, E. Mycorrhiza: the ultimate in reciprocal parasitism? *BioScience* 22:577–583. 1972.
28. Hanstein, J. Die Scheitzelzellgruppe im Vegetationspunkt der Phanerogamen. *Festschr. Niederrhein. Ges. Natur- u. Heilkunde* 1868:109–134. 1868.
29. Hanstein, J. Die Entwickelung des *Keimes der Monokotylen und der Dikotylen. Bot. Abhandl.* 1:1–112. 1870.
30. Harkes, P. A. A. Structure and dynamics of the root cap of *Avena sativa* L. *Acta Bot. Neerl.* 22:321–328. 1973.
31. Harrison-Murray, R. S., and D. T. Clarkson. Relationships between structural development and the absorption of ions by the root system of *Cucurbita pepo. Planta* 114:1–16. 1973.
32. Hayward, H. E. *The structure of economic plants.* New York, Macmillan. 1938.
33. Iversen, T.-H., and P. Larsen. Movement of amyloplasts in the statocytes of geotropically stimulated roots. The preinversion effect. *Physiol. Plant.* 28:172–181. 1973.
34. Jenny, H., and K. Grossenbacher. Root-

soil boundary zones as seen in the electron microscope. *Proc. Soil Sci. Soc. Amer.* 27:273–277. 1963.

35. Jensen, W. A., and L. G. Kavaljian. An analysis of cell morphology and the periodicity of division in the root tip of *Allium cepa. Amer. J. Bot.* 45:365–372. 1958.
36. Juniper, B. E., and R. M. Roberts. Polysaccharide synthesis and the fine structure of root cap cells. *J. Roy. Micr. Soc.* 85:63–72. 1966.
37. Karas, I., and M. E. McCully. Further studies of the histology of lateral root development in *Zea mays. Protoplasma* 77:243–269. 1973.
38. Leiser, A. T. A mucilaginous root sheath in Ericaceae. *Amer. J. Bot.* 55:391–398. 1968.
39. Libbenga, K. R., and P. A. A. Harkness. Initial proliferation of cortical cells in the formation of root nodules in *Pisum sativum* L. *Planta* 114:17–28. 1973.
40. Luhan, M. Das Abschlussgewebe der Wurzeln unserer Alpenpflanzen. *Ber. Deut. Bot. Ges.* 68:87–92. 1955.
41. Luhan, M. Neues zur Anatomie der Alpenpflanzen. *Ber. Deut. Bot. Ges.* 72: 262–267. 1959.
42. Morré, D. J., D. D. Jones, and H. H. Mollenhauer. Golgi apparatus mediated polysaccharide secretion by outer root cap cells of *Zea mays.* I. Kinetics and secretory pathway. *Planta* 74:286–30l. 1967.
43. Nagahashi, G., W. W. Thomson, and R. T. Leonard. The Casparian strip as a barrier to the movement of lanthanum in corn roots. *Science* 183:670–671. 1974.
44. Northcote, D. H., and J. D. Pickett-Heaps. A function of the golgi apparatus in polysaccharide synthesis and transport in the root-cap cells of wheat. *Biochem. J.* 98:159–167. 1966.
45. Pate, J. S., B. E. S. Gunning, and L. G. Briarty. Ultrastructure and functioning of the transport system of the leguminous root nodule. *Planta* 85:11–34. 1969.
46. Peirson, D. R., and E. B. Dumbroff. Demonstration of a complete Casparian strip in *Avena* and *Ipomoea* by a fluorescent staining technique. *Can. J. Bot.* 47: 1869–1871. 1969.
47. Perbal, G. L'action des statolithes dans la réponse géotropique des racines de *Lens culinaris. Planta* 116:153–171. 1974.
48. Peterson, R. L. Differentiation and maturation of primary tissues in white mustard root tips. *Can. J. Bot.* 45:319–331. 1967.
49. Pilet, P. E. Géoperception et géoréaction racinaires. *Physiol. Vég.* 10:347–367. 1972.
50. Pillai, A., and S. K. Pillai. Air spaces in the roots of some monocotyledons. *Proc. Ind. Acad. Sci.* 55:296–301. 1962.
51. Popham, R. A. Levels of tissue differentiation in primary roots of *Pisum sativum. Amer. J. Bot.* 42:529–540. 1955.
52. Raju, M. V. S., R. T. Coupland, and T. A. Steves. On the occurrence of root buds on perennial plants in Saskatchewan. *Can. J. Bot.* 44:33–37. 1966.
53. Reyneke, W. F., and H. P. Van Der Schijff. The anatomy of contractile roots in *Eucomis* l'Hérit. *Ann Bot.* 38:977–982. 1974.
54. Richardson, S. D. The influence of rooting medium on the sturcture and development of the root cap in seedlings of *Acer saccharinum* L. *New Phytol.* 54:336–337. 1955.
55. Riedhart, J. M., and A. T. Guard. On the anatomy of the roots of apple seedlings. *Bot. Gaz.* 118:191–194. 1957.
56. Riopel, J. L., and T. A. Steeves. Studies

on the roots of *Musa acuminata*. I. The anatomy and development of main roots. *Ann. Bot.* 28:475– 490. 1964.

57. Robards, A. W., S. M. Jackson, D. T. Clarkson, and J. Sanderson. The structure of barley roots in relation to the transport of ions into the stele. *Protoplasma* 77:291–311. 1973.
58. Rosene, H. F. The water absorptive capacity of winter rye root-hairs. *New Phytol.* 54:95–97. 1955.
59. Row, H. C., and J. R. Reeder. Root hair development as evidence of relationships among genera of Gramineae. *Amer. J. Bot.* 44:596–601. 1957.
60. Safir, G. R., J. S. Boyer, and J. W. Gerdemann. Nutrient status and mycorrhizal enhancement of water transport in soybean. *Plant Physiol.* 49:700–703. 1972.
61. Schüepp, O. Untersuchungen über Wachstum und Formwechsel von Vegetationspunkten. *Jb. Wiss. Bot.* 57: 17–79. 1917.
62. Scott, T. K. Auxins and roots. *Ann. Rev. Plant Physiol.* 23:235–258. 1972.
63. Seago, J. L. Developmental anatomy in roots of *Ipomoea purpurea*. I. Radicle and primary root. *Amer. J. Bot.* 58:604–615. 1971. II. Initiation and development of secondary roots. *Amer. J. Bot.* 60:607–618. 1973.
64. Shen-Miller, J., and R. R. Hinchman. Gravity sensing in plants: a critique of the statolith theory. *BioScience* 24:643–651. 1974.
65. Street, H. E. Physiology of root growth. *Ann. Rev. Plant Physiol.* 17:315–344. 1966.
66. Street, H. E., and E. H. Roberts. Factors controlling meristematic activity in excised roots. I. Experiments showing the operation of internal factors. *Physiol. Plant.* 5:498–509. 1952.
67. Tanaka, Y., and F. W. Woods. Root and root hair growth in relation to supply and internal mobility of calcium. *Bot. Gaz.* 133:29–34. 1972.
68. Tanton, T. W., and S. H. Crowdy. Water pathways in higher plants. II. Water pathways in roots. *J. Exp. Bot.* 23:600–618. 1972.
69. Tepfer, D. A., and H. T. Bonnett. The role of phytochrome in the geotropic behavior of roots of *Convolvulus arvensis*. *Planta* 106:311–324. 1972.
70. Tomlinson, P. B. *Anatomy of the monocotyledons.* II. *Palmae.* III. *Commelinales—Zingiberales.* Oxford, Clarendon Press. 1961 and 1969.
71. Volkmann, D. Amyloplasten und Endomembranen. Das Geoperzeptionssystem der Primärwurzel. *Protoplasma* 79:159–183. 1974.
72. Voronin, N. S. Ob evolyutsii kornei rastenij. [On evolution of roots of plants.] *Byul. Moskov. Obshch. Isp. Prirody, Otd. Biol.* 61:47–58. 1956.
73. Webster, P. L., and H. D. Langenauer. Experimental control of the activity of the quiescent center in excised root tips of *Zea mays*. *Planta* 112:91–100. 1973.
74. Wilcox, H. Primary organization of active and dormant roots of noble fir. *Abies procera*. *Amer. J. Bot.* 41:812–821. 1954.
75. Wilkins, M. B. Geotropism. *Ann. Rev. Plant Physiol.* 17:379–408. 1966.
76. Wilkinson, R. E. Adventitious shoots on saltcedar roots. *Bot. Gaz.* 127:103–104. 1966.
77. Wilson, K., and J. N. Honey. Root contraction in *Hyacinthus orientalis*. *Ann. Bot.* 30:47–61. 1966.

15

The Root: Secondary State of Growth and Adventitious Roots

The secondary growth in roots, as in stems, consists of the formation of secondary vascular tissues from a vascular cambium and of periderm from a phellogen. Secondary growth is characteristic of roots of gymnosperms and occurs in various amounts in most dicotyledons. As was mentioned in chapter 14 the roots of monocotyledons commonly lack secondary growth. The secondary growth of roots may show peculiarities in relation to functional specialization.

COMMON TYPE OF SECONDARY GROWTH

The vascular cambium is initiated by divisions of those procambial cells that remain undifferentiated between the primary phloem and the primary xylem (fig. 15.1,*A–D*). Thus, in the beginning, the cambium has the form of strips, the number of which depends on the type of root (fig. 15.1,*C*). There are two strips in a diarch root, three in a triarch root, and so forth. Subsequently, the pericyclic cells located outside the xylem ridges also become active as a cambium, and then the cambium completely encircles the xylem core. This early cambium has the same outline as the xylem; in cross sections it is oval in diarch roots, triangular in triarch roots, and many-angled in polyarch roots. The cambium located on the inner face of the phloem begins to function earlier than the pericyclic part of the cambium. By formation of secondary xylem opposite the phloem the cambium is displaced outward, and eventually its circumference becomes circular in cross section (fig. 15.1,*E*).

The cambium produces phloem and xylem cells (fig. 15.2) by periclinal divisions and increases in circumference by anticlinal divisions. The cambium arising on the inner face of the phloem produces conducting elements and associated cells of the xylem and phloem (fig. 15.2,*A*). In some roots the cambium originating in the pericycle produces ray parenchyma (figs. 15.2,*B*, and 15.3,*B*,*C*). Rays also appear in the other parts of the secondary tissues (fig. 15.2,*C*), but those origin-

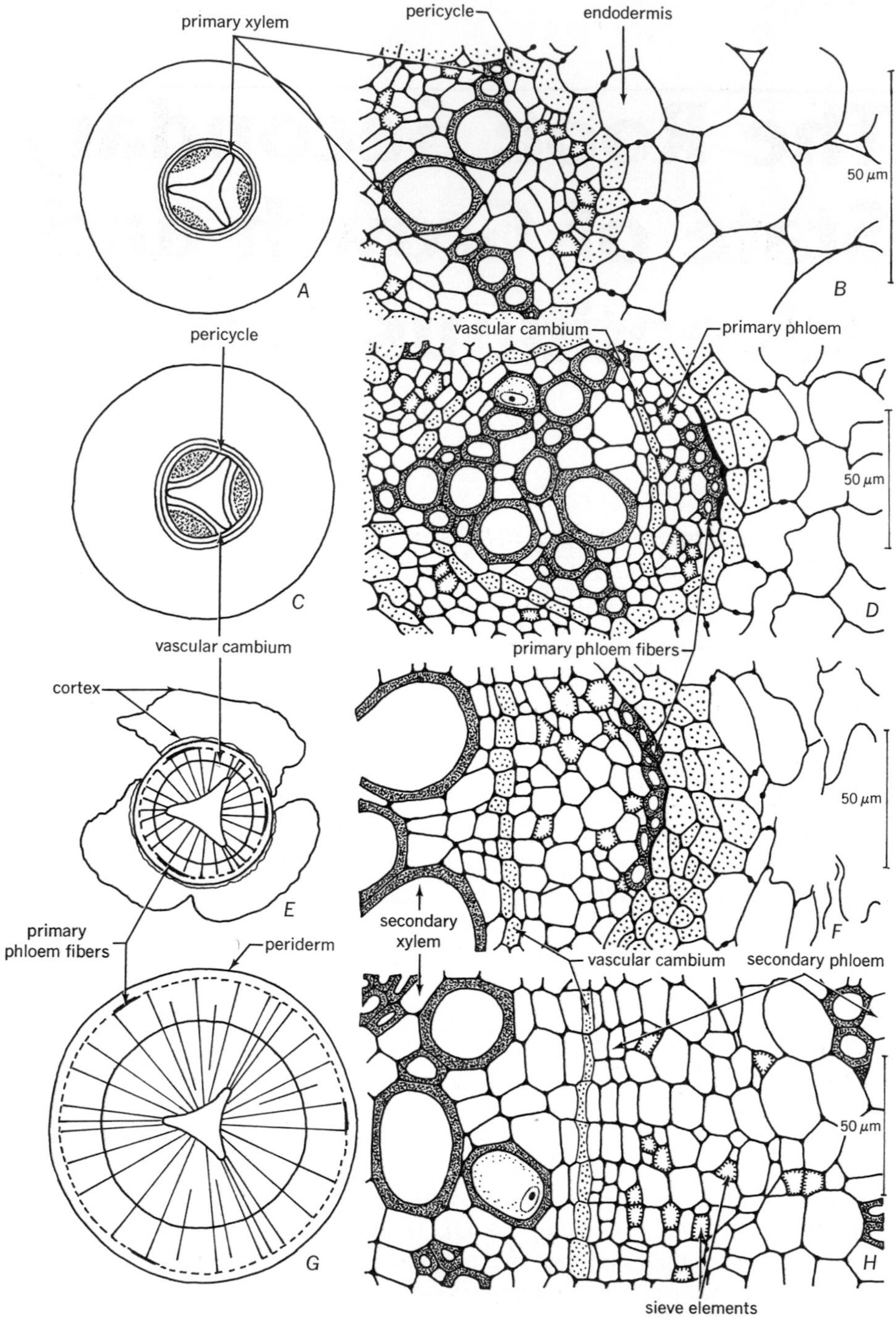

Figure 15.1 Diagrams and detailed drawings of cross sections of root of alfalfa (*Medicago sativa*) in different stages of development. *A*, *B*, primary stage of growth. *C*, *D*, initiation of vascular cambium. *E*, *F*, secondary growth of vascular cylinder, cell division in pericycle, and rupture of cortex. *G*,*H*, secondary growth is established.

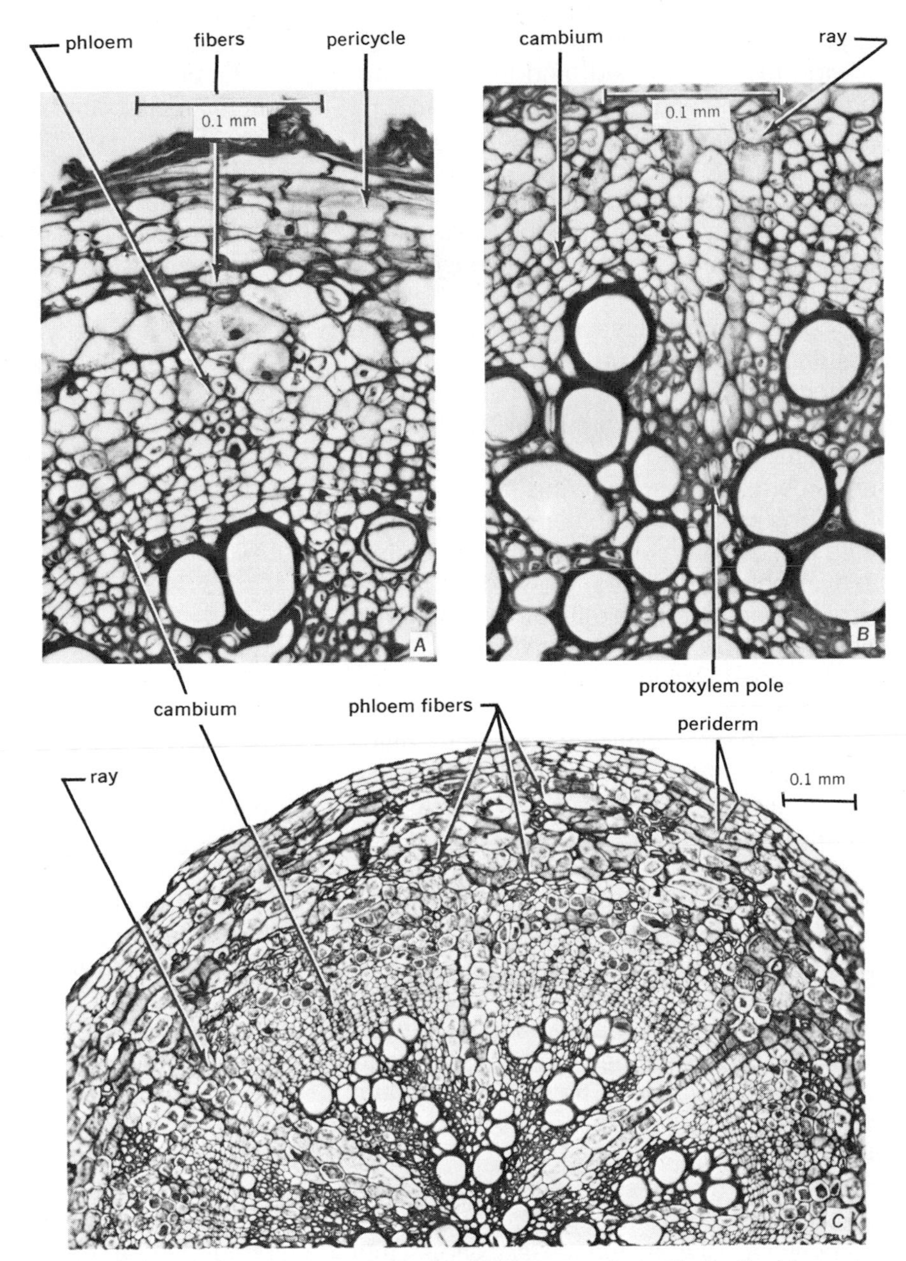

Figure 15.2 Cross sections of root of alfalfa (*Medicago sativa*) with details of secondary growth. *A*, *B*, early stage of secondary growth similar to that in fig. 15.1, *E*, *F*. *A*, arrangement of tissues along radius through phloem; cortex (outside pericycle) is collapsed. *B*, region opposite a protoxylem pole. *C*, root in advanced stage of secondary growth similar to that in fig. 15.1, *G*, *H*.

ating in the pericycle opposite the xylem ridges are frequently the widest. In some roots no wide rays are formed, and the xylem appears rather homogeneous (fig. 15.3,*A*,D).

The formation of periderm follows the initiation of secondary vascular growth. The pericyclic cells undergo periclinal and anticlinal divisions (figs. 15.1,*D*,*F*, and 15.2,*A*). The periclinal divisions cause an increase of the number of pericyclic layers in radial extent. The combined increase in thickness of the vascular tissues and of the pericycle forces the cortex outward. The cortex does not undergo an increase in circumference but becomes ruptured and is cast off together with the epidermis and endodermis (fig. 15.1,*E*). A phellogen arises in the outer part of the pericycle and forms phellem toward the outside. It may produce phelloderm toward the inside, but such phelloderm is difficult to distinguish from the pericycle which proliferated before the phellogen was initiated.

In perennial roots the activity of the vascular cambium continues through many years. The phellogen also continues its activity but may become replaced by phellogens arising at greater depths in the root. If such development occurs, the root, like a stem, has a rhytidome.

HERBACEOUS DICOTYLEDON

Secondary growth in a herbaceous dicotyledon may be exemplified by the root of *Medicago sativa*, alfalfa.[10, 19] The secondary xylem contains vessels of various diameters, mostly with scalariformly and reticulately pitted secondary walls. The vessels are accompanied by fibers and parenchyma cells. Wide rays of parenchyma divide the axial xylem into sectors (fig. 15.2,*C*). During the secondary growth the primary xylem becomes considerably modified by dilatation growth of the primary xylem parenchyma. The files of primary tracheary elements are broken and partly crushed.

The phloem contains sieve tubes with companion cells, fibers, and parenchyma cells (fig. 15.1,*H*). The wide rays of the xylem are continuous through the cambium with similar rays in the phloem (fig. 15.2,*C*). The outer phloem contains only fibers and storage parenchyma; the old sieve tubes are crushed. The phloem merges imperceptibly with the pericyclic parenchyma beneath the periderm, except where fibers are present. Cork derived from the phellogen forms the protective tissue.

The amount of secondary growth varies in different herbaceous dicotyledons as do also the structure of tissues and the distinctness of periderm (fig. 15.3).

WOODY SPECIES

The organization of secondary vascular tissues in roots of woody species resembles that just outlined for the root of alfalfa.[5] Usually the roots of trees have a larger proportion of elements with lignified secondary walls (fig. 15.4), but roots of herbaceous plants also may become strongly sclerified (fig. 15.3,*A*). The roots of gymnosperms (fig. 15.4,*A*) have a type of secondary growth similar to that in trees of dicotyledons (fig. 15.4,*B*), except that their conducting elements are phylogenetically less advanced (chapters 8 and 9).

Roots and stems of trees show histologic differences, particularly in the wood. As compared with stems, the roots have a smaller proportion of elements with lignified secondary walls in bark and wood and, correspondingly, a higher proportion of parenchyma tissue.[7] In a *Platanus* species, for example, the root has higher and wider vascular rays than the stem.[21] The root wood of *Platanus* also has wider vessels than the stem

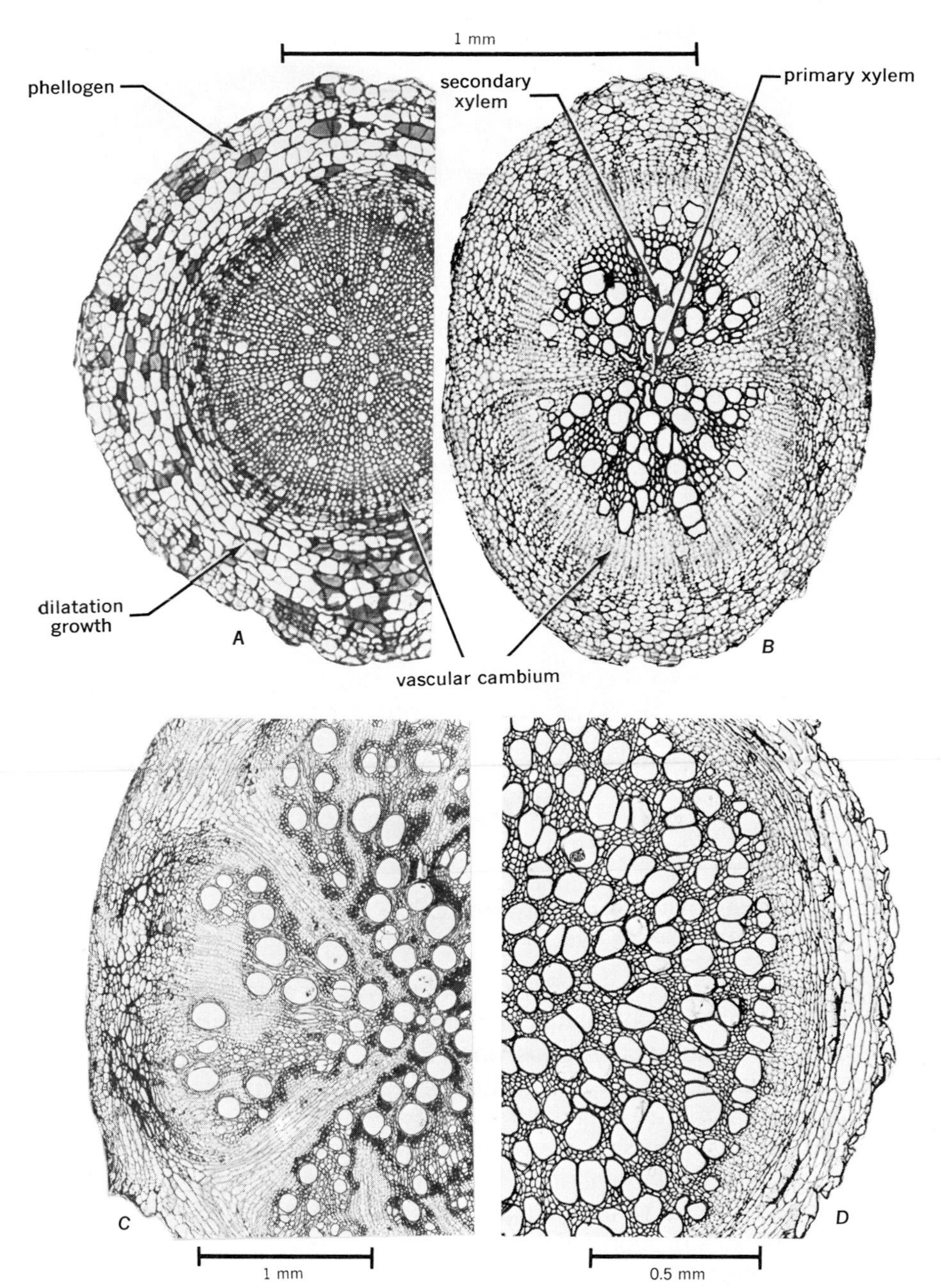

Figure 15.3 Cross sections of roots of herbaceous plants in secondary state of growth. *A*, tomato (*Lycopersicon esculentum*). *B*, cabbage (*Brassica oleracea*). *C*, pumpkin (*Cucurbita sp.*). *D*, potato (*Solanum tuberosum*).

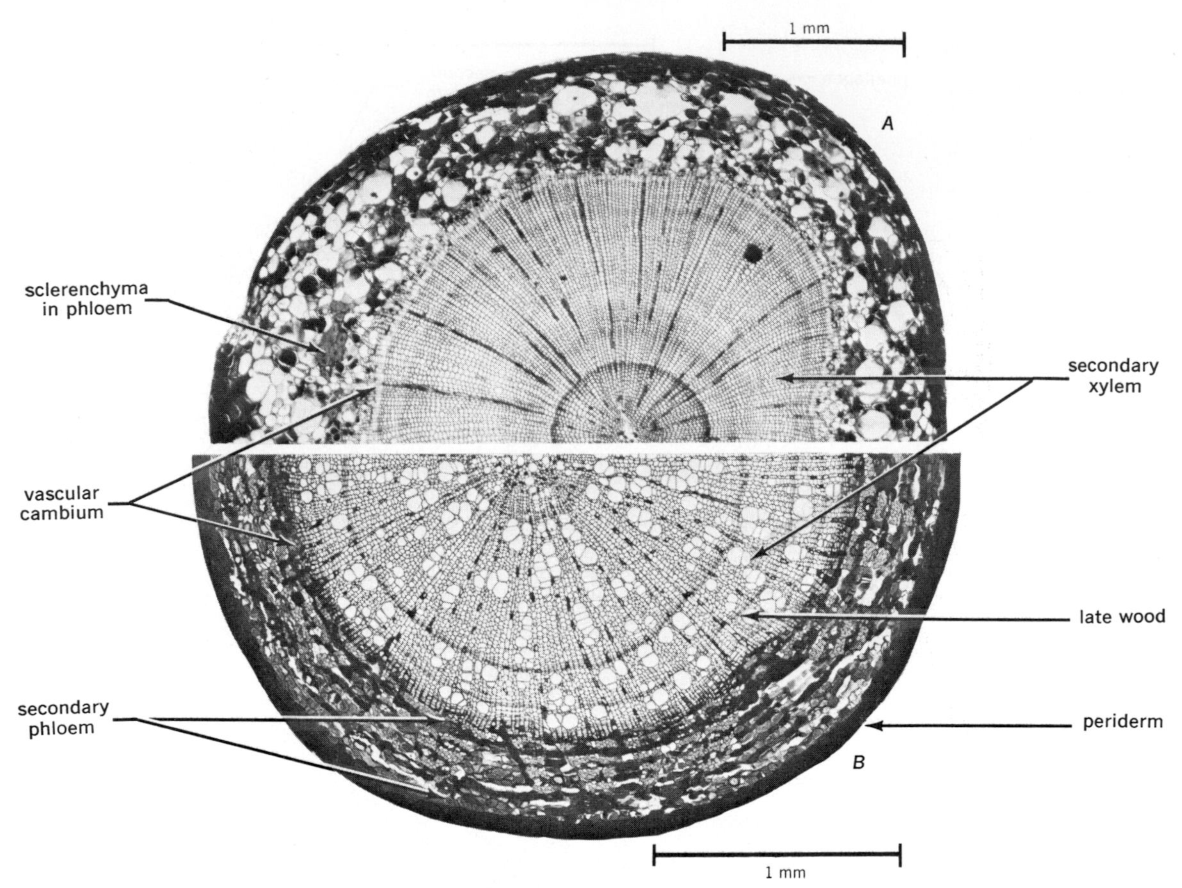

Figure 15.4 Cross sections of roots of woody species in secondary state of growth. *A*, *Abies*, fir. *B*, *Tilia*. (After Esau, *Plant Anatomy*, 2nd ed. John Wiley & Sons, 1965.)

wood and these are so uniformly distributed that the delimitation of the annual increments is obscured. The study of *Platanus* wood indicates, moreover, that the wood of the root is phylogenetically more primitive than that of the stem in having longer parenchyma strands, longer vessel members with a larger number of bars in the scalariform perforation plates, as well as a proportionally lower value for the elongation of fibers as related to the length of the vessel members. This observation conflicts with the data on the evolution of the primary xylem in monocotyledons in which the specialization of tracheary elements advances from the root toward the aerial parts (chapter 8).

The histologic differences between the secondary tissues of stems and roots are determined to a large extent by differences in the environment in which the two parts of the plant body develop. If roots of dicotyledon or gymnosperm trees are exposed to light and air, the wood that develops after the exposure assumes most of the characteristics of the wood in stems.[7, 14]

An important horticultural and ecological aspect of secondary growth in roots is the occurrence of natural root grafting between different trees of the same species, a phenomenon that is widespread among dicotyledons and gymnosperms in the tropics and in the temperate zone.[2] Two roots of the same tree also develop graft unions. Where roots come in contact with each other they become united through secondary growth. The grafting of roots establishes a continuity of the vascular system between the two graft partners, which is demonstrated by movement of experimentally introduced dyes, poisons, and radioactive substances from one tree to another.[2, 13] In a given stand of trees, a large number of individuals become interconnected, and if some trees are cut down, the stumps remain alive for a long time.[13, 18] The grafting of roots is one of the effective means of transmission of infectious diseases from one tree to another.

The development of natural root grafts was studied in aerial roots of *Ficus globosa*.[15] These roots have epidermal hairs which fuse with the hairs of another root when two roots come in contact with one another (fig. 15.5,*A*). As the roots increase in thickness by secondary growth, they come into closer contact and press against each other. Rays near the contact region proliferate and produce parenchyma tissue. The epidermis and the compressed cortex and phloem of both roots are crushed and obliterated, and the xylem is invaded by the proliferating parenchyma (fig.

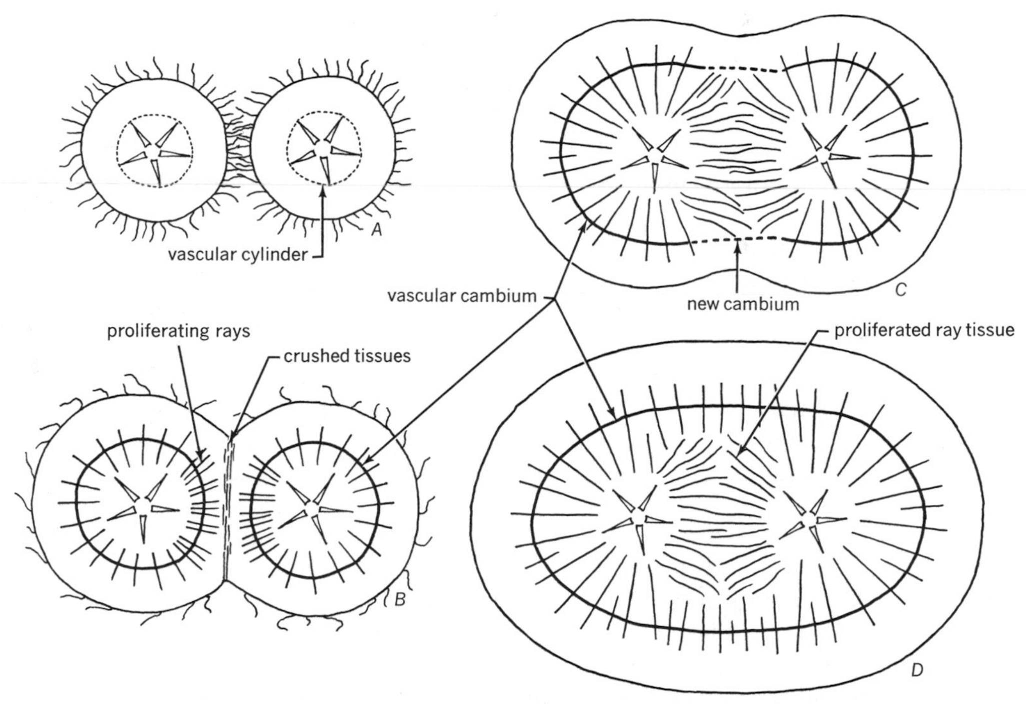

Figure 15.5 Diagrams illustrating development of graft union between two aerial roots of *Ficus globosa*. Transections. *A*, roots growing close together show fusion of epidermal hairs. *B*, enlargement of roots by secondary growth brings them closer together. In region of contact, tissues are crushed and rays begin to proliferate. *C*, new cambium arises in parenchyma derived from proliferated rays and connects the cambia of the two roots. *D*, the two roots are completely united. (Constructed from data in Rao.[15])

15.5,*B*). The proliferated parenchyma of the two roots merges and produces a cambium in continuity with the cambia of the two roots (new cambium in fig. 15.5,*C*). The now continuous cambial cylinder forms vascular tissues as though the two roots were one root (fig. 15.5,*D*).

VARIATIONS IN SECONDARY GROWTH

Herbaceous dicotyledons often have a limited amount of secondary growth, which may be associated with characteristic features. In *Actaea,* for example, the conducting part of the secondary vascular tissues appears in discrete strands separated from each other by wide rays of large-celled parenchyma. These rays originate in the pericycle opposite the protoxylem ridges. The divisions extending these rays radially occur in line with the vascular cambium between the xylem and the phloem. The dividing cells in the rays may be interpreted accordingly as part of the cambium, although they are not readily identifiable as such in sections.

Actaea, Convolvulus, and some other herbaceous plants with roots having a limited amount of secondary growth have a superficial periderm and therefore retain their cortex. The endodermis may increase in circumference by radial cell division and tangential cell enlargement, like the rest of the cortex (*Actaea*), or it may be crushed (*Convolvulus*). In roots of *Citrus sinesis* the periderm is first formed beneath the epidermis; later a deeper periderm arises in the pericycle. In certain families (Rosaceae, Myrtaceae, Onagraceae, Hypericaceae), a protective tissue called polyderm is formed by the pericycle (chapter 12). The products of the initiating layer of polyderm are not ordinary cork cells but consist of rows of nonsuberized parenchyma cells alternating with rows of suberized cells resembling endodermal cells.

STORAGE ROOTS

Several variations in secondary structure occur in connection with the development of storage roots (usually a combination of root and hypocotyl). In roots of Apiaceae, as those of fennel (*Foeniculum*[3]) or carrot (*Daucus*[4]), the secondary growth is of the ordinary kind, but parenchyma predominates in the xylem and phloem. In the beet (*Beta*[10]), however, the main increase in thickness results from the so-called anomalous type of growth (fig. 15.6). A series of supernumerary cambia arranged approximately concentrically arise outside the normal vascular core. The cambial cells are derived from cells of pericycle and phloem and produce several increments of vascular tissues, each composed of storage parenchyma and strands of xylem and phloem separated from one another by wide radial panels of parenchyma (fig. 15.6,*B*).

Another complex type of anomalous growth occurs in the fleshy adventitious roots of *Ipomoea batatas,* the sweet potato.[10] The xylem contains a large proportion of parenchyma and arises in the usual manner. Cambia develop in the parenchyma around individual vessels or vessel groups and produce a few tracheary elements toward the vessels, a few sieve tubes and laticifers away from the vessels, and a considerable number of storage parenchyma cells in both directions (fig. 15.7,*A*,*B*). Thus, phloem elements appear within the part of the root that originally differentiated as xylem. A cambium in normal position separates the xylem from the phloem in normal position, and a periderm of pericyclic origin occurs on the periphery (fig. 15.7,*B*). In the fleshy taproots and stems of some Brassicaceae (turnip, radish, kohlrabi, rutabaga, and others) parenchyma of the xylem and pith (if present) proliferates, and subsequently cambia and vascular tissues arise in this parenchyma[10] (fig. 15.7,*C*,*D*).

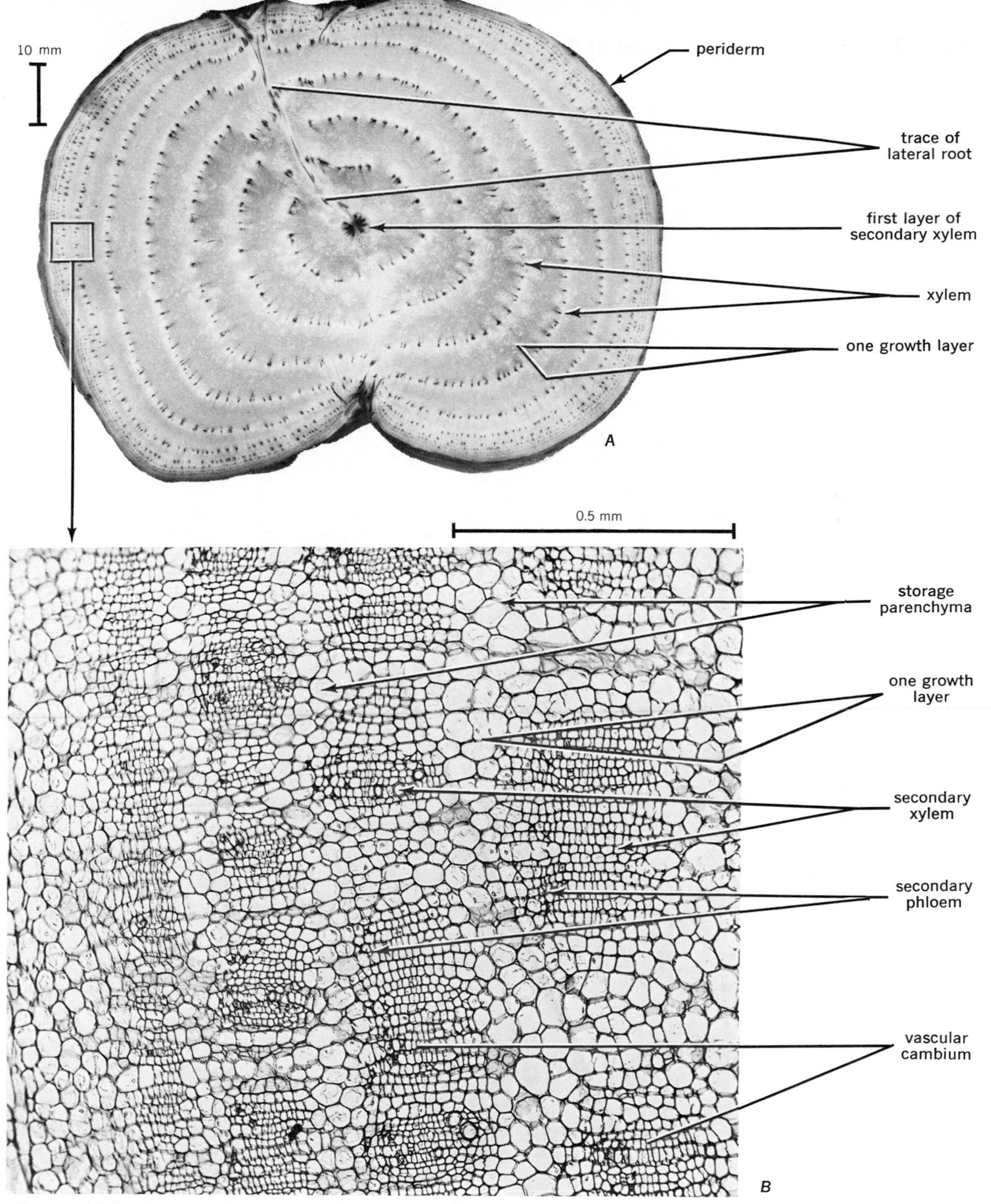

Figure 15.6 Cross sections of root of sugar beet (*Beta vulgaris*) illustrating anomalous secondary growth resulting from formation of many cambial layers outside the ordinary cambium, each of which gives rise to xylem and phloem cells together with storage parenchyma. (From E. Artschwager, *J. Agr. Res.* 33:143–176, 1926.)

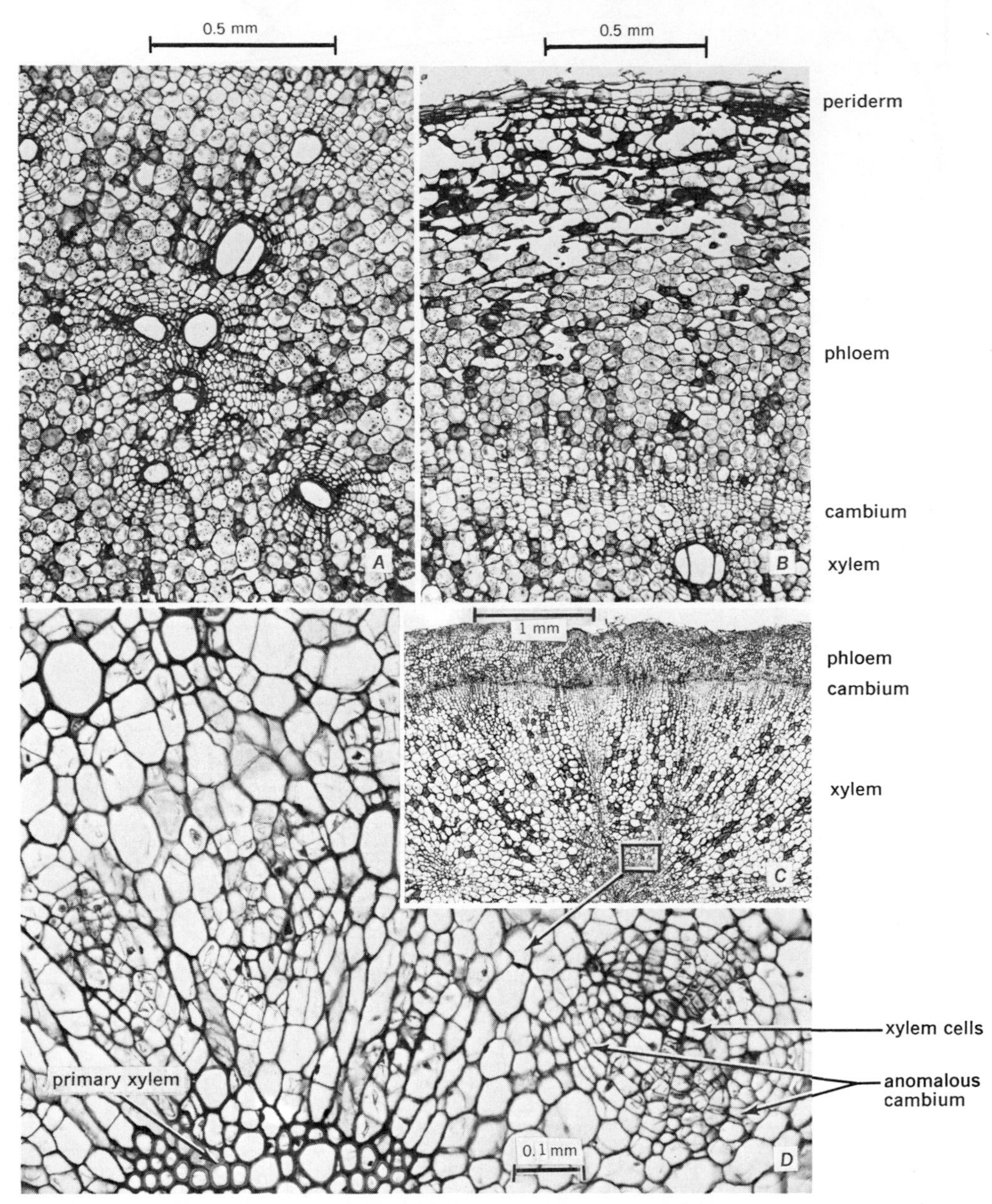

Figure 15.7 Cross sections illustrating anomalous secondary growth in storage roots. *A*, *B*, sweet potato (*Ipomoea batatas*). Anomalous cambium around vessels of secondary xylem in *A*; normal cambium in *B*. *C*, *D*, radish (*Raphanus sativus*) with anomalous cambium in secondary xylem.

The common character of all the fleshy storage organs derived from hypocotyls, roots, and sometimes stems (kohlrabi, *Brassica caulorapa*) is the possession of abundant storage parenchyma permeated by vascular tissues. This close association between conducting and storage tissues is attained by various modifications in the form of secondary growth.

PHYSIOLOGIC ASPECTS OF SECONDARY GROWTH IN ROOTS

The observations that excised roots consistently fail to undergo secondary growth without special treatments indicate that the development of vascular cambium in roots depends on some factor or factors in the shoot.[24] In conformity with this interpretation, studies on the effect of photoperiod on root growth are showing that the thickening of storage roots characteristic of some crop plants depends on the relative length of day and night to which the shoot is exposed. This relation is regarded as evidence that under the proper photoperiodic stimulus the shoot synthesizes substances that are transported to the roots and there induce and sustain cambial activity. In experiments with excised roots, which received various combinations of nutrients through an agar medium, these substances have been identified as sugar, vitamins, and hormones. The combinations of substances required by the roots are not necessarily the same for different species. For secondary growth in pea roots, for example, sucrose and a relatively high concentration of auxin must be present in the medium, whereas radish roots require, additionally, a low concentration of cytokinin.[23, 25]

The dependence of the root upon the shoot for the stimulation of cambial activity is perceived also in the pattern of the seasonal cambial activity characteristic of trees periodically undergoing cessation of growth. As was pointed out in chapter 10, the initiation of cambial activity in the shoot in the spring is apparently stimulated by growth substances synthesized in the reactivating buds. Numerous studies have shown that the stimulus is propagated from the buds downward through the stem to the roots and to their tips.[7] Thus, the spring reactivation of the cambium in the root (if the root becomes completely dormant) lags behind that in the stem. The cessation of secondary growth in the fall also proceeds basipetally in the stem, and it appears to occur later in the root than in the stem.[6] But, judged by studies on species of *Pinus*,[7] in the root itself the termination of cambial activity progresses basipetally as in the shoot, that is, from the tip to the base of the root.

ADVENTITIOUS ROOTS

The term adventitious root has somewhat varied meanings. In this book it is used broadly to designate roots that arise on aerial plant parts, on underground stems, and on more or less old root parts, especially those having undergone secondary growth. Adventitious roots may develop in intact plants growing under natural conditions, or arise in connection with infections by disease agents (fig. 15.8), or after experimental surgery or other injury. They may develop on excised plant parts or in tissue cultures. Some adventitious roots develop from preformed dormant primordia that require an additional stimulus to resume growth. Sometimes the distinction between adventitious and lateral roots is not sharp. One can use as a guide a definition of lateral roots as roots that arise in a typically acropetal succession on such other roots that are still in the primary state of growth. The

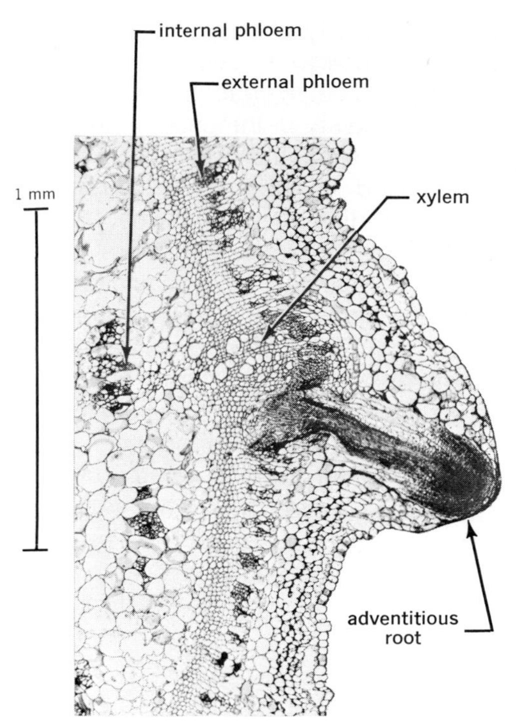

Figure 15.8 Cross section of stem of aster yellows-infected tomato. Adventitious root originated in parenchyma of external phloem. External and internal phloem is degenerating in response to mycoplasmal infection. (Slide, courtesy of E. A. Rasa.)

roots producing laterals may be taproots or adventitious roots or laterals of various orders on the two kinds of roots.

Adventitious roots are widely distributed in all vascular plants and are formed in many locations on the plant.[1,8] They may occur at nodes in association with axillary shoots (e.g., tillering in grasses) but may also be independent of axillary buds and may develop on internodes. In some species excised leaves readily form adventitious roots (*Begonia, Peperomia, Sedum*). The development of adventitious roots is important in the propagation of plants by means of stem cuttings or excised leaves. Adventitious roots constitute the main root system in lower vascular plants, monocotyledons, and dicotyledons that are propagated by means of rhizomes and runners, and in water plants, saprophytes, and parasites. Monocotyledons are particularly well known for their elaborate adventitious root systems. A few weeks after germination a cereal plant can produce 50 meters of roots composed of a dozen or more adventitious roots bearing some 2000 lateral roots.[16] A monocotyledon is usually characterized as having the adventitious roots early replacing the seedling taproot,[22] but apparently the taproot does not necessarily die off when the adventitious roots develop.[11,12]

The origin and development of adventitious roots resemble those of lateral roots: they usually have an endogenous origin and arise close to the vascular tissues (fig. 15.8), and they grow through tissues located outside the point of origin. During this growth the adventitious root originating in a relatively old stem may find an obstacle in the form of a sheath of sclerenchyma on the periphery of the vascular cylinder, which deflects the root from its normally radial course.[20,22]

In young stems of dicotyledons and gymnosperms the adventitious roots commonly arise in the interfascicular parenchyma and, in older stems, in the vascular ray near the cambium. Thus the new root appears close to both xylem and phloem.[17,20] When adventitious roots are formed in cuttings they may originate in the callus tissue that is often formed at the base of cuttings. If it is stated that an adventitious root arises in the pericycle in a stem of a gymnosperm or an angiosperm, the developmental history reveals that the tissue is primary phloem (chapter 16).

Adventitious and the so-called regenerated lateral roots are of great importance in the growth of trees. Although consisting initially of a taproot and its laterals, tree roots are frequently injured during their progress through the soil or are pruned at tree planting time in cultural practice so that, during further growth, the original roots are replaced by roots arising near the injured surfaces ("replacement roots"). The origin of replacement roots has been well illustrated by a study of pruned roots of *Abies nobilis,* the noble fir.[26] Wound healing and callus formation (chapter 17) occur on the pruned surfaces of roots. Subsequently replacement roots arise, in part from the undisturbed tissues beneath the wound surface, in part from derivatives of the callus. Younger pruned roots produce replacement roots in the same positions where lateral roots arise, that is, opposite the protoxylem ridges, and they are initiated in the pericycle or in the pericyclic callus. If the root is pruned after some secondary tissues have been formed, the replacement roots arise in various positions around the circumference of the vascular cylinder and are initiated in the vascular cambium or in the cambium that has developed in the callus.

The adventitious root primordia are initiated by divisions of parenchyma cells—callus cells or other parenchyma cells—resembling the divisions initiating the lateral roots in the pericycle of young roots. Before the adventitious root emerges from the stem or root it differentiates an apical meristem, a rootcap, and the beginning of the vascular cylinder and cortex. When vascular elements differentiate in the adventitious root the callus cells or other parenchyma cells located at the proximal end of the primordium differentiate into vascular elements and provide a connection with corresponding elements of the initiating organ.

The phenomenon of adventitious root formation has been widely explored in connection with research on growth substances. In cuttings that are naturally able to regenerate roots, applied auxins increase the number of developing adventitious roots, whereas applied gibberellins reduce their number. To elucidate these antagonistic effects at the cytologic level, the action of the two substances was investigated by the use of root primordia in different stages of development on an intact willow plant.[9] The action of auxin was studied by the effect of inhibition of its transport by triiodobenzoic acid. The gibberellic acid was added to the nutrient solution into which the plant roots were placed. The study showed that older primordia were less dependent on basipetally transported auxin than primordia just being initiated. Gibberellic acid caused a reduction in the number of cells in growing primordia but had little effect on the initiation process. The concluding inference was that gibberellic acid blocks the action of auxin in some process of root formation that occurs after the root is initiated.

REFERENCES

1. Baranova, E. A. *Zakonomernosti obrazovaniya pridatochnykh kornej u rastenij* [Laws of formation of adventitious roots in plants.] *Trudy Glav. Bot. Sada* 2:168–193. 1951.
2. Bormann, F. H. Root grafting and non-competitive relationship between trees. Pp. 237–246. In: *Tree growth.* T. T. Kozlowski, ed. New York, Ronald Press Company. 1962.
3. Bruch, H. Beiträge zur Morphologie und Entwicklungsgeschichte der Fenchelwurzel (*Foeniculum vulgare* Mill.) *Beitr. Biol. Pflanz.* 32:1–26. 1955.
4. Esau, K. Developmental anatomy of the fleshy storage organ of *Daucus carota. Hilgardia* 13:175–226. 1940.
5. Esau, K. Vascular differentiation in the pear root. *Hilgardia* 15:299–311. 1943.

6. Esau, K. *Plant anatomy.* 2nd ed. New York, John Wiley & Sons. 1965.
7. Fayle, D. C. F. Radial growth in tree roots. *Fac. Forest., Univ. Toronto, Tech. Rep.* 9:1–183. 1968.
8. Guttenberg, H. von. Der primäre Bau der Angiospermenwurzel. *Handbuch der Pflanzenanatomie.* Band 8. Teil 5. 1968.
9. Haissig, B. E. Meristematic activity during adventitious root primordium development. Influences of endogenous auxin and applied gibberellic acid. *Plant Physiol.* 49:886–892. 1972.
10. Hayward, H. E. *The structure of economic plants.* New York, The Macmillan Company. 1938.
11. Kausche, W. Die Primärwurzel von *Zea mays* L. *Planta* 73:328–332. 1967.
12. Kausche, W. Lebensdauer der Primärwurzel von Monokotylen. *Naturwiss.* 54:475. 1967.
13. Miller, L., and F. W. Woods. Root grafting in loblolly pine. *Bot. Gaz.* 126:252–255. 1965.
14. Morrison, T. M. Comparative histology of secondary xylem in buried and exposed roots of dicotyledonous trees. *Phytomorphology* 3:427–430. 1953.
15. Rao, A. N. Developmental anatomy of natural root grafts in *Ficus globosa. Aust. J. Bot.* 14:269–276. 1966.
16. Russell, R. S. Root systems and plant nutrition—some new approaches. *Endeavour* 29:60–66. 1970.
17. Satoo, S. Origin and development of adventitious roots in layered branches of 4 species of conifers. *J. Jap. Forestry Soc.* 37:314–316. 1955.
18. Schultz, R. P. Intraspecific root grafting in slash pine. *Bot. Gaz.* 133:26–29. 1972.
19. Simonds, A. O. Histological studies of the development of the root and crown of alfalfa. *Iowa State Col. J. Sci.* 9:641–659. 1935.
20. Stangler, B. B. Origin and development of adventitious roots in stem cuttings of chrysanthemum, carnation, and rose. *N.Y. Agr. Exp. Sta. Mem.* 342. 1956.
21. Süss, H., and W. R. Müller-Stoll. Zur Anatomie des Ast-, Stamm- und Wurzelholzes von *Platanus* x *acerifolia* (Ait.) Willd. *Österr. Bot. Z.* 121:227–249. 1973.
22. Tomlinson, P. B. *Anatomy of the monocotyledons.* II. *Palmae.* Oxford, Clarendon Press. 1961.
23. Torrey, J. G. Cellular patterns in developing roots. *Symp. Soc. Exp. Biol.* 17:285–314. 1963.
24. Torrey, J. G. *Development of flowering plants.* New York, The Macmillan Company. 1967.
25. Torrey, J. G., and R. S. Loomis. Auxin-cytokinin control of secondary vascular tissue formation in isolated roots of *Raphanus. Amer. J. Bot.* 54:1098–1106. 1967.
26. Wilcox, H. Regeneration of injured root systems in noble fir. *Bot. Gaz.* 116:221–234. 1955.

16

The Stem: Primary State of Growth

EXTERNAL MORPHOLOGY

The close association of the stem with the leaves makes the aerial part of the plant axis structurally more complex than the root. The term shoot, which refers to the stem and leaves as one system, serves to express this association.

In contrast to the root, the shoot has nodes and internodes, with one or more leaves attached at each node. Depending on the degree of development of the internodes, the shoot assumes different aspects. It may be an elongated structure with easily recognizable nodes and internodes, or it may be a condensed structure without discernible internodes and with leaves crowded in a rosette or a bulb. Other features that add to the variation in the external aspect of the shoot are the arrangement of leaves, their manner of insertion, development or nondevelopment of axillary buds into lateral shoots, and the level at which branching, if present, occurs. Differences are also associated with the type of growth and habitat of the shoot, that is, whether the shoot grows in air or underground (rhizome, tuber, corm, bulb), in water or on land, and whether it is upright, or climbing, or creeping.

PRIMARY STRUCTURE

The stem, like the root, consists of three tissue systems, the dermal, the fundamental, and the fascicular or vascular (fig. 16.1). The variations in the primary structure in stems of different species and the larger taxa are based chiefly on differences in the relative distribution of the fundamental and vascular tissues. In the conifers and dicotyledons the vascular system of the internode commonly appears as a hollow cylinder delimiting an outer and an inner region of ground tissue, the cortex and the pith, respectively. The subdivisions of the vascular system, the vascular bundles, are separated from each other by more or less wide panels of ground parenchyma—the interfascicular parenchyma—that interconnects the pith and the cortex

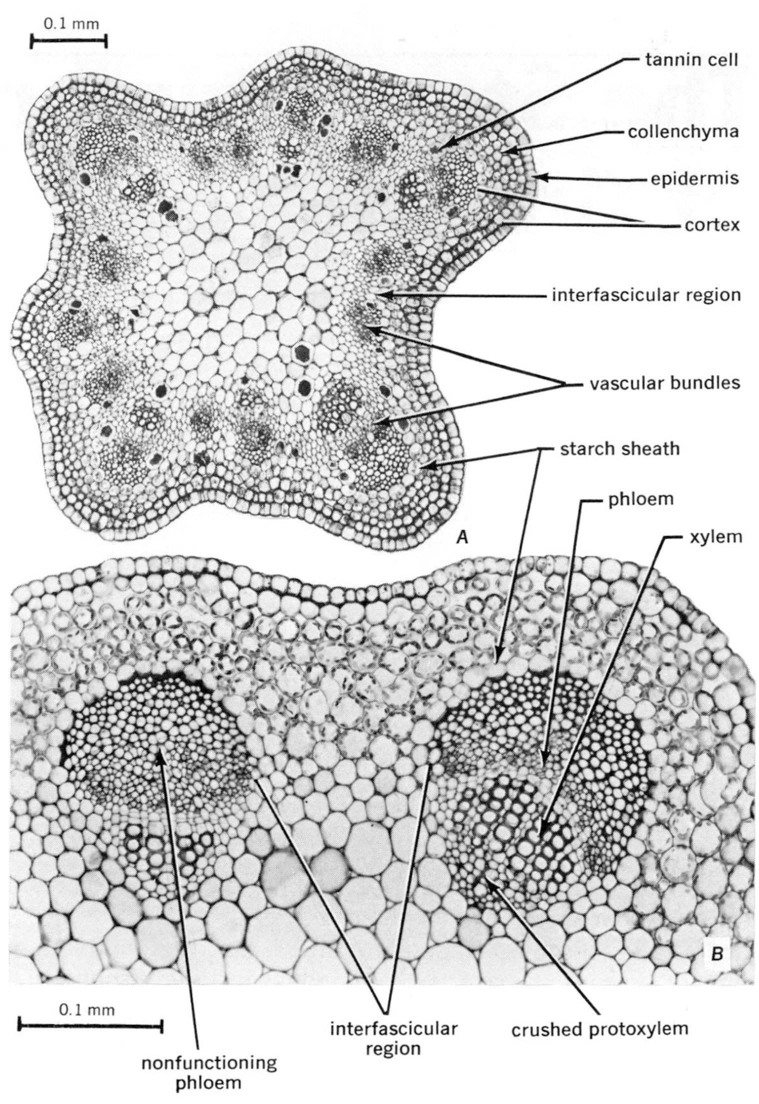

Figure 16.1 Cross sections of herbaceous dicotyledom stems in primary state of growth. *A, Lotus corniculatus,* complete section. *B, Trifolium hybridum,* clover, partial section with two vascular bundles. (*A,* from H. W. Hansen, *Iowa State Coll. Jour. Sci.* 27:563–600, 1953; *B,* from J. E. Sass, *Botanical Microtechnique,* 3rd ed. Iowa State College Press, 1958.)

(fig. 16.1). This tissue is called interfascicular because it occurs between the bundles, or fascicles. A panel of interfascicular parenchyma is often called the medullary or pith ray. If the interfascicular regions are narrow, one or two cell layers in width, the recognition of the individual units of the vascular system requires a developmental study.[2]

Stems of many ferns, some herbaceous dicotyledons, and most monocotyledons have a complex arrangement of vascular tissues. As seen in transections, the bundles may occur in more than one ring (chapter 17) or may appear scattered throughout the cross section (fig. 16.2,*A*). The delimitation of the ground tissue into cortex and pith is less precise or does not exist when the vascular bundles do not form a ring in cross sections of internodes.

Epidermis

The principal features of the epidermis of aerial plant parts are described in chapter 7. Stomata constitute a less prominent epidermal component in the stem than in the leaf. The stem epidermis commonly consists of one layer of cells and has a cuticle and cutinized walls. It is a living tissue capable of mitotic activity, an important characteristic in view of the stresses to which the tissue is subjected during the primary and secondary increase in thickness of the stem. The epidermal cells respond to these stresses by tangential enlargement and radial divisions. The persistence of mitotic activity in the stem epidermis is particularly impressive in species with long delayed periderm formation (chapter 12).

Cortex and pith

The cortex of stems contains parenchyma, usually with chloroplasts (fig. 16.1,*B*). Intercellular spaces are prominent but sometimes are largely restricted to the median part of the cortex. In many aquatic angiosperms the cortex develops as an aerenchyma with a system of large intercellular spaces (chapter 17). The peripheral part of the cortex frequently contains collenchyma (fig. 16.1,*A*), in strands or in a more or less continuous layer. In some plants, notably grasses, sclerenchyma rather than collenchyma develops as the primary supporting tissue in the outer region of the stem (fig. 16.2,*B*). The conifers typically have no special strengthening tissue in the cortex.

As was shown in chapter 14, the innermost layer of the cortex of roots of vascular plants has special wall characteristics and is called endodermis. The stems of conifers and angiosperms commonly lack a morphologically differentiated endodermis. In young stems, the innermost layer or layers may contain abundant starch and thus be recognized as a starch sheath (fig. 16.1). Some dicotyledons, however, do develop casparian strips in the innermost cortical layer of the stem,[57] and many lower vascular plants have a clearly differentiated stem endodermis.

When no starch accumulates and no special wall characteristics develop in the innermost cortical layer, the delimitation of the cortex from the vascular region may be problematic. But whether or not the vascular region is structurally demarcated from the cortex, it is surrounded by a physiological boundary resulting from chemical interaction between materials derived from vascular tissues and those present in the cortex. Under appropriate conditions, chemical reactions may lead to the development of a casparian strip or deposition of starch or both together.[52] Broadly defined, the term *endodermis,* and *endodermoid* as an adjective, are applicable to the boundary in any of its manifestations.[12]

The pith is commonly composed of parenchyma, which may contain chloroplasts. In many stems the central part of the pith is destroyed during growth. Frequently this destruction occurs only in the internodes, whereas the nodes retain their pith (nodal diaphragms). Sometimes series of horizontal

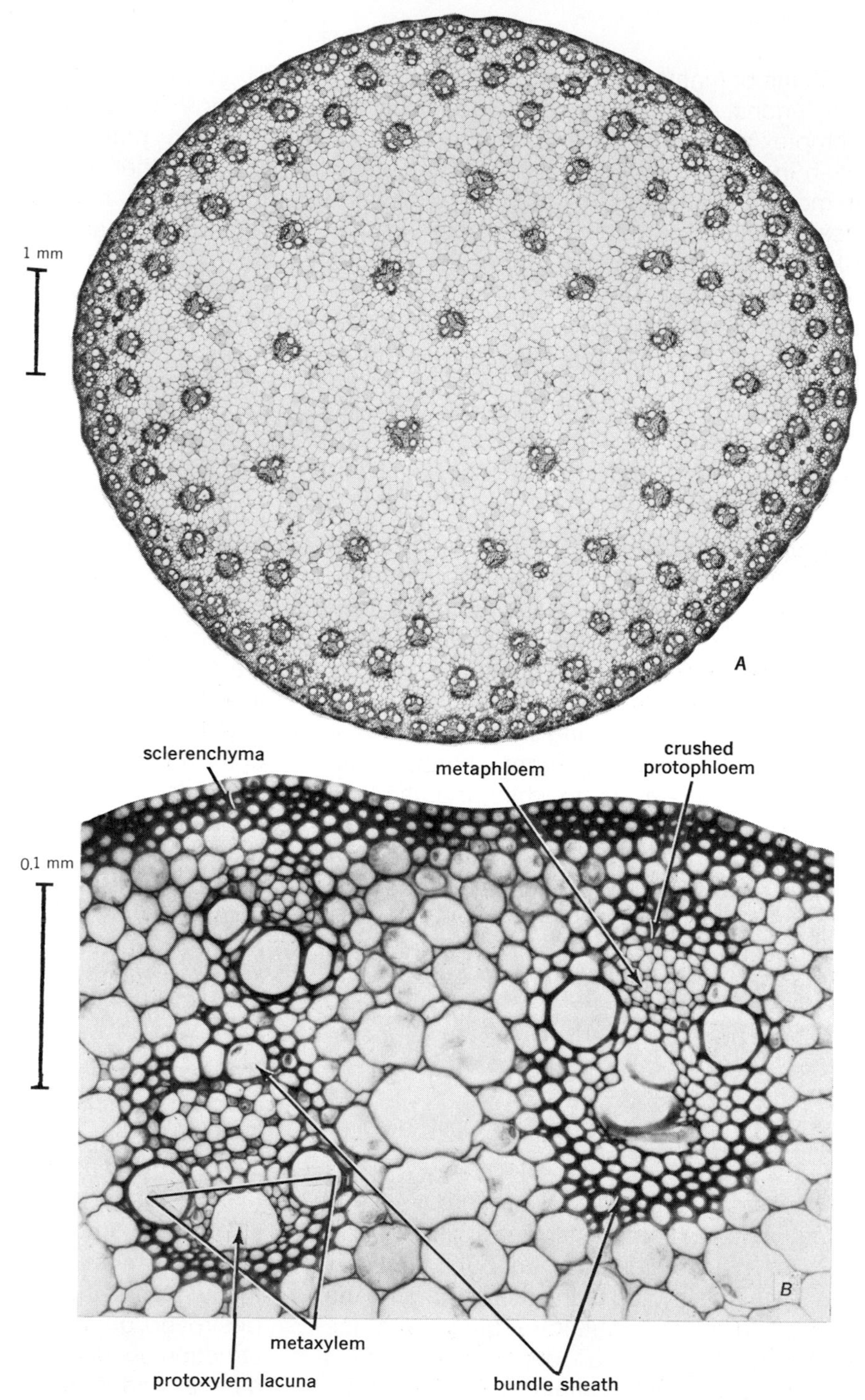

Figure 16.2 Vascular system in stem of *Zea mays* (corn) shown in transections. *A*, vascular bundles distributed throughout section but are more numerous near the periphery. *B*, details in vascular bundles.

plates of pith tissue also remain in the internodes (*Juglans, Pterocarya*). The pith has prominent intercellular spaces, at least in the central part. The peripheral part may be distinct from the inner in having compactly arranged small cells and greater longevity. Since the pith is also referred to as the medulla, the distinct peripheral zone of the pith is called *perimedullary zone* or *medullary sheath* (chapter 17).

Both cortex and pith may contain various idioblasts, including cells with crystals and other ergastic contents, and sclereids. If the plant has laticifers, such may be present in pith and cortex.

Vascular system

VASCULAR BUNDLES

The primary vascular system of seed plants consists of strands variable in size and degree of distinctness. Discrete individual strands are commonly referred to as vascular bundles. The phloem and xylem show variations in their relative position in vascular bundles. The prevalent arrangement is *collateral,* in which the phloem occurs on one side (abaxial) of the xylem (figs. 16.1,*B*, and 16.2,*B*). In certain ferns and in some dicotyledons (Apocynaceae, Asclepiadaceae, Convolvulaceae, Cucurbitaceae, Solanaceae, and certain tribes of Asteraceae) one part of the phloem occurs on the outer side and another on the inner side of the xylem. This arrangement is called *bicollateral* (fig. 16.9,*A*) and the two complements of phloem are referred to as the *external* (abaxial) and the *internal* (adaxial) phloem.

The vascular bundles may be also *concentric,*[12] in which either the phloem surrounds the xylem (*amphicribral* bundles) or the xylem surrounds the phloem (*amphivasal* bundles). The amphivasal bundles appear to be phylogenetically rather specialized and are found in certain positions in stems of some dicotyledons (medullary bundles in *Rheum, Rumex, Mesembryanthemum, Begonia*) and monocotyledons (Araceae, Liliaceae, Juncaceae, Cyperaceae). Amphicribral bundles are most common in ferns but are found also in angiosperms. Small bundles in flowers, fruits, and ovules may be amphicribral (chapter 20).

LEAF ARRANGEMENT AND VASCULAR ORGANIZATION

The patterns formed by the vascular strands in the stem of a seed plant reflect the close relation between stem and leaves. At each node, one or more vascular bundles diverge from the cylinder of strands in the stem toward the leaf or leaves attached at that node. The extensions from the vascular system in the stem toward the leaves are referred to as *leaf traces.* A leaf trace extends from its connection with a bundle in the stem to the insertion of the leaf. A single leaf may have one or more leaf traces (one in figs. 16.3 and 16.4, three in fig. 16.6). The traces also vary in length as measured by the number of internodes they traverse before they diverge toward the leaf (less than 1 internode in the conifers, 1 to 35 in different species of dicotyledons[10]).

The leaf arrangement, or *phyllotaxy* (also *phyllotaxis*), shows many variations, but basically there are three main types: the whorled, with several leaves at each node (the decussate type with two opposite leaves at each node may be regarded as a subtype of the whorled, fig. 16.27); the distichous, or two-ranked, with leaves single at each node but disposed in two opposite ranks (fig. 16.5); and the alternate, or helical (fig. 16.3,*A,C*).

Students of phyllotaxy are concerned with

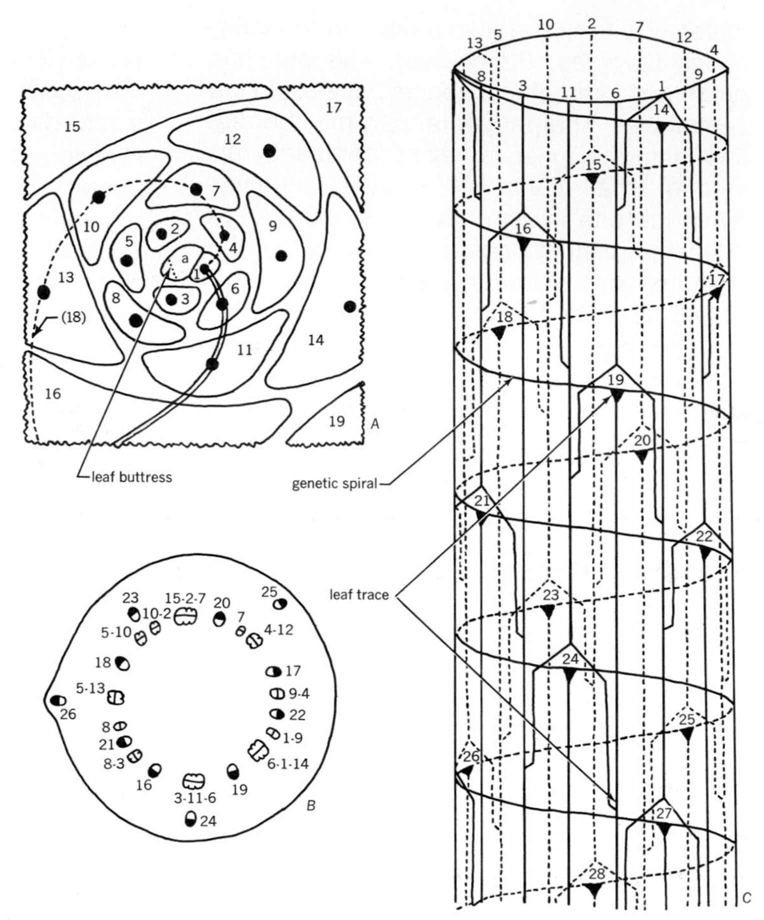

Figure 16.3 Diagrams of primary vascular system of *Hectorella caespitosa* in transections (*A, B*) and in three-dimensional view (*C*). The numbers relate leaf traces and sympodia to the leaves with which they are connected. *A,* stem apex (*a*) and leaves 1–19. Black circles, midveins. The two curved lines (one broken, the other double) indicate a pair of contact parastichies. *B,* stem cut near node of leaf 23. Leaf traces are distinguished from sympodial stem bundles by blackened phloem regions. *C,* diagram of interconnections of leaf traces and stem bundles. Black triangles, leaf bases. (Adapted from Skipworth.[46])

the meaning of the specific leaf arrangements and use various features to analyse the patterns in mathematical terms.[8,55] The classical method is to use the angle of divergence between two successive leaves. Examples of divergence are $\frac{1}{2}$, $\frac{1}{3}$, $\frac{2}{5}$, $\frac{3}{8}$, $\frac{5}{13}$, (the Fibonacci series; see the Glossary) and other fractions of the circumference that approach 0.382 (angle of divergence of 137.5°). Thus, one speaks of $\frac{1}{2}$, $\frac{1}{3}$, $\frac{2}{5}$, and so forth, phyllotaxy. In a plant with alternate leaf arrangement, a line drawn from leaf to leaf connecting leaves in the order of their origin at the shoot apex, is a helix, or *parastichy,* which constitutes the "genetic spiral" of the classical botanists.

The fractional expression of the angle of di-

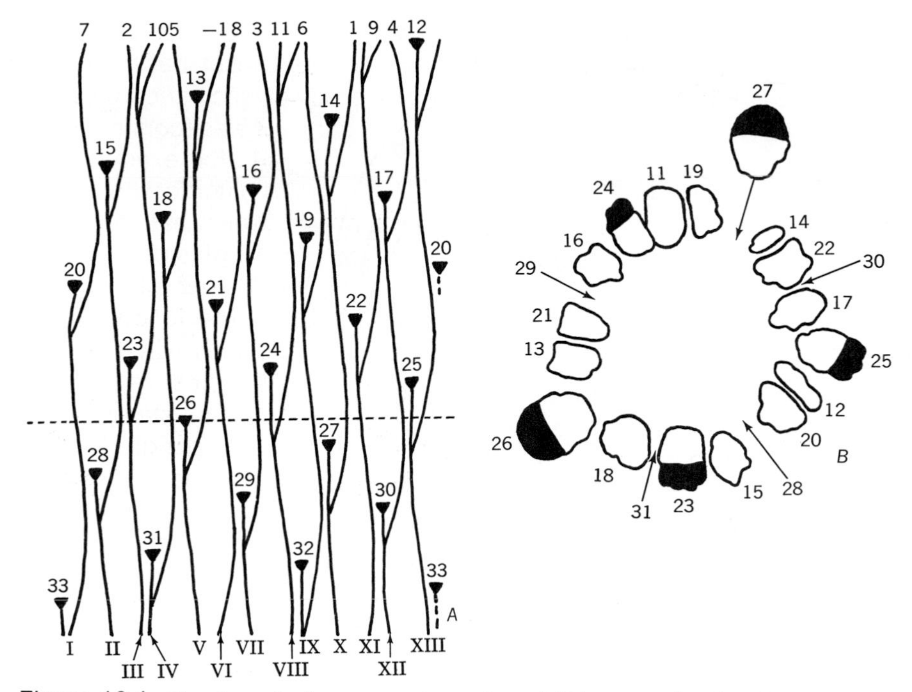

Figure 16.4 Structure of primary vascular system of *Abies concolor. A,* outer perspective of vascular system shown as though split open. The Roman numerals indicate the 13 sympodia, the Arabic numerals, leaf traces. The black triangles represent leaf bases. The number −1 refers to a leaf that was not yet formed at the shoot apex. *B,* transverse section of vascular cylinder at level indicated by the broken line in *A.* Leaf traces are differentiated from sympodial stem bundles by blackened phloem regions. (Adapted from Namboodiri and Beck.[33])

vergence tells something about the distribution of leaves along the genetic spiral. In $\frac{5}{13}$ phyllotaxy, for example, five windings about the axis include 13 leaves, with leaves *n* and *n* plus 13 located one above the other (fig. 16.3,*C,* leaves 13 and 26, 14 and 27, 15 and 28, etc.).

In addition to the genetic-spiral parastichy, systems of other parastichies can be projected on a given shoot, some flatter, others steeper, some winding clockwise, other counterclockwise. Plants with decussate and distichous phyllotaxies have straight-line relationships between superimposed leaves. Such leaf series are called *orthostichies.*

In figure 16.3, illustrating a dicotyledon (*Hectorella*[46]) with alternate leaf arrangement, the genetic spiral represents a flat counterclockwise (as seen from above) parastichy (fig. 16.3,*C*). Somewhat steeper counterclockwise parastichies can be projected through leaf insertions at intervals of 3 internodes, for example, 1-4-7-10-13, and still steeper clockwise parastichies through leaf insertions at intervals of 5 internodes, for example, 1-6-11-16-21.

Instead of referring to internodes one can speak of the *plastochron* to indicate the intervals between leaves of a given parastichy. This term refers to the time interval between

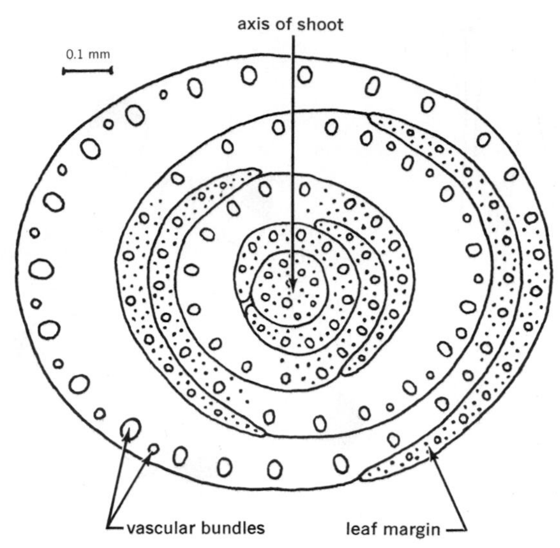

Figure 16.5 Transection of shoot tip of *Zea mays* (corn) showing distichous arrangement of leaves. Each leaf encircles the stem. All but the youngest leaf have overlapping margins. Stippling indicates young, intensively growing regions.

the inception of two successive leaves at the apex. Thus one can say that in figure 16.3,*A*, the leaves occur at plastochronic intervals of 1 (genetic spiral), 2, 3, 5, and 13 in the different parastichy systems.

The phyllotactic patterns are often studied in transverse sections of shoot tips where the young leaves are not yet separated by elongated internodes. In such sections it is possible to recognize the parastichies along which the leaves are proximal to one another when they are initiated at the apex. These parastichies are called *contact* parastichies.[6] In figure 16.3,*A*, two sets of contact parastichies are indicated by curved lines drawn through one parastichy in each set. One set consists of 5 parastichies passing through leaves 5 plastochrons apart, the other includes 3 parastichies with plastochronic intervals of 3 between successive leaves in one parastichy.

Certain of various parastichies in a given shoot pass through leaves that have direct vascular connections with one another. In the *Abies* (conifer) vascular system in figure 16.4, *A*, for example, 13 parastichies of leaves coincide with 13 longitudinal bundles each of which has vascular connections with leaves 13 plastochrons apart (e.g., leaves 33-20-7, 28-15-2, etc.).

The longitudinal bundles from which leaf traces diverge are referred to as *stem bundles, axial bundles,* or *cauline* (from *caulis,* stalk) *bundles.* The combination of stem bundles with the leaf traces that diverge from them are called *sympodia.* There are 13 sympodia in the *Abies* shoot (fig. 16.4,*A*).

The 13 sympodia in figure 16.4,*A*, are not interconnected with one another by vascular tissue but constitute independent units of the vascular system. This kind of vascular system is called *open.* The *Hectorella* stem in figure 16.2,*C*, also has 13 sympodia, but since each leaf is connected with two sympodia no free sympodia are present, and the interconnected vascular system is called *closed.*

Because of their continuity with the leaves through the leaf traces, the sympodia can be numerically related to the leaves that receive the leaf traces from them. In the cross sections of stems in figures 16.3,*B*, and 16.4,*B*, the vascular bundles are given the same numbers as the leaves with which they are connected. The bundles with the phloem shown in black are leaf traces. The other bundles are stem bundles sectioned below the level of divergence of those leaf traces the numbers of which they bear.

The scheme of analyzing the vascular system of stem in terms of the numbers of the leaves with which the units of the system are associated does not imply that the stem has no vascular tissue of its own. It indicates, however, that the cauline vascular system is organized in relation to the leaves. Moreover, if stem and leaf have common phylogenetic origin,[17] there is no fundamental distinction between leaf traces and stem bundles. The pertinent terms have a descriptive, topographic meaning.

Experiments involving removal of leaf primordia in ferns and angiosperms[13] yielded results supporting the concept that the usual form of the vascular system of shoots is determined by the developmental interaction between leaf and stem. If leaves are permitted to develop, connections with these leaves are formed through leaf traces. If leaf development is suppressed, the vascular system in the stem forms a continuous column without leaf traces and hence resembles those in plant axes in which leaves are absent or are small and widely spaced on the axis (many rhizomes, corms, scapes, inflorescence axes, roots).

Leaf traces and stem bundles show structural differences. In transections of stems, the leaf traces are clearly circumscribed and have a larger amount of xylem than the cauline bundles. As seen in longitudinal views, the leaf traces increase in size above their divergence from the stem bundles[10] (fig. 16.6). Another notable feature of the leaf traces in representatives of diverse taxa is the occurrence of vascular parenchyma cells differentiated as transfer cells with a profusion of wall ingrowths.[21] These cells are particularly conspicuous in the xylem of the trace but are less well developed in the phloem. The association of transfer cells with the diverging leaf traces suggests a specialization for an intensive lateral transfer of solutes strategically located in relation to the apical meristem in a seedling and to axillary buds in adult plants.

In the monocotyledons, the vascular connection between leaves and stem is highly complex because each leaf has many leaf traces and, accordingly, the stem has many individual strands (fig. 16.2,*A*). The numerous vascular bundles seen in the leaves in a cross section of a shoot tip of *Zea mays* (fig. 16.5), for example, are joined to the stem as separate strands and are connected with equally numerous strands distributed throughout the entire cross section.

The vascular systems of stems are commonly studied by recording the position of vascular strands in drawings made from serial transections of the shoot and by reconstructing the three-dimensional organization from these drawings. This method of analysis is complex enough with regard to shoots having one or a few traces to a leaf; but it is not well suited for monocotyledons because of the multiplicity of strands in their shoots. In the coconut palm, for example, some 20,000 vascular strands appear in the central region of stem transection and tens of thousands of fibrous strands (which are also connected with the leaves) in the peripheral region.[61]

A substantial progress toward understanding the monocotyledonous vascular systems was made through the introduction of structural analysis by means of frame-by-frame cinematography of individual transverse sections of stems through a microscope. The film images are projected on graph paper so that the measurements of bundle displacements in the successive sections can be plotted. The method permits a quick analysis of the three-dimensional structure.[58–61] However, the communication of the resulting data to persons not directly involved in the study remains a problem and is done by means of idealized and simplified diagrams. Two such diagrams in figure 16.7 depict parts of the vascular system of the small palm *Rhapis excelsa.*

The leaf traces in Rhapis are derived from a small number of large strands and a large number of small strands in the axis. The smaller strands form more leaf traces than do the larger ones (fig. 16.7,*A*). The strands in

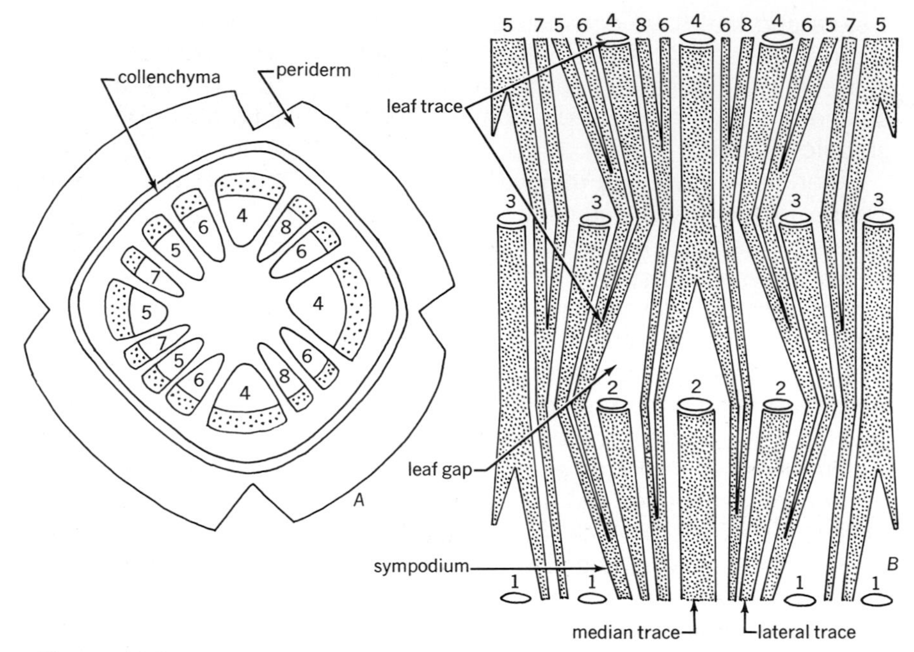

Figure 16.6 Diagrams illustrating vascular system in a dicotyledon stem (*Ulmus*). *A*, cross section. *B*, longitudinal view showing the vascular cylinder as though cut through median leaf trace 5 and spread out in one plane. The numbers in both views indicate leaf traces. The cross section in *A* corresponds with the topmost view in *B*. (After E. Smithson, *Proc. Leeds Phil. Soc.* 6:211–220, 1954.)

the stem exhibit a longitudinal course along a shallow helix (not so indicated in fig. 16.7) and occur at various distances from the periphery of the stem at different levels of the shoot. The larger bundles appear closer to the center than the smaller bundles. At intervals, a strand is sharply bent toward a leaf insertion and divided into branches (fig. 16.7). One of the two large branches constitutes a leaf trace, the other is continued as a vertical stem bundle. Several small bundles serve as traces to an axillary inflorescence and other small bundles form connections (bridges) with neighboring bundles (fig. 16.7,*B*).

The study of the vascular system in large monocotyledons has revealed that these plants have two systems of bundles, an inner and an outer.[61] In the inner system, the leaf traces are connected with continuous vertical bundles, in the outer system with bundles that end blindly in the cortex. This difference is a result of the relatively late differentiation of the outer bundles.

Some monocotyledons have nodal plates with networks of vascular bundles (e.g., festucoid grasses,[23,37] *Zea*[28]). These moncotyledons have either "scattered" vascular bundles (fig. 16.2,*A*) or bundles localized near the periphery and seen as a ring in a cross section of a stem (chapter 17).

LEAF GAPS

In the nodal region, where a leaf trace is bent away from the vascular cylinder in the

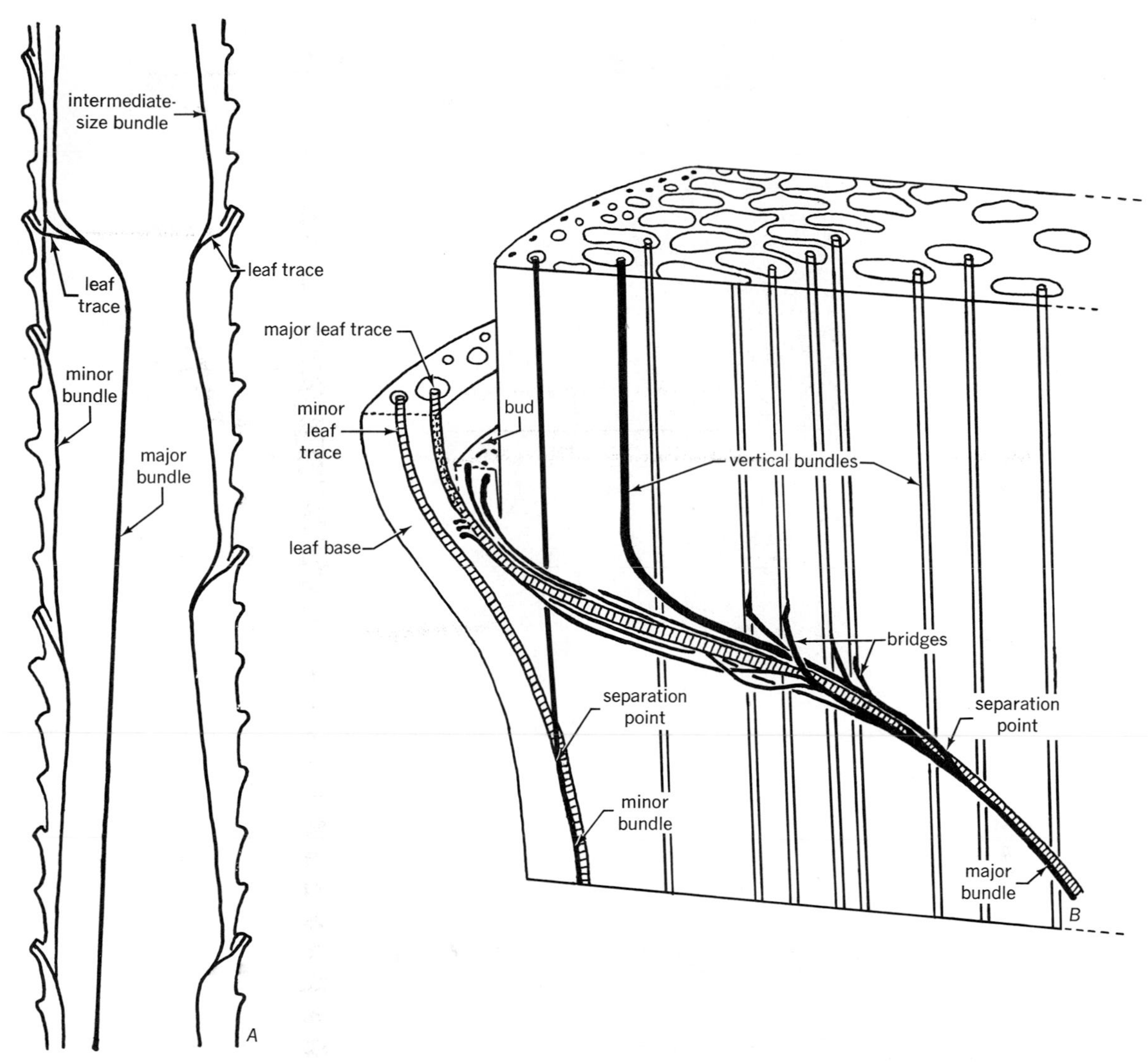

Figure 16.7 Diagrams of vascular system of *Rhapis excelsa,* a palm. *A,* three sizes of bundles shown in their vertical course. Leaf traces diverge from vertical bundles. Major bundle occurs deeper in the axis than the smaller bundles. The stem axis is foreshortened four times in relation to stem diameter. *B,* details of separation between leaf traces and vertical bundles. Bridges are interconnections between bundles. (Adapted from Zimmermann and Tomlinson.[58])

stem toward the leaf base, a region of parenchyma occurs in the cylinder (fig. 16.4,*B*, see leaf traces 25–28). This region, which in transverse sections appears like a rather wide interfascicular region, is called *leaf gap.* A leaf gap thus is a parenchymatous region in the vascular cylinder of the stem located opposite (adaxially to) the upper part of a leaf trace, approximately at the level of leaf insertion.

Leaf gaps vary in extent, laterally and vertically. If the vascular cylinder has many ax-

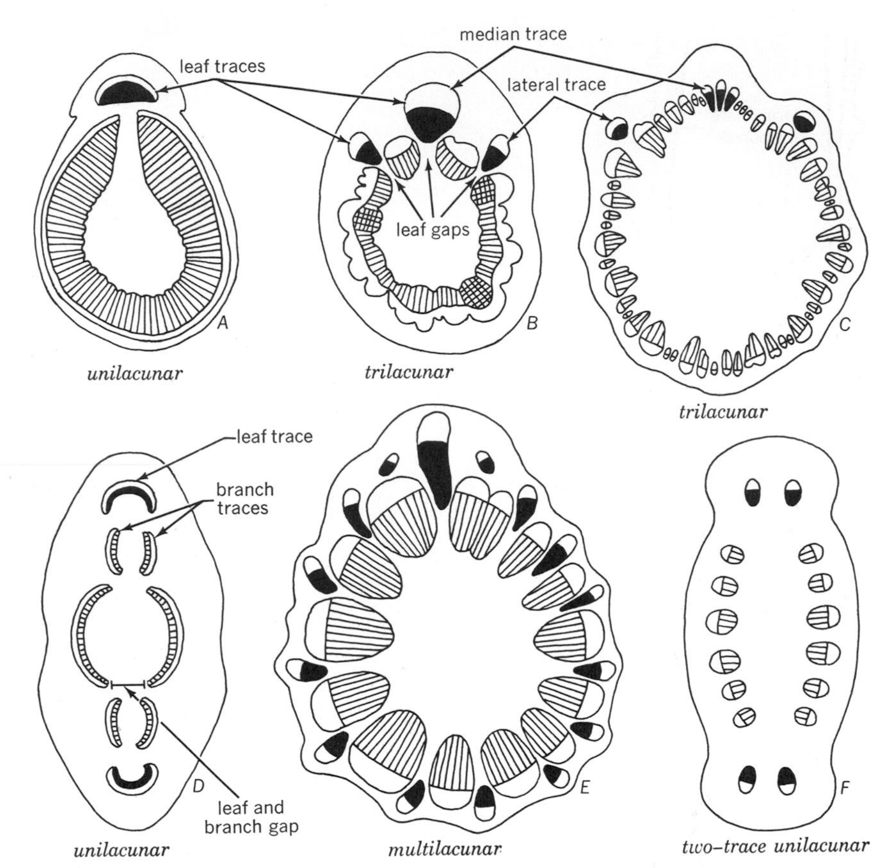

Figure 16.8 Cross sections of stems with different types of nodal structure. Leaf traces are indicated by blackened xylem regions. *A, Spiraea. B, Salix. C, Brassica. D, Veronica. E, Rumex. F, Clerodendron.* (After Esau, *Plant Anatomy,* 2nd ed. John Wiley & Sons, 1965.)

ially elongated interfascicular regions, the gaps merge with these (fig. 16.6,*B*). In such stems the delimitation of the gap can be made only arbitrarily.

The numbers of leaf traces and leaf gaps vary in different plants (fig. 16.8) and may vary in the same plant at different levels. Hence the nodal structure has different forms, which are designated by special terms. The term gap is replaced by that of lacuna, and the nodes are said to be *unilacunar, trilacunar,* and *multilacunar* depending on whether a given leaf is associated with one, three, or several lacunae at the node. If more than one leaf is inserted at a node, the node is characterized with reference to one leaf (fig. 16.8,*D*). If a leaf has more than one leaf trace and is confronted by three or more gaps, the gap (and the associated trace) that occurs in the median position with reference to the leaf is called median and the other lateral (figs. 16.6,*B*, and 16.8,*B*,*C*). A given trace may consist of more than one bundle (fig. 16.8,*C*, median trace).

In most vascular plants only one trace is related to each gap. A deviating type of nodal structure is represented by the *two-trace unilacunar* condition (fig. 16.8,*F*.). The two traces are branches from different sympodia and are therefore considered to be separate traces. The original interpretation of the two-trace unilacunar node as precursory to the other nodal types[5] is proposed to be replaced by the concept that the single-trace unilacunar condition is more primitive.[33,36]

BRANCH TRACES AND BRANCH GAPS

Buds commonly develop in the axils of leaves so that, in addition to leaf traces, the vascular bundles that connect the main stem with the branch may be recognized in the nodal region. These strands are called branch traces (figs. 16.8,*D*, and 16.9,*B*). Actually the branch traces are leaf traces, namely, leaf traces of the first leaves on a branch called prophylls (fig. 16.9,*A*). In conifers and dicotyledons two prophylls occur opposite one another oriented so that a plane bisecting both prophylls would be parallel with the plane of the subtending leaf. Two traces, one to each prophyll and each composed of one or more bundles, connect the bud with the main axis. In monocotyledons the side shoot also has two traces, although there is only one prophyll (sometimes interpreted as a double structure) at the base of the axillary shoot.

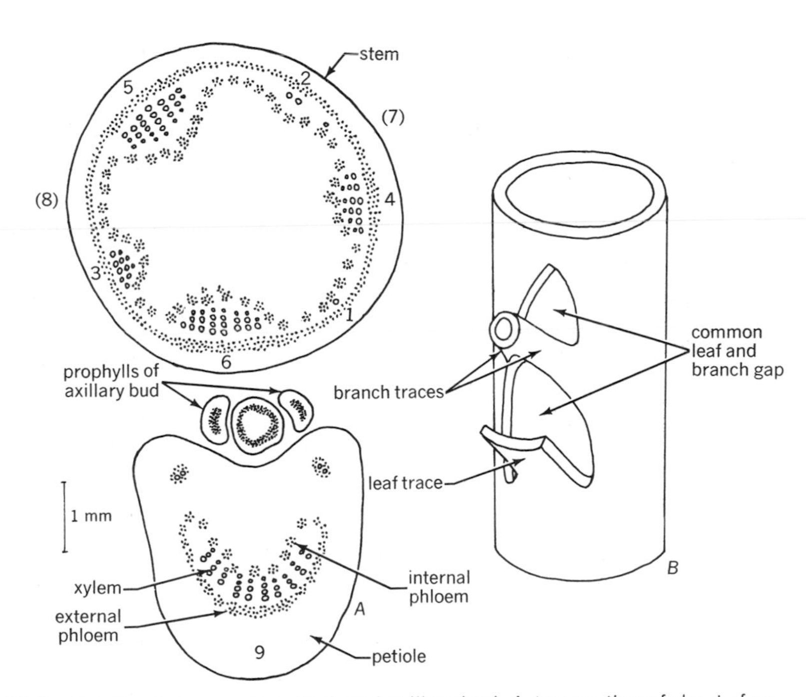

Figure 16.9 Relation between stem, leaf, and axillary bud. *A*, transection of shoot of potato (*Solanum tuberosum*) made below the node of the sixth leaf from the apex. Leaves 7 and 8 are omitted. Leaf 9 subtends the axillary bud shown. The numbers 1–6 in the stem section relate the vascular bundles (delimited as such only by the xylem) to the leaves with which they are connected. The petiole of leaf 9 illustrates a bicollateral vascular bundle, with external and internal phloem close to each other at margins. *B*, diagram showing relation of branch traces and a leaf trace to the vascular system in main stem.

The branch traces usually diverge from the main stem right and left of the median trace of the leaf subtending the branch (or the single trace if no laterals are present) so that a branch and the subtending leaf have a common gap (fig. 16.9,*B*). The branch traces extend through variable distances in the main axis and at some level are connected with the vascular system of this axis.

THE CONCEPT OF THE STELE

A well-known concept pertaining to the phylogeny of the form of the primary vascular system in the axis is that of the stele. The stelar concept was introduced[53] for the purpose of analyzing the homology and evolution of axial structure in different plant taxa. The stele, meaning a column, was defined as the core of the plant axis (stem and root) including the vascular system and all the interfascicular regions, gaps, pith (if present), and some fundamental tissue on the periphery of the vascular system, the pericycle. The plant axis was thus envisioned as consisting of a central column surrounded by the cortex, with the epidermis forming the surface layer.

The stelar concept has proved to be useful in comparative and phylogenetic studies of vascular plants. In time, the concept and the associated terminology changed and the term stele has come to mean the vascular system rather than the combination of vascular and ground tissue in the central column of the axis. In fact, it is difficult to apply the original concept of the stele meaningfully to the stem of seed plants because of the usual absence of a clear structural delimitation between the cortex and the vascular region in the stem of these taxa. In the original concept of the stele, a delimitation of the latter by an endodermis and a pericycle is considered an important evidence of the reality of the stele as a morphological entity.

In some stems of the dicotyledons a continuous or nearly continuous cylinder of fibers occurs on the periphery of the primary vascular cylinder. These fibers may arise from the same meristem as the phloem (*Pelargonium*) or may originate outside the phloem (*Aristolochia, Cucurbita*), but inside of the starch sheath, which is the innermost layer of the cortex in seed plants.[3] Thus, in some instances, a tissue of nonphloic origin occurs between the vascular tissue and the cortex. It was precisely such tissue for which the term pericycle was coined originally, then extended to all stems, despite the fact that in most stems of seed plants the phloem is in contact with the cortex, and the peripheral fibers are primary phloem fibers. If the fibers arise inside the starch sheath but outside the phloem they may be given the topographic term of *perivascular fibers.*

The classification of steles into types is based mainly on the relative distribution of the vascular and nonvascular tissues as seen in the axis in primary state of development. In the simplest type of stele, which is also considered to be the most primitive phylogenetically, the vascular tissue forms a solid column. This is the *protostele.* In a protostele the phloem may surround the xylem in a relatively uniform layer, or the two vascular tissues may intermingle in the form of strands or plates. Protosteles are found in Lycophyta and Sphenophyta, but they occur also in the lower parts of ferns and in the stems of some water plants in the angiosperms. The vascular cylinder of roots of seed plants is classified as a protostele.

The second form of stele is the *siphonostele,* or tubular stele, in which the vascular tissue surrounds a nonvascular core, the pith. The siphonostele and its variations are characteristic of Lycophyta, Sphenophyta, and Pterophyta. Some of these variations are the *ectophloic siphonostele,* with the phloem tissue appearing only on the outside of the xylem, and the *amphiphloic siphonostele,* with the phloem present on both the outer and the inner sides of the xylem cylinder. In its simplest form the siphonostele has no leaf gaps. In some ferns the leaf gaps are relatively short vertically, and since interfascicular regions are absent the vascular tissue forms a continuous ring in a cross section of an internode. In other ferns the leaf gaps are vertically elongated and overlap in the internodes so that the vascular cylinder appears dissected into strands, each with the phloem surrounding the xylem (amphicribral concentric vascular bundles). Such modification is called a *dictyostele.*

The stele of the gymnosperms and angiosperms, comprising a system of strands and interfascicular regions, is called *eustele.* When the vascular system consists of widely dispersed bundles, as in some monocotyledons, the stele is termed *atactostele* (fig. 16.2).

For a long time, the eustele of seed plants was considered to have evolved from the type of siphonostele found in ferns (filicinean type) through dissection into strands, notably by lengthening and overlapping of leaf gaps.[17] Results of extensive comparative studies of vascular systems of conifers and dicotyledons, combined with surveys of the fossil record, suggest, however, that the vascular systems of seed plants has been derived not from a filicinean type of siphonostele but directly from a more ancient protostelic condition, the kind found in pteridophytic gymnosperms.[1,33,47] The probable sequence of this evolution is illustrated in figure 16.10 by the use of examples of steles of pteridophytic progymnosperms (fig. 16.10,*A–C*) and a living conifer (fig. 16.10,*D*).

In the primitive condition, traces to appendages diverge radially from the outer surface of the protostelic column, without leaving gaps (fig. 16.10,*A*). The stele is dissected into longitudinal columns with consequent appearance of a pith (fig. 16.10,*B*). (Evolutionary dissection here means an ontogenetic replacement of a certain amount of vascular tissue by parenchyma). Further dissection of the columns of vascular tissue results in the appearance of numerous discrete strands (fig. 16.10,*C,D*), which in their relation with the leaf traces are homologous with the stem bundles in the sympodia of living conifers. Thus, according to the proposed concept of stelar evolution, leaf-gap formation was not involved in the dissection of the vascular cylinder in the ancestors of seed plants and, therefore, the term leaf gap has only a descriptive meaning with reference to these taxa.

Although the sequence in figure 16.10 ends with the conifers, it leads also to the dicotyledons, and probably to the monocotyledons as well, since the stelar structure is basically the same in the conifers and the angiosperms.[10]

DEVELOPMENT

Shoot apex

The shoot apex is the seat of the apical meristem and its derivative meristematic tissue that together lay the foundation of the primary plant body. The stem with its nodes and internodes, leaves, axillary buds and the resulting lateral shoots, and later the reproductive structures, arise through the activity of apical meristems. Depending on the kind of structures produced at the shoot apex one distinguishes between vegetative and reproductive (or floral) shoot apices and apical meristems.

The shoot apex produces lateral organs

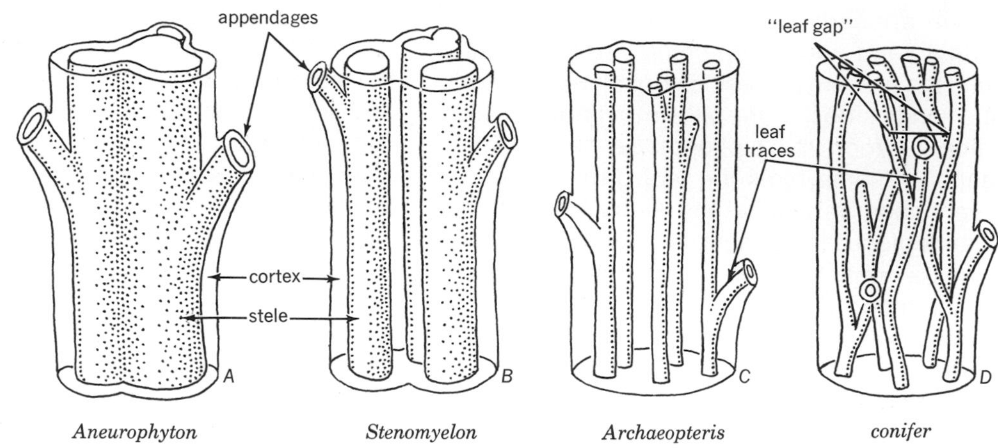

Figure 16.10 Diagrams illustrating probable trend of evolution of primary vascular system of progymnosperms and gymnosperms. *A*, three-ribbed solid protostele; trace of appendage (leaf or leaf precursor) diverges radially and leaves no gap. *B*, three-stranded vascular system with pith; trace divergence as in *A*. *C*, as *B*, but strands are more numerous and the pith larger. *D*, axial sympodial strands are more numerous that in *C* and have an undulating course as in conifers with helical phyllotaxy. Interfascicular region above the diverging leaf trace resembles a leaf gap. (Adapted from Namboodiri and Beck.[33])

and, therefore, the structure and activity of the apical meristem of the shoot must be considered in relation to the origin of lateral organs, especially the leaves. As was discussed in chapter 14, the root apex produces no lateral organs and also differs from the shoot apex in having a rootcap.

In seed plants, the apical meristem of the first shoot is organized in the embryo before or after the appearance of the cotyledon or cotyledons.[20,35,43] According to a common concept, an apical meristem consists (1) of certain cells, the initials, that are the source of all body cells and (2) of the derivatives of these initials, cells that are actively dividing but by their position and planes of division foreshadow the tissue regions subjacent to the apical meristem.

The initials and their most recent derivatives together may be regarded as the least determined part of an apical meristem. The term *promeristem* or *protomeristem*[25] is the classic way to designate this part of the apex. The term may be applied in the same sense to the root apex, but with respect to that organ promeristem has acquired a different meaning in the formulation of the quiescent-center concept (chapter 14).

The number of cells interpreted as initials, their spatial relation to each other and to their derivatives show variations in different taxa.[19,38] Studies of these aspects of apical meristems in shoots have led to a characterization of these meristems in terms of their geometric and developmental organization. With a progressive refinement of observational and experimental techniques investigators are learning to correlate biochemical events with morphological and developmental features of the apical meristems, including the changes related to the periodic initiation of leaf primordia (plastochronic changes), to seasonal periodicity, and to the passage of the apical meristem from the vegetative to the reproductive stage (chapter 20).

The apical meristem of the shoot, like that

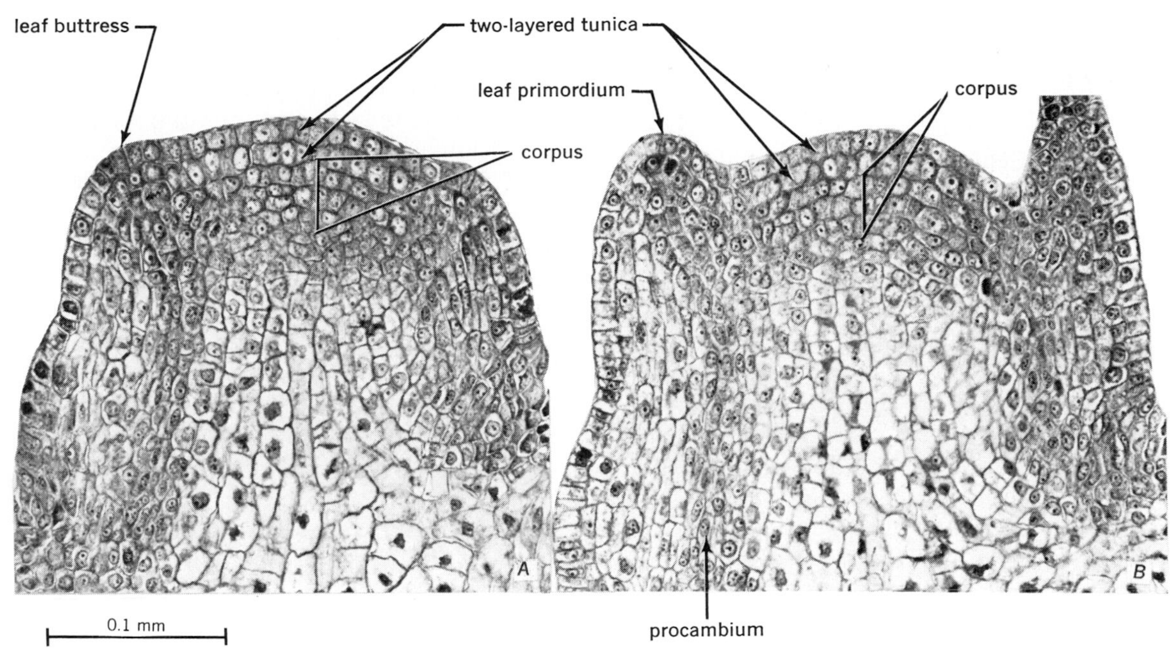

Figure 16.11 Longitudinal sections of shoot apices of potato (*Solanum tuberosum*) showing tunica-corpus organization of apical meristem and two stages in the initiation of a leaf primordium; leaf buttress stage in *A*, and beginning of upward growth in *B*. (From Sussex.[49])

of the root, merges with the differentiating tissues. Beneath the apical meristem, tissue regions are progressively differentiated through changes of cells in size, degree of vacuolation, and rate and orientation of mitoses (fig. 16.11). Still farther from the apex phenomena of organogenesis (e.g., initiation of leaf primordia) and histogenesis (differentiation of tissues) become dominant. The peripheral region, from which originate the foliar primordia, epidermis, cortex, and vascular tissues, becomes distinguishable from the future pith. In the peripheral region, the cells remain meristematic (relatively small in size and not strongly vacuolated) longer than in the pith region. The vascular tissues are initiated in the form of procambium, the cells of which assume a rather narrow elongated form because of predominance of longitudinal divisions (fig. 16.11). They become distinct from the less elongated, wider, and more vacuolated cells of the ground meristem, the precursor of the ground tissue. In the meantime, the protoderm progressively acquires the specific characteristics of the epidermis. Thus, increasingly farther from the apical meristem the three tissue systems, the epidermal, the vascular, and the fundamental, are differentiated, first in the form of their meristematic precursors (sometimes called "primary meristems"), protoderm, procambium, and ground meristem, later as mature tissues. Meristematic activity and differentiation phenomena overlap, and tissue maturation does not coincide precisely in the different tissue systems.

In the following, the variations in the organization of shoot apices and the seemingly conflicting interpretations of the functioning of apical meristems are reviewed. A description of apical meristems of lower vascular plants, interpreted as having a single initial cell, precedes that of the shoot apices of seed plants.

APICAL MERISTEMS WITH APICAL CELLS

Structurally, the simplest apical meristems are those in which a single large initial cell, the *apical cell,* dominates the distal cell group and appears to be the source of all body cells in the shoot (fig. 16.12). The single cell is sometimes lenticular but more commonly pyramidal in shape. Typically, it divides parallel to its faces (except the outermost) so that the resulting derivatives have an orderly arrangement indicating their ontogenetic relation to the apical cell (fig. 16.12,*A*), even after they themselves divide and produce generations of cells (fig. 16.12,*B*).

The solitary apical cell is not only relatively large but is also conspicuously vacuolated. The nearest derivatives are also strongly vacuolated, but, as they divide, smaller cells with denser protoplasts are eventually produced. This change in cytological appearance is noted mainly along the periphery of the shoot apex where leaf primordia originate; the pith meristem (if present) is more highly vacuolated.

The occurrence of apical meristems with single initial cells in lower vascular plants is often deduced from single median longitudinal sections, with considerable reliance on occasional division figures in the uppermost cell and the arrangement of the presumed derivatives. Comparative surveys of fern apices, combining anatomic studies with culture techniques and observations on surface growth in live apices, indicate that less emphasis on the uniqueness of the apical cell, at least in the ferns, is warranted.[31] In this taxon, the subsurface cells, which are derived from the surface initials, function as a promeristem and provide most of the cells that after further divisions give rise to the internal tissues of the shoot.

TUNICA-CORPUS ORGANIZATION

In the gymnosperms and angiosperms groups of apical initials occur in the apical meristem. In the angiosperms, the cells in the distal part of the apex are arranged in layers (stratified meristem; figs. 16.11 and 16.13), an organization suggesting that two or more layers grow independently from one another, and that two or more superimposed tiers of initials are present. Morphologically, the ini-

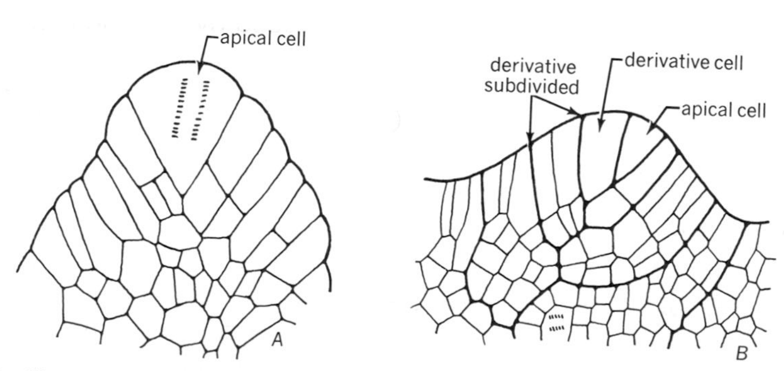

Figure 16.12 Shoot apices with apical cells. *A, Equisetum hyemale;* apical cell in division. *B, Polypodium peroussum;* subdivided derivatives of the apical cell indicated by slightly thicker walls. (From photomicrographs on pp. 81 and 299 in D. W. Bierhorst, *Morphology of vascular plants,* New York, Macmillan, 1971.)

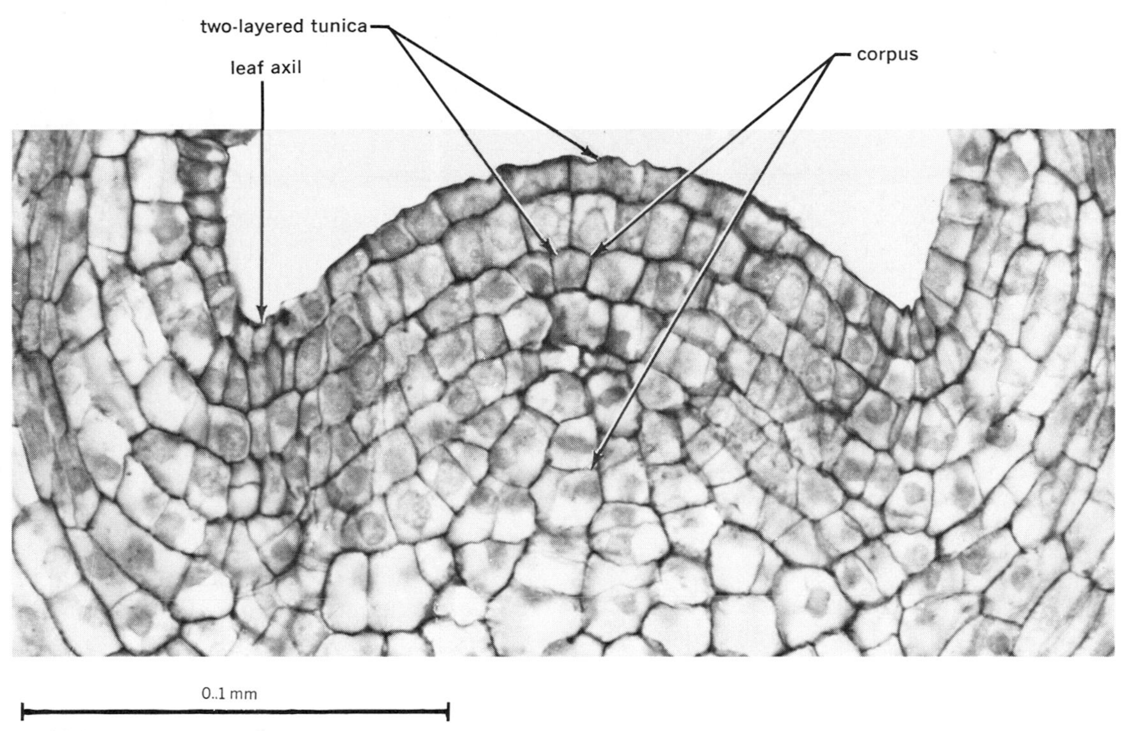

Figure 16.13 Longitudinal section of shoot apex of *Coleus* with a two-layered tunica. See figure 16.21 for bud in leaf axil at a later plastochron stage.

tials are usually not distinguishable from their immediate derivatives.

The formulation of the idea of separate tiers of initials is embodied in the *tunica-corpus* concept of apical organization introduced by Schmidt.[44] It states that the initial region of the apical meristem consists of (1) the tunica, one or more peripheral layers of cells which divide in planes perpendicular to the surface of the meristem (anticlinal divisions), and (2) the corpus, a body of cells several layers deep in which the cells divide in various planes (figs. 16.11, 16.13, and 16.14). Thus, the corpus is a core of cells that adds bulk to the apical meristem by increase in volume, and the tunica is a mantle of one or more cell layers that maintain their continuity over the enlarging core by surface growth. Of course, the tunica and the corpus do not increase in extent and volume indefinitely, for, as they form new cells, the older cells become incorporated in the shoot regions below the apical meristem.

The corpus and each layer of tunica are visualized as having their own initials. In the tunica the initials are disposed in the median axial position. By anticlinal divisions these cells form progenies of new cells some of which remain at the apex as initials, others function as derivatives which, by subsequent divisions, contribute cells to the peripheral part of the shoot (fig. 16.14,*B*). The initials of the corpus appear beneath those of the tunica. By periclinal divisions (parallel with the apical surface) these initials give derivatives to the corpus below, the cells of which

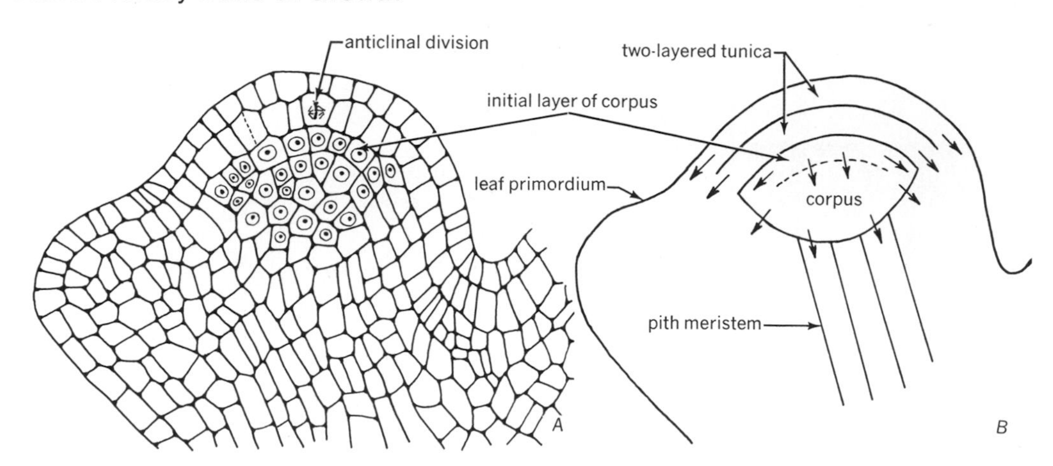

Figure 16.14 Shoot apex of *Pisum* (pea). Cellular details in *A*, interpretative diagram in *B*. The pith meristem does not show typical rib-meristem form of growth.

divide in various planes. Cells produced by divisions in the corpus are added to the center of the axis, that is, to the pith meristem, and commonly also to the peripheral region (fig. 16.14).

The initials of the corpus may form a well defined layer in contrast to the less orderly arranged cells in the mass of the corpus (fig. 16.14). When this pattern is present the delimitation between the tunica and the corpus is difficult (figs. 16.11,*A*, and 16.13), but if shoot apices are collected in different stages of development, the uppermost layer of the corpus will be found undergoing periodic periclinal divisions. After such a division a second orderly layer appears temporarily in the corpus. The outer corpus is then described as being stratified (figs. 16.11,*B*, and 16.13[49]).

Some investigators include the unstable parallel layers of the corpus in the tunica and state that the tunica fluctuates in the number of layers. The term *mantle* has been proposed for the less rigidly defined tunica and the term *core* for the body of cells covered by the mantle.[39]

The number of tunica layers varies in angiosperms. More than half of the species studied among the dicotyledons have a two-layered tunica. The reports of higher numbers, four and five, are subject to the qualification that some workers include the innermost parallel layer or layers in the tunica, others in the corpus. One and two are common numbers of tunica layers in the monocotyledons (fig. 16.15).

With some exceptions (*Araucaria*, *Ephedra*), the gymnosperms do not show a tunica-corpus organization in the shoot apex;[19] that is, they do not have stable surface layers dividing only anticlinally. The outermost layer of the apical meristem undergoes periclinal and anticlinal divisions and contributes cells to the peripheral and interior tissues of the shoot (figs. 16.16 and 16.17). The surface cells located in the median position in the apical meristem are interpreted as initials.

The view that the layers in the apical meristem of plants with the tunica-corpus organization are relatively independent is supported by observations on periclinal cytochimeras. Such chimeras are plants in which one or more layers parallel to the surface of the plant body (hence *periclinal* chimeras)

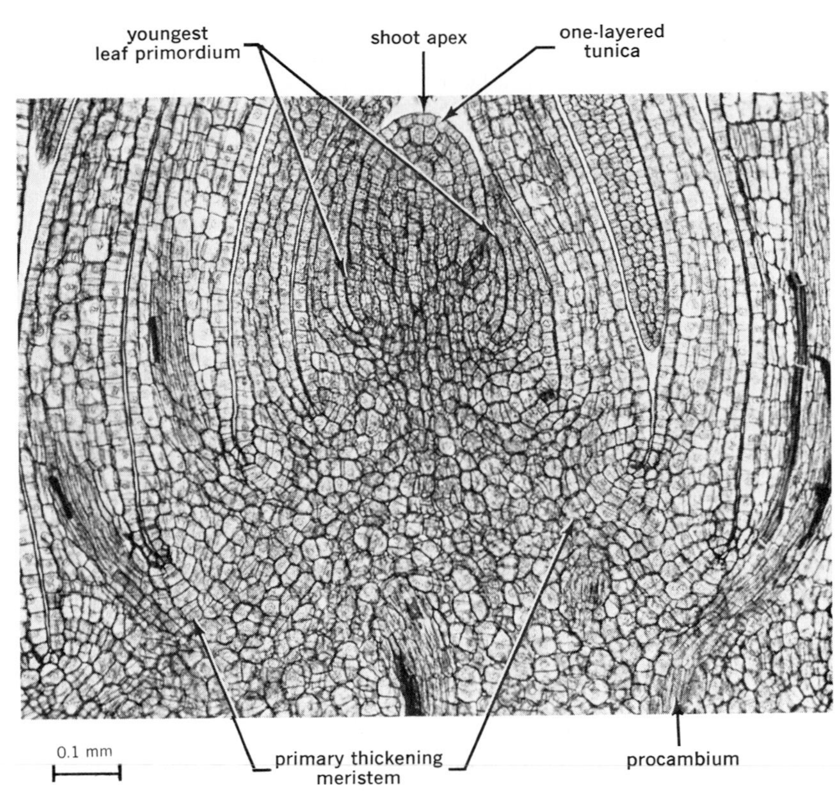

Figure 16.15 Longitudinal section of shoot apex of *Zea mays* (corn) with a one-layered tunica. Parts of each leaf occur on both sides of the axis because leaves encircle the stem in their growth. See figure 16.5.

show heritable characteristics different from those in adjacent layers. The differences serve as markers that may be followed through continuous cell lineages to similar differences in the cell layers of the apical meristem. Some chimeras have combinations of layers with diploid and polyploid nuclei. Polyploidization of nuclei may be induced by treating shoot apices with colchicine. As a result, one or another layer in the apical meristem becomes populated with polyploid nuclei and the change is propagated through the progeny of the layer in the differentiating plant body.[9] Periclinal chimeras are also available among mutants with defective, colorless plastids. As in the nuclear chimeras, the deviating feature—the defective plastids in this instance—can be traced on cell-to-cell basis between the apical meristem and the mature tissues.[48] Most ot the plants studied by reference to cytochimeras are dicotyledons with a two-layered tunica. In these plants, periclinal cytochimeras have clearly revealed the existence of three independent layers (two layers of tunica and one of corpus initials) in the apical meristem.

The results obtained with cytochimeras also support the basic premise of the tunica-corpus concept that the differentiation of the various regions of the plant is not predetermined in the organization of the apical meristem even if its layers are independent. True,

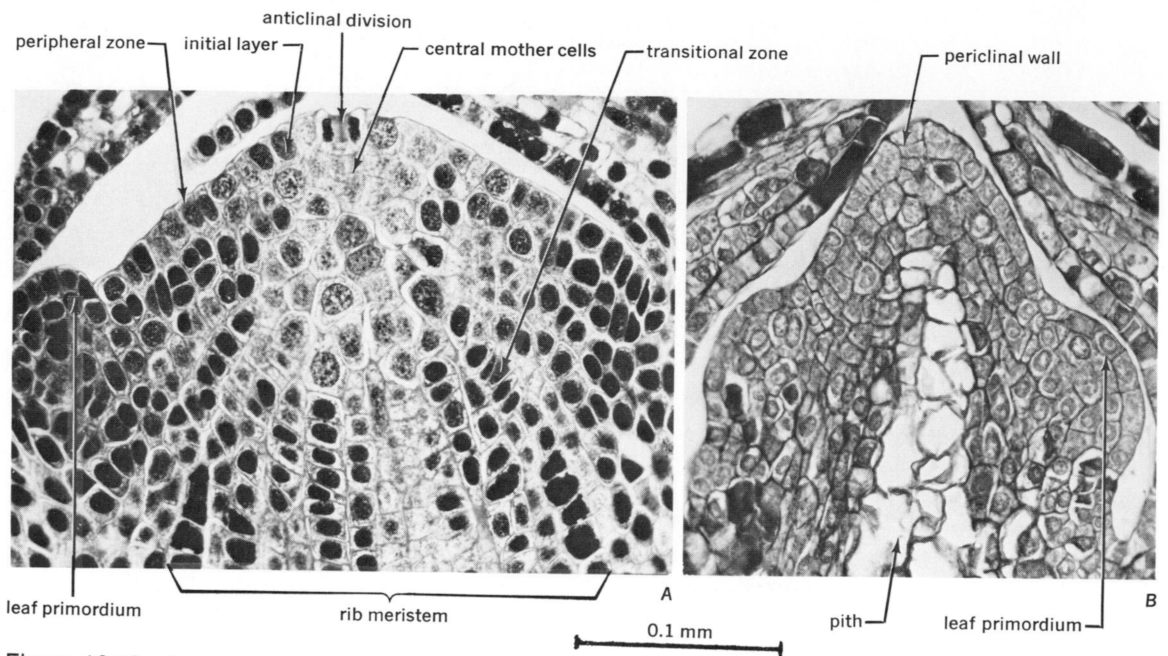

Figure 16.16 Shoot apices of conifers in longitudinal sections. *A, Pinus strobus,* pine. *B, Cupressus macrocarpa,* Monterey cypress. In *A,* the transitional zone merges with the rib meristem (file meristem). In *B,* the mother cells are not well differentiated and the pith is vacuolated close to the apex. (Slides, courtesy of: A. R. Spurr, *A;* A. S. Foster, *B.*)

the epidermis arises consistently from the outermost layer of the meristem, but the destinies of the derivatives of the deeper layers are not fixed. The derivatives of the second tunica layer divide anticlinally and periclinally and contribute various amounts to the subepidermal part of the peripheral region. The derivatives of the third layer also divide variously and produce the central region and variable amounts of peripheral tissues.

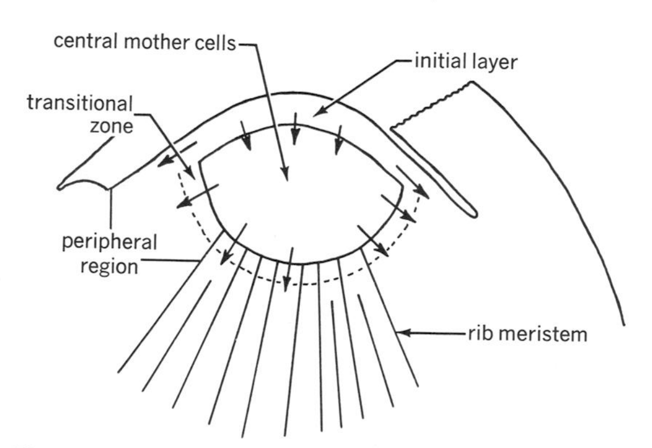

Figure 16.17 Diagram of a longitudinal section of a gymosperm shoot apex (*Pinus,* fig. 16.16, *A*). Arrows indicate the direction in which cells are contributed by the regions in the apical meristem.

The premise about the lack of predestination of tissue regions in the apical meristem distinguishes the tunica-corpus concept from the classic *histogen theory* of Hanstein,[22] according to which the epidermis, the cortex, and the vascular cylinder have their own precursory meristems, the dermatogen, the periblem, and the plerome, each with its own initials in the shoot or root (chapter 14) apex.

CYTO-HISTOLOGICAL ZONATION

The term cyto-histological zonation (see the Glossary) refers to the differentiation of regions with distinctive cytological characteristics in the shoot apex. Whereas the tunica-corpus concept helps our understanding of the cell arrangement and growth in the apical meristem, the recognition of regions of cells varied in character among the nearest derivatives of the apical meristem reveals the relation of this meristem to the tissue and organ differentiation in the shoot.[27] Zonation based on cell differentiation in the shoot apex was first described for *Ginkgo*[16] and has since been recognized in other gymnosperms and most angiosperms.[19] Because of the research on zonation, emphasis did not remain centered on the least determined part of the shoot apex (promeristem) but was extended to the derivative regions of the apical meristem. The unavoidable result was that the concept of the apical meristem was broadened and the distinction between the terms shoot apex (apical meristem and derivative meristematic regions) and apical meristem became less clear.[7]

The concept of cyto-histological zonation is discussed in the following with reference to shoot apices of *Ginkgo*[16] and *Pinus* (figs. 16.16,*A,* and 16.17). The apical meristem in these two genera includes a group of initials along the apical surface and their lateral and subjacent derivatives. The latter form the *central mother cell zone.* This entire distal group of cells is conspicuously vacuolated, a feature associated with a relatively low rate of mitotic activity. Moreover, the central mother cells often have thickened and distinctly pitted primary walls.

The apical meristem is surrounded by the peripheral region, or *peripheral meristem* (often inappropriately called flank meristem; see the Glossary), and beneath the central mother cells is the *pith meristem.* The peripheral meristem originates in part from the lateral derivatives of the apical initials, in part from the central mother cells. The pith meristem is formed by divisions along the periphery of the central mother cells in a layer of cells called *transitional zone.* At the height of its activity it resembles a cambium. The degree of discreteness of the transitional zone apparently depends on the vigor of growth since it may vary in the same plant in relation to seasonal or plastochronic periodicity.

The peripheral zone has dense, deeply staining protoplasts and is active mitotically. A particularly intense activity in localized positions results in the formation of leaf primordia (fig. 16.16,*B*). The peripheral zone is also concerned with the elongation of the shoot (anticlinal divisions) and increase in width (periclinal divisions). The cells in the pith meristem, which are considerably vacuolated, divide transversely so that the derivatives of individual cells form vertical files. A meristem showing this pattern of growth is called *file meristem* or *rib meristem.* In figure 16.16,*A*, the file meristem is actually continuing the pattern of growth established in the transitional zone. Some vertical divisions occur also and result in an increase in the number of files. In some gymnosperms the central group of vacuolated cells and the file meristem are poorly differentiated and the pith appears only a few cell layers beneath the surface of the apical meristem (fig. 16.16,*B*).

INQUIRIES INTO THE IDENTITY OF APICAL INITIALS

As was reviewed in chapter 14, the center of the apical meristem in a root undergoes a reduction in meristematic activity as the root passes beyond the initial stages of development. Meristematic acitvity is then transferred to the nearest derivatives of the original initials. The root exhibits a quiescent center in the middle of the apical meristem.

A similar view regarding the distribution of

meristematic activity in the shoot apex is advocated by the French cytologists.[4,35] According to this view, the distal axial cells (*zone axiale*) are relatively inert and the real initiating regions are the peripheral and subterminal zones where the stem tissues and leaf primordia arise. After the apical structure is organized in the embryonic or postembryonic growth, the distal group of cells becomes the waiting meristem (*méristème d'attente*), for it stays in a quiescent state until the reproductive stage is reached and meristematic activity is resumed in the distal cells. During the vegetative stage meristematic activity is centered in the initiating ring (*anneau initial*), corresponding to the peripheral zone, and in the pith meristem.

The French concept of apical growth in the shoot has served as a considerable stimulant for further research on apical meristems.[7,19,35] Counts of mitoses in different regions of the shoot apex (fig. 16.18), feeding shoot tips with radioactively labeled compounds to detect the location and synthesis of protein, DNA, and RNA, histochemical tests, experimental manipulations, and tracing of cell patterns in fixed and living shoot apices are providing data that, in essence, corroborate the postulate of the relative infrequency of mitotic activity in the distal zone. This conclusion, however, has not lead to a universal abandonment of the concept that the most distal cells are the ultimate source of all body cells in the shoot. Considering the geometry of the apex, one can deduce *a priori* that, in view of the exponential growth of the derivatives of the apical meristem, a few divisions in the distal cells would result in the propagation of any distinctive genome (sum total of genes) characteristics of these cells through large poplulations of cells. Periclinal chimeras are often cited as providing evidence of periodic additions of cells from the distal zone the progenies of which can be recognized in mature parts of the shoot. In this sense, the distal apical cells can be interpreted as the initials.

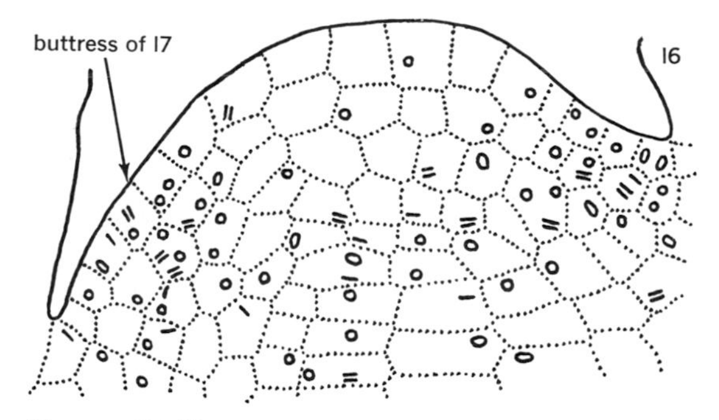

Figure 16.18 Longitudinal section of shoot apex of *Chrysanthemum segetum* from a plant in which the buttress of the seventh leaf (*l*) was emerging. The markings in the cells indicate nuclear division figures. These represent a sum of divisions observed in apices of ten different plants all in the same stage of development. The markers are: circle, prophase; one dash, metaphase; two dashes, anaphase; oval, telophase. (Adapted from A. Lance, *Ann. Sci. Nat., Bot. Sér.* 11, 18:91–421, 1957.)

The relation between the initials and the immediate derivatives in the apical meristem is flexible. A cell functions as an initial not because of any inherent properties but because of its position. (See similar concept of initials in the vascular cambium; chapter 10.) At the time of division of an initial it is impossible to predict which of the two daughter cells will "inherit" the function of the initial and which will become the derivative (e.g., fig. 16.12,*A*). It is also known that a given initial may be replaced by a cell which through prior history would be classified as a derivative of an initial.

As Newman[34] has stated, no cells are permanent initials. Instead, products of successive divisions constituting a "continuing meristematic residue" function as initials. This concept is used in Newman's classification of apical meristems designed for all groups of vascular plants. Three types of apical meristems are recognized: (1) monopodial, as in ferns (fig. 16.12,*B*)—the residue is in the superficial layer and any kind of division contributes to growth in length and breath; (2) simplex, as in gymnosperms (fig. 16.17)—the residue is in a single superficial layer and both anticlinal and periclinal divisions are needed for bulk growth; (3) duplex, as in angiosperms (fig. 16.14,*B*)—the residue occurs in at least two surface layers with two contrasting modes of growth, anticlinal divisions near the surface, divisions in at least two planes deeper in the apical meristem.

Origin of leaves

When a leaf primordium is initiated in the peripheral region of the shoot apex, localized cell divisions cause the formation of a protrusion (the so-called *leaf buttress*) on the side of the axis (fig. 16.11). In shoots with a helical leaf arrangement (figs. 16.3 and 16.26), the divisions alternate in different sectors around the circumference of the apical meristem and the resulting periodic enlargement of the apex, as seen from above, is asymmetric. In shoots with a decussate leaf arrangment (figs. 16.19 and 16.27), the enlargement is symmetrical because the intensified meristematic activity occurs simultaneously on opposite sides. Thus, the initiation of leaves causes periodic changes in the size and form of the shoot apex.

As illustrated for a shoot with decussate phyllotaxy, the increase in size of the apex reaches its maximum just before the pair of leaf primordia emerge (fig. 16.19,*A*). The shoot apex is in *maximal-area phase* of plastochronic growth. As the leaf primordia become elevated, the apical meristem is decreased in width (fig. 16.19,*B–F*). The apex enters the *minimal-area phase* of plastochronic growth. Before a pair of new primordia is formed the apex returns to the maximal phase. The extension now occurs perpendicular to the longest diameter of the preceding maximal phase, but the enlargement of the apical meristem is evident also between the members of the pair of leaves the growth of which had previously caused the reduction in the size of the apex (fig. 16.19,*E*).

The relation between the growing leaf primordium and the apical meristem varies greatly in different species. Figure 16.19 illustrates one extreme in which the apical meristem almost vanishes between the enlarging leaf primordia (fig. 16.19,*D*). In other species, the apical meristem is affected much less (fig. 16.11), and in species in which the apical meristem is elevated considerably above the organogenic region the apex does not undergo plastochronic changes in size (fig. 16.15). But whatever may be the relation between the peripheral region where the leaves originate and the distal region, the peripheral region is the main contributor to the growth of the leaf primordium. Accordingly, this region shows higher counts of mitosis (fig. 16.18), contains more RNA, and demonstrates a larger number of labeled nuclei in tests for incorporation of ^{3}H-thymidine (synthesis of DNA) than do the other regions of the shoot apex, especially the distal axial cell group.

The histologic details of the origin of leaves have been obtained for many plants in different taxa. In gymnosperms and dicotyledons the divisions initiating leaf primordia commonly occur in the second or third layer from the surface (fig. 16.11,*A*). Periclinal divisions add cells toward the periphery and are largely responsible for the lateral protrusion of the primordium. Anticlinal divisions occur in the surface layer and also accompany the periclinal divisions in the deeper layers. In monocotyledons, leaf primordia are frequently initiated by periclinal divisions in the surface layer.

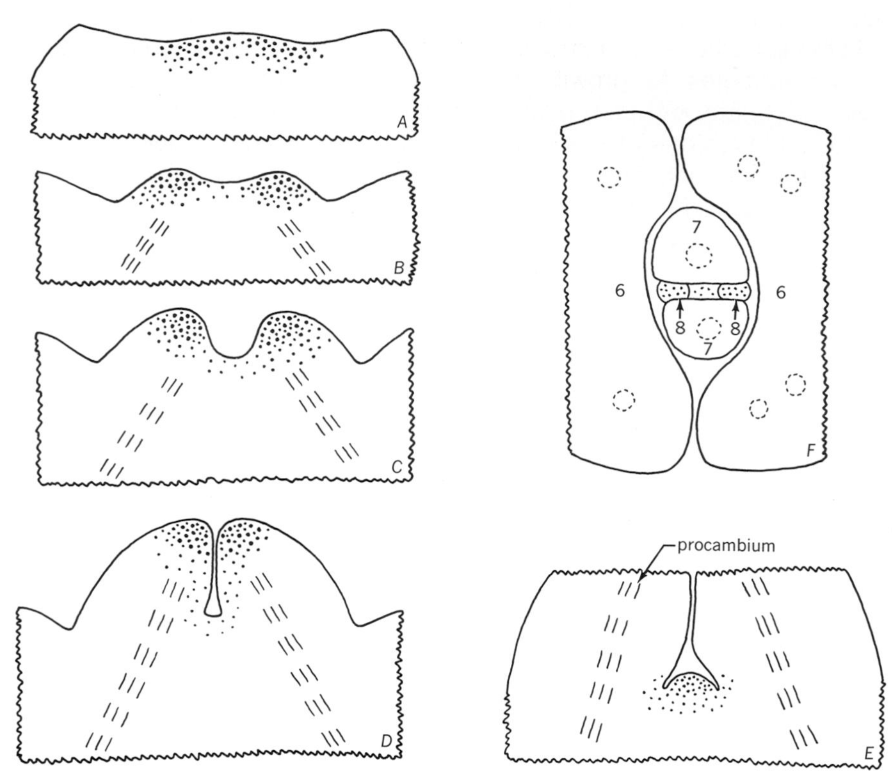

Figure 16.19 Outlines of developing leaf primordia of *Kalanchoë* from longitudinal (*A–E*) and transverse (*F*) sections of shoots sampled during the initiation and development of the eighth leaf pair. *A,* after plastochron 7; apex in maximal phase. *B,* early plastochron 8; leaf pair 8 has been initiated. *C,* leaves of pair 8 somewhat elongated. *D,* mid-phase of plastochron 8; apex in minimal phase. *E,* early plastochron 9; the primordia of pair 9 alternate with those of pair 8 and therefore do not appear in the plane of figure *E*; the enlarging apex between the two primordia 8 is visible. *F,* early plastochron 8, phase similar to that in *B.* (From photomicrographs in D. B. Stein and O. L. Stein, *Amer. J. Bot.* 47:132–140, 1960.)

Derivatives of either tunica or corpus undergo divisions leading to the formation of a primordium. If the tunica is deep, the entire primordium may originate from its derivatives. Otherwise the leaf tissue may be traced to both tunica and corpus (fig. 16.13).

Leaf primordia arise in positions around the circumference of the apical meristem that are correlated with the phyllotaxy of the shoot. The causal relations of the orderly initiation of leaves have interested botanists for a long time. One view is that the new leaf is initiated in a locus—"the first available space"—that is removed from inhibitions emanating from the apical meristem and the most recently formed leaf primordia.[54] Some authors propose that the creation of the available space depends on physiological events at the apex,[7] but it may well be that physical constraints also are an important element in the growth of leaves at the apex.[55] Furthermore, the arrangement of leaves is correlated with the architecture of the vascular system in the stem so that the spatial relation of the leaves to one another is part of an overall pattern in shoot organization. The developmental relation between the leaves and the leaf traces in the stem suggests that the events determining the specific leaf arrangements are not necessarily limited to the apical region.[13]

Origin of branches

In the seed plants, branches commonly originate as buds in the axils of leaves. The axillary buds arise at variable plastochronic distances from the apical meristem, most frequently in the axil of second or third leaf from the apex. At these levels, the tissue in the leaf axil usually has a typical meristematic appearance (figs. 16.13 and 16.20,*A*) and the bud is formed in continuity with the apical meristem. With further growth of the parent shoot, vacuolation occurs above the axillary meristem and the latter becomes separated from the apical meristem (figs. 16.20,*B,* and 16.21). If the axillary meristem is initiated during later plastochrons, it may be separated from the apex by vacuolated cells from the start and is then characterized as a *detached meristem.*

When a bud is formed, periclinal and anticlinal divisions occur in a variable number of cell layers in the leaf axil, the bud meristem is elevated above the surface, and the apical meristem of the bud is organized (fig. 16.20,*B*). In many plants, orderly divisions occur along the basal and lateral limits of the incipient bud and form a zone of parallel curving layers referred to as the *shell zone* (fig. 16.20,*A*) because of its shell-like shape. It has been suggested that the specific patterns of cell division that give rise to buds may be determined by the direction and magnitude of stresses in the region where the bud is initiated (fig. 16.13) and that the development of the shell zone is an early response to the stresses generated in the axil of a growing leaf.[30] The shell zone is possibly responsible for the upward growth of the bud, just as the transitional zone in a terminal shoot apex (fig.

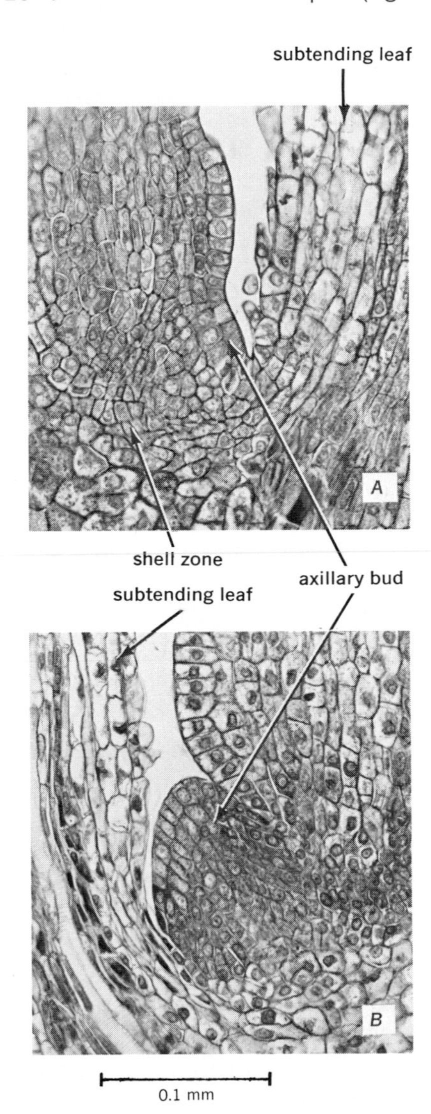

Figure 16.20 Origin of axillary buds in potato (*Solanum tuberosum*). Longitudinal sections of nodes showing an earlier (*A*) and a later (*B*) stage of bud development. (From Sussex.[49])

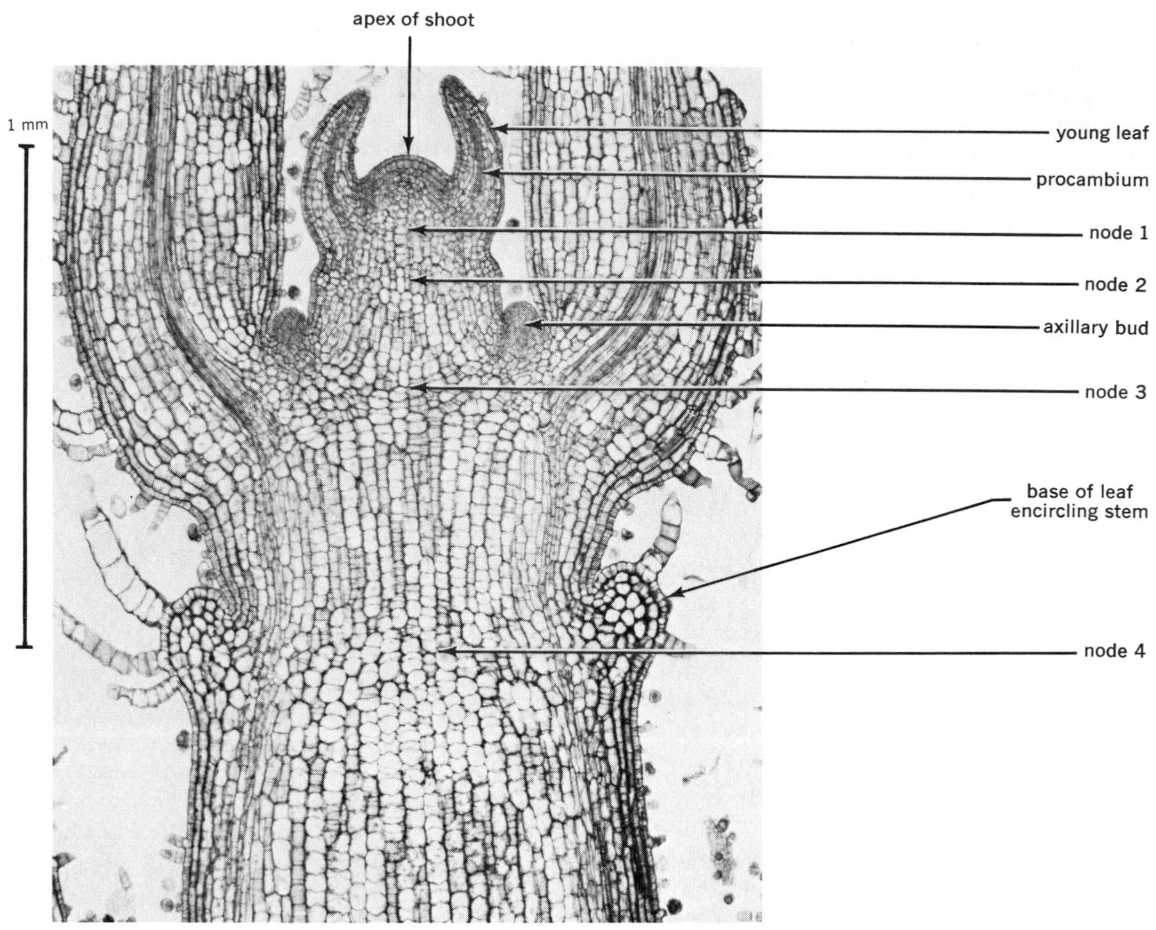

Figure 16.21 Longitudinal section of shoot of *Coleus.* The pith illustrates result of rib-meristem form of growth in length.

16.16,*A*) appears to be associated with an active upward thrust of the apical meristem. If the axillary bud is not dormant, its upward growth is followed by the initiation of leaf primordia beginning with the prophylls (fig. 16.9,*A*).

Shoots may develop from adventitious buds. Such buds arise with no direct relation to the apical meristem. Adventitious buds may develop on roots, stems, hypocotyls, and leaves. They originate in callus tissue of cuttings or near wounds, in the vascular cambium or on the periphery of the vascular cylinder. The epidermis may produce adventitious buds. Depending on the depth of the initiating tissue, the buds may have an exogenous or an endogenous origin.[40] If the adventitious buds arise in mature tissues, their initiation involves the phenomenon of dedifferentiation, that is, change of mature into meristematic tissue.

Factors determining the initiation and development of buds, both axillary and adventitious, are apparently numerous, and their interrelationship is complex. Buds may remain dormant because of suppression of their growth by the terminal shoot (apical dominance). There may also be a competition among the axillary buds. Distribution of growth substances and competition for nutrients are two of the most important factors in the formation of buds,[32,51,56] and at different stages of bud development different combinations of factors are required.

Primary growth of the stem

The amount and pattern of primary growth has a determining effect upon the form of the shoot, particularly in herbaceous plants. In the stem, primary growth is expressed in the elongation and the widening of the axis below the apical meristem. In ordinary leafy shoots, elongation occurs chiefly in the internodes. At the apex, leaves appear at close levels to one another so that nodes and internodes do not exist as separate regions. Later, rapid growth in length between the leaf insertion zones, or nodes, gives rise to the internodes (figs. 16.21 and 16.22,*A*). This elongation is based on the file-meristem type of growth forming longitudinal files of cells in the cortex and pith meristems by repeated transverse divisions (fig. 16.21). Later, cell enlargement also occurs and eventually replaces cell division. Most of the linear extension of an internode may result from growth by cell elongation[26] but the final relative length of an internode may be determined by cell number rather than cell length.[29] The rate of internodal elongation changes from level to level in a developing plant[11] so that the mature plant has internodes of variable length (fig. 16.22,*B*). During the rosette stage of growth characteristic of some plants, internodal elongation does not occur and the leaves remain crowded.

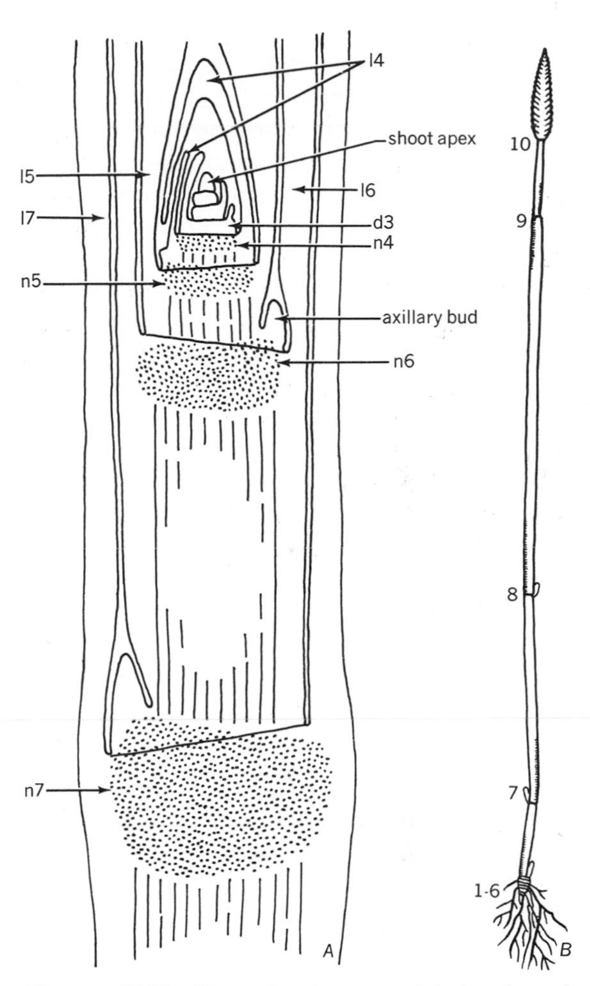

Figure 16.22 Elongation in monocotyledon (grass) stems. *A*, median longitudinal section crossing the plane of leaves (*l*) on rhizome of *Agropyron repens*. Shows relationship of developing nodes (*n*) and internodes (below each *n*, marked with vertical lines) to discs of insertion (one marked *d3*) of leaf primordia. Each leaf appears on both sides of the axis because of encircling growth. Compare with fig. 16.5. Leaves and nodes are numbered from the apex downward. *B*, stem of wheat (*Triticum aestivum*) with leaves removed. Shows crowded internodes (1–6) at base and elongated internodes above. The nodes are numbered from the oldest to the youngest. Inflorescence at the apex. (Adapted from Hitch and Sharman.[23])

In angiosperms, and particularly the monocotyledons, the extent of the internodal elongation is of primary importance in establishing the characteristic morphology of a species.[24] The meristematic activity causing the elongation of an internode may be rather uniform throughout the internode, or it may occur as a wave progressing from the base of the internode acropetally,[18] or it may be largely restricted to the base of the internode.[26] A localized meristematic region in the elongating internode is called an *intercalary meristem;* it is intercalated between regions of tissues more advanced in development. An intercalary meristem region may retain a capacity for growth long after the elongation of the particular internode is completed. In grasses, for example, resumption of growth at the base of an internode (and the surrounding leaf sheath) enable lodged plants to lift their culms from the ground by growth in the intercalary meristem region on the side turned toward the ground.[12] The region involved in this growth is called a joint or pulvinus.

Vascular tissue differentiates in the intercalary meristematic zone but the mature vascular elements are affected by the elongation of the ground tissue. Tracheary elements are destroyed by stretching and their function is taken over by newly differentiating elements or by lacunae appearing in their place.[14] The mature sieve elements are also destroyed. In *Lolium perenne*[15] and wheat[37] the crushed sieve elements are not immediately replaced (in hours in wheat, in 1 to 2 days in *Lolium*) by new cells so that a constriction or even complete blockage of the food conduit occurs and possibly results in diversion of the assimilate to the meristem.

The growth in thickness of the axis involves periclinal divisions and cell enlargement in both the pith and the cortex. Primary thickening varies in detail in different plants. It is usually moderate in species with secondary growth (fig. 16.23). Herbaceous dicotyledons of specialized types, such as rosette and succulent types,[42] and many monocotyledons have massive primary growth. This growth usually occurs so close to the apical meristem that the latter appears inserted on a shallow cone (fig. 16.24), on a flat plateau, or even in a depression. In dicotyledons, the primary thickening growth is emphasized in the pith (medullary thickening) or in the cortex (cortical thickening), or is dispersed

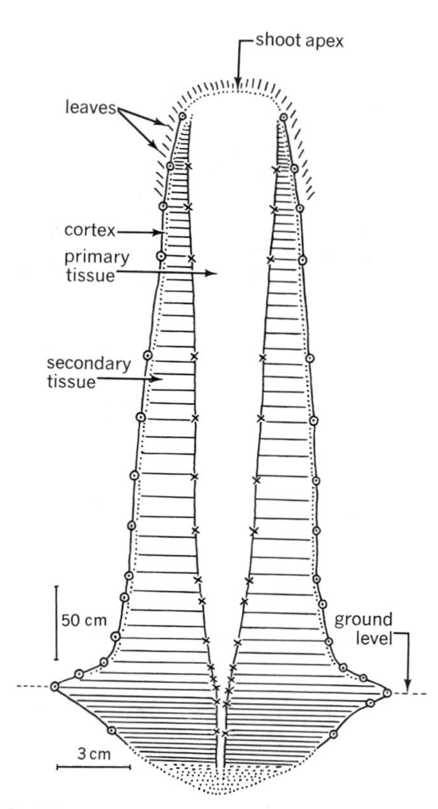

Figure 16.23 Diagrammatic representation of axis, in longitudinal section, of a young *Cordyline australis* (Agavaceae) tree, a representative of a large monocotyledon. Plotted from measurements on a specimen about 4 meters high to the leafy crown. The vertical axis is foreshortened to 16 times in relation to stem diameter. The primary axis is obconical and mechanical stability is provided by early secondary growth. (Adapted from Tomlinson and Esler.[50])

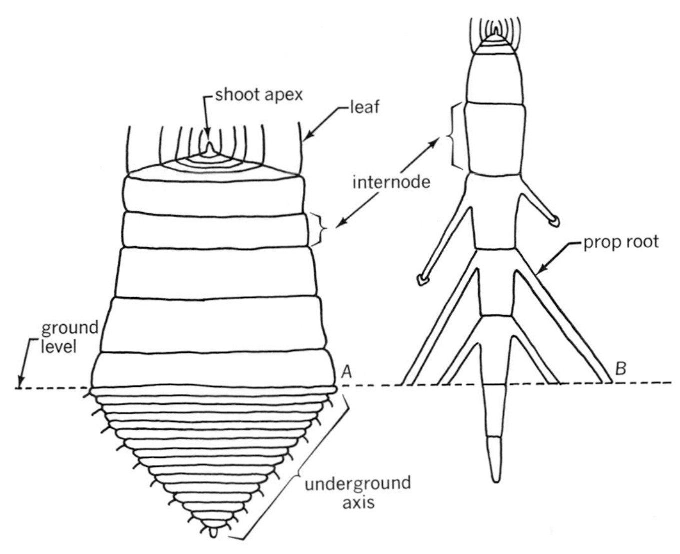

Figure 16.24 Large monocotyledons during establishment growth. *A*, type found in most palms: early underground stem part widens progressively but forms short internodes; growth in height by means of longer internodes occurs after the diameter of adult stem is reached (at ground level). *B*, type found in *Pandanus* and a few palms: growth in height begins early; the widening stem is supported by adventitious prop roots. (Adapted from Zimmermann and Tomlinson.[60])

throughout the axis. In some monocotyledons thickening growth is dominant in a relatively narrow region near the periphery of the axis and because of its conspicuousness the region is called *primary thickening meristem.* Its cells are arranged in orderly anticlinal series (fig. 16.15).

During primary thickening the stem assumes an obconical form because the later formed internodes become wider than the older ones. If this kind of growth would continue indefinitely an unstable axis would result. In most dicotyledons and some monocotyledons, secondary growth stabilizes the axis by increasing the thickness of the axis beginning with the base (fig. 16.23). Most monocotyledons, however, lack secondary growth and, if they become arborescent, must produce a stable axis by other means.[50,60] In the majority of the arborescent monocotyledons, including most palms, growth in height of the seedling remains minimal while, during the establishment growth, an obconical underground axis develops. Growth in height begins after the size of the crown and the diameter of the stem reach those of the adult plant (fig. 16.24,*A*). In a few palms establishment growth is associated with growth in height through internodal elongation. The resulting axis is stabilized by a development of supporting prop roots (fig. 16.24,*B*).

Vascular differentiation

Developmental studies of the primary vascular system clearly show that the topographic organization of the differentiating system is a small replica of that of the mature system. This feature is brought out in figure

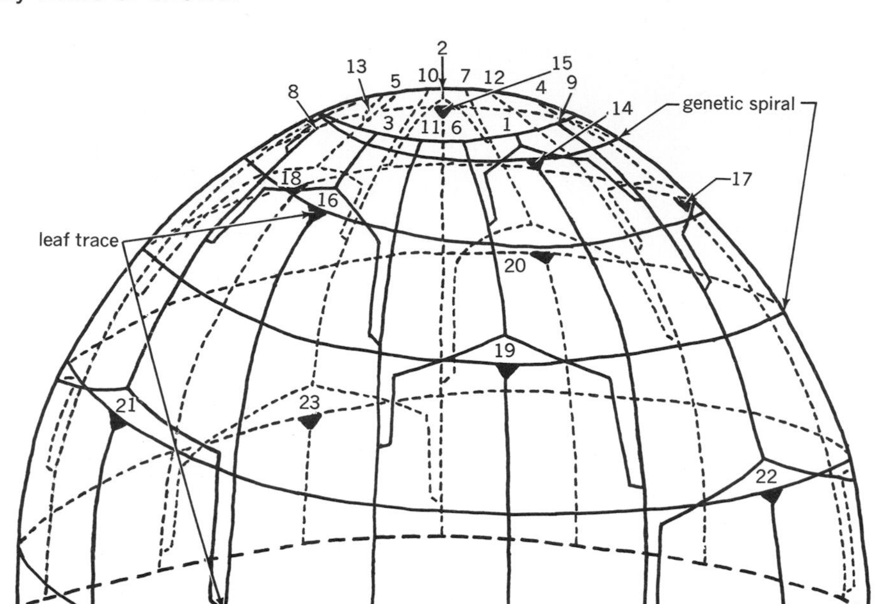

Figure 16.25 Diagrammatic representation of the primary vascular system of *Hectorella caespitosa* near the apex of the shoot. The numbers relate leaf traces to leaves. Compare with fig. 16.3. (Adapted from Skipworth.[46])

16.25 in which the previously analysed mature vascular system of the dicotyledon *Hectorella* (fig. 16.3,*C*) has been projected onto a conical form representing, diagrammatically, the tip of the shoot. The lengths of the component bundles and the distances between the coils of the "genetic spiral" have been appropriately reduced in the upward direction in order to indicate that the upper internodes had not yet completed their elongation.

The principles of differentiation of the primary vascular system in seed plants are reviewed in the following paragraphs by the use of some specific examples.

ORIGIN OF PROCAMBIUM

The differentiation of procambium is illustrated in figure 16.26 in successive cross sections of a shoot of *Linum* (flax). Complementary views of longitudinal sections of dicotyledon (potato) shoots of similar age appear in figure 16.11. In both plants, the vascular system of the stem forms a single cylinder of leaf traces and stem bundles. No vascular differentiation is detectable above the level of leaf initiation (figs. 16.11 and 16.26,*A*). Farther below, where leaf primordia come into view, some vacuolation is present (light stippling and no stippling in fig. 16.26,*B*). It indicates a beginning of differentiation of cortex and pith. Increased vacuolation in cortex and pith delimits an approximately circular zone as seen in transections (fig. 16.26,*C*). For descriptive purposes, this meristematic zone is called *residual meristem.*[13] It appears as though it were a residium of the apical meristem retained among

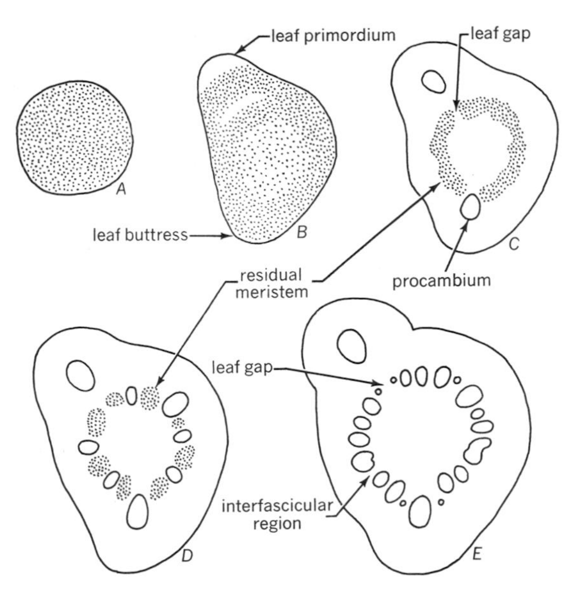

Figure 16.26 Cross sections of flax stem taken at different levels beginning with apical meristem (*A*) and ending with node in which all primary tissues are delimited (*E*). Stages in differentiation of vascular system: vacuolation of cortex (starts in leaf primordium in *B*) and pith; the resulting blocking out of prospective vascular region as residual meristem associated with procambial strands; and completion of procambial differentiation within the residual meristem.

tissues more advanced in differentiation. Beneath the insertion of leaf primordia, longitudinal divisions, not followed by an appreciable increase in cell width, produce somewhat elongated cells of the first procambium in the residual meristem (figs. 16.11 and 16.21).

At successively later stages of development more and more procambial strands differentiate from the residual meristem (fig. 16.26,*D*). Whereas the first procambial strands are traces of the nearest (oldest) leaf primordia (fig. 16.26,*C*), those appearing later are stem bundles and extensions of traces of higher (younger) leaves (fig. 16.26,*D*). After all vascular bundles characteristic of the given stem level have differentiated, the remaining residual meristem differentiates into interfascicular parenchyma (fig. 16.26,*E*). At the nodes, some of the residual meristem becomes leaf-gap parenchyma. The gaps are usually discernible in earlier stages of stem development, that is, at higher levels of the shoot, than are the interfascicular regions (fig. 16.26,*C*).

The sequence described implies an initial procambial continuity between the sites of leaf primordia and the vascular cylinder of the axis. As the leaf primordia are elevated by growth above their buttresses, procambium differentiates within the primordia in continuity with the procambium of the previously defined leaf traces (figs. 16.11 and 16.19). The differentiation of procambium in successively higher internodes and its continuation in the developing leaf primordia is referred to as *acropetal differentiation of procambium.*

Continuous acropetal differentiation of procambium in the leaf traces and stem bundles has been recorded in a number of conifers and dicotyledons.[10,13,41] In the monocotyledons, the pattern may deviate from that common in conifers and dicotyledons. In the investigated grasses (wheat and *Dactylis glomerata*) the earlier procambial leaf traces originate in the stem at the level of insertion of the leaf primordium, the later traces successively higher in the leaf itself. From the point of origin, the various traces differentiate basipetally in the stem, where they become connected with older vascular bundles, and acropetally in the developing leaf.[23,37,45]

ORIGIN OF PHLOEM AND XYLEM

Procambial differentiation is followed by that of vascular elements, first in the older, then in the younger procambial strands at the given level of the stem. The earlier the bundles begin to differentiate, the larger they are at maturity so that in a given transection of

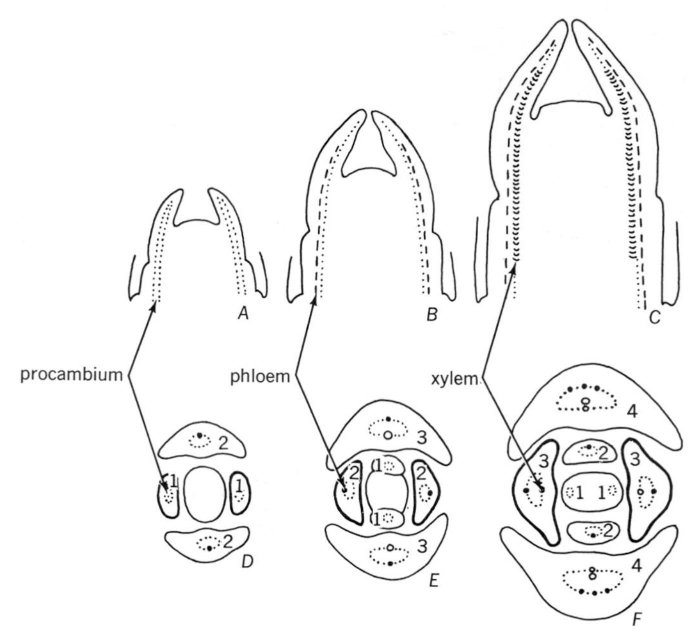

Figure 16.27 Diagrams illustrating sequence in vascular differentiation, as seen in longitudinal (*A–C*) and cross (*D–F*) sections of a shoot with decussate leaf arrangement (*Lonicera,* honeysuckle): acropetal and continuous differentiation of procambium and phloem; discontinuous initiation and bidirectional differentiation of xylem. The first phloem elements mature before the first xylem elements.

stem the bundle system has a heterogeneous appearance during development and at maturity.

In collateral vascular bundles, the first phloem appears in the outer part and the xylem in the inner part of a procambial strand (chapter 11). The subsequent differentiation of the phloem is centripetal; that is, new phloem elements appear closer to the center of the stem. The xylem differentiates in the opposite direction, that is, centrifugally, and is called *endarch xylem.* The term means that protoxylem elements occupy the innermost (with regard to the center of the axis) position in the xylem bundle. (Compare with exarch xylem in the root; chapter 14.)

In the longitudinal course of differentiation the first phloem and the first xylem usually show contrasting developmental sequences in both gymnosperms and dicotyledons.[10,13,41] The phloem follows the procambium rather closely in its acropetal course of differentiation (fig. 16.27,*A,B*). The xylem commonly differentiates first in the basal part of the leaf or below it somewhere in the leaf traces and then progresses acropetally toward the leaf and basipetally in the axis (figs. 8.8 and 16.27,*C*) where a connection with the xylem of an older bundle is established. In a given leaf trace, the first phloem appears before the first xylem (fig. 16.27). If the plant has internal phloem, the latter begins to differentiate later than the external phloem.

The initial development of xylem and phloem in wheat[37] (*Triticum aestivum*) contrasts with that in the dicotyledons and gymnosperms. Although in a given leaf trace the phloem appears 2 to 3 days before the xylem, the later course of differentiation is similar for the two tissues. Both appear first at sites isolated from differentiated vascular tissues and develop from here bidirectionally down the stem and up the leaf primordia.

VASCULAR CONNECTION BETWEEN BUD AND STEM

The course of procambial development in bud traces depends on the character of the bud.[13] An axillary bud that is initiated close to the apical meristem and soon develops into a branch becomes connected with the vascular system of the stem by branch traces which are equivalent to leaf traces in their developmental relation to the bundles in the stem. A bud, the development of which is delayed (e.g., by apical dominance of the main shoot), is initially separated from the vascular cylinder of the stem by vacuolated parenchyma. When a bud of this kind begins to grow, either spontaneously or after removal of the shoot tip (the source of apical dominance), it establishes a direct vascular connection with the nearest vascular strands in the central cylinder through the vacuolated parenchyma.

REFERENCES

1. Beck, C. B. The appearance of gymnospermous structure. *Biol. Rev.* 45:379–400. 1970.
2. Benzing, D. H. Developmental patterns in stem primary xylem of woody Ranales. I. Species with unilacunar nodes. *Amer. J. Bot.* 54:805–813. 1967. II. Species with trilacunar and multilacunar nodes. *Amer. J. Bot.* 54:813–820. 1967.
3. Blyth, A. Origin of primary extraxylary stem fibers in dicotyledons. *Univ. Calif. Pubs. Bot.* 30:145–232. 1958.
4. Buvat, R. Le méristème apical de la tige. *Ann. Biol.* 31:595–656. 1955.
5. Carlquist, S. *Comparative plant anatomy.* New York, Holt, Rinehart and Winston. 1961.
6. Church, A. H. *On the interpretation of phenomena of phyllotaxis.* London, Oxford University Press. 1920.
7. Cutter, E. G. Recent experimental studies of the shoot apex and shoot morphogenesis. *Bot. Rev.* 31:7–113. 1965.
8. Cutter, E. G. *Plant anatomy: experiment and interpretation.* Part 2. *Organs.* London, Edward Arnold. 1971.
9. Dermen, H. Periclinal cytochimeras and origin of tissues in stem and leaf of peach. *Amer. J. Bot.* 40:154–168. 1953.
10. Devadas, C., and C. B. Beck. Development and morphology of stelar components in the stems of some members of the Leguminosae and Rosaceae. *Amer. J. Bot.* 58:432–446. 1971. Comparative morphology of the primary vascular systems in some species of Rosaceae and Leguminosae. *Amer. J. Bot.* 59:557–567. 1972.
11. Enright, A. M. and B. G. Cumbie. Stem anatomy and internodal development in *Phaseolus vulgaris. Amer. J. Bot.* 60:915–922. 1973.
12. Esau, K. *Plant anatomy.* 2nd ed. New York, John Wiley & Sons. 1965.
13. Esau, K. *Vascular differentiation in plants.* New York, Holt, Rinehart and Winston. 1965.
14. Evans, P. S. Intercalary growth in the aerial shoot of *Eleocharis acuta.* R. Br. Prodr. I. Structure of the growing zone. *Ann. Bot.* 29:205–217. 1965.
15. Forde, B. J. Differentiation and continuity of the phloem in the leaf intercalary meristem of *Lolium perenne. Amer. J. Bot.* 52:953–961. 1965.
16. Foster, A. S. Structure and growth of the

shoot apex in *Ginkgo biloba*. *Bull. Torrey Bot. Club* 65:531–556. 1938.

17. Foster, A. S., and E. M. Gifford, Jr. *Comparative morphology of vascular plants*. 2nd ed. San Francisco, California, W. H. Freeman and Company. 1974.
18. Garrison, R. The growth of internode in *Helianthus*. *Bot. Gaz.* 134:246–255. 1973.
19. Gifford, E. M., Jr., and G. E. Corson, Jr. The shoot apex in seed plants. *Bot. Rev.* 37:143–229. 1971.
20. Gregory, R. A., and J. A. Romberger. The shoot apical ontogeny of the *Picea abies* seedling. I. Anatomy, apical dome diameter, and plastochron duration. *Amer. J. Bot.* 59:587–597. 1972.
21. Gunning, B. E. S., J. S. Pate, and L. W. Green. Transfer cells in the vascular system of stems: taxonomy, association with nodes, and structure. *Protoplasma* 71:147–171. 1970.
22. Hanstein, J. Die Scheitelzellgruppe im Vegetationspunkt der Phanerogamen. *Festschr. Niederrhein. Ges. Natur-u. Heilkunde* 1868:109–134. 1868.
23. Hitch, P. A., and B. C. Sharman. The vascular pattern of festucoid grass axes, with particular reference to nodal plexi. *Bot. Gaz.* 132:38–56. 1971.
24. Holttum, R. E. Growth-habits of monocotyledons, variations on a theme. *Phytomorphology* 5:399–413. 1955.
25. Jackson, B. D. *A glossary of botanic terms*. 4th ed. New York, Hafner Publishing Co. 1953.
26. Kaufman, P. B., S. J. Cassel, and P. A. Adams. On nature of intercalary growth and cellular differentiation in internodes of *Avena sativa*. *Amer. J. Bot.* 126:1–13. 1965.
27. Kondrat'eva, E. A. O stroenii verkhushki vegetativnogo pobega pokrytosemennykh. [Concerning structure of vegetative shoot apex in angiosperms.] *Leningrad Univ. Vest. Ser. Biol. Geog. i Geolog.* 10:3–15. 1955.
28. Kumazawa, M. Studies on the vascular course in maize plant. *Phytomorphology* 11:128–139. 1961.
29. Lam, O. C. III, and C. L. Brown. Shoot growth and histogenesis of *Liquidambar styraciflua* L. under different photoperiods. *Bot. Gaz.* 135:149–154. 1974.
30. Lintilhac, P. M. Differentiation,organogenesis, and the tectonics of cell wall orientation. III. Theoretical consideration of cell wall mechanics. *Amer. J. Bot.* 61:230–237. 1974.
31. McAlpin, B. W., and R. A. White. Shoot organization in the Filicales: the promeristem. *Amer. J. Bot.* 61:562–579. 1974.
32. McIntyre, G. I. Environmental control of apical dominance in *Phaseolus vulgaris*. *Can. J. Bot.* 51:293–299. 1973.
33. Namboodiri, K. K., and C. B. Beck. A comparative study of the primary vascular system of conifers. I. Genera with helical phyllotaxis. *Amer. J. Bot.* 55:447–457. 1968. II. Genera with opposite and whorled phyllotaxis. *Amer. J. Bot.* 55:458–463. 1968. III. Stelar evolution in gymnosperms. *Amer. J. Bot.* 55:464–472. 1968.
34. Newman, I. W. Pattern in the meristems of vascular plants. III. Pursuing the patterns in the apical meristems where no cell is a permanent cell. *J. Linn. Soc. London, Bot.* 59:185–214. 1965.
35. Nougarède, A. Experimental cytology of the shoot apical cells during vegetative growth and flowering. *Internatl. Rev. Cytol.* 21:203–351. 1967.
36. Pant, D. D., and B. Mehra. Nodal anatomy in retrospect. *Phytomorphology* 14:384–387. 1964.
37. Patrick, J. W. Vascular system of the

stem of the wheat plant. I. Mature state. *Aust. J. Bot.* 20:49–63. 1972. II. Development. *Aust. J. Bot.* 20:65–78. 1972.

38. Popham, R. A. Principal types of vegetative shoot apex organization in vascular plants. *Ohio J. Sci.* 51:249–270. 1951.
39. Popham, R. A., and A. P. Chan. Zonation in the vegetative stem tip of *Chrysanthemum morifolium* Bailey. *Amer. J. Bot.* 37:476–484. 1950.
40. Priestley, J. H., and C. F. Swingle. Vegetative propagation from the standpoint of plant anatomy. *U. S. Dept. Agr. Tech. Bull.* 151. 1929.
41. Pulawska, Z. Correlations in the development of the leaves and leaf traces in the shoot of *Actinidia arguta* Planch. *Acta. Soc. Bot. Polon.* 34:697–712. 1965.
42. Rauh, W., and F. Rappert. Über das Vorkommen und die Histogenese von Scheitelgruben bei krautigen Dikotylen, mit besonderer Berücksichtigung der Ganz- und Halbrosettenpflanzen. *Planta* 43:325–360. 1954.
43. Saint-Côme, R. Applications des techniques histoautoradiographiques et des méthodes statistiques à l'étude du fonctionnement apical chez le *Coleus blumei* Benth. *Rev. Gén. Bot.* 73:241–324. 1966.
44. Schmidt, A. Histologische Studien an phanerogamen Vegetationspunkten. *Bot. Arch.* 8:345–404. 1924.
45. Sharman, B. C., and P. A. Hitch. Initiation of procambial strands in leaf primordia of bread wheat, *Triticum aestivum. Ann. Bot.* 31:229–243. 1967.
46. Skipworth, J. P. The primary vascular system and phyllotaxis in *Hectorella caespitosa* Hook. f. *New Zeal. J. Sci.* 5:253–258. 1962.
47. Slade, B. F. Stelar evolution in vascular plants. *New Phytol.* 70:879–884. 1971.
48. Stewart, R. N., P. Semeniuk, and H. Dermen. Competition and accommodation between apical layers and their derivatives in the ontogeny of chimeral shoots of *Pelargonium* X *Hortorum. Amer. J. Bot.* 61:54–67. 1974.
49. Sussex, I. M. Morphogenesis in *Solanum tuberosum* L.: apical structure and developmental pattern of the juvenile shoot. *Phytomorphology* 5:253–273. 1955.
50. Tomlinson, P. B., and A. E. Esler. Establishment growth in woody monocotyledons native to New Zealand. *New Zeal. J. Bot.* 11:627–644. 1973.
51. Tucker, D. J., and T. A. Mansfield. Apical dominance in *Xanthium strumarium;* a discussion in relation to current hypotheses of correlative inhibition. *J. Exp. Bot.* 24:731–740. 1973.
52. Van Fleet, D. S. The cell forms, and their common substance reactions, in the parenchyma-vascular boundary. *Bull. Torrey Bot. Club* 77:340–353. 1950.
53. Van Tieghem, P., and H. Douliot. Sur la polystélie. *Ann. Sci. Nat., Bot. Sér.* 7. 3:275–322. 1886.
54. Wetmore, R. H. Growth and development in the shoot system of plants. *Soc. Devel. and Growth, Symposium* 14: 173–190. 1956.
55. Williams, R. F. *The shoot apex and leaf growth: a study in quantitative biology.* London-New York, Cambridge University Press. 1975.
56. Yun, K.-B., and J. M. Naylor. Regulation of cell reproduction in bud meristems of *Tradescantia paludosa. Can. J. Bot.* 51:1137–1145. 1973.
57. Ziegenspeck, H. Vorkommen und Bedeutung von Endodermen und Endodermoiden bei oberirdischen Organen der Phanerogamen im Lichte der Fluoroskopie. *Mikroskopie* 7:202–208. 1952.

58. Zimmermann, M. H., and P. B. Tomlinson. Anatomy of the palm *Rhapis excelsa,* I. Mature vegetative axis. *J. Arnold Arb.* 46:160–178. 1965.
59. Zimmermann, M. H., and P. B. Tomlinson. Vascular construction and development in the aerial stem of *Prionium* (Juncaceae). *Amer. J. Bot.* 55:1100–1109. 1968.
60. Zimmermann, M. H., and P. B. Tomlinson. The vascular system in the axis of *Dracaena fragrans* (Agavaceae), 1. Distribution and development of primary strands. *J. Arnold Arb.* 50:370–383. 1969. 2. Distribution and development of secondary vascular tissue. *J. Arnold Arb.* 51:478–491. 1970.
61. Zimmermann, M. H., and P. B. Tomlinson. The vascular system of monocotyledonous stems. *Bot. Gaz.* 133:141–155. 1972.

17

The Stem: Secondary State of Growth and Structural Types

SECONDARY GROWTH

Secondary growth resulting from the activity of the vascular cambium (chapter 10) increases the amount of vascular tissues in stems, beginning with the part of the shoot or seedling axis that has ceased to elongate. It contributes to the thickness of the axis but not to its length. Secondary growth occurs chiefly in the main stem and its branches but may be observed in limited amounts in leaves, particularly the petioles and midribs. It is characteristic of gymnosperms and woody dicotyledons but is also found in herbaceous dicotyledons. Some herbaceous dicotyledons and most monocotyledons have no secondary thickening. The secondary growth that does occur in monocotyledons is of a special type. The term secondary growth includes also the formation of periderm from the phellogen (chapter 12).

Establishment and extent of vascular cambium

The vascular cambium arises in part from the procambium within the vascular bundles and in part from the interfascicular parenchyma (fig. 17.1). The parts of the vascular cambium arising in the two positions are called *fascicular* and *interfascicular cambium,* respectively. In those stems in which the leaf traces and stem bundles are laterally almost contiguous, the interfascicular cambium is difficult to delimit from the fascicular cambium, except in the leaf gaps.

At the beginning of cambial development, divisions initiating the cambium within the vascular bundles frequently precede those appearing in the interfascicular regions. If these regions are wide, their initial cambial divisions may start next to the fascicular cambium and spread tangentially. This observa-

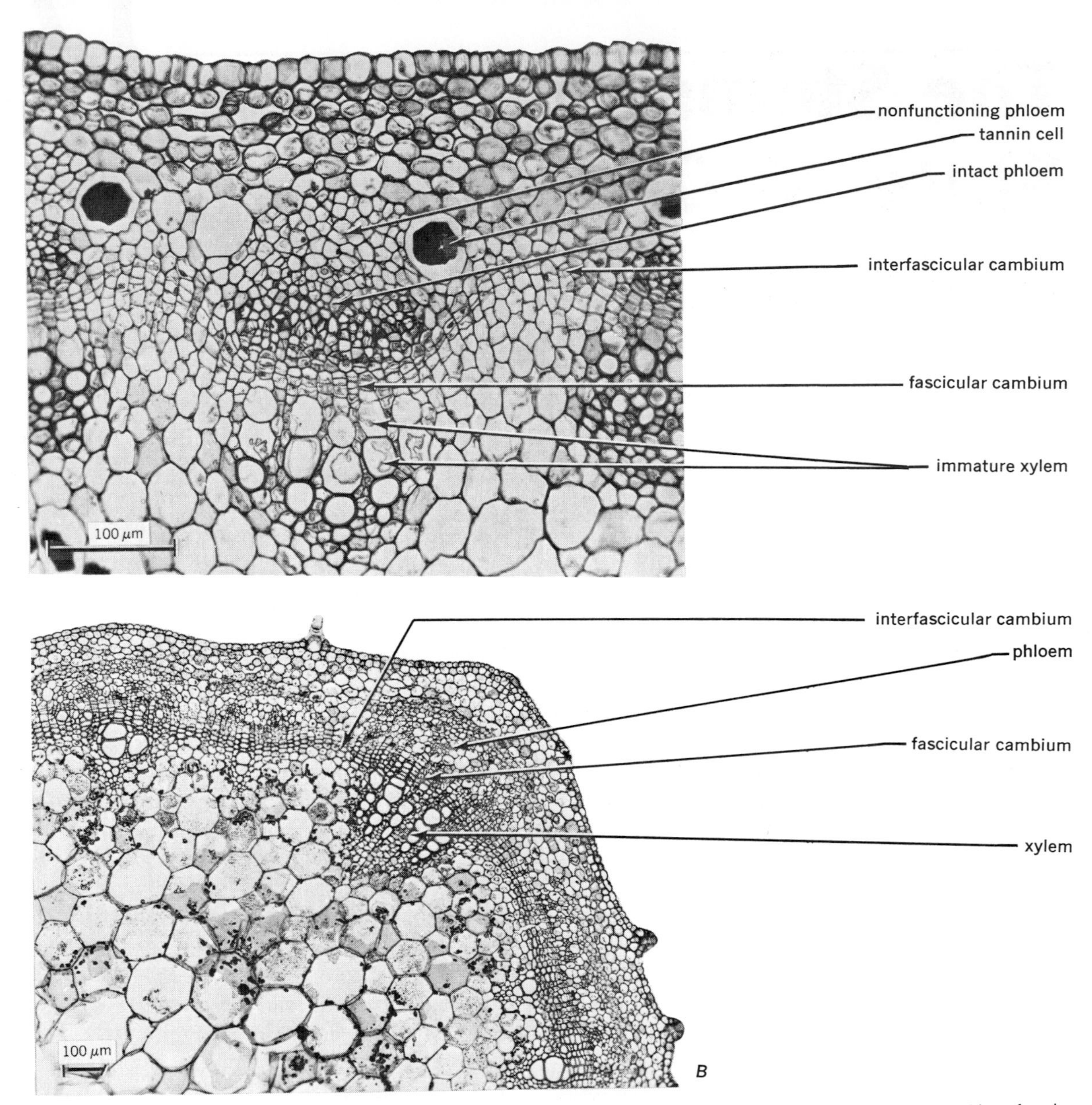

Figure 17.1 Cross sections of stems showing an earlier (*A*) and a later (*B*) stage in activity of fascicular and interfascicular cambia. *A*, *Lotus corniculatus*. *B*, *Medicago sativa*, alfalfa. (Courtesy of J. E. Sass. *B*, from J. E. Sass, *Botanical Microtechnique*, 3rd ed. The Iowa State College Press, 1958.)

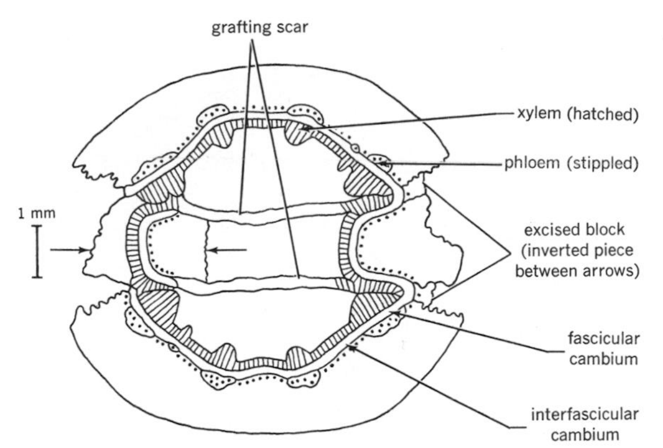

Figure 17.2 Transection of hypocotyl of *Ricinus communis* in which a block of tissue was excised by means of a rectangular punch, then reinserted after a fragment, including the interfascicular tissue on one side of the block, was cut out and inverted (between unlabeled arrows). At the time of the operation the plant was 7 days old and had no secondary growth. The section shown was made 10 days after the inversion. It shows results of early secondary growth. Cambial continuity was established between the stem and the grafted block to the right. The cambium of the inverted piece (left) arose in isolation and produced xylem to the outside and phloem to the inside of the stem. (From photomicrograph in Siebers.[25])

tion has led to the assumption that the development of the interfascicular cambium depends on some stimulus from the vascular bundles. Experiments designed to test this concept by the use of interfascicular regions in *Ricinus* hypocotyl provided no evidence of such stimulation.[25] In one type of experiment, blocks of hypocotyl, which were excised and replaced in reverse, showed no adaptation to the new position but produced interfascicular cambium at the normal site, even though this site was not in line with the fascicular cambium (fig. 17.2). Moreover, the cambium in the inverted block produced xylem and phloem in inverse direction topographically, hence, it exhibited the same kind of radial polarity that it would have had in uninverted position. In another type of experiment, strips of interfascicular tissue separated from adjacent vascular tissue by means of radially inserted razor blades or excised and cultured in a basal medium showed a normal development of cambium and vascular tissues. These results were interpreted to mean that the future cambial cells in the interfascicular regions were determined as such and had the radial polarity impressed on them many days before cambial activity started at the particular level of the hypocotyl.

In the common type of secondary growth the vascular cambium becomes a complete cylinder and produces continuous cylinders of secondary phloem and xylem (figs. 17.3 to 17.5), each with its axial and radial systems of cells (chapters 8 to 11), the axial cells derived from the fusiform cambial initials, the radial from the ray initials (chapter 10).

The establishment of the cambium, as seen in stem transections, does not occur uniformly around the circumference of the axis. As was explained in chapter 16, the vascular bundles at a given level of a stem in primary state of growth are in different developmental stages because of their association with leaves of different ages. Hence, some bundles enter the secondary-growth stage earlier than others. The transition from primary to secondary growth in stems of *Populus deltoides*[16] was found occurring in the internode associated with the first mature leaf below the apex, and the first cambial divisions appeared in the vascular bundles above the divergence of leaf traces of the next older leaf. From this site, the initiation of cambial activity progressed acropetally in the stem along the parastichy (helix) connecting the leaves in the order of their origin at the shoot apex.

In branches developing from axillary buds, the vascular cambium appears in the same manner as in the main axis, and the cambia of branches and trunk form a continuous sheet. The stem cambium is also continuous with that of the root. The cambium appears before

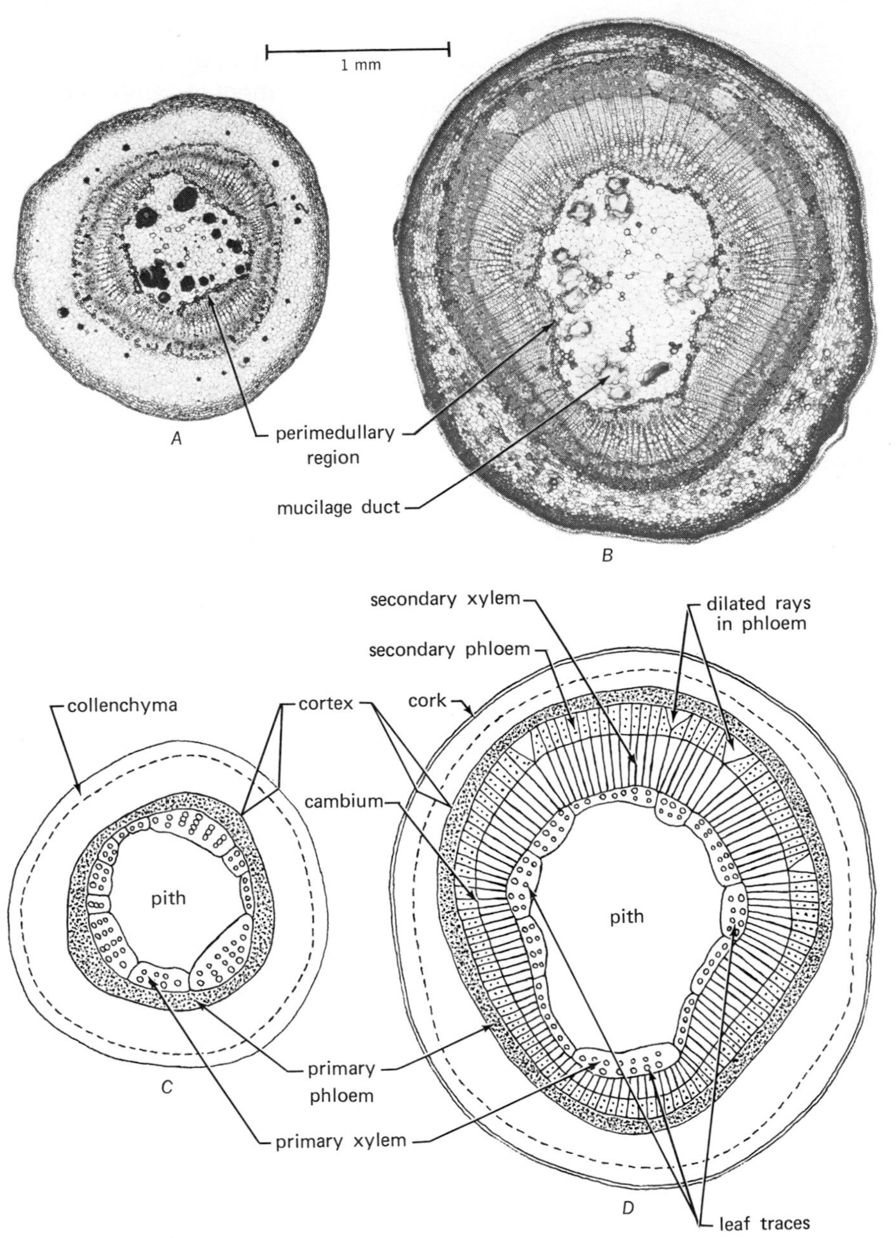

Figure 17.3 Cross sections of *Tilia* stem made before (*A, C*) and after (*B, D*) secondary growth was initiated. (From K. Esau, *Plant Anatomy,* 2nd ed. John Wiley & Sons, 1965.)

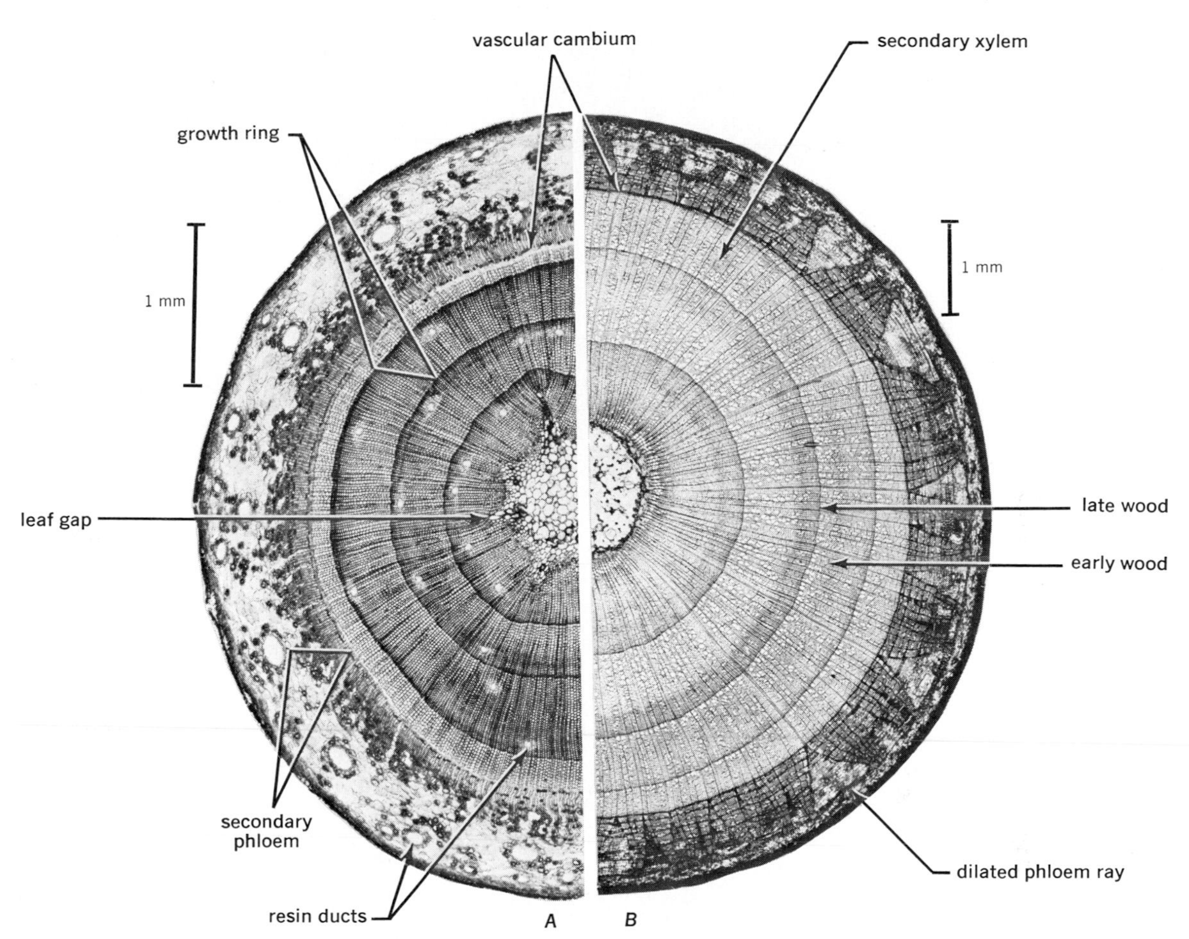

Figure 17.4 Cross sections of stems of *Pinus* (*A*) and *Tilia* (*B*), each with periderm and several increments of secondary vascular tissues. (After K. Esau, *Plant Anatomy,* 2nd ed. John Wiley & Sons, 1965.)

the end of the first year of growth of a shoot, after the internodes have completed their elongation. Because of the basic continuity of the cambium throughout the tree, vascular tissues formed in the oldest part of the tree are directly connected, through successively younger parts, with the first-year secondary growth in the new shoots of that year. Furthermore, the secondary tissues at the base of a young shoot are connected with the primary growth in the upper levels of the same shoot. The continuity between the primary and secondary growth in a shoot can be demonstrated by introducing an eosin solution through a cut in a leaf and seeing it pass through the primary xylem of the leaf traces into that of the stem bundles and finally into the secondary xylem.[29]

Effect of secondary growth on tissues formed earlier

The interpolation of secondary vascular tissues between the primary phloem and the

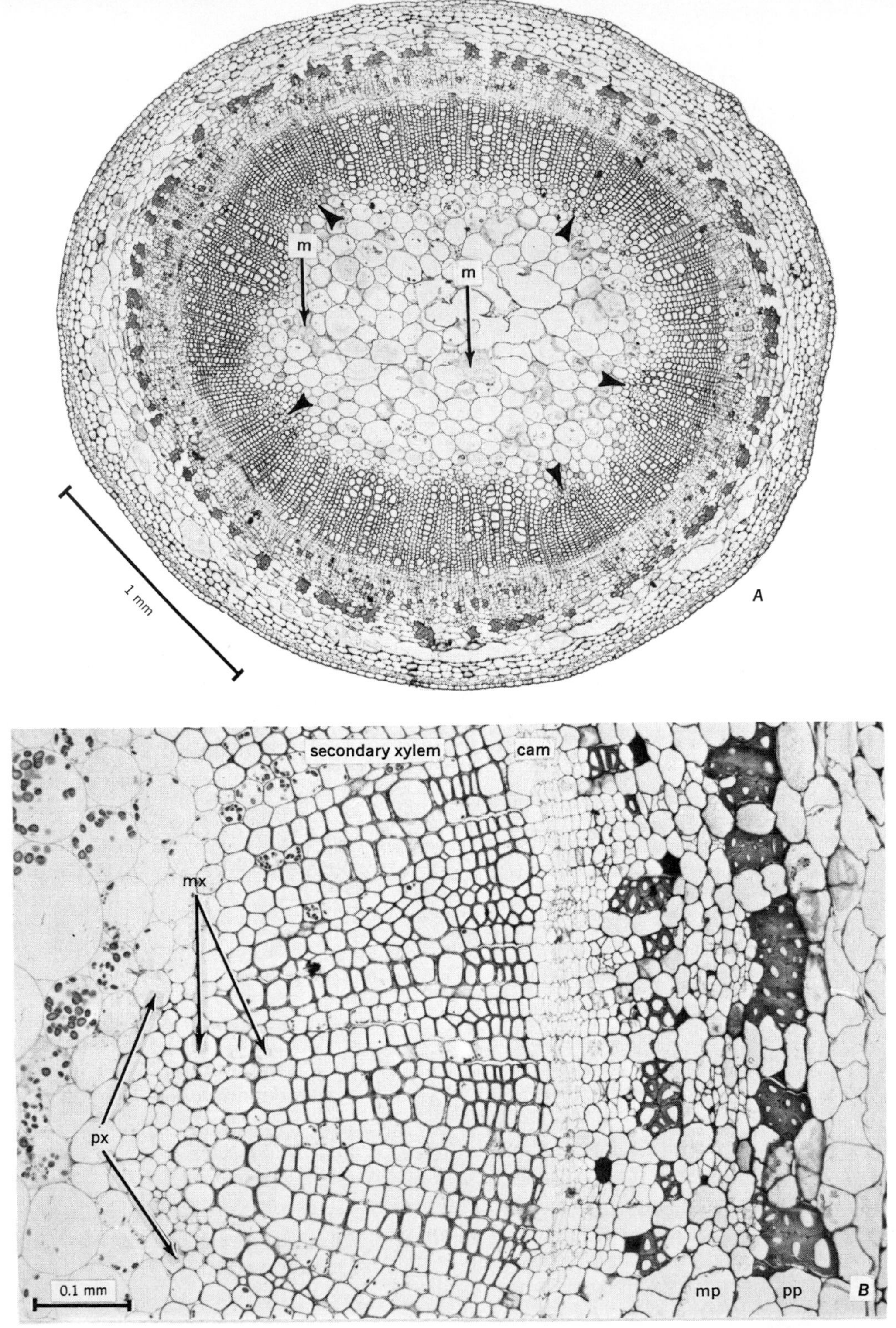

Figure 17.5 *A*, transection of stem of *Hibiscus cannabinus* in secondary state of growth. The most prominent protoxylem regions are indicated by arrowheads, mucilage cells by *m*. *B*, transection of part of same stem. Details: *cam*, cambium; *mp*, metaphloem; *mx*, metaxylem; *pp*, protophloem; *px*, protoxylem. Protophloem is represented by fibers. Secondary phloem, with fibers, between metaphloem and cambium. (From Esau and Morrow.[12])

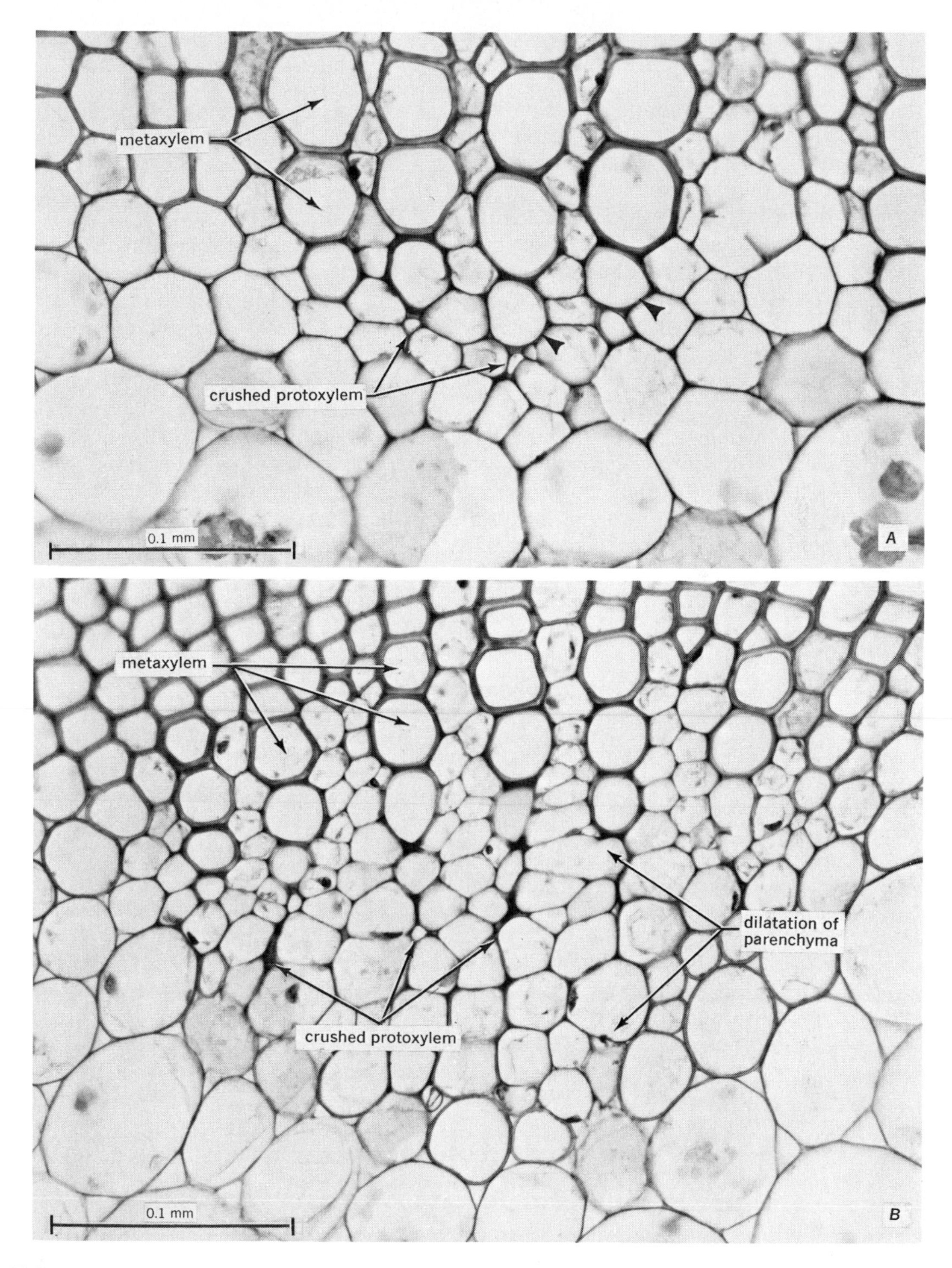

Figure 17.6 Wedges of primary xylem in transection from *Hibiscus cannabinus* stem as in fig. 17.5. *A*, some protoxylem elements are partly crushed, others are intact but stretched longitudinally as indicated by lack of secondary wall along part of circumference (arrowheads). *B*, tracheary elements of protoxylem partly or completely crushed, parenchyma cells remaining intact and undergoing dilatation growth. (From Esau and Morrow.[12])

primary xylem brings about considerable changes in the interior of the stem, especially with regard to tissues located outside the cambium. The pith and the primary xylem are covered by the secondary xylem and the primary conducting elements cease to function. The protoxylem becomes completely crushed (fig. 17.5,*B*, and 17.6) but the associated parenchyma may remain alive for many years. Sometimes the pith is deformed by an inward pressure of the enlarging secondary body.

The primary phloem is pushed outward (fig. 17.3). It becomes nonconducting and the protophloem develops fibers in many dicotyledons (fig. 17.5). The thin-walled cells collapse or are crushed. The cortex may persist for many years. It increases in circumference through cell expansion in periclinal direction and cell divisions in the anticlinal plane. The epidermis also may persist by undergoing cell enlargement and cell division in accordance with the increase in circumference of the stem.

With continued secondary growth, the secondary phloem also is subjected to pressure from the inside produced by the enlarging woody cylinder. Depending on the structure of this phloem it is more or less changed in its aspect. When fibers are abundant and occur in tangential bands (fig. 17.7) the old nonconducting phloem may not be crushed. In other species with or without fibers, large masses of cells (sieve elements and the closely associated cells) are completely crushed. The accommodation to the increasing circumference occurs by cell division in phloem parenchyma and rays. Sometimes this growth is

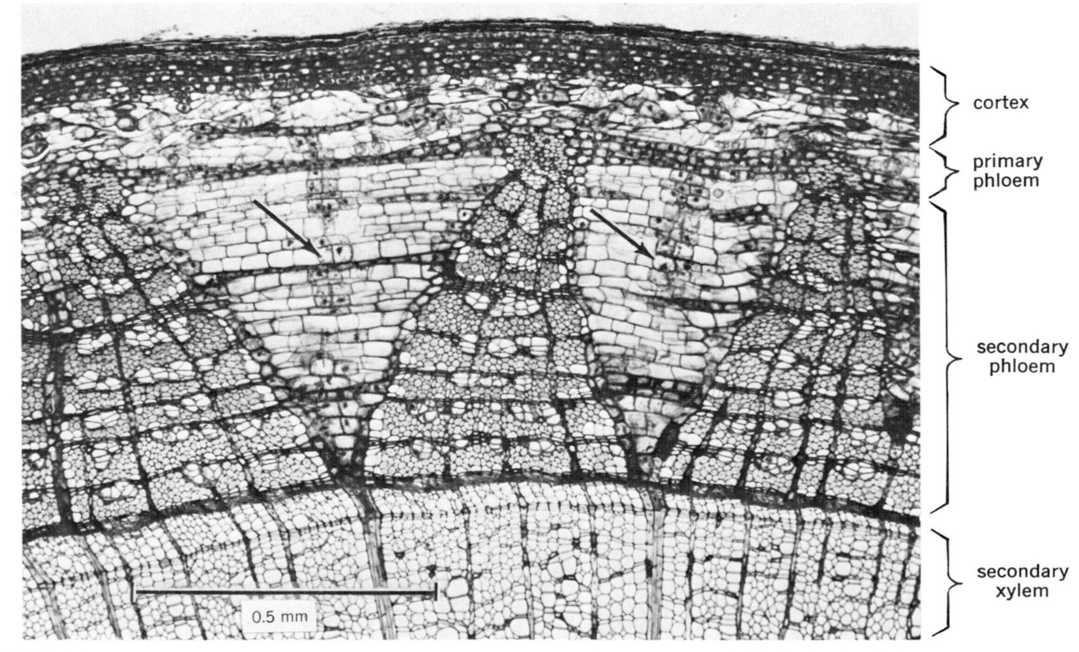

Figure 17.7 Transection of part of four-year-old *Tilia* stem. Two rays are much dilated. Arrows indicate recently divided cells in the rays. In axial system, fiber bands alternate with bands of sieve elements and associated cells. (From Esau.[10])

restricted to certain rays, which become dilated ("flaring") toward the periphery (figs. 17.3, 17.4,*B*, and 17.7).

In many species, periderm is formed during secondary growth and may remain in superficial position for some years (chapter 12). If rhytidome formation occurs, the primary tissues, and later also successive layers of secondary phloem, are gradually cut off. This elimination of peripheral layers periodically relieves the stress resulting from the increase in circumference of the axis.

EFFECT ON LEAF GAPS AND LEAF TRACES

When the interfascicular cambium appears in the gap parenchyma, the pertinent divisions and subsequent production of xylem and phloem occur first along the margins of the gap and progress toward its middle (fig. 17.8). If the gap is wide, 2 or more years may elapse before the cambium forms through the entire expanse of the gap. The gap, therefore, decreases in width stepwise in the successive increments of the xylem (fig. 17.8,*C*,*D*).

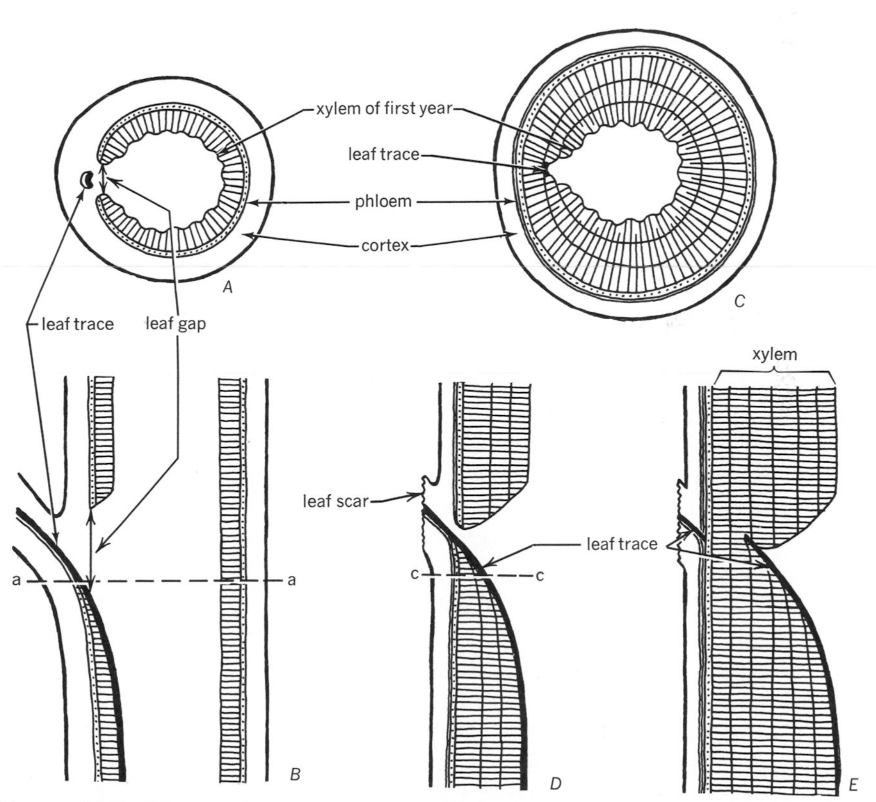

Figure 17.8 Diagrams illustrating effect of secondary growth on leaf traces and leaf gaps in cross (*A, C*) and longitudinal (*B, D, E*) sections of stems made through the nodal region. The sequence is: gap open (*A, B*); gap partly restricted by secondary vascular tissues (*D*); gap closed and leaf trace ruptured (*C, E*). (From K. Esau, *Plant Anatomy*, 2nd ed. John Wiley & Sons, 1965.)

When the process of xylem formation in front of the gap is completed, the gap is said to be closed.

The leaf trace confronting the gap undergoes substantial changes (fig. 17.8). In deciduous species it becomes severed from the leaf bundle at the end of the first season. The lower part of the trace is within the cylinder of stem bundles, and before the leaf falls, it develops a fascicular cambium in line with that of the other vascular bundles in the stem. The upper part, however, is directed outward and ends at the leaf scar. This stublike end has little or no cambial activity. If this end of the trace makes a small angle with the vascular cylinder, the vascular tissues formed above it in the gap region eventually cause a rupture of the trace, and the severed end is carried outward (fig. 17.8,*D*,*E*). If the end has a nearly horizontal orientation, it becomes embedded in the secondary tissue without rupture.

Wound healing and grafting

Secondary growth and cambial activity are often involved in wound-healing phenomena and in formation of union between stock and scion during grafting. Injuries in leaves, young twigs, and other plant parts not having secondary growth usually result first in the formation of a cicatrice, that is, collapse of dead cells and deposition of substances that appear to protect the surface from desiccation and from outside injuries. Subsequently, a periderm develops from living cells underlying the cicatrice. When branches or trunks having secondary growth are injured, the formation of periderm is preceded by the development of callus, a parenchyma tissue resulting from a proliferation of various cells near the surface of the wound. Callus also provides the tissue through which cambial continuity is restored if it was severed by the wound.[24]

The establishment of a graft union is based on phenomena similar to those associated with wound healing. Breakdown products of dead cells on the cut surfaces of stock and scion form a necrotic layer, the contact layer,[4, 8] which corresponds to the cicatrice in wound healing. Intact cells next to the contact layer enlarge, divide, and form callus tissue which fills the space left between the stock and the scion when the two were put together. Eventually the cambia of the two graft partners become continuous across the callus by a change of callus cells into cambial cells (fig. 17.9). The matching of the cambia when the scion is attached to the stock facilitates the establishment of cambial continuity and the successful junction of the subsequently formed secondary vascular tissues. The callus arises from various living cells in the vascular region, among which phloem ray cells and immature xylem ray cells may be particularly active.

The establishment of graft unions involves many difficulties, and the causes of success or failure of grafts are incompletely known. Sometimes failure results from technical causes, for example, a poor matching of cambia of stock and scion. The beneficial effect of the removal of wood from a bud scion[22] is another example of the importance of technique. Peculiar normal structure is sometimes regarded as an obstacle to a successful union, but apparently refinement of methods may overcome this difficulty. Thus, successful grafts have been obtained even in monocotyledons that have no cambial activity in normal development.[15,19] The most seriously limiting factor, however is not the nature of the union but the genetically determined incompatibility, which brings about an antagonistic interaction between the stock and the scion.[21]

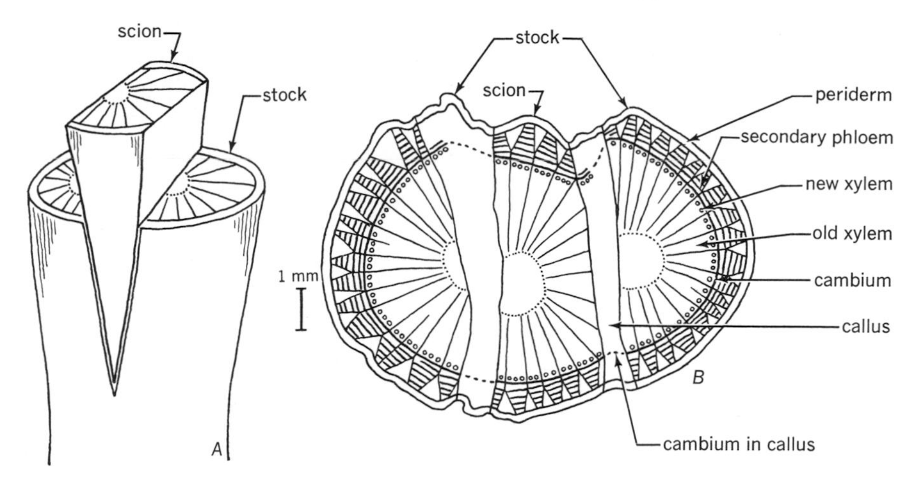

Figure 17.9 Illustration of stock and scion union in a cleft graft. *A*, diagram of a cleft graft at the level where the wedge of scion tissue is inserted in a cleft in the stem of the stock. *B*, transection of a cleft graft in stem of *Hibiscus*. The periderm has become continuous between the stock and the scion. Within the callus, cambial divisions have connected the cambia of the two graft partners. The new cambium has started producing vascular tissues. (*B*, from a photomicrograph in Sharples and Gunnery.[24])

TYPES OF STEMS

Stems differ in primary and secondary structure so that it is sometimes useful to distinguish between types of stems. It is customary to speak of woody and herbaceous stems, of vines, of monocotyledonous stems, and of stems with anomalous secondary growth.

Conifer

The stem of pine is used here as an example of a woody conifer stem. In the primary state the stem has discrete vascular bundles—the leaf traces and stem bundles—separated from one another by relatively narrow interfascicular regions. The vascular cambium, composed of fascicular and interfascicular parts, forms a continuous cylinder of secondary xylem and secondary phloem (fig. 17.4,*A*). (The structure of these tissues is dealt with in chapters 9 and 11.) Opposite the gaps, secondary tissues are formed gradually so that the gap parenchyma projects into the earlier part of the secondary xylem. The primary xylem of the original vascular bundles may be recognized next to the pith, but the primary phloem is completely obliterated. When the crushed primary phloem is still evident, a demarkation between the phloem and the cortex may be made. Otherwise the boundary is obscure because the primary phloem forms no fibers. The cortex contains resin ducts which become considerably enlarged tangentially as the stem increases in circumference. The initial periderm arises beneath the epidermis and is not replaced by deeper periderms for many years.

Woody dicotyledon

In most of the arborescent dicotyledons the interfascicular regions are narrow (*Salix,*

Prunus, Quercus) to very narrow (*Tilia*). In these species the secondary tissues form a continuous cylinder, that is, a cylinder not interrupted by wide rays.

Tilia (Tiliaceae; figs. 17.3, 17.4,*B*, and 17.7) illustrates a number of features common to the stem of a woody dicotyledon. On the inner edge of the continuous secondary xylem the primary xylem has a slightly uneven outline around the pith and can be delimited from the secondary xylem only approximately. The secondary xylem has a somewhat denser appearance than the primary and contains, in the axial system, vessel elements, tracheids, fibers, and xylem parenchyma cells in banded paratracheal arrangement. Wide and narrow rays are present. The secondary phloem has a characteristic appearance because of the dilatation of some of the rays and because of the alternation of bands of fibers and bands containing sieve tubes, companion cells, and parenchyma cells. The initial periderm arises beneath the epidermis and persists for many years. The cortex is retained during this time. The cortex may be easily delimited from the phloem because the latter contains fibers in its peripheral part (protophloem fibers), as well as in deeper layers (secondary phloem fibers). The pith is parenchymatic and contains mucilage cells or cavities. The outer part of the pith long remains active as a storage tissue. This is the medullary sheath or perimedullary region (fig. 17.3,*A*,*B*).

Herbaceous dicotyledon

Many herbaceous dicotyledons have secondary growth of the ordinary type and their stems resemble those of woody dicotyledons of comparable age. *Hibiscus cannabinus* (Malvaceae) provides an example of this kind of herbaceous stem[12] (fig. 17.5). The stem maintains its epidermis during early secondary growth. In older stems, periderm with lenticels arises in the epidermis. One or two layers of cortex beneath the epidermis contain chloroplasts. These layers are followed successively by two to three layers of collenchyma, parenchyma with mucilage cells, and additional parenchyma. The primary phloem contains fibers next to the cortex (protophloem fibers). Fibers also develop in the secondary phloem, but not in the metaphloem (fig. 17.5,*B*). Vascular cambium separates the phloem from the secondary xylem, which forms a compact cylinder. The rays in the secondary vascular tissues are at first uniseriate. Later, multiseriate rays also appear. Moreover, many rays become dilated in the outer phloem as the stem grows older. The pith is parenchymatic but contains mucilage cells. Starch and crystals may be present in pith, cortex, rays, and axial parenchyma. The border between the xylem and pith is uneven because the primary xylem strands protrude to various distances into the pith (fig. 17.5,*A*). The largest primary xylem bundles contain protoxylem in which the tracheary elements become crushed (fig. 17.6).

The stem of *Pelargonium* (Geraniaceae) has narrow interfascicular regions so that the vascular bundles are compactly arranged in the vascular cylinder (fig. 17.10,*A*). The secondary tissues form a continuous cylinder. The vascular region is surrounded by several rows of primary phloem fibers with lignified secondary walls. These fibers arise in the procambium, most of them in association with sieve tubes, some between the phloem strands.[2] In an older stem, the epidermis is replaced by periderm of subepidermal origin (chapter 12). The cortex and the pith are parenchymatic.

In *Helianthus* (Asteraceae), the interfascicular regions are wider than in *Pelargonium* (fig. 17.10,*B*). Interfascicular cambium and a

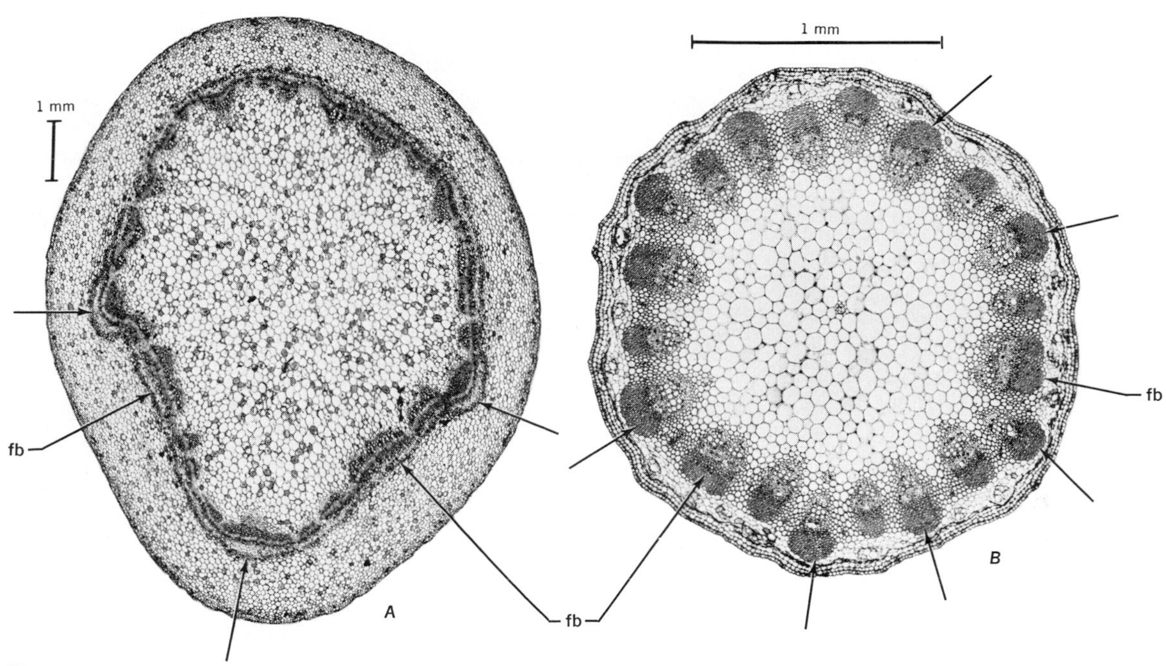

Figure 17.10 Transverse sections of herbaceous dicotyledonous stems in primary state of growth. The arrows indicate leaf traces. At *fb*, partly fused bundles. *A*, *Perlargonium;* narrow interfascicular regions. *B*, *Helianthus;* wide interfascicular regions. (*A*, from K. Esau, *Plant Anatomy*, 2nd ed. John Wiley & Sons, 1965.)

continuous cylinder of vascular tissues are formed at lower levels of the stem. At upper levels, no secondary growth occurs and the interfascicular parenchyma becomes sclerified. Small phloem bundles may be present in the interfascicular regions and in the pith. Primary phloem fibers appear as massive bundle caps in transections. Periderm is commonly absent, and the epidermal cells increase in size tangentially and divide anticlinally. Stems of some Asteraceae have an endodermis with casparian strips.

In the stem of *Medicago,* alfalfa, the vascular bundles are clearly separated from one another by relatively wide interfascicular regions (fig. 17.1,*B*). Some secondary growth occurs at the base of the stem, but the interfascicular cambium produces a small amount of phloem and mostly sclerenchyma cells on the xylem side. This type of secondary growth, which is found in many herbaceous Fabaceae, is combined with histological peculiarities of the secondary xylem and distinguishes the herbaceous stem structure from that of the related woody species. Hence, the commonly expressed generalization that a herbaceous stem resembles the stems of related woody species does not have a universal validity.[5, 6]

The stem of *Coleus* (Lamiaceae) also has a distinctive type of secondary growth. The xylem regions containing tracheary elements and most of the phloem are formed in the fascicular regions, sclerenchyma and a small number of sieve tube-containing strands in the interfascicular regions (fig. 17.11).

Some herbaceous dicotyledons produce no secondary tissues and their vascular

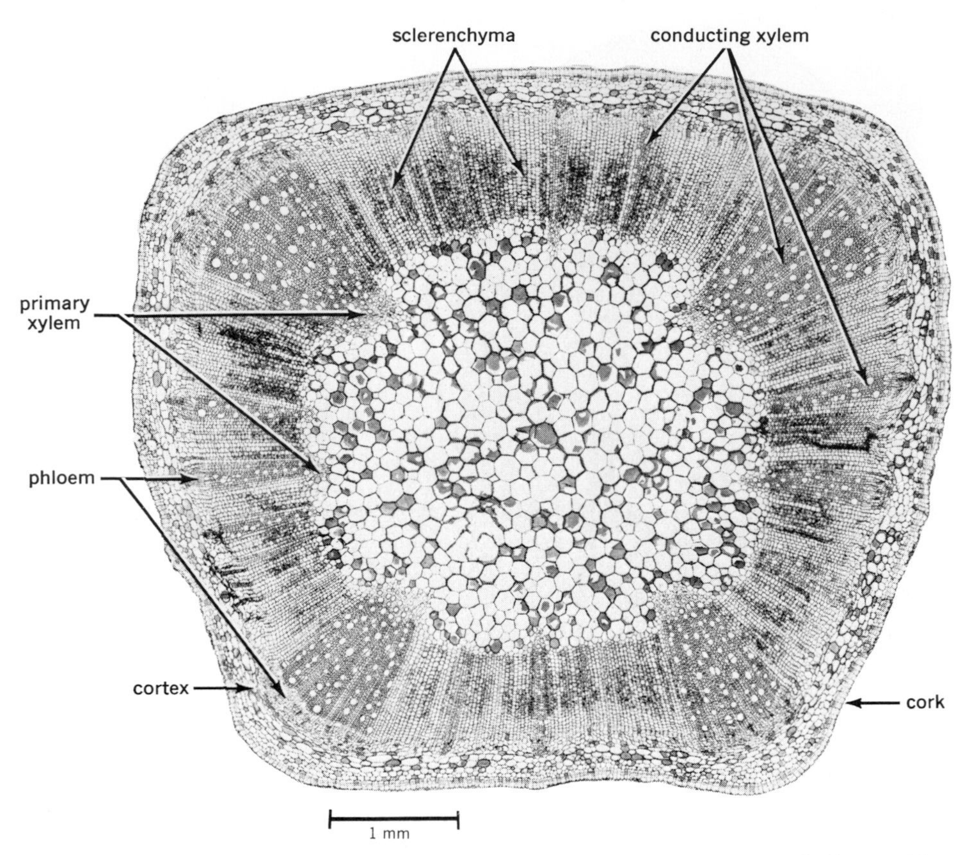

Figure 17.11 Transverse section of *Coleus* (herbaceous dicotyledon) stem in secondary state of growth. The cambium has formed xylem and phloem in the fascicular regions, sclerenchyma mainly in the interfascicular regions. The plant has a decussate phyllotaxy with a pair of leaves at each node.

system may resemble those of the monocotyledons in having widely spaced vascular bundles of the "closed" type (*Ranunculus,* Ranunculaceae, fig. 17.12,*A*; *Nelumbo,* Nelumbonaceae).

All the stems described thus far have a collateral arrangement of xylem and phloem. In the Solanaceae, for example, tomato, potato, and tobacco, internal phloem is present. When secondary growth occurs in these stems, the cambium appears only between the outer phloem and the xylem.[27] *Cucurbita* (Cucurbitaceae) also has internal phloem and its habit may be classified as that of a herbaceous as well as of a vine type of plant. The bicollateral vascular bundles appear in two series, an outer and an inner (fig. 17.12, *B*). They are embedded in ground parenchyma. The inner part of the pith breaks down in early stages of primary growth. The vascular system and the associated ground parenchyma are enclosed in a cylinder of perivascular fibers with lignified secondary walls and living contents. The cortex is composed of parenchyma and collenchyma. A starch sheath occurs immediately outside the perivascular fibers. Secondary growth is limited to the vascular bundles, specifically to

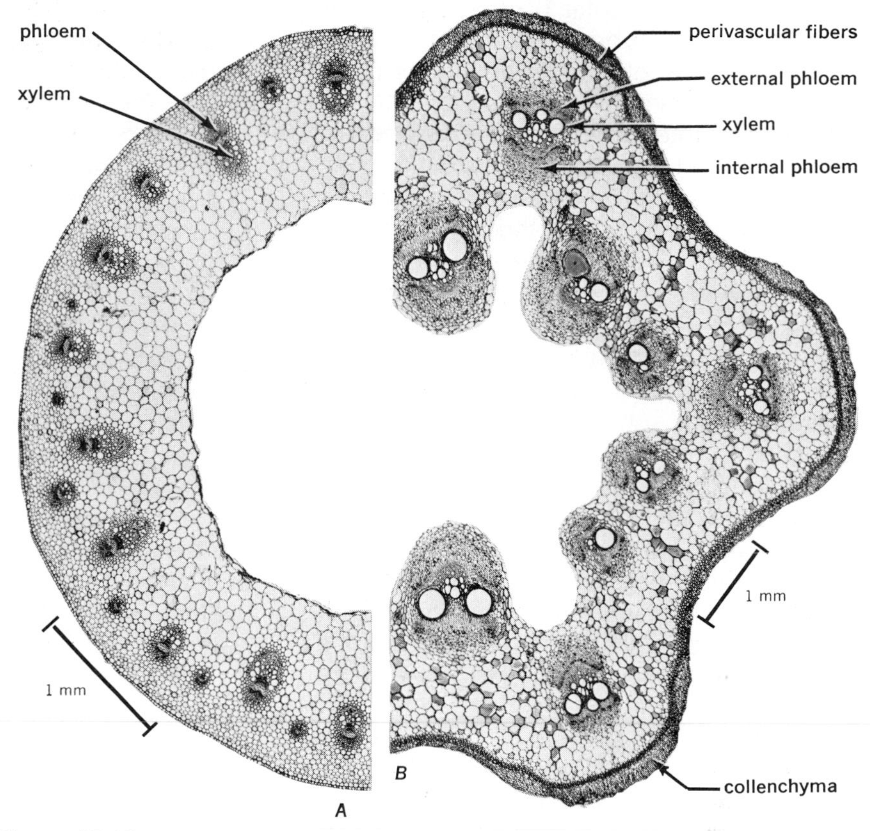

Figure 17.12 Cross sections of herbaceous dicotyledonous stems. *A*, *Ranunculus*. *B*, *Cucurbita*, a vine. (After K. Esau, *Plant Anatomy*, 2nd ed. John Wiley & Sons, 1965.)

the region between the outer phloem and the xylem. Some Cucurbitaceae have rather extensive secondary growth, including a radial extension of the interfascicular regions and formation of periderm.

Herbaceous stems may become more or less modified with reference to their principal function. An example of conspicuous modification is found in the tuber of the potato.[13] The primary vascular system is more homogeneous in appearance than that of the aerial stem because the leaves are scalelike and their traces are small. The internodes remain short. A small amount of secondary growth occurs. The main mass of the storage tissue is derived from the perimedullary region containing the internal phloem. The pith, including the perimedullary region, is voluminous, and the internal phloem is dispersed more widely than it is in the aerial stem or in the rhizome.

Dicotyledonous vine

A common characteristic of vine types of stems are wide rays that make the secondary body appear dissected. The stem of the grapevine, *Vitis* (chapter 12), may serve as an example.[9] In the primary state the vascular system consists of strands of various sizes. The vascular cambium originates in the fascicular and interfascicular regions and becomes continuous. The interfascicular cambium forms parenchyma so that wide rays are formed in continuity with the interfascicular regions. New wide rays arise from time to time within the wedgelike blocks of the vertical system. These rays are not continuous with the interfascicular regions, but their development maintains the dissected appearance of the secondary body during its increase in circumference. The primary phloem (protophloem) develops fibers after the tissue ceases to function in conduction. Fibers arranged in tangential bands occur in the secondary phloem. The cortex is composed of collenchyma and parenchyma, both with chloroplasts. The innermost layer of the cortex is a starch sheath. The pith is composed of parenchyma.

The initial periderm in *Vitis* arises not immediately beneath the epidermis but deeper (chapter 12). It first appears within the primary phloem beneath the primary phloem fibers and is derived from metaphloem parenchyma cells. The phellogen becomes continuous between the vascular bundles by differentiating also from the interfascicular parenchyma. The outer part of the phloem and the cortex are sloughed off in one continuous piece ("ring bark"). The sequent periderms arise in progressively deeper layers of the secondary phloem.

The separation of the vascular tissues into wedgelike blocks is present also in *Aristolochia* (fig. 17.13). This plant shows some additional peculiarities so that it is sometimes classified as having anomalous secondary growth. In the primary state, the widely spaced collateral vascular bundles surround a parenchymatic pith. The vascular system and the ground parenchyma in which it is embedded is surrounded by a cylinder of perivascular fibers. The phloem contains no fibers. The cortex consists of parenchyma and collenchyma. The innermost layer of the cortex located next to the perivascular fibers is a starch sheath in young stems. During secondary growth the individual strands are extended radially into wedge-shaped blocks. Cell division occurs in an ill-defined interfascicular cambium. This cambium produces rays composed of parenchyma similar to that of the primary interfascicular regions. As the vascular cylinder increases in circumference, the cylinder of perivascular fibers is ruptured, mostly in front of the rays (fig. 17.13). Adjacent parenchyma cells invade the breaks by intrusive growth[10] and may differentiate into sclereids. The pith and the interfascicular regions are partly crushed. Periderm develops in the collenchyma beneath the epidermis. It is initiated as isolated vertical strips extending from node to node and takes several years to spread over the entire surface. The cork is layered because of an alternation of radially unextended cells with cells that are larger in radial dimension. A considerable amount of phelloderm is produced.

Dicotyledons with anomalous secondary growth

The term anomalous secondary growth is used to indicate the forms of cambial activity that deviate from that commonly found in conifers and in woody dicotyledons of the temperate regions. The forms of anomalous secondary growth vary considerably and intergrade with normal forms.[20] In some plants with anomalous growth the vascular cambium occurs in normal position, but the secondary body shows an unusual distribution of xylem and phloem. In *Leptadenia* (Asclepiada-

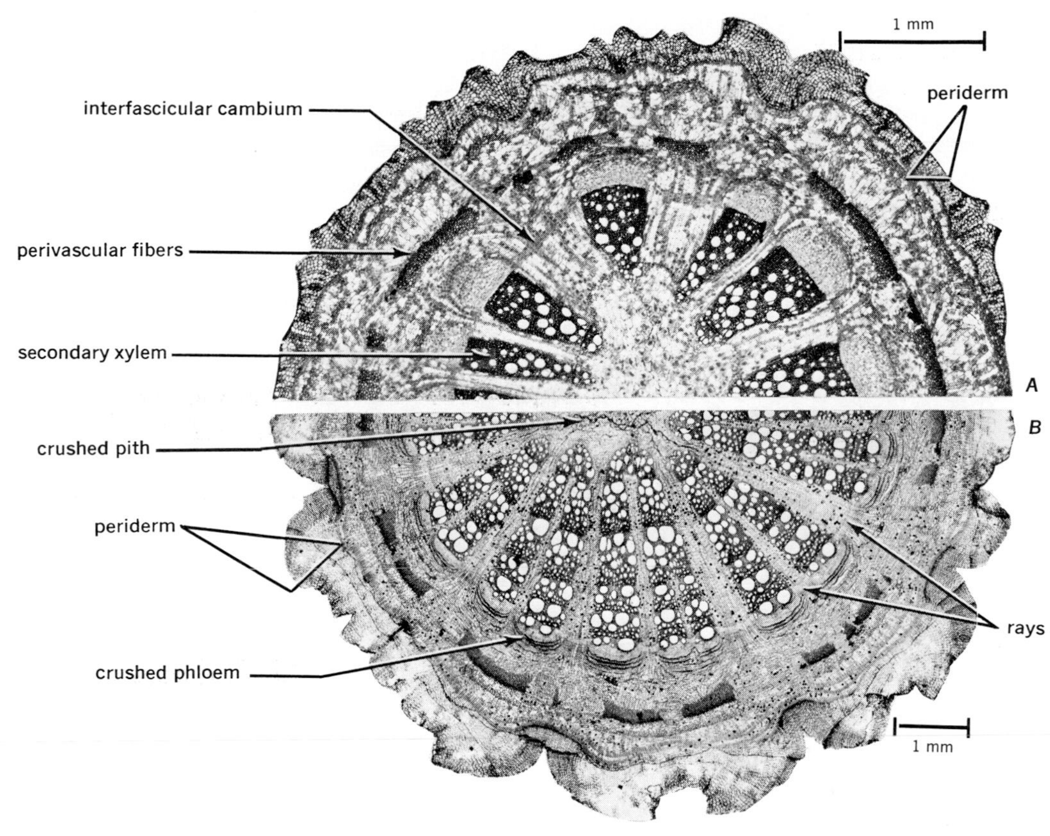

Figure 17.13 Cross sections of stem of *Aristolochia,* a vine, in an earlier (*A*) and a later (*B*) stage of growth. (After K. Esau, *Plant Anatomy,* 2nd ed. John Wiley & Sons, 1965.)

ceae), *Strychnos* (Loganiaceae), *Thunbergia* (Acanthaceae)[17] phloem is formed not only toward the outside but from time to time also toward the inside. Thus strands of secondary phloem, which anastomose with one another, become embedded in the secondary xylem (fig. 17.14,*A*). In the Amaranthaceae, Chenopodiaceae, Menispermaceae, and Nyctaginaceae series of vascular cambia arise successively farther outward from the center of the stem, each producing xylem toward the inside and phloem toward the outside (fig. 17.14,*B*). (Cf. sugar beet root in chapter 15.) Phloem and xylem form strands embedded in parenchyma tissue sometimes called conjunctive tissue (fig. 17.15). This tissue results from cambial activity between the vascular strands resembling interfascicular cambial activity but having a limited duration.

The initiation of secondary growth with multiple cambia was studied in *Bougainvillea spectabilis* (Nyctaginaceae).[26] This species has no normal cambium so that the primary vascular strands remain discrete. A complete cylinder of meristem arises outside these bundles. It produces vascular bundles and conjunctive tissue like a cambium. Some of the bundles differentiate among the inner derivatives of this cambium, others are formed in a way common for dicotyledons, the phloem

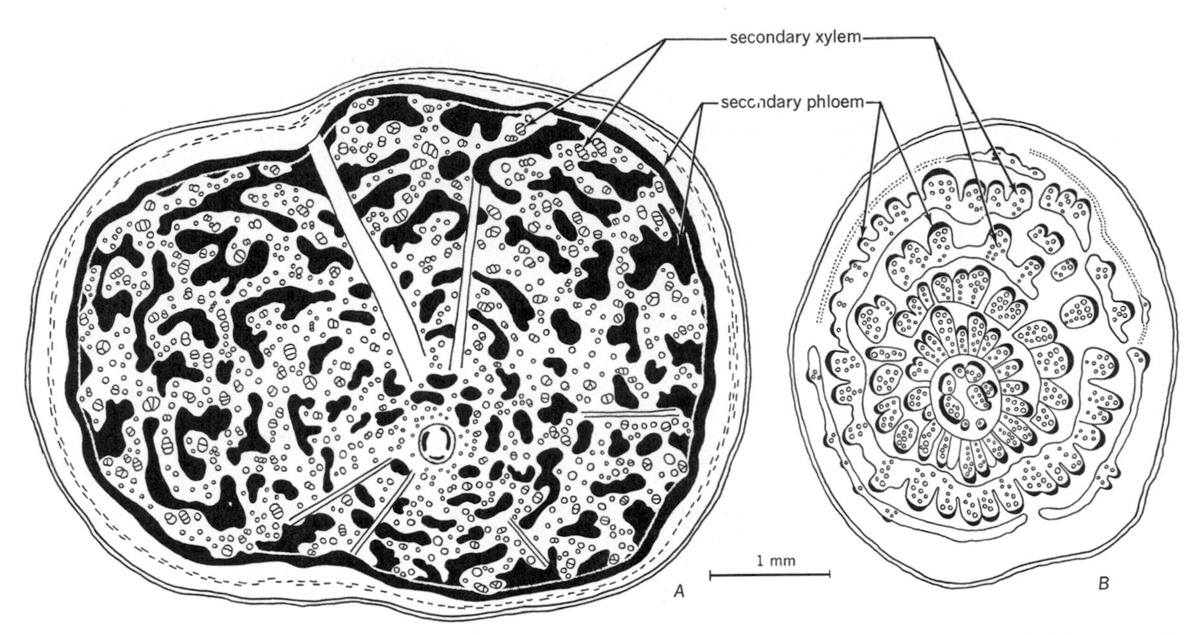

Figure 17.14 Transections of stems with anomalous secondary growth. *A*, *Leptadenia spartium*, Asclepidaceae, with secondary phloem strands embedded in secondary xylem (included phloem). *B*, *Boerhaavia diffusa*, Nyctaginaceae, with successive increments of secondary vascular tissues, each composed of xylem, phloem, and parenchyma. Each increment arises from a separate cambial layer. (From K. Esau, *Plant Anatomy*, 2nd ed. John Wiley & Sons, 1965.)

arising among the outer derivatives, the xylem among the inner derivatives of the cambium. In the second case, a new sheet of cambium appears outside the phloem.[11]

Representatives of the Bignoniaceae illustrate yet another type of anomalous growth[7] (fig. 17.16). After the usual type of cambial cylinder is formed at the end of primary growth, four strips of cambium cease to produce xylem but continue to give off derivatives to the phloem side. Thus, two kinds of cambium become established, one showing bidirectional activity, the other unidirectional activity. The anomalously formed panel of phloem becomes deeply embedded in the xylem, for the normal cambium produces xylem on both sides of a phloem panel. The centrifugally produced phloem is pushed outward and shears against the stationary xylem. As the stem increases in circumference additional strips of cambium change from bidirectional to unidirectional cell production. Some of the new phloem panels develop next to the earlier ones, others appear in new regions of the xylem. Each phloem panel is flanked by wide rays and retains its original tangential diameter, but the groups of panels increase in width toward the outside through the addition of new panels to each group (fig. 17.16). The xylem is thus repeatedly fissured by the insertion of new phloem panels. In old stems, it is further broken up by dilatation growth of xylem and pith parenchyma.

Dilatation of parenchyma plays an important part in producing the anomalous stem structure in *Bauhinia* (Caesalpiniaceae). The xylem becomes fissured and new vascular bundles arise in the dilatation tissue. Band-shaped stems in certain species of this genus

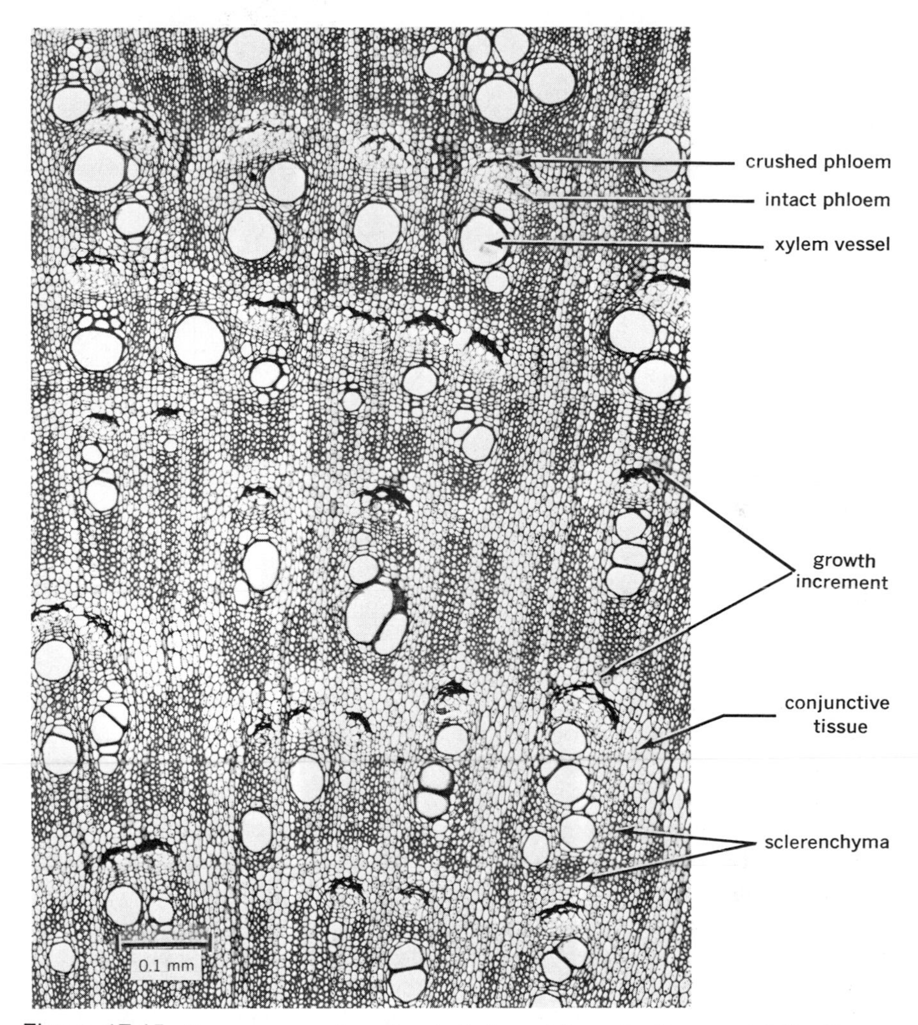

Figure 17.15 Transverse section of part of stem of *Bougainvillea spectabilis*. Shows several increments of anomalous secondary growth each containing vascular bundles embedded in conjunctive ground tissue. Compare with fig. 17.14, *B*. (From Esau and Cheadle[11].)

result from a retardation of cambial activity on two sides and an acceleration on two other sides.[1]

Monocotyledons

The stems (culms) of the Poaceae, seen in transections, have widely spaced vascular bundles not restricted to one circle. The bundles are either in two circles (fig. 17.17; *Avena, Hordeum, Secale, Triticum, Oryza*) or scattered throughout the section (*Bambusa, Saccharum, Sorghum, Zea;* chapter 16). In grasses with a circular arrangement of bundles a continuous cylinder of sclerenchyma occurs close to the periphery. The

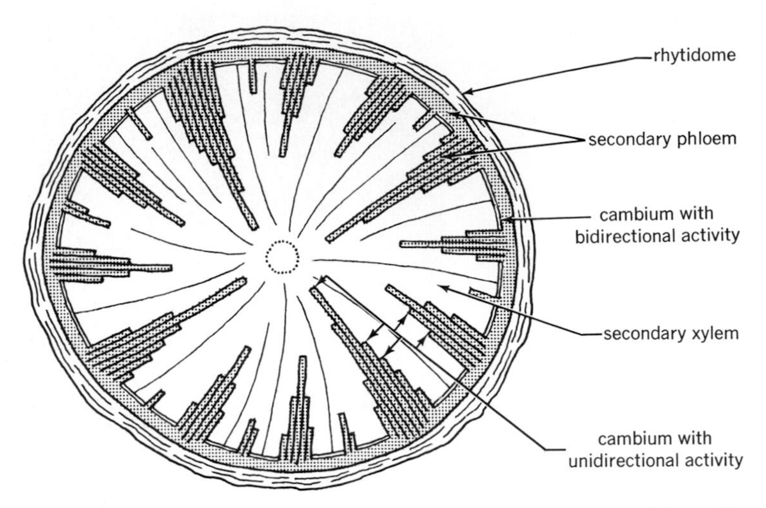

Figure 17.16 Diagram of transection of stem with anomalous secondary growth found in Bignoniaceae. The secondary xylem is fissured by insertion of panels of phloem. Each panel has its own cambium which produces cells only in the outward direction. The xylem and the peripheral layer of phloem are produced by cambium giving off cells in two directions.

outer smaller bundles are embedded in this sclerenchyma. Fiber strands occur between the small bundles and the epidermis, and strands of chlorenchyma alternate with the fiber strands. Stomata occur in the epidermis adjoining the chlorenchyma. The pith often breaks down, except at the nodes, in grass stems that have the bundles arranged in circles. In stems with scattered bundles no peripheral fibrous layer develops, but the subepidermal and deeper-lying parenchyma may be sclerified.[18] In both kinds of stems the vascular bundles are entirely primary and are enclosed in sheaths of sclerenchyma.

Monocotyledons other than the Poaceae also have vascular bundles scattered or in rings near the periphery, as seen in stem transections. In *Tradescantia* (Commelinaceae), the central cylinder with scattered bundles is limited externally by a 1-4 layered sclerotic cylinder and the vascular bundles are interconnected at the nodes by nodal networks of vascular tissue.[14]

Condensation of vascular tissue characteristic of hydrophytes is common in aquatic monocotyledons.[23] In *Potamogeton* (Potamogetonaceae), for example, a wide cortex consisting of aerenchyma encloses a compact vascular cylinder delimited by a small-celled endodermis (fig. 17.18). Variable amounts of pith tissue occur in different species of the genus. There are eight vascular bundles, the two largest oriented in the median plane of the distichously arranged leaves. The large bundle with phloem on both sides of the xylem in figure 17.18,*B,* above, is a composite of three leaf traces which are free at higher levels of the same stem. The bundles marked *m* and *l* in the same figure are a median and two lateral traces to a leaf attached to the node at a level above that of the section shown. The xylem of the vascular bundles is represented by lacunae which have replaced the tracheary elements torn during the extension growth of the internode. The phloem, in contrast, is well developed

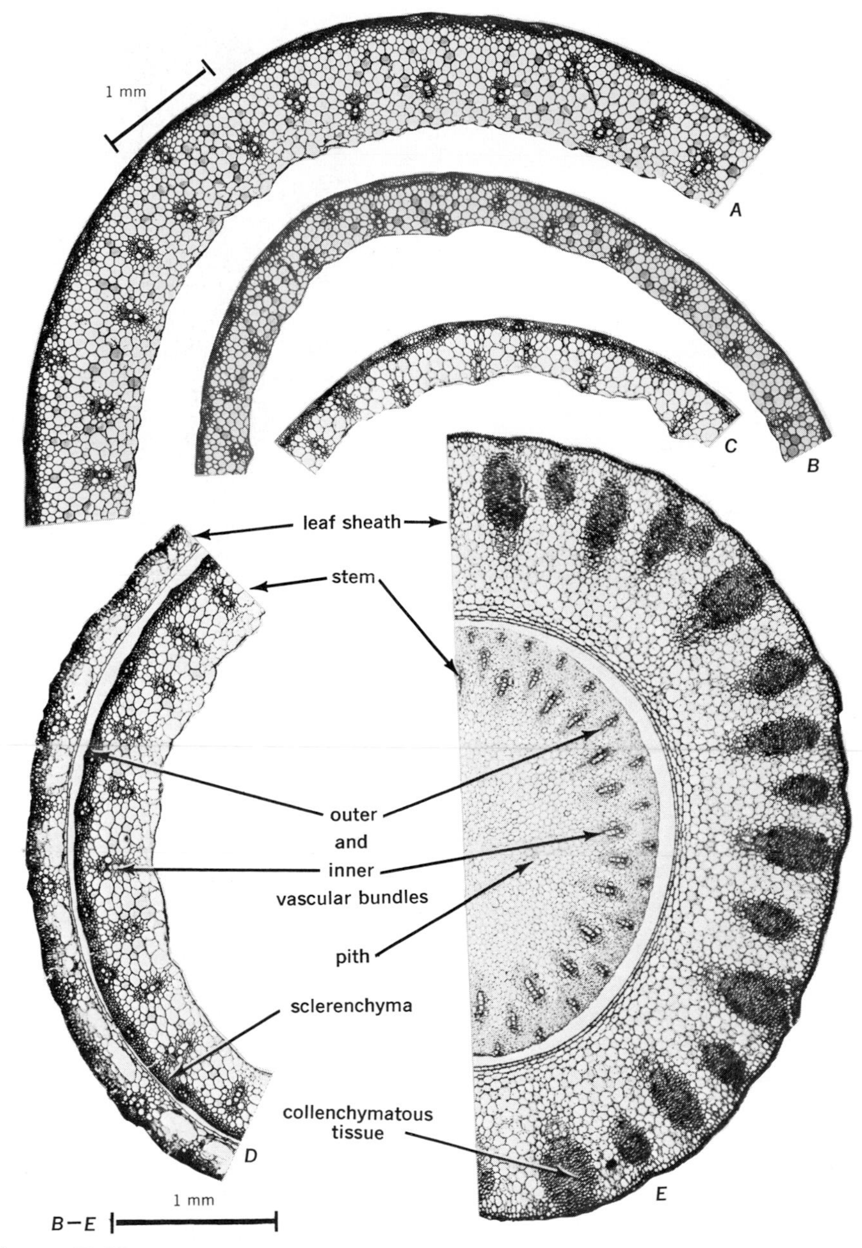

Figure 17.17 Stems of grasses in cross sections. *A*, *Avena*, oat. *B*, *Hordeum*, barley. C, *Secale*, rye. *D*, *E*, *Triticum*, wheat, both showing stem and leaf sheath as they appear in middle of internode (*D*) and near node (*E*).

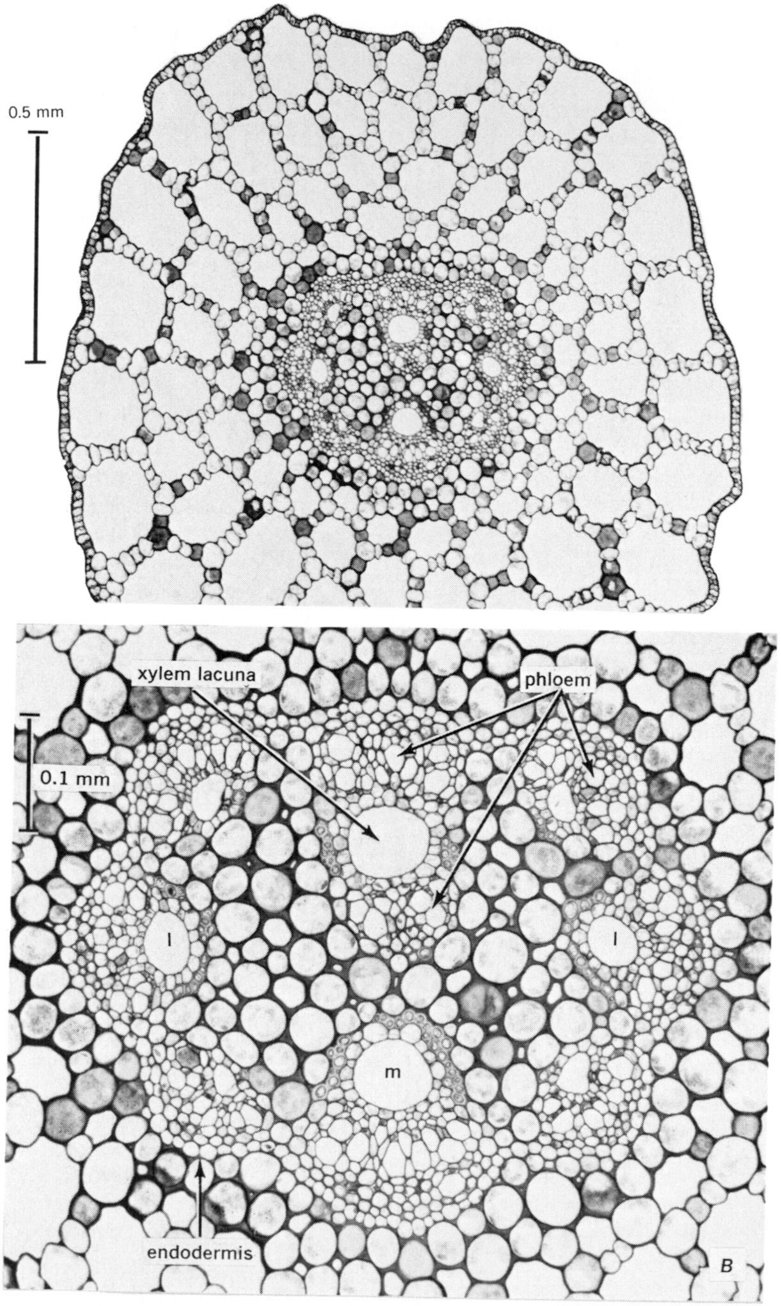

Figure 17.18 Transections of stem of *Potamogeton*, an aquatic monocotyledon. *A*, cortex consisting of aerenchyma encloses a compact central cylinder. *B*, enlarged view of central cylinder. The three traces of the nearest leaf above are marked *m* (median) and *l* (lateral).

and consists of wide sieve elements, distinct companion cells, and phloem parenchyma cells.

SECONDARY GROWTH

Monocotyledons usually lack secondary growth from a vascular cambium but may develop thick stems (e.g., palms) by a thickening growth resulting from division and enlargement of ground parenchyma cells. This growth is called *diffuse secondary growth.*[28] Secondary growth by means of a special kind of cambium occurs in herbaceous and woody Liliflorae (*Agave, Aloe, Cordyline, Dracaena, Sansevieria, Yucca*) and other taxa.[3,30] This cambium is continuous with the primary thickening meristem (chapter 16) if the latter is discernible, but it functions in the part of the stem that has completed elongation. The cambium arises in the parenchyma outside the primary vascular bundles and produces secondary vascular bundles and parenchyma toward the inside and a small amount of parenchyma toward the outside (fig. 17.19). In the development of the vascular bundles, individual cells derived from the cambium divide longitudinally; then two or three of the resulting cells repeat the longitudinal divisions. The products of the final divisions differentiate into vascular elements and associated sclerenchyma cells. Vertically, many tiers of cells combine to form a vascular bundle. The secondary vascular

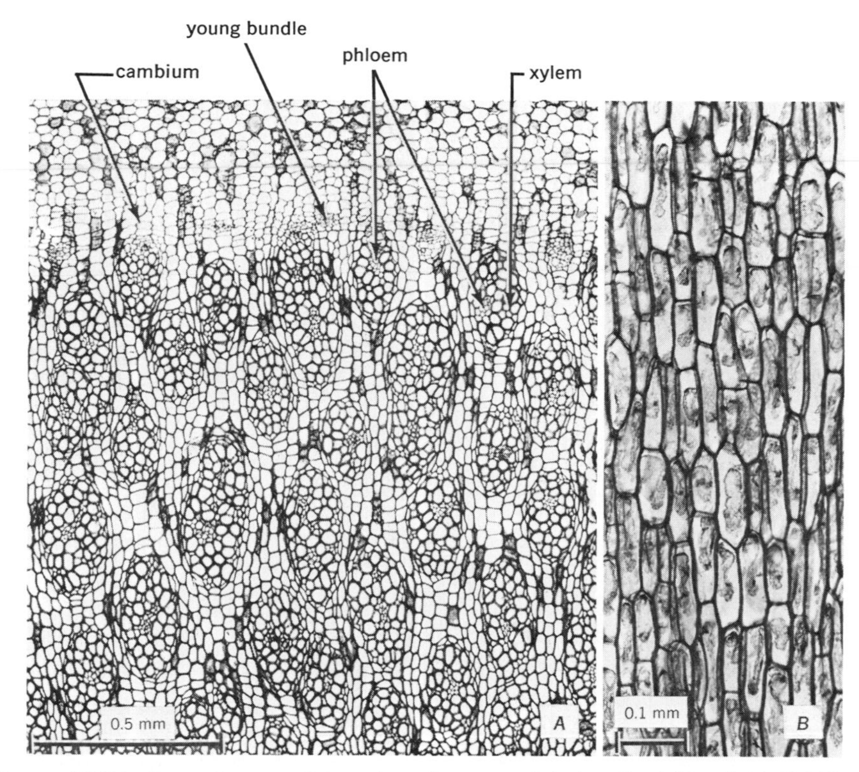

Figure 17.19 Secondary growth in a stem of a monocotyledon, *Cordyline*. *A*, cross section of secondary vascular tissues. *B*, tangential section of cambium. (From V. I. Cheadle: *A*, *Amer. Jour. Bot.* 30:484–490, 1943; *B*, *Bot. Gaz.* 98:535–555, 1937.)

bundles may be collateral or amphivasal and occur in more or less definite radial files. Some monocotyledons form the type of periderm found in dicotyledons (*Aloe, Cocos, Roystonea*); others have the so-called storied cork as a protective tissue (chapter 12).

REFERENCES

1. Basson, P. W., and D. W. Bierhorst. An analysis of differential lateral growth in the stem of *Bauhinia surinamensis*. *Bull. Torrey Bot. Club.* 94:404–411. 1967.
2. Blyth, A. Origin of primary extraxylary stem fibers in dicotyledons. *Univ. Calif. Pubs. Bot.* 30:145–232. 1958.
3. Cheadle, V. I. Secondary growth by means of a thickening ring in certain monocotyledons. *Bot. Gaz. 98:535*–555. 1937.
4. Copes, D. Graft union formation in Douglas-fir. *Amer. J. Bot.* 56:285–289. 1969.
5. Cumbie, B. G. Anatomical studies in the Leguminosae. *Trop. Woods* 113:1–47. 1960.
6. Cumbie, B. G., and D. Mertz. Xylem anatomy in *Sophora* (Leguminosae) in relation to habit. *Amer. J. Bot.* 49:33–40. 1962.
7. Dobbins, D. R. Studies of the anomalous cambial activity in *Doxantha unguis-cati* (Bignoniaceae). II. A case of differential production of secondary tissues. *Amer. J. Bot.* 58:697–705. 1971.
8. Dormling, I. Anatomical and histological examination of the union of scion and stock in grafts of Scots pine (*Pinus sylvestris* L.) and Norway spruce (*Picea abies* (L.) Karst.) *Studia Forestalia Suecica* No. 13, 136 p. 1963.
9. Esau, K. Phloem structure in the grapevine, and its seasonal changes. *Hilgardia* 18:217–296. 1948.
10. Esau, K. On the anatomy of the woody plant. Pp. 35–50. In: *Cellular ultrastructure of woody plants.* W. A. Côté, Jr., ed. Syracuse University Press. 1965.
11. Esau, K., and V. I. Cheadle. Secondary growth in *Bougainvillea. Ann. Bot.* 33:807–819. 1969.
12. Esau, K., and I. B. Morrow. Spatial relation between xylem and phloem in the stem of *Hibiscus cannabinus* L. (Malvaceae). *Bot. J. Linn. Soc.* 68:43–50. 1974.
13. Hayward, H. E. *The structure of economic plants.* New York, The Macmillan Company. 1938.
14. Heyser, W. Das Phloem von *Tradescantia albiflora.* I. Lichtmikroskopische Untersuchungen. *Flora* 159:286–309. 1970.
15. Krenke, N. P. *Wundkompensation, Transplantation und Chimären bei Pflanzen.* Berlin, Julius Springer. 1933.
16. Larson, P. R., and J. G. Isebrands. Anatomy of the primary-secondary transition zone in stems of *Populus deltoides. Wood Sci. Tech.* 8:11–26. 1974.
17. Mullenders, W. L'origine du phloème interxylémien chez *Stylidium et Thunbergia.* Étude anatomique. *Cellule* 51: 5–48. 1947.
18. Murdy, W. H. The strengthening system in the stem of maize. *Ann. Missouri Bot. Gard.* 47:205–226. 1960.
19. Muzik, T. J. Role of parenchyma in graft union in vanilla orchid. *Science* 127:82. 1958.
20. Philipson, W. R., J. M. Ward, and B. G. Butterfield. *The vascular cambium. Its development and activity.* London, Chapman and Hall. 1971.
21. Roberts, R. H. Theoretical aspects of graftage. *Bot. Rev.* 15:423–463. 1949.
22. Scaramuzzi, F. Le basi istogenetiche

dell'innesto "ad occhio." Ricerche sul pesco. [Researches on the histogenic process in bud-union of peach trees.] *Ann. Sper. Agr.* 6:517–537. 1952.

23. Schenk, H. Vergleichende Anatomie der submersen Gewächse. *Bibl. Bot. Heft* 1, 67 p. 1886.
24. Sharples, A., and H. Gunnery. Callus formation in *Hibiscus Rosa-sinensis* L. and *Hevea brasiliensis* Müll. Arg. *Ann. Bot.* 47:827–840. 1933.
25. Siebers, A. M. Initiation of radial polarity in the interfascicular cambium of *Ricinus communis* L. *Acta Bot. Neerl.* 20:211–220. 1971. Differentiation of isolated interfascicular tissue of *Ricinus communis* L. *Acta Bot. Neerl.* 20:343–355. 1971.
26. Stevenson, D. W., and R. A. Popham. Ontogeny of the primary thickening meristem in seedlings of *Bougainvillea spectabilis*. *Amer. J. Bot.* 60:1–9. 1973.
27. Thompson, N. P., and C. Heimsch. Stem anatomy and aspects of development in tomato. *Amer. J. Bot.* 51:7–19. 1964.
28. Tomlinson, P. B. *Anatomy of the monocotyledons.* II. *Palmae.* Oxford, Clarendon Press. 1961.
29. Wareing, P. F., and D. L. Roberts. Photoperiodic control of cambial activity in *Robinia pseudoacacia* L. *New Phytol.* 55:356–366. 1956.
30. Zimmermann, M. H., and P. B. Tomlinson. The vascular system in the axis of *Dracaena fragrans* (Agavaceae), 2. Distribution and development of secondary vascular tissue. *J. Arnold Arb.* 51:478–491. 1970.

18

The Leaf: Basic Structure and Development

MORPHOLOGY

The leaf, in the wide sense of the term, is highly variable in both structure and function. The foliage leaf usually shows its specialization as a photosynthetic organ by the expanded flat form of its blade, the lamina. The blade may be attached to the stem by the petiole, or the leaf may be without a petiole (sessile leaf). If the base of either a sessile or a petiolate leaf encircles the stem, it is referred to as a sheathing base. Sometimes the sheathing part of the leaf is developed into a conspicuous leaf sheath. Plants with multilacunar nodes characteristically have sheathing bases. Outgrowths at the base of the leaf, the stipules, are often present in leaves associated with trilacunar nodes. A simple leaf has one blade; a compound leaf has two or more leaflet blades attached to a common axis, the rachis. The leaflets also may be compound.

Two other common leaf forms are the cotyledons, the first leaves on the plant (chapters 23 and 24), and the cataphylls. The cataphylls are represented by various protective and storage bracts or scales. They are simpler than the foliage leaves in shape and histology. The first bracts on a lateral shoot are called prophylls (chapter 16). The prophylls may be followed by foliage leaves or by a succession of other bracts intergrading with the foliage leaves.

HISTOLOGY OF ANGIOSPERM LEAF

Like the root and the stem the leaf consists of dermal, vascular, and ground tissue systems (fig. 18.1). Since the leaf commonly has no secondary growth—sometimes in a limited amount in petioles and large veins—the epidermis persists as the dermal system. Bud scales, however, may develop periderm.

Epidermis

The main features of the leaf epidermis are compact arrangement of cells and the pres-

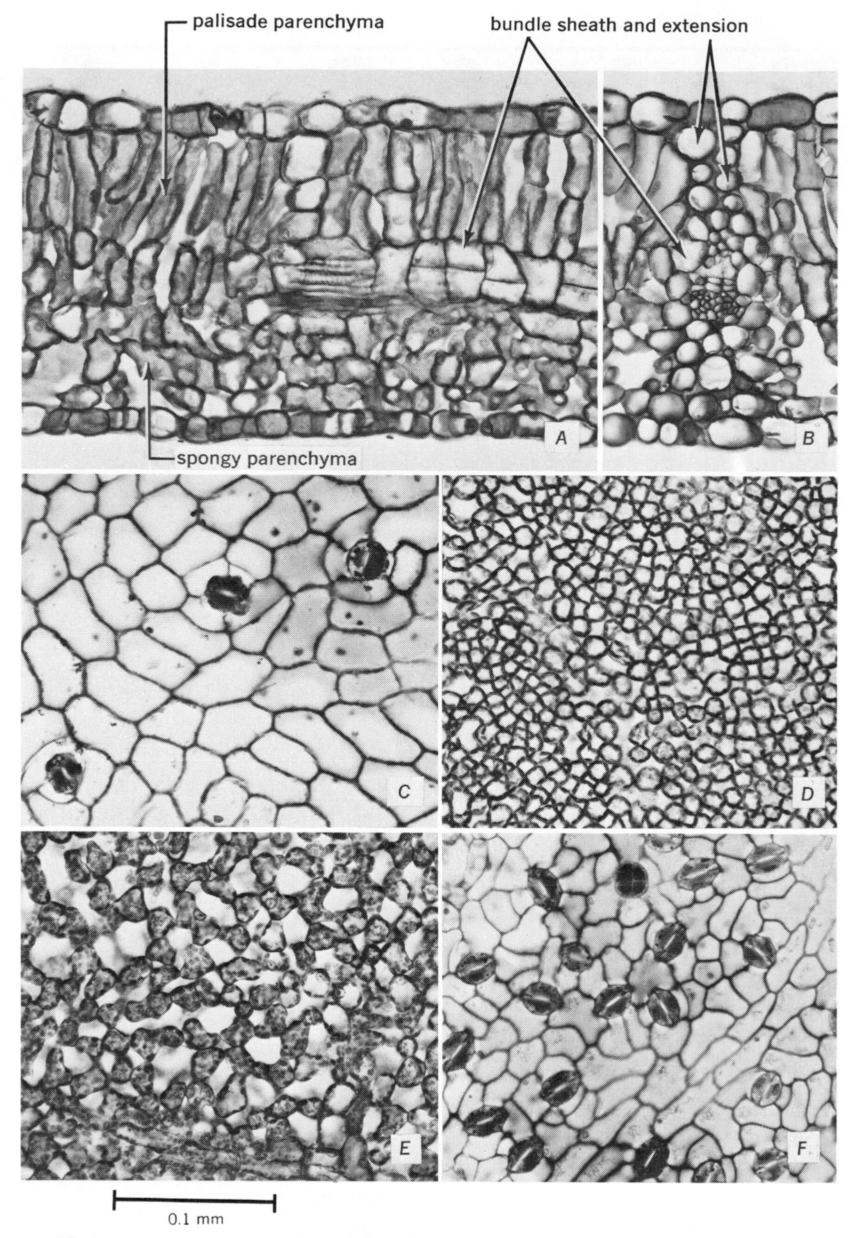

Figure 18.1 Structure of lilac (*Syringa*) leaf. *A, B,* cross sections of mesophyll with a small vascular bundle embedded in it. *C–F,* paradermal sections showing upper epidermis (*C*), palisade parenchyma (*D*), spongy parenchyma (*E*), and lower epidermis (*F*).

ence of cuticle and stomata (chapter 7). The stomata may occur on both sides of the leaf (amphistomatic leaf; fig. 18.1) or only on one side, either on the upper (epistomatic leaf) or, most commonly, on the lower side (hypostomatic leaf). In the broad leaves of dicotyledons the stomata are scattered (fig. 18.1,*F*). In the narrow elongated leaves characteristic of monocotyledons and conifers, the stomata occur in rows parallel with the long axis of the leaf. The stomata may be on the same level as the other epidermal cells, or they may be located above the surface of the epidermis (raised stomata), or below it (sunken stomata). Several stomata may appear in a depression called stomatal crypt. Raised stomata are associated with a hydrophytic habitat providing a large supply of available water, sunken stomata with a xerophytic habitat characterized by a low supply of available water. Although this relation is not strict, experiments show that a raised position of stomata may be induced by enclosing developing leaves in vapor-saturated atmosphere.[1] This observation is used to explain the frequently raised position of stomata that are located in stomatal crypts: the crypts, usually protected by trichomes, probably contain a moist atmosphere.

Mesophyll

The main part of the ground tissue of a leaf blade is the mesophyll containing many chloroplasts and a large volume of intercellular spaces. The mesophyll may be relatively homogeneous or may be differentiated into palisade parenchyma and spongy parenchyma (fig. 18.1). The palisade parenchyma consists of cells elongated perpendicular to the surface of the blade. Although the palisade tissue appears more compact than the spongy tissue (fig. 18.1,*A*) a considerable part of the long sides of the palisade cells is exposed to the intercellular air (figs. 18.1,*D,* and 18.2). Leaves may have one or more rows of palisade parenchyma. In plants of the temperate regions characterized by abundant available water in the soil (mesophytic habitat), the palisade parenchyma is usually located on the upper (adaxial, or "ventral") side of the blade, the spongy parenchyma on the lower (abaxial, or "dorsal"). A leaf with such structure is called *bifacial* or *dorsiventral.* If the palisade tissue occurs on both sides of the leaf, as is common in leaves of xerophytic habitats (chapter 19), the leaf is called *unifacial, isobilateral,* or *isolateral.*

The spongy parenchyma consists of cells of various shapes, frequently irregular, with branches extending from one cell to the other. Limitation of such connections to the ends of the branches gives the spongy parenchyma an appearance of a three-dimensional net, with the meshes enclosing the intercellular spaces (figs. 18.1,*E,* and 18.2). With regard to the connections between adjacent cells, the spongy parenchyma tissue has a dominantly horizontal continuity parallel with the surface of the leaf, whereas the palisade parenchyma is continuous mainly in the direction perpendicular to the surface of the leaf.

The loose structure of the mesophyll is responsible for a large total surface area between the cells and the internal air, a surface which many times exceeds that between the epidermis and the external air; and, of the two kinds of mesophyll tissue, the palisade parenchyma has a larger internal free surface than the spongy parenchyma.[6]

Vascular system

The vascular system of the leaf is distributed throughout the blade and thus shows a close spatial relation to the mesophyll. The

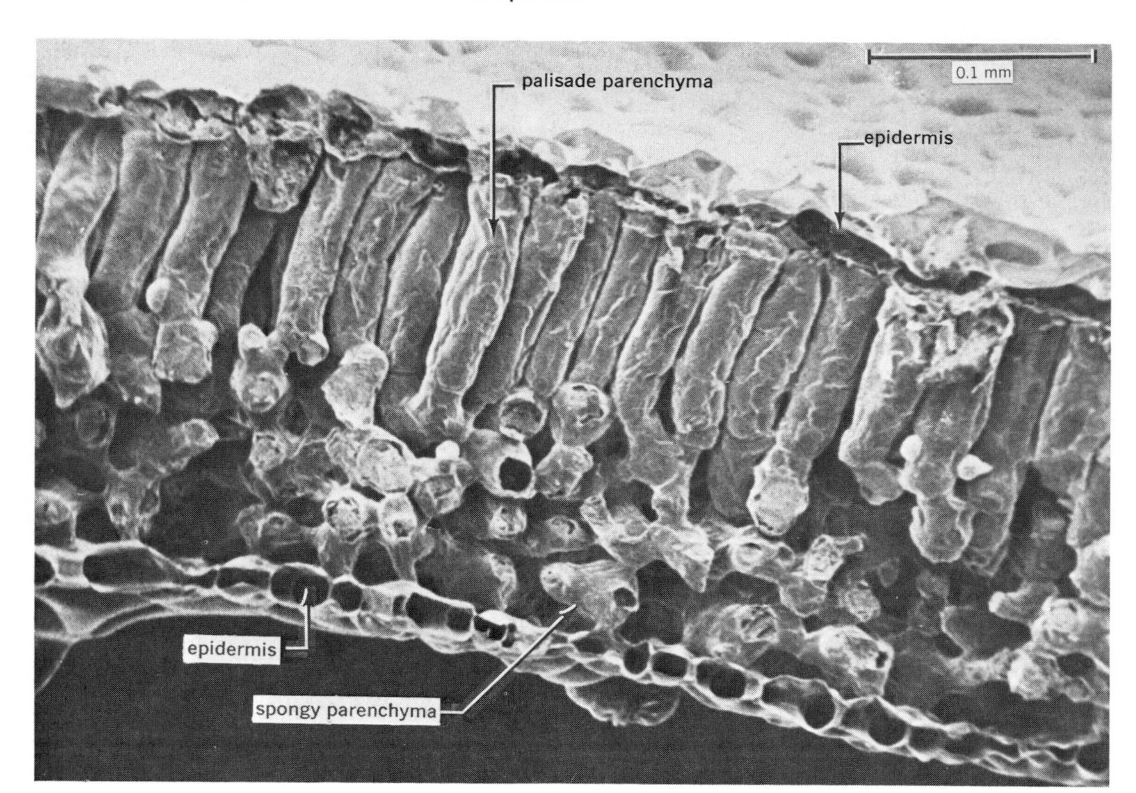

Figure 18.2 Scanning electron microscope micrograph of fresh freeze-fractured bean (*Phaseolus vulgaris*) leaf. (From B. Bole and E. Parsons, *J. Microscopy* 98:91–97, 1973.)

vascular strands form an interconnected system in the median plane of the blade parallel with the surface of the leaf. The vascular bundles in the leaf are commonly called *veins,* and the pattern formed by these veins, *venation.* As seen with the unaided eye, the venation appears in two main patterns, the reticulate, or netted, and the parallel. Reticulate venation may be described as a branching system with successively thinner veins diverging as branches from the thicker veins (fig. 18.3). In the parallel-veined leaves, strands of relatively uniform size are oriented longitudinally or nearly so (fig. 18.4,*A*,*B*). The designation "parallel" is an approximation, since the bundles converge and join at the apex of the leaf and are not truly parallel even in the widest part of the leaf. Netted venation is most common in dicotyledons, parallel venation in monocotyledons.

Leaves with reticulate venation often have the largest vein, the midvein, along the median longitudinal axis of the leaf (fig. 18.3). The midvein is connected laterally with somewhat smaller lateral veins. Each of these veins is connected with still smaller veins, from which other small veins diverge. In dicotyledons, the number of orders of veins as determined by their size and point of divergence from another vein ranges from two to more than five.[21] The ultimate branchings form meshes delimiting small areas, or *areoles,* of mesophyll (fig. 18.5). Freely terminating vein endings may or may not extend

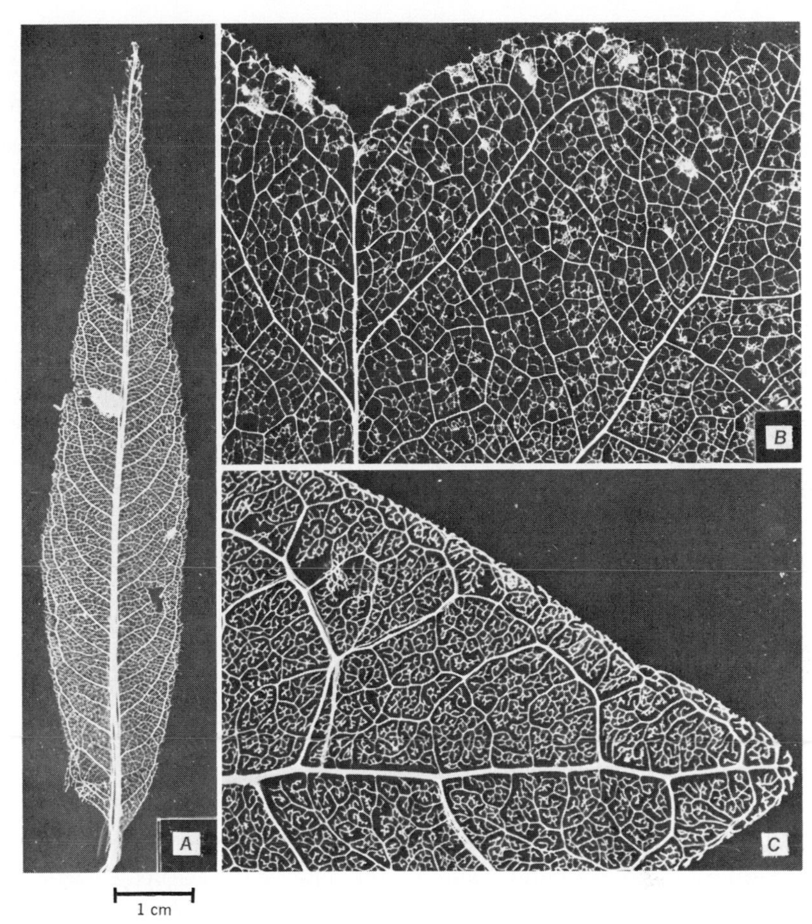

Figure 18.3 Leaf "skeletons" showing reticulate type of venation. *A,* willow (*Salix*). *B,* tulip tree (*Liriodendron*). *C,* privet (*Ligustrum*). (From R. T. Whittenberger and J. Naghski, *Amer. J. Bot.* 35:719–722, 1948.)

into the areoles.[12,40] In the venation pattern designated as open dichotomous[13] closed meshes are absent.

In the parallel-veined leaves the main veins may vary in size, with the smaller and larger veins alternating. These longitudinal veins are interconnected by considerably smaller veins, the commissural bundles, which may appear singly as simple cross connections[2] or as a complex network.[40] Thus, at the microscopic level, the parallel-veined leaves also have a reticulate arrangement of vascular bundles (fig. 18.4,*A*,*B*).

The number and arrangement of the vascular bundles in the petiole and midvein vary widely (figs. 18.4,*C*,*D*, and 18.6). The lateral veins usually consist of single strands in which the vascular tissues diminish in amount from the first-order lateral veins toward the ultimate ramifications. In the bundle ends, xylem elements frequently extend farther than the phloem elements, but in

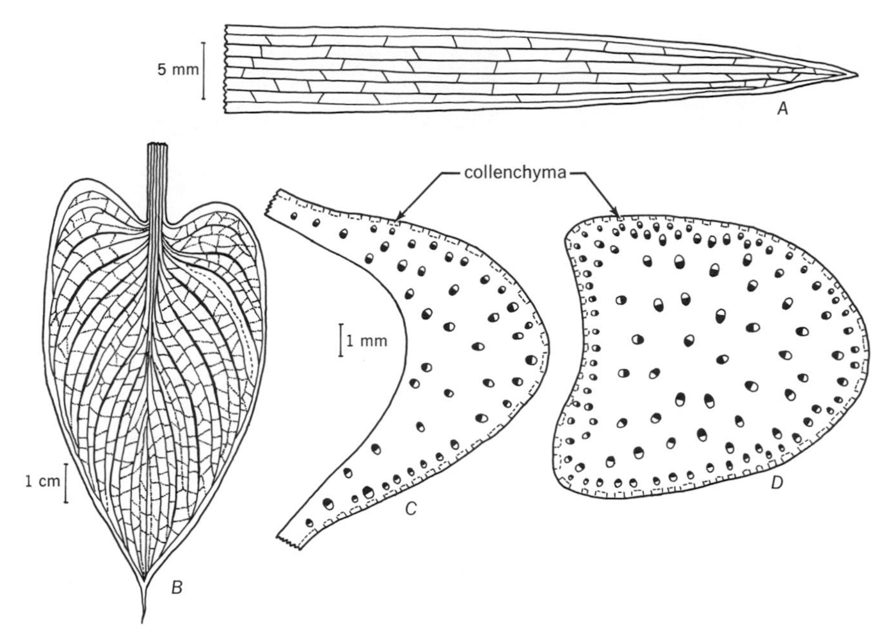

Figure 18.4 Vascular system in monocotyledon leaves. *A,* venation in a grass leaf. *B,* venation in *Zantedeschia leaf. C, D,* arrangement of vascular bundles as seen in cross sections of midvein (*C*) and petiole (*D*) of *Zantedeschia* leaf. (After K. Esau, *Plant Anatomy,* 2nd ed. John Wiley & Sons, 1965.)

some plants the phloem accompanies the xylem to the ends of veins (fig. 18.7). The xylem of vein endings usually consists of short tracheids (fig. 18.7,*A*), the phloem of short narrow sieve elements and enlarged companion cells (fig. 18.7,*B*). Depending on the plant group the vascular bundles are collateral or bicollateral, although in leaves having internal (adaxial) phloem this phloem may not extend into the smaller veins. When the bundles are collateral, the xylem occurs on the adaxial side of the leaf, the phloem on the abaxial.

In the dicotyledons, the smaller veins are embedded in the mesophyll (figs. 18.1,*B*, and 18.8,*B*, arrows) but the larger veins are enclosed in ground tissue that is not differentiated as mesophyll and has relatively few chloroplasts. The tissue associated with the larger veins rises above the surface of the leaf and forms ribs, most commonly on the abaxial side of the blade (fig. 18.8). Collenchyma or sclerenchyma may be present beneath the epidermis of the vein rib, on one or both sides of the vein. The Latin word for rib, *costa,* is used in the term *intercostal areas* referring to the leaf tissue between the ribs.

The small vascular bundles located in the mesophyll are enclosed in one or more layers of compactly arranged cells forming the *bundle sheath* (fig. 18.1,*B*). Bundle sheaths may be parenchymatic or sclerenchymatic, and the two kinds of sheaths may be combined. Suberization of cell walls in bundle sheaths recorded in several species indicates that these cells might function as an endodermis.[31] Bundle sheaths extend to the

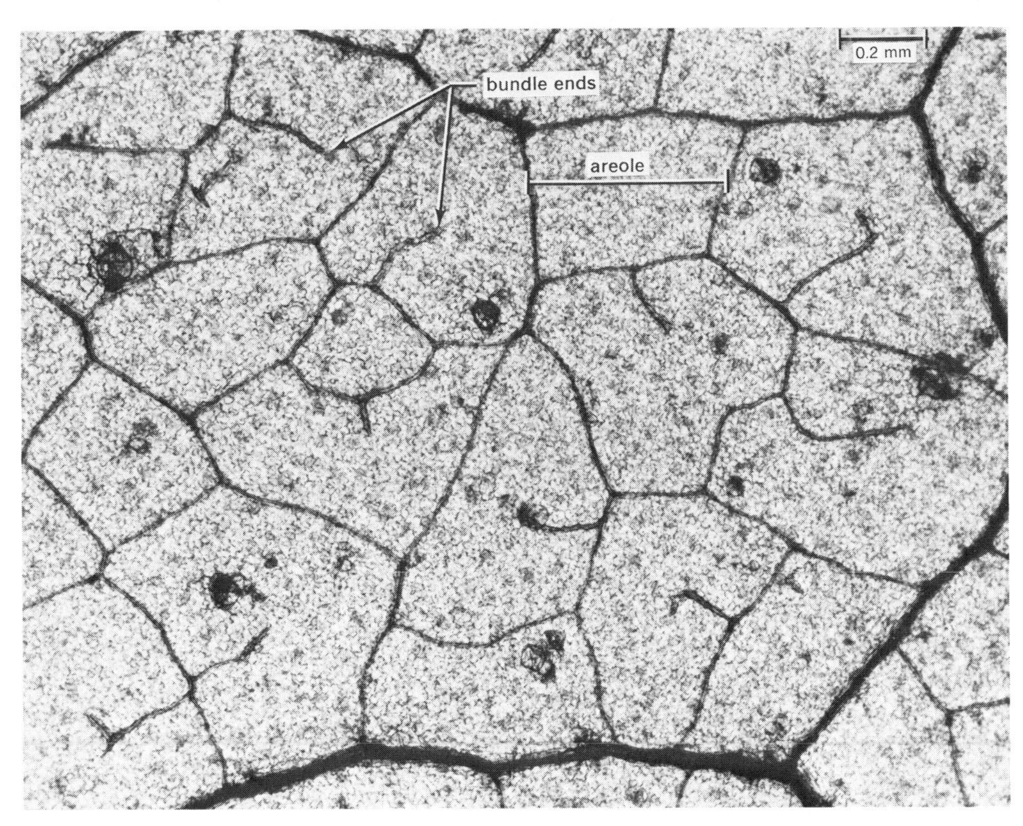

Figure 18.5 Piece of tobacco (*Nicotiana tabacum*) leaf partially cleared to show minor veins. Free bundle ends occur in some of the areoles.

ends of bundles (fig. 18.7) so that the vascular tissue is rarely directly exposed to intercellular spaces. An exception is found in those hydathodes in which the xylem elements at bundle ends release water into the intercellular spaces (chapter 13). In many dicotyledons the bundle sheaths are connected with the epidermis by panels of cells resembling the bundle-sheath cells (fig. 18.1,*B*). These panels are called *bundle-sheath extensions.*[50] The term is applied only to the smaller vascular bundles embedded in the mesophyll; it does not refer to the vein-rib tissue associated with the larger veins. Bundle sheaths and bundle-sheath extensions occur in the monocotyledons also. In some taxa they consist of sclerenchyma.

Leaves show a significant correlation between the character of the vascular system and those structural features of the nonvascular tissues that may have an influence upon conduction. Among the nonvascular tissues the epidermis and the spongy mesophyll, both tissues in which cells have extensive lateral contacts, may be assumed to be better adapted for lateral conduction than the palisade mesophyll with its dominant cell connection in the abaxial-adaxial direction. In conformity with this concept, the ratio of palisade tissue to spongy tissue is closely re-

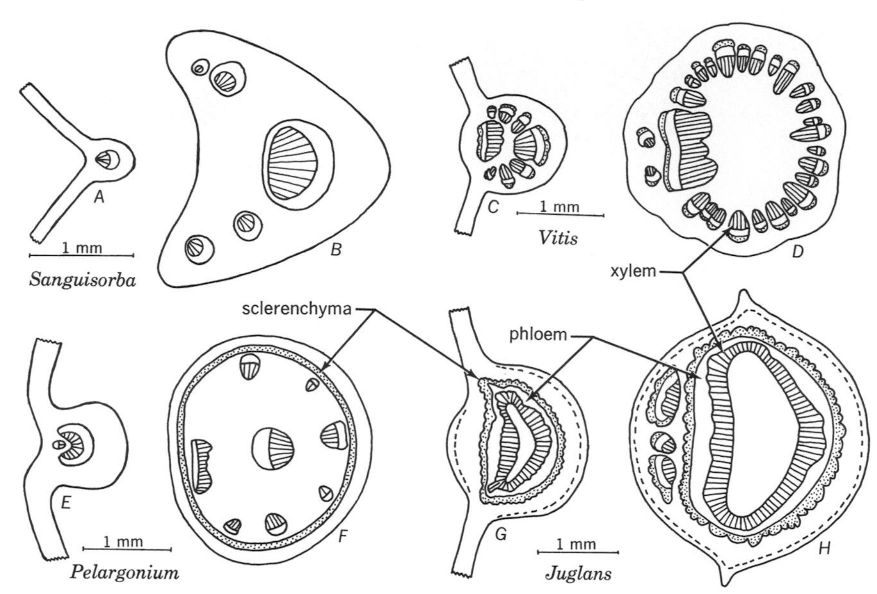

Figure 18.6 Arrangement of vascular tissues in midveins (*A*, *C*, *E*, *G*) and petioles (*B*, *D*, *F*, and *H*) of dicotyledon leaves.

lated to vein spacing: the greater this ratio, the closer is the vein spacing.[37,50] Then, there is evidence that parenchymatic bundle-sheath extensions conduct water toward the epidermis, where a lateral spread occurs. Accordingly, when sheath extensions are present the venation is less dense than when they are absent.

MINOR VEINS AND TRANSLOCATION OF SOLUTES.

The small veins embedded in the mesophyll are referred to as minor veins (fig. 18.9) although they play a major role in transport of water and food. They distribute the transpiration stream through the mesophyll[4,45] and serve as starting points for the uptake of the products of photosynthesis and their translocation out of the leaf.

The outstanding characteristic of minor veins is the prominence of vascular parenchyma cells, particularly of those in the phloem. Most of these cells have dense protoplasts (fig. 18.10) and numerous plasmodesmata of the branched type (chapter 11) connecting them with sieve elements. The density of the cytoplasm is detectable even when the cells have large vacuoles (fig. 18.9). In some species the dense cells extend to the bundle ends where they are not associated with sieve elements. The dense cells are interpreted as companion cells or as cells functionally equivalent to companion cells when they have no ontogenetic relation to sieve elements. The less dense cells are phloem parenchyma cells.

Fischer[9] called the dense cells *intermediary cells* to express the concept that these cells mediate between the mesophyll and the sieve elements in the transfer of the photosynthate. The term intermediary is now applied to the less dense cells as well because they too are involved in the process of

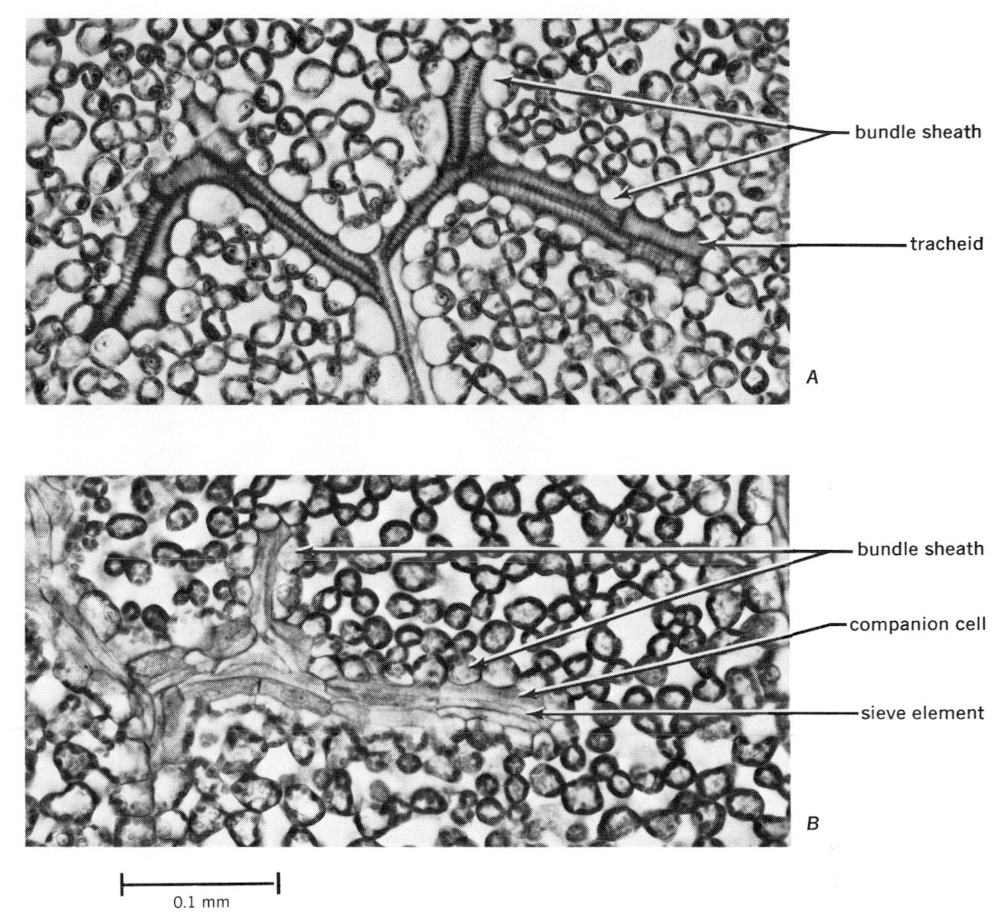

Figure 18.7 Bundle ends in lilac (*Syringa*) leaf as seen in paradermal sections. *A*, section through terminal tracheids of xylem. *B*, section through terminal sieve elements and companion cells of phloem.

loading of the minor veins with the photosynthate.

In a number of herbaceous dicotyledons the intermediary cells develop wall ingrowths with the corresponding increase in surface area of the plasmalemma (fig. 18.10) and are then called transfer cells[33,34] (chapters 5 and 13). It is significant that ingrowths develop in the vein cells when the differentiating leaf is converted from a sink to a source of carbohydrates and begins to export the latter.[34]

Two distributional patterns of wall protuberances are recognized in the transfer cells in the phloem of minor veins (fig. 18.10). In the *a* type of cell, which is the dense companion cell, the ingrowths occur on all walls but are less prolific where plasmodesmata connect the cell with the sieve element. In the *b* type of cell, the phloem parenchyma cell, the protuberances develop preferentially on wall areas next to or closest to sieve elements and their companion cells.

Intermediary cells with or without wall ingrowths are visualized as being concerned

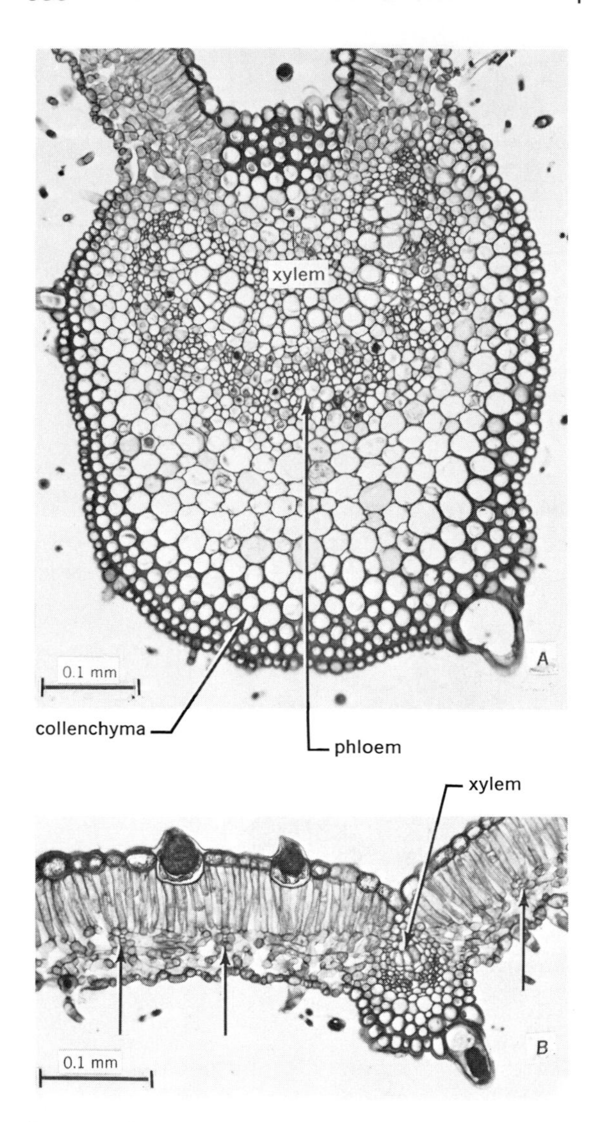

Figure 18.8 Structure of midvein (*A*) and small vein and mesophyll (*B*) of hemp (*Cannabis*) leaf. Unlabeled arrows in *B* indicate minor veins in mesophyll.

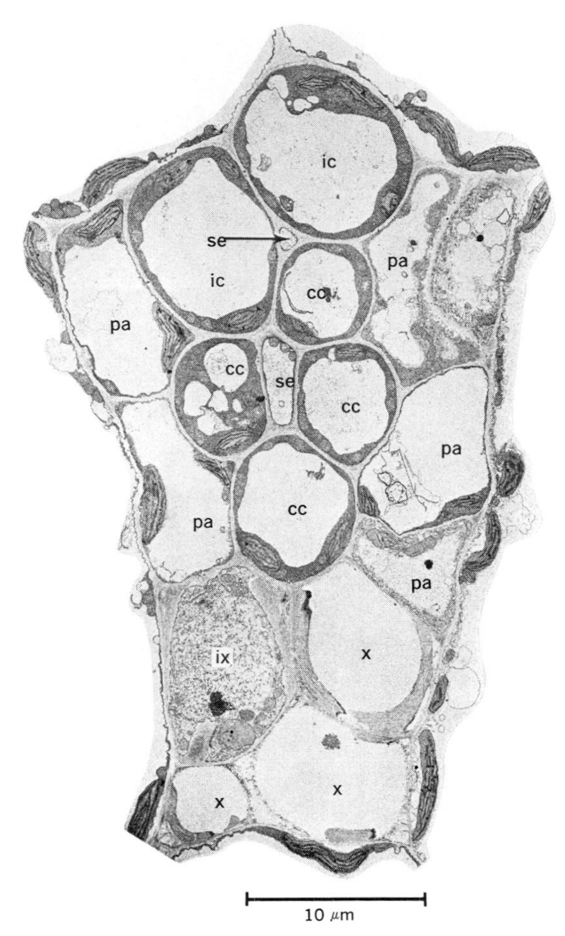

Figure 18.9 Electron micrograph of minor vein from sugar beet (*Beta vulgaris*) leaf. The sieve elements (*se*) are narrower than the associated cells: companion cells (*cc*), parenchyma cells (*pa*), and intermediary cells (*ic*) that resemble companion cells. Three xylem elements are mature (*x*), one immature (*ix*). (From K. Esau, *New Phytologist* 71:161–168, 1972.)

with retrieval of solutes and their transfer to the sieve elements, whether the solutes originate through photosynthesis or are brought to the leaf by the transpiration stream. Both *symplastic* (confined to the cytoplasm) and *apoplastic* (occurring in the cell walls, which constitute the *free space* together with the intercellular spaces) movements are apparently conceivable.[3,18,34] In figure 18.11, the continuous lines indicate the symplastic pathway, the dotted lines the apoplastic pathway, and in both the arrowheads give the direction of movement. Solutes arriving in the

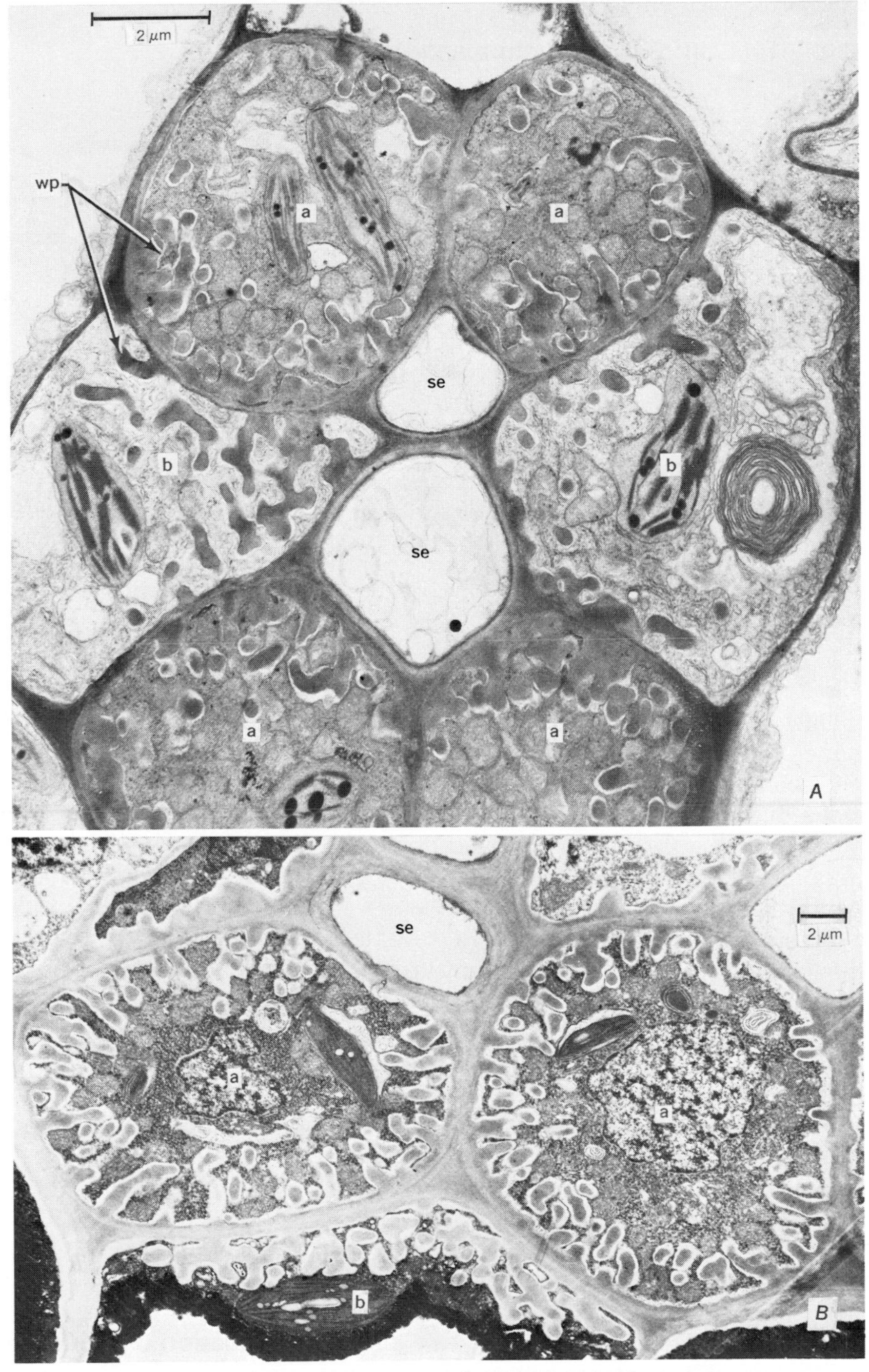

Figure 18.10 Electron micrographs of cell groups from minor veins in leaves of, *A, Senecio vulgaris* (Asteraceae), and, *B, Armeria corsica* (Plumbaginaceae). The narrow sieve elements (*se*) are associated with intermediary cells differentiated as transfer cells with wall protuberances (*wp*). The transfer cells are of the *a*-type (companion cells) and the *b*-type (parenchyma cells). (From Pate and Gunning.[33])

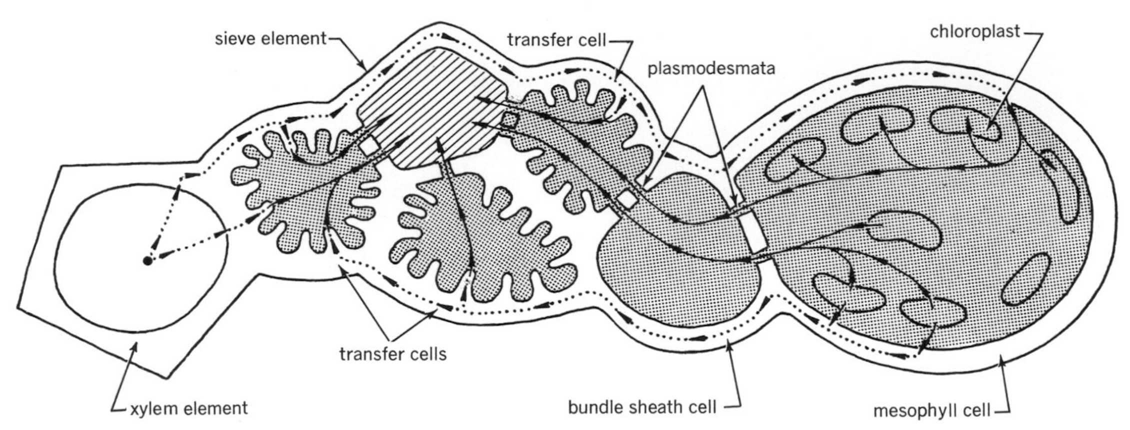

Figure 18.11 Diagram of a group of cells from a minor vein of a dicotyledon illustrating the movement of solutes during loading of a sieve element. Continuous lines, symplastic routes; dotted lines, apoplastic routes; arrowheads indicate direction of movement. The cells adjacent to the sieve element are shown as transfer cells with wall ingrowths. They could be intermediary cells without the ingrowths: see fig. 18.9. (Constructed from data in Pate and Gunning.[34])

xylem (dot) may be carried various distances through the cell wall before they enter the protoplasts of parenchymatic cells and are transported symplastically to the sieve elements. The solutes generated by photosynthesis may move through the symplast to the sieve element or leak into the cell wall and combine apoplastic and symplastic movements on their way to the sieve element.

Autoradiographic studies have revealed that labeled photosynthate is rapidly taken up by minor veins in exporting leaves. Solute accumulation and the resulting osmotic concentration reach higher levels in the loaded veins than in the surrounding mesophyll.[18] Hence, loading of the veins appears to occur against the positive gradient and is therefore regarded as an active process. The cytology and localization of intermediary cells in the minor veins are consistent with the concept that these cells are carrying out the energy-requiring transfer of carbohydrates to the conduit in the phloem.

Loading of the phloem with sugar is associated with entry of water across the plasmalemma from the free-space solution. The resulting rise of hydrostatic pressure in the sieve element is a major factor in the long-distance transport in the phloem. Thus, the loading of the minor veins generates the motive force of translocation. The movement in the phloem, however, depends also on the presence of *sink* regions where the osmotic pressure is lower than in the leaf veins and solutes are removed from the conduit. Companion cells and phloem parenchyma cells are actively involved in the retrieval of carbohydrates from the sieve element for use in growth and storage. The relation between the intermediary cells and the sieve elements in minor veins is merely one of the more striking expressions of the functional dependence of the enucleate conduit upon cells with complete and metabolically active protoplasts.

DEVELOPMENT

Initiation of leaf primordia

A leaf is initiated by cell divisions in the peripheral region of the shoot apex, more or less far below the distal zone of the apical

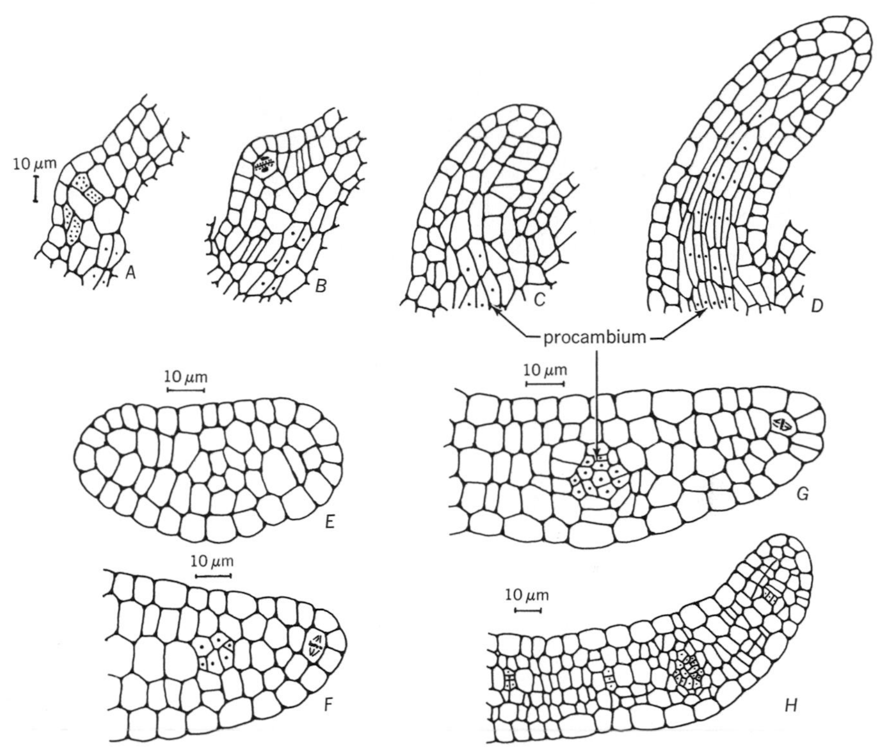

Figure 18.12 Origin of leaf primordium and blade in flax (*Linum*). *A–D,* longitudinal sections; *E–H,* cross sections, *A, B,* emergence of primordium through enlargement and periclinal divisions of subsurface cells. *C, D,* upward growth of primordium. *E,* primordium before initiation of blade. *F–H,* growth of blade: products of marginal meristematic activity form a plate meristem in which intercalary anticlinal divisons predominate. Procambial cells are indicated by dots. (After G. Girolami, *Amer. J. Bot.* 41:264–273, 1954.)

meristem (chapter 16). The location of these divisions in relation to the sites of the previously formed leaves is determined by the phyllotaxy of the plant. In many angiosperms, the first divisions occur in one or more layers beneath the protoderm and are periclinal to the surface of the shoot apex (fig. 18.12,*A,B*). The periclinal divisions are soon followed by anticlinal divisions in the protoderm and the layers beneath it. In some taxa, the protoderm cells also divide perclinally[27,32] (fig. 18.13). The initial divisions give rise to a lateral protrusion, often called leaf buttress,[11] on the shoot apex (figs. 18.12,*A,B*; 18.13,*B–D*; and 18.14,*A*).

Growth in length and width

The leaf primoridum extends upward from the initial protrusion as a bladeless conical or peglike protuberance referred to as the leaf axis (figs. 18.14,*B,F*, and 18.15,*A*). The primordium is dorsiventral as it emerges.[20] In subsequent stages, the primordium increases in height (length) and width and, by growing more actively on the abaxial side than on the adaxial side, develops a curvature toward the shoot apex (fig. 18.14,*C–E*).

The increase in width of the leaf includes the spread of the leaf base laterally so that the apical meristem is encircled to a greater or

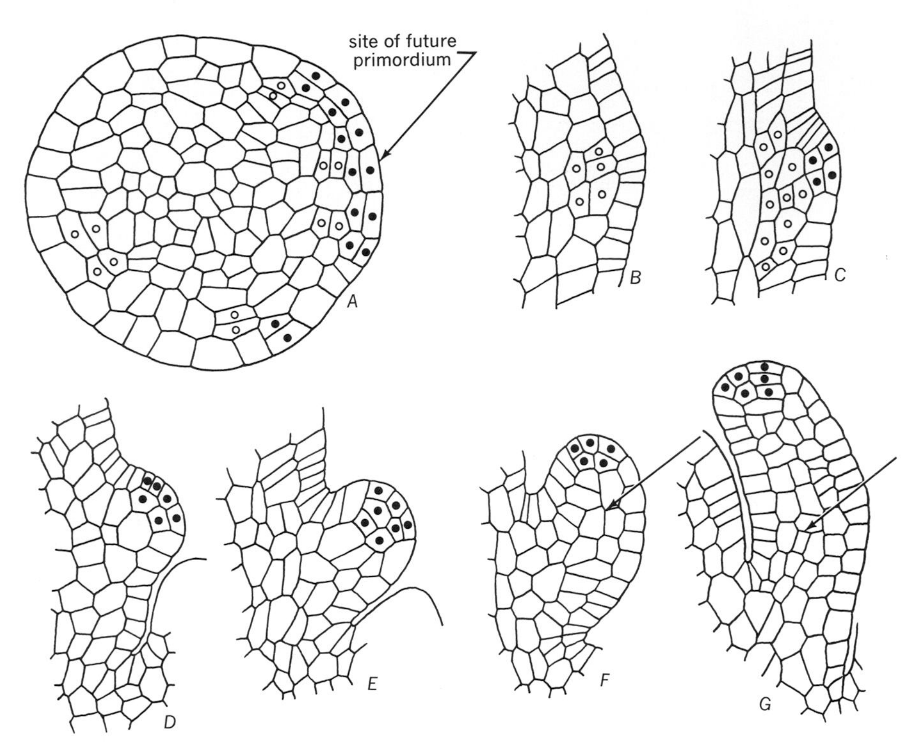

Figure 18.13 Origin of leaf in barley (*Hordeum*). *A,* transverse section of shoot at level of origin of a leaf primordium: periclinal divisions have occurred in the protoderm (dots) and in the subsurface layer (circles). Some distance from the site of primordium, several periclinal divisions have initiated the circumferential growth of the future leaf base. *B–G,* longitudinal sections of leaf primordia in successive stages of development. Periclinal divisions in subsurface layer (*B,* circles) are followed by similar divisions in the protoderm (*C,* dots). The elongating primordium (*D–G*) shows additional periclinal divisions at its apex. At arrows in *F* and *G,* increase in number of layers in leaf. (Redrawn from Klaus.[27])

lesser extent depending on the degree to which the mature leaf ensheaths the stem (fig. 18.14,*C*). The encircling part of the leaf base may give rise to stipules (fig. 18.14,*D*,*E*) or to a sheath as in the monocotyledons.[25]

The growth concerned with the lateral extension of the primordium becomes localized along two margins of the leaf axis and the primordium differentiates above the base into a midrib and two panels of the leaf blade (fig. 18.15,*B*). Lateral veins become blocked out in the form of procambial strands as the blade continues to widen and, in petiolate leaves, a petiole develops between the blade and the leaf base (fig. 18.15,*C*). The growing blade may spread sideways from the midrib or bend toward the adaxial side of the leaf (fig. 18.15,*D*) so that it encloses the apical meristem and the younger leaf primordia. The midrib often increases in thickness on its adaxial side by orderly periclinal divisions within a region constituting the *adaxial meristem* (fig. 18.15,*D*). The petiole also increases in thickness by means of an adaxial meristem.

The early growth of a leaf is commonly di-

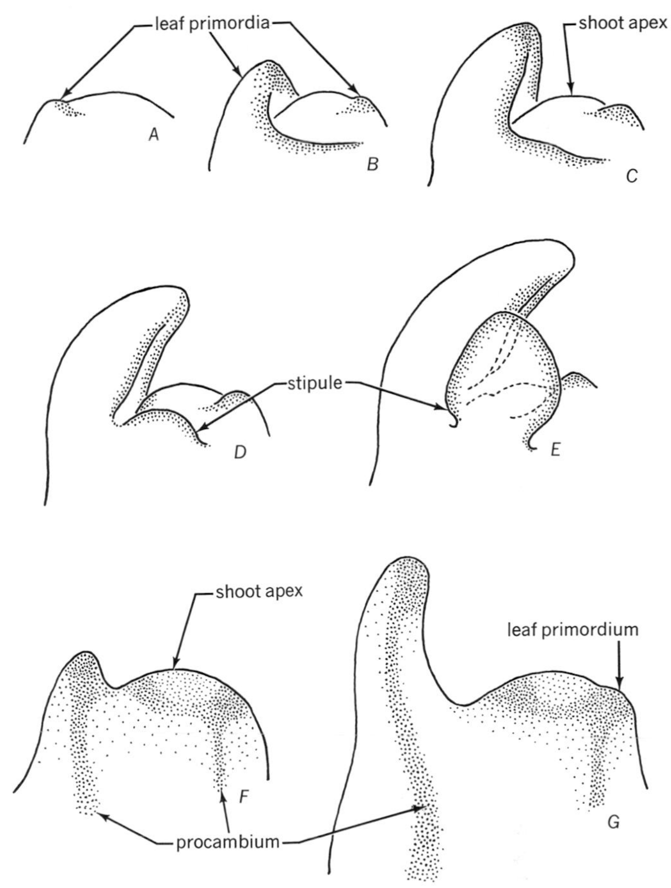

Figure 18.14 Development of a dicotyledon (*Oenothera biennis*) leaf schematically represented in three-dimensional views (*A–E*) and in longitudinal profiles (*F–G*). The leaf arises as a dorsiventral protuberance (*A*) which grows in height and width according to a pattern determined by the distribution of meristematic activity (stippling in *A–E*). The longitudinal profiles show the relation between the development of primordia and of procambium. In *F,* right, the procambium is shown to be present before the leaf primordium has emerged as a protuberance. The region in the shoot apex where primordia arise is more active meristematically (heavy stippling in *F* and *G*) than the most distal region (light stippling at apex). (Redrawn from Hagemann.[20])

vided into apical and marginal growth, the first concerned with the elongation of the primordium, the second with the lateral extension that produces the two panels of the blade. Thus, the growing primordium has an *apical meristem* at the tip and two *marginal meristems* along opposite sides of the axis (fig. 18.15). Each meristem is at least two cells in depth, for typically in dicotyledons the outermost layer, or protoderm, grows by anticlinal divisions independently from the subjacent layers. The internal tissues are derived from the layer or layers beneath the protoderm (fig. 18.12). In some taxa, notably

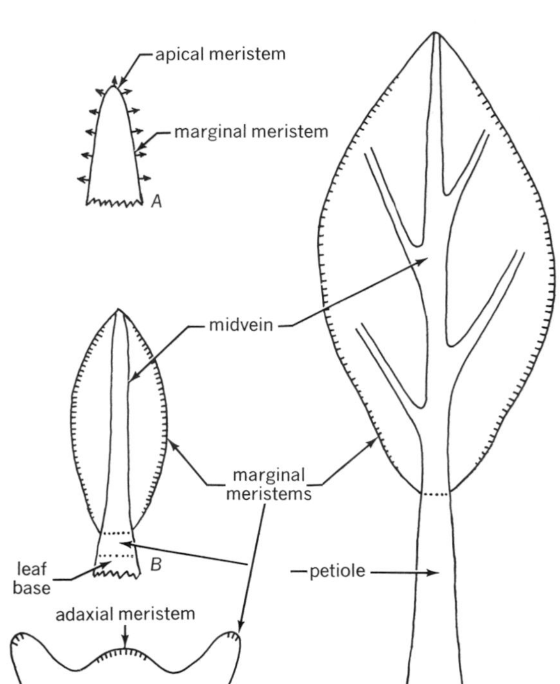

Figure 18.15 Diagrams interpreting growth of a dicotyledon leaf. *A*, undifferentiated leaf primordium (leaf axis) without blade and petiole; arrows indicate direction of growth. *B, C*, two stages of: growth of blade initiated by marginal meristems; differentiation of major veins; and elongation of petiole. *D*, cross section of blade showing position of adaxial and marginal meristems.

the monocotyledons, the outermost layer may add cells to the internal tissue by periclinal divisions (fig. 18.13).

When the concept of apical and marginal growth of the leaf primordium was formulated[11] the two kinds of meristems were pictured as having distinct initials. Thus, the protoderm would be derived from apical and marginal initials dividing anticlinally, the inner tissues from subapical and submarginal initials alternately dividing anticlinally and periclinally in a sequence characteristic of the species. The concept of leaf meristems with initials was developed by observing cell patterns in sectioned leaves and visualizing ontogenetic cell lineages by reference to these patterns. Occasionally seen mitoses and their orientation in cells occupying sites appropriate for the initials (fig. 18.12,*B*,*F*,*G*) were given much weight in constructing the diagrams (fig. 18.16) depicting the relation between cell layers and the presumed initials.

Later studies on distribution of mitoses in early growth of leaves indicate, however, that the concept of distinct initials in the apical and marginal meristems is not universally applicable. The cells in subapical and submarginal position, in particular, are not as precisely related to the inner cell rows as was originally visualized[5,17,30,43] (fig. 18.17). Observations on chimeral plants (chapter 16) also showed that the ontogenetic relation between the submarginal cells and cell layers in the blade may be highly variable. In a number of stable natural chimeras with variegated leaves, having three independent layers (LI, LII, and LIII) at the apex, the inner tissues of the leaf may be derived from at least two layers of the apical meristem, LII and LIII, and the contributions from these layers may vary in amount and distribution not only between species or different plants but also between different leaves on the same plant.[44] If, for example, the chimeral plant has the genotype G-W-G (green-white-green) at the apex, the W (LII) and G (LIII) kinds of cells occur in the leaf in various combinations depending on the frequency, duration, and distribution of divisions among the derivatives of the W and G cells (fig. 18.18). Patterns of this kind cannot be interpreted as results of an orderly origin of tissues from subapical and submarginal initials, especially not from a single layer of such initials as is usually depicted in the diagrams (fig. 18.16).

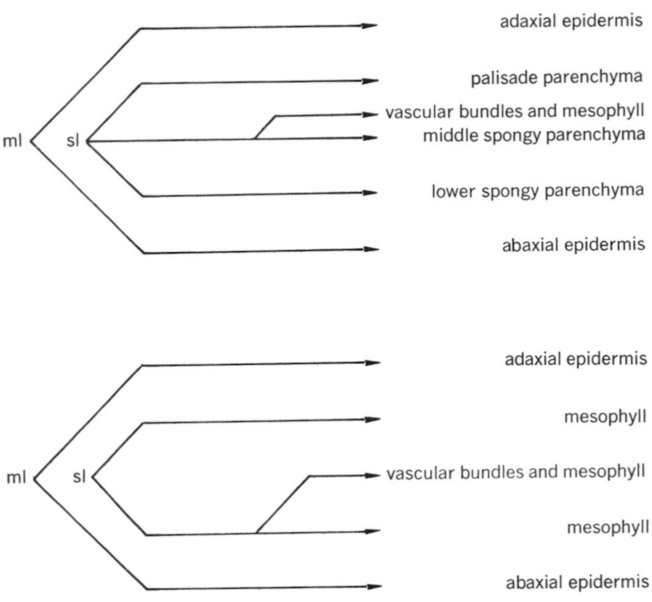

Figure 18.16 Diagrams illustrating two patterns of possible ontogenetic relation between marginal meristem and derivative cell layers of the plate meristem forming the blade. Marginal layer (*ml*) produces the protoderm. Submarginal layer (*sl*) gives rise to the mesophyll and vascular tissue. The sites of the hypothetical marginal and submarginal initials would be at *ml* and *sl*, respectively.

The derivatives of the cells at the apex and along the margins of the leaf divide in their turn. These *intercalary divisions,* which are distributed at random, make the major contribution to the increase in size of the leaf. Intercalary growth follows that at the apex and margins so closely that in overall views of developing leaves the intercalary mitotic activity cannot be delimited from apical and marginal activity[30] (fig. 18.17). The randomness in the distribution of mitoses in growing leaves has been demonstrated in whole leaves of several dicotyledons that were cleared and stained with the Feulgen DNA reagent, a technique revealing dividing nuclei[16,24,46] (fig. 18.19).

The intercalary meristematic activity ceases earlier at the apex than at the base (fig. 18.19) so that the leaf tissues differentiate and mature in the basipetal direction. Moreover, cell divisions may be less numerous in the apical region than farther below, a difference that results in greater width of the leaf toward the base. The basipetal order of maturation is particularly clear in the long narrow monocotyledon leaves in which growth becomes conspicuously localized at the bases of leaf blade and sheath zones referred to as intercalary meristems (chapter 16.)

Growth of the blade

In many species the mesophyll between the veins is remarkably uniform in thickness (chapter 19). This feature results from the establishment of a certain rather limited number of layers of cells close to the margin (figs. 18.12,*F–H*; 18.16) and the subsequent growth

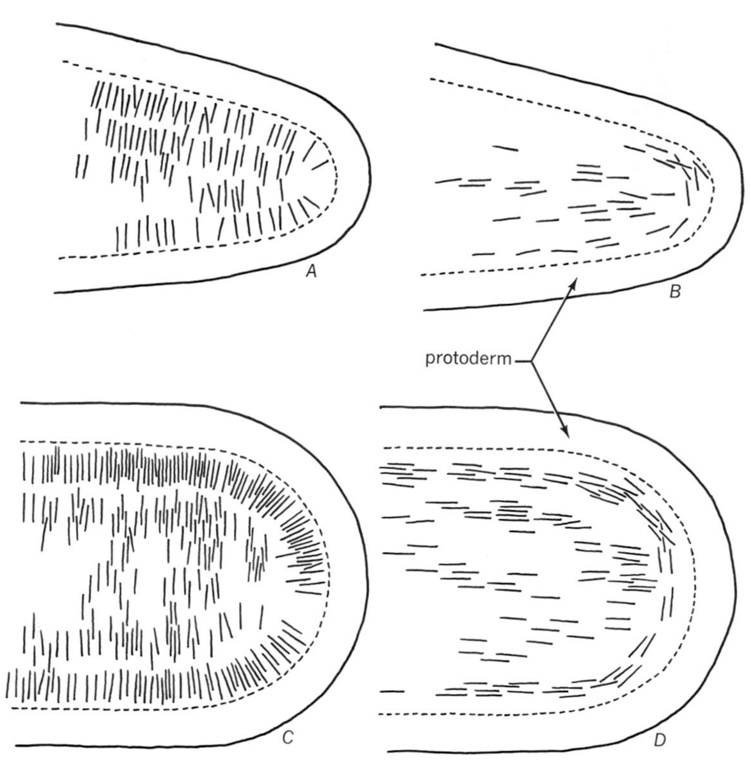

Figure 18.17 Diagrams illustrating meristematic activity at margins of *Lupinus* leaflets of two sizes: *A*, *B*, 600 μm long; *C*, *D*, 8500 μm long. The short lines indicate equatorial planes or cell plates of dividing cells. The division figures were recorded in serial sections and assembled: anticlinal divisions in *A* and *C*, periclinal divisions in *B* and *D*. Periclinal divisions at margin establish the number of layers in the leaf. Anticlinal divisions extend the layers (plate-meristem activity). (Redrawn from Fuchs.[17])

of the layers by anticlinal divisions. The leaf appears to be composed of several self-perpetuating layers of cells. A meristematic tissue consisting of parallel layers of cells dividing anticlinally is called *plate meristem.* The divisions in this meristem constitute a major part of the intercalary growth by means of which the leaf reaches its mature size. Some studies indicate that plate meristem activity may bring about a hundredfold increase in surface area of the leaf without an increase in the number of mesophyll layers.[41] In some plants (*Trifolium,*[5] *Xanthium*[30]), however, periclinal divisions may occur in the plate meristem at a considerable distance from the margin and cause an increase in thickness of the blade.

The plate meristem type of growth is disturbed in the regions where vascular bundles differentiate. Anticlinal, periclinal, and oblique divisions give origin to procambium and to tissues associated with the vascular bundles such as bundle sheaths or ground tissue of a vein rib (figs. 18.12,*F–H*, and 18.24,*A*,*B*).

The establishment of the characteristic layering in the leaf blade is interpreted as the function of the marginal meristem.[17,30] In the

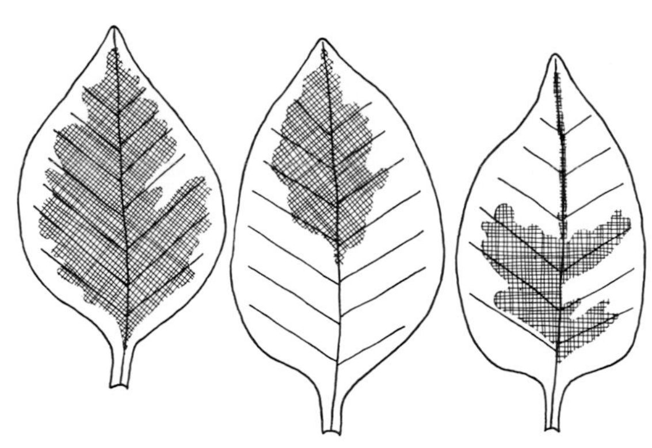

Figure 18.18 Variegated leaves of tobacco (*Nicotiana tabacum*) from plastidial chimeral shoots with G-W-G genetic differences in the layers of the apical meristem: L I, potentially green plastids (G); L II, potentially colorless plastids (W); L III, potentially green plastids (G). The dark areas in the figure indicate green cells contributed by the L III layer. The relative contribution of L II (W) and L III (G) are highly variable in the three leaves. (Adapted from a photograph in Stewart and Dermen.[44])

leaf of *Xanthium,*[30] the marginal meristem is estimated to be extending through about four cell diameters from the margin. Its outermost layer (protoderm) shows predominantly anticlinal divisions. Beneath the protoderm different proportions of three kinds of divisions, anticlinal, periclinal, and oblique, are taking place in the marginal and plate meristems (fig. 18.17). The high proportion of anticlinal divisions in the plate meristem (fig. 18.20) is associated with an intensive lateral extension of the leaf. Duration of mitotic activity was calculated to be at least 18 days in the marginal meristem and about 23 days in the plate meristem.

Variations in developmental pattern

Leaves are more variable in form than the two other vegetative organs, root and stem. They may be radially symmetrical like stems (*Hakea*), tubular (the insectiverous *Nepenthes*), or sword-shaped (numerous monocotyledons). Many plants have compound leaves with leaflets borne on a rachis (fig. 18.21,*A*). Leaves may be either tangentially or radially flattened with reference to the stem (fig. 18.22).

Despite their morphologic differences the leaves show merely variations in the basic pattern of development reviewed in the preceding sections. Limited marginal growth is responsible for the small width of the blade in many angiosperms and most conifers.[32] If the leaf is compound, marginal growth is divided into localized centers from each of which a separate leaflet develops (fig. 18.21). The leaflets then pass through developmental stages similar to those shown by simple leaves.

In the radially flattened unifacial leaves, which have a radial rather than a bifacial distribution of photosynthetic and vascular tissues (fig. 18.22,*C*, contrasted with 18.22,*A*), growth is distributed differently from that in bifacial leaves. Marginal meristem activity is suppressed, whereas the adaxial meristematic activity is accentuated and brings about the radial extension of the leaf.[25] This recognition of a common principle in the development of bifacial and unifacial leaves is clarifying the homologies between monocotyledon and dicotyledon leaves.[26]

One of the highly problematic phenomena in leaf development is the segmenting of palm leaves. As exemplified by the coconut leaf, the blade of which is segmented into pinnae (leaflets), the lamina develops on the rachis as two marginal panels facing the shoot apex. The lamina panels are at first smooth but soon develop corrugations (fig. 18.23,*A*). In sections perpendicular to the corrugations each panel of the blade appears as a series of interconnected zig-zag folds, or plications (fig. 18.23,*B*). Later, a separation occurs in each abaxial fold, whereas vascular

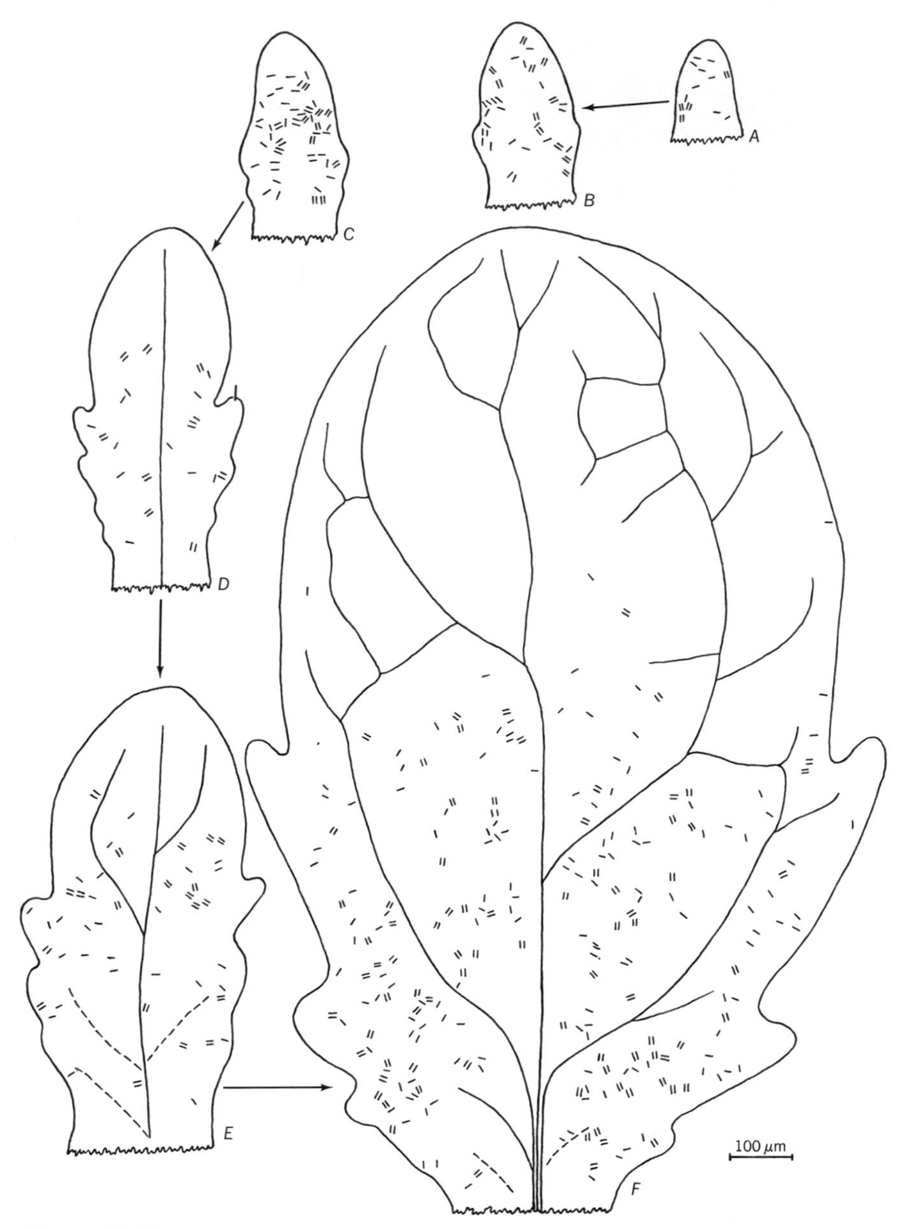

Figure 18.19 Distribution of mitoses in third pair of leaves above the cotyledons in *Paulownia tomentosa*. (Only one leaf of a given pair is shown.) Small lines represent divisions: single lines, metaphases; paired lines, anaphases and early telophases. The orientation of the lines indicates the direction of the divisions with reference to the length and width of the leaf. Leaves *A* (140 μm) and *B* (275 μm) are entirely meristematic. In leaf *C* (320 μm), mitotic activity is fading out at the apex. In subsequent stages, the cessation of meristematic activity advances basipetally (*D–F*). The differentiation of lateral veins also shows a basipetal progression. (Redrawn from Jeune.[24])

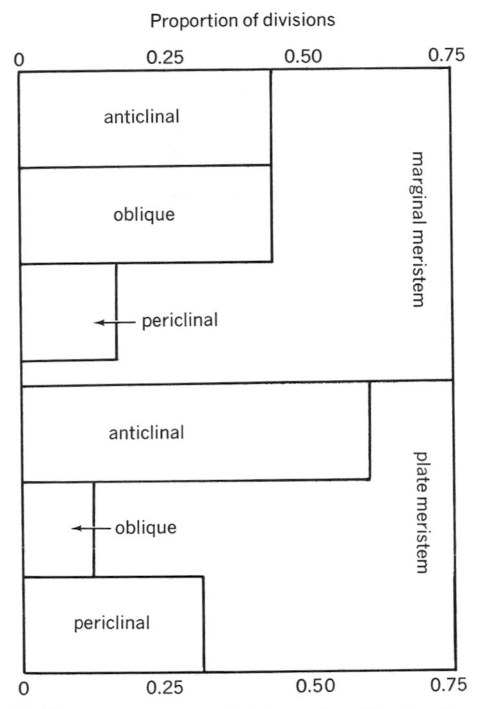

Figure 18.20 Proportion of divisions classified by the angular position of cell plate with reference to the surface of the leaf of *Xanthium*. The histogram is based on a record of metaphase, anaphase, and telophase figures seen in the marginal regions of several leaves of known plastochron age. The dividing line between the marginal and plate meristems was placed at about four cell diameters from the margin. The number of divisions represented under marginal meristem is 111, under plate meristem, 171. In both regions, the cell plates are preferentially oriented in particular directions and the pattern of orientation differs between the two regions. (Adapted from Maksymowych.[30])

tissue develops in each adaxial fold, now the midrib of an individual leaflet (fig. 18.23,*C*,*D*). The marginal strip of the lamina, which is not involved in the development of the plications (fig. 18.23,*A*) becomes detached at leaf maturation. Extension of the rachis separates the leaflets from one another (fig. 18.23,*D*).

The mechanism of formation of the folds in the developing palm leaf is a controversial issue. Some investigators visualize differential marginal growth, followed by intercalary growth, producing alternate ridges and furrows, a combination giving the effect of folds.[36] Others promote the concept that the folds appear as a result of splitting of the tissue, that is, separation of adjacent cell files along the middle lamella of the cell wall.[47] It is likely that both, differential growth and separation between cell layers, are involved in this complex type of leaf development.

If leaf ontogeny varies during different stages of development of a given plant, so that one can distinguish between juvenile and adult leaf forms (chapter 19), the phenomenon is called *heteroblasty* (as contrasted with *homoblasty* when leaves are all alike).

Differentiation of mesophyll

In a bifacial dicotyledon leaf the differentiation of mesophyll begins with an anticlinal elongation of the future palisade cells accompanied by anticlinal divisions (fig. 18.24,*A*). The spongy parenchyma cells also divide anticlinally, but less frequently than the palisade cells. Commonly, they remain approximately isodiametric in form during these divisions. While the divisions are still in progress in the palisade tissue, the adjacent epidermal cells cease dividing and enlarge, particularly in the plane parallel with the surface of the leaf (paradermal plane). As was observed in *Xanthium* leaf, the expansion of epidermal cells lasts longer and proceeds at a higher rate than that of the palisade cells. Thus, eventually, several palisade cells are found to be attached to one epidermal cell (fig. 18.24,*C*). Usually divisions continue longest in the palisade tissue and, accordingly, in *Xanthium* leaf, DNA synthesis in the palisade layer was found to be lasting 3.5 days longer and occurring at a significantly higher rate than in the epidermis and spongy meso-

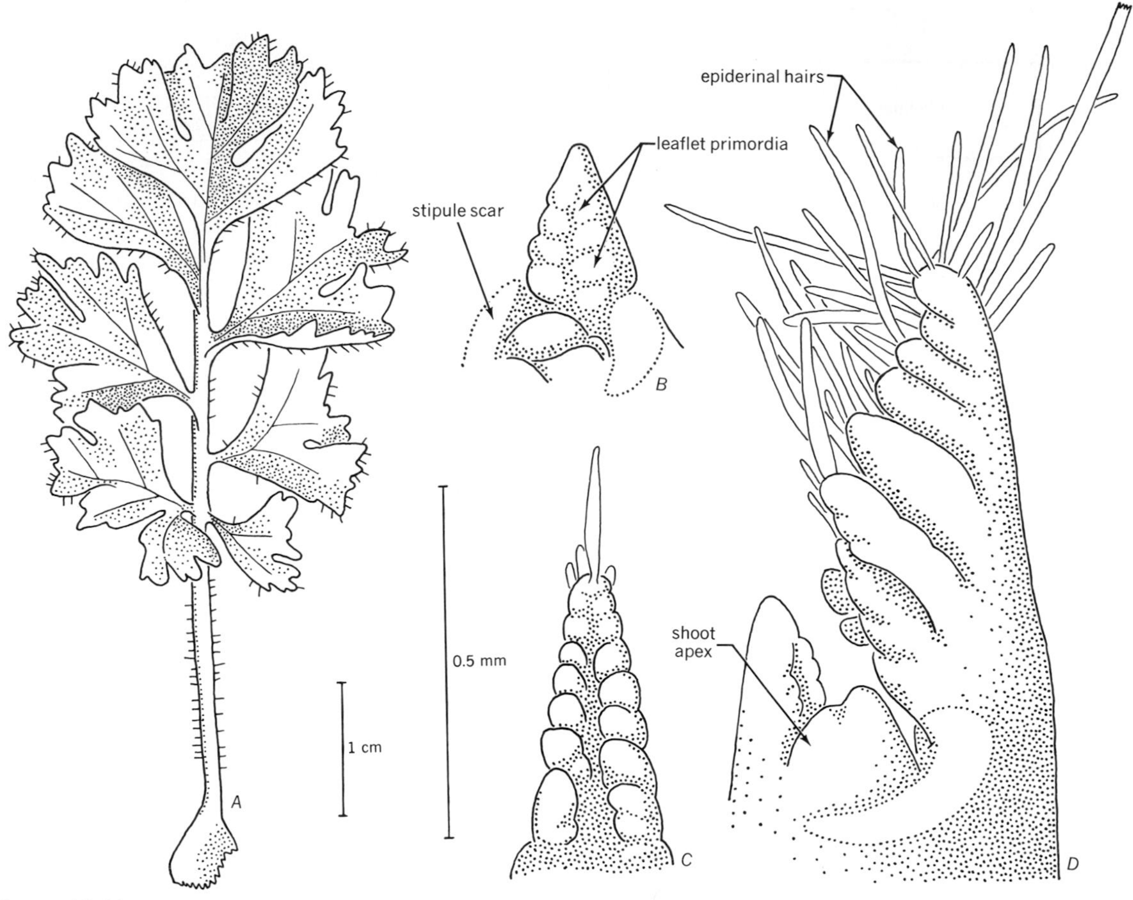

Figure 18.21 Development of leaf in *Pelargonium alternans*. *A*, mature leaf consisting of a rachis and lobed leaflets. *B–D*, three stages in leaflet development. The leaflets arise in marginal position on the primordial axis (*B*) and extend in the adaxial direction toward the shoot apex (*C, D*). Epidermal hairs are prominently displayed in the young leaves (*C, D*). The stipules were cut away during preparation of the specimens. (Redrawn from Hagemann.[20])

phyll.[30] After the divisions are completed a separation of the palisade cells from one another occurs along the anticlinal walls (fig. 18.24,*C*). Formation of intercellular spaces in the spongy parenchyma precedes that in the palisade tissue (fig. 18.24,*B*). The separation of the spongy parenchyma cells is combined with localized growth of cells resulting, in many species, in the development of branched cells. The stomata differentiate concomitantly with or after the development of intercellular spaces in the mesophyll. The expansion of the leaf in surface area is associated with an increase in number and size of chloroplasts and in the amount of chlorophyll.[30,39]

Development of vascular tissues

Vascular development in a leaf of a dicotyledon begins with the differentiation of pro-

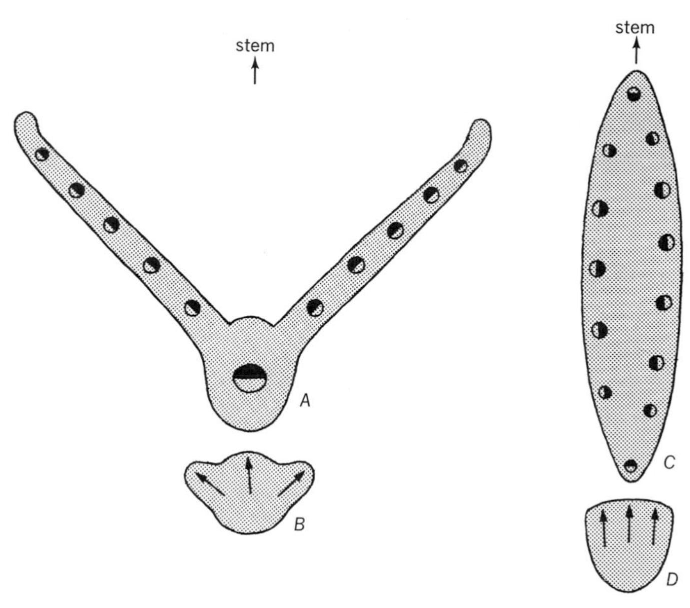

Figure 18.22 Diagrams contrasting transectional profiles of a bifacial (*A, B*) and a unifacial (*C, D*) leaf blades. In an early stage of development, the bifacial leaf shows a differentiation between marginal and adaxial directions of growth (*B*); the bifacial leaf shows no marginal growth but extends in the adaxial direction.

cambium in the future midvein, a phenomenon that may be discerned while the leaf is still in an early primordial stage (fig. 18.14,*F*,*G*). This procambium differentiates acropetally in continuity with the leaf-trace procambium in the axis. The lateral veins of the first order develop from the midrib toward the margins as the leaf expands laterally.[40] Lateral veins of various orders originate among the derivatives of the marginal meristems (fig. 18.12). The larger lateral veins are initiated earlier and closer to the marginal meristems than are the smaller ones. The smallest veins appear among the larger veins first near the apex of the leaf, then successively farther down in consonance with the basipetal maturation of the leaf. Fewer cells are involved in the initiation of the smaller veins than in that of the larger. The smallest veins may be uniseriate in origin; that is they may arise from cell series that are only one cell in diameter.[40] The differentiation of procambium in the blade is usually a continuous process in the sense that successively formed procambial strands arise in continuity with those formed earlier.[40] Formation of new vascular bundles occurs during the entire stage of intercalary growth of the leaf in successively more vacuolated tissue.[12,41]

Xylem differentiation, which can be conveniently studied in cleared leaves, follows a somewhat varied course depending on the manner of growth of the blade.[7] Most of the veins commonly exhibit a continuous progressive maturation of tracheary elements from the larger to the smaller veins, but discontinuities in the process may occur in veins of various categories. The final maturation of xylem, which occurs in the smallest veins, follows the initiation of the last procambium from the apex of the leaf downward.

A method to study phloem differentiation in

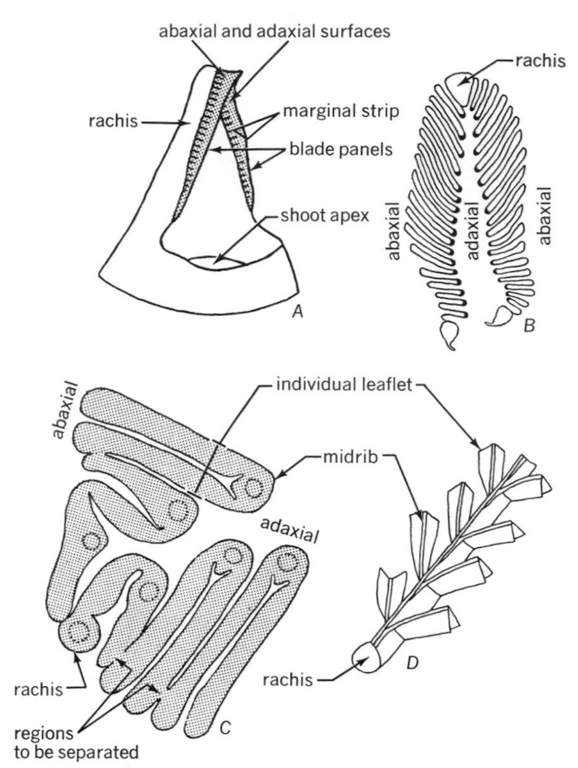

Figure 18.23 Interpretation of leaflet formation in a palm, *Cocos nucifera*. *A,* young leaf differentiated into: leaf base encircling the shoot apex (not shown as such); rachis; and blade panels, which are initiated by marginal growth along two sides of the rachis. The short parallel lines in the panels indicate corrugations. *B,* section through the blade panels approximately tangential to the rachis except that the tip of the rachis is included. Older stage than in *A.* The corrugations appear as two series of zig-zag folds. *C,* higher magnification of a few folds. Each adaxial fold is revealed as a midrib of a leaflet. Each abaxial fold constitutes a junction between two adjacent leaflets, which eventually breaks. *D,* portion of leaf consisting of rachis and bases of leaflets. (*A–C,* constructed from data in Periasamy;[36] *D,* adapted from Tomlinson.[47])

cleared leaf blades has also been developed.[14,15] Its application to a study of *Lupinus* leaflets of various ages has shown that phloem differentiation occurs acropetally and continuously in the median vein and progresses into the larger lateral veins in continuity with the midvein. The phloem of the successively smaller veins likewise develops continuously from the larger to the smaller veins. The interconnections between veins are sometimes established by bidirectional differentiation of phloem in the connecting bundles.

In the monocotyledons, vascular differentiation in leaf blades has been investigated in three grasses, *Zea,*[42] *Lolium,*[10] and *Triticum.*[35] The procambium of the larger veins differentiates acropetally, that of the smaller veins, alternating with the larger, and of the transverse interconnections progresses basipetally. A similar sequence of xylem and phloem differentiation, that is, an earlier acropetal and later basipetal progression, was recorded in all three grasses (chapter 16).

The development of the vascular tissues is correlated with that of other tissues in bringing about the physiological maturation of the leaf. Each leaf progresses from a heterotrophic to an autotrophic state, importing photosynthates during its rapid expansion and exporting photosynthates when its own capacity for photosynthesis is built up.[23,48] The initial export occurs from the early-maturing lamina tip. It is directed away from the leaf, while the still developing lower part of the leaf depends on outside import for nutrients. The import to developing leaves occurs from the older exporting leaves, which supply photosynthates to specific regions of the lamina of the young leaf in accordance with the phyllotactically determined vascular connections between the exporting and the importing leaves.[28] The basipetal development of stomata, minor veins, and intercellular spaces is correlated with a basipetal maturation of the photosynthetic and gaseous exchange systems and a basipetal development of the capacity to export photosynthates.

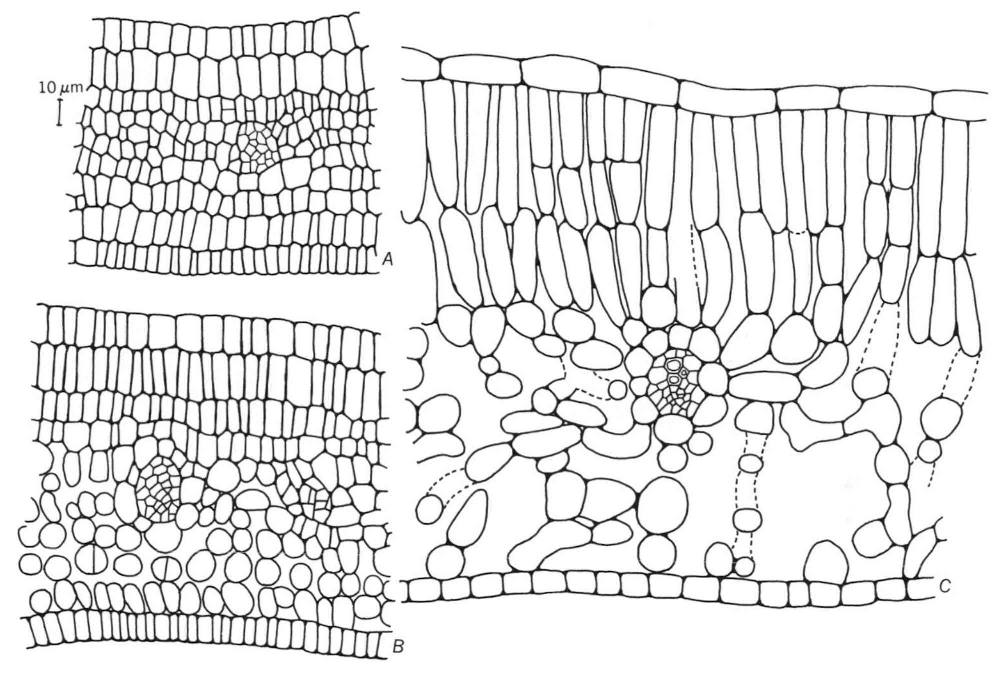

Figure 18.24 Differentiation of mesophyll as seen in cross sections of *Pyrus* leaf. *A,* leaf still compact. *B,* spongy parenchyma with conspicuous intercellular spaces; palisade cells in characteristic orderly arrangement. *C, mature leaf.*

ABSCISSION

The active separation of a leaf from a branch, without injury to the branch, is called leaf abscission. The seasonal defoliation of trees results from leaf abscission, but various injuries also may cause leaf fall. Plant parts other than leaves are shed by abscission likewise. Abscission is an adaptation that serves to remove senile leaves, ripe fruits, and flowers that did not set, and is a means of self-pruning when excess of shoots is present. In connection with the increasing mechanization of agriculture, regulation of abscission in such practices as controlled defoliation, thinning of fruits, and timing of fruit fall is becoming indispensable.

Leaf abscission is commonly prepared near the base of the petiole by cytological and biochemical changes in cells along which the leaf eventually separates from the branch (or a leaflet from a rachis). The tissue region concerned is referred to as *abscission region or abscission zone* (fig. 18.25). Two layers are discernible in the abscission zone, an *abscission,* or *separation, layer,* through which the detachment occurs, and a *protective layer,* which at leaf fall protects from desiccation and invasion by parasites the surface that becomes exposed.

In most leaves, flowers, fruits, and some stems,[22] the preparation of the abscission zone occurs during ontogeny; but it may take place directly in response to conditions that provoke abscission. The abscission zone is often structurally weak in having reduced amounts of sclerified tissue and a vascular system that is condensed in the center in-

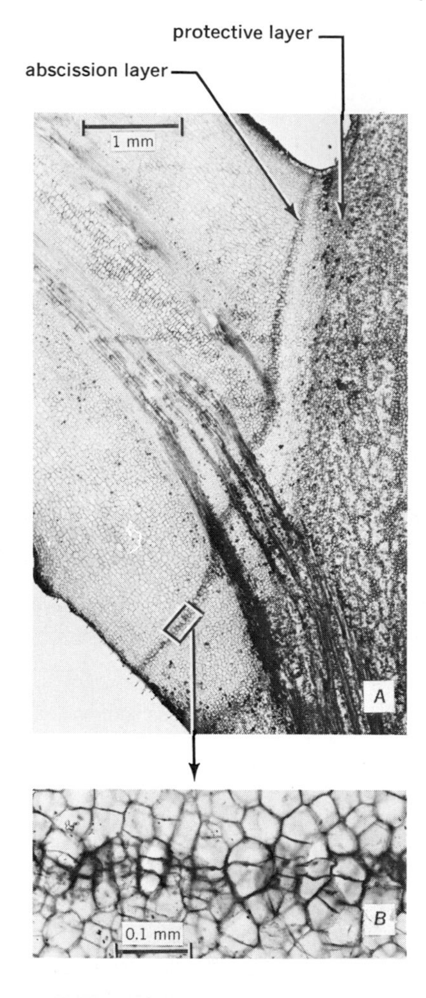

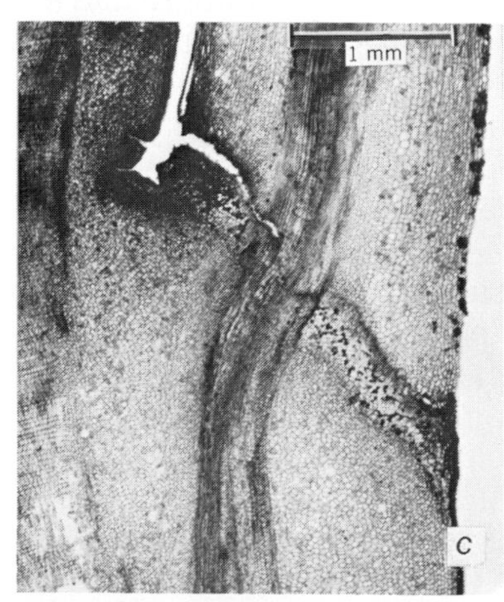

stead of distributed near the periphery. This kind of abscission zone occurs, for example, at the junction of petiole and pulvinus in some leaves.

Several cellular processes have been associated with abscission, not necessarily all in the same species.[19,29,49] Cell division may precede the actual separation (fig. 18.25,*B*), the new cell walls being specifically affected by the subsequent degradation phenomena. Cell enlargement also has been observed in the process. If it occurs in the proximal region, it can introduce shear forces across cell walls in the separation layer. A senescence of cells in the distal region, resulting in the mobilization of nutrients from the leaf into the proximal region, appears to be required in abscission in some species. Lignification of cell walls in the distal region may be one of the aspects of senescence.[38] Tylose formation in tracheary elements and callose deposition in sieve elements and parenchyma cells may occur in advance of abscission. Finally, the central event in abscission, the enzymic degradation of cell walls, causes the actual separation. The wall changes include loss of cementing effectiveness of the middle lamella, partly by removal of calcium, hydrolysis of cellulose walls themselves, and rupture of the sclerfied tracheary elements.

The protective layer, or cicatrice, is formed through deposition of protective substances, such as suberin and wound gum, in the cell walls and the intercellular spaces. In woody species, the protective layer is sooner or later replaced by periderm that develops beneath the protective layer in continuity with the periderm elsewhere in the stem.

Figure 18.25 Abscission zone in walnut, *Juglans (A, B)*, and cherry, *Prunus* (*C*), as seen in longitudinal sections through leaf bases. Cell division has occurred in the abscission layer of walnut (*A, B*), whereas in cherry (*C*) the cells of this layer have begun to separate from each other. (After K. Esau, *Plant Anatomy* 2nd ed. John Wiley & Sons, 1965.)

The shedding of leaves is not necessarily associated with dissolution phenomena. In most monocotyledons and in some herbaceous dicotyledons physical stresses appear to bring about the separation of leaves. Mechanical breakage but no chemical changes have been observed in the shedding of needles in *Picea.*[8]

Much research is conducted on the regulatory agents in abscission.[19,29] The two best known agents are auxin and ethylene. The effects of other natural plant hormones, gibberellin, cytokinin, and abscisic acid, are less well understood.[29] Auxin deters abscission when it is applied after the abscission zone has developed but has not yet undergone any structural weakening. Auxin can also inhibit the differentiation of the abscission zone. Ethylene appears to be the major agent driving abscission to completion. The need for enzyme synthesis in abscission, especially of cell-wall degradation and respiration enzymes, is generally recognized. Peroxidase brings about an increased synthesis of ethylene, and phosphatase is associated with the aging phenomena.[19,38]

REFERENCES

1. Aykin, S. Hygromorphic stomata in xeromorphic plants. *Istanbul Univ. Rev. Fac. Sci. Ser. B. Sci. Nat.* 18:75–90. 1952.
2. Blackman, E. The morphology and development of cross veins in the leaves of bread wheat (*Triticum aestivum* L.). *Ann. Bot.* 35:653–665. 1971.
3. Cataldo, D. A. Vein loading: the role of the symplast in intercellular transport of carbohydrate between the mesophyll and minor veins of tobacco leaves. *Plant Physiol.* 53:912–917. 1974.
4. Crowdy, S. H., and T. W. Tanton. Water pathways in higher plants. I. Free space in wheat leaves. *J. Exp. Bot.* 21: 102–111. 1970.
5. Denne, M. P. Leaf development in *Trifolium repens. Bot. Gaz.* 127:202–210. 1966.
6. Esau, K. *Plant anatomy.* 2nd ed. New York, John Wiley & Sons. 1965.
7. Esau, K. *Vascular differentiation in plants.* New York, Holt, Rinehart and Winston. 1965.
8. Facey, V. Abscission of leaves in *Picea glauca* (Moench.) Voss and *Abies balsamea* L. *Proc. North Dakota Acad. Sci.* 10:38–43. 1956.
9. Fischer, A. *Untersuchungen über das Siebröhren-System der Cucurbitaceen.* Berlin, Gebrüder Borntraeger. 1884.
10. Forde, B. J. Differentiation and continuity of the phloem in the leaf intercalary meristem of *Lolium perenne. Amer. J. Bot.* 52:953–961. 1965.
11. Foster, A. S. Leaf differentiation in angiosperms. *Bot. Rev.* 2:349–372. 1936.
12. Foster, A. S. Foliar venation in angiosperms from an ontogenetic standpoint. *Amer. J. Bot.* 39:752–766. 1952.
13. Foster, A. S. The morphology and relationships of *Circaeaster. J. Arnold Arb.* 44:299–327. 1963.
14. Fuchs, C. Recherches ontogéniques sur le phloème de la foliole du *Lupinus albus* L. *C. R. Acad. Sci. Paris D.* 262:91–94. 1966.
15. Fuchs, C. Recherches ontogéniques sur le phloème des nervures d'ordre supérieur a 2 dans la foliole du *Lupinus albus* L. *C. R. Acad. Sci. Paris D* 262: 752–755. 1966.
16. Fuchs, C. Observations sur l'extension en largeur du limbe foliaire du *Lupinus albus* L. *C. R. Acad. Sci. Paris D* 263:1212–1215. 1966.
17. Fuchs, C. Localisation des divisions dans le méristème marginal des feuilles

des *Lupinus albus* L. *C. R. Acad. Sci. Paris D* 267:722–725. 1968.

18. Geiger, D. R., R. T. Giaquinta, S. A. Sovonick, and R. J. Fellows. Solute distribution in sugar beet leaves in relation to phloem loading and translocation. *Plant Physiol.* 52:585–589. 1973.
19. Hagemann, P. Histochemische Muster beim Blattfall. *Ber. Schweiz. Bot. Ges.* 81:97–138. 1971.
20. Hagemann, W. Studien zur Entwicklungsgeschichte der Angiospermenblätter. *Bot. Jb.* 90:297–413. 1970.
21. Hickey, L. J. Classification of the architecture of dicotyledonous leaves. *Amer. J. Bot.* 60:17–33. 1973.
22. Höster, H. R., W. Liese, and P. Böttcher. Untersuchungen zur Morphologie und Histologie der Zweigabwürfe von *Populus* "Robusta." *Forstwiss. Centralbl.* 87:321–384. 1968.
23. Isebrands, J. G., and P. R. Larson. Anatomical changes during leaf ontogeny in *Populus deltoides*. *Amer. J. Bot.* 60:199–208. 1973.
24. Jeune, B. Observations et expérimentation sur les feuilles juvéniles du *Paulownia tomentosa* H. Bn. *Bull Soc. Bot. France* 119:215–230. 1972.
25. Kaplan, D. R. Comparative foliar histogenesis in *Acorus calamus* and its bearing on the phyllode theory of monocotyledonous leaves. *Amer. J. Bot.* 57:331–361. 1970.
26. Kaplan, D. R. The monocotyledons: their evolution and comparative biology. VII. The problem of leaf morphology and evolution in the monocotyledons. *Quart. Rev. Biol.* 48:437–457. 1973.
27. Klaus, H. Ontogenetische und histogenetische Untersuchungen an der Gerste (*Hordeum distichon* L.) *Bot. Jb.* 85: 45–79. 1966.
28. Larson, P. R., and R. E. Dickson. Distribution of imported ^{14}C in developing leaves of eastern cottonwood according to phyllotaxy. *Planta* 111:95–112. 1973.
29. Leopold, A. C. Physiological processes involved in abscission. *Hort. Sci.* 6: 376–378. 1971.
30. Maksymowych, R. *Analysis of leaf development.* Cambridge University Press. 1973.
31. O'Brien, T. B., and D. J. Carr. A suberised layer in the cell walls of the bundle sheath of grasses. *Aust. J. Biol. Sci.* 23:275–287. 1970.
32. Owens, J. N. Initiation and development of leaves in Douglas fir. *Can J. Bot.* 46:271–278. 1968.
33. Pate, J. S., and B. E. S. Gunning. Vascular transfer cells in angiosperm leaves. A taxonomic and morphological survey. *Protoplasma* 68:135–156. 1969.
34. Pate, J. S., and B. E. S, Gunning. Transfer cells. *Ann. Rev. Plant Physiol.* 23:173–196. 1972.
35. Patrick, J. W. Vascular system of the stem of wheat plant. II. Development. *Aust. J. Bot.* 20:65–78. 1972.
36. Periasamy, K. Morphological and ontogenetic studies in palms—I. Development of the plicate condition in the palm-leaf. *Phytomorphology* 12:54–64. 1962. II. Growth pattern of the leaves of *Cocos nucifera* and *Borassus flabellifer* after the initiation of plications. *Aust. J. Bot.* 13:225–234. 1965.
37. Philpott, J. A blade tissue study of leaves of forty-seven species of *Ficus*. *Bot. Gaz.* 115:15–35. 1953.
38. Poovaiah, B. W. Formation of callose and lignin during leaf abscission. *Amer. J. Bot.* 61:829–834. 1974.
39. Possingham, J. V., and W. Saurer. Changes in chloroplast number per cell during leaf development in spinach. *Planta* 86:186–194. 1969.

40. Pray, T. R. Foliar venation of angiosperms. I. Mature venation of *Liriodendron. Amer J. Bot.* 41:663–670. 1954. II. Histogenesis of the venation of *Liriodendron. Amer. J. Bot.* 42:18–27. 1955. III. Pattern and histology of the venation of *Hosta. Amer J. Bot.* 42:611–618. 1955. IV. Histogenesis of the venation of *Hosta. Amer. J. Bot.* 42:698–706. 1955.
41. Schneider, R. Histogenetische Untersuchungen über den Bau der Laubblätter, insbesondere ihres Mesophylls. *Österr. Bot. Z.* 99:253–285. 1952.
42. Sharman, B. C. Developmental anatomy of the shoot of *Zea mays* L. *Ann. Bot.* 6:245–282. 1942.
43. Shushan, S., and M. A. Johnson. The shoot apex and leaf of *Dianthus caryophyllus* L. *Bull. Torrey Bot. Club* 82: 266–283. 1955.
44. Stewart, R. N., and H. Dermen. A new leaf: flexibility in ontogenesis shown by analysis of the contribution of derivatives of the shoot apical layers to leaves on periclinally chimeral plants. *Amer. J. Bot.* 62:935–947. 1975.
45. Tanton, T. W., and S. H. Crowdy. Water pathways in higher plants. III. The transpiration stream within leaves. *J. Exp. Bot.* 23:619–628. 1972.
46. Thomasson M. Quelque observations sur la répartition des zones de croissance de la feuille du *Jasminum nudiflorum* Lindl. *Candollea* 25:297–340. 1970.
47. Tomlinson, P. B. *Anatomy of the monocotyledons.* II. *Palmae.* Oxford Clarendon Press. 1961.
48. Turgeon, R., and J. A. Webb. Leaf development and phloem transport in *Cucurbita pepo. Planta* 113:179–191. 1973.
49. Webster, B. D. A morphogenetic study of leaf abscission in *Phaseolus. Amer. J. Bot.* 57:443–451. 1970.
50. Wylie, R. B. The bundle sheath extension in leaves of dicotyledons. *Amer. J. Bot.* 39:645–651. 1952.

19

The Leaf: Variations in Structure

LEAF STRUCTURE AND ENVIRONMENT

Plants growing in different habitats show structural differences that are commonly interpreted as evolutionary adaptations to the conditions of the specific habitat. But different plants growing in the same ecologic niche overcome the adverse conditions of the particular environment in different ways. In a habitat deficient in water, for example, some plants develop features protecting the aerial parts from excessive loss of water, others form underground water storage organs or develop roots reaching great depths, and still others control the problem by having a short life span restricted to the time when water supply is most abundant. Thus, one character may serve as well as another in enabling the plant to become adapted to a given environmental condition and, consequently, plants can reach the same overall level of adaptation by different character combinations.[7,37]

Availability of water is an especially important factor affecting the form and structure of plants. On the basis of their water relations plants are usually classified as xerophytes, mesophytes, and hydrophytes. The xerophytes are adapted to dry habitats; mesophytes require abundant available soil water and a relatively humid atmosphere; and hydrophytes (or hygrophytes) depend on a large supply of moisture or grow partly or completely submerged in water. The structural features typical of plants of the various habitats or plants having such features are referred to as xeromorphic, mesomorphic, or hydromorphic, respectively. The peculiarities distinguishing plants of the various habitats are most striking in leaves.

The features commonly interpreted as characteristic of mesophytes and hydrophytes are pointed out in the descriptions of the various examples of leaves later in the chapter. The xeromorphic characters, however, are given much attention in the literature and are, therefore, reviewed separately in some detail.

XEROMORPHY

One of the most prevalent characteristics of xeromorphic leaves is the high ratio of volume to surface; that is, the leaves are small

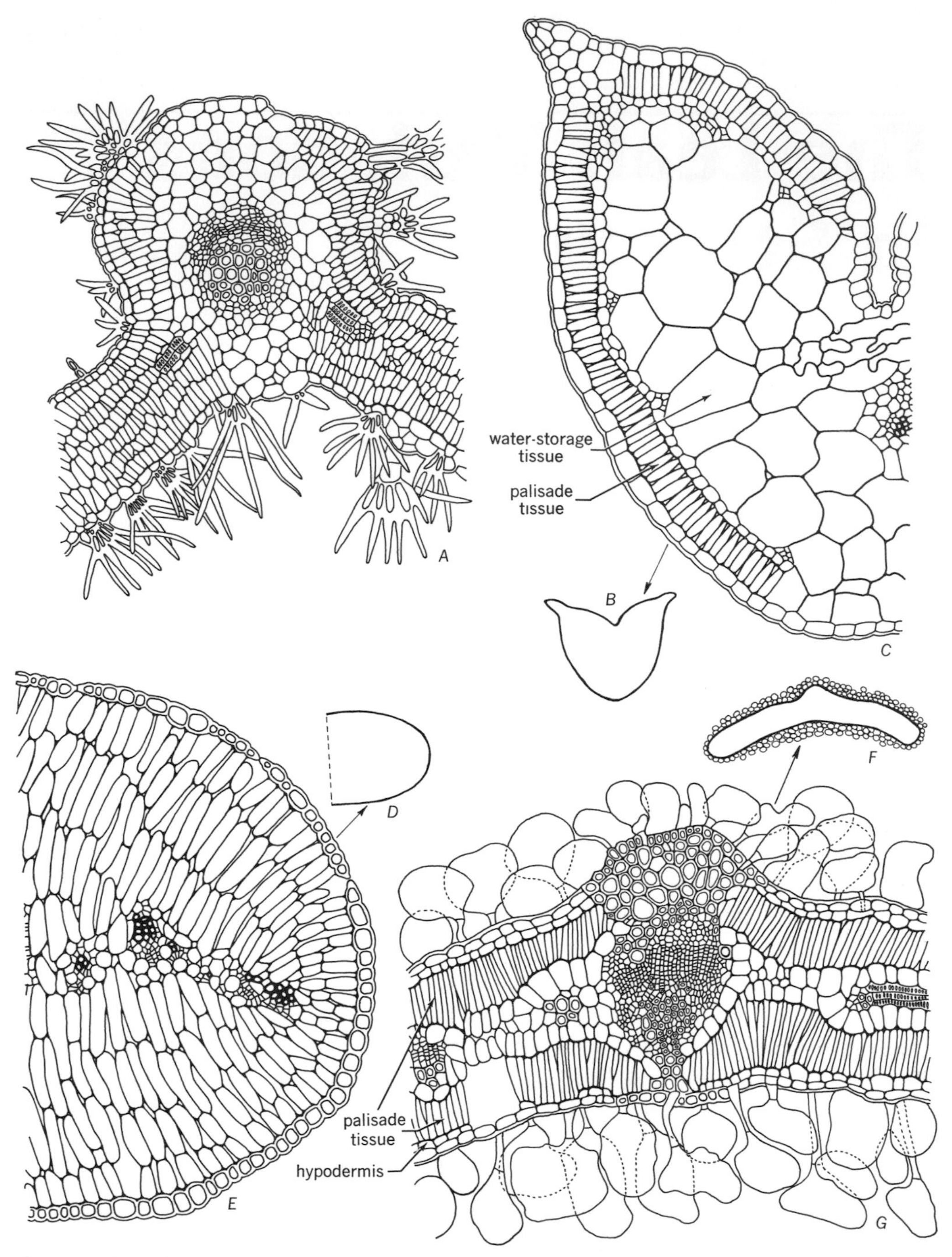

Figure 19.1 Cross sections of leaves showing various xeromorphic features. *A, Sphaeralcea incana,* entire mesophyll differentiated as palisade tissue; also trichomes. *B, C, Salsola kali,* succulent leaf with a large-celled water tissue enclosed by a single layer of palisade parenchyma. Some water cells to the right are shown in a wilted condition—response to depletion of water supply. *D, E, Greggia camporum,* low ratio of surface to volume, and the entire mesophyll is differentiated as palisade tissue. *F, G, Atriplex canescens,* vesicular hairs and isobilateral mesophyll. (Redrawn from Shields.[41])

and compact.[41,45] This character is associated with distinct internal characteristics such as thick mesophyll, with the palisade tissue more strongly developed than the spongy parenchyma, or present alone (fig. 19.1,*A*,*E*,*G*); small intercellular-space volume; compact network of veins and the related low frequency of bundle-sheath extensions; high stomatal frequency; and sometimes small cells.

A xerophytic flora may also have a high proportion of representatives with leaves having a hypodermis,[51] a tissue with few or no chloroplasts (fig. 19.1,*G*). Mechanical strengthening of leaves through abundant development of sclerenchyma, common among xerophytes, is thought to reduce the injurious effect of wilting,[45] and, accordingly, much sclerenchyma is found in plants of habitats with continuous or periodic dryness, such as the hot deserts.[47] Thick cell walls, especially in the epidermis (fig. 19.1,*E*), and thick cuticles are often recorded in xerophytic plants, but generally cuticle thickness is variable. Stomata may occur in cavities, stomatal crypts (*Nerium oleander*), or in grooves (Ericales[23]), lined with epidermal hairs. Some xerophytes are succulent plants with their own peculiar histologic features, especially the presence of a water-storage tissue (fig. 19.1,*C*) and paucity of vascular tissue.

Trichomes are abundant in many xerophytes (fig. 19.1,*A*,*G*), and, if the same pubescent species has mesophytic and xerophytic forms, the latter usually have a denser covering of hairs. The effect of environment on trichome development was studied in an aquatic plant, *Polygonum amphibium*. Lateral buds developing in air gave rise to densely pubescent shoots, whereas those developing in contact with water or under water-saturated soil remained glabrous throughout the season[33]. Studies on the effect of trichomes upon loss of water have given variable results, but it is likely that sometimes the trichomes have a role in insulating the mesophyll from excessive heat.[5]

Xeromorphic features (and other ecotypic features) show variable degrees of constancy, but they may be well fixed genetically in a given species. Environmental factors, however, may induce a degree of xeromorphy in normally mesomorphic leaves or intensify the xeromorphic characters in xerophytes.[41,47] The deficiency of moisture is only one such factor. Nutrient deficiencies and cold may induce stronger expression of xeromorphy than lack of moisture.[45] Succulence, for example, is increased when nitrogen is deficient. It may also develop in shore plants exposed to a spray of sea water[8] (fig. 19.2,*A*,*B*), apparently under the effect of sodium chloride.[6]

Another important formative factor is light. Leaves developing in light of high intensity show a higher degree of xeromorphy than those protected from light. This developmental reaction is the basis for the differentiation between sun and shade leaves. It has frequently been observed that leaves developing in direct sunlight are smaller but thicker and have a more strongly differentiated palisade tissue than leaves developing in the shade[50] (fig. 19.2,*C*–*E*).

LEAF MORPHOLOGY AND POSITION ON PLANT

Progressively developing leaves on a plant show more or less pronounced differences in morphology. The leaves undergo a heteroblastic development and hence reveal a structural gradient in the plant. Changes in leaf form during the ontogeny of a plant often occur in seedling stages so that there is a difference between juvenile and adult forms. In plants with compound leaves, those in the seedling may be entire (*Fragaria, Fraxinus*); or, on the contrary, the early leaves may be compound, the later ones simple by suppression of leaflet development (*Acacia*). In *Del-*

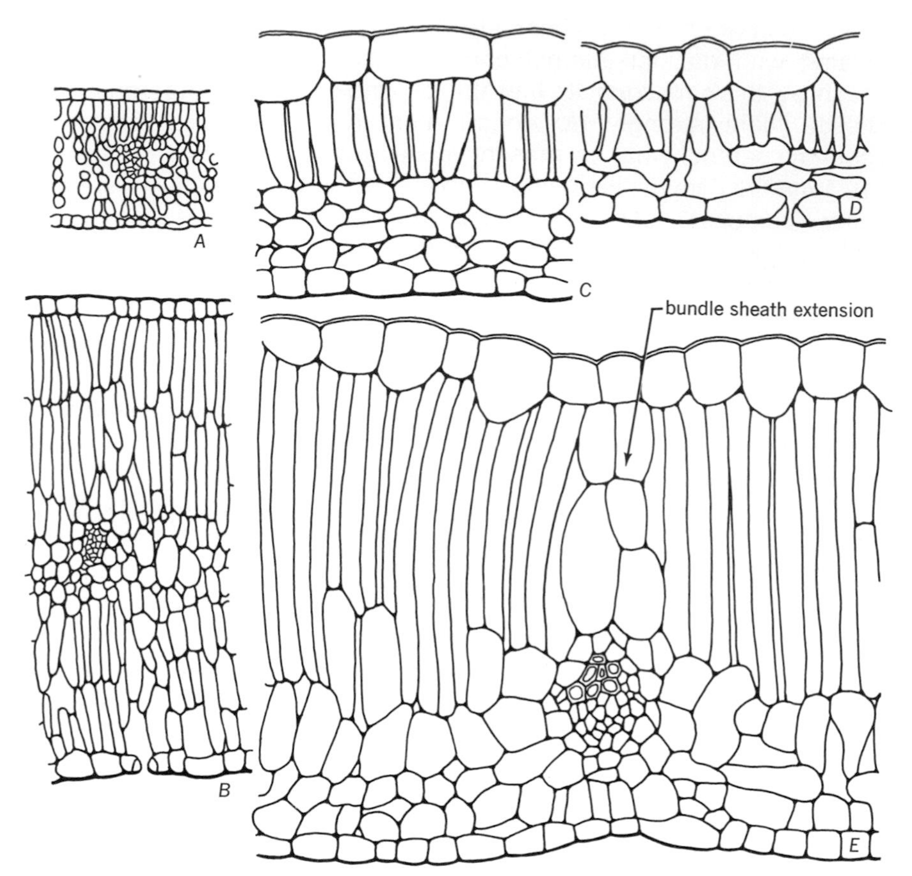

Figure 19.2 Effect of environmental factors on leaf structure. *A, B, Baccharis halimifolia,* leaf blades in cross sections showing normal leaf (*A*) and a succulent leaf (*B*), both taken from the same plant, except that *B* was derived from the side exposed to ocean spray. (Redrawn from Boyce.[8]) *C–E, Acer platanoides,* leaf blades in cross sections, all from one tree, showing effect of light upon structure of mesophyll: *C,* from interior of crown, moderately shaded; *D,* from deep interior, strongly shaded; *E,* from a sunny location. (Redrawn from Wylie.[50])

phinium, the number of segments of the divided lamina increases with the age of the plant, and in *Ulmus,* the number of marginal teeth increases. In some species, heteroblastic development results in a change in the nature of the lateral appendage. The juvenile leaves have the form of normal foliage leaves, whereas the later leaves are modified as phyllodes (*Oxalis*), spines (*Ulex*), or tendrils (*Lathyrus aphaca*).

In perennial woody plants, the first leaves on a new shoot are the bud scales, or cataphylls, which are quite distinct from the foliage leaves succeeding them, although forms transitional between the two kinds of leaves may be produced. Still another change in leaf form occurs during the transition from vegetative to reproductive stage of growth when the foliage leaves are succeeded by bracts subtending inflorescence branches and the floral structures.

The causal factors in positional changes in

leaves appear to be varied and complex but, in general, heteroblastic development is interpreted as a response of the apical meristem to a changing physiological condition in the plant.[1] External factors, such as daylength, temperature, availability of water and nutrients affect the physiologic state of the plant so that production of bud scales is a seasonal event and the appearance of floral bracts is correlated with the photoperiodic induction of the reproductive stage.

In *Ipomoea,* in which the juvenile leaves are entire and the adult leaves lobed, sugars were found to have an accelerating effect on the development of the adult form, whereas casein hydrolysate retarded the phenomenon. Certain growth retardants also promoted formation of the adult leaf . The author of the study[34] recalled that a hypothesis about the controlling effect of carbohydrate nutrition in heteroblastic development was proposed early in the century by Goebel.[19]

Heteroblastic leaf development is affected by endogenous factors such as hormone levels, for example those of gibberellinlike substances,[16] and by some undetermined influences of older leaves upon younger leaves. In ivy (*Hedera helix*), in which the differences between the juvenile and adult forms of shoots are expressed not only in leaf morphology but also in the growth habit of the plant, the mature form can be induced to produce juvenile shoots by application of gibberellic acid.[38]

Internally, earlier leaves have a simpler, less differentiated structure than the later-formed leaves. The difference may be especially pronounced in the development of palisade parenchyma. In later leaves, the palisade tissue may undergo more anticlinal divisions, contain more cell layers, and have longer cells than in the earlier leaves.[40] In Douglas fir, differentiation between cataphylls and foliage leaves occurs during the early enlargement of the primordia, and the cataphylls develop a simple mesophyll without a distinction between palisade and spongy parenchyma.[36] The structural modification of leaves at successively higher levels is sometimes described as an increase in xeromorphy.[1]

DICOTYLEDON LEAVES

VARIATIONS IN MESOPHYLL STRUCTURE

Many herbaceous dicotyledons have leaves with a relatively undifferentiated mesophyll. The palisade tissue is absent or weakly developed; the intercellular volume is large; the leaf is often thin; the epidermis bears a thin cuticle; and the stomata are more or less raised. When strongly expressed such features characterize hydromorphic leaves, but they are also found, in varying degrees, in herbaceous plants growing in conditions of more moderate amounts of available moisture. Examples of leaves with relatively undifferentiated mesophyll are those of *Pisum sativum, Linum usitatissimum,* and *Lactuca sativa.* In the sugar beet (*Beta vulgaris*) the cell form in the mesophyll is associated with leaf thickness. In thin leaves the mesophyll consists of short rounded cells, in thick leaves most of the cells are elongated.

A thin loosely organized mesophyll with a single row of palisade cells is found in *Ipomoea batatas, Pastinaca sativa* (fig. 19.3,*A*), *Raphanus sativus, Solanum tuberosum,* and *Lycopersicon esculentum.* The similarly constructed leaves of *Cannabis sativa* (chapter 18) and *Humulus lupulus* have cystolith-containing cells in the epidermis and numerous trichomes, glandular and nonglandular. In alfalfa (*Medicago sativa*) the palisade consists of two rows of rather short cells. In another genus of the Fabaceae, the soybean (*Glycine*), the mesophyll cells occupying the interveinal space consist of long

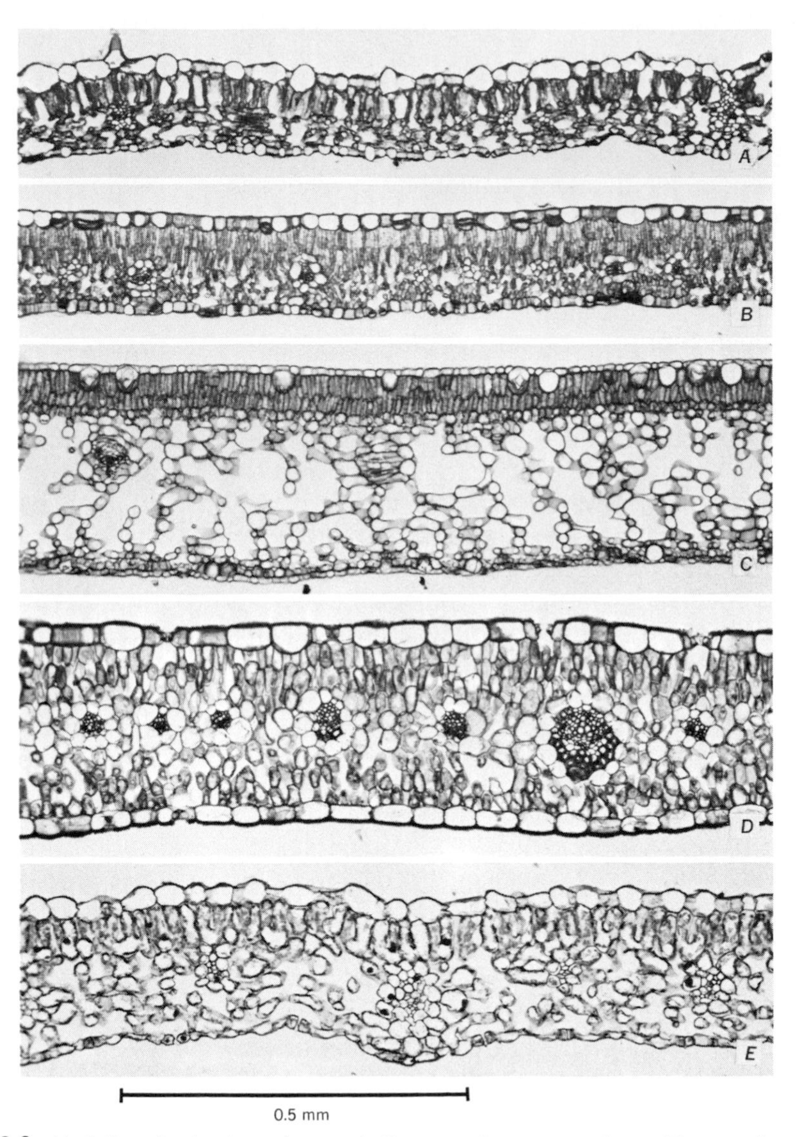

Figure 19.3 Variations in structure of mesophyll as seen in cross sections of leaves. *A*, parsnip (*Pastinaca sativa*). *B*, peach (*Prunus persica*). *C*, lemon (*Citrus limon*). *D*, *Dianthus*. *E*, *Lilium*.

multiarmed cells which are connected to the bundle sheath cells at the level of the phloem.[49] The thin leaves of *Gossypium* (cotton) have long palisade cells that occupy approximately one-third to one-half of the blade thickness. The cotton leaf has lysigenous glands in the mesophyll and pitlike nectaries with club-shaped papillae on the abaxial surface of the main vein ribs.

Various shrubby and woody species furnish

examples of leaves with well-differentiated palisade parenchyma on the adaxial side of the leaf, that is, typical dorsiventral mesomorphic leaves (fig. 19.3,*B*; *Vitis, Syringa, Ligustrum, Pyrus*), as well as leaves with various combinations of xeromorphic features. The leaves of *Citrus* are thick and leathery and have a thick cuticle with wax layers. The compact palisade contrasts strikingly with the loose thick spongy parenchyma (fig. 19.3,*C*). Lysigenous cavities occur in the mesophyll (chapter 13). *Ficus* leaves have a chlorophyll-free hypodermis derived from the epidermis (multiple epidermis). They also contain cystoliths in the epidermis (chapter 13) and laticifers in the mesophyll.

The isobilateral mesophyll with palisade parenchyma on both sides of the leaf (a xeromorphic character) is exemplified by *Artemisia, Atriplex* (fig. 19.1,*G*), *Chrysothamnus, Sarcobatus,* and many other genera of various families.[32] A modification of the isobilateral form is the centric[32] found in leaves that are very narrow or close to cylindrical. In such leaves the abaxial and adaxial palisade cells form an almost continuous layer (fig. 19.1,*C*). A nearly isobilateral leaf may also result when the palisade and the spongy tissues are not clearly differentiated (fig. 19.3,*D*).

In *Salsola,* the palisade tissue surrounds a parenchyma of large colorless cells interpreted as water storage tissue (fig. 19.1,*C*). Such tissue may occur also in median position in flat leaves[41] (*Haplopappus spinulosus*) and, outside the mesophyll, as hypodermis or multiple epidermis. In the fleshy leaves of *Peperomia* the multiple epidermis may be fifteen layers of cells in depth, exceeding the thickness of the mesophyll.[30]

The leaves of aquatics vary in relation to their growth conditions. Some have partially or completely submerged leaves; others have floating leaves; and still others combinations of floating and submerged leaves. The submerged leaves are commonly highly dissected and thus contrast with the less dissected or entire leaves growing above the water on the same plant. The formation of dissected leaves is a direct response to environmental conditions and can be induced experimentally by submerging shoots or by changing daylight conditions.[14] Common features of hydromorphic leaves are large air spaces (fig. 19.6,*A*), small amount of sclerenchyma, and a weak development of the vascular system.

SUPPORTING TISSUE

In the dicotyledons the supporting tissue in leaves may be collenchyma or sclerenchyma, and the vascular bundles as such also contribute to the support of the blade. The collenchyma occurs along the larger veins, on one or both sides (chapter 18), and the nonconducting part of the xylem and phloem may have collenchymatously thickened cell walls. Sclerenchyma occurs in the form of "bundle caps" (in transections), bundle sheaths, and bundle-sheath extensions composed of fibrous cells, and as sclereids in the mesophyll. Examples of vascular bundles accompanied by sclerenchyma are found in the Boraginaceae, Caryophyllaceae (fig. 19.3,*D*), Lamiaceae, Lauraceae, all families of Fabales, Monimiaceae, Proteaceae, some Rosaceae, and Sterculiaceae.[32]

PETIOLE

The petiole of dicotyledon leaves contains the same tissues as the stem, often in similar arrangement. The epidermis has some stomata and the ground tissue contains chloroplasts. Collenchyma or sclerenchyma occur as supporting tissues. The vascular

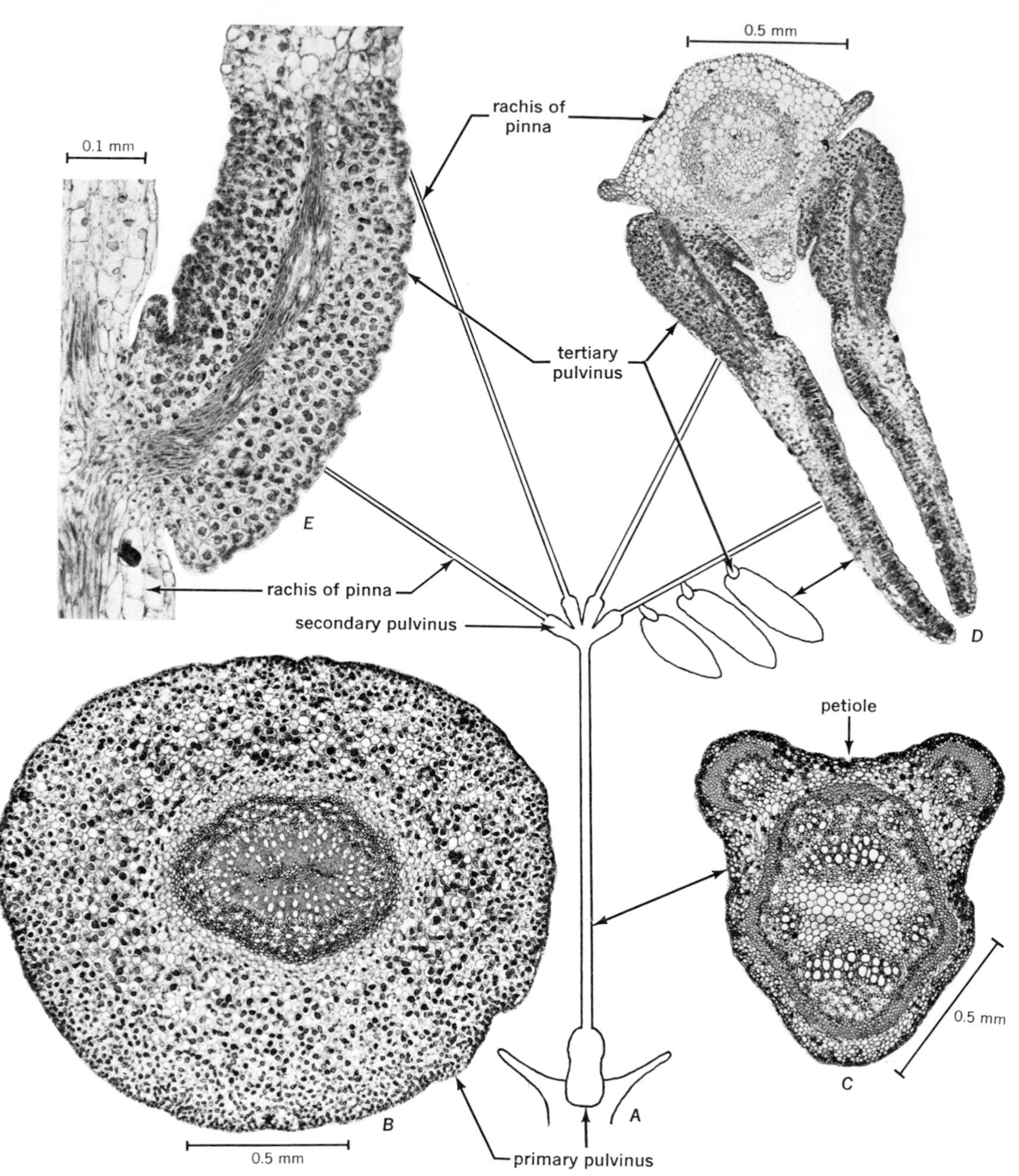

Figure 19.4 Structure of pulvini in *Mimosa pudica*. *A*, diagram showing distribution of pulvini. Upon stimulation leaflets "close" by action of tertiary pulvini at their bases, pinnae approach each other by action of secondary pulvini at bases of rachises, leaf bends down at primary pulvinus. *B*, transection of primary pulvinus; parenchyma at its lower (abaxial) side contracts upon stimulation. *C*, transection of petiole above pulvinus. In *B* and *C*, the adaxial (upper) side is above. *D*, transection through rachis of pinna and two leaflets in "closed" condition. *E*, longitudinal section of tertiary pulvinus in stimulated state: the adaxial side (left) is contracted. (*B* and *C* from K. Esau, *Ann. Bot.* 34:505–515, 1970.)

tissue shows a great variety of arrangements (chapter 18).

Certain plants have jointlike thickenings, the pulvini, at the base of petioles and also at the bases of the petiolules in compound leaves (fig. 19.4). The pulvini are involved in movements of leaves which occur in response to a variety of endogenous and exogenous stimuli. The best known pulvini are those of the Fabales.[9,39,48] The leaflets of the compound leaves of various Fabales move away from each other in the light and approach each other in the dark (nyctinastic movement). This "opening" and "closing" of leaves is a circadian phenomenon and persists even when plants are kept continuously in light or in darkness. The well-known movements of leaves of *Mimosa pudica* in response to touch or other external stimuli are also mediated by the activity of pulvini.[48]

The pulvinus is somewhat swollen and its surface wrinkled. The vascular system is concentrated in the center even though above and below it the vascular tissue is disposed close to the periphery of the petiole (fig. 19.4). The largest volume is occupied by thin-walled parenchyma with intercellular spaces. Stomata are few or absent. Trichomes may be present.

The movement is associated with changes in turgor and the concomitant contractions and expansions of the ground-parenchyma cells on the opposite sides of the pulvinus. The leaflets close when the adaxial cells of the pulvinus contract, open when the abaxial cells on the opposite sides of the pulvinus. The leaflets close when the adaxial cells of stimulation. As was shown for *Albizzia,* the turgor changes in the pulvinus are caused by a flux of potassium. Elemental analyses and microprobe studies have shown that potassium moves from the contracting cells into the expanding cells in measurable amounts sufficient to exert a significant osmotic effect.[39] The flux of potassium is the basis for movement of leaves whether these are controlled by phytochrome or by an endogenous rhythm. The relationship agrees with the concept that oscillatory alterations in membrane permeability have a key role in leaf movements.[13]

MONOCOTYLEDON LEAVES

The leaves of the monocotyledons vary in form and structure, and some resemble those of the dicotyledons. Monocotyledon leaves may have petioles and blades (*Canna, Zantedeschia, Hosta*), but the majority are differentiated into blade and sheath, and the blade is relatively narrow. The venation is typically parallel.

The anatomic structure ranges from hydromorphic to extreme xeromorphic. Hydrophytes in the monocotyledons show the same basic features as those in the dicotyledons, especially with regard to the abundance of aerenchyma. In the Butomaceae, at least 80 percent of plant volume is occupied by intercellular spaces.[46] An example of a dorsiventral leaf with the palisade parenchyma on the adaxial side is found in *Lilium* (fig. 19.3,*E*). The dorsiventral leaf of banana (*Musa sapientum*) is thick and has several layers of palisade and a wide region of spongy parenchyma with large lacunae.[43] The rigid leaves of *Carex* (fig. 19.5) have prominently developed sclerenchyma and, in the mesophyll, air cavities containing large thin-walled cells.

Many monocotyledons have partly unifacial leaves resulting from replacement of marginal growth by adaxial growth in the leaf primordium stage (chapter 18). The *Iris* leaf, for example, has a unifacial blade flattened not parallel with the tangent of the axis but perpendicular to it. Its vascular bundles appear partly in one file, partly in two files close together, and half of the bundles are oriented

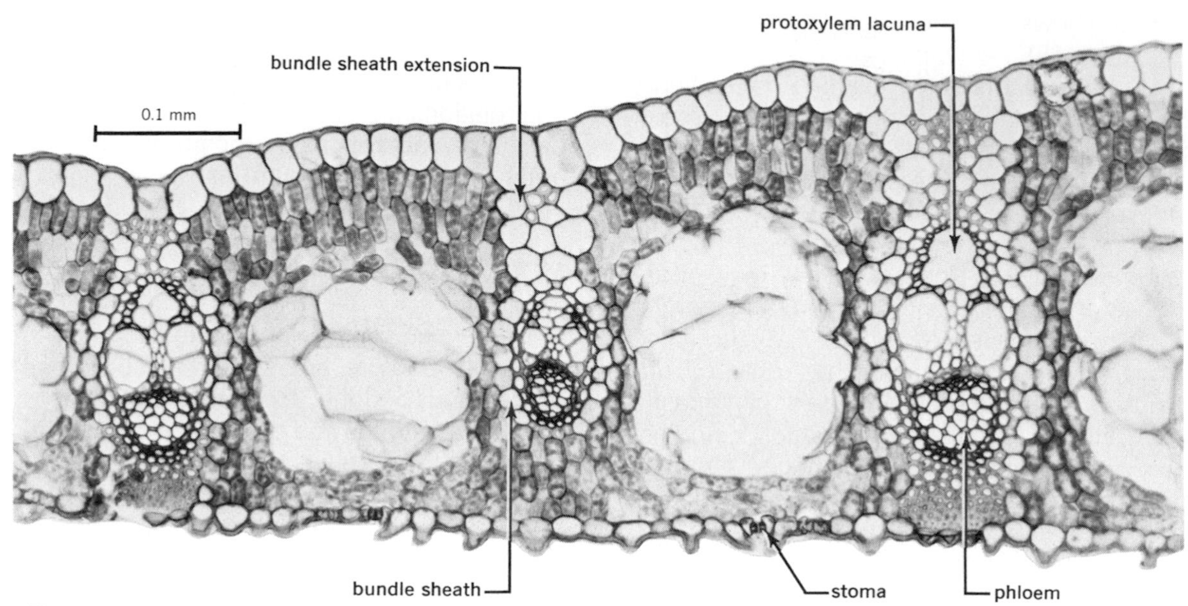

Figure 19.5 Cross section of *Carex* leaf illustrating air cavities containing large thin-walled cells without chloroplasts. (Slide courtesy of J. E. Sass.)

with their xylem to one side, the other half to the other side (fig. 19.6,*D*). The sheath of this leaf develops by means of marginal meristematic activity and its bundles appear in the usual position in the two halves of the leaf (fig. 19.6,*E*). Certain species of *Allium* have tubular leaves (fig. 19.6,*B,C*). The palisade tissue appears next to the epidermis around the entire circumference, and beneath it is the spongy parenchyma. The center of the leaf is occupied by a cavity surrounded by remnants of parenchyma cells that initially occupied the region of the cavity.

Numerous monocotyledon leaves develop large amounts of sclerenchyma, which in some species serves as an important source of commercial hard leaf fibers (chapter 6). The fibers are associated with the vascular bundles or appear as independent strands.

GRASS LEAVES

The grass leaf typically consists of a more or less narrow blade and a sheath enclosing the stem. Auricules and a ligule commonly occur between the blade and the sheath. Vascular bundles of different sizes alternate rather regularly with one another (figs. 19.7, and 19.8,*A,B*) and are interconnected by small transverse commissural strands (chapter 18). The median bundle may be the largest (fig. 19.8,*A*) or the median part of the blade is thickened on the adaxial side.[15]

The mesophyll of grasses shows, as a rule, no distinct differentiation into palisade and spongy parenchyma, although sometimes the cell rows beneath both epidermal layers are more regularly arranged than the rest of the mesophyll. In some grasses, especially in the Eragrostoideae and Panicoideae, the mesophyll cells surround the vascular bundles in an orderly manner, each cell oriented with its longer diameter at right angles to the bundle so that in transverse sections the mesophyll cells appear to radiate from the bundles (figs. 19.7, and 19.8,*A,B*). The mesophyll cells may be typically parenchymatic (Pooideae), somewhat lobed (Bambusoideae), or deeply

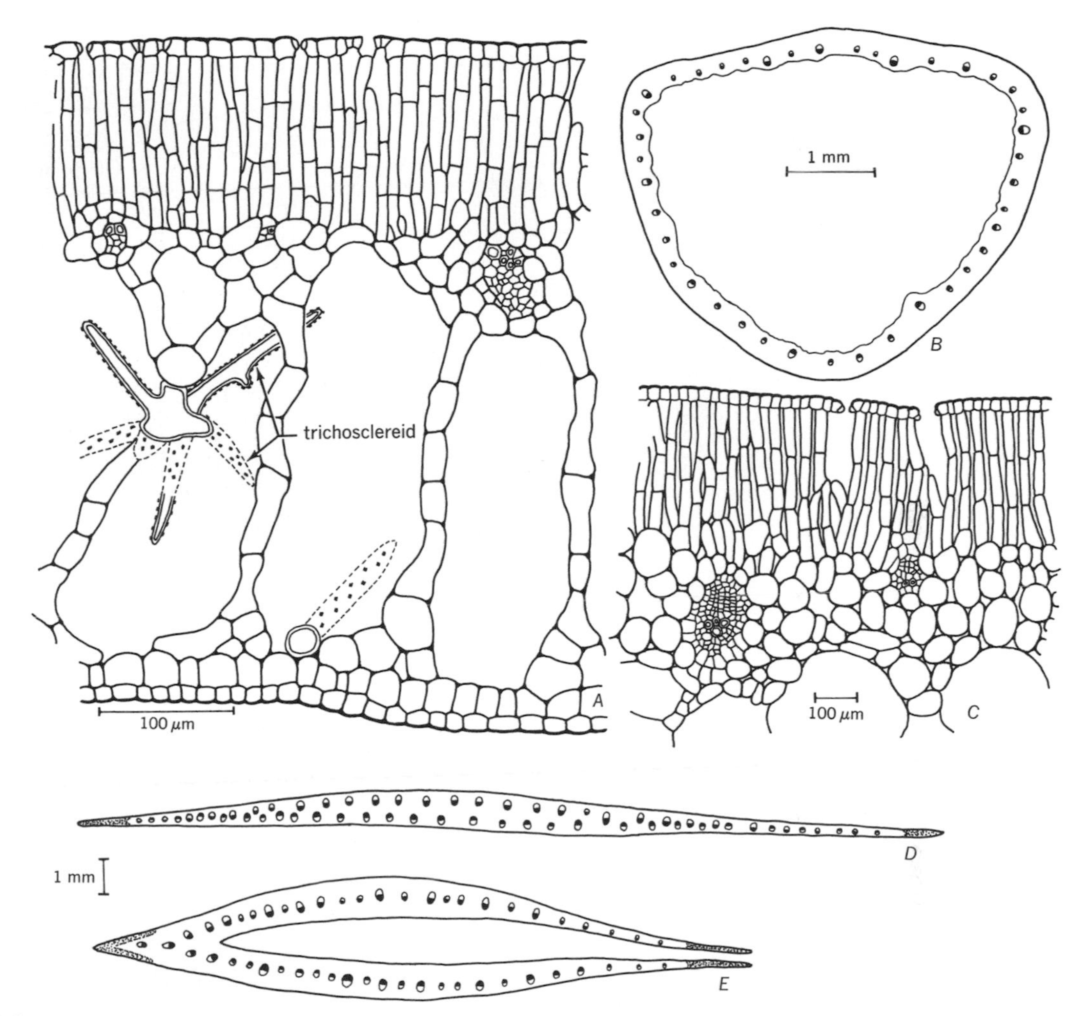

Figure 19.6 Variations in leaf structure as seen in cross sections. *A*, floating leaf of *Nymphaea*, a dicotyledon water plant. The air spaces are on the side in contact with water. *B–E*, monocotyledons: *B*, *C*, hollow tubular leaf of onion (*Allium cepa*); *D*, *E*, *Iris* leaf cut through unifacial blade (*D*) and bifacial sheath (*E*).

lobed, or branched (some Eragrostoideae and Panicoideae).[17]

The epidermis of grasses contains a variety of cells (chapter 7). The ground mass comprises narrow, elongated cells, which often have strongly undulate anticlinal walls. The narrow guard cells of the stomata are associated with subsidiary cells. Silica cells, cork cells, and trichomes may be present.

Enlarged epidermal cells with thin anticlinal walls, referred to as bulliform cells (fig. 19.8), are often described as cells participating in involution and folding movements of grass leaves. In a number of xeric grasses, enlarged epidermal cells line adaxial grooves between the vein ribs and are continuous with similarly enlarged mesophyll cells, the hinge cells (fig. 19.7; colorless cells according to Metcalfe[31]). During excessive loss of water, the bulliform cells, or the hinge

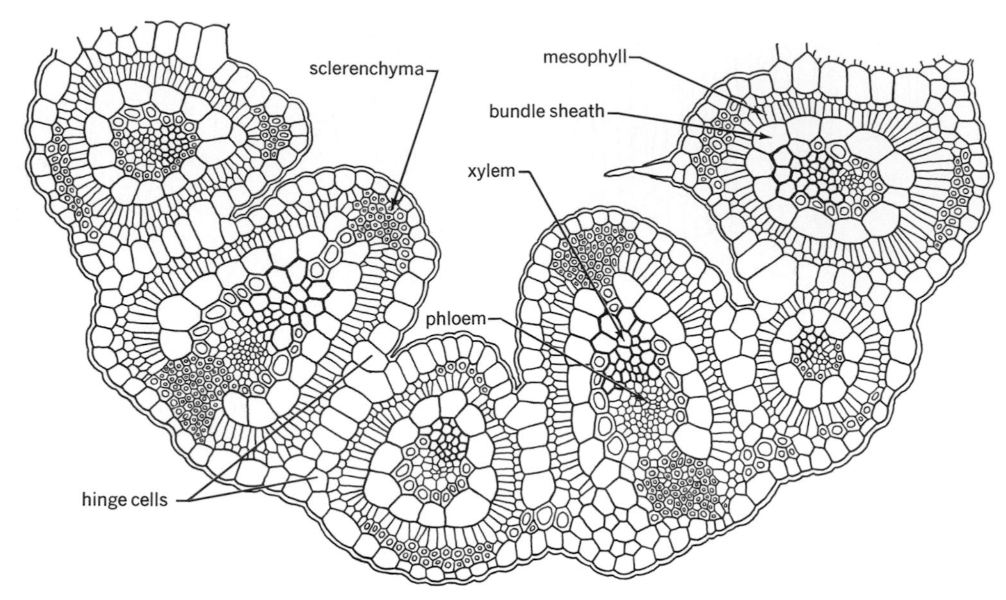

Figure 19.7 Transverse section from leaf of *Bouteloua breviseta* illustrating structure associated with C_4 photosynthetic cycle. A naturally involute leaf. (Redrawn from Shields.[42])

cells, or both types in conjunction, become flaccid and enable the leaf to fold or to roll (fig. 19.9). But the shrinkage of the various large thin-walled cells is only one contributing factor in the involution phenomenon, for leaves without such cells likewise respond by rolling to loss of moisture. Differential shrinkage of other tissues, distribution of sclerenchyma, and cohesive forces among tissues also contribute to rolling and folding of leaves.[42] The unrolling of leaves as they emerge from the bud (from *A* to *B* in fig. 19.9) is a growth phenomenon based on enlargement of bulliform and mesophyll cells.

The bundle sheaths of grasses show variations that are significant taxonomically[10,12] and as indicators of the type of photosynthesis characteristic of the species. In the Pooideae (or Festucoideae) grasses (figs. 19.8,*C*, and 19.10,*B*,*C*), two bundle sheaths are usually present, an inner *mestome* sheath with thickened walls and small, little-differentiated plastids, and an outer parenchyma sheath with thin walls and somewhat larger plastids but smaller than those in the mesophyll. The mestome sheath may have a suberized lamella in its walls and thus be analogous to an endodermis.[12,35] In the Eragrostoideae (figs. 19.7 and 19.11), the parenchyma sheath is well developed, has thicker walls than the mesophyll , and contains large plastids with grana. An inner more or less thick-walled sheath is commonly present. In the Panicoideae (figs. 19.8,*A*,*B*, and 19.10,*A*), the inner sheath is usually absent, whereas the single parenchyma sheath consists of large cells with walls thicker than those of the mesophyll cells. The sheath chloroplasts are as large or larger than the mesophyll chloroplasts, but have few or no grana.

Grass leaves have strongly developed sclerenchyma. Commonly fibers appear in longitudinal plates extending from the larger vascular bundles to the epidermis. The larg-

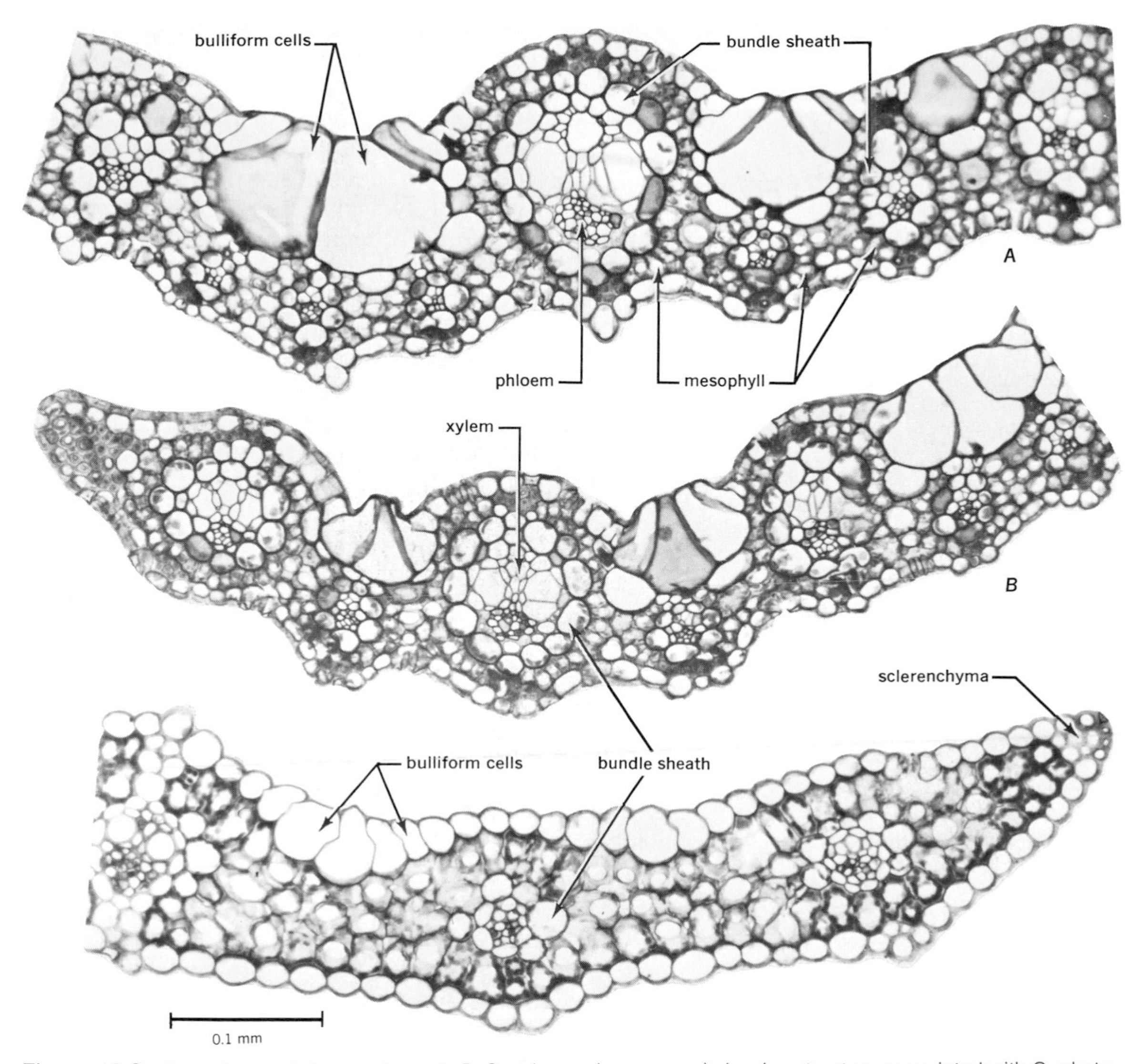

Figure 19.8 Grass leaves in transections. *A, B, Saccharum* (sugar cane) showing structure associated with C_4 photosynthetic cycle. *C, Avena* (oat), a C_3 plant. Double bundle sheaths as in wheat (fig. 19.10, *B, C*). Spatial association between mesophyll and vascular bundles closer in sugar cane than in oat. (Slides, courtesy of J. E. Sass.)

est vascular bundles may be enclosed in fibers as well and be associated with plates of fibers on both sides (fig. 19.10,*C*). Smaller bundles may be connected with only one fiber plate. In some species fibers occur in subepidermal strands or plates having no contact with the vascular bundles (fig. 19.10,*B*). The leaf margins have fibers (fig. 19.8, *C*), and so does the epidermis in some species.

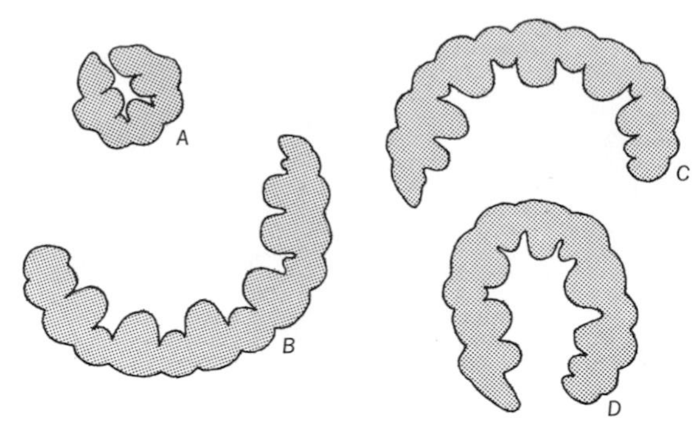

Figure 19.9 Variations in involution of leaf of *Bouteloua breviseta* shown in transverse sections. *A*, a young, still completely involuted leaf. *B*, open mature leaf naturally partially involuted (see fig. 19.7). *C*, section from the same leaf as in *B* but left to dry in air for 48 hours before it was fixed. Involution more pronounced than in *B*. *D*, mature leaf wilted naturally on the plant. Curvature greater than in *C*. (After Shields.[42])

GRASS LEAF STRUCTURE AND TYPE OF PHOTOSYNTHESIS

After the discovery of the C_4 or Hatch-Slack pathway of photosynthesis in sugar cane[27] comparative grass leaf anatomy became the object of intensive investigation in relation to photosynthesis. The most common photosynthetic cycle, the C_3 or Calvin-Benson pathway, is characterized by the three-carbon compound 3-phosphoglyceric acid as the first product of photosynthesis. In the C_4 cycle, four-carbon acids are produced in the initial stages of CO_2 fixation. The C_4 plants utilize CO_2 more efficiently than the C_3 plants and, accordingly, show a low CO_2 release resulting from the light-dependent respiration, or photorespiration.

The C_4 cycle is characteristic of plants that require relatively high temperatures for growth. In the angiosperms, this cycle has been recorded in representatives of some ten families (Amaranthaceae, Aizoaceae, Chenopodiaceae, Asteraceae, Cyperaceae, Euphorbiaceae, Poaceae, Nyctaginaceae, Portulacaceae, and Zygophyllaceae[27]). About half of the species of the Poaceae are included among the C_4 plants.[44] The C_4 plants are of tropical origin and occur widely in xerophytic environments. Since so few angiosperms are specialized for the C_4 photosynthesis cycle the C_4 condition is considered to be of more recent origin than the C_3 condition.[11] The discovery of the C_4 pathway and subsequent studies, mainly in grasses, have greatly advanced research in photosynthesis and brought an increased awareness of the interrelationship between metabolic activities and physiology, anatomy, ecology, evolution, and economic factors such as plant productivity and plant competition.[2]

The C_3 and C_4 plants show anatomical and ultrastructural differences with a high degree of consistency, a phenomenon much explored in the Poaceae.[12,27] The structure of parenchymatic bundle sheath is particularly important in distinguishing C_3 and C_4 grasses. In C_3 plants, this sheath has few organelles and rather small chloroplasts so that at low magnifications the cells appear empty and clear in striking contrast to the chloroplast-rich mesophyll (figs. 19.8,*C*, and 19.10,*B*,*C*). In C_4 plants, the sheath has a high content of organelles, especially mitochondria and microbodies,[3] and its chloroplasts are deep green and large, commonly larger than the mesophyll chloroplasts (figs. 19.10, *A*, and 19.11). The sheath cell chloroplasts may be localized next to the outer tangential wall or on the opposite side of the cell[25] (figs. 19.10,*A*, and 19.11). The walls of the sheath are somewhat thickened. When the vascular bundles of C_4 plants are enzymically isolated from the mesophyll the sheath is found to be tightly adhering to the vascular tissue, with its cells arranged in a double helix[4] (fig. 19.12,*A*,*B*).

The bundle sheath chloroplasts in C_4

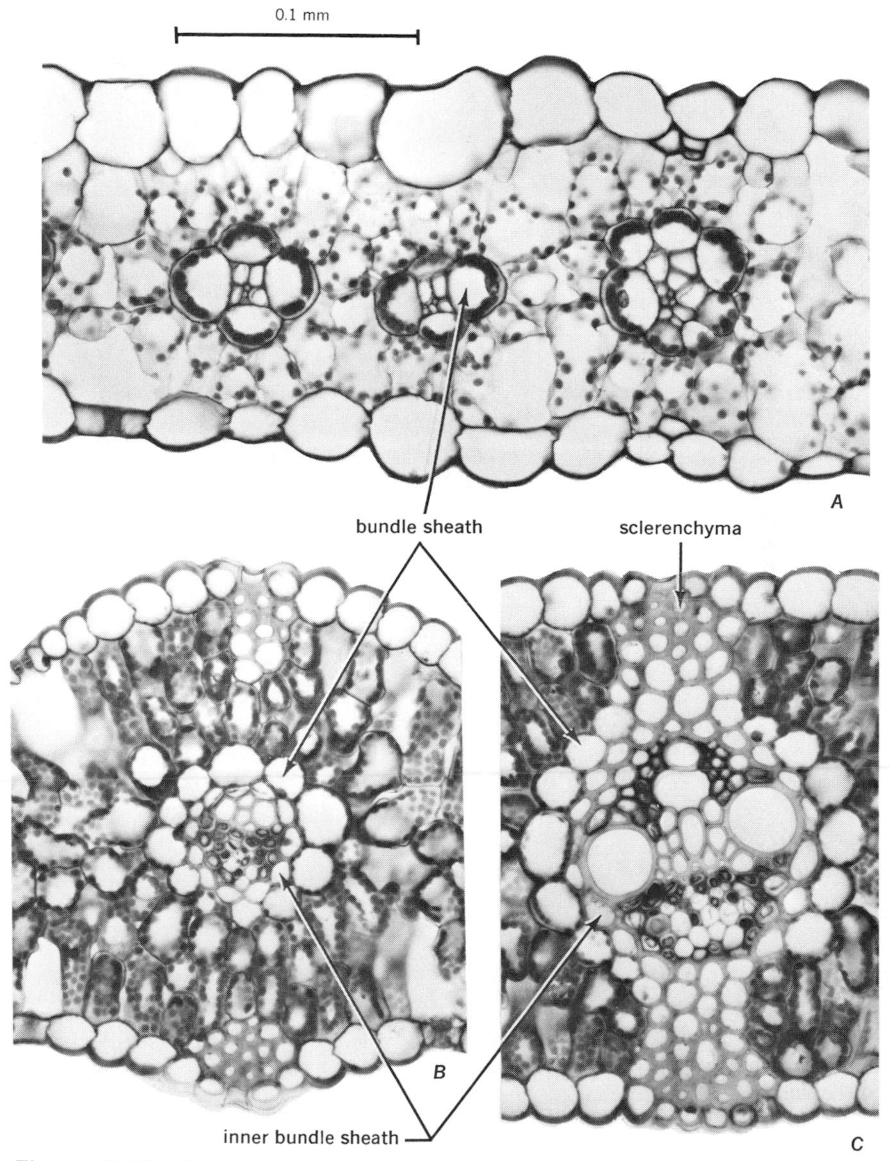

Figure 19.10 Grass leaves in transections. *A, Zea mays,* a C_4 plant. The bundle sheath has larger plastids than the mesophyll. The sheath plastids are localized next to the wall farthest from the vascular tissues. *B, C, Triticum,* a C_3 plant, showing a minor (*B*) and a large (*C*) vein. Two bundle sheaths are present, the inner having smaller cells than the outer. Few small chloroplasts in the outer sheath.

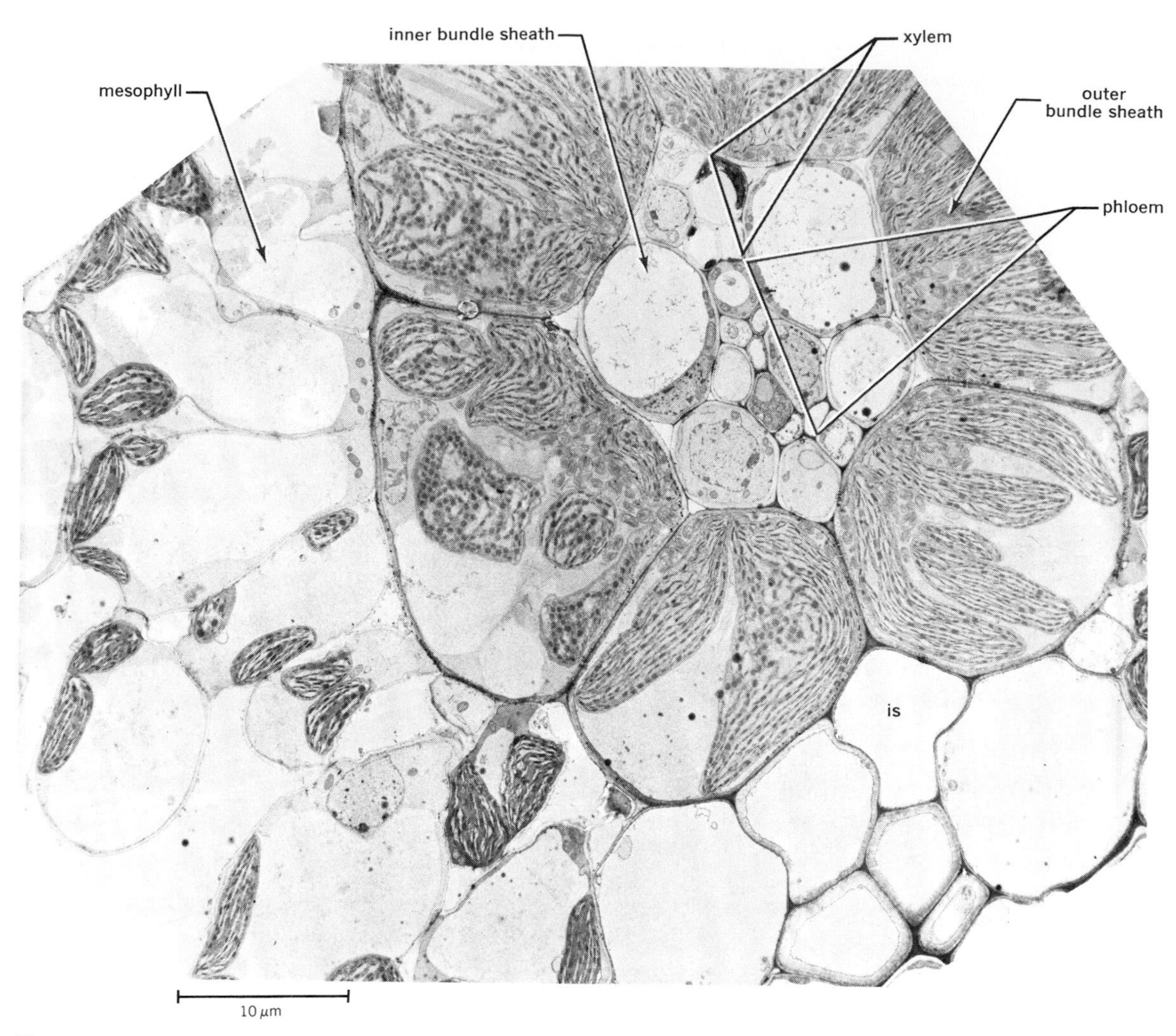

Figure 19.11 Electron micrograph of transection of leaf of *Cynodon dactylon* (Bermuda grass), a C_4 plant. The prominent outer bundle sheath has much larger chloroplasts than the mesophyll. The chloroplasts are aggregated next to the walls adjoining the incomplete inner sheath. There are intercellular spaces between the sheaths and between the inner sheath and the vascular tissue. (From Black and Mollenhauer.[3])

plants vary in certain characteristics. They may form larger and more numerous starch grains than the mesophyll chloroplasts and, in contrast to the latter, may show a reduced grana development or none . In general, the degree of distinction between the sheath chloroplasts and those of the mesophyll varies in different C_4 plants.[3,25] Comparative biochemical studies on C_4 plants suggest that the variation in morphology and localization of the sheath chloroplasts are correlated with different enzymic activities.[21]

A rather consistent character of chloroplasts in both sheath and mesophyll of C_4

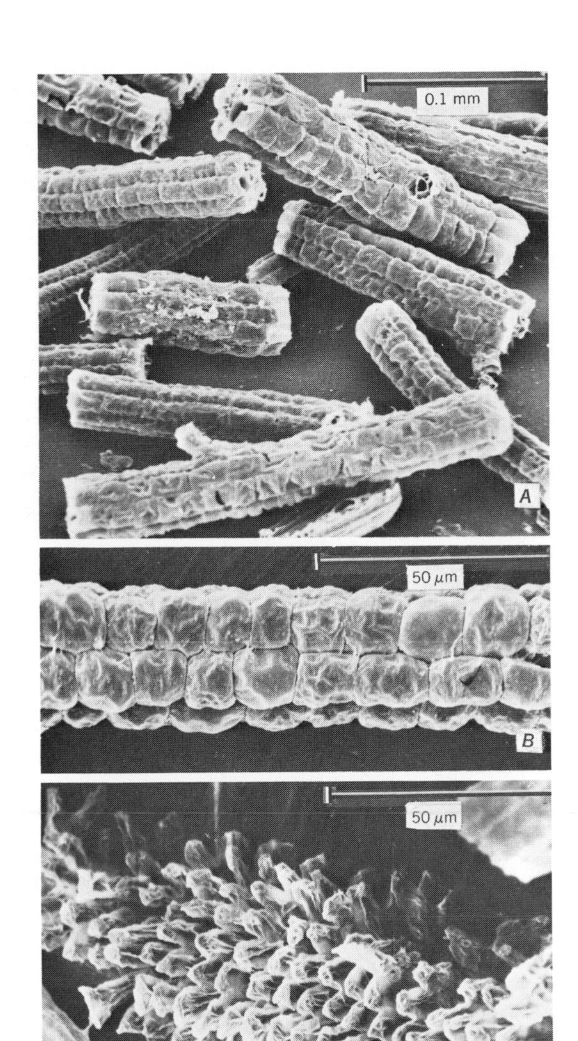

Figure 19.12 Scanning electron micrographs of isolated tissues from C_4 grass leaves. *A,* bundle sheaths with enclosed vascular tissues from a leaf of *Pennisetum purpureum. B,* a single bundle sheath with enclosed vascular tissue from a leaf of *Digitaria decumbens* (crab grass). *C,* vein of *Cynodon dactylon* (Bermuda grass) leaf with mesophyll having remained attached to the bundle sheath enclosing the vascular tissue. (From Black, Campbell, Chen, and Dittrich.[4])

plants is the peripheral reticulum, a system of anastomosing tudules contiguous with the inner membrane of the chloroplast envelope. Since this reticulum has also been observed in some C_3 plants its relation to C_4 photosynthesis is uncertain.[27]

A commonly mentioned anatomic feature of C_4 plants is the orderly arrangement of mesophyll cells with reference to the bundle sheath, the two together forming concentric layers around the vascular bundle as seen in transection (figs. 19.7 and 19.8,*A*,*B*). In contrast to the sheath cells the mesophyll cells have intercellular spaces among them. The loose arrangement of the mesophyll cells is clearly revealed in scanning electron microscope micrographs of bundles isolated from leaves but with the mesophyll still attached to their sheaths[4] (fig. 19.12,*C*).

The concentric layering of mesophyll and bundle sheath seen in cross sections of C_4 grass leaves varies from such an extreme form as in *Buchloe*[27] or *Bouteloua* (fig. 19.7) to a barely discernible form as in *Zea* (fig. 19.10,*A*). But despite the variations, the spatial association between the mesophyll and the vascular bundles appears to be closer in C_4 plants than in the C_3 plants (figs. 19.8,*A*,*B*, and 19.10,*A* as contrasted with fig. 19.8,*C*). The C_4 grass leaf is often described as an assemblage of rodlike units composed of vascular bundles and associated mesophyll (figs. 19.7, 19.8,*A*,*B*, and 19.12,*C*)

In observing the concentric arrangement of the parenchymatic layers around the vascular bundles of certain grasses and sedges Haberlandt[22] compared the mesophyll layer with a wreath (Kranz), saying that its tubular cells appeared to radiate from the sheath. In the literature dealing with the characteristics of the C_4 plants, the designation kranz in "kranz type" includes both the mesophyll and the sheath and the term kranz syndrome (also C_4 syndrome) refers to the collection of structural and physiological features of C_4 plants.[11,27]

The specialized anatomy of C_4 plants furnishes an outstanding example of correlation between structure and function in plants. But the assumption that a plant revealing one aspect of the C_4 syndrome may be expected to possess the whole syndrome[11] must be qualified in view of the evidence that species intermediate between those showing C_3 and C_4 features do occur.[26]

GYMNOSPERM LEAVES

The gymnosperm leaves are less variable in structure than those of the angiosperms and are rather unresponsive to environmental conditions. Most gymnosperms are evergreen. The well-known exceptions are *Ginkgo, Larix,* and *Taxodium.* The leaves of conifers, which constitute the largest number of species among the gymnosperms, have been studied most frequently. The pine leaf, therefore, is described first, and those of other conifers and other gymnosperms are treated comparatively.

Pine needles originate on dwarf branches (short shoots), singly or most commonly in groups of two to several. Depending on this number the transectional form varies (figs. 19.13,*A*, and 19.14,*A*) from approximately oval to triangular. The needle has a thick-walled epidermis with a heavy cuticle and deeply sunken stomata with overarching subsidiary cells (chapter 7). The stomata occur on all sides and are in vertical rows. A sclerified fibrous hypodermis occurs beneath the epidermis, except under the rows of stomata. The mesophyll consists of parenchyma cells with vertical ridges protruding into the lumen of the cell (fig. 19.13). The ridges develop as invaginations composed of the wall layer to one side the middle lamella.[24] An interwall space separates the two halves of the ridge and gives the wall invagination the appearance of a fold ("plicate mesophyll"). The mesophyll is not differentiated into palisade and spongy parenchyma. Resin ducts occur in the mesophyll.

The vascular tissue usually forms one bundle or two bundles side by side and occurs in a central position in the needle. The xylem is on the adaxial, the phloem on the abaxial side. The xylem consists of protoxylem and metaxylem. The latter shows orderly radial seriation of cells, with rows of xylem parenchyma cells alternating with those of the tracheids. Possibly a small amount of secondary growth occurs after the primary extension growth of the needle is terminated but the bulk of the xylem is metaxylem.

The vascular bundle is surrounded by a peculiar tissue known as *transfusion tissue.* It is composed of tracheids and parenchyma cells (fig. 19.13,*B*,*C*). The tracheids that occur next to the vascular bundle are elongated; those farther away have the same shape as the parenchyma cells. The walls of the tracheids, though provided with secondary thickenings, are relatively thin and lightly lignified and bear bordered pits. The tracheids usually appear somewhat deformed, probably because of the pressure of the associated living parenchyma cells. Some densely cytoplasmic cells interpreted as albuminous cells occur next to the phloem (fig. 19.13,*C*).

The vascular bundle and the associated transfusion tissue are surrounded by a thick-walled endodermis. No intercellular spaces are present in the endodermis and the tissues enclosed by it. The endodermis is often described as having casparian strips in early stages of development and a suberin lamella later.[28] In the mature state the endodermis has

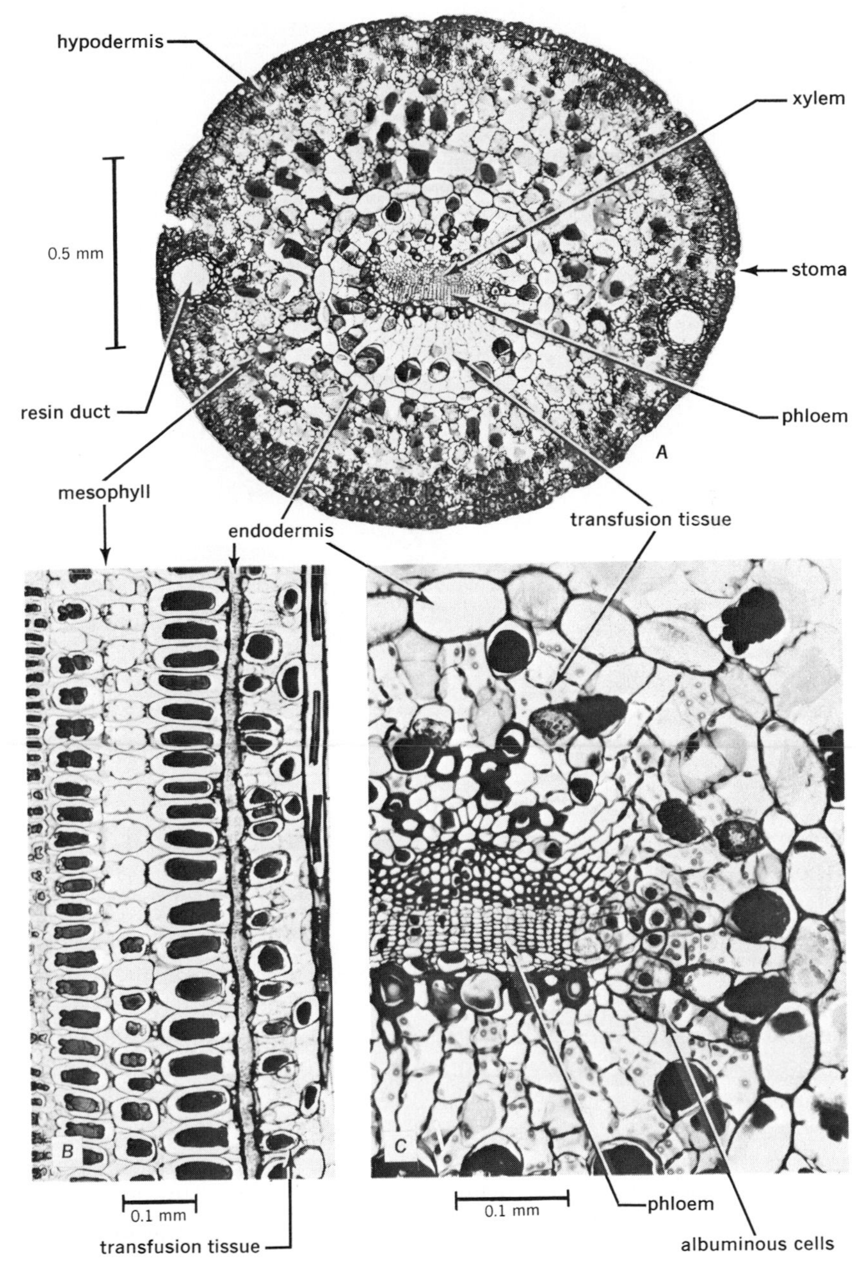

Figure 19.13 Structure of pine (*Pinus monophylla*) leaf. *A*, cross section of entire needle. *B*, longitudinal section through mesophyll and transfusion tissue. *C*, cross section of part of vascular bundle, transfusion tissue, and endodermis. Circles in transfusion cells are bordered pits in tracheid walls.

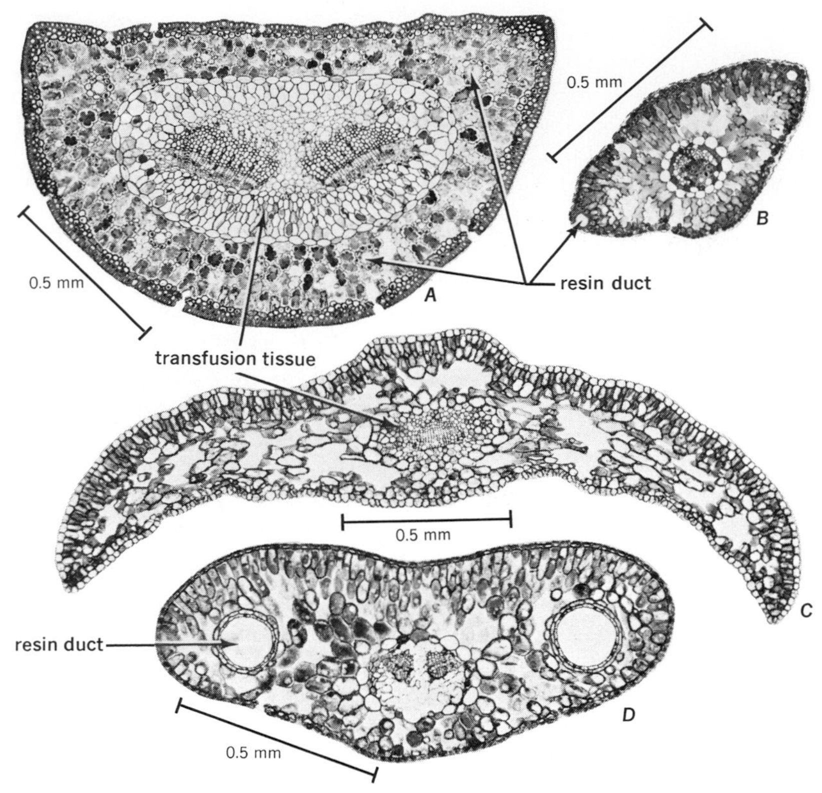

Figure 19.14 Cross sections of various conifer leaves. *A*, pine (*Pinus nigra*). *B*, larch (*Larix*). *C*, ground-hemlock (*Taxus canadensis*). *D*, fir (*Abies*).

secondary lignified walls with the suberin confined to the anticlinal walls.

The structrual features just described are found in many other conifers, usually with quantitative differences, but are absent in some.[28] The leaves of the conifers may be single veined and scalelike (Taxodiaceae, Cupressaceae, Podocarpaceae), or single veined and needlelike (*Abies, Larix, Picea, Pinus*), or broadly ovate, with many veins (Araucariaceae). Still other forms may be found.

The sclerified hypodermis is up to five layers in thickness in *Araucaria* but is entirely absent in some conifers (*Taxus*, fig. 19.14,*C; Torreya*). Most conifers do not have a plicate mesophyll. Palisade and spongy parenchyma are differentiated in such genera as *Abies* (fig. 19.14,*D*), *Cunninghamia, Dacrydium, Sequoia, Taxus* (fig. 19.14,*C*), *Torreya;* and in *Araucaria* and *Podocarpus* the palisade occurs on both sides of the leaf.

The boundary between the vascular region and the mesophyll is not equally clear in conifer leaves.[28] The Pinaceae (*Abies*, fig. 19.14,*D; Larix*, fig. 19.14,*B; Picea, Pinus*, fig. 19.14,*A*) have a distinctly differentiated endodermis. In species of *Taxus* (fig. 19.14,*C*), in *Sequoia sempervirens, Metasequoia glyptostroboides, Juniperus communis,* and *Arau-*

caria excelsa only a parenchymatous sheath is present.

The transfusion tissue is characteristic of the conifers, but it varies in amount and arrangement.[18,28] It occurs right and left from the vascular bundle in such genera as *Cunninghamia, Cupressus, Juniperus, Thuja, Torreya, Sequoia,* and *Taxus* (fig. 19.14,*C*); as an arc about the xylem in *Araucaria, Dammara, Sciadopitys;* right and left from the vascular bundle but most abundantly next to the phloem in *Larix* and in species of *Abies* and *Cedrus;* and completely surrounding the vascular bundle in *Pinus* (fig. 19.14,*A*). The transfusion tracheids may have reticulate or pitted secondary walls. The tissue is often considered to be concerned with the translocation of water and food materials between the vascular bundle and the mesophyll.

In addition to the transfusion tissue associated with the vascular bundle a so-called accessory transfusion tissue has been described in *Podocarpus*[20] and *Dacrydium.*[29] This tissue consists of thick-walled elongated cells extending outward from the bundle sheath into the mesophyll. It is not in contact with the transfusion tissue next to the vascular bundle.

The arrangement and number of resin ducts are variable even in the same species. Among the single-veined genera only *Taxus* (fig. 19.14,*C*) lacks resin ducts. As a basic pattern, the Cupressineae, the Taxineae (other than *Taxus*), *Sequoia, Podocarpus,* and most species of *Tsuga* have one resin duct located between the vein and the lower epidermis; the Abietineae, except *Tsuga,* have two resin ducts, right and left from the bundle (fig. 19.14,*D*); the many-veined genera (*Araucaria*) have one duct between each two bundles. In addition to the basic duct, variable numbers of accessory ones may be present. In *Pinus* two lateral resin ducts occur almost invariably (figs. 19.13,*A,* and 19.14,*A*). The others vary in number and distribution.

To illustrate leaves of gymnosperms other than conifers those of *Cycas* and *Ginkgo* are described. The large leaves of the Cycadales are compound, and their broad single-veined pinnae are stiff and rigid. The leaf of *Cycas revoluta* (fig. 19.15,*A*) has a thick cuticle, a thick-walled epidermis, and sunken stomata on the abaxial side. The mesophyll is composed of palisade and spongy parenchyma. A layer or two of hypodermal sclerenchyma occur on the adaxial side. An endodermis with walls thickened next to the vascular tissue is present. The xylem shows a primitive arrangement in that the protoxylem is toward the abaxial, the metaxylem toward the adaxial side. A layer of parenchyma envelops the protoxylem, and some secondary xylem is present next to the phloem. A few transfusion tracheids occur on both sides of the metaxylem. Accessory transfusion tissue composed of elongated parenchyma cells and tracheids with bordered pits is present.[28] Both kinds of cells appear to be in contact with intercellular spaces.

The *Ginkgo* (fig. 19.15,*B*) leaf is wide at the apex and narrow at the base. It has numerous dichotomously branching veins. The epidermis has relatively thin walls, and hypodermal sclerenchyma is absent. The guard cells are slightly depressed and occur only on the abaxial side. There is a single palisade layer of rather short lobed cells and below it is the spongy parenchyma. Each of the numerous vascular bundles has a uniseriate lignified endodermis. Tannins are often abundant in the endodermal sheath, especially in the small bundles. A few transfusion tracheids flank the xylem of each bundle. The vascular bundles alternate with mucilage ducts.

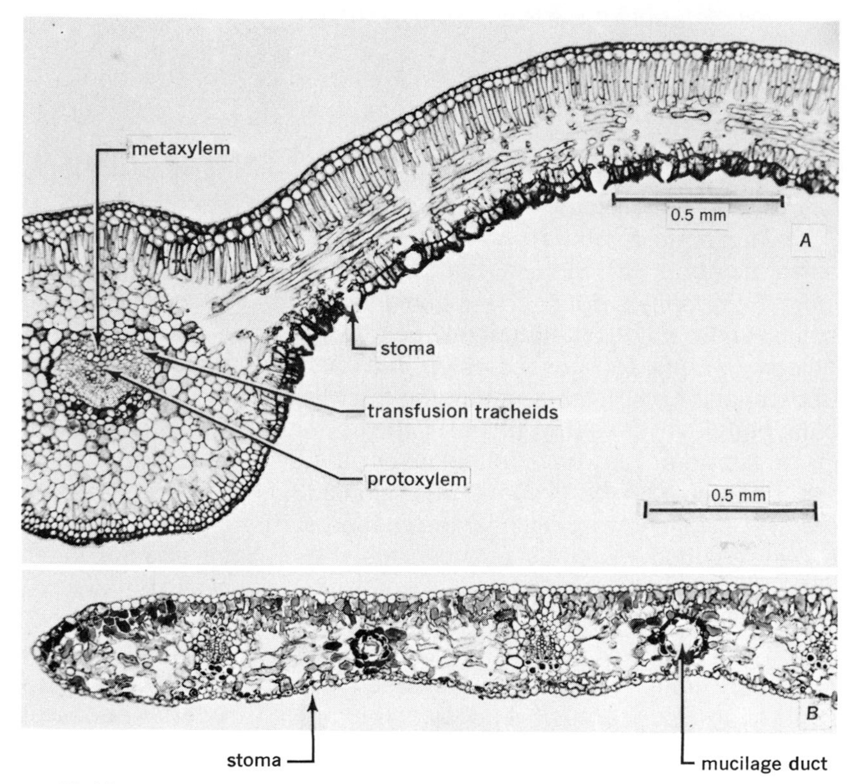

Figure 19.15 Cross sections of gymnosperm nonconifer leaves. *A,* cycad (*Cycas revoluta*). *B,* maidenhair tree (*Ginkgo biloba*).

REFERENCES

1. Allsopp, A. Shoot morphogenesis. *Ann. Rev. Plant Physiol.* 15:225–254. 1964.
2. Black, C. C., Jr. Photosynthetic carbon fixation in relation to net CO_2 uptake. *Ann. Rev. Plant Physiol.* 24:253–286. 1973.
3. Black, C. C., and H. H. Mollenhauer. Structure and distribution of chloroplasts and other organelles in leaves with various rates of photosynthesis. *Plant Physiol.* 47:15–23. 1971.
4. Black, C. C., W. H. Campbell, T. M. Chen, and P. Dittrich. The Monocotyledons: their evolution and comparative biology. III. Pathways of carbon dioxide metabolism related to net carbon dioxide assimilation by monocotyledons. *Quart. Rev. Biol.* 48:299–313. 1973.
5. Black, R. F. The leaf anatomy of Australian members of the genus *Atriplex.* I *Atriplex vesicaria* Heward and *A. nummularia* Lindl. *Aust. J. Bot.* 2:269–286. 1954.
6. Black, R. F. Effect of sodium chloride on leaf succulence and area of *Atriplex hastata* L. *Aust. J. Bot.* 6:306–321. 1958.
7. Böcher, T. W., and O. B. Lyshede. Anatomical studies in xerophytic apophyllous plants. II. Additional species

from South American shrub steppes. *Kon. Danske Videnskab. Selskab Biol. Skr.* 18, 4, 137 p. 1972.

8. Boyce, S. G. The salt spray community. *Ecological Monographs* 24:29–67. 1954.
9. Brown, H. S., and F. T. Addicott. The anatomy of experimental leaflet abscission in *Phaseolus vulgaris*. *Amer. J. Bot.* 37:650–656. 1950.
10. Brown, W. V. Leaf anatomy in grass systematics. *Bot. Gaz.* 119:170–178. 1958.
11. Brown, W. V., and B. N. Smith. Grass evolution, the kranz syndrome, $^{13}C/^{12}C$ ratios, and continental drift. *Nature* 239: 345–346. 1972.
12. Carolin, R. C., S. W. L. Jacobs, and M. Vesk. The structure of the cells of the mesophyll and parenchymatous bundle sheath of the Gramineae. *Bot. J. Linn. Soc.* 66:259–275. 1973.
13. Cumming, B. G., and E. Wagner. Rhythmic processes in plants. *Ann. Rev. Plant Physiol.* 19:381–416. 1968.
14. Cutter, E. G. *Plant anatomy: experiment and interpretation.* Part 2. *Organs.* London, Edward Arnold. 1971.
15. Esau, K. *Plant anatomy.* 2nd ed. New York, John Wiley & Sons. 1965.
16. Feldman, L. J., and E. G. Cutter. Regulation of leaf form in *Centauria solstitialis* L. I. Leaf development on whole plants in sterile culture. *Bot. Gaz.* 131:31–39. 1970. II. The developmental potentialities of excised leaf primordia in sterile culture. *Bot. Gaz.* 131:39–49. 1970.
17. Freier, F. Las células clorenchimáticas del mesófilo de las Gramíneas. *Rev. Argentina Agron.* 26:1–16. 1959.
18. Gathy, P. Les feuilles de *Larix.* Étude anatomique. *Cellule* 56:331–353. 1954.
19. Goebel, K. *Einleitung in die experimentelle Morphologie der Pflanzen.* Leipzig, Teubner. 1908.
20. Griffith, M. M. Foliar ontogeny in *Podocarpus macrophyllus,* with special reference to the transfusion tissue. *Amer. J. Bot.* 44:705–715. 1957.
21. Gutierrez, M., V. E. Gracen, and G. E. Edwards. Biochemical and cytological relationships in C_4 plants. *Planta* 119: 279–300. 1974.
22. Haberlandt, G. F. J. *Physiologische Pflanzenanatomie.* 3te Aufl. Leipzig, Wilhelm Engelmann.
23. Hagerup, O. The morphology and systematics of the leaves in Ericales. *Phytomorphology* 3:459–464. 1953.
24. Harris, W. M. Ultrastructural observations on the mesophyll cells of pine leaves. *Can. J. Bot.* 49:1107–1109. 1971.
25. Johnson, Sister C., and W. V. Brown. Grass leaf ultrastructural variations. *Amer. J. Bot.* 60:727–735. 1973.
26. Kennedy, R. A., and W. M. Laetsch. Plant species intermediate for C_3, C_4 photosynthesis. *Science* 184:1087–1089. 1974.
27. Laetsch, W. M. The C_4 syndrome: a structural analysis. *Ann. Rev. Plant Physiol.* 25:27–52. 1974.
28. Lederer, B. Vergleichende Untersuchungen über das Transfusionsgewebe einiger rezenter Gymnospermen. *Bot. Stud.* No. 4:1–42. 1955.
29. Lee, C. L. The anatomy and ontogeny of the leaf of *Dacrydium taxoides*. *Amer. J. Bot.* 39:393–398. 1952.
30. Linsbauer, K. Die Epidermis. In: *Handbuch der Pflanzenanatomie.* K. Linsbauer, ed. Band 4. Lief. 27. 1930.
31. Metcalfe, C. R. *Anatomy of the monocotyledons.* I. *Gramineae.* Oxford, Clarendon Press. 1960.
32. Metcalfe, C. R. and L. Chalk. *Anatomy of the dicotyledons.* 2 vols. Oxford, Clarendon Press. 1950.
33. Mitchell, R. S. Variation in the *Po-*

lygonum amphibium complex and its taxonomic significance. *Univ. Calif. Pubs. Bot.* 45:1–65. 1968.

34. Njoku, E. The effect of sugars and applied chemicals on heteroblastic development in *Ipomoea purpurea* grown in aseptic culture. *Amer. J. Bot.* 58:61–64. 1971.
35. O'Brien, T. B., and D. J. Carr. A suberized layer in the cell walls of the bundle sheath of grasses. *Aust. J. Biol. Sci.* 23:275–287. 1970.
36. Owens, J. N. Initiation and development of leaves in Douglas fir. *Can. J. Bot.* 46:271–278. 1968.
37. Pyykkö, M. The leaf anatomy of East Patagonian xeromorphic plants. *Ann. Bot. Fenn.* 3:453–622. 1966.
38. Rogler, C. E., and M. E. Dahmus. Gibberellic acid-induced phase change in *Hedera helix* as studied by deoxyribonucleic acid-ribonucleic acid hybridization. *Plant Physiol.* 54:88–94. 1974.
39. Satter, R. L., and A. W. Galston. Potassium flux: a common feature of *Albizzia* leaflet movement controlled by phytochrome or endogenous rhythm. *Science* 174:518–519. 1971.
40. Schneider, R. Histogenetische Untersuchungen über den Bau der Laubblätter, insbesondere ihres Mesophylls. *Österr. Bot. Z.* 99:253–285. 1952.
41. Shields, L. M. Leaf xeromorphy as related to physiological and structural influences. *Bot. Rev.* 16:399–447. 1950.
42. Shields, L. M. The involution mechanism in leaves of certain xeric grasses. *Phytomorphology* 1:225–241. 1951.
43. Skutch, A. F. Anatomy of leaf of banana, *Musa sapientum* L., var. hort. Gros Michel. *Bot. Gaz.* 84:337–391. 1927.
44. Smith, B. N., and W. V. Brown. The kranz syndrome in the Gramineae as indicated by carbon isotope ratios. *Amer. J. Bot.* 60:505–513. 1973.
45. Stalfelt, M. G. Morphologie und Anatomie des Blattes als Transpirationsorgan. *Handb. Pflanzenphysiol.* 3:324–341. 1956.
46. Stant, M. Y. Anatomy of the Butomaceae. *J. Linn. Soc. London, Bot.* 60:31–60. 1967.
47. Vasilevskaya, V. K. *Formirovanie lista zasukhoustojchivykh rastenij.* [Formation of leaves of drought-resistant plants.] Akad. Nauk Turkmen SSR. 183 pp. 1954.
48. Weintraub, M. Leaf movements in *Mimosa pudica* L. *New Phytol.* 50:357–382. 1952.
49. Weston, G. O., and D. D. Cass. Observations on the development of the paraveinal mesophyll of soybean leaves. *Bot. Gaz.* 134:232–235. 1973.
50. Wylie, R. B. Differences in foliar organization among leaves from four locations in the crown of an isolated tree (*Acer platanoides*). *Proc. Iowa Acad. Sci.* 56:189–198. 1949.
51. Wylie, R. B. Leaf organization of some woody dicotyledons from New Zealand. *Amer. J. Bot.* 41:186–191. 1954.

20

The Flower: Structure and Development

CONCEPT

The flower has been the object of many studies with regard to its structure, organography, and ontogeny, but there is no general agreement about the concept of the flower.[1,2,5,10,16,31,57] One of the obstacles to a full comprehension of the nature of the flower is the inadequacy of the fossil record. Botanists are therefore inclined to rely on comparative studies of flowers in extant taxa and to use the data to theorize about the relation of the flower and its parts to the other plant organs and to the reproductive structures in nonflowering taxa.

The classical theory on the nature of the flower, which was much influenced by the largely philosophical ideas of Goethe,[2] homologizes the flower with the vegetative shoot and the individual parts of the flower with the leaves, or *phyllomes.* The concept that leaves of the type found in ferns, gymnosperms, and angiosperms (megaphylls) have evolved from branch systems[6] (see also telome theory[19]) has given rise to the speculation that, in their parallel evolution, leaves and floral parts diverged before the leaflike form emerged. Nevertheless, in their initiation, ontogeny, and basic vascular organization the parts of the less specialized flowers show considerable similarity to leaves, whereas those of the more specialized flowers can often be related to leaflike structures by transitional forms. These aspects of floral structure explain the continued adherence of many botanists to the theory of the shootlike nature of the flower.

Certain workers, however, emphasize the deviating forms and propose that at least some of the floral organs are not related to leaves and that the concept of the flower requires a fundamental revision.[53] In the following, the flower is described in agreement with the theory of homology between the vegetative shoot and the flower, with due regard to the circumstance that the floral features of certain taxa are not readily explained by the theory .

STRUCTURE

Flowers, flower parts, and their arrangement

The flower is an assemblage of sterile and fertile, or reproductive, parts (organs) borne on an axis, the *receptacle* (fig. 20.5). The sterile parts are the *sepals* forming the *calyx* and the *petals* forming the *corolla.* Calyx and corolla together are called *perianth.* If the perianth is not differentiated into calyx and corolla, the individual members are called *tepals.* The reproductive parts are the *stamens* (microsporophylls) and *carpels* (megasporophylls). The stamens constitute the *androecium,* the carpels the *gynoecium.*

The arrangement of the floral parts on the axis and the relation of parts to each other are highly variable and the variations are of particular concern in taxonomic and phylogenetic studies of flowers. If the flower is regarded as a modified shoot, the differences in floral structure may be interpreted as deviations of various degrees from the basic shoot form; and, in this sense, the greater the deviation, the more highly specialized is the flower.

The vegetative shoot is characterized by indeterminate growth. In contrast, the flower shows determinate growth, for its apical meristem ceases to be active after it has produced all the floral parts. The more highly specialized flowers have a shorter growth period and produce a shorter axis and a smaller more definite number of floral parts than the more primitive flowers. Further indications of increasing specialization are: *whorled* arrangement of parts instead of *spiral* or *helical;* a whorl growing as a unit (*cohesion* of members of a whorl); two or more different whorls growing as a unit (*adnation* of one whorl to another); bilateral symmetry (*zygomorphy*) instead of radial symmetry (*actinomorphy*); and inferior ovary (*epigyny*) instead of superior ovary (*hypogyny*). The flower may lack certain parts. If the gynoecium or the androecium is absent, the flower is termed unisexual, or imperfect. The two kinds of unisexual flowers are the carpellate (or pistillate) and the staminate.

Flowers are grouped into inflorescences or they occur singly in a terminal position on the axis. Inflorescences are branch systems bearing flowers and more or less modified foliar organs, frequently in the form of bracts. Inflorescences are of many different types which develop according to specific ontogenetic sequences.[19]

The presence or absence of one or more floral parts, the size, pigmentation, and mutual arrangements of the parts are responsible for the existence of a great variety of floral types. There is an increasing recognition that the evolutionary interpretation of floral typology must take into account the adaptive aspects of floral structure, particularly in relation to pollination, fruit and seed dispersal,[10,18] and protection of reproductive structures from predators.[63] With regard to pollination, two large ecologic groups of plants have evolved, plants pollinated by abiotic agencies (wind, water, rain) and characterized by unshowy flowers, and plants pollinated by biotic agents (insects, birds, and other animals) and having showy flowers[38] with a variety of nectaries (chapter 13).

Among plants pollinated by animals diversification in floral typology can be related to the sensory perception of the different classes of pollinators. The selective pressure by the exploiters-pollinators explains also some specializations in inflorescences. Often the visual effect of the flowers is increased by their assemblage into dense showy clusters. At high levels of specialization, the inflorescences resemble large solitary flowers in shape and color, as for example, the flower heads of the Asteraceae, the cyathia of the Euphorbiaceae, and the umbels of the Apiaceae. In some condensed inflorescences the showy part is represented by the involucre (*Poinsettia, Cornus florida*). The concept of directional trend of floral evolution occurring under the selective pressure of pollinators seems to be supported by the available fossil record.[39] This trend generally leads from simple, actinomorphic flowers at lower levels toward complex, zygomorphic flowers at upper levels.

Sepal and petal

The sepals and petals resemble leaves in their internal structure. They consist of ground parenchyma, a more or less branched vascular system, and an epidermis (fig. 20.1). Crystal-containing cells, laticifers, tannin cells, and other idioblasts may be present. Starch is often formed in young petals. Green sepals contain chloroplasts but rarely show a differentiation into palisade and spongy parenchyma. The color of petals, which plays an important role in making the flowers attractive to biotic pollinators, results from pigments in chromoplasts (carotenoids) and in the cell sap (flavonoids, mainly anthocyanins) and from various modifying conditions such as, for example, acidity of cell sap.[48]

A study of flavonoid pigments in *Impatiens balsamina* revealed a contrast between the pigmentation in petals and that in sepals and in the vegetative parts.[21] This difference was

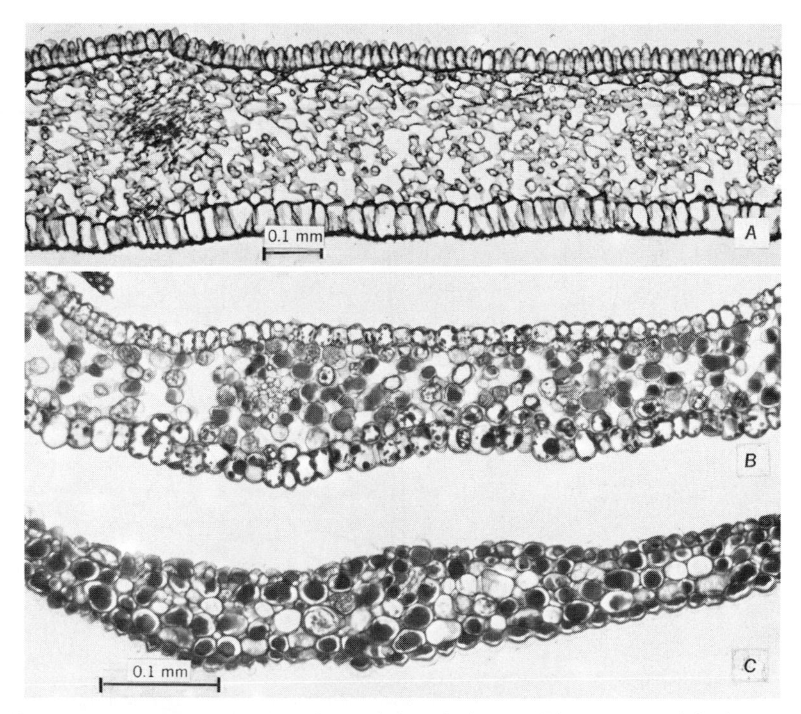

Figure 20.1 Cross sections of petal of rose (*A*) and of petal (*B*) and sepal (*C*) of *Cassiope*. (Slides courtesy of, *A*: A. T. Guard; *B, C*: B. F. Palser.)

interpreted as indicating that in the petals, in which the pigment has a specific function in connection with attracting the pollinating insects, selective forces operating during the evolution of the species have led to a development of specialized pigmentation. In *Rudbeckia hirta,* the petal bases contain flavonol glucosides, which absorb ultraviolet light and make the petal bases distinguishable as "nectar guides" to appropriate insect pollinators.[59] In the Brassicaceae, which also are pollinated by insects, the flowers show a variety of ultraviolet reflectance patterns. These patterns differentiate closely related taxa and may be of diagnostic value in taxonomy.[27] The epidermal cells of petals often contain volatile oils that impart the characteristic fragrance to flowers.

In some plants the anticlinal walls of the epidermis in petals are wavy or bear internal ridges. The outer walls may be convex or papillate (fig. 20.1,*A*), especially on the adaxial side. The epidermis of both sepals and petals may have stomata and trichomes. The stomata on the petals either resemble those on the foliage leaves or are incompletely differentiated.

Stamen

The most common type of stamen in angiosperms consists of an *anther* divided into *pollen sacs,* or microsporangia, and borne on a thin single-veined stalk, or *filament* (fig. 20.2,*A*). Each pollen sac includes wall layers and a locule in which the microspores are produced. The majority of angiosperms have tetrasporangiate anthers, that is, the type with two locules in each of the two lobes (figs. 20.5

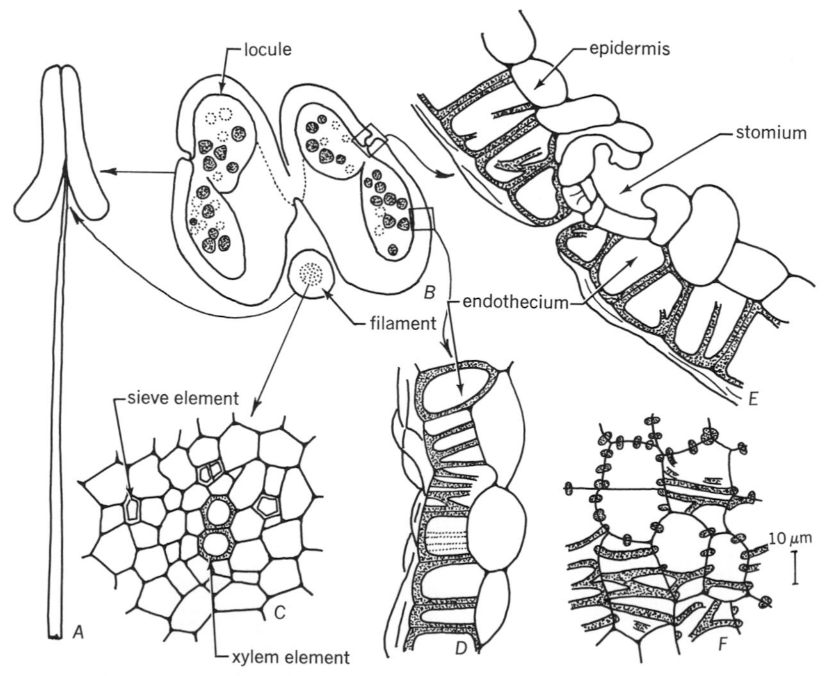

Figure 20.2 Stamen of *Prunus* (*A*) and its parts: cross sections of anther (*B*), vascular bundle of filament (*C*), anther wall (*D, E*); and endothecium in face and sectional views of the secondary cell wall (*F*).

and 20.7), and a few have bisporangiate anthers with one locule in each half of the anther (chapter 21). At maturity, before the anthers dehisce, the partitions between the anther locules may break down (fig. 20.2,*B*) in such a way that a tetrasporangiate anther appears as though it were bilocular and a bisporangiate as though it were unilocular. Some botanists accept such secondary modifications as diagnostic features and falsely characterize tetrasporangiate and bisporangiate anthers with broken-down partitions as bilocular and unilocular, respectively.[11]

Among the more primitive dicotyledons certain taxa have leaflike three-veined stamens bearing microsporangia on the adaxial surface between the midvein and the lateral veins. The single-veined stamens are regarded as products of evolutionary reduction in width of the leaflike stamens with a concomitant decrease in the number of veins.[19] According to an extensive survey, 95 percent of angiosperms have single-veined stamens.[64]

The filament is relatively simple in structure. Parenchyma surrounds the vascular bundle, which may be amphicribral (fig. 20.2,*B,C*). The cutinized epidermis may have trichomes, and on both filament and anther, stomata may be present. The vascular bundle traverses the entire filament and ends blindly either at the base of the anther or in the *connective tissue* located between the two anther halves.

Anthers commonly dehisce; that is, they open spontaneously. In many species, dehiscence is preceded by the previously mentioned breakdown of the partitions between locules of the same anther half (fig. 20.2,*B*). Later, the outer tissue in this region, which is sometimes represented by a single epidermal layer (fig. 20.2,*E*), breaks also, and the pollen is released through an often slitlike break (*stomium*). The subepidermal wall layer in the anther, the *endothecium,* which bears secondary wall thickenings in the form of strips (fig. 20.2,*D–F*), appears to promote the rupture of the stomium by differential shrinkage when the anther dries up. In some species, the stomium is a pore formed on the side or at the apex of the anther lobe.

Gynoecium

The morphology of the gynoecium and the pertinent terminlogy are eliciting more debate than those regarding any other part of the flower.[16] Some of the outstanding controversies concern (1) the nature of the carpel, that is, whether the carpel is a phyllome, axial structure, or a special kind of organ not related to others in the flower (*sui generis*); (2) the meaning of the so-called congenital union (or fusion) of carpels assumed to be present when a gynoecium arises as a unit during ontogeny; and (3) the delimitation of the carpel or carpels in flowers with inferior ovaries. The terms ovary and ovule are deemed inappropriate by some botanists because the definitions of these terms in botany do not agree with those of the same terms in zoology.[66]

THE CARPEL

The carpel is classically interpreted as the basic unit of the gynoecium. A flower may have one carpel or more than one. If two or more carpels are present, they may be free from one another (*apocarpous* gynoecium; fig. 22.1,*A,C*) or they may be united (*syncarpous* gynoecium; fig. 22.1,*B*). A gynoecium with a single carpel is classified as apocarpous.

An old established term used with regard to the gynoecium is *pistil.* It refers to a single carpel in an apocarpous gynoecium (simple pistil) or to an entire syncarpous gynoecium (compound pistil). Some botanists advocate the abandonment of the term pistil[16] but, as defined above, the term continues to be useful.

As a foliar structure, the carpel is described as folded lengthwise in such a way that the adaxial surface is enclosed and the margins are more or less completely united. According to the prevalent view, the folding is *conduplicate,* with margins remaining flat (fig. 20.3,*A,* and 20.6,*E*). Evolutionary changes in this type of carpel are thought to have led to a reduction of the marginal areas (fig. 20.3,*B*). An older concept implies that the folding includes an *involution* of the margins so that the suture is lined with the abaxial surface (fig. 20.3,*C*). In mature state, the distinction between a conduplicate carpel with reduced margins and an involute carpel is difficult to make. Probably both types of folding of carpels occur depending on the taxon.[16]

The union of carpels in a syncarpous gynoecium follows two basic plans: the carpels are joined either in a folded condition, abaxial surface to abaxial surface (fig. 20.4,*C*), or in an unfolded or partially folded condition, margin to margin (fig. 20.4, *D*). The first type of arrangement results in a bilocular or multilocular gynoecium (fig. 20.4,*A*,*C*), the second, in a gynoecium with one locule (fig. 20.4,*D*). Secondary modifications may bring about variations in the basic plans. One example of such modification is found in the Brassicaceae in which two carpels are united by their margins but the gynoecium is not unilocular because a partition is formed by an outgrowth from the placentae at carpel margins (fig. 20.3,*D*).

The carpel of an apocarpous gynoecium or the entire syncarpous gynoecium is commonly differentiated into the lower fertile part, the *ovary,* and the upper sterile part, the *style* (fig. 20.5,*A,* and 20.7,*B*). A more or less extensive peripheral part of the style is differen-

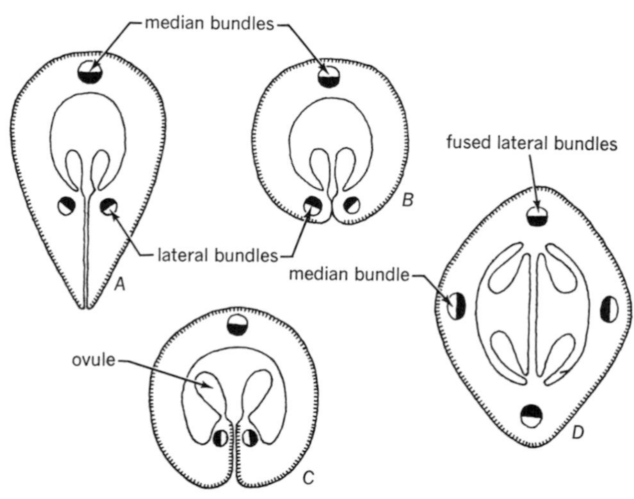

Figure 20.3 Diagrams of cross sections of ovaries composed of one (*A–C*) and two (*D*) carpels. *A,* conduplicately folded carpel with laminar placentation. *B,* conduplicately folded carpel with reduced margins. *C,* carpel involuted in folding. *D,* Brassicaceae type of ovary with a partition derived from the parietal placentae. (The median bundle is commonly referred to as dorsal bundle, the lateral as ventral bundle.)

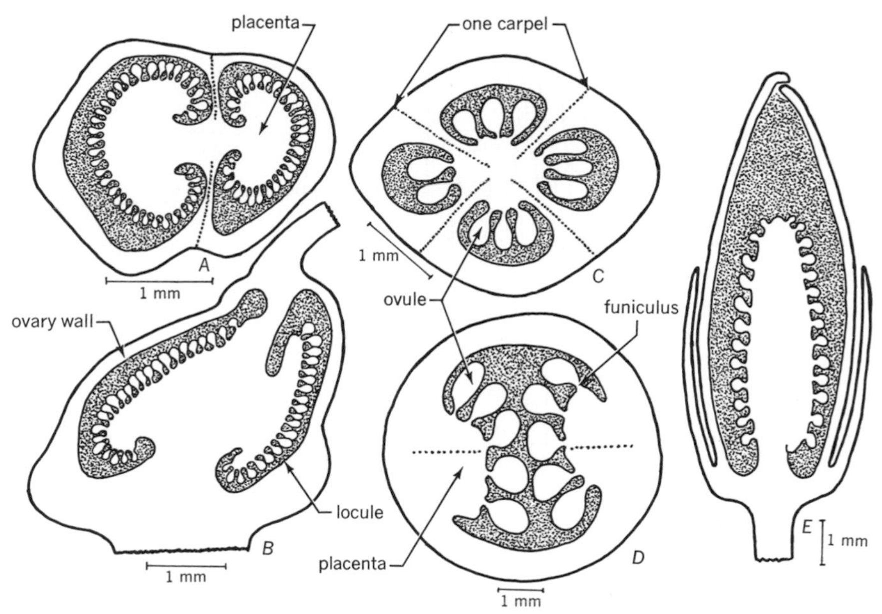

Figure 20.4 Placentation in ovaries in cross (*A, C, D*) and longitudinal (*B, E*) sections. *A, B, Antirrhinum,* axile, two carpels. *C, Fuchsia,* axile, four carpels. *D, Ribes,* parietal, two carpels. *E, Dodecatheon,* free central, five carpels.

tiated as a *stigma,* usually in the upper region of the style. If there is no extended structure that may be called style, the stigma is described as sessile, that is, sessile on the ovary. In the less specialized angiosperms the carpels appear as folded styleless structures with the stigmatic tissue covering the unsealed margins (fig. 20.13,*B*)

OVARY

Within the ovary one distinguishes the ovary wall, the locule (cavity) or locules, and, in a multilocular ovary, the partitions. The ovules are borne on certain regions of the ovary wall located on its inner (adaxial) side. An ovule-bearing region constitutes the *placenta.* Each carpel has two placentae. In a given carpel, the placentae occur either close to the margins or at some distance from them. On this basis, one distinguishes between *marginal* (fig. 20.3,*C*) and *laminar* (fig. 20.3,*A*) placentations. The latter is considered to be the more primitive. A placenta may be a conspicuous outgrowth, sometimes almost occluding the lumen of the ovarian cavity (fig. 20.4,*A*,*B*).

The position of the placentae in an ovary is related to the method of union of carpels.[50] The placentae of marginally joined carpels are said to be inserted on the ovary wall and the placentation is called *parietal* (fig. 20.4,*D*). If the carpels are joined in a folded condition, the ovary is bilocular or multilocular and the placentae occur in the center of the ovary where the carpel margins meet (fig. 20.4,*A*,*C*). This kind of placentation is called *axile.* Various modifications of these two basic placentations occur in different taxa. The partitions in a multilocular ovary may disappear in terms of either phylogeny or ontogeny,[16] so that a *free central* placentation

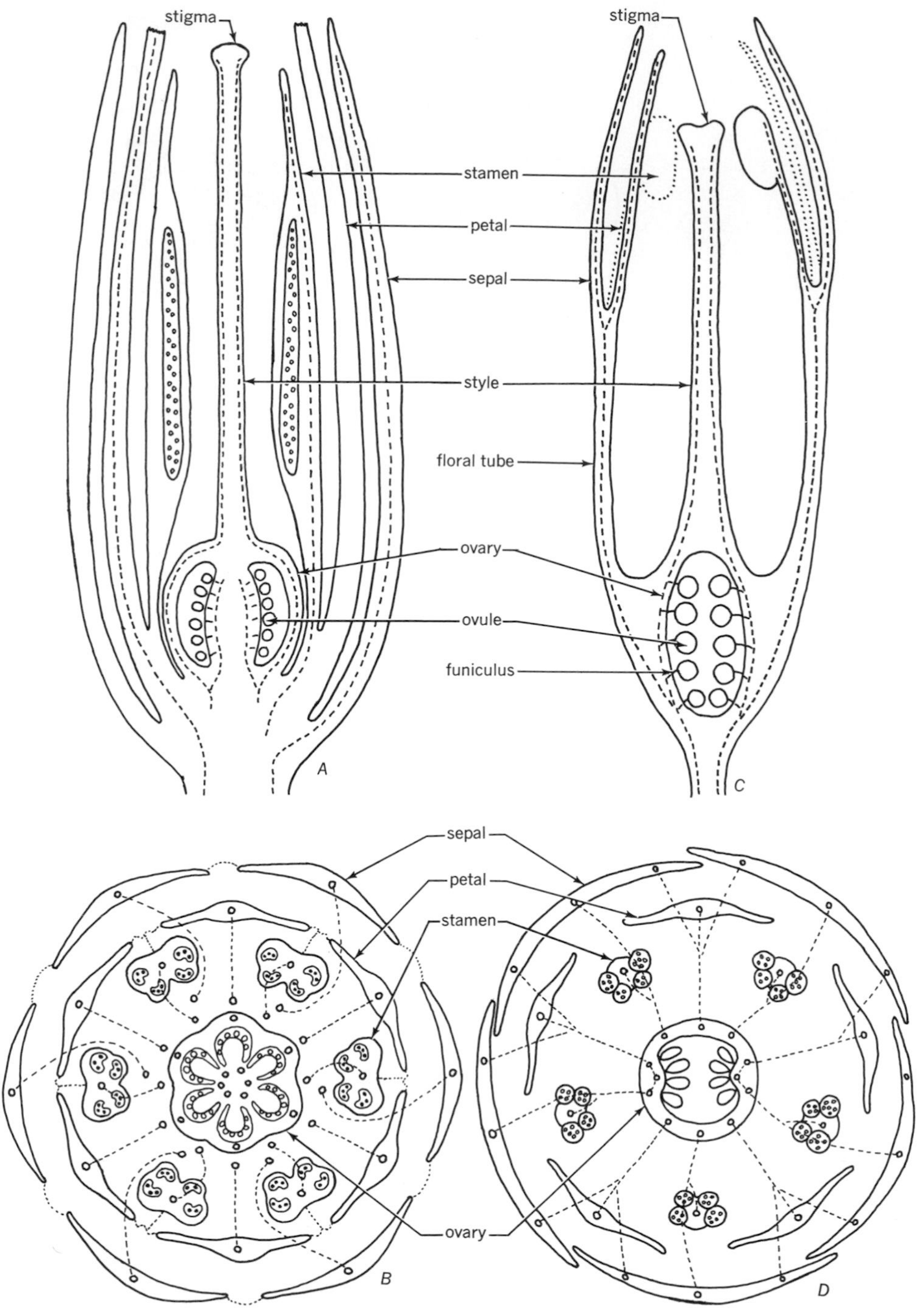

Figure 20.5 Diagrams of flowers in longitudinal (*A, C*) and cross (*B, D*) sections. *A, B, Lycopersicon,* horticultural multicarpellate variety of tomato; hypogynous, axile placentation. *C, D, Ribes;* epigynous, parietal placentation. Broken lines indicate course of vascular bundles and their interconnections.

results (fig. 20.4,*E*). The placenta in a unilocular ovary may occur at the base and the placentation is then called *basal.*

The understanding of placentation is materially aided by examination of the vascularization of the carpel. Most commonly the carpel has three veins, one median ("dorsal") and two lateral ("ventral;" fig. 20.3,*A–C,* and 20.6,*E,F*), and the vascular supply of the ovules is commonly derived from the lateral bundles (fig. 20.6,*D*). In an ovary with axile placentation the lateral bundles appear in the center of the ovary (fig. 20.5,*B*), with the phloem turned inward (adaxially), the xylem outward (abaxially). The lateral bundles of two adjacent carpellary margins—of the same carpel (figs. 20.3,*B*, and 20.7,*A*) or of two different carpels (fig. 20.3,*D*, and 20.5,*B*), depending on placentation—may be more or less completely united.

The interpretation of the placentae as carpellary in origin can be readily accepted for many representatives of angiosperms. In some, however, the relation between carpels and ovules is obscure. It has been suggested, therefore, that the placentae may be of axial origin, at least in some groups of plants.[43] Such placentae are supposed to occur on the receptacle that constitutes the bottom of the ovary (e.g., Poaceae[4]) or is prolonged upward in the center of the ovary (some representatives with central placentation). This view is contested by investigators who insist that despite mature appearances the placentae originate from the carpels.[17]

The delimitation of the carpels and the interpretation of origin of placentae is even more difficult in epigynous flowers (fig. 20.5,*C,D*). In such flowers the ovary is visualized as embedded in extracarpellary tissue derived either from the receptacle[37] or from the floral tube , that is, fused bases of sepals, petals, and stamens.[15] Some authors consider that the nature of the extracarpellary tissue varies in different groups of angiosperms.[30] There is also the problem of recognizing whether the carpels actually line the extracarpellary tissue or whether they are limited to the upper part of the gynoecium, that is, the part that forms the top of the ovary and bears the styles.[37] If the latter concept is accepted, the placentae would not be derived from the carpels. The nature of the cup enclosing the gynoecium but not joined to it in *perigynous* flowers (fig. 20.7) is similarly subject to various interpretations.

When the flower is fully expanded (*anthesis*) the ovary is not highly differentiated histologically. Its wall consists mainly of parenchyma penetrated by vascular bundles, although in some taxa the gynoecium may develop sclerified layers and accumulate crystals and tannins as protective devices.[63] The outer epidermis is cuticularized and may have stomata. The ovary undergoes its major histologic differentiation when it develops into the fruit.

STYLE AND STIGMA

The style is an upward prolongation of the carpel (fig. 20.6,*F*). In syncarpous gynoecia the style, if single, is derived from all the carpels composing the gynoecium (fig. 20.5,*A,C*). The carpels may be incompletely united, with the style a single compound structure at the base and a multiple structure at the top; or there may be as many stylar units (stylar branches, or stylodes) as there are carpels in the syncarpous ovary. The styles and stylodes may be solid or have a canal in the center. In most angiosperms the styles are solid. The mature stigma provides an environment suitable for the germination of the pollen grains and is called receptive when it reaches mature state. Receptive stigmas may be covered with secreted material ("wet stigmas") or they may lack such

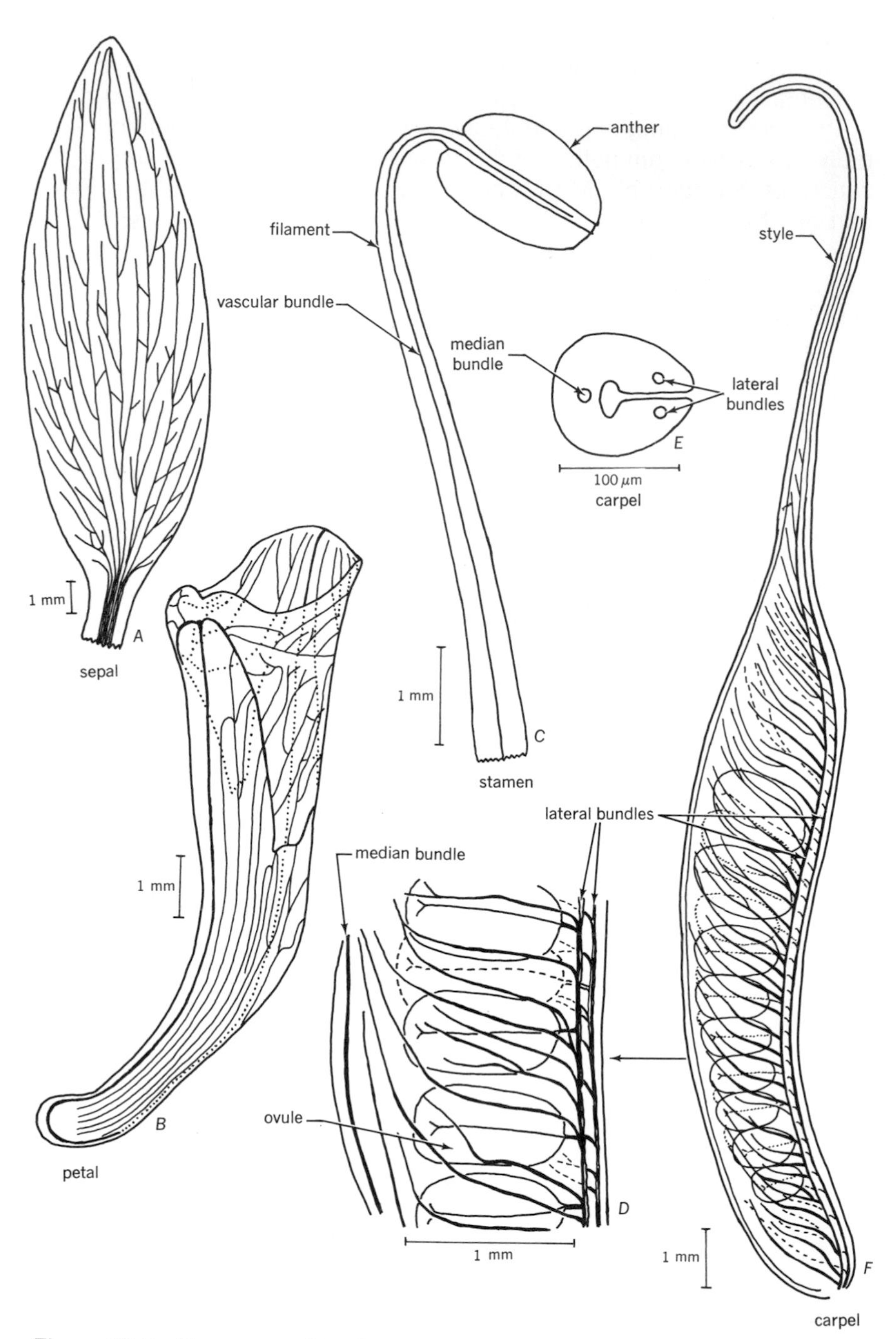

Figure 20.6 Flower parts of *Aquilegia*. Longitudinal views of sepal (*A*), petal (*B*), stamen (*C*), and carpel (*D, F*), and cross section of carpel (*E*). (Redrawn from Tepfer.[58])

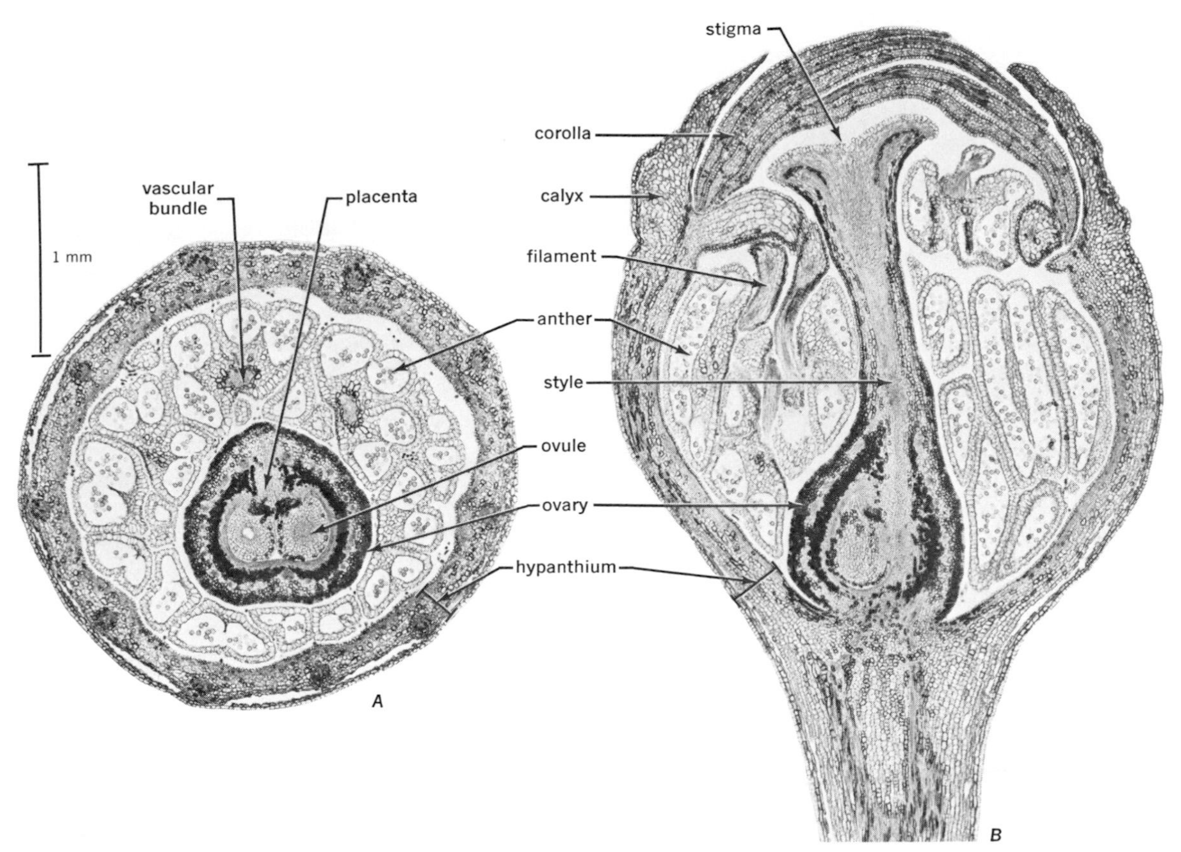

Figure 20.7 The flower of choke-cherry (*Prunus virginiana*) in transverse (*A*) and longitudinal (*B*) sections. The floral cup (hypanthium) extending above the ovary is not united with the latter. The flower is perigynous. The single carpel encloses two ovules. The stamens are crowded in the hypanthium because the flower was not yet open.

material ("dry stigmas").[29] The wet type of stigma is essentially a glandular tissue. The epidermal cells of the stigma are commonly elongated into papillae (fig. 20.8), short hairs, or long branched hairs (fig. 20.14). The stigmatic tissue is connected with the ovule in the ovarian cavity by the pollen *transmitting tissue,* which serves as a path for the growing pollen tube and as a source of nutrients.[36] In styles having canals, the transmitting tissue lines the canal. In solid styles the transmitting tissue forms one or more strands embedded in the ground tissue or associated with the vascular bundles.

The glandular stigma resembles a nectary in structure and function. The epidermis and the subepidermal layers produce the secretion that eventually forms a film over the epidermal walls. The secretion, analyzed in a number of angiosperms,[40] contains little or no free sugar but consists principally of lipid and phenolic compounds (anthocyanins, flavonoids, cinnamic acids). The lipids probably correspond to the waxy component of the epidermal wall and serve to prevent loss of water. The phenolics occur as glycosides and esters and their hydrolysis may provide the sugars necessary for pollen germination.

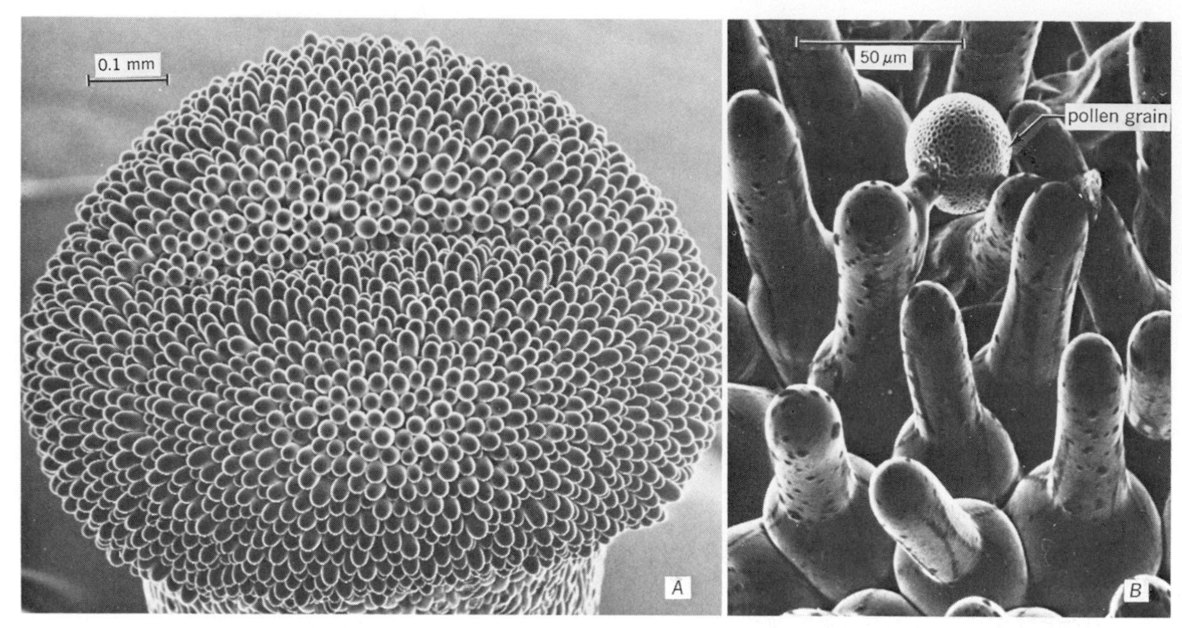

Figure 20.8 Stereoscan pictures of stigma of *Brassica oleracea,* kale. *A,* immature stigma from a bud 5 mm long; the papillae are closely appressed to one another. *B,* mature papillae have swollen bases; a pollen tube from a pollen grain is attached to a papilla. (From Okendon;[47] the photographs were taken at Long Ashton Research Station, Bristol, by courtesy of Dr. W. E. Thomas.)

The phenolics could have other functions as well, such as protection from insects, inhibition of infectious diseases, and stimulation or inhibition of pollen germination, perhaps in relation to the phenomena of compatibility and incompatibility in pollination. Studies of pollination in *Brassica* with the scanning electron microscope (fig.20.8) indicate that the incompatibility barrier is not a physical one but must involve an incompatibility substance.[47] In a survey of some eighty angiosperm families,[41] the surfaces of the stigmatic papillae were found bearing an external protein pellicle. The authors propose that the hydrophilic protein pellicle may function in the capture and hydration of the pollen grains and also may be the site of the recognition reactions involved in incompatibility responses.

As recorded in *Raphanus* flowers, which have glandular stigmas with papillae, the pollen tube penetrates the cuticle of a stigmatic papilla[14] before it begins to elongate. In the nonglandular stigma of cotton, the pollen tube grows along the surface of a stigmatic hair.[29] In both types of flower, the pollen tube eventually reaches the transmitting tissue of the style. In cotton, which has a solid style, this tissue consists of thick-walled cells arranged in vertical files. The layered wall is rich in pectic substances and the pollen tube grows within one of the layers. In some other species with solid styles, the transmitting tissue consists of cells surrounded by an intercellular substance pectic in nature and mucilaginous in consistency. During ontogeny of the style, this substance appears in the intercellular spaces as the original contact between cells is broken.[16,52] The pollen tube grows through the mucilaginous substance.

In styles with canals, the transmitting tissue is the glandular lining of the canal. The pollen tube grows along this tissue or deeper in the lining, apparently without penetrating the cells. In lily pistils, the walls of the transmitting tissue facing the canal have ingrowths of the type recorded in transfer cells.[51] The lily pistil has been used for study of the possible regulation of the direction of growth of the pollen tube by substances produced by the gynoecium.[51] The results showed that the transmitting tissue attracts pollen tubes growing on agar and that, in the pistil, the chemotropic activity of the transmitting tissue and its exudate begins distally at anthesis and progresses basipetally in advance of the growing pollen tube.

Vascular system

The vascular system of the flower has been studied extensively with regard to taxonomic and phyletic relationships among the angiosperms and to the morphologic nature of the flower and its parts.[15,16,46,49,65] The use of the vascularization pattern in interpretations of the flower is considered to be justifiable because, with regard to the evolutionary specialization, the vascular system appears to be slower to change (more conservative) than the organs that it serves and consequently may reveal former boundaries, numbers, and categories of organs in flowers in which reduction, crowding, and various types of union of parts obscures the identity and the interrelationships of the organs.[46] At the same time, the value of vascular structure as an evidence of phyletic relationships varies with the taxon and therefore the data on vascular organization must be correlated with those of other characters of the flower.

In relatively unspecialized flowers the vascular organization is readily interpreted as being comparable to that of the vegetative shoot. The vascular system of the floral axis can be described as a cylindrical complex of sympodial bundles from which the traces to the floral parts are derived. At successive levels certain bundles assume an oblique course and become part of the vascular supply of the floral parts inserted at that level. Because of the common shortness of internodes in flowers and various modifications in the interrelationships of floral parts, the bundles often branch and combine with each other in a more irregular manner than they do in a vegetative shoot.[45,56]

The number of traces of the different floral parts varies in the same flower. The usual pattern is as follows (fig. 20.5, broken lines, and 20.6). Each sepal has as many traces as the foliage leaf of the same plant. Each petal in a dicotyledon has one trace; each perianth member (tepal) in a monocotyledon, one to many. Within the sepal and petal the vascular bundles form more or less complex systems resembling those in foliage leaves (fig. 20.6,*A*,*B*). The specialized type of stamen has one trace, which is continued as a single bundle in the filament and anther (fig. 20.6,*C*). The carpel has three traces (fig. 20.3,*A*–*C*), and their prolongations within the carpel may have branches (fig. 20.6,*D*,*F*). The ovules are supplied by branches from carpellary bundles, usually the two lateral ones (fig. 20.6,*D*). Carpellary vascular bundles are continuous through the style (fig.20.5,*A*,*C*). As was mentioned regarding the carpels, the vascular bundles of floral parts may be fused when these parts are united.

In some epigynous flowers certain bundles show an inverted orientation of xylem and phloem (fig. 20.9). Such arrangement is inter-

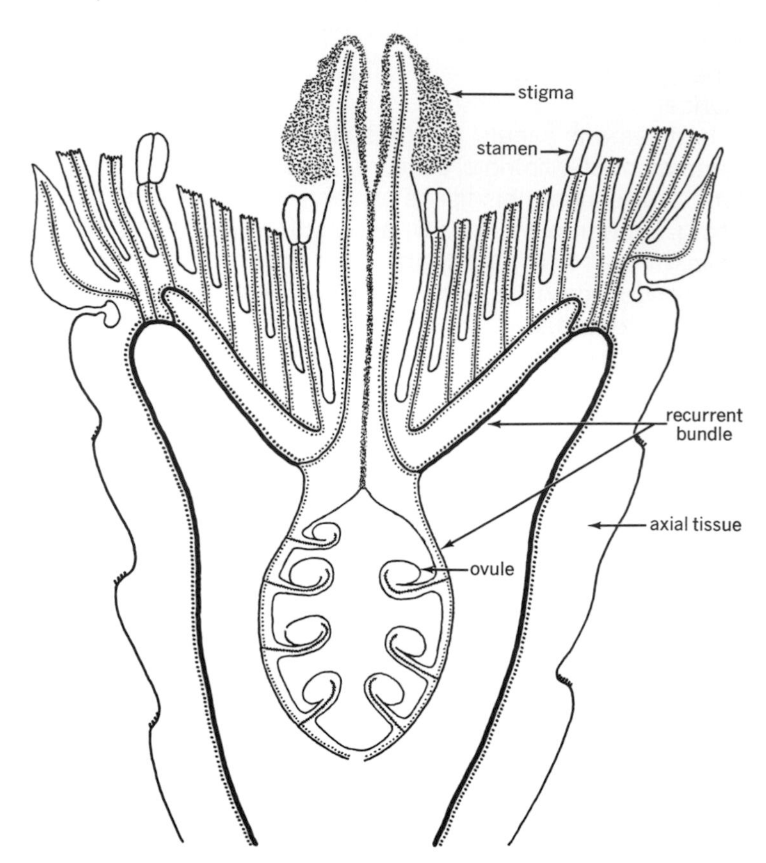

Figure 20.9 Diagram of a longitudinal section of flower of *Opuntia dillenii*. Epigynous, ovary embedded in axial (receptacular) tissue. Recurrent bundles have phloem (indicated by dots) on the adaxial side. The bundles are amphicribral in upper parts of the flower. (Redrawn from Tiagi.[60])

preted as an indication that the ovary is embedded in receptacular (axial) tissue, whereas absence of inverted bundles is regarded as evidence that the extracarpellary tissue is appendicular in nature (floral tube).

The difference between the two kinds of epigynous flowers can be visualized as a result of contrasting morphogenetic events in their development.[30] In a flower interpreted as having a floral tube enclosing the ovarian cavity , the meristematically active floral apex becomes increasingly more concave, for the peripheral region of it becomes elevated as the successive floral organs are initiated. An upward zonal growth of the united bases of the perianth members accentuates the concavity. Because of the uniform upward growth of the tissue around the apical meristem the vascular bundles have a direct acropetal course in the tissue enclosing the ovarian cavity and in the tube above, with the traces to the individual organs diverging from them at the appropriate levels (fig. 20.5,*C,D*).

In a flower in which the ovary is assumed to

be embedded in axial tissue,[60] the floral organs arise, as usual, at the periphery of the apical mound. Concomitantly, the axis expands radially and then becomes cup-shaped by growing upward around the periphery. The developing vascular tissue is accommodated to the curvature of the cup so that the parts of the bundles located underneath the bases of stamens and perianth members assume an inclined course (fig. 20.9). This part of the vascular system constitutes the recurrent bundles in which the phloem appears on the side opposite to that in the vertical bundles.

DEVELOPMENT

When the plant reaches the reproductive stage of development some or all of the apical meristems on its shoot cease to initiate foliage leaves and begin to produce floral parts according to a sequence characteristic of the species, with a variable number of bracts intervening between the leaves and the flower. In this process, the apical meristems pass from indeterminate to determinate growth because the formation of the flower is usually the final event in the activity of a given apical meristem. In annual plants, the advent of the reproductive stage means also the approaching completion of the entire life cycle. In perennials, flowering is repeated a variable number of times depending on the longevity of the plant.

A flower may arise at the apex of the main shoot or at those of the lateral branches, or in both places. In many species, the change to flowering involves the formation of an inflorescence, a process not readily distinguishable histologically from the formation of a flower because, in an inflorescence, the two kinds of events merge in a continuous chain of successively modified morphologic and physiologic states.

An often observed phenomenon during the initiation of the reproductive stage is a sudden and rapid elongation of the axis. Such growth is particularly striking in plants having rosette-like habit during the vegetative stage as, for example, grasses and bulbous plants[4,42]. The elongated axis produces either a single flower or an inflorescence. When flowers are borne on branched inflorescences an accelerated production of axillary buds indicates the approaching flowering (fig. 20.15,*E*).

Induction of flowering

Floral initiation is controlled by external factors but only within the limits of reactivity of the plant to a given environment.[26] Many plants have characteristic responses to length of day (photoperiod) and to temperature and enter the reproductive stage under specific combinations of these two factors. Depending on their response to the length of day, the plants are identified as long-day, short-day, and neutral-day plants. The endogenous circadian rhythm is apparently implicated in the photoperiodic time measurement by plants. Furthermore, the plant must undergo a certain amount of development before it reaches the "ripeness-to-flower" stage and responds to the required photoperiodic condition. Many plants require an exposure to low temperature before they can produce flowers. In the practice of vernalization, germinating seed of such plants is given a cold treatment in order to hasten subsequent flowering.

In a plant ready to bloom, the effect of the appropriate photoperiod, mediated by phytochrome, results in the synthesis of a transmissible factor referred to as the floral stimulus. According to some studies, the stimulus is formed in leaves, including the cotyledons, and is transported to the apical meristems where it evokes changes committing the plant to subsequent flower formation. During the evocation stage, a rise in RNA synthesis has been observed as the earliest change in the apical meristems in several species.[7,54] A related phenomenon is the increase in amount of total protein and the associated formation of new ribosomes.[28] A rise in the mitotic index is another early event following the arrival of the floral stimulus in the meristem.[32]

Quantitative ultrastructural studies reveal a number of cytological changes in evoked meristems.[23,24] The earliest detectable change is a replacement of the large vacuoles carried over from the vegetative apices by numerous small vacuoles.The number of mitochondria increases parallel with an intensification of succinic hydrogenase activity, a reflection of a rise in cellular respiratory activity. An increase in nucleolar size accompanied by marked vacuolation is a relatively late event. Of particular interest is the increase in the degree of dispersal of the chromatin in enlarged nuclei. The dispersed chromatin/condensed chromatin ratio is thus higher in the evoked meristem than in the vegetative. It is known from studies of plant and animal cells that dispersed chromatin is far more active than the condensed chromatin in suppporting the transcription of genetic information from DNA.[24]

After the evocation stage, DNA synthesis is stimulated, and further rise in mitotic activity occurs. Both processes are involved in the production of cells that will give rise to flower primordia. Thus the plant enters the stage of floral morphogenesis. Whereas ontogenetic reversion of reproductive meristems to a vegetative state may occur during the evocation stage, with the establishment of the morphogenetic stage flower production becomes almost inevitable.[7]

Floral meristem

When the apical meristem enters the reproductive stage it undergoes more or less conspicuous morphologic changes (fig. 20.10). These changes are related, at least in part, to the altered mode of production of lateral appendages after the indeterminate growth of the vegetative stage ceases. During the latter, the apical meristem grows upward and widens before a new foliar plastochron begins. During the development of the flower, in contrast, the area of the apical meristem gradually diminishes as the successive floral parts arise (fig. 20.10,*B*,*C*). In some flowers a certain amount of apical meristem remains after the carpels are initiated (fig. 20.13,*G*), but the tissue ceases to be active; in others, the carpels or the ovules appear to arise from the terminal part of the apical meristem (fig. 20.10,*D*). Depending on the type of flower, its organs, or at least some of them, originate acropetally in helical sequence[62] or the organs of a given kind arise approximately at one level in one whorl or two closely succeeding one another (fig. 20.13). The whorled condition is considered to be the more specialized.

Small depth and a comparatively broad expanse of the meristematic tissue are common histologic features of the floral meristem. The broad apex is occupied by a mantle of meristematic cells covering a vacuolated core of ground tissue no longer concerned with upward growth. These features may be encountered in meristems of single flowers (fig. 20.10,*B*) and in those of inflorescences (fig. 20.11). The tunica and corpus organization may or may not be identifiable in the floral

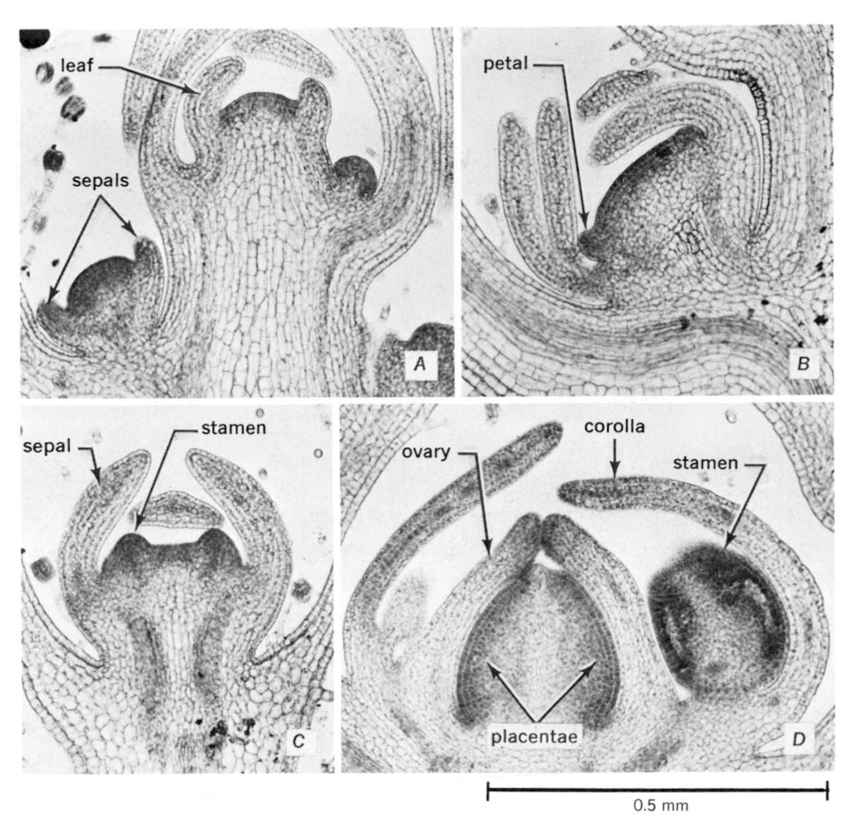

Figure 20.10 Early stages of flower development in *Antirrhinum,* snapdragon. Longitudinal sections. *A,* vegetative (above) and floral (right and left below vegetative) apices. *B,* flower with sepals and primordia of petals. (Petals are initiated as discrete units, later intercalary growth at their bases forms a sympetalous corolla.) *C,* flower cut in plane exposing stamen primordia. *D,* flower with gynoecium; carpels enclose massive placentae but are not yet prolonged into a style. (Part of young style in fig. 20.4, *B.*)

meristem. If it is present, the number of tunica layers may be the same as in the vegetative apex of the same species or may be smaller or larger.[20]

The initiation of floral organs involves both an increase in mitotic activity and a change in distribution of this activity. The appearance of a uniform meristematic mantle zone is a result of effacement of the difference between the less active distal zone and the more active peripheral zone seen in vegetative apices. Mitotic activity and the concentration of DNA become more uniformly distributed than they are during leaf formation.[20]

Origin and development of floral parts

HISTOGENESIS

Floral organs are initiated like foliage leaves by periclinal divisions of cells located more or less deeply beneath the protoderm[58,61] or also in the protoderm itself (often

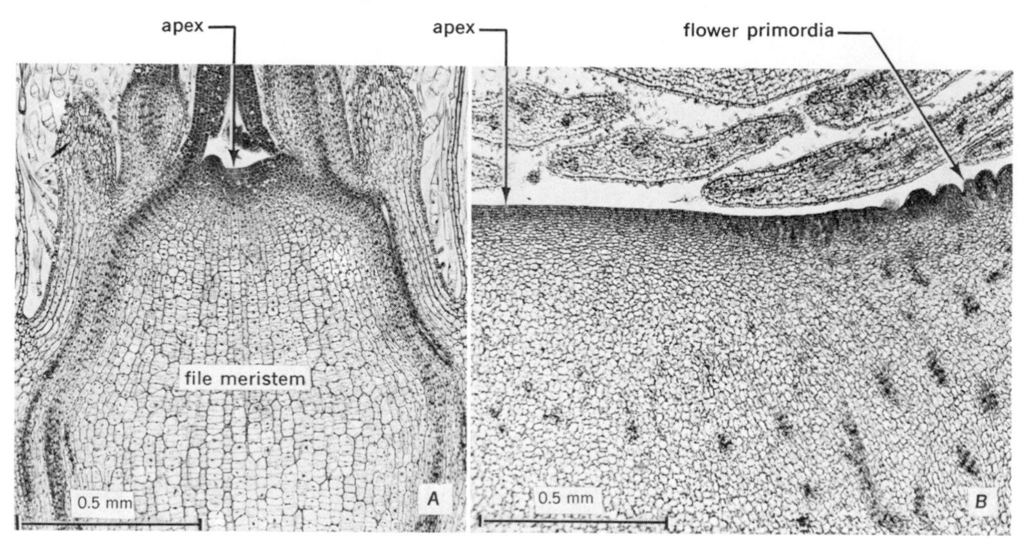

Figure 20.11 *Helianthus annuus,* sunflower, vegetative shoot apex (*A*) and less than half of floral apex (*B*). (*A,* from K. Esau, *Amer. J. Bot.* 32:18–29, 1945. *B*, slide courtesy of A. T. Guard.)

in monocotyledons[4,33]). The depth of the site of these divisions may be the same as in the origin of leaves in the same species or may be different; moreover, the depth may vary for the different parts of the same flower.

The location of the initial divisions is often used to interpret homologies in floral organs. Shallow divisions are regarded as evidence that the organ is leaflike, deeper divisions, as sometimes observed in the origin of stamens,[12] are assumed to denote axial nature. The value of such evidence is questioned in studies of cytochimeras (chapters 16 and 18) which make possible the identification of the origin of a tissue from specific layers of the apical meristem.[13] Such studies indicate a high degree of flexibility in the number of cells in a given organ derived from the first three layers of the apex.

The initial periclinal divisions are followed by others, including anticlinal, and the primordium becomes a protuberance (fig. 20.10,*B*). Growth in width and length soon reveals the dorsiventral form of the protuberance as in leaf primordia. The staminal filament is an exception, for it remains narrow. Because floral organs are often comparatively small and succeed one another at close intervals, a buttresslike prominence may not be identifiable at their initiation.

The sepals resemble foliage leaves most closely in their initiation and development. The growth pattern of the petals is more or less similar to that of the leaves. The stamens arise as stout short structures (fig. 20.10,*C*,*D*), the filament developing subsequently by intercalary growth. The elongation of the filament in *Cleome hassleriana* was found to be under control of the anther, which was the source of auxin (IAA) inducing the elongation.[34,35] If the perianth parts and stamens show cohesion or adnation, the union of parts may be congenital or ontogenetic or a combination of the two methods. If congenital union is present, the product of the presumed union develops as one structure by zonal intercalary growth.[30] When cell lineages were followed in the congenitally united floral tube in a cytochimeral peach (*Prunus;* another species in fig. 20.7) the tube, in agreement with the concept of congenital union, was found lacking any trace of demarcations suggesting component parts but showed continuous concentric layers of derivatives from the three outermost layers of the apical meristem.[13] The compound nature of such floral tube may be deduced from an observation in *Downingia* in which the sepal primordia arise separately and the tube is formed by growth of their common base.[30]

The development of the gynoecium varies in detail in relation to the union of carpels with one another and with other floral parts. If the carpels are not united, the individual carpel primordium is initiated at one point by periclinal divisions in the apex. This point becomes the median part of the carpel from which divisions are propagated in the adaxial direction and lead to the formation of a horseshoe-shaped or a circular welt (figs. 20.12,*A–C,* and 20.13,*G,H*). Upward growth, comparable to marginal growth in a leaf, elevates the carpel as a saclike structure. In some species, intercalary elongation at the base of the carpel gives rise to a congenitally united basal part (fig. 20.12,*D,E*).

In syncarpous ovaries the carpels arise as individual primordia or jointly as a unit structure in which the delimitation of the individual carpels is obscured. Various degrees of congenital and ontogenetic unions are exhibited by the developing gynoecium in different species.[22,44] The ontogenetic union may be so firm that the suture is not identifiable in the mature state. Such union involves enlargement and interpenetration of epidermal cells,

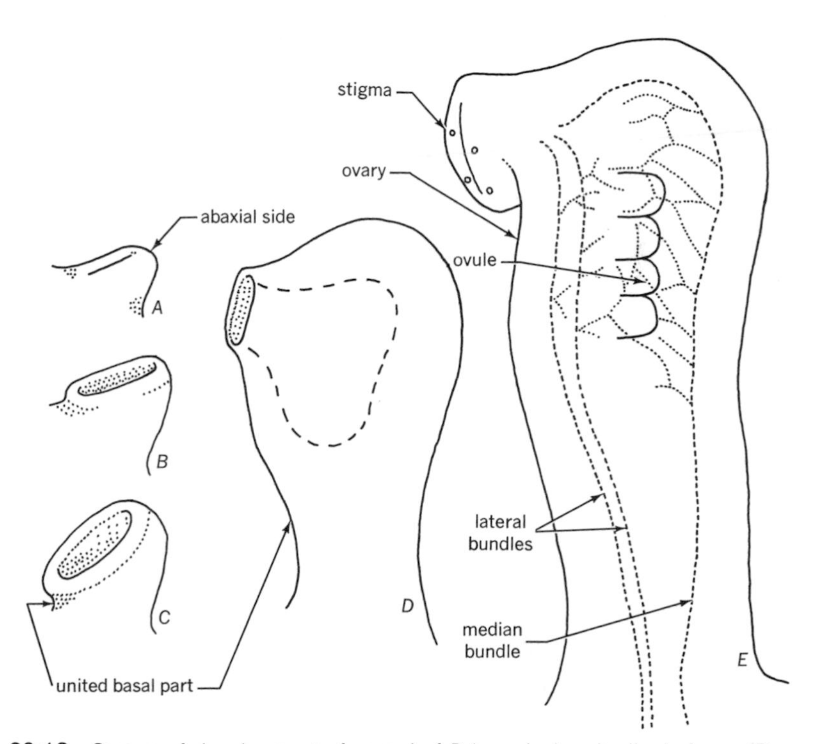

Figure 20.12 Stages of development of carpel of *Drimys* in longitudinal views. (Redrawn from Tucker.[61])

Figure 20.13 Development of flower of *Butomus umbellatus* shown in surface views of partly dissected flowers. *A*, entire flower. *B*, mature gynoecium. *C–F*, stages of development of the successive floral parts including the inception of carpels. *G*, *H*, two subsequent stages of carpel development. Details *c*1, carpel in outer whorl; *c*2, carpel in inner whorl; *f*, floral apex; *it*, inner tepal; *ot*, outer tepal; *s*, stamen. (From Singh and Sattler.[55] Reproduced by permission of the National Research Council of Canada from the *Canadian Journal of Botany*, Volume 52, 1974. pp. 223–230.)

sometimes accompanied by division of the epidermal cells.[22] Ontogenetic union of carpel margins in the cytochimeral peach mentioned previously was identifiable by the arrangement of the derivatives from two outermost apical layers in bands radially oriented with reference to the suture.[13] Depending on the degree of union of carpels the style of a syncarpous pistil grows out from the ovary either as a unit structure or as stylodal prolongations of the individual carpels, free or partly united.

ORGANOGENESIS

In contrast to the repetitive pattern in foliage leaf formation, the morphogenetic pattern in floral development changes from one set of floral parts to the next one. The difference in the form between the successive types of floral parts is evident from early ontogeny. Figure 20.13 illustrates this feature for the flower of *Butomus umbellatus* (a monocotyledon) by means of low-power photographs of partly dissected flowers in different stages of development.[55] The 6 tepals, 3 outer and 3 inner, arise as lateral protrusions that soon assume a dorsiventral shape (*C–F*). The 9 stamens, 6 in pairs opposite the outer tepals and 3 single ones opposite the inner tepals, appear as bumps that represent the anthers. The two-lobed condition of the future anthers is slightly indicated in *F*. The 6 carpels arise in two whorls (*F*), one opposite the outer

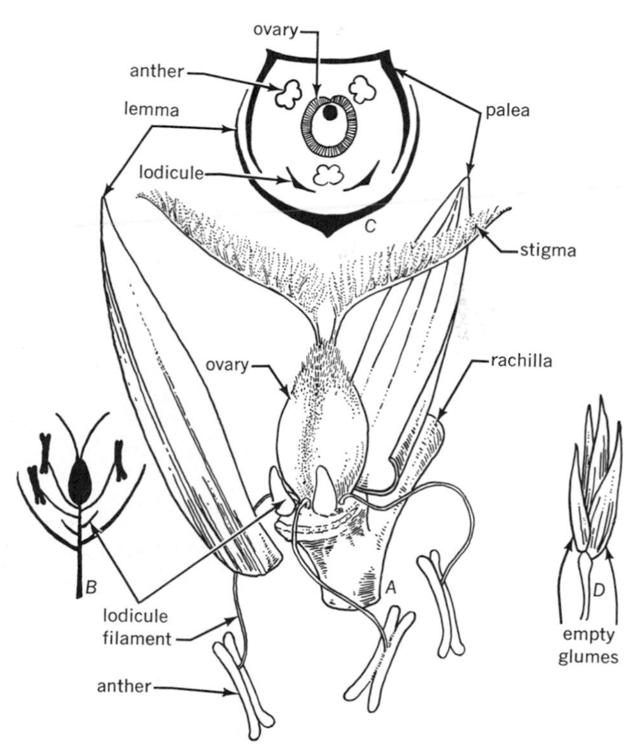

Figure 20.14 The grass flower. *A,* partly dissected flower at anthesis. *B,* longitudinal and, *C,* transverse diagrams of flower. *D,* spikelet. Lodicules are small scales outside the stamens. (From K. Esau, *Plant Anatomy,* 2nd ed. John Wiley & Sons, Inc. 1965; originally redrawn from A. M. Johnson, *Taxonomy of the Flowering Plants,* Appleton-Century-Crofts, Inc. 1931.)

tepals (*c*1) and the other opposite the inner tepals (*c*2). Starting as rounded bumps on the side of the apex the carpels become horseshoe-shaped (*G*) and then leaflike by marginal growth. With upward growth they assume the form of conduplicately folded leaves (*H*). Their adaxial margins do not fuse but a sessile stigma develops on the distal marginal surface (*B*). Some residual tissue from the apical meristem is retained above the carpels.

The sequential changes in the pattern of organ formation has been well illustrated in studies of the development of inflorescences and flowers of grasses.[3,4,8,9,33] The mature structure of a grass spikelet and flower is ex-

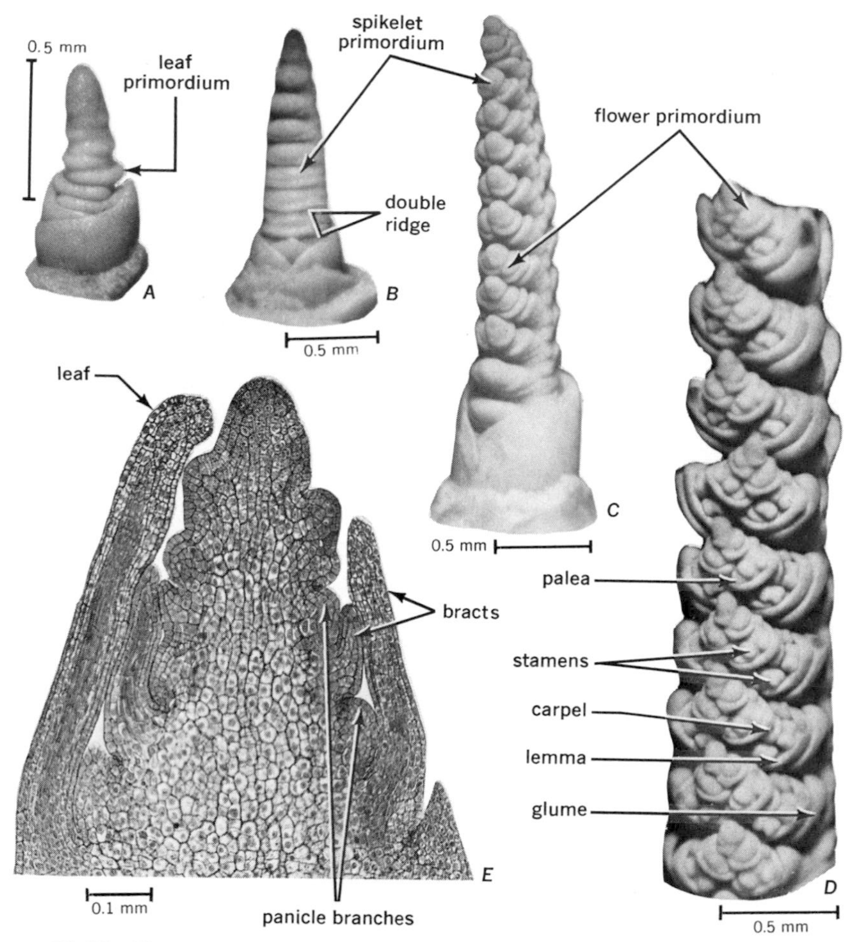

Figure 20.15 Floral initiation in grasses. *A–D, Triticum* (wheat). *A*, vegetative apex with leaf primordia, somewhat elongated in advance of spike formation. *B*, young spike with double ridges, each composed of spikelet primordium and subtending leaf primordium. *C*, spike with spikelets developing first flower primordia. *D*, part of spike with spikelets having several florets each. *E*, longitudinal section of shoot apex of *Bromus* at transition to flowering stage; initiation of panicle branches. (*A–D*, from Barnard.[3] *E*, from J. E. Sass and J. Skogman, *Iowa State Coll. J. Sci.* 25:513–519, 1951.)

plained in figure 20.14 and 20.16,*F,G*. During development, the rapidly elongating inflorescence of a spike type forms first a series of semicircular ridges, the leaf primordia (fig. 20.15,*A*). At a later stage, double ridges appear (fig. 20.15,*B*). The doubleness results from initiation of spikelet primordia in the axils of the leaf primordia. The latter develop less and less strongly toward the apex of the spike and are overtopped by the developing spikelets. In a panicle type of inflorescence, branches of one or more orders arise in the

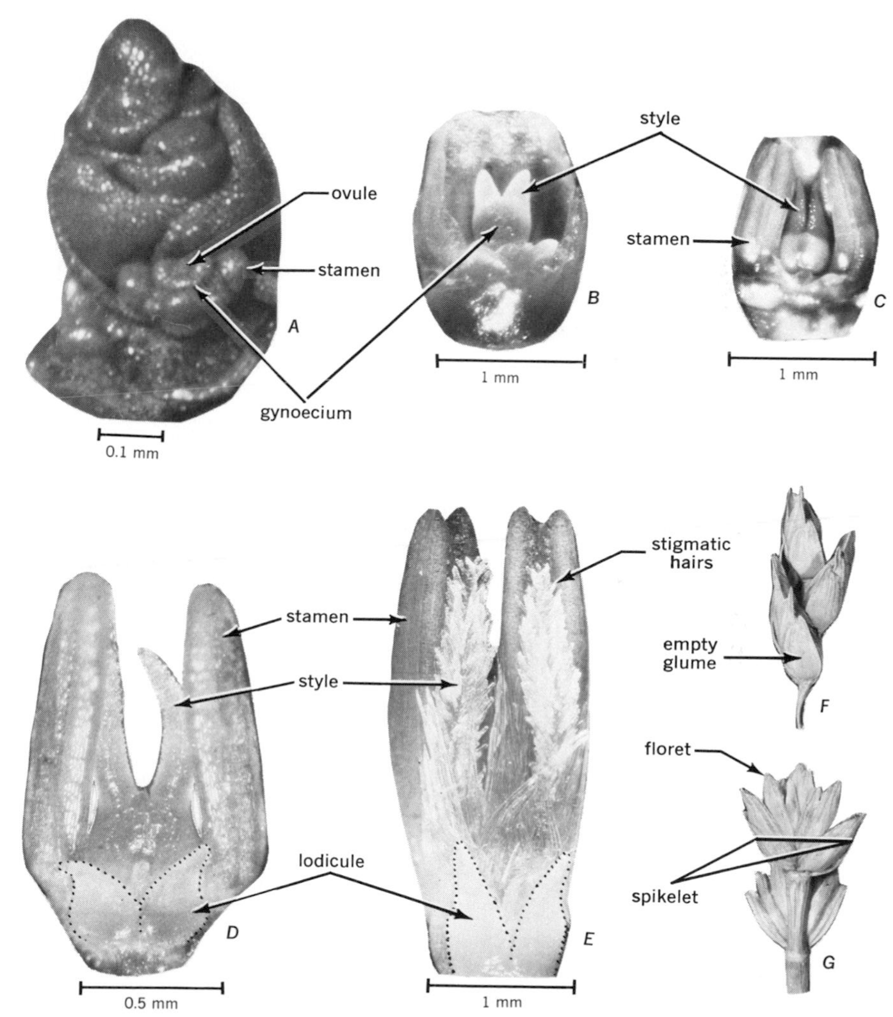

Figure 20.16 Florets of Cyperaceae (*A*) and Poaceae (*B–G*). *A*, spikelet of *Scirpus* with spiral arrangement of glumes and flower primordia; gynoecium still open, ovule exposed. *B, D, E*, florets of *Avena* (oat) showing gynoecium in three stages of development. *C*, part of floret of *Triticum* (wheat) with young gynoecium. *F, G*, pieces of wheat spikes showing spikelets in side view (*F*) and abaxial (*G*, above) views. (*A*, from Barnard;[4] *B–G*, from Bonnett: *C*;[8] *B, D, E*;[9] *F, G*, courtesy of O. T. Bonnett.)

axils of bracts (fig. 20.15,*E*) and spikelets arise on the branches.

Within the spikelets, the glumes, florets, and floral organs develop in an acropetal sequence (figs. 20.15,*C,D,* and 20.16,*A*). The gynoecium of the Poaceae and Cyperaceae is often interpreted as syncarpous, consisting of two or three carpels. But it arises at the apex as a single ringlike structure surrounding the emerging single ovule[4] (fig. 20.16,*A*). In grasses, two styles arise above the ovary (fig. 20.16,*B–D*) and develop stigmatic hairs (fig. 20.16,*E*).

The morphological and functional dissimilarities among the different floral parts suggest that not one but a succession of appropriate physiologic states are involved in the differentiation of a flower. This concept is supported by experiments in which organ regeneration in surgically halved apices was used to determine the timing of the establishment of irreversible patterns during organogenesis.[25] It was found that the median distal tissue of the meristem gradually loses its potential for regeneration of a specific kind of organ, but that different plants show marked differences in the time at which the floral meristem is committed to the formation of a particular kind of organ. A sequential gene activation resulting in changes in the enzyme protein pattern is discussed as the possible mechanism regulating the process.

REFERENCES

1. Andrews, H. N. Early seed plants. *Science* 142:925–931. 1963.
2. Arber, A. *The natural philosophy of plant form.* Cambridge, Cambridge University Press. 1950.
3. Barnard, C. Histogenesis of the inflorescence and flower of *Triticum aestivum* L. *Aust. J. Bot.* 3:1–20. 1955.
4. Barnard, C. Floral histogenesis in the monocotyledons. I. The Gramineae. *Aust. J. Bot.* 5:1–20. 1957. II. The Cyperaceae. *Aust. J. Bot.* 5:115–128. 1957.
5. Barnard, C. The interpretation of the angiosperm flower. *Aust. J. Sci.* 24:64–72. 1961.
6. Beck, C. B. On the anatomy and morphology of lateral branch systems of *Archaeopteris. Amer. J. Bot.* 58:758–784. 1971.
7. Bernier, G. Structural and metabolic changes in the shoot apex in transition to flowering. *Can. J. Bot.* 49:803–819. 1971.
8. Bonnett, O. T. The development of the wheat spike. *J. Agr. Res.* 53:445–451. 1936.
9. Bonnett, O. T. The development of the oat panicle. *J. Agr. Res.* 54:927–931. 1937.
10. Carlquist, S. Toward acceptable evolutionary interpretations of floral anatomy. *Phytomorphology* 19:332–362. 1969.
11. Davis, G. L. *Systematic embryology of the angiosperms.* New York, John Wiley & Sons. 1966.
12. Dengler, N. G. Ontogeny of the vegetative and floral apex of *Calycanthus occidentalis. Can. J. Bot.* 50:1349–1356. 1972.
13. Dermen, H., and R. N. Stewart. Ontogenetic study of floral organs of peach (*Prunus persica*) utilizing cytochimeral plants. *Amer. J. Bot.* 60:283–291. 1973.
14. Dickinson, H. G., and D. Lewis. Cytochemical and ultrastructural differences between intraspecific compatible and incompatible pollinations in *Raphanus. Proc. Roy. Soc. London B.* 183:21–38. 1973.
15. Douglas, G. E. The inferior ovary. *Bot. Rev.* 10:125–186. 1944. II. *Bot. Rev.* 23:1–46. 1957.

16. Eames, A. J. *Morphology of the angiosperms.* New York, McGraw-Hill. 1961.
17. Eckardt, T. Vergleichende Studie über die morphologischen Beziehungen zwischen Fruchtblatt, Samenanlage und Blütenachse bei einigen Angiospermen. *Neue Hefte zur Morphologie* No. 3:1–91. 1957.
18. Faegri, K., and L. van der Pijl. *The principles of pollination ecology.* 2nd ed. Oxford, Pergamon Press. 1971.
19. Foster, A. S., and E. M. Gifford, Jr. *Comparative morphology of vascular plants.* 2nd ed. San Francisco, Freeman and Company. 1974.
20. Gifford, E. M., Jr., and G. E. Corson, Jr. The shoot apex in seed plants. *Bot. Rev.* 37:143–229. 1971.
21. Hagen, C. W., Jr. The differentiation of pigmentation in flower parts. I. The flavonoid pigments of *Impatiens balsamina,* genotype $11HHP^rP^r$, and their distribution within the plant. *Amer. J. Bot.* 53:46–54. 1966. II. Changes in pigments during development of buds in *Impatiens balsamina genotype* $11HHP^rP^r$. *Amer. J. Bot.* 53:54–60. 1966.
22. Hartl, D. Morphologische Studien am Pistill der Scrophulariaceen. *Österr. Bot. Z.* 103:185–242. 1956.
23. Havelange, A., and G. Bernier. Descriptive and quantitative study of ultrastructural changes in the apical meristem of mustard in transition to flowering. I. The cell and nucleus. *J. Cell Sci.* 15:633–644. 1974.
24. Havelange, A., G. Bernier, and A. Jacqmard. Descriptive and quantitative study of ultrastructural changes in the apical meristem of mustard in transition to flowering. II. The cytoplasm, mitochondria and proplastids. *J. Cell Sci.* 16:421–432. 1974.
25. Hicks, G. S., and I. M. Sussex. Organ regeneration in sterile culture after median bisection of the flower primordia of *Nicotiana tabacum. Bot. Gaz.* 132: 350–363. 1971.
26. Hillman, W. S. *The physiology of flowering.* New York, Holt, Rinehart and Winston. 1962.
27. Horovitz, A., and Y. Cohen. Ultraviolet reflectance characteristics in flowers of crucifers. *Amer. J. Bot.* 59:706–713. 1972.
28. Jacqmard, A., J. P. Miksche, and G. Bernier. Quantitative study of nucleic acids and proteins in the shoot apex of *Sinapis alba* during transition from the vegetative to the reproductive condition. *Amer. J. Bot.* 59:714–721. 1972.
29. Jensen, W. A., and D. B. Fisher. Cotton embryogenesis: the tissues of the stigma and style and their relation to the pollen tube. *Planta* 84:97–121. 1969.
30. Kaplan, D. R. Floral morphology, organogenesis and interpretation of the inferior ovary in *Downingia bacigalupii. Amer. J. Bot.* 54:1274–1290. 1967.
31. Kaussmann, B. *Pflanzenanatomie.* Jena, Gustav Fischer. 1963.
32. King, R. W. Timing in *Chenopodium rubrum* of export of the floral stimulus from the cotyledons and its action on the apex. *Can. J. Bot.* 50:697–702. 1972.
33. Klaus, H. Ontogenetische und histogenetische Untersuchungen an der Gerste (*Hordeum distichon* L.) *Bot. Jb.* 85:45–79. 1966.
34. Koevenig, J. L. Floral development and stamen filament elongation in *Cleome hassleriana. Amer. J. Bot.* 60:122–235. 1973.
35. Koevenig, J. L., and D. Sillix. Movement of IAA in sections from spider flower (*Cleome hassleriana*) stamen filaments. *Amer. J. Bot.* 60:231–235. 1973.
36. Kroh, M., H. Miki-Hirosige, W. Rosen,

and F. Loewus. Incorporation of label into pollen tube walls from myoinositol-labeled *Lilium longiflorum* pistils. *Plant Physiol.* 45:92–94. 1970.

37. Leins, P. Das Karpell im ober-und unterständigen Gynoeceum. *Ber. Deut. Bot. Ges.* 85:291–294. 1972.
38. Leppik, E. E. Morphogenic classification of flower types. *Phytomorphology* 18:451–466. 1968.
39. Leppik, E. E. Paleontological evidence on the morphogenic development of flower types. *Phytomorphology* 21:164–174. 1971.
40. Martin, F. W., and J. L. Brewbaker. The nature of stigmatic exudate and its role in pollen germination. Pp. 262–272. In: *Pollen development and physiology.* J. Heslop-Harrison, ed. London, Butterworth. 1971.
41. Mattsson, O., R. B. Knox, J. Heslop-Harrison, and Y. Heslop-Harrison. Protein pellicle of stigmatic papillae as a probable recognition site in incompatibility reactions. *Nature* 247:298–300. 1974.
42. Mitrakos, K., L. Price, and H. Tzanni. The growth pattern of the flowering shoot of *Urginea maritima* L. (Liliaceae). *Amer. J. Bot.* 61:920–924. 1974.
43. Moeliono, B. M. *Caulinary or carpellary placentation among dicotyledons.* 2 vols. Assen, The Netherlands, Royal Van Gorcum Ltd. 1971.
44. Morf, E. Vergleichend-morphologische Untersuchungen am Gynoëceum der Saxifragaceen. *Ber. Schweiz. Bot. Ges.* 60:516–590. 1950.
45. Moseley, M. F., Jr. Morphological studies of the Nymphaeaceae. III. The floral anatomy of *Nuphar. Phytomorphology* 15:54–84. 1965.
46. Moseley, M. F., Jr. The value of the vascular system in the study of the flower. *Phytomorphology* 17:159–164. 1967.
47. Okendon, D. J. Pollen tube growth and the site of the incompatibility reaction in *Brassica oleracea. New Phytol.* 71:519–522. 1972.
48. Paech, K. Colour development in flowers. *Ann. Rev. Plant Physiol.* 6: 273–298. 1955.
49. Puri, V. The role of floral anatomy in the solution of morphological problems. *Bot. Rev.* 17:471–553. 1951.
50. Puri, V. Placentation in agiosperms. *Bot. Rev.* 18:603–651. 1952.
51. Rosen, W. G. Pollen tube growth and fine structure. Pp. 177–185. Pistil-pollen interactions in *Lilium.* Pp. 239–254. In: *Pollen development and physiology.* J. Heslop-Harrison, ed. London, Butterworth. 1971.
52. Sassen, M. M. A. The stylar transmitting tissue. *Acta Bot. Neerl.* 23:99–108. 1974.
53. Sattler, R. A new approach to gynoecial morphology. *Phytomorphology* 24:22–35. 1974.
54. Seidlova, F. Changes in growth and RNA synthesis in shoot apical meristem of *Chenopodium rubrum* as a consequence of photoperiodic induction. *Z. Pflanzenphysiol.* 73:394–404. 1974.
55. Singh, V., and R. Sattler. Floral development of *Butomus umbellatus. Can. J. Bot.* 52:223–230. 1974.
56. Sporne, K. R. Some aspects of floral vascular systems. *Proc. Linn. Soc. London B.* 169:75–84. 1958.
57. Takhtajan, A. *Die Evolution der Angiospermen.* Jena, Gustav Fischer. 1959.
58. Tepfer, S. S. Floral anatomy and ontogeny in *Aquilegia formosa* var. *truncata* and *Ranunculus repens. Univ. Calif. Pubs. Bot.* 25:513–648. 1953.

59. Thompson, W. R., J. Meinwald, D. Aneshansley, and T. Eisner. Flavonoids: pigments responsible for ultraviolet absorption in nectar guide of flower. *Science* 177:528–530. 1972.
60. Tiagi, Y. D. Studies in floral morphology. II. Vascular anatomy of the flower of certain species of the Cactaceae. *J. Indian Bot. Soc.* 34:408–428. 1955.
61. Tucker, S. C. Ontogeny of the inflorescence and the flower of *Drimys winteri* var. *chilensis*. *Univ. Calif. Pubs. Bot.* 30:257–336. 1959.
62. Tucker, S. C. Phyllotaxis and vascular organization of the carpels in *Michelia fuscata*. *Amer. J. Bot.* 48:60–71. 1961.
63. Uhl, N. W., and H. E. Moore, Jr. The protection of pollen and ovules in palms. *Principes* 17:111–149. 1973.
64. Wilson, C. L. The telome theory and the origin of the stamen. *Amer. J. Bot.* 29:759–764. 1942.
65. Wilson, C. L., and T. Just. The morphology of the flower. *Bot Rev.* 5:97–131. 1939.
66. Wolff, E. T. The terminology of plant reproduction. *Plant Sci. Bull.* 18:14–15. 1972.

21

The Flower: Reproductive Cycle

The reproductive cycle in the angiosperms, which is the subject of this chapter, is characterized by an extremely short haploid (gametophytic) phase which does not involve the development of nutritionally independent gametophytes. The development of the female gametophyte (embryo sac), the formation of the female gamete (egg), fertilization, and the early stages of development of the new sporophyte (embryo) take place on the parent sporophyte. The male gametophytic phase begins with the formation of the pollen grain within the anther of the parent sporophyte and continues with the development of the pollen tube on the stigma. In a common sequence of female gametogenesis, the megaspore is separated from the female gamete by three mitotic divisions, and in the male gametogenesis only two divisions intervene between the formation of the microspore (uninucleate pollen grain) and the male gametes (two sperms in the pollen grain or in the pollen tube).

The short duration of the gametophytic phases and their localization in the same structures that are concerned with sporogenesis (anthers and ovules) explain why the terms androecium and gynoecium, meaning male and female organs, were adopted for stamens and carpels, which actually function as sporophylls.

MICROSPOROGENESIS

Microsporangium and microspores

The microsporophyll, or stamen, bears the microsporangia, or pollen sacs, the function of which is to produce the microspores, or pollen grains. The microsporangium consists of the sporogenous tissue (later the microspores) within the pollen sac locule and of the special wall layers surrounding the sporogenous tissue and ontogenetically related to it. The development of the anther on the stamen and its final structural features are determined by the development of microsporangia and the eventual release of pollen.

The course in the ontogeny of microsporangia and their walls is consistent in large taxa. The sequence found in most dicotyledons may be exemplified by the cotton (*Gossypium arboreum*) anther.[38] The anther is bilobed and bisporangiate (fig. 21.1,*A*). In a still meristematic lobe, periclinal divisions occur in the first layer beneath the protoderm (fig. 21.1,*B*,*C*). This layer is called *archesporial,* for the inner derivatives of it become the *primary sporogenous cells* (fig. 21.1,*D*,*E*). The outer derivatives of the archesporial layer, which constitute the *primary parietal layer,* also undergo periclinal divisions forming two *secondary parietal layers.* The outer of these two layers divides periclinally so that, in total, three parietal layers are

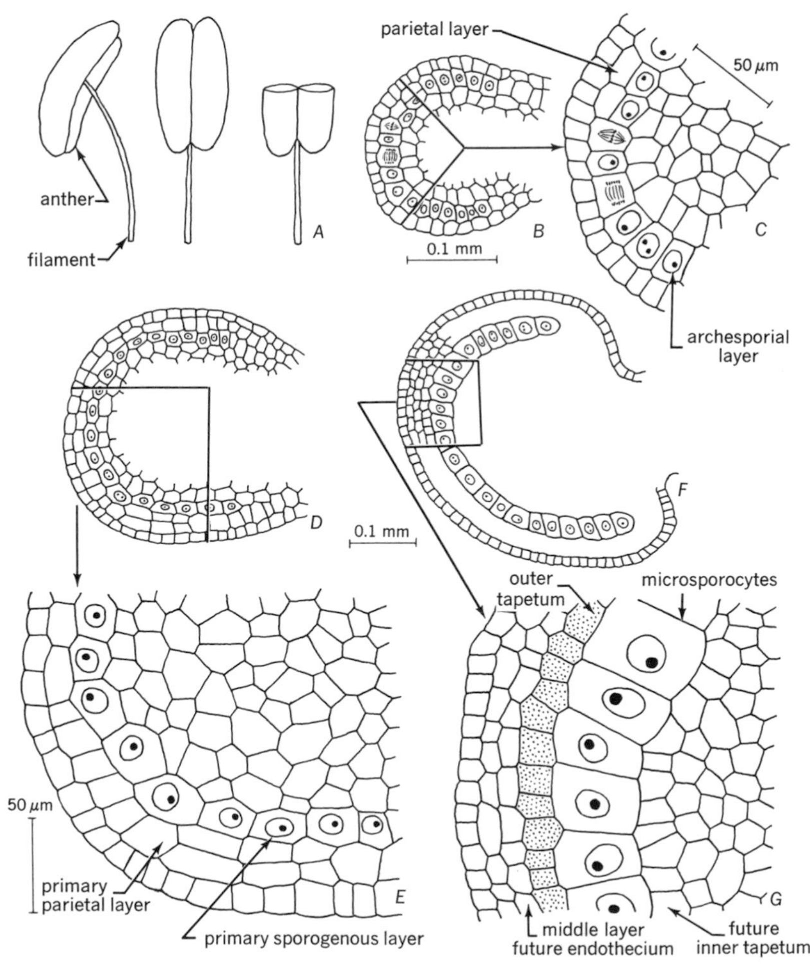

Figure 21.1 Pollen development in cotton (*Gossypium arboreum*). *A*, stamens showing bilobed bisporangiate structure. *B–G*, partial cross sections of microsporangia. *B*, *C*, archesporial stage. *D*, *E*, after formation of primary parietal layer and sporogenous layer. *F*, *G*, after formation of all wall layers and of microsporocytes. (*B–G*, adapted from Joshi, Wadhwani, and Johri.[38])

formed (fig. 21.1,*F*,*G*): the future *endothecium* beneath the epidermis, the *middle layer,* and the innermost *tapetum.* In some dicotyledons, both secondary parietal layers divide periclinally each contributing a middle layer, a sequence that is considered to be primitive.[10] In the monocotyledons, the inner secondary parietal layer divides periclinally to give rise to the middle layer and the tapetum, whereas the outer directly differentiates into the endothecium.

While the wall layers are formed the primary sporogenous cells divide or function directly as spore mother cells but after a considerable enlargement (fig. 21.1,*F*,*G*). The cells undergo meiosis resulting in the production of haploid *microspores.* Thus the spore mother cells are *microsporocytes.* They are also called *pollen mother cells* because the pollen grains in the uninucleate stage are the microspores.

Initially the microsporocytes are compactly arranged (fig. 21.1,*G*) and are interconnected by plasmodesmata. During the early meiotic prophase, a massive deposit of a callose-type polymer, β-1,3-glucan, appears outside the plasmalemmas, the original walls disintegrate, and the microsporocytes round off.[30] Instead of plasmodesmata, wide cytoplasmic bridges interconnect the microsporocytes through the callose, apparently a common phenomenon in angiosperms. The channels in the callose may be more than 1.5 μm in diameter and, in a scanning microscope, appear as holes lined with plasmalemma or filled with coagulated cytoplasm.[63,64] The connections between the microsporocytes are low-resistance junctions as demonstrated by electrophysiological techniques.[53]

One can visualize that the development of the channels through the callose wall converts the entire mass of microsporocytes in a microsporangium into a coenocyte through which a rapid transport and distribution of nutrients and growth substances occur.[26] The continuity of the microsporocytes is thought to account for the closely synchronous meiosis within a microsporangium frequently observed in angiosperms. No intersporocytic connections and no synchronous divisions were recorded in microsporangia of a few gymnosperms examined in this regard.[55] The connections between the angiosperm microsporocytes disappear before meiosis II so that the tetrads of spores resulting from the meiotic divisions are completely isolated from other tetrads of the same locule.

Microsporogenesis follows two patterns determined by the timing of cytokinesis in relation to meiosis. In the *simultaneous* cytokinesis, which is known to be characteristic of 186 families of angiosperms,[10] no wall is formed after meiosis I, although a phragmoplast usually appears (fig. 21.2,*A*). After meiosis II, phragmoplasts are formed between all four nuclei (fig. 21.2,*B*,*E*). The role of these phragmoplasts is not as well defined as in somatic cytokinesis, but their appearance is followed by an aggregation and fusion of vesicles midway between adjacent nuclei and an invagination of the plasmalemma[15,62] (fig. 21.2,*C*,*F*). The members of the tetrad become isolated from one another by a callose wall which is continuous with the callose surrounding the entire tetrad (fig. 21.2,*D*,*G*). In the *successive cytokinesis,* walls appear after the first and the second divisions.[42] This type of microsporogenesis was found in 40 families, most of them monocotyledons. The prevailing arrangements of the members of the tetrads are tetrahedral (fig. 21.2,*G*) or tetragonal (isobilateral; fig. 21.2,*D*), depending on the orientation of the meiotic spindle axes and of the related cleavage planes[59] (fig. 21.2). As determined in *Triticum,* the polarity governing the symmetry of microsporal arrangement is established during the premeiotic prophase.[16]

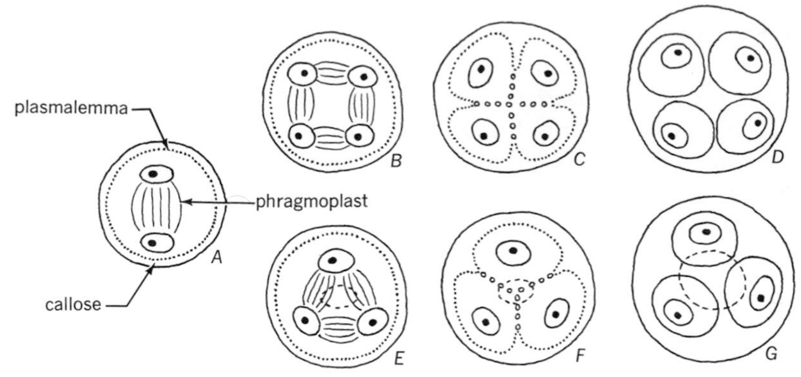

Figure 21.2 Diagrams of microsporogenesis leading to the formation of tetrads in tetragonal (*B–D*) and tetrahedral (*E–G*) arrangements. *A*, meiosis I. *B, E*, meiosis II. *C, F*, callose wall formation between microspores. *D, G*, tetrads of microspores embedded in callose wall at beginning of pollen wall formation.

Wall layers in the microsporangium

The tapetum is a specialized layer the development of which is synchronized with the events occurring in the adjacent sporogenous tissue.[21] It has a dual origin. Externally to the microsporangium the tapetal cells are derived from the primary parietal layer, whereas internally, metamorphosed cells of the connective tissue complete the tapetum around the circumference of the locule (fig. 21.3,*A*,*B*). Tapetal cells have dense cytoplasm before microsporogenesis and may become polyploid or multinucleate. The tapetum completely invests the locule and is considered to have a nutritional function by transferring food materials to the differentiating pollen grains either directly or after metabolizing them.[7,21]

Two main types of tapetum are recognized, the *secretory* (or *glandular*) and the *amoeboid* (or *plasmodial*). The secretory tapetum has been recorded in more angiosperm families than the amoeboid.[10] When the meiocytes are in prophase, the secretory tapetum develops an increased amount of endoplasmic reticulum and dictyosome-derived vesicles. After meiosis, the tapetum begins to break down. Its walls are lysed and the disintegrating cytoplasm is released into the locule where its condensed remains, called tryphine, are deposited on the pollen grains as an external coat.[14] Lipid is a major component of tryphine.

In the amoeboid tapetum, the protoplasts are intact when the walls are lysed and they intrude among the developing pollen grains (fig. 21.3,*A*,*C–F*). Fusion of the free protoplasts along the periphery of the locule gives rise to the so-called periplasmodium (fig. 21.3,*E*). In *Tradescantia,* the dissolution of the tapetal walls occurs before meiosis and the tapetal plasmodium infiltrates among the meiocytes.[43] The plasmodium continues to be associated with the developing pollen grains until these are mature. During the desiccation of the anther, shortly before anthesis, the plasmodium is dehydrated and deposited as debris (tryphine) on the surfaces of the pollen grains.

The number of middle layers in angiosperm anthers depends on the type of wall formation, but usually only one layer is formed in both dicotyledons and monocotyledons.[10] Al-

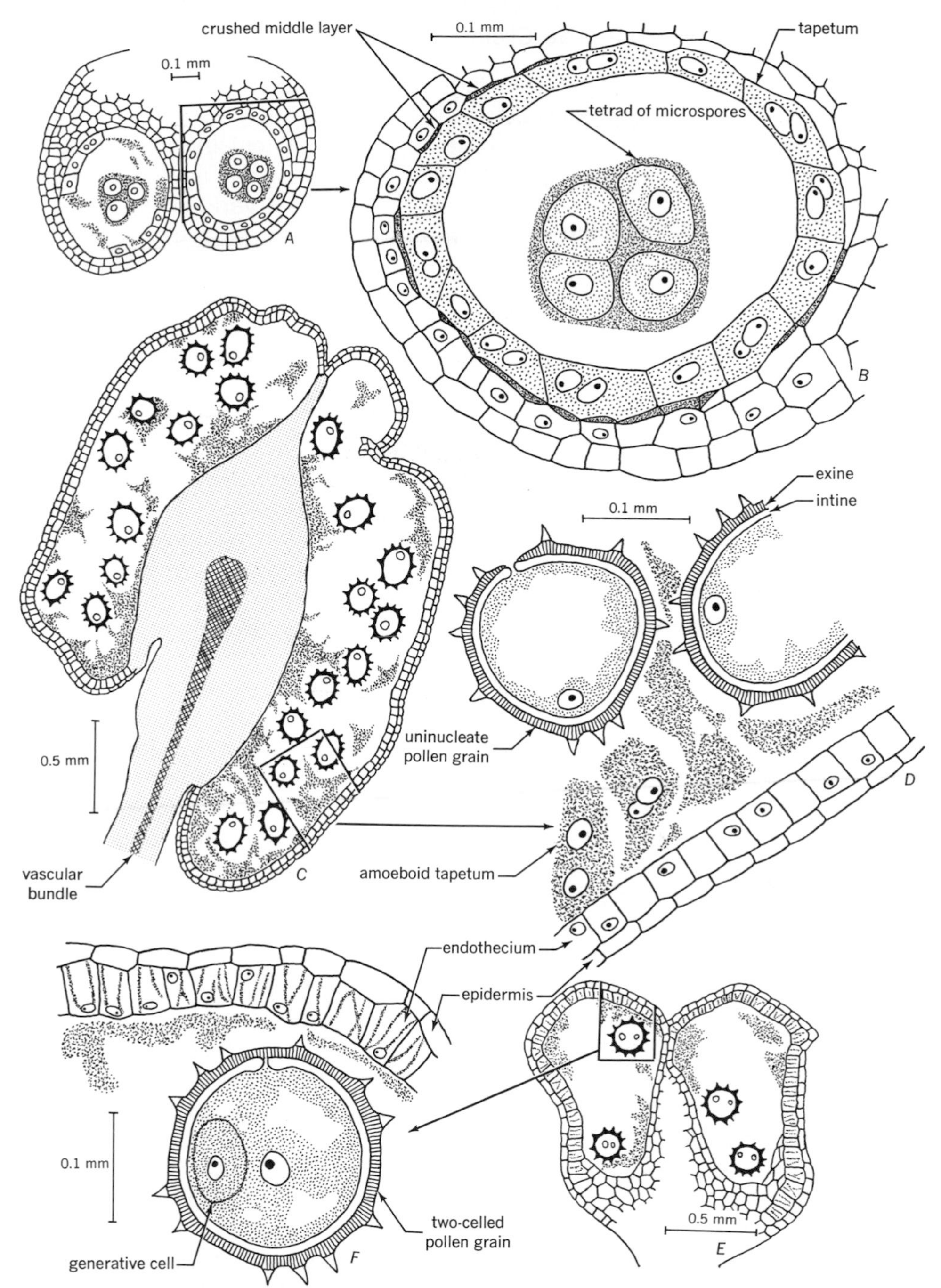

Figure 21.3 Pollen development in cotton (*Gossypium arboreum*). Partial cross (*A*, *B*) and oblique longitudinal (*C*–*F*) sections of microsporangia. *A*, *B*, tetrads in callose wall at beginning of exine formation. *C*, *D*, uninucleate pollen grains with exine and intine; tapetum in plasmodial stage. *E*, *F*, mature binucleate two-celled pollen grains; tapetum forms a periplasmodium. (Adapted from Joshi, Wadhwani, and Johri.[38])

most invariably, the middle layer degenerates early. It is then crushed between the tapetum and the immature endothecium and eventually absorbed by the adjacent cells (fig. 21.3,*B*).

The endothecium is a typical wall layer in the angiosperm anther except in those in which dehiscence occurs by means of pores. The outstanding characteristic of the endothecium are the secondary wall thickenings, commonly in the form of strips or bands (chapter 20). These thickenings are located on the anticlinal and the inner tangential walls. On the latter, the thickening may be even rather than in bands. Connective tissue cells on the inner side of the microsporangium may also develop the bandlike secondary thickenings (*Lilium*). In the region of the *stomium,* that is, the opening that appears when anthers release the pollen, endothecial thickenings do not develop. Because of the type of wall thickening the endothecium is often called the fibrous layer.

The endothecium becomes mature rather late in the development of the anther. It has no secondary thickenings during meiosis and wall development in the pollen grains (fig. 21.3,*A–D*). The cells increase in radial extent and rapidly develop the wall thickenings shortly before the pollen is shed (fig. 21.3,*E*,*F*). It has been suggested that the timing in the differentiation of the endothecium is governed by the requirement for proper distribution of nutrients during the different stages of microsporogenesis.[22] The presence of the relatively rigid endothecium during microsporogenesis also would hinder the increase in volume occurring in the developing sporogenous tissue. The wall thickenings in the endothecium are sometimes interpreted as being lignified or suberized.[20] In *Chenopodium,* the thickenings have no lignin and consist of the oriented α-cellulose.[22]

POLLEN

The outstanding feature of the mature pollen grain is its wall, which is sculptured in a great variety of patterns but at the same time shows a high degree of consistency in organization, chemistry, and development. The multiplicity of studies on the pollen wall has resulted in the development of a complex terminology with numerous synonyms. One of the terminologies is used below.[27] The wall typically consists of an *exine* and an *intine* (fig. 21.7). The exine is either a uniform sheath or is subdivided into the outer *sexine* and the inner *nexine.* The sexine is the sculptured part. It is attached to the nexine by means of rods, the *bacula,* which may be united into a *tectum* above (fig. 21.4,*A*) or remain free (fig. 21.4,*B*). Bacula are absent and the exine appears amorphous in primitive angiosperm pollen.[60]

The pollen grain has apertures (pores), that is, limited thin-walled areas in the exine (fig. 21.7) through which the pollen tube usually emerges during germination and which allows the grain to change in volume during changes in humidity.[59] The apertures may be rounded (porate pollen) or furrowlike (colpate pollen), and their number varies. The basic numbers in angiosperms are one (monocolpate pollen) in monocotyledons and many Ranales and three (tricolpate pollen) in the majority of dicotyledons.[20]

The most distinctive chemical component of the pollen wall is the *sporopollenin* which consists of oxidative polymers of carotenoids and carotenoid esters.[52] Sporopollenin is remarkably resistant to various chemicals, high temperature, and to agents of natural decay of organic matter and is thought to be mainly responsible for the preservation of pollen in fossil plant deposits that have lost most other traces of biological structure. Silicon, the

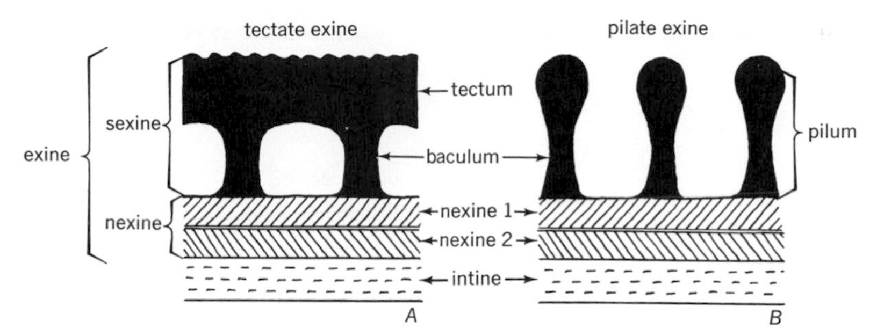

Figure 21.4 Diagrams interpreting organization of pollen wall. The nomenclature is based on the publications of G. Erdtman. (Redrawn from Heslop-Harrison.[25])

presence of which was discovered in the pollen exine of certain dicotyledons, may add to the resistance of pollen to geological weathering and biological decay.[8] The relation of sporopollenin to the other wall constituents is best explained by reference to pollen wall ontogeny.[25,27]

Pollen wall development begins while the tetrads are still enclosed in callose. Initially, the individual microspores are without walls and their plasmalemmas are in contact with the callose (fig. 21.3,*B*). Studies on several species have shown that endoplasmic reticulum cisternae become applied to regions beneath the plasmalemma where eventually this membrane forms flattened protrusions (fig. 21.5,*A*,*B*). These regions mark the sites of future apertures. Where the endoplasmic reticulum is absent, the first wall, the *primexine,* is deposited outside the plasmalemma. The primexine is composed of cellulose and is finely fibrillar. Almost immediately upon its deposition, the primexine becomes radially traversed by rods, the *probacula* (fig. 21.5,*C*), suspected of being composed of lipid and protein. More material is added to the probacula and they become interconnected at their bases and, in the tectate type of exine (fig. 21.4,*A*), also at their tops (fig. 21.5,*C*). The resulting netlike product clearly foreshadows the organization of the mature exine. The patterned component becomes increasingly electron dense and resistant to acetolysis, for *protosporopollenin* becomes associated with it.

While the microspores are still enclosed in callose[44] or after the spores are released from the tetrad by dissolution of the callose wall, they increase in size and new exine material (sporopollenin) is deposited. The exine assumes its mature organization into sexine, bacula, and nexine 1 and nexine 2, the latter added last to the exine. Exine material extends to the aperture regions where in the meantime a layer of intine appears (fig. 21.5,*D*). The intine increases in thickness and spreads beyond the aperture regions, and the plasmalemma ceases to bulge (fig. 21.5,*E*,*F*). The intine is composed of pectin and cellulose and shows layering. In the final period of pollen maturation coating substances derived from the tapetum are applied to the pollen.

Pollen walls carry proteins in all species of angiosperms examined in this regard.[28] Among these proteins are many enzymes and allergenic types of proteins causing hayfever (e.g., ragweed, *Ambrosia,* pollen). The proteins have different origins and may be

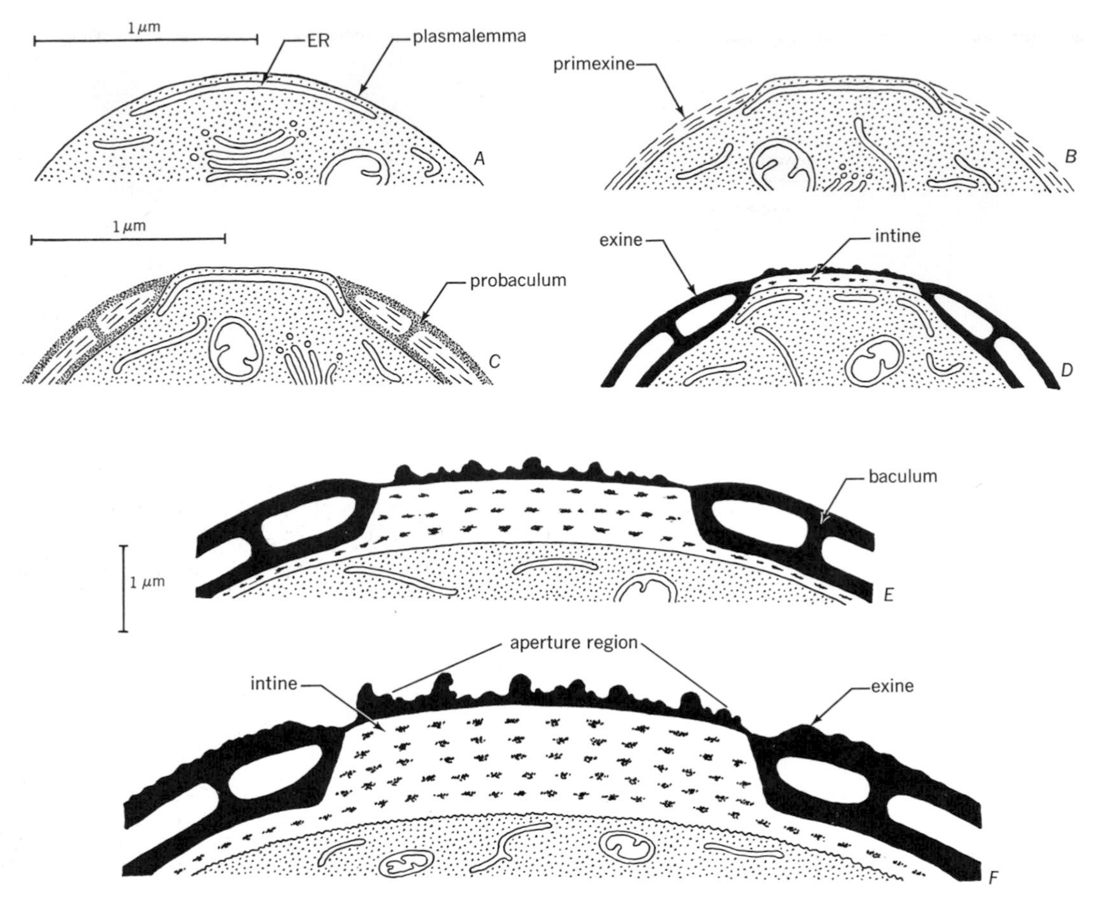

Figure 21.5 Development of pollen wall in *Silene pendula* shown in partial sections. *A–C,* stages within the callose wall of the microsporocyte. *D–E,* development after release from the callose wall. (Adapted from Heslop-Harrison.[25])

responsible for certain differences in the incompatibility reactions observable in pollination studies.

MALE GAMETOPHYTE

Gametogenesis

Before the pollen is shed, a mitotic division into *vegetative* and *generative* nuclei gives rise to the two-celled gametophyte. The generative cell may immediately divide mitotically into two *sperms,* or male gametes, or the second mitotic division may occur after the pollen grain germinates. Among the angiosperms surveyed, the pollen is shed in a two-celled state in 192 families, in a three-celled state in 115 families. In still others the character is variable.[10]

Ultrastructural research serves to clarify the nature of the gametophytic cells, especially with regard to their delimitation from one another. The mitotic division resulting in the formation of the generative nucleus is followed by cytokinesis resembling that in somatic cells.[18] The division is polarized in that the premitotic microspore nucleus is displaced toward the wall on the side of the grain away from the germination furrow or furrows[17,49] (fig. 21.6,*A*). The generative nucleus is given off toward the wall, and cytokinesis (fig. 21.6,*B*) results in a hemispherical wall enclosing the nucleus and the accompanying small amount of cytoplasm. This wall is connected to the intine and is lined with the plasmalemmas of the vegetative and generative cells (figs. 21.6,*C* and 21.7). In several species the new wall was found to stain a clear blue with lacmoid and to show intense yellow fluorescence after treatment with dilute aniline blue, reactions identifying callose.[23,44] The wall appears to have plasmodesmata in *Linum*[56] (fig. 21.6,*C*).

The generative cell becomes detached from the pollen wall and then rounds off so that the contact area with the intine gradually decreases (fig. 21.6,*D*). The free cell becomes surrounded by the cytoplasm of the vegetative cell. In some species, the wall of the generative cell disappears and the cell is delimited only by two plasmalemmas[44,56] (fig. 21.6,*E*). In other species, the wall remains, reacts histochemically as a carbohydrate, and has plasmodesmata.[37]

As is known from work with the light microscope, the two gametic nuclei are produced by mitosis of the generative nucleus.[42] Electron microscopic studies are confirming the concept that the two sperms are cells. In the sugar beet[29] (fig. 21.8) and barley,[4] in which the sperms are present in the mature pollen grain, they are individual elongated ellipsoidal cells bound by two unit membranes, the two plasmalemmas, one belonging to the gamete, the other to the vegetative cell. A narrow space occurs between the two membranes.

Between the microspore and the mature gametophyte stages, the structure of the pollen grain undergoes certain changes. The components of the microspore comprise a large nucleus, a large vacuole, and various cytoplasmic organelles, although the endoplasmic reticulum is relatively sparse (fig. 21.6,*A*). Amyloplasts are present, with large amounts of starch in some species. In *Impatiens,* starch is used up in young microspores during wall formation in the tetrad stage.[19] Lipid globules may be abundant. While the wall is still growing, dictyosomes producing vesicles are conspicuous.

After the first mitosis, the large vacuole, which remains in the vegetative cell, breaks up into smaller units and the vacuolar volume decreases. The organelles of the vegetative cell increase in number. The endoplasmic reticulum forms voluminous stacks in some species[37,41,49] (fig. 21.7). Starch and lipid may increase greatly in amount, apparently as reserves accumulating before germination.[19]

The generative cell is less rich in organelles than the vegetative cell, and this characteristic is passed to the sperm cells. A special point of interest is whether the gametes contain plastids that could possibly be transmitted to the egg cell during fertilization. In some species, occasional plastids were seen in the generative cell and the gametes.[4,29,37,56,57] In the sugar beet and barley numerous microtubules, arranged parallel with the long axis of the cell, were observed in the gametes. It has been suggested that the microtubules are involved in the control of shape of the sperm cells[4] and of that of the generative cell.[49]

Pollen tube

The pollen normally germinates on the stigma but most of the information on pollen

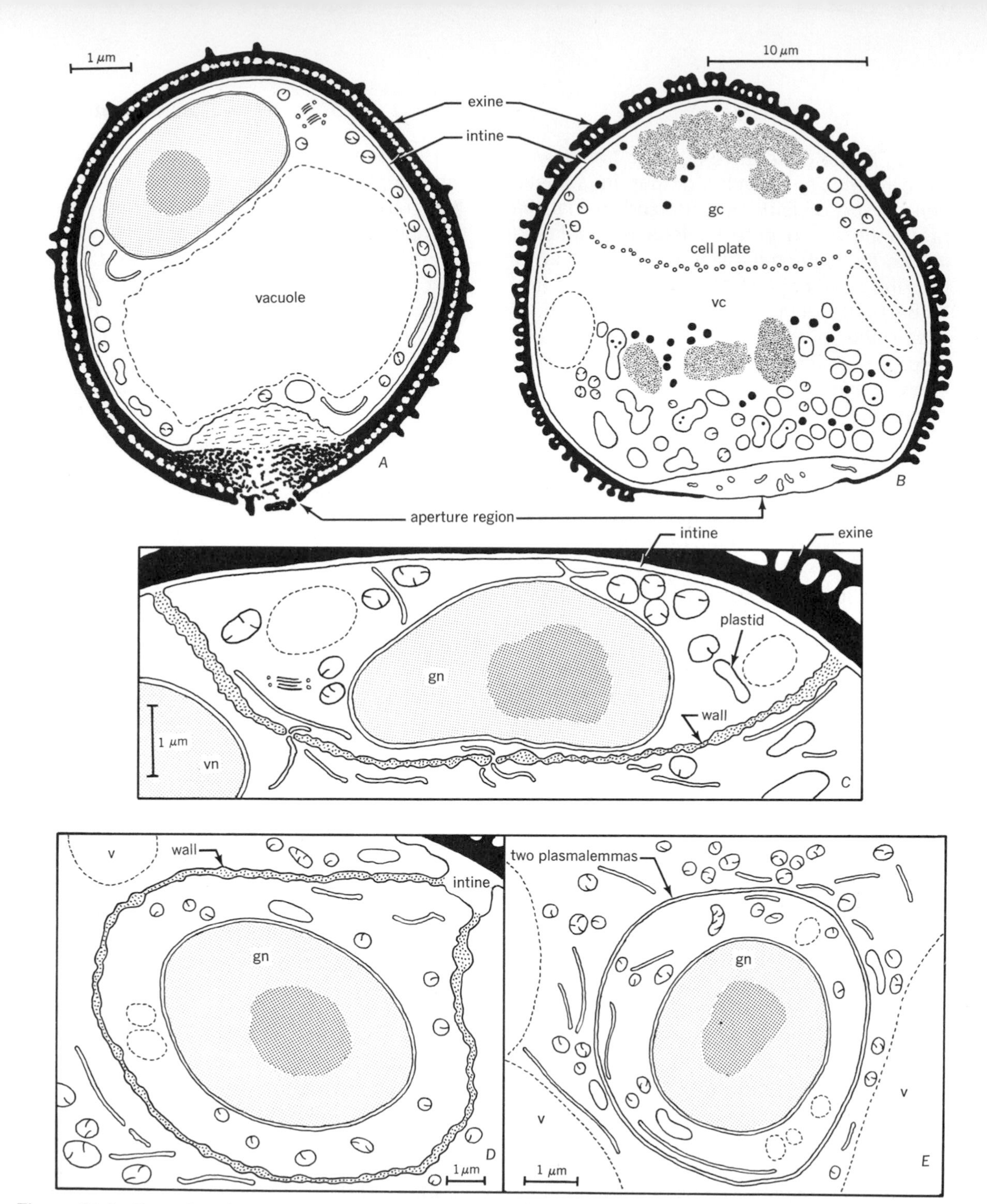

Figure 21.6 Formation of generative cell in a pollen grain. Assemblage of drawings made from electron micrographs illustrating pollen of: *A, Jasione montana* (Dunbar[17]); *B, Haemanthus katharinae* (Sanger and Jackson[49]); *C, D, E, Linum usitatissimum* (Vazart[56]). The stages are: *A,* before mitosis; the nucleus is located opposite the aperture region. *B,* cytokinesis after mitosis; developing curved cell plate separates the generative cell (*gc*) from the vegetative cell (*vc*); the nuclei of the two cells are in early telophase. *C,* generative cell delimited by a curved wall which is continuous with the intine of the pollen grain. *D,* generative cell separating from the pollen wall; a limited connection with the intine is still present. *E,* generative cell within the cytoplasm of vegetative cell; it is delimited by two plasmalemmas. Details: *gc,* generative cell; *gn,* generative nucleus; *v,* vacuole; *vc,* vegetative cell; *vn,* vegetative nucleus.

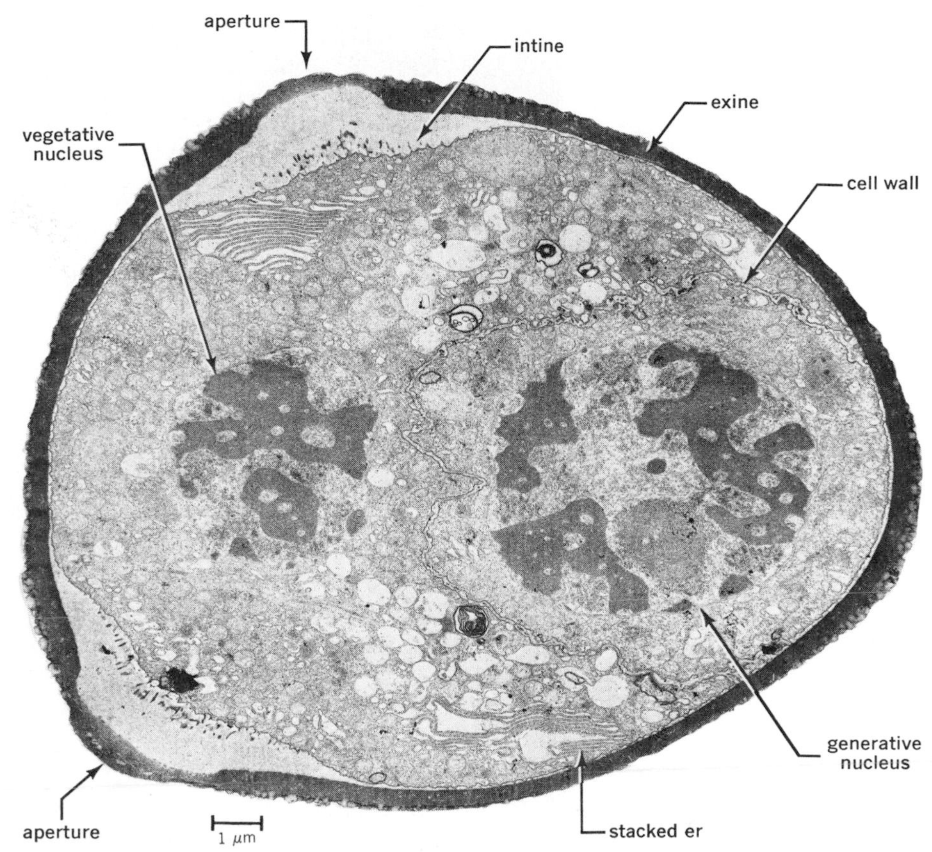

Figure 21.7 Electron micrograph of two-celled pollen grain of *Montropa uniflora*. Generative nucleus and associated cytoplasm (generative cell) delimited by a wall which is continuous with the thin intine of the pollen grain. Vegetative cell partly surrounds the generative cell. (From Lutz and Sjolund.[41])

tube growth is derived from studies of pollen germinated in cultures (fig. 21.9). Germination occurs soon after the pollen is placed on an appropriate medium (fig. 21.9,*B*; within an hour on cotton stigma[36]). The pollen tube grows rapidly, often several mm/hr in vitro,[47] 2 mm/hr in cotton style.[36] Its growth in culture is highly polarized, for it is restricted to a small, clearly defined region of about 3 to 5 μm in length at the tip of the pollen tube. Whether this type of growth occurs in the style as well is not known.[48] Water uptake and activation or synthesis of enzymes are presumed to be the factors initiating germination. Before the pollen tube emerges, protein synthesis starts in the pollen grain, but no dependence of germination on RNA synthesis has been observed. While the pollen tube is growing in vitro, polysomes increase in number and both protein and RNA synthesis are demonstrable.

The wall of the pollen tube may be thin or rather thick. Its microfibrillar organization is more highly oriented in the more mature regions than at the tip. The microfibrils are commonly interpreted as being cellulosic, but in the pollen tube of *Lilium longiflorum*,

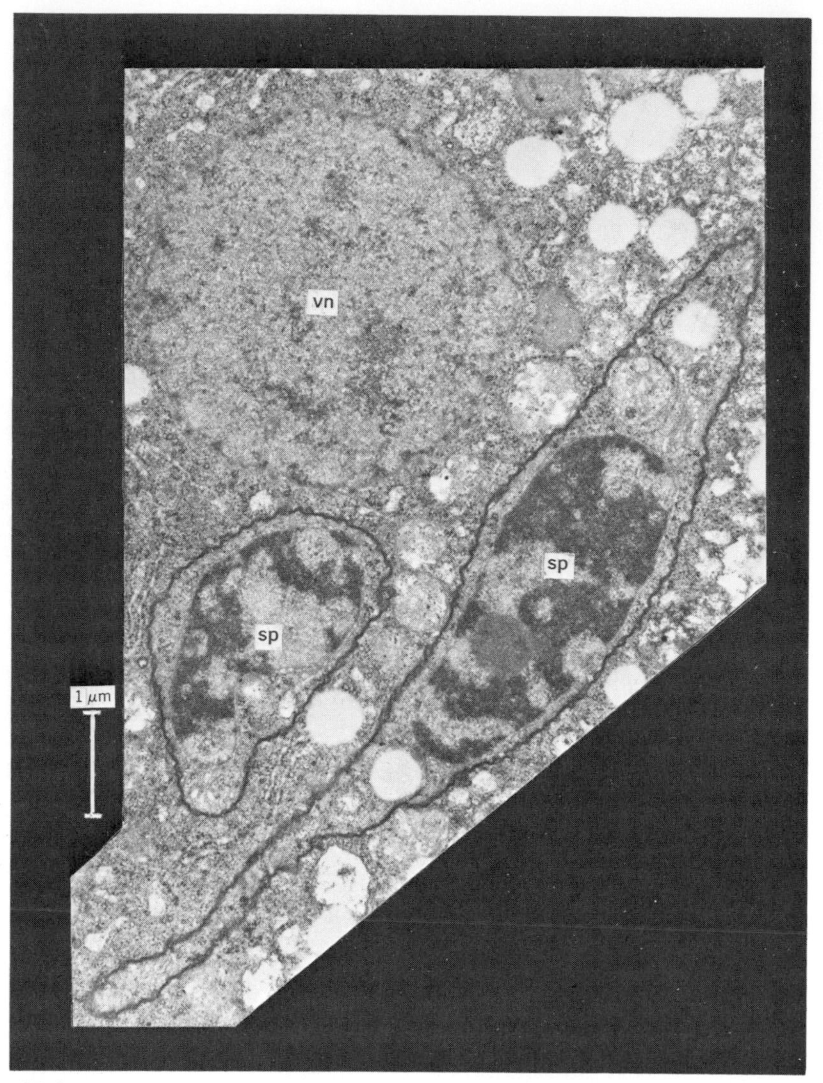

Figure 21.8 Electron micrograph of sperm cells (*sp*) and vegetative nucleus (*vn*) from a three-celled pollen grain of sugar beet (*Beta vulgaris*). The sperms are positioned at right angles to one another. They are delimited by two membranes interpreted as two plasmalemmas, the outer derived from the vegetative cell, the inner from the sperm cell. (From Hoefert.[29])

the wall consists predominantly of a fibrillar alkali-resistant crystalline β-1,3-polyglucan which occurs together with a minor amount of the common cellulosic β-1,4-glucans.[24]

The relation of the pollen tube to the intine that occurs in the aperture region has not been fully explored. As seen with the light microscope, the pollen tube appears to penetrate and push aside with ease the plastic layer of intine in the aperture.[1] Electron mi-

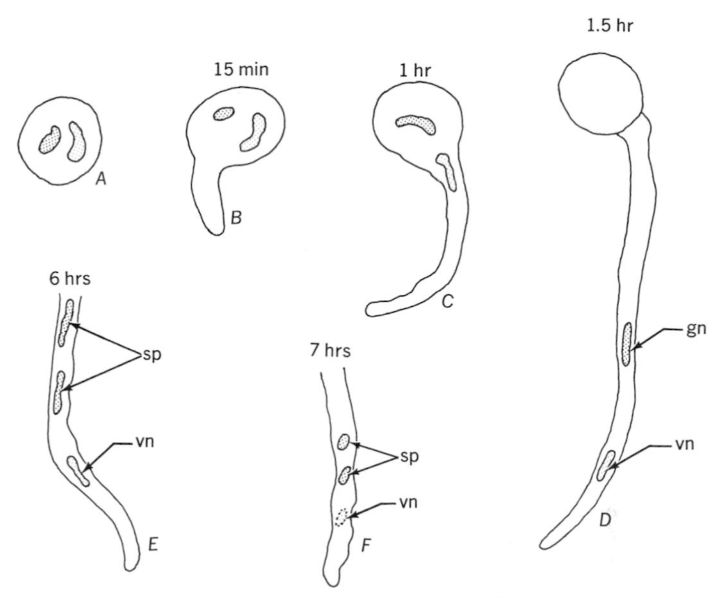

Figure 21.9 Germination of pollen grain of *Scilla* (monocotyledon) in vitro. Medium: 2% agar, 7% cane sugar, traces of sterile yeast. *A*, resting binucleate pollen grain, *B–E*, stages of germination at times after placement on medium indicated above the drawings. Only tips of the pollen tubes in *E* and *F*. The nuclei are compact at end of germination in *F*. Details: *gn*, generative nucleus; *sp*, sperm nucleus; *vn*, vegetative nucleus. (Adapted from R. A. Brink, *Amer. J. Bot.* 11:351–364, 1924.)

croscopy reveals that, upon germination of pollen, the intine becomes continuous with the pollen tube wall.[9,13]

The cytoplasm of the growing pollen tube commonly accumulates at the tip, with some differentiation between the apical and subapical regions. The vegetative nucleus, the sperms (or the generative nucleus), the cytoplasmic organelles, and small vesicles, all transferred from the pollen grain, occur in the subapical region (fig. 21.9,*D*,*E*). The apical end is devoid of organelles but contains numerous vesicles. The vesicles are derived from dictyosomes and probably also from endoplasmic reticulum and, at least some of them, are concerned with wall synthesis.[54] In many species, the older parts of the elongating pollen tube are successively sealed off by plugs of callose.

MEGASPOROGENESIS

Ovule

The ovule developing from the placenta of the ovary (chapter 20) is the site of formation of megaspores and development of the embryo sac (female gametophyte). The ovule commonly consists of the following principal parts (fig. 21.10): *nucellus,* the central body with vegetative cells enclosing the sporogenous cells; one or two *integuments* (thus, *unitegmic* and *bitegmic* ovules) enclosing the nucellus; *funiculus* (fig. 21.10,*C*), the stalk connecting the ovule with the placenta. The region where the nucellus, the integuments, and the funiculus merge is called the *chalaza,* a not clearly defined region.

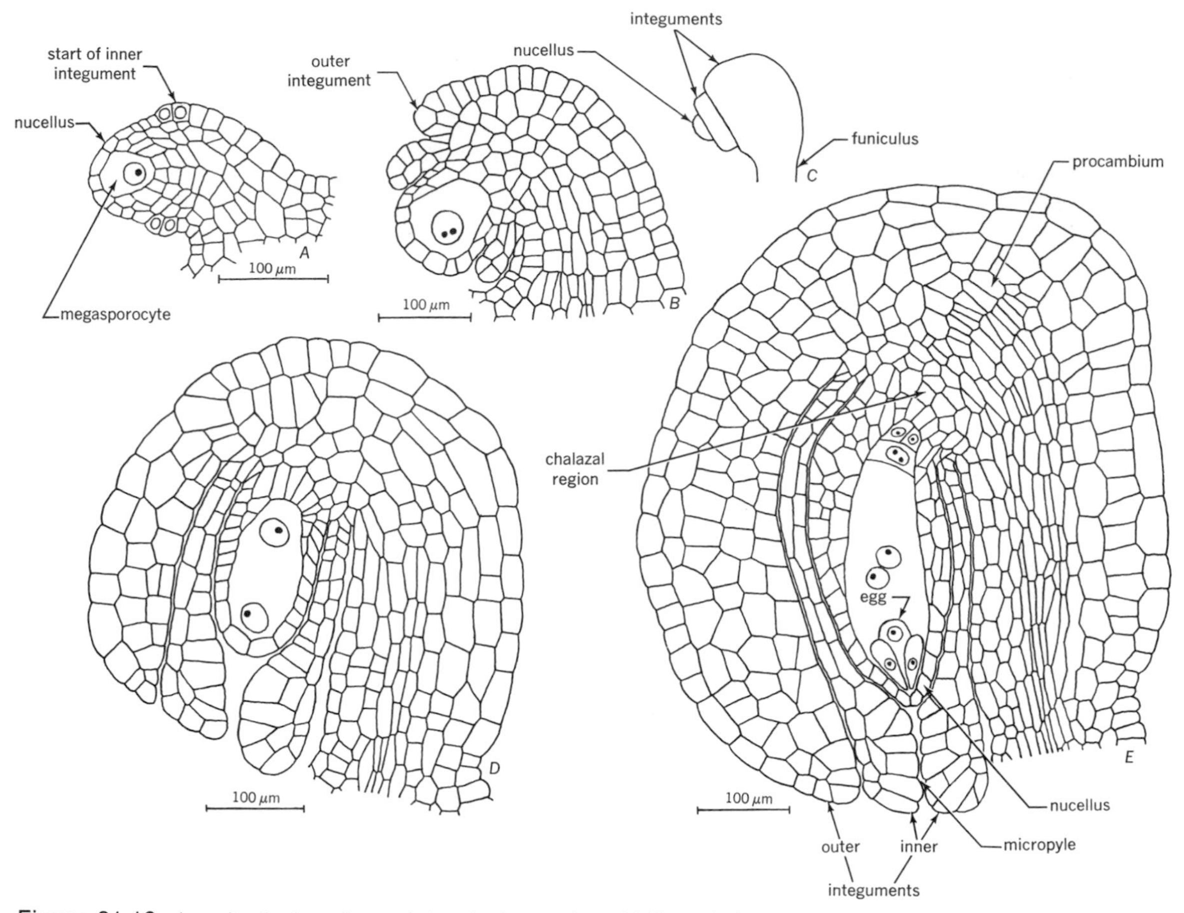

Figure 21.10 Longitudinal sections of developing ovules of *Lilium tigrinum.* As the integuments develop the ovule curves and assumes its anatropous form. (*A, B, D, E,* adapted from Bouman.[3])

The parts of the ovule become gradually delimited as the reproductive cells pass through megasporogenesis and embryo sac development. As exemplified by the ovule of lily[3] (fig. 21.10), the ovule emerges on the placenta as a conical protuberance with the first sporogenous cell, the *archesporial cell,* becoming visible almost immediately. The two integuments are initiated by periclinal divisions in the epidermis, first the inner integument, then the outer. Their appearance delimits the nucellus. The integuments grow like ringlike welts gradually enclosing the nucellus. They complete development by the time the embryo sac is ready for fertilization. An opening, the *micropyle,* remains where the inner integument arches over the nucellus. One or both integuments may be involved in the formation of the micropyle[10] (figs. 21.10,*E*, and 21.14).

The development of the highly vacuolate thin-walled embryo sac surrounded by actively dividing integuments tightly investing the nucellus (fig. 21.10,*E*) poses the question as to how the embryo sac retains its shape and does not become crushed. A study relating the orientation of cell walls in the cotton ovule with the distribution of tension and compression stresses in a two-dimensional plastic model of the ovule suggests that the integuments serve to protect the developing embryo sac by insuring that all compressive forces circumvent it.[39,40]

During the development of integuments the ovule may remain upright (*atropous* or *orthotropous* ovule) or become inverted (*anatropous* ovule; fig. 21.10). Between these two basic types of ovule are many other variously curved and variously named forms of ovules.[10,42,50] The ovules also vary in the size of the nucellus. An ovule with a large nucellus is called *crassinucellate,* one with a small nucellus, *tenuinucellate.*

When the female gametophyte is ready for fertilization the ovule, like the ovary, is still relatively undifferentiated histologically. The vascular system, connected with that of the placenta through the funiculus, extends to the chalaza, usually as a single strand. But it may be more elaborate and may develop also in one or both integuments and become branched. The ground tissue is parenchymatic and the epidermis bears a cuticle. Because of the commonly epidermal origin of the integuments three cuticular layers may be distinguished: the outer, on the outside of the outer integument and the funiculus; the median, double in origin, between the two integuments; and the inner, also double, between the inner integument and the nucellus.

Parts of the ovule are disorganized during the development of the embryo sac and embryo. The vegetative tissue of the nucellus may be partly or entirely resorbed. In the latter instance, the embryo sac comes in contact with the inner epidermis of the adjacent integument. This epidermis commonly differentiates into the *integumentary tapetum,* or *endothelium,* a layer of deeply staining cells with an abundance of endoplasmic reticulum. A connection of this tapetum with the nutrition of the embryo is assumed because of the disintegration of the ovule tissue next to the tapetum and the persistence of the tapetum to seed maturity. This assumption is yet to be correlated with the characteristics of the walls of the tapetum and that of the embryo sac. No plasmodesmata have been seen between the embryo sac and the endothelium[12] and a double cuticle occurs between the two.[58] The wall of the embryo sac, however, may bear wall ingrowths indicating transfer activity.[45]

Megaspores

Megaspores result from the meiotic division of the megaspore mother cell, or *megasporocyte.* In crassinucellate condition, the archesporial cell undergoes a periclinal division and produces a parietal cell and the megasporocyte. In tenuinucellate ovules, the archesporial cell functions as the megasporocyte directly[10] (fig. 21.10,*A*,*B*). The common sequence of divisions of the latter (*Polygonum* type) is illustrated for *Solanum* in figure 21.11,*A–E*.[61] The megasporocyte may have rather dense cytoplasm or be vacuolated to various degrees. The cell undergoes two meiotic divisions resulting in the formation of a linear tetrad of haploid megaspores. The chalazal megaspore enlarges in preparation for the first mitotic division of the gametophyte. The other three megaspores degenerate. Fluorescence microscopy indicates that a temporary deposition of callose occurs in megasporogenesis, as well as in microsporogenesis.[46] Callose envelops the megaspore preparing for division and also appears in the walls separating the degenerating megaspores. All this callose soon disappears.

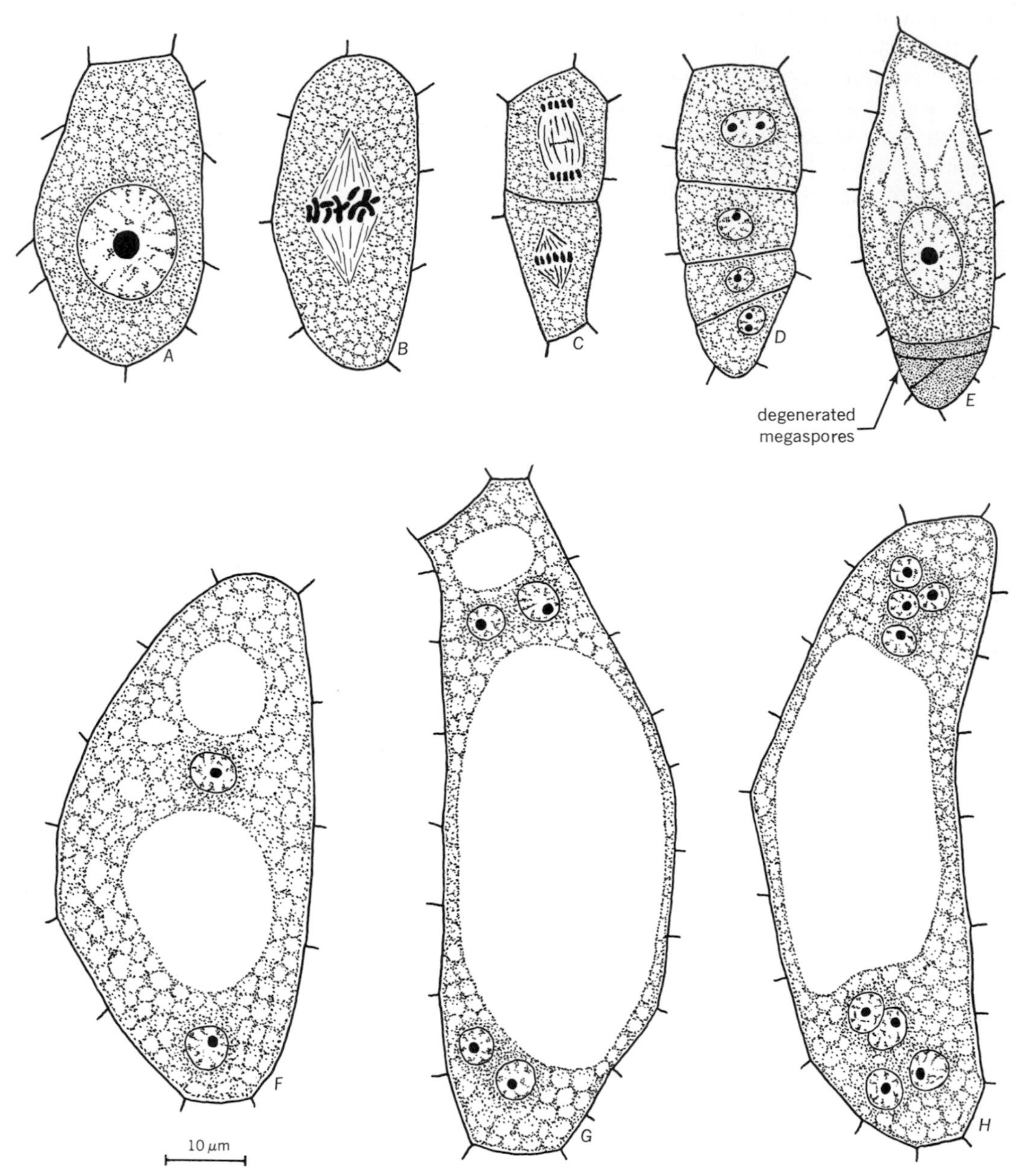

Figure 21.11 Megasporogenesis and embryo sac development in *Solanum demissum*. *A*, megasporocyte. *B*, *C*, meiosis. *D*, four megaspores. *E*, functional megaspore and three degenerated ones. *F–H*, embryo sacs with two, four, and eight nuclei. (Redrawn from Walker.[61])

FEMALE GAMETOPHYTE

Mitotic divisions occur in three generations of nuclei of the gametophyte so that an eight-nucleate embryo sac is formed (fig. 21.11,*E–H*). During these divisions the former megaspore cell enlarges and becomes much vacuolated. The eight-nucleate cell is organized into the seven-celled embryo sac through delimitation by cell walls of six of the nuclei and associated cytoplasm (fig. 21.12,*A–C*). The three cells at the micropylar end constitute the *egg apparatus* which is composed of the *egg* and two *synergids.* At the opposite end of the embryo sac are three *antipodal cells.* Between the two groups of cells is the large *central cell* containing two *polar nuclei* derived one from each of the two groups of four nuclei. The polar nuclei may fuse before fertilization and form the diploid *secondary endosperm nucleus.*[10] As observed in cotton,[32] the fusing nuclei are first joined by means of endoplasmic reticulum cisternae extending from the nuclear envelopes. In this step, the outer membranes of the two nuclear envelopes become continuous. Later, the inner nuclear membranes come in contact and fuse, establishing bridges. Enlargement and fusion of the bridges completes the union of the two nuclei.

There are numerous variations in angiosperm embryo sac formation.[10,20,42] One of the deviating types (*Fritillaria* type; also found in *Lilium,* fig. 21.10) is compared with the *Polygonum* type (common type as in *Solanum,* figs. 21.11 and 21.12) in figure 21.13. The distinctive features in *Fritillaria* are: (1) no walls are formed between the megaspore nuclei and (2) all four nuclei are involved in embryo sac development. The chromosomes of three of these nuclei combine during mitosis so that the resulting two nuclei are triploid. In the eight-nucleate embryo sac, four nuclei are triploid and four are haploid. The polar nuclei fuse into a tetraploid secondary endosperm nucleus. The *Polygonum* type of embryo sac (fig. 21.13, upper series) is called *monosporic* because it is derived from one of four nuclei resulting from meiosis. The *Fritillaria* type of embryo sac (fig. 21.13, lower series) is called *tetrasporic* because it contains derivatives from all four nuclei resulting from meiosis. There is also a *bisporic* embryo sac type[42] (*Allium*).

The components of the embryo sac show considerable cytologic differentiation from one another, a feature emphasized by electron microscope investigations. The antipodals have extremely variable developmental histories in angiosperms. In *Zea mays,*[12] these cells proliferate into a many-celled ephemeral tissue as in other grasses. The cells have wall ingrowths where they border on the nucellus and their richness in appropriate organelles indicates a potentiality for high rate of respiration and synthetic activity. The central cell has two large polar nuclei or one fusion nucleus and a large vacuolar volume. Dictyosomes are prominent among the organelles. Lipid bodies and plastids with starch have been recorded. Wall ingrowths may be present. The cell appears metabolically active, possibly partly in relation to its long-lasting growth in size.[32]

The synergids are the most complex cells in the embryo sac and have a vital role in the fertilization process[5,35] Their large vacuoles occur in the chalazal end, the nuclei closer to the micropylar end (fig. 21.12,*B*). Some form of *filiform apparatus* consisting of more or less elaborate system of wall ingrowths occurs at the micropylar end (fig. 21.15,*A,D*). Structurally this system is loose and porous and appears to be rich in pectic substances and hemicelluloses. The cytoplasm associated with the apparatus is rich in organelles. Each synergid is enclosed by a partial

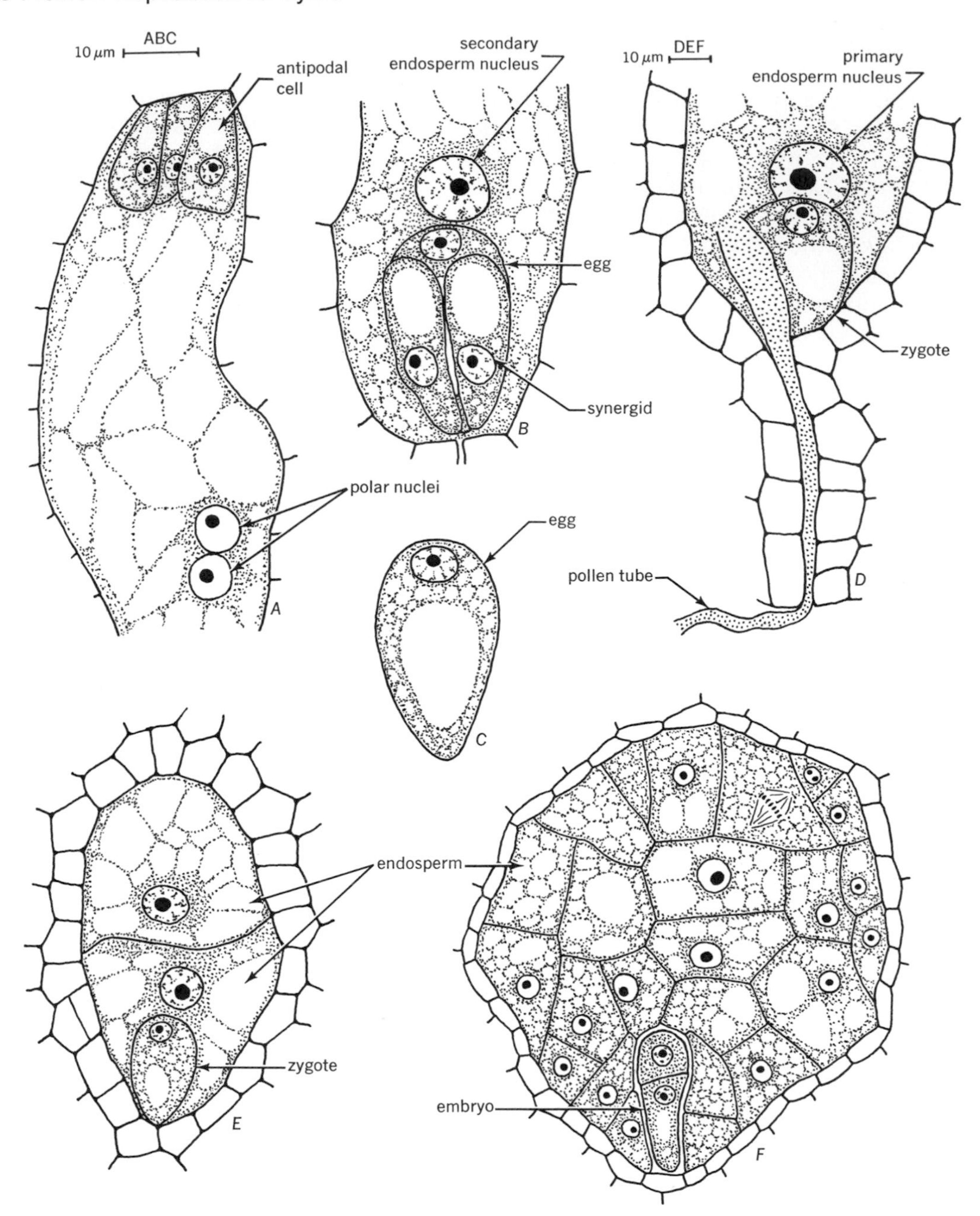

Figure 21.12 Embryo sac of *Solanum demissum*. *A, B, C,* parts of the same eight-nucleate, seven-celled embryo sac; the egg is drawn separately in *C* to show the contrast between egg and synergids in localization of vacuoles and nuclei. *D,* remnant of pollen tube after fertilization; degenerated synergids are not shown. *E,* zygote and two-celled endosperm. *F,* two-celled embryo and many-celled endosperm. (Redrawn from Walker.[61])

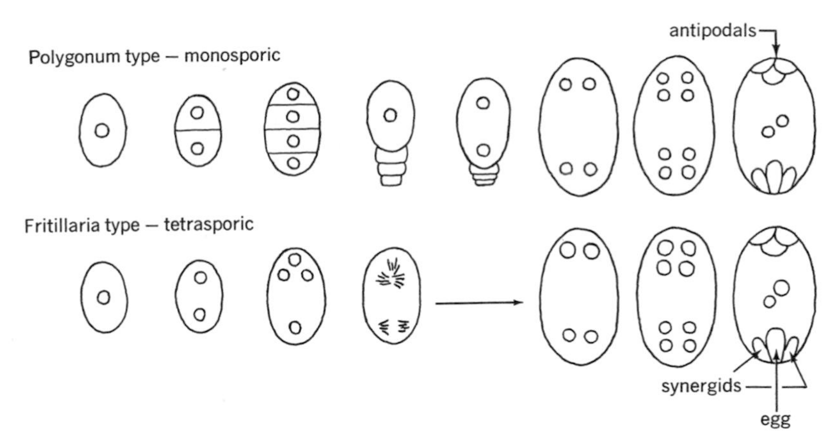

Figure 21.13 Diagrams illustrating monosporic and tetrasporic types of megasporogenesis and embryo sac development. In *Polygonum,* the embryo sac is derived from one megaspore. In *Fritillaria,* all four nuclei formed in meiosis participate in the formation of embryo sac. The larger circles in the embryo sacs of *Fritillaria* indicate triploid nuclei.

wall limited to two-thirds of the cell at the micropylar end. At the chalazal end, only a plasmalemma separates the synergid from the central cell. In the cotton embryo sac, the synergid wall is composed of cellulose, hemicelluloses, and pectin.[31]

The eggs of angiosperms vary greatly in size and structure but usually they are highly vacuolated cells. The large vacuole is commonly located at the micropylar end, the nucleus closer to the opposite end[12] (fig. 21.12,*B*,*C*). In cotton, however, the vacuole extends axially through the whole cell.[32] In *Zea* and cotton, the egg has a partial wall at the micropylar end, as do the synergids, with the plasmalemma forming the boundary at the opposite end. In *Capsella,*[51] the wall extends over the entire cell but has numerous discontinuities at the chalazal end. Electron microscopic observations suggest that eggs may vary in degree of metabolic activity. An egg associated with synergids appears to be a rather inactive cell as judged by the amount and state of the organelles.[12,32] In a species of *Plumbago,*[6] the egg is not accompanied by synergids, has a filiform apparatus at its micropylar end, and its cytologic structure indicates that the cell has a relatively intense metabolic activity.

FERTILIZATION

The pollen tube passing through the style eventually reaches the ovarian cavity. Here, it proceeds in the transmitting tissue lining the ovary wall and the placenta, sometimes also the funiculus (*Allium*). The pollen tube comes in contact with the ovule and enters the embryo sac through the micropyle (figs. 21.12,*D*, and 21.14,*A*), or sometimes through the chalazal tissue. If the ovule is crassinucellate, nucellar tissue intervenes between the micropyle and the embryo sac and must be penetrated by the pollen tube[38] (fig. 21.14,*B*). In the cotton ovule, a column of cells degenerate in the nucellus and thus a path is prepared for the passage of the pollen tube.[34]

In several species examined ultrastructurally,[35] the pollen tube was found entering one of the synergids through the filiform apparatus and discharging its contents within

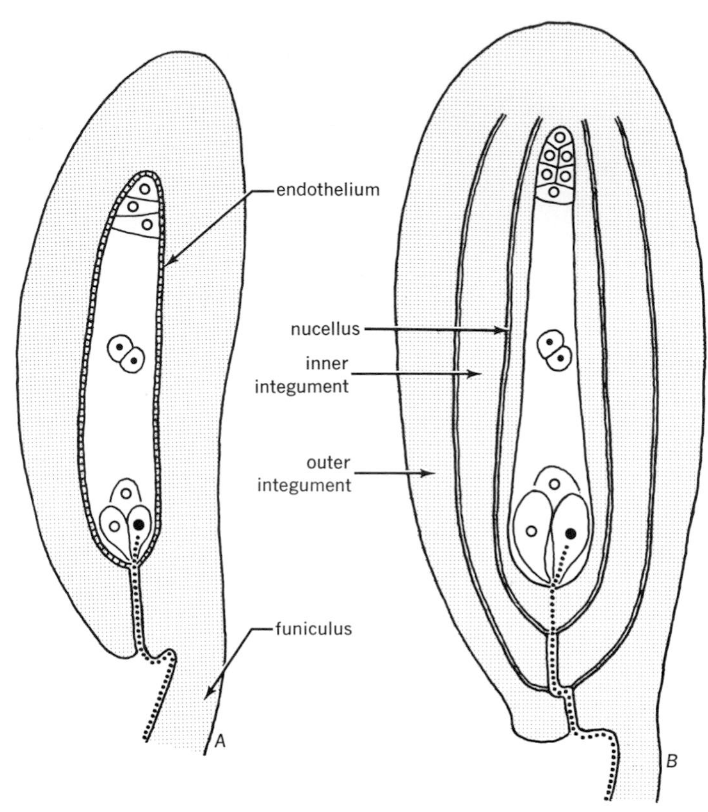

Figure 21.14 Outline drawings of longitudinal sections of ovules showing contrasting paths of pollen tubes (dotted lines) related to contrasting structures of the ovules. *A*, unitegmic, tenuinucellate; pollen tube reaches synergid through a simple micropyle. *B*, bitegmic, crassinucellate; pollen tube reaches synergid by passing through a double micropyle and nucellus. (*B*, cotton, adapted from Joshi, Wadhwani, and Johri.[38])

the cell. This course of events may be detectable with the light microscope also.[2] The synergid receiving the pollen tube partly degenerates before the pollen tube arrives, a phenomenon possibly associated with a release of chemotropic substances that guide the pollen tube into the altered synergid. The second synergid usually degenerates later, but in wheat it often breaks down at the same time as the first (fig. 21.15,*A*). The process of pollen tube discharge varies in detail. The common features are that the pollen tube stops growing part way in the synergid, bursts open, and releases its contents. The vegetative nucleus and that of the synergid degenerate and become the so-called X-bodies. The plasmalemma of the degenerated synergid disappears. One of the sperms comes in contact with the plasmalemma of the egg, the other with that of the central cell. The hypothesis has been advanced that the plasmalemmas of the sperms fuse with those of the two cells after which the sperm nuclei enter the cells.[35] In angio-

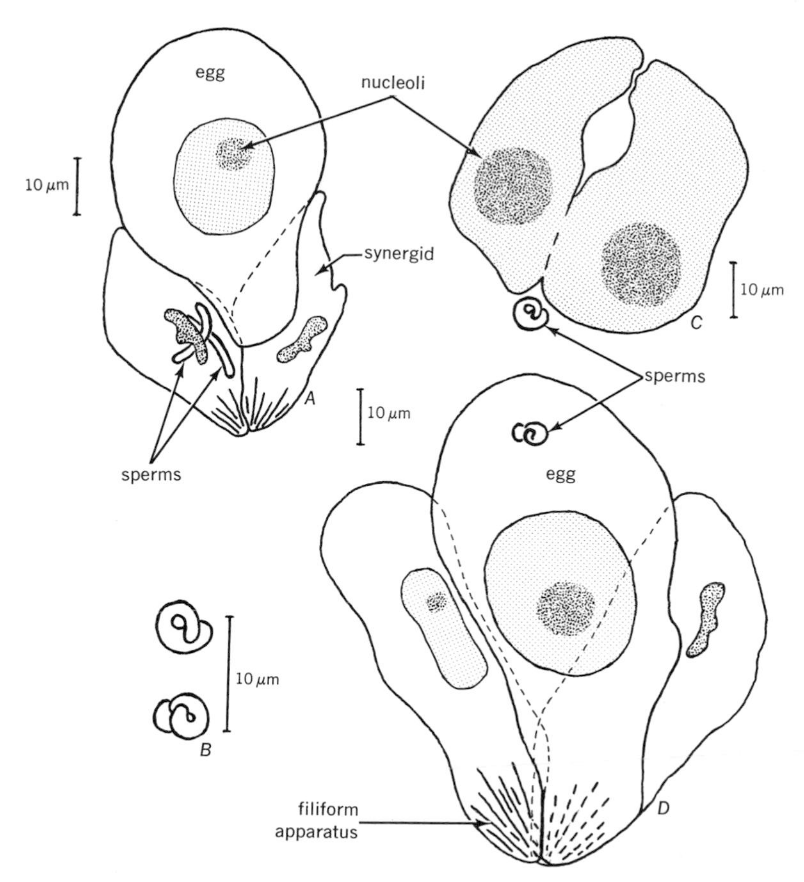

Figure 21.15 Fertilization in wheat. *A,* slender, elongated sperm cells have entered one of the synergids; both synergids have degenerated nuclei. *B,* coiled sperms. *C,* fusing polars, and, *D,* egg apparatus from one embryo sac; one of the coiled sperms is next to the polar nuclei, the other has entered the egg cell. (Adapted from Batygina.[2])

sperms, whole sperms or cell organelles of the sperms have not been seen in the egg or the central cell in electron microscope studies. In light microscope views of fertilization in wheat,[2] the sperms are seen becoming tightly coiled when they leave the synergid (fig. 21.15,*B*) and in that form they are also detectable within the egg and fusing with the polars (fig. 21.15,*C*,*D*).

The fusion of the sperm nuclei with (1) the egg nucleus and (2) the polar nuclei or the product of their fusion (double fertilization) occurs as described for the fusion of the polar nuclei. The outer membranes of the nuclear envelopes fuse first, then also the inner membranes, and thus the continuity between the contents of the fusing nuclei is established. The nucleoli fuse after the union of the nuclear membranes is accomplished.[11] The fertilized egg becomes the zygote, the product of the nuclear fusion in the central cell (triple fusion), the *primary endosperm nucleus.*

Zygote

As compared with the egg, the zygote undergoes profound cytologic changes. Its wall is completed at the chalazal end, the vacuolar volume is decreased, ribosomes are increased in number and form large polysomes, and starch accumulates in the plastids. The cell appears to become supplied with materials and information necessary for the forthcoming divisions. In cotton, the reduction in vacuolar content is associated with a reduction in size of the zygote to about half of that of the egg.[33] Zygotes of other species examined ultrastructurally show only minor shrinkage (chapter 24). The first division of the zygote (fig. 21.12,*F*) frequently occurs after the division of the primary endosperm nucleus that initiates the formation of the endosperm (fig. 21.12,*E*).

REFERENCES

1. Bailey, I. W. Some useful techniques in the study and interpretation of pollen morphology. *J. Arnold Arb.* 41:141–148. 1960.
2. Batygina, T. B. *Embriologiya pshenitsy.* [*Embryology of wheat.*] Leningrad, Kolos. 1974.
3. Bouman, F. The application of tegumentary studies to taxonomic and phylogenetic problems. *Ber. Deut. Bot. Ges.* 84:169–177. 1971.
4. Cass, D. D. An ultrastructural and Nomarski-interference study of the sperms of barley. *Can. J. Bot.* 51:601–605. 1973.
5. Cass, D. D., and W. A. Jensen. Fertilization in barley. *Amer. J. Bot.* 57:62–70. 1970.
6. Cass, D. D., and I. Karas. Ultrastructural organization of the egg of *Plumbago zeylanica. Protoplasma* 81:49–62. 1974.
7. Christensen, J. E., and H. T. Horner, Jr Pollen pore development and its spatial orientation during microsporogenesis in the grass *Sorghum bicolor. Amer. J. Bot.* 61:604–623. 1974.
8. Crang, R. E., and G. May. Evidence for silicon as a prevalent elemental component in pollen wall structure. *Can. J. Bot.* 52:2171–2174. 1974.
9. Crang, R. E., and G. B. Miles. An electron microscope study of germinating *Lychnis alba* pollen. *Amer. J. Bot.* 56:398–405. 1969.
10. Davis, G. L. *Systematic embryology of the angiosperms.* New York, John Wiley & Sons. 1966.
11. Deschamps, R. d'A. Étude ultrastructurale de la double fécondation chez le *Linum catharticum* L. *C. R. Acad. Sci. Paris D.* 279:263–265. 1974.
12. Diboll, A. G., and D. A. Larson. An electron microscopic study of the mature megagametophyte in *Zea mays. Amer. J. Bot.* 53:391–402. 1966.
13. Dickinson, H. G., and D. Lewis. Cytochemical and ultrastructural differences between intraspecific compatible and incompatible pollinations in *Raphanus. Proc. Roy. Soc. London B.* 183:21–38. 1973.
14. Dickinson, H. G., and D. Lewis. The formation of the tryphine coating the pollen grains of *Raphanus,* and its properties relating to the self-incompatibility system. *Proc. Roy. Soc. London B.* 184:149–165. 1973.
15. Dietrich, J. Le cloisonnement simultané des cellules-mères polliniques de deux Dicotylédones: *Paeonia tenuifolia* (Renonculacées) et *Campanula rapunculoides* (Campanulacées). *C. R. Acad. Sci. Paris D.* 276:509–512. 1973.
16. Dover, G. A. The organization and polarity of pollen mother cells of *Triticum aestivum. J. Cell Sci.* 11:699–711. 1972.

17. Dunbar, A. Pollen ontogeny in some species of Campanulaceae. A study by electron microscopy. *Bot. Notiser* 126: 277–315. 1973.
18. Dupuis, F. Formation de la paroi de séparation entre la cellule génératrice et la cellule végétative dans le pollen d'*Impatiens balsamina* L. *Bull. Soc. Bot. France* 119:41–50. 1973.
19. Dupuis, F. Evolution of the plastidial system during the microsporogenesis in *Impatiens balsamina* L. Pp. 65–71. In: *Fertilization in higher plants.* H. F. Linskens, ed. Amsterdam, North-Holland Publishing Company. 1974.
20. Eames, A. J. *Morphology of angiosperms.* New York, McGraw-Hill. 1961.
21. Echlin, P. The role of the tapetum during microsporogenesis of angiosperms. Pp. 41–61. In: *Pollen development and physiology.* J. Heslop-Harrison, ed. London, Butterworth. 1971.
22. Fossard, R. A. De. Development and histochemistry of the endothecium in the anthers of in vitro grown *Chenopodium rubrum* L. *Bot. Gaz.* 130:10–22. 1969.
23. Górska-Brylass, A. The "callose stage" of the generative cells in pollen grains. *Grana Palynol.* 10:21–30. 1970.
24. Herth, W., W. W. Franke, H. Bittiger, A. Kuppel, and G. Keilich. Alkali-resistant fibrils of β-1,3- and β-1,4-glucans: structural polysaccharides in the pollen tube wall of *Lilium longiflorum. Cytobiologie* 9:344–367. 1974.
25. Heslop-Harrison, J. Ultrastructural aspects of differentiation in sporogenous tissue. *Symp. Soc. Exp. Biol.* No. 17. *Cell Differentiation.* Pp. 315–340. 1963.
26. Heslop-Harrison, J. Cytoplasmic connexions between angiosperm meiocytes. *Ann Bot.* 30:221–230. 1966.
27. Heslop-Harrison, J. The pollen wall: structure and development. Pp. 75–98. In: *Pollen development and physiology.* J. Heslop-Harrison, ed. London, Butterworth. 1971.
28. Heslop-Harrison, J., Y. Heslop-Harrison, R. B. Knox, and B. Howlett. Pollen-wall proteins: "gametophytic" and "sporophytic" fractions in the pollen walls of Malvaceae. *Ann. Bot.* 37:403–412. 1973.
29. Hoefert, L. L. Fine structure of sperm cells in pollen grains in *Beta. Protoplasma* 68:237–240. 1969.
30. Horner, H. T., and M. A. Rogers. A comparative light and electron microscopic study of microsporogenesis in male-fertile and cytoplasmic male-sterile pepper (*Capsicum annuum*). *Can. J. Bot.* 52:435–441. 1974.
31. Jensen, W. A. The ultrastructure and histochemistry of the synergids of cotton. *Amer. J. Bot.* 52:238–256. 1965.
32. Jensen, W. A. The ultrastructure and composition of the egg and central cell of cotton. *Amer. J. Bot.* 52:781–797. 1965.
33. Jensen, W. A. Cotton embryogenesis: the zygote. *Planta* 79:346–366. 1968.
34. Jensen, W. A. Cotton embryogenesis: pollen tube development in the nucellus. *Can. J. Bot.* 47:383–385. 1969.
35. Jensen, W. A. Fertilization in flowering plants. *BioScience* 23:21–27. 1973.
36. Jensen, W. A., and D. B. Fisher. Cotton embryogenesis: the pollen tube in stigma and style. *Protoplasma* 69:215–235. 1970.
37. Jensen, W. A., M. Ashton, and L. R. Heckard. Ultrastructural studies of the pollen of subtribe Castilleiinae, family Scrophulariaceae. *Bot. Gaz.* 135:210–218. 1974.
38. Joshi, P. C., A. M. Wadhwani, and B. M. Johri. Morphological and embryological studies of *Gossypium* L. *Proc. Natl . Inst. Sci. India* 33:37–93. 1967.
39. Lintilhac, P. M. Differentiation, organo-

genesis, and the tectonics of cell wall orientation. II. Separation of stresses in a two-dimensional model. *Amer. J. Bot.* 61:135–140. 1974.

40. Lintilhac, P. M., and W. A. Jensen. Differentiation, organogenesis, and the tectonics of cell wall orientation. I. Preliminary observations on the development of the ovule in cotton. *Amer. J. Bot.* 61:129–134. 1974.
41. Lutz, R. W., and R. D. Sjolund. Development of the generative cell wall in *Monotropa uniflora* L. pollen. *Plant Physiol.* 52:498–500. 1973.
42. Maheshwari, P. *An introduction to the embryology of angiosperms.* New York, McGraw-Hill. 1950.
43. Mepham, R. H., and G. R. Lane. Formation and development of the tapetal periplasmodium in *Tradescantia bracteata. Protoplasma* 68:175–192. 1969.
44. Mepham R. H., and G. R. Lane. Observations on the fine structure of developing microspores of *Tradescantia bracteata. Protoplasma* 70:1–20. 1970.
45. Newcomb, W., and T. A. Steeves. *Helianthus annuus* embryogenesis: embryo sac wall projections before and after fertilization. *Bot. Gaz.* 132:367–371. 1971.
46. Rodkiewicz, B. Callose in cell walls during megasporogenesis in angiosperms. *Planta* 93:39–47. 1970.
47. Rosen, W. G. Ultrastructure and physiology of pollen. *Ann. Rev. Plant Physiol.* 19:435–462. 1968.
48. Rosen, W. G. Pollen tube growth and fine structure. Pp. 177–185. Pistil-pollen interactions in *Lilium.* Pp. 239–254. In: *Pollen development and physiology.* J. Heslop-Harrison, ed. London, Butterworth. 1971.
49. Sanger, J. M., and W. T. Jackson. Fine structure study of pollen development in *Haemanthus katharinae* Baker. I. Formation of vegetative and generative cells. *J. Cell Sci.* 8:289–301. 1971. II. Microtubules and elongation of the generative cells. *J. Cell Sci.* 8:303–315. 1971. III. Changes in organelles during development of the vegetative cell. *J. Cell Sci.* 8:317–329. 1971.
50. Savchenko. M. I. *Morfologiya semyapochki pokrytosemennykh rastenij.* [Morphology of the ovule in angiosperms.] Leningrad, Nauka. 1973.
51. Schulz, Sister R., and W. A. Jensen. Capsella embryogenesis: the egg, zygote, and young embryo. *Amer. J. Bot.* 55:807–819. 1968.
52. Shaw, G. The chemistry of sporopollenin. Pp. 305–348. In: *Sporopollenin.* J. Brooks, P. R. Grant, M. D. Muir, P. van Gijzel, and G. Shaw, eds. London, Academic Press. 1971.
53. Spitzer, N. C. Low resistance connections between cells in the developing anther of the lily. *J. Cell Biol.* 45:565–575. 1970.
54. VanDerWoude, W. J., D. J. Morré, and C. E. Bracker. Isolation and characterization of secretory vesicles in germinated pollen of *Lilium longiflorum. J. Cell Sci.* 8:331–351. 1971.
55. Vasil, I. K. The new biology of pollen. *Naturwiss.* 60:247–253. 1973.
56. Vazart, B. Structure et évolution de la cellule génératrice du Lin, *Linum usitatissimum* L., au cours des premiers stades de la maturation du pollen. *Rev. Cytol. Biol. Vég.* 32:101–114. 1969.
57. Vazart, J. Aspects infrastructuraux de la reproduction sexuée chez le Lin. Derniers stades de la différenciation du pollen. Structure inframicroscopique de la cellule génératice et des gamètes. *Rev. Cytol. Biol. Vég.* 33:289–310. 1970.
58. Vazart, B., and J. Vazart. Infrastructure

de l'ovule de Lin, *Linum usitatissimum* L. L'assise jaquette ou endothélium. *C. R. Acad. Sci. Paris* 261:2927–2930. 1965.

59. Walker, J. W. Aperture evolution in the pollen of primitive angiosperms. *Amer. J. Bot.* 61:1112–1137. 1974.
60. Walker, J. W., and J. J. Skvarla. Primitively columellaless pollen: a new concept in the evolutionary morphology of angiosperms. *Science* 187:445–447. 1975.
61. Walker, R. I. Cytological and embryological studies in *Solanum,* section Tuberarium. *Bull. Torrey Bot. Club* 82:87–101. 1955.
62. Waterkeyn, L. Les parois microsporocytaires de nature callosique chez *Helleborus* et *Tradescantia. Cellule* 62:225–255. 1962.
63. Whelan, E. D. P. Discontinuities in the callose wall, intermeiocyte connections, and cytomixis in angiosperm meiocytes. *Can. J. Bot.* 52:1219–1224. 1974.
64. Whelan, E. D. P., G. H. Haggis, and E. J. Ford. Scanning electron microscopy of the callose wall and intermeiocyte connection in angiosperms. *Can. J. Bot.* 52:1215–1218. 1974.

22

The Fruit

CONCEPT AND CLASSIFICATION

Fertilization of the egg evokes embryo development and seed formation. Concomitantly, the flower undergoes changes leading to the development of the fruit in which the gynoecium becomes the sole or the basic component. The perianth and the stamens usually wither and fall and later, after pollination, the style dries up except in species in which it functions in fruit dispersal (various Ranunculaceae, some Rosaceae[33]). The ovary, however, increases in size and undergoes a variety of histologic modifications, some or all of which result in the development of devices that facilitate seed dispersal. In many taxa, accessory, or noncarpellary, tissue becomes associated with the ovary and may dominate over the carpellary tissue in the mature fruit. Within the ovary itself, the ovary wall, the partitions in syncarpous gynoecia, and the placentae participate in fruit formation to various degrees.

Fruits may develop without fertilization and seed development. This phenomenon, known as *parthenocarpy,* is widespread, especially in species with large numbers of ovules per fruit, such as banana, fig, melon, pineapple, and tomato. Parthenocarpy may occur without pollination (citrus, pepper, pumpkin, tomato) or it may require the stimulation of pollination (orchid). Seedless fruits may also develop as a result of abortion of embryos (cherry, grape, peach). In a strict definition of parthenocarpy,[32] such fruits are not parthenocarpic because fertilization is involved in initiating seed development.

Fruit and seed colaborate in various ways in functioning as devices of plant dispersal. In the more primitive taxa, the seed develops features that make it independent of the fruit in utilizing the various agencies of seed dispersal, but in the more advanced angiosperms, the fruit is the dominant entity of dispersal and displaces the seed as such.[33] Thus, the functional, and consequently also the morphological, relation between the fruit and the seed is highly variable and the ecological role of the fruit has many aspects. This variability of the fruit is the source of difficulties botanists experience in defining the fruit and in establishing systems of classification.

Strictly defined, the fruit is the matured ovary. A more acceptable, broader definition regards the fruit as a derivative of the gynoecium and whatever extracarpellary parts it may be united with in the fruiting stage. Examples of extracarpellary tissues in the fruit are receptacle in the strawberry, calyx in the mulberry, bracts in the pineapple, and floral tube or receptacle in fruits derived from epigynous flowers such as the apple and the pumpkin.

Morphologic fruit classifications usually relate the fruit to the type of flower and gynoecium from which it develops, with emphasis on the relationship of the carpels to one another and to the other floral parts. In a common classification of this kind, the following major types of fruit are distinguished. The *simple fruit,* a product of a single pistil which may consist of one carpel or of two or more united carpels (bean pod, tomato, plum); *aggregate fruit,* a fruit formed from an apocarpous gynoecium, each carpel retaining its identity in mature state (raspberry, strawberry); *multiple fruit,* a fruit derived from an inflorescence, that is, from combined gynoecia of many flowers (mulberry, pineapple). If any of these fruits contain extracarpellary tissue, they are additionally called *accessory fruits.*[4] Thus, an apple is a simple accessory fruit; strawberry, an aggregate accessory fruit; mulberry, a multiple accessory fruit. Accessory fruits are also termed, rather inaptly, false or spurious fruits.

If one considers the fruit to be the product of the total gynoecium and any other floral parts that may be joined with the gynoecium congenitally or ontogenetically, the terms accessory, false, and spurious as applied to fruits become obsolete. Four characters suffice for a foundation of an inclusive morphologic classification of fruits according to the broadened definition of the fruit:[54] (1) *aggregate fruit,* carpels not united with one another; (2) *unit fruit,* carpels united; (3) *free fruit,* from a superior ovary; (4) *cup fruit,* from an inferior ovary embedded in a "cup" of noncarpellary tissue or from a superior ovary associated in fruit with a cuplike hypanthium. Combinations of these four characters provide a classification into four major fruit types: (1) *aggregate free fruit,* fruit derived from an apocarpous hypogynous flower (fig. 22.1,*A*); *unit free fruit,* from a syncarpous hypogynous flower (fig. 22.1,*B*); (3) *aggregate cup fruit,* from an apocarpous perigynous flower (fig. 22.1,*C*); and (4) *unit cup fruit,* from a syncarpous epigynous flower (fig. 22.1,*D*).

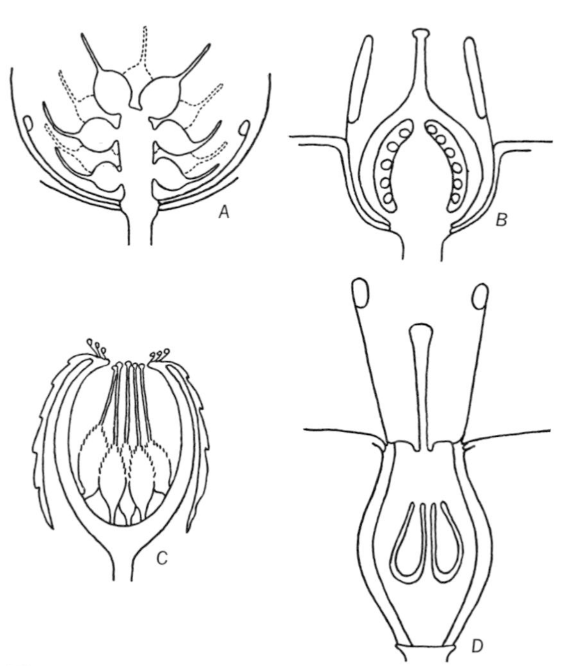

Figure 22.1 Diagrams of longitudinal sections of flowers from which fruit types named in parentheses are derived according to Winkler's[54] classification. *A, Ranunculus,* apocarpous, hypogynous (aggregate free fruit). *B, Solanum,* syncarpous, hypogynous (unit free fruit). *C, Rosa,* apocarpous, perigynous (aggregate cup fruit). *D, Cornus,* syncarpous, epigynous (unit cup fruit). (From K. Esau, *Plant Anatomy* 2nd ed. John Wiley & Sons, 1965.)

An individual carpel in an aggregate fruit forms a *fruitlet.*[55] The four types of fruit are further subdivided on the basis of combinations of many features among which the arrangement and type of union of carpels and the nature of the fruit wall and its dehiscence or nondehiscence are especially important.[2]

The subdivision of fruits into types according to the structure of flowers from which they develop, is one of several morphologic fruit classifications none of which has found a general acceptance. One of the weaknesses ascribed to morphologic classifications is that they largely neglect the functional modifications of fruits. But formulating terms for all such modifications would result in a chaotic and unmanageable proliferation of categories.[33] Morphologic classifications continue to be useful for organizing simple descriptions of fruits and for correlating the fruit and the flower developmentally.

THE FRUIT WALL

The fruit wall is commonly used to categorize fruits in descriptive anatomic and morphologic classifications. This wall is referred to as the *pericarp,* a term that, in a strict sense, means the matured ovary wall. If the term fruit is intended to denote a matured gynoecium with or without accessory tissue, the term pericarp may be extended to include the extracarpellary tissue if it is associated with the ovary in fruit.[2] To make the relation between the flower and the fruit clear, in this chapter the term pericarp is used only when the origin of the tissue from the ovary wall needs to be indicated, and the term fruit wall denotes either the pericarp in the narrow sense or the combination of pericarp and extracarpellary tissue. In some fruits derived from epigynous flowers the pericarp proper is distinguishable from the associated tissue, in others no such division is discernible. As was mentioned in chapter 20, some botanists have advanced the concept that carpellary tissue is not associated with the ovarian cavity in epigynous flowers.

The fruit wall may be more or less highly differentiated, and frequently the pericarp proper shows two or three distinct layers. If such layers are recognizable, they are referred to, beginning with the outermost layer, as *exocarp* (or *epicarp*), *mesocarp,* and *endocarp.* These terms are commonly used for purpose of description without relation to the ontogenetic origin of the layers. The exocarp, for example, sometimes denotes the epidermis alone, sometimes the epidermis together with some subjacent tissue. Investigators often disagree on the delimitation of the pericarp layers in the same kind of fruit. In this chapter, the pericarp layers are listed only if they are mentioned by the authors whose work is reviewed.

The main types of fruits that may be distinguished on the basis of fruit wall histology are the *dry fruits,* which are subdivided into *dehiscent* and *indehiscent,* and the *fleshy fruits.*

FRUIT TYPES

Dry fruit

DEHISCENT FRUIT

Dehiscent fruit walls commonly occur in fruits containing several seeds. A dehiscent fruit may develop from a single carpel (follicle, legume) or from more than one carpel (capsule, silique).

The method of dehiscence is highly variable.[18] If the ovary is derived from a single carpel, the break may occur longitudinally through (1) the suture resulting from the union

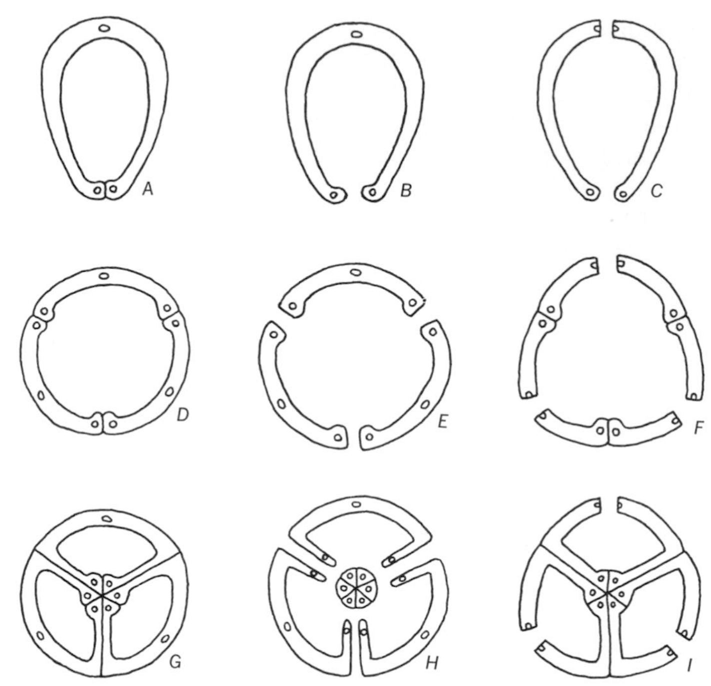

Figure 22.2 Diagrams of cross sections of dry dehiscent fruits illustrating three types of fruits, still unopened (*A, D, G*), and examples of dehiscence of such fruits (*B–C, E–F, H–I*). *A,* single carpel. *D,* three carpels, parietal placentation. *G,* three carpels, axile placentation. The median and two lateral bundles are indicated in each carpel. The median bundles are shown split in two in *C, F, I.* Other details in text. (Adapted from Kaden.[18])

of carpel margins; (2) the back of the carpel; (3) the suture and the back simultaneously (fig. 22.2,*A–C*). In syncarpous fruits with parietal placentation, dehiscence may occur through the suture between two carpels or through the backs of carpels (fig. 22.2,*D–F*). In syncarpous fruits with axile placentation (fig. 22.2,*G*), the separation along the line of union of contiguous carpels, that is, through the septae (*septicidal dehiscence*), may be combined with a breaking away from the central column of tissue (fig. 22.2,*H*). Dehiscence through the backs of carpels opens the individual locules (*loculicidal dehiscence;* fig. 22.2,*I*). In the fruit types mentioned above, the longitudinal dehiscence may also occur without reference to the suture or the median vein.[18] In some fruits, dehiscence occurs through a horizontal circular region involving all carpels (*circumscissile dehiscence*) or through pores (*poricidal dehiscence*).

A well-known example of a dry dehiscent fruit is the legume found in many Fabaceae. (The legumes and a variety of other dry fruits are commonly called by the inexact but convenient term pod.) The legume is derived from a superior ovary which embodies a single carpel. It dehisces along the suture of carpel margins and along the median vein (figs. 22.2,*C,* and 22.4,*B*).

As shown for the *Glycine* (soybean; fig. 22.3) pod, the pericarp layers from outside in are: exocarp composed of the outer epi-

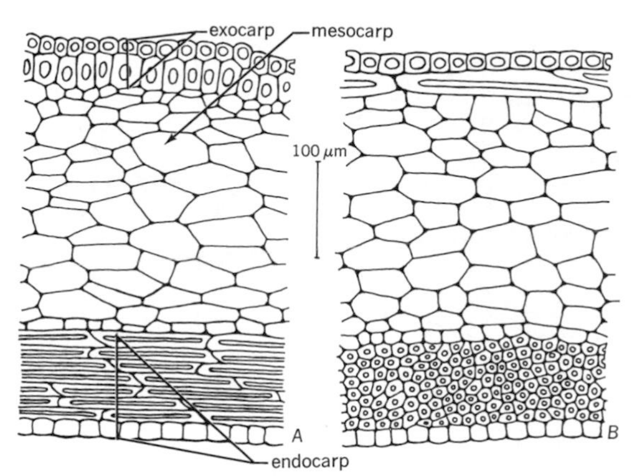

Figure 22.3 Pericarp of soybean (*Glycine*) legume. *A*, cross section; *B*, oblique longitudinal section. Sclerified cells occur in exocarp and endocarp. Vascular bundles are not shown. (Adapted from Monsi.[29])

dermis and hypodermis, both with thickened cell walls; mesocarp parenchyma; endocarp including several layers of sclerenchyma cells and the inner epidermis.[29] The cells in the hypodermis and the sclerenchyma are elongated but the long axes of the two kinds of cells are oriented in opposite planes. As a result, the outer and the inner layers of the pericarp shrink in different directions and the developing stresses promote the opening of the dehydrated mature fruit. Another feature found to be associated with the differential shrinkage of the pericarp in some legumes is the dissimilar orientation of cell wall microfibrils in different layers of the sclerenchyma, in low-pitched helices in some, in steeply pitched helices in others. Tensions developing during dehydration in some legumes are so great that the separated valves of the pod become twisted.

In *Phaseolus* (bean; fig. 22.4,*A,B*) pod, the epidermis and the hypodermis, which are physically closely associated (the "skin"), is the source of recurrent texture defects in canned and frozen green beans, for it may slough and leave a ragged surface.[37] The parenchyma beneath the hypodermis and extending to the sclerenchyma contains chloroplasts with starch granules. This parenchyma encloses a network of small vascular bundles near the sclerenchyma (fig. 22.4,*A*) interconnecting the median and the lateral bundles. The lignified sclerenchyma associated with the vascular bundles, particularly with the large median and lateral bundles, is the cause of "stringiness" in some varieties of green beans.

In *Glycine,* the sclerenchuma of the endocarp is delimited on the inside by a single layer of epidermis (fig. 22.3). In *Phaseolus,* the inner epidermis undergoes numerous periclinal divisions and becomes a massive layer of usually nonphotosynthetic inner parenchyma (fig. 22.4,*A*). When this parenchyma is still soft and translucent, the green bean pod is in prime state for use as a green vegetable. At that time, the outer parenchyma appears more mature, is less compact, and has thicker walls. In some legumes (*Pisum, Vicia faba*), the inner parenchyma produces hairlike extensions intruding deep into the locule. During this process, the radial series of cells separate from one another next to the locule and assume the form of multicellular hairs after the component cells elongate. This tissue is interpreted as a device to maintain favorable moisture conditions within the immature pod.[20]

The capsule, another type of dehiscent fruit, is composed of more than one carpel and dehisces in various ways (fig. 22.2,*D–I*). In the capsule of *Linum usitatissiumum,* the outer part of the pericarp consists of rigid, highly lignified cells, the inner part of parenchyma cells. In the capsule of *Nicotiana tabacum,* the distribution of the thick-walled and thin-walled cells is reversed.

A special type of capsule is the silique which is characteristic of many Brassicaceae. According to a favored interpretation,

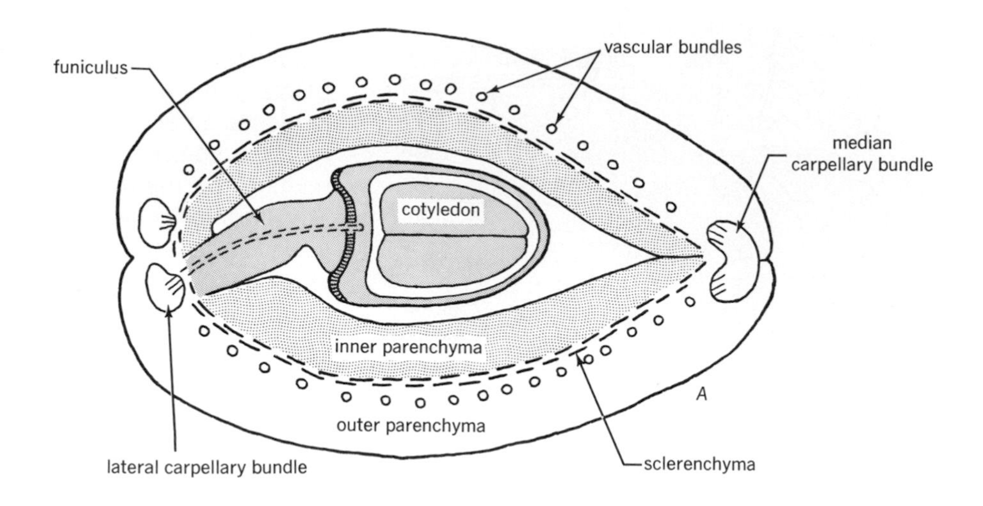

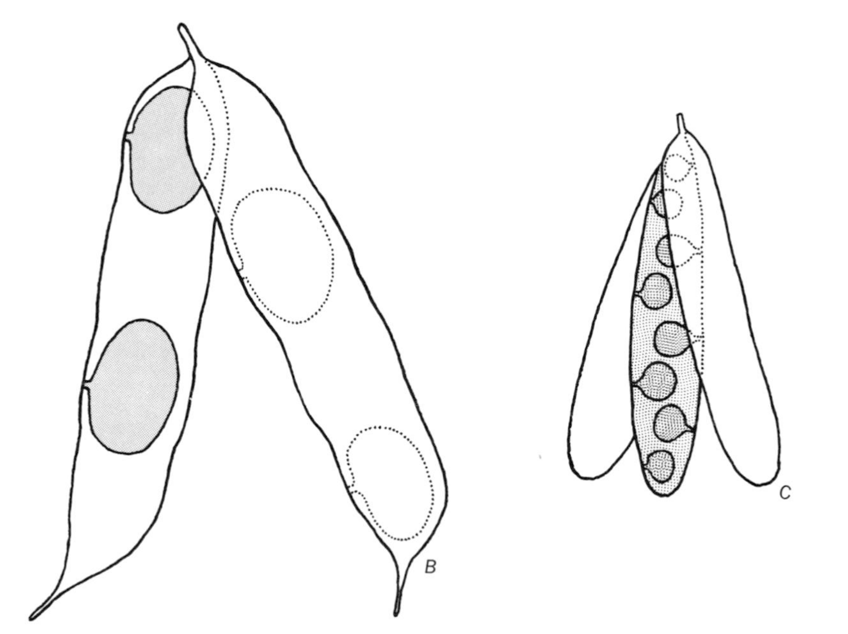

Figure 22.4 Diagrams of examples of legume (*Phaseolus*, *A*, *B*) and silique (*Brassica*, *C*). *A*, cross section of legume at level of the cotyledons in embryo. *B*, single carpel with dehiscence along the median and lateral bundles. *C*, two carpels with false partition holding the seeds after dehiscence.

the silique consists of two carpels joined margin to margin and a false partition that develops from the marginal placentae and divides the locule in two (chapter 20). The exocarp and mesocarp have thin walls and the endocarp is sclerenchymatic. A prominent rib occurs along the juncture of the carpels. At maturity, the carpels separate along the suture leaving the seeds attached to the rib which forms a frame (replum) around the partition (fig. 22.4,*C*).

INDEHISCENT FRUIT

An indehiscent fruit usually originates from an ovary in which only one seed develops, though more than one ovule may be present. The pericarp of an indehiscent fruit often resembles a seed coat in structure. The actual seed coat in such fruits may become obliterated to a considerable degree (achene of the Asteraceae; fig. 22.5) or fused with the pericarp (caryopsis of the Poaceae; fig. 22.6).

The achene of the Asteraceae develops from an inferior ovary and by this definition is a cypsella.[17] The extracarpellary tissue is interpreted as the floral tube but there is no differentiation of the fruit wall into pericarp and accessory tissue. The seed coat, derived from a single integument, is disorganized and compressed except for the outer epidermis which develops thick walls. The fruit wall is also largely disorganized and is reduced to an outer sclerified tissue and some parenchyma[10] (fig. 22.5,*D*).

The fruit of grasses, the caryopsis, has been extensively studied.[41] The covering layers of the wheat caryopsis are composed of the pericarp and the remains of the seed coat. In the pericarp the layers from the outside in are (fig. 22.6): outer epidermis covered with a cuticle; one or more layers of

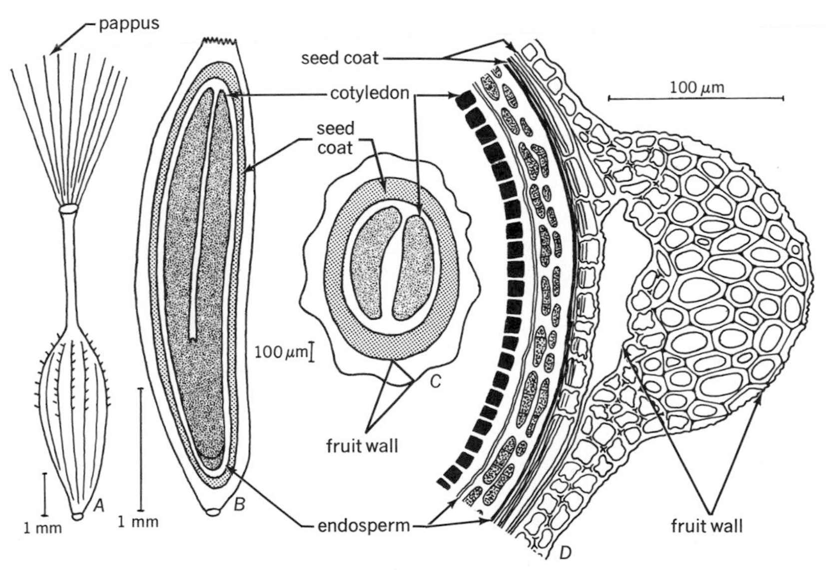

Figure 22.5 Achene (cypsela type) of *Lactuca sativa* (lettuce). *A,* entire fruit with pappus. *B, C,* diagrams of longitudinal (*B*) and transverse (*C*) sections. *D,* part of fruit coat with subjacent layers, in transection. (Adapted from K. Esau, *Plant Anatomy* 2nd ed. John Wiley & Sons, 1965.)

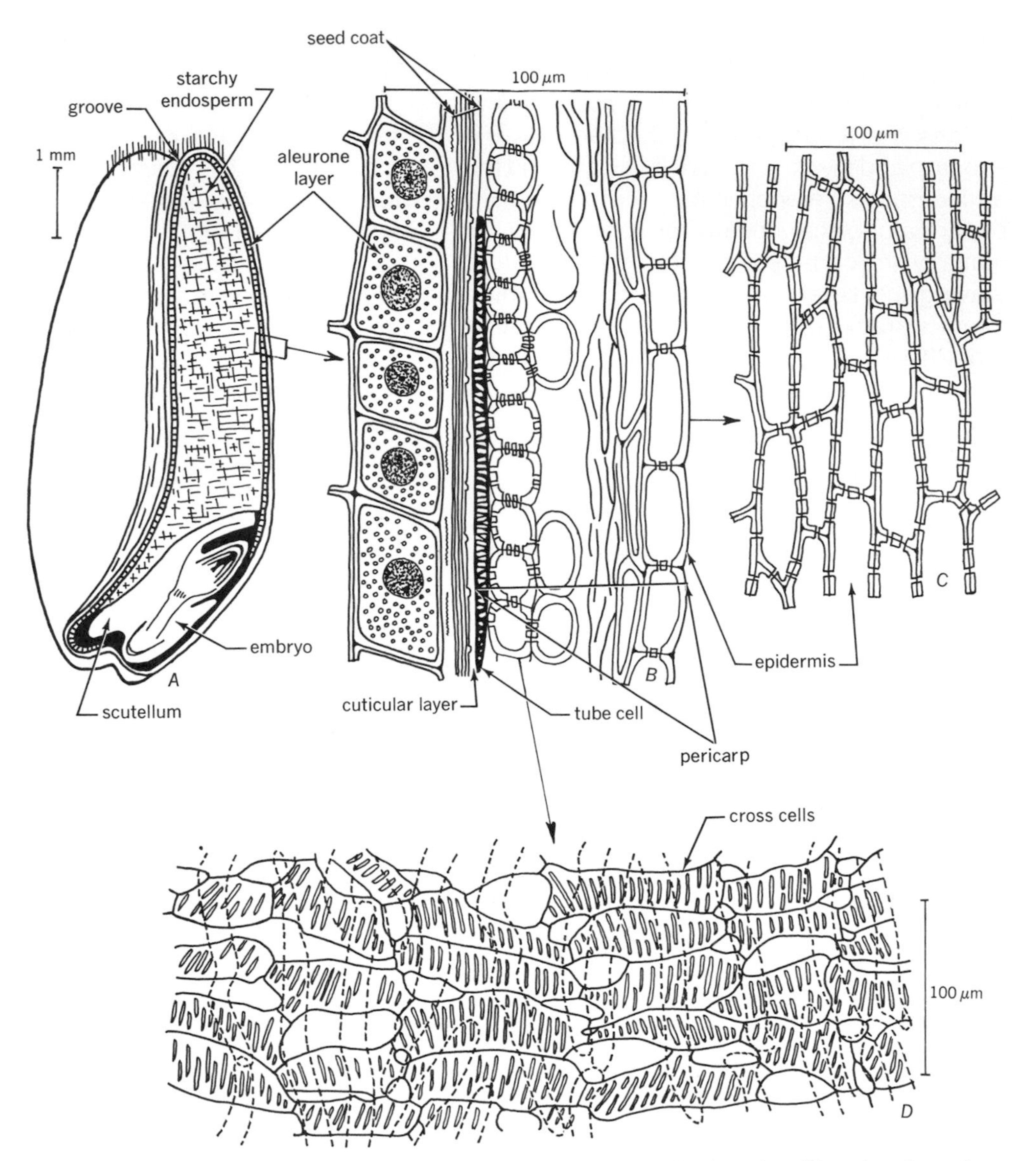

Figure 22.6 Caryopsis (*A*) of wheat (*Triticum*) and parts of its pericarp in longitudinal section (*B*) and surface views (*C, D*). (*A, B,* from K. Esau, *Plant Anatomy* 2nd ed. John Wiley and Sons, 1965. *C, D,* drawn from photomicrographs in Bradbury, MacMasters, and Cull.[6])

parenchyma, partly compressed; partly resorbed parenchyma; cross cells, elongated transversely to the long axis of the grain and having thick lignified walls; and remains of the inner epidermis in the form of lignified cells elongated parallel with the long axis of the grain (tube cells). In the development of the seed coat, the outer integument disintegrates, and the inner becomes altered and compressed. The compressed inner integument contains pigment, gives a positive reaction for fatty materials, and is covered by a cuticle on both sides.[6]

Within the seed coat lies the endosperm which forms about 83 percent of the fruit and contains starch and proteins. The outermost endosperm layer is the aleurone layer which contains lipid and protein reserves and produces enzymes essential for initiating germination (chapter 23). The aleurone layer surrounds the starchy endosperm and embryo. The bran of wheat, which is removed in the milling of white flour, forms about 14 percent of the grain and includes the pericarp, remnants of the nucellus and integuments, and the aleurone layer.[5] The embryo is also separated ("wheat germ") because the oil it contains reduces the keeping quality of the flour.

Among the proteins of the wheat flour, the water soluble glutens govern the varietal differences in breadmaking. The glycolipids (complexes of carbohydrates and lipids) are also essential for producing bread of acceptable textural quality because of their interaction with gluten proteins. Studies of this interaction have led to the improvement of the processes of enriching wheat flour with proteins.[34]

The degree of modification of the seed coat and pericarp during the development of the caryopsis shows a wide range of variation. In the grain of *Zea*,[22] the outer pericarp consists of cells with thick pitted walls and is much compressed. The central pericarp disintegrates. The inner pericarp remains thin walled and is variously stretched, torn, and compressed. The integuments disintegrate completely. A cuticular layer occurs between the thick-walled nucellar epidermis and the pericarp. *Sorghum* shows less collapse in the pericarp than many other cereals.[42]

A feature requiring consideration with regard to water absorption by the germinating grain is the development of cuticular layers in the seed coat. Such layers are derived from the integuments and the nucellar epidermis, although the inner pericarp layer may also contribute fatty materials. In the mature grain the cuticular layers are essentially fused into one layer closely associated with the remnants of the seed coat (fig. 22.6,*B*). In the wheat grain, the cutinized seed coat is interrupted in the groove region where a strand of pigmented tissue is inserted between the longitudinal vascular bundle and the endosperm.[57] It has been suggested that during germination water passes through the cell walls of the pigment strand.[23]

In millet (*Echinochloa utilis*), the gap in the cutinized seed coat is small and occurs at the base of the caryopsis.[58] The gap contains two kinds of cells, nucellar cells and cells similar to those of the pigment strand in wheat. The aleurone cells lying opposite the nucellar cells are differentiated as transfer cells provided with wall ingrowths (fig. 22.7). Similar cells, and in the same position in the caryopsis, were depicted in *Zea*[22] and identified as aleurone transfer cells in *Setaria*.[40] Possibly these cells aid in the transfer of solutes to the developing endosperm and embryo through the narrow gap in the seed coat. In grasses with grooves the gap is much extended along the pigment strand and aleurone cells do not have wall ingrowths.[58]

The type of fruit called schizocarp separates into one-seeded units, the mericarps, each a derivative of one carpel. Although the schizocarp is mentioned here under indehiscent fruits, the separation into mericarps has the same relation to seed dispersal as the release of seeds by dehiscence of a fruit. In fact, some botanists regard the schizocarp as intermediate between dehiscent and indehiscent dry fruits.[19]

As an example of a schizocarp, figure 22.8 illustrates the fruit (cremocarp) of *Carum*, a genus in the Apiaceae. The fruit is derived

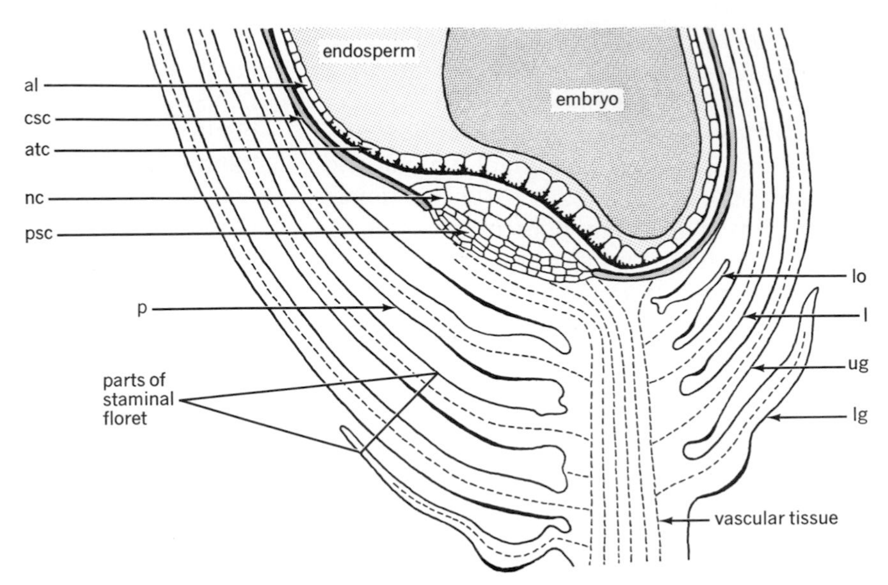

Figure 22.7 Part of median longitudinal section of a mature spikelet of millet (*Echinochloa utilis*) showing the region where the cutinized seed coat is interrupted and the aleurone cells have wall ingrowths indicating transfer cell function. Details are: *al*, aleurone layer; *atc*, aleurone transfer cells; *csc*, cutinized seed coat; *l*, lemma; *lg*, lower glume; *lo*, lodicule; *nc*, nucellar cells; *p*, palea; *psc*, pigment-strand type of cells; *ug*, upper glume. (Adapted from Zee and O'Brien.[58])

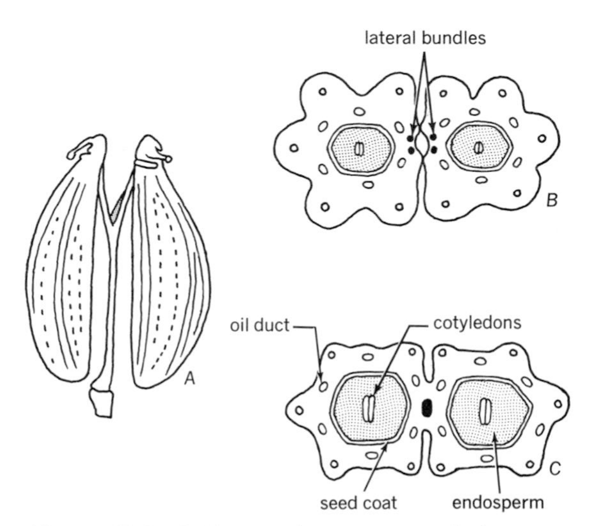

Figure 22.8 Schizocarp (cremocarp type) of caraway (*Carum carvi*, Apiaceae). *A*, fruit partly dehisced into two mericarps. *B*, *C*, transections of fruit at two levels before dehiscence. Lateral carpellary bundles are shown in black. (Adapted from Kaden.[19])

from an inferior ovary and shows no distinction between a pericarp and accessory tissue. The mericarps of *Carum* separate in such a way that the united lateral bundles and associated tissue form a column forked at the top (fig. 22.8,*A*). This feature is reflected in the arrangement of the lateral vascular bundles into one unit at lower levels (fig. 22.8,*C*) and two paired units at higher levels (fig. 22.8,*B*). Each mericarp has five ribs in which vascular bundles are located. Six oil ducts alternate with the vascular bundles. In some fruits of the Apiaceae no column is formed during their separation into mericarps.

Fleshy fruit

The fleshy fruits, like the dry fruits, may be derived from monocarpellate or multicarpellate gynoecia. Their walls may consist of the pericarp or of pericarp fused with extracarpellary tissues. The inner or the outer part of the fruit wall or the entire fruit wall may become fleshy by differentiating into soft or succulent parenchyma. Parts other than the wall, such as placentae and partitions in multilocular ovaries, may become fleshy.

The fleshy-fruit character is considered to be relatively new from the evolutionary aspect.[33] According to this concept, in earlier stages of the evolutionary process, the fleshy part was an outgrowth of the inner layer of the pericarp (pulpa) penetrating among the seeds in the locule. Later, the entire pericarp became a fleshy storage tissue attractive to animals acting as agents of seed dispersal.

FRUIT WITH A RIND

Among the fruits in which a rind confines the fleshy tissue to the inner part of the fruit are citrus, banana, and fruits of the Cucurbitaceae. The citrus fruit,[43] called hesperidium, develops from a superior ovary and has about ten carpels (fig. 22.9,*A,B*). The placentation is axile. The exocarp, or flavedo (yellow tissue), consists of a cuticle-covered outer epidermis and compact subepidermal parenchyma with oil glands and crystal-containing cells. The mesocarp, or albedo (white tissue), consists of parenchyma with large intercellular spaces and cells with armlike extensions (aerenchyma). A vascular network occurs in this tissue. The endocarp is composed of the inner epidermis and a few layers of compact parenchyma. The partitions between the locules are extensions of endocarp and mesocarp tissues. In some kinds of citrus, the segments (carpels) in the mature fruit are easily separated along the internal layer of parenchyma in the partitions. The "peel" of a mature fruit consists of the exocarp and all but the innermost layer of the mesocarp.

The endocarp produces the juice sacs that completely fill the carpel locules (fig. 22.9,*D*). The juice sacs are multicellular club-shaped structures with long slender stalks. At maturity, an epidermis bearing a cuticle with wax secretions[11] encloses large, highly vacuolated cells containing the juice. The juice sacs arise from the epidermis of the endocarp but the subepidermal layer may add cells also. The sacs are initiated as small protuberances, mainly on the abaxial side of the carpels (fig. 22.9,*C*), and elongate by apical growth.A massive meristem in subterminal position forms the bulk of the juice sac. Development of the juice sacs begins at the time when the flower opens and continues until the style withers.

The chloroplasts responsible for the green color of immature fruits occur in the parenchyma of the exocarp. They become changed into chromoplasts as the fruit matures. In the Valencia orange, which develops the maximum orange color in the winter months in southern California, a reversion to green color occurs in the spring. The change involves a restoration of chloroplasts from chromoplasts, including a reformation of grana.[51]

The pepo of the Cucurbitaceae develops from an inferior ovary. No dividing line is discernible between carpellary and extracarpellary tissues (fig. 22.10). The arrangement of carpels in Cucurbitaceae is a subject of considerable controversy, particularly with regard to the type of placentation.[27] The developing seeds are embedded in parenchyma tissue and no locules as such are delimited (figs. 22.10 and 22.11,*A*). The placentae appear as though they were in parietal position, but the presence of typically inverted lateral carpellary bundles in the center of the ovary[27] contradicts this impression. Moreover, a gentle teasing apart of the tissue occluding a locule reveals that the ovules are free from the fruit wall and are

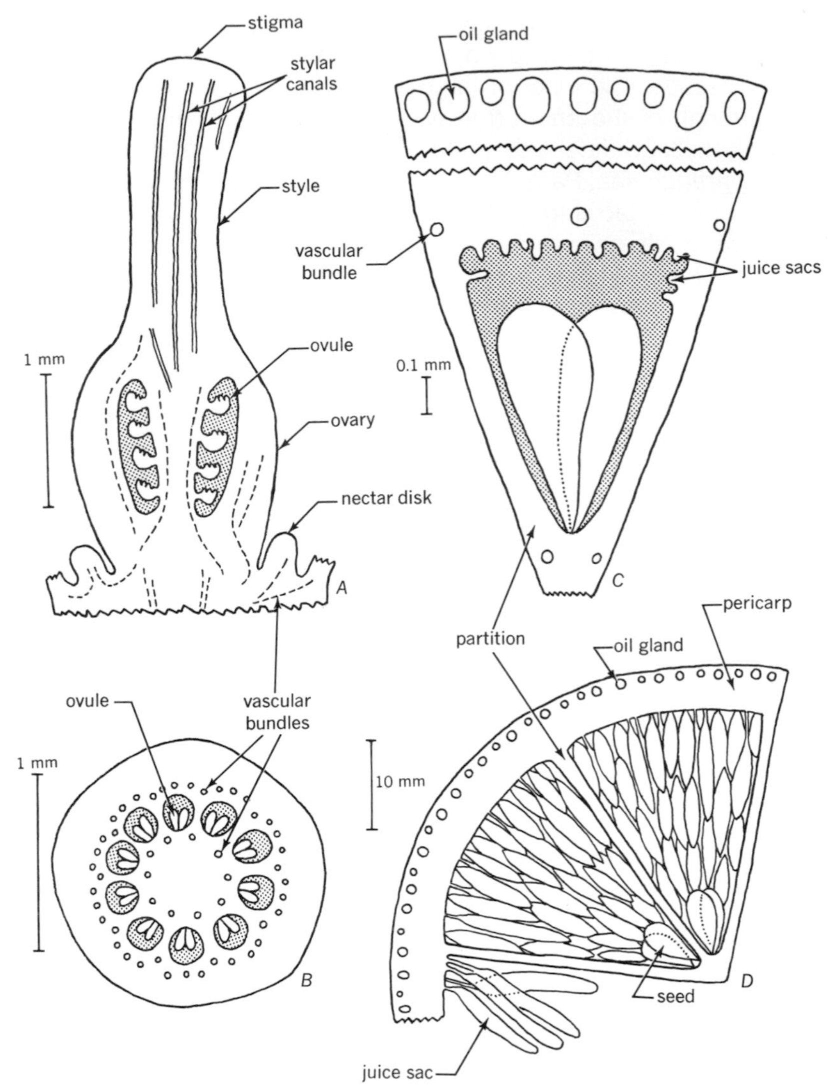

Figure 22.9 Citrus fruit. *A*, *B*, young ovary from flower of *Citrus aurantifolia* (lime) in longitudinal (*A*) and transverse (*B*) sections. *C*, part of transection of young fruit of *C. sinensis* (orange) including one carpel. Juice sacs in early stage of growth. *D*, diagram of orange fruit including two carpels and part of a third. Juice sacs fill the locules. (*A–C*, constructed from photomicrographs in Schneider.[43])

borne on an incurved panel of tissue (fig.22.11,*B*).

An interpretation of carpel arrangement, which agrees with data on the development of the ovary in a cucurbit,[15] is given in figure 22.11. The diagram in *C* depicts the form and arrangement of the carpels as they would appear without occlusion of the locules and without association with extracarpellary tissue. The margins of the carpels are incurved first centripetally and thus form the partitions between the locules. A second

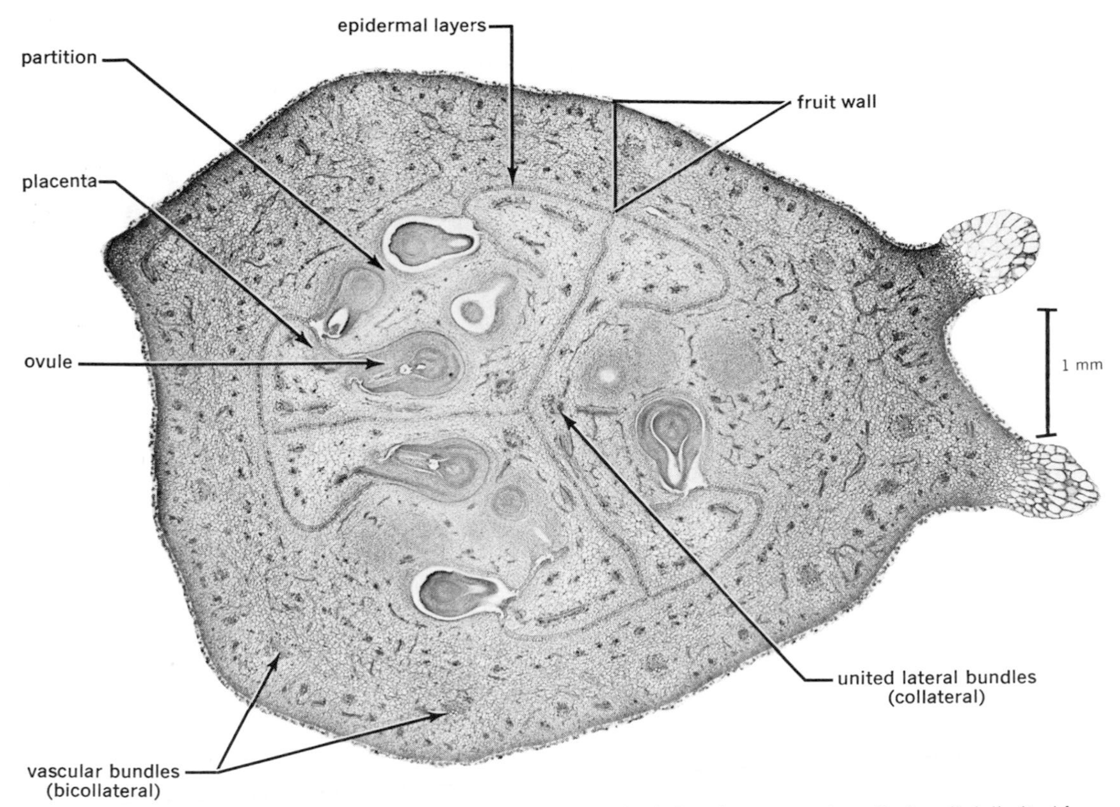

Figure 22.10 Transverse section of fruit of *Cucumis sativus* (cucumber). Seed not yet mature. Fruit wall delimited from placentae by epidermal layers. Epidermal layers also between margins of each carpel. These layers converge in the center. The inverted lateral bundles of contiguous carpels are united.

curvature carries the carpellary margins centrifugally so that each locule is divided. The placentae also are curved and extend centripetally (figs. 22.10 and 22.11,*A*). Double epidermal layers occur between the carpellary margins along the second curvature and between the placentae and the backs of the carpels (fig. 22.10; solid lines in fig. 22.11,*A*). No line of union is visible between the margins of contiguous carpels along the first curvature of the carpels (dashed lines in fig. 22.11,*A*). The tissue occluding the locules is of carpellary origin. It becomes part of the central edible fleshy tissue in certain Cucurbitaceae.

A survey of many species of cucurbitaceous fruits has revealed a number of common broad histologic features and variations in details.[27] A uniseriate outer epidermis is covered with a cuticle and has stomata. The subepidermal tissue varies in width and may consist of parenchyma (green, yellow, or colorless) and is sometimes collenchymatic. Fibers and phloem strands may be present. In some taxa, a continuous or discontinuous layer of sclereids occurs at greater depth. Beneath this sclerenchyma is a parenchyma which, at maturity extends to the center of the fruit (cucumber, watermelon) or is torn and replaced by a central cavity (*Cucurbita maxima*, muskmelon). Whatever parenchyma is retained may be juicy, or starchy, or permeated with intercellular

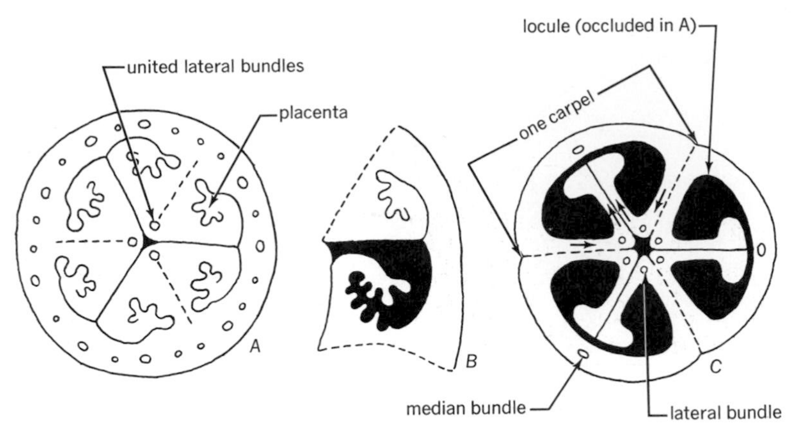

Figure 22.11 *A,* transection of ovary of *Citrullus vulgaris* (watermelon). The dashed lines indicate regions of union of carpels, solid lines delimit margins of carpels within each carpel and the placentae from the fruit wall (cf. fig. 22.10). The lateral bundles of contiguous carpels are united (cf. fig. 22.10). *B,* transection of one carpel with one placenta freed from the surrounding tissue. *C,* a diagram interpreting the arrangement of carpels in a cucurbitaceous ovary. For clarity, the occlusion of the locules and the union of lateral bundles are not shown. (*A, B,* from photographs of cleared sections in L. K. Mann, *Bot. Gaz.* 105:257–262, 1943.)

spaces. The inner epidermis of the fruit wall may form a membranous covering on the seeds. The vascular system consists mainly of bicollateral vascular bundles. Some of these are carpellary bundles, others belong to the extracarpellary part, which appears to be appendicular in origin[27] (floral tube).

The external coloring of cucurbitaceous fruits depends on plastids in the subepidermal parenchyma. The plastids are chloroplasts in green fruits, chromoplasts in yellow fruits. In a variety of *Cucurbita pepo,* a spontaneous regreening may occur as a result of reversion of yellow chromoplasts to chloroplasts.[9] The coloring of the internal flesh also is determined by chromoplast differentiation. In watermelons, the pink and red pigments occur in the form of needles and plates.[27]

The banana fruit (*Musa acuminata*) arises as a superior ovary and may produce seed or develop parthenocarpically.[28] Three carpels are arranged according to the axile placentation pattern (fig. 22.12,*A,B*). The seeded and parthenocarpic fruits are similar in structure at beginning of flowering. Later, the ovules of the parthenocarpic fruits degenerate and the locules are occuled with pulp which originates from the pericarp and the partitions (fig. 22.12,*A*). The pulp is rich in starch. In the seeded variety, the mature seeds almost fill the locules, and very little pulp is produced (fig. 22.12,*B*).

Numerous vertical vascular bundles, accompanied by laticifers, are embedded in the parenchyma of the banana pericarp. Internally from this region occur, first, a zone of aerenchyma and, second, a zone with horizontally oriented vascular bundles. This system is connected with the vertical vascular bundles of the pericarp and with the lateral carpellary bundles in the center of the fruit (fig. 22.12,*A,B*). The aerenchyma is probably responsible for the ease with which the peel can be removed from the fruit.

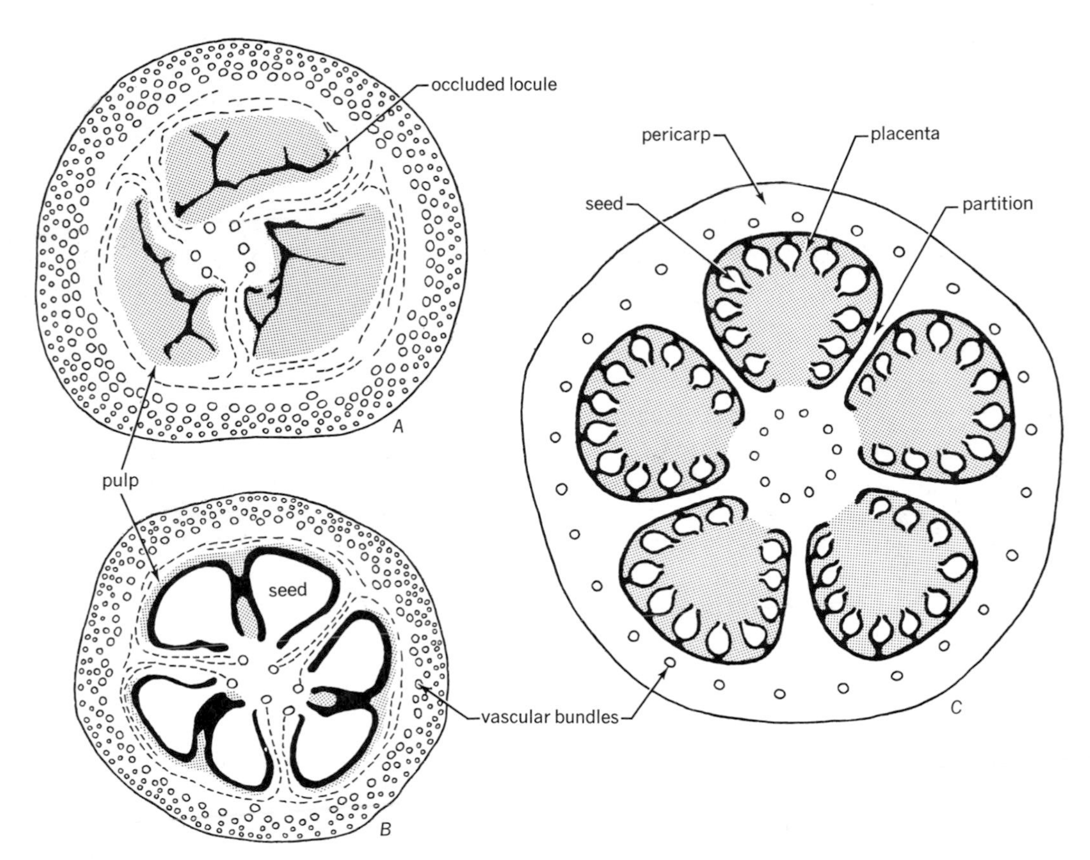

Figure 22.12 Transections of fruits of *Musa acuminata* (banana, *A*, *B*) and *Lycopersicon esculentum* (*C*). *A*, parthenocarpic fruit with ovarian cavities occluded by pulp derived from pericarp and partitions. *B*, seeded fruit with ovarian cavities occupied by seeds and small amount of pulp. *C*, the placentae have occluded the locules and have embedded the ovules. The tissue among the ovules is gelatinous in mature fruit. In all drawings solid black indicates locular spaces. (*A*, *B*, adapted from Mohan Ram, Ram, and Steward.[28])

FRUIT WITHOUT A RIND

A fruit consisting of fleshy tissue throughout and having no rind is well illustrated by the berry of *Lycopersicon esculentum* (Solanaceae), the tomato. The typical number of carpels in the Solanaceae is two, but the cultivated tomato other than the cherry variety shows a variable larger number of carpels. The fleshy tissue includes the pericarp, the partitions, and the large placentae. The fruit has axile placentation but the placentae fill much of the locular space. Placental tissue also invades the spaces among the ovules (fig. 22.12,*C*). Toward maturity, this part of the placental tissue becomes degraded and assumes a gelatinous consistency. Cellulase may be one of the enzymes involved in this process.[13]

The change in the color of tomato fruit during ripening occurs through a transformation of chloroplasts into chromoplasts. This process is actuated by light and involves absorption of red light by phytochrome, ethylene production, and other biochemical events.[21] Two kinds of chromoplasts develop.[24] One, located in the gelatinous tissue, contains globular inclusions with β-carotene. The other, located in the pericarp, includes voluminous paracrystalline sheets of lycopene in addition to the globules with β-carotene.

A drupe, as in *Prunus,* is a fleshy fruit originating from a single carpel in a perigynous flower (chapter 20). The ventral suture between the carpel margins is detectable as an indentation in various *Prunus* species. In some peach and nectarine varieties, the indentation is particularly pronounced in the pit and may be associated

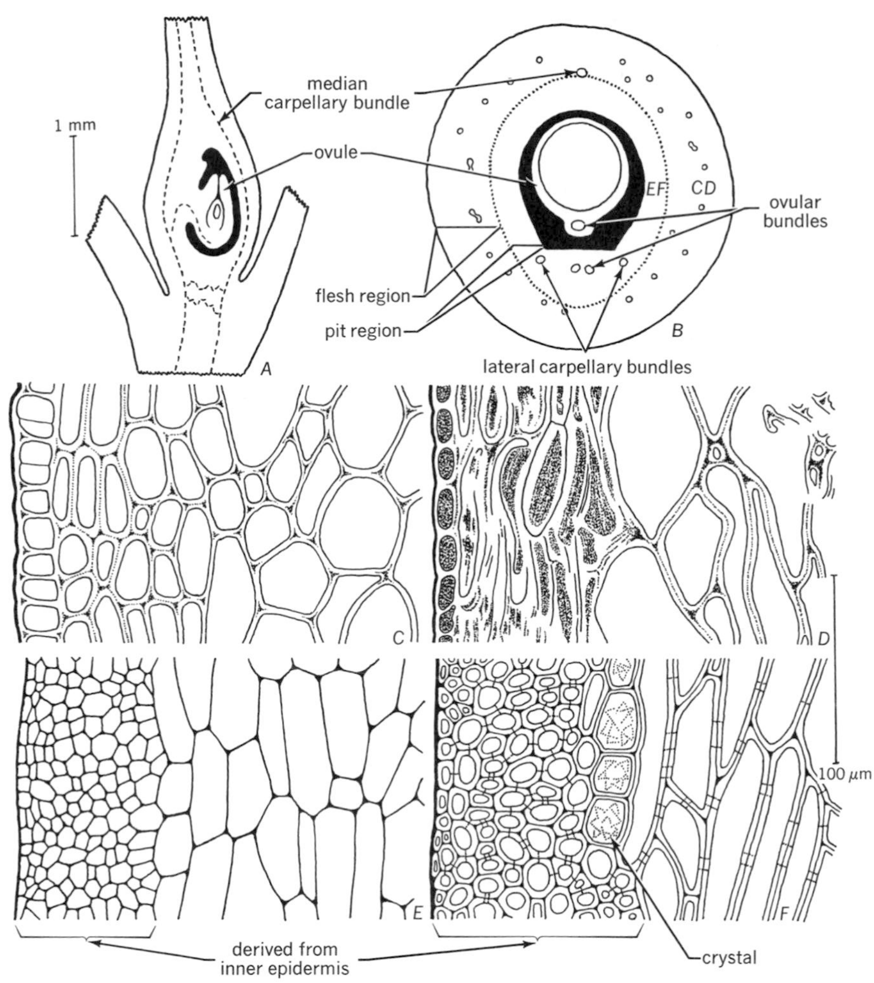

Figure 22.13 Fruit of *Prunus domestica* (prune). *A,* longitudinal and *B,* transverse sections of the ovary. Ovarian cavity in solid black. *C, D,* outer part of pericarp 6 weeks after full bloom (*C*) and 2 weeks after abscission. Disorganized protoplasts and partly collapsed cell walls in *D. E, F,* inner part of endocarp. *E,* 6 weeks after full bloom, at completion of cell division. *F,* 8 weeks after full bloom, at beginning of lignification. *C–F* from transverse sections. (From photomicrographs in Sterling.[48])

with an inadequate adhesion along the suture which results in the "split pit" phenomenon.[39]

The pericarp may be divided into a relatively thin exocarp composed of the epidermis and subepidermal collenchyma, a fleshy mesocarp, and a stony endocarp (pit wall; fig. 22.13). Fine fibrils of wax on the surface of the cuticularized epidermis cause the characteristic bloom on plum fruits.[45] In peaches, the mesocarp shows chemical and histological differences between the "melting-fleshed" and the canning "cling" types of fruit. In the former, cell walls decrease in thickness and the cells disorganize as the fruit ripens. In the fruit of prune, the stony endocarp is derived from three regions of the ovary wall.[48] The inner epidermis forms a multiseriate layer of vertically elongated sclereids; the next outer region is a multiseriate layer of tangentially elongated sclereids (fig. 22.13,*E,F*); and the two to four layers still farther outward differentiate into isodiametric sclereids. Vascular bundles occur in the fleshy mesocarp and in the stony endocarp (fig. 22.13,*B*).

The individual drupelets of the raspberry (*Rubus*) are assembled into an aggregate fruit on an elevated receptacle (fig. 22.14,*A*). The individual drupelet has a stony endocarp composed of elongated curved sclereids oriented differently in the different layers[36] (fig. 22.14,*B,C*). The parenchymatic mesocarp is a succulent pulp. The exocarp bears epidermal hairs that hold the drupelets together at maturity.[36]

The fruit of *Pyrus* (*P. malus,* apple; *P. communis,* pear), the pome, arises from an inferior ovary, and the extracarpellary part composes the bulk of the fruit flesh. According to one concept, this tissue is appendicular in origin

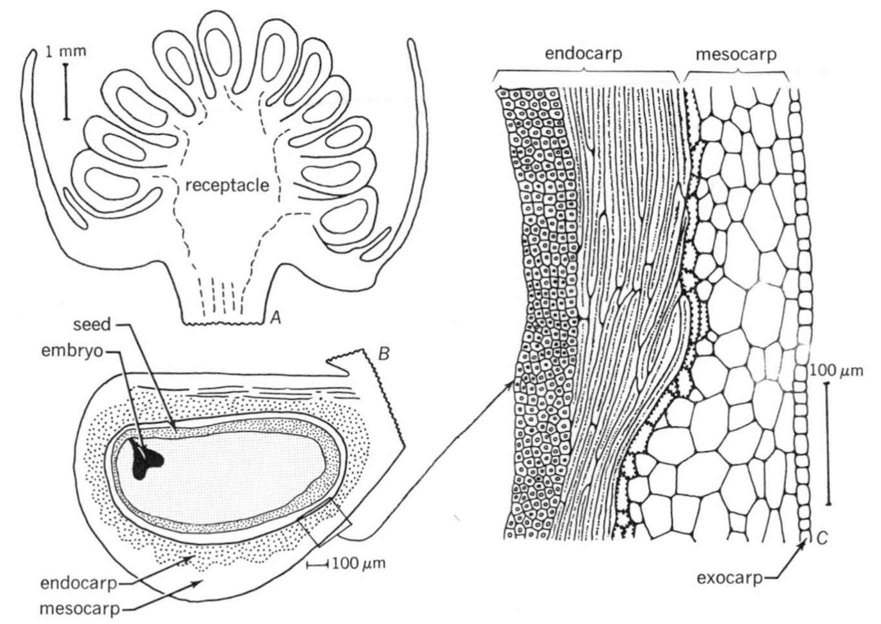

Figure 22.14 Fruit of *Rubus strigosus* (raspberry). All longitudinal sections. *A,* entire immature fruit: aggregation of drupelets on an enlarged receptacle. *B,* single drupelet and, *C,* fragment if its pericarp. The sclereids in the two parts of the endocarp are oriented in opposite planes with their long axes. (From photomicrographs in Reeve.[36])

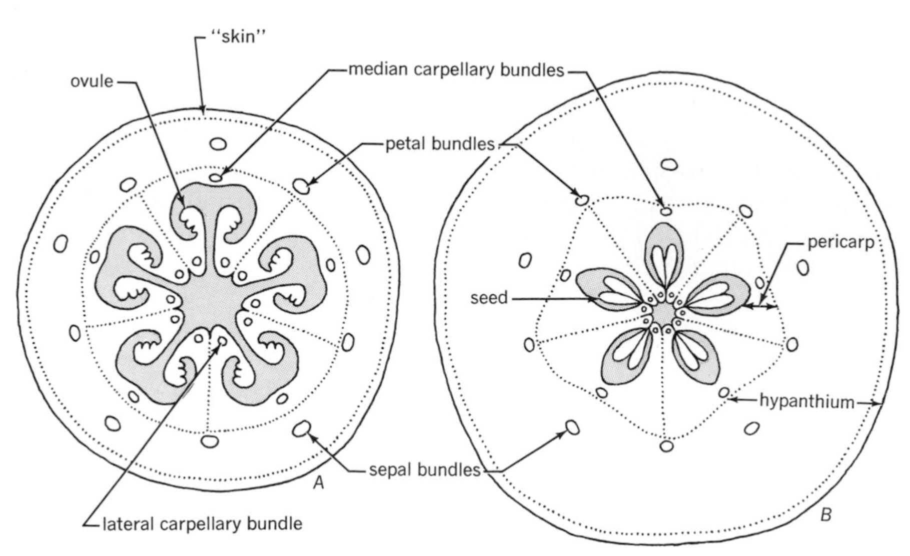

Figure 22.15 Fruit of *Pyrus malus* (apple) in cross sections. *A*, immature, is shown at higher magnification than *B*, mature. Shading indicates open space. The labeling is in accord with the interpretation of the extracarpellary tissue as a hypanthium. (Based on photomicrographs in Fisk and Millington.[12])

(floral tube or hypanthium). Other botanists interpret the accessory tissue as part of the receptacle and call the outer fleshy part of the fruit cortex. The following description of the pome fruit (apple and pear) and the labeling in the pertinent illustrations are based on the concept of appendicular origin of the extracarpellary tissue.

The boundary between the floral tube and the ovary (the core of the fruit) includes the median bundles of the carpels but excludes the vascular bundles assigned to the floral tube (fig. 22.15,*B*). In the mature fruit, a thin layer of compactly arranged cells indicates the presumed boundary.[12] The ovary consists of five carpels united to form axile placentation. The margins of the contiguous carpels, which jointly form the partitions are not individually identifiable except near the center of the core where each partition is slightly indented and shows two lateral carpellary bundles (fig. 22.16). The two margins of one carpel are not joined in the early stages of growth but come in contact with one another later and appear to join (figs. 22.15 and 22.16). An opening may remain in the center of the core or be closed by the carpel margins.

The characteristics of the tissues of the apple, beginning from the outside, are the following. The outer epidermis is covered with a cuticle that increases in thickness as the fruit grows. Wax on the surface of the cuticle occurs in overlapping platelets.[45] Stomata and trichomes occur in young fruits. Lenticels later replace the stomata. In "russeting" apples patches of cork develop in the outer layers of the fruit. The subepidermal tissue is compact and has thickened walls. It forms, together with the epidermis, the relatively tough "skin" of the fruit (fig. 22.15; also in pear, fig. 22.16). Anthocyanins occur in the skin of the red varieties of apple.[16] Chloroplasts which change into chromoplasts with carotenoids during the ripening process, are

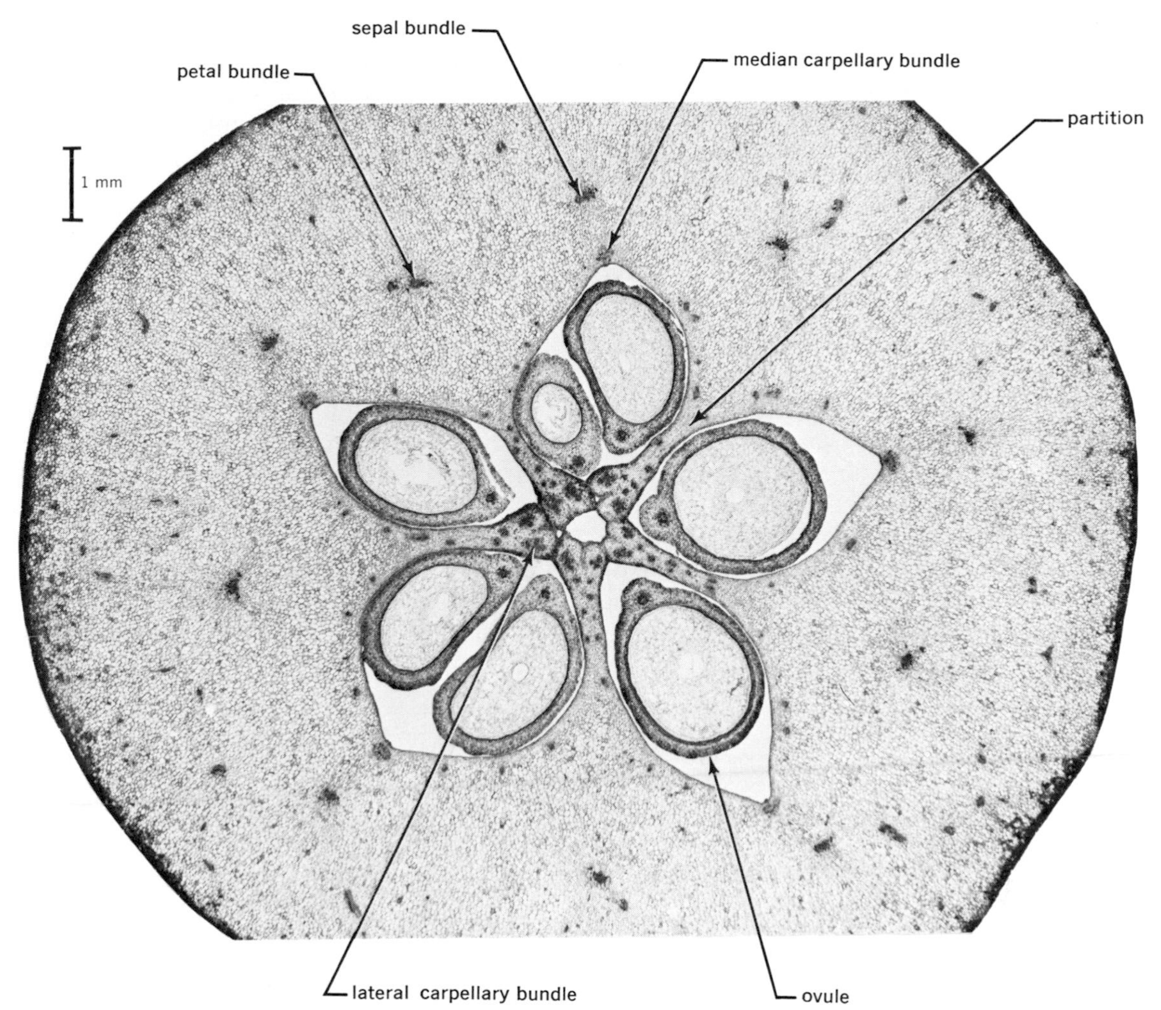

Figure 22.16 Photomicrograph of transverse section of young fruit of *Pyrus* (pear). Labeling is in accord with the appendicular theory of origin of extracarpellary tissue. Stone cells were just beginning to differentiate.

also located in the peripheral tissues of the fruit. According to an ultrastructural study,[8] the plastids of the epidermis and the subepidermal tissue accumulate ferritin and membrane-bound proteinaceous inclusions.

The bulk of the extracarpellary tissue is parenchyma permeated with intercellular spaces. The "mealy" condition of this tissue in overripe fruits results from further separation of cells and enlargement of air spaces. The main vascular bundles of the floral tube are located in the parenchyma (fig. 22.15; also in pear, fig. 22.16). Branches from these bundles form an anastomosing system throughout the flesh.

The core of the pome consists of two kinds of tissue, parenchyma in which the median and lateral carpellary bundles are located and cartilaginous tissue lining the locules and composed of sclereids. Both these tissues are considered to be the pericarp, with the endocarp restricted to the cartilaginous tissue. Each locule has two seeds, sometimes more.

In the pear, sclereids constitute a characteristic tissue element of the fruit flesh. The sclereids occur in clusters visible with slight magnification especially if stained for lignin with phloroglucinol and hydrochloric acid. According to a developmental study of *Pyrus communis,*[14] sclereid differentiation begins two weeks after anthesis and continues into the twentieth week of fruit growth. The process starts in the deeper regions of the flesh and advances outward. The sclereid primordia enlarge conspicuously in advance of the surrounding cells and often divide one or two times. They stop growing and deposit the secondary wall which becomes heavily lignified. The surrounding cells, which are still growing at this time, show concentric divisions about the sclereid clusters and elongate radially with respect to them (fig. 22.17).

FRUIT GROWTH

Studies on fruit growth are centered on fleshy fruits because of their horticultural importance. Since the majority of fruits are not fleshy, the information on fruit growth is restricted in scope.[25]

The overall growth of fleshy fruits follows two patterns. One group of fruits (avocado, banana, citrus, melon, pome, tomato) have a simple sigmoid growth curve indicating an initial exponential increase in size followed

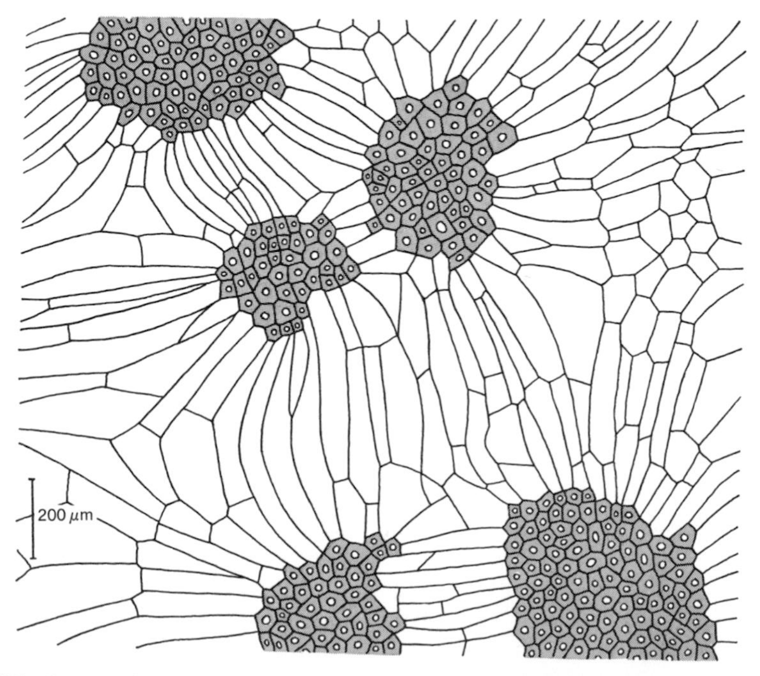

Figure 22.17 Groups of sclereids in transection of pericarp of *Pyrus nivalis* (pear). (From photomicrograph in Staritsky.[47])

by a slowing down in a sigmoid pattern. Another group of fruits (grape, fig, olive, stone fruits) show a cyclical development and have a double sigmoid growth curve.[4,25]

The two processes in the increase in size of fruit, cell division (combined with cell enlargement before each mitosis) and cell enlargement occur sequentially to a great extent. Commonly, the early enlargement depends on cell multiplication starting before anthesis and continuing after fruit set. This stage is gradually replaced by that of cell expansion, which occupies the longest period of growth. Such pattern was recorded, for example, for citrus [52] and pome[16] fruits. Avocado constitutes an apparent exception in that it shows cell multiplication throughout the entire growth period.[3] The duration and timing of the two stages of growth vary in different fruits. The ultimate size of the fruit may be correlated with cell size, as in cherries, or with cell number, as in apples.[25]

In many fruits, the development of intercellular spaces in the fleshy parenchyma contributes much to the increase in volume. In apples, for instance, 25 percent of the volume is occupied by air spaces and, therefore, the last half of growth in volume is more rapid than the concomitant increase in weight. In contrast, the grape berry shows a greater increase in weight than in volume in the later period of growth.[25]

The composition of a fruit of tissues of different genomes with different numbers of chromosomes (2*n* parent sporophyte, 2*n* embryo possibly of distinct genome, 3*n* endosperm) introduces a complicating factor into the growth pattern. At different periods of fruit development preferential growth is variously distributed among the components of the fruit, the pericarp, the seed coat, the endosperm, and the embryo. Thus in stone fruits, the initial enlargment occurs in the sporophytic part of the ovule. Subsequently, growth becomes largely confined to the endosperm and embryo and is completed with the expansion of the mesocarp.[39] The different components of the fruit may have different types of growth curves. In the bean, the entire fruit has a simple sigmoid growth curve, the seed components, a double growth curve.[7]

The heterogeneity in fruit structure is differently expressed in fruits of different taxa. The fruit in apple and strawberry is composed mostly of parental tissue, in the grass fruit of endosperm, and in the achene of Asteraceae of embryo. Thus, centers of growth vary in different types of fruit.

The basic feature of morphogenesis—polarized distribtuion of cell division and cell enlargement—has been recorded in growing fruits[30,46] (fig. 22.18). Changes in the direction of growth bring about changes in the length/diameter ratio which serves as a parameter in allometric studies (comparison of growth rate of one part of organ or organism with that of another) of form development.[44] In a variety of pear, the length/diameter ratio

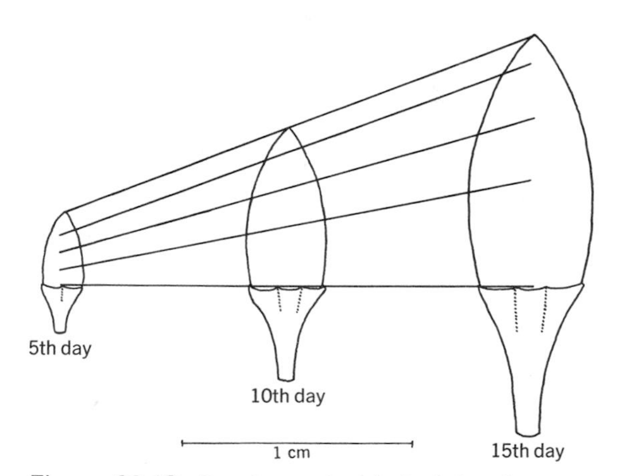

Figure 22.18 Development of fruit of *Capsicum annuum* var. *minimum* (red pepper). Ink marks were placed at equal distances on the fruit 5 days after anthesis. Lines indicate the unequal increase in distance between the marks in the two successively older fruits. Growth was most rapid at the base of the fruit. (After Munting.[30])

was 1.1 at anthesis, 2.7 fourteen days later, and 1.3 in harvest-ripe fruit 133 days after anthesis.[1] The fruit of a variety of *Capsicum annuum* showed a continuous increase in the ratio of length to diameter, starting with 1.3 at anthesis and ending with 2.2 at maturity. This elongation was most rapid at the base of the fruit[30] (fig. 22.18).

Many factors, both internal and external, regulate fruit development. One of the internal factors is the development of seed.[26,47] The discovery that treatment with auxin paste replaces seeds in inducing a strawberry fruit to grow[31] has stimulated intensive research on growth substances in relation to fruit development.[25] Studies on parthenocarpic fruits have suggested that young ovules are not absent from such fruits but are actally crucial for their development. Although seedless varieties have generally a lower level of auxins and gibberellins than the seeded, in the developing fruits the ovules probably exert their influence by supplying the growth substances at the critical time of flowering.[32]

On the whole, the relation of growth substances to fruit growth is highly complex. All categories of plant hormones—auxins, gibberellins, cytokinins, inhibitors, and the ethylene—influence fruit development. Since the fruit is a composite organ, its various components grow in a predetermined sequence in which different factors become limiting at different stages.[32] There may also be conflicts with the vegetative development when a hormonal stimulus is decisive in fruit development if provided at the right time.

An example of a relation of a hormone to a specific stage of fruit developlent is the activation by ethylene of phenomena leading to ripening of the fruit. One of these phenomena is the relatviely sudden rise of respiration called *respiration climacteric.* The alteration of respiratory activity is now known to be dependent on available levels of ethylene. Thus, an enhancement of biosynthesis of ethylene rather than the rise in respiration is the primary event in the transition from growth to senescence of the fruit. Increased respiration is one of many secondary events occuring at that period. Some others are increase in RNA and protein synthesis and change in cell permeability. Rhodes,[38] therefore, defines the term climacteric as a period in the ontogeny of certain fruits (most fleshy fruits) during which numerous biochemical changes, which set off the transition from growth to senescence and ripening in the fruits, are initiated by the autocatalytic production of ethylene.

FRUIT ABSCISSION

Causal factors in fruit abscission, including action of growth substances, have been intensively investigated by plant physiologists, for the tendency of many kinds of fruit to drop prematurely is a widespread horticultural problem.[25] Fruits abscise in different stages of development. At maturity, they do so either with seed enclosed (indehiscent fruits; commonly fleshy fruits) or after a release of seeds (dehiscent fruits).

The location of the abscission zone varies in different fruits, and a given fruit may have more than one abscission zone. Pome fruits usually separate at the base of the pedicel. In early fruit fall, plums are shed with the stalks attached, later without the stalks. In cherries, the first separation occurs at the base of the fruit, the second at the base of the stalk (preferred in commercial harvesting), and the third at the base of the spur that bore the fruit cluster. In citrus, the separation layer is formed beneath the ovary where the vascular bundles diverge from the receptacle into the carpels.[43]

The degree of differentiation of the abscission zone varies in different fruits.[10] In the separation layer of mature apples, cells in several tiers increase in size, secondary walls in sclerenchyma lose their anisotropic qualities, and the middle lamellae, primary walls, and much of the secondary wall thickening dissolve. Vessels and fibers are ruptured. In citrus,[53] large quantities of starch accumulate in the absicission region and pectin disappears, with the consequent weakening of cell walls. Preharvest abscission in citrus causes considerable loss to growers and, to some extent, is controllable by application of 2,4-dichlorophenoxyacetic acid (2,4-D). This treatment prevents the hydrolysis of pectic materials.

In *Prunus cerasus* (sour cherry) abscission layer formation between the fruit and the pedicel appears 12 to 15 days before the fruit matures.[50] The layer, composed of five to eight rows of cells, becomes apparent because of a low affinity for the hematoxylin stain. During abscission, cells separate without rupture of walls but later collapse. *Prunus avium* (sweet cherry) shows no defined abscission layer at the base of the fruit.

Much work has been done in relation to efficiency of cherry fruit removal by means of harvesting machines (shaking of trees). In studies of effect of growth substances on abscission in this fruit, a device is used to measure the force required to pull the fruit off the stalk. In early stages of growth, this force increases in relation to tissue differentiation in stalk and fruit, but at the end of this development, when the separation layer is formed, a rapid decline in fruit-removal force takes place.[56] At this time, the fruit becomes sensitive to exogenous application of ethylene, which promotes the loss in resistance to breaking.

Histochemical localization of enzymes in the abscission zones of maturing cherry fruits showed that dehydrogenase, acid phosphatase, and peroxidase activities were particularly high at the juncture of receptacle and fruit tissue.[35] Synthesis of protein and RNA also was found to be localized in the abscission zone of this fruit.[49] Application of cycloheximide, an inhibitor of RNA and protein synthesis, delayed abscission. Synthesis of RNA and protein thus appears to be causally related to abscission, probably as a requirement for the synthesis of wall-degrading enzymes such as cellulase. The synthesis of this enzyme is enhanced by ethylene and abscisic acid,[35] both of which are known to promote the separation process.

REFERENCES

1. Bain, J. M. Some morphological, anatomical, and physiological changes in the pear fruit (*Pyrus communis* var. Williams bon Chrétien) during development and following harvest. *Aust. J. Bot.* 9:99–123. 1961.
2. Bauman-Bodenheim, M. G. Prinzipien eines Fruchtsystems der Angiospermen. *Ber. Schweiz. Bot. Ges.* 64:94–112. 1954.
3. Biale, J. B., and R. E. Young. The avocado pear. Chap. 1, pp. 1–63. In *The biochemistry of fruits and their products.* A. C. Hulme, ed. Vol. 2. London and New York, Academic Press. 1971.
4. Bollard, E. G. The physiology and nutrition of developing fruits. Chap. 14, pp. 387–425. In: *The biochemistry of fruits and their products.* A. C. Hulme, ed. Vol. 1. London and New York, Academic Press. 1970.
5. Bradbury, D., I. M. Cull, and M. M. MacMasters. Structure of the mature wheat kernel. I. Gross anatomy and relationships of parts. *Cereal Chem.* 33:329–342. 1956.

6. Bradbury, D., M. M. MacMasters, and I. M. Cull. Structure of mature wheat kernel. II. Microscopic structure of pericarp, seed coat, and other coverings of the endosperm and germ of hard red winter wheat. *Cereal Chem.* 33:342–360. 1956.
7. Carr, D. J., and K. G. M. Skene. Diauxic growth curves of seeds with reference to French beans. *Aust. J. Biol. Sci.* 14:1–12. 1961.
8. Catesson, A.-M. Évolution des plastes de pomme au cours de la maturation du fruit. Modifications ultrastructurales et accumulation de ferritin. *J. Microscopie* 9:949–974. 1974.
9. Devidé, Z., and N. Ljubesić. The reversion of chromoplasts to chloroplasts in pumpkin fruits. *Z. Pflanzenphysiol.* 73:296–306. 1974.
10. Esau, K. *Plant anatomy.* 2nd ed. New York, John Wiley & Sons. 1965.
11. Fahn, A., I. Shomer, and I. Ben-Gera. Occurrence and structure of epicuticular wax on juice vesicles of citrus fruits. *Ann. Bot.* 38:869–872. 1974.
12. Fisk, E. L., and W. F. Millington. *Atlas of plant morphology. Portfolio II: Photomicrographs of flower, fruit and seed.* Minneapolis, Minnesota, Burgess Publishing Company. 1962.
13. Hall, C. B. Cellulase activity in tomato fruits according to portion and maturity. *Bot. Gaz.* 125:156–157. 1964.
14. Häuptli, F. Die Sklereidendifferenzierung in *Pyrus communis.* Morphologische, anatomische und histochemische Untersuchungen. *Ber. Schweiz. Bot. Ges.* 81:273–318. 1971.
15. Hayward, H. E. *The structure of economic plants.* New York, Macmillan. 1938.
16. Hulme, A. C., and M. J. C. Rhodes. Pome fruits. Chap. 10, pp. 333–373. In: *The biochemistry of fruits and their products.* A. C. Hulme, ed. Vol. 2. London and New York, Academic Press. 1971.
17. Jackson, B. D. *A glossary of botanic terms.* 4th ed. New York, Hafner Publishing Co. 1953.
18. Kaden, N. N. Tipy prodol'nogo vskryvaniya plodov. [Types of longitudinal dehiscence in fruits.] *Bot. Zh.* 47:495–505. 1962.
19. Kaden, N. N. K voprosu o drobnykh plodakh. [To the question of schisocarpic fruits.] *Bot. Zh.* 49:966–973. 1964.
20. Kaniewski, K. Hairs in the loculus of the broad-bean (*Vicia faba* L.) fruit. *Bull. Acad. Polon. Sci.* Cl. V 41:585–594. 1968.
21. Khudari, A. K. The ripening of tomatoes. *Amer. Sci.* 60:696–707. 1972.
22. Kiesselbach, T. A., and E. R. Walker. Structure of certain specialized tissues in the kernel of corn. *Amer. J. Bot.* 39:561–569. 1952.
23. Krauss, L. Etwicklungsgeschichte der Früchte von *Hordeum, Triticum, Bromus* und *Poa* mit besonderer Berücksichtigung ihrer Samenschalen. *Jb. Wiss. Bot.* 77:733–808. 1933.
24. Laval-Martin, D. La maturation du fruit de tomate cerise: mise en évidence, par cryodécapage, de l'évolution des chloroplastes en deux types de chromoplastes. *Protoplasma* 82:33–59. 1974.
25. Leopold, A. C. *Plant growth and development.* New York, McGraw-Hill. 1964.
26. Luckwill, L. C. Factors controlling the growth and form of fruits. *J. Linn. Soc. London, Bot.* 56:294–302. 1959.
27. Matienko, B. T. *Sravnitel'naya anatomiya i ul'trastruktura plodov tykvennykh. [Comparative anatomy and ultrastructure of cucurbitaceous fruits.*] Kishinev, Kartya Moldovenyaske. 1969.
28. Mohan Ram, H. Y., M. Ram, and F. C.

Steward. Growth and development of the banana plant. 3. A. The origin of the inflorescence and the development of the flowers. B. The structure and development of the fruit. *Ann. Bot.* 26:657-673. 1962.

29. Monsi, M. Untersuchungen über den Mechanismus der Schleuderbewegung der Sojabohnen-Hülse. *Jap. J. Bot.* 12: 437–474. 1943.
30. Munting, A. J. Development of flower and fruit of *Capsicum annuum* L. *Acta Bot. Neerl.* 23:415–432. 1974.
31. Nitsch, J. P. Growth and morphogenesis of the strawberry as related to auxin. *Amer. J. Bot.* 37:211–215. 1950.
32. Nitsch, J. P. Hormonal factors in growth and development. Chapter 15, pp. 428–472. In: *The biochemistry of fruits and their products.* A. C. Hulme, ed. Vol. 1. London and New York, Academic Press. 1970.
33. Pijl, L. van der. *Principles of dispersal in higher plants.* 2nd. ed. Berlin-Heidelberg-New York, Springer-Verlag. 1972.
34. Pomeranz, Y. From wheat to bread: a biochemical study. *Amer. Sci.* 61:683–691. 1973.
35. Poovaiah, B. W., H. P. Rasmussen, and M. J. Bukovac. Histochemical localization of enzymes in the abscission zones of maturing sour and sweet cherry fruit. *J. Amer. Soc. Hort. Sci.* 98:16–18. 1973.
36. Reeve, R. M. Fruit histogenesis in *Rubus strigosus.* I. Outer epidermis, parenchyma, and receptacle. *Amer. J. Bot.* 41:152–160. 1954. II. Endocarp tissues. *Amer. J. Bot.* 41:173–181. 1954.
37. Reeve, R. M., and M. S. Brown. Histological development of the green bean pod as related to culinary texture. I. Early stages of pod development. 2. Structure and composition at edible maturity. *J. Food Sci.* 33:321–331. 1968.
38. Rhodes, M. J. C. The climacteric and ripening of fruits. Chap. 17, pp. 521–533. In: *The biochemistry of fruits and their products.* A. C. Hulme, ed. Vol. 1. London and New York, Academic Press. 1970.
39. Romani, R. J., and W. G. Jennings. Stone fruits. Chap. 12, pp. 411–436. In: *The biochemistry of fruits and their products.* A. C. Hulme, ed. Vol. 2. London and New York, Academic Press. 1971.
40. Rost, T. L., and N. R. Lersten. Transfer aleurone cells in *Setaria lutescens* (Gramineae). *Protoplasma* 71:403–408. 1970.
41. Rost, T. L., and N. R. Lersten. A synopsis and selected bibliography of grass caryopsis anatomy and fine structure. *Iowa State J. Res.* 48:47–87. 1973.
42. Sanders, E. H. Developmental morphology of the kernel in grain sorghum. *Cereal Chem.* 32:12–25. 1955.
43. Schneider, H. The anatomy of citrus. Chap. 1, pp. 1-85. In: *The citrus industry.* W. Reuther, L. D. Batchelor, and H. J. Weber, eds. Berkeley, California, University of California, Division of Agricultural Sciences. 1968.
44. Sinnott, E. W. *Plant morphogenesis.* New York, McGraw-Hill. 1960.
45. Skene, D. S. The fine structure of apple, pear, and plum fruit surfaces, their changes during ripening, and their response to polishing. *Ann. Bot.* 27: 581–587. 1963.
46. Skene, D. S. The distribution of growth and cell division in the fruit of Cox's Orange Pippin. *Ann. Bot.* 30:493–512. 1966.
47. Staritsky, G. The morphogenesis of the inflorescence, flower and fruit of *Pyrus nivalis* Jacquin var. orientalis Terpó. *Med. Landbouwhogesch. Wageningen* 70:1–91. 1970.
48. Sterling, C. Developmental anatomy of

the fruit of *Prunus domestica* L. *Bull. Torrey Bot. Club* 80:457–477. 1953.
49. Stösser, R. Localization of RNA and protein synthesis in the developing abscission layer in fruit of *Prunus cerasus* L. *Z. Pflanzenphysiol.* 64:328–334. 1971.
50. Stösser, R., H. P. Rasmussen, and M. J. Bukovac. A histological study of abscission layer formation in cherry fruits during maturation. *J. Amer. Soc. Hort. Sci.* 94:239–243. 1969.
51. Thomson, W. W., L. N. Lewis, and C. W. Coggins. The reversion of chromoplasts to chloroplasts in Valencia oranges. *Cytologia* 32:117–124. 1967.
52. Ting, S. V., and J. A. Attaway. Citrus fruits. Chap. 3, pp. 107–169. In: *The biochemistry of fruits and their products.* A. C. Hulme, ed. Vol. 2. London and New York, Academic Press. 1971.
53. Wilson, W. C., and C. H. Hendershott. Anatomical and histological studies of abscission of oranges. *Proc. Amer. Soc. Hort. Sci.* 92:203–210. 1968.
54. Winkler, H. Versuch eines "natürlichen" Systems der Früchte. *Beitr. Biol. Pflanz.* 26:201–220. 1939.
55. Winkler, H. Zur Einigung und Weiterführung in der Frage des Fruchtsystems. *Beitr. Biol. Pflanz.* 27:92–130. 1940.
56. Wittenbach, V. A., and M. J. Bukovac. Cherry fruit abscission. Evidence for time of initiation and the involvement of ethylene. *Plant Physiol.* 54:494–498. 1974.
57. Zee, S.-Y., and T. P. O'Brien. Studies on the ontogeny of the pigment strand in the caryopsis of wheat. *Aust. J. Biol. Sci.* 23:1153–1171. 1970.
58. Zee, S.-Y., and T. P. O'Brien. Aleurone transfer cells and other structural features of the spikelet of millet. *Aust. J. Biol. Sci.* 24:391–395. 1971.

23

The Seed

CONCEPT AND MORPHOLOGY

The seed is the seat of partial development of the new sporophyte, the embryo, and it thus plays a major role in providing the continuity between successive generations of seed plants. The seed of angiosperms, which is the topic of this chapter, develops from the ovule as a consequence of double fertilization. In the mature seed, the embryo is protected by the surrounding seed coat and supplied with stored nutrients. Seeds have a greater chance of surviving and of producing new generations than do spores. The capacity to produce seeds made the seed plants, particularly the angiosperms, dominant over the spore-bearing plants during the recent geologic times.[23]

Seeds are an important source of food for animals and man. Among the angiosperms, the Poaceae provide more food seeds than any other plant family, and the Fabaceae are the second most important family in this respect. Seeds have many other uses beside serving for food. They may be, for example, the source of beverages (coffee, cocoa, beer), medicines, fibers (cotton), and industrial oils.[54]

A true seed is a matured ovule containing the embryo and stored nutrients, with the integument or integuments differentiated as the protective seed coat, or *testa* (fig. 23.1). Plants have other units of dissemination (the "diaspores"[16]) resembling seeds in the functional sense. Among these diaspores are certain one-seeded fruits such as the achene of the Asteraceae, the caryopsis of the Poaceae, the mericarp of the Apiaceae, and others.

The nutrient reserves in the seed maintain the sporophyte emerging from the germinating seed until it becomes a photosynthetically active organism. The storage of food reserves is, therefore, one of the primary functions of the seed. In some seeds, the storage occurs mainly outside the embryo, in the *endosperm* or the *perisperm*. The endosperm is a derivative of the fused polar and sperm nuclei in the central cell of the embryo sac and the perisperm is sporophyte nucellar tissue. In many dicotyledons, however, the

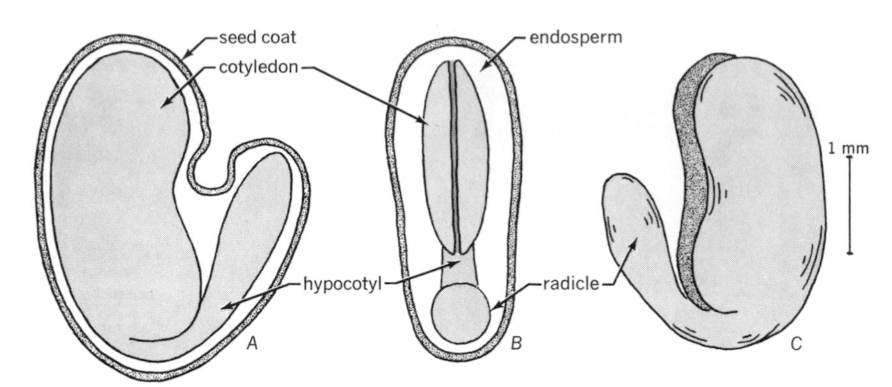

Figure 23.1 Diagrams of the seed of *Crotalaria intermedia* (Fabaceae) in longitudinal sections made parallel with (*A*) and perpendicular to (*B*) the plane of cotyledons, and of an embryo dissected from the seed (*C*). (Adapted from Miller.[31])

two storage tissues are transient and are absorbed more or less completely by the developing embryo before the seed becomes dormant. The food is then stored in the body of the embryo, notably the cotyledons.

The various structural details of the ovule are preserved to various degrees during the transformation of the ovule into the seed. The embryo alone or the embryo and the endosperm come to occupy the largest volume of the seed, whereas the integuments, in developing into the seed coat, usually undergo a considerable reduction in thickness and a partial disorganization. The micropyle may remain visible as an occluded pore or may be obliterated. The funiculus, whole or in part, abscises from the ovule and leaves a scar, the *hilum.* In anatropous ovules the part of the funiculus that is adnate to the ovule remains recognizable as a longitudinal ridge, the *raphe,* on one side of the seed.

The variability in morphology of angiosperm seed and the relative constancy of seed structure in narrow taxonomic units permit the use of seed characteristics in taxonomic studies.[29,33,54] Yet, comparative morphology of seeds is a rather neglected field despite the importance of seed identification in seed testing and crop improvement work.[18]

In external topography, the important features of seeds are shape, size, seed-coat surface, placement of the hilum, and presence or absence of such structures as *aril* (outgrowth from the funiculus), *caruncle* (integumentary protuberance near the micropyle), or *elaiosome* (oily appendage used as food by ants). Scanning electron microscopy is proving to be an especially suitable tool for study of seed surface.[7,58] In internal structure, the anatomy of the seed coat appears to be of greatest value in determining taxonomic relationships.[31,35,54,55,60] In developmental studies used to characterize seeds, the position of the ovule (whether upright or inverted in various ways) is an important feature.[11,48] In his comprehensive seed classification based on internal structure, Martin[29] selected the embryo as the principal diagnostic feature and used its shape, size, and position to divide the seeds into twelve types grouped into three divisions (fig. 23.2).

SEED DEVELOPMENT

Double fertilization provides the stimulus that sets off the developmental events culminating in the formation of the seed. Growth and

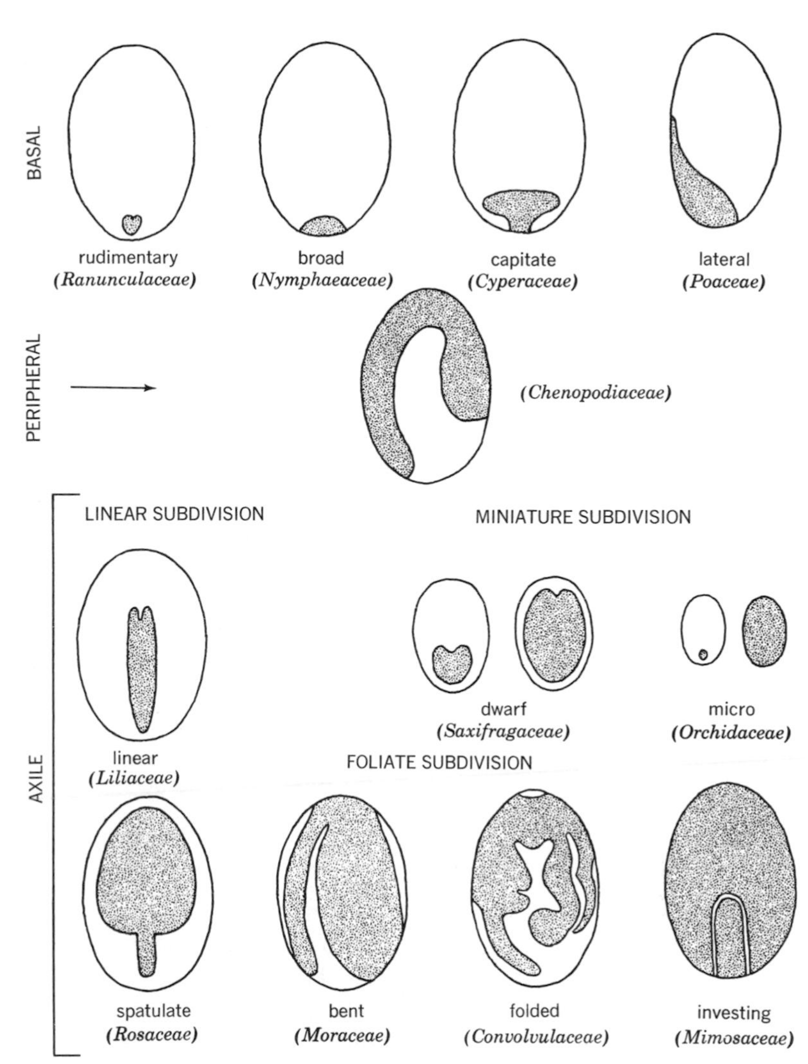

Figure 23.2 Diagrams illustrating typology of seeds based on size, shape, and position of embryo as seen in longitudinal sections of mature seeds. Classification by Martin[29] developed after a survey of 1287 genera of gymnosperms and angiosperms. Examples of families in which the particular type is common are given in parentheses. Shading indicates the embryo. (Later, Martin,[29] p. 529, merged the BASAL with the PERIPHERAL.) (Adapted from Martin.[29])

differentiation of the ovule, embryo sac, endosperm, and embryo occur in a series of interdependent stages and follow a characteristic sequence. The comparative volume changes of the ovule, endosperm, and embryo in a developing pea seed shows the basic pattern of this integrated development.[28] After fertilization, growth of the ovule is soon followed by that of the endosperm (fig. 23.3). The increase in the amount of endosperm is asso-

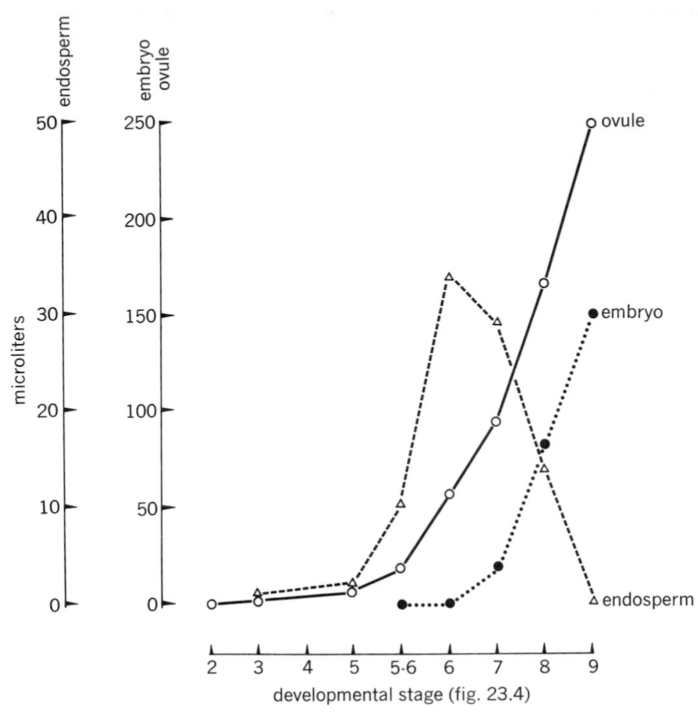

Figure 23.3 Graph illustrating relative volume changes of the entire ovule, endosperm, and embryo during seed development in the pea (*Pisum sativum*). (Compare with fig. 23.4.) Ovule and embryo volume were determined by liquid displacement in tubes of various diameters. The volume of the mostly noncellular endosperm was determined by withdrawing the liquid contents into calibrated micropipettes. (Adapted from Marinos.[28])

ciated with the enlargement of the embryo sac (fig. 23.4). The embryo is the last to show a measurable increase in volume but it begins to grow rapidly after the endosperm reaches maximum volume. The final number of cells is reached by the pea embryo before half of the seed formation is completed, a feature found in other dicotyledons also.[12] The later development is a result of cell enlargement. Embryo growth in the pea occurs at the expense of the endosperm so that the volume of the latter decreases and the embryo almost fills the embryo sac before dormancy is reached (figs. 23.3 and 23.4). The suspensor is disintegrated at this time.[8]

Uptake of materials by the embryo from the endosperm has been demonstrated experimentally for *Aesculus:* ^{14}C substrates injected into the embryo sac became localized in the cotyledons.[24] The growing embryo sac forms a sink into which water and soluble materials are drawn from the adjacent ovular tissues which in their turn are supplied with water, photosynthates, and nitrogen from the vegetative plant parts through the vascular tissues in the funiculus.[39] The transfer of food to the embryo sac involves a more or less extensive digestion of ovular tissues.

The developing endosperm and embryo contain relatively large amounts of hormones.[14,24] In the growing fruits of the pea, a close relationship was observed between

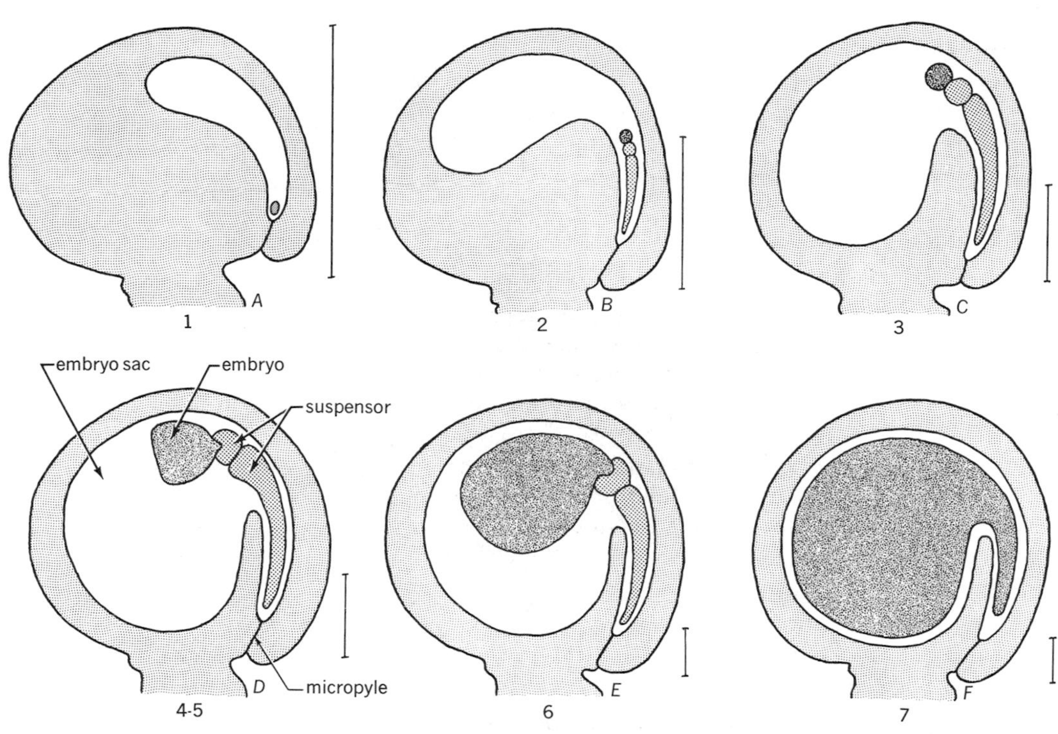

Figure 23.4 Diagrams illustrating seed development in the pea (*Pisum sativum*). The diagrams are drawn at different magnifications to emphasize the growth relation between embryo, endosperm (profiles of embryo sac), and entire ovule. In the embryo, a distinction between suspensor and embryo proper is made in *B–E*. No suspensor is present in later stages (*F*). The numbers under the diagrams indicate certain stages of development after fertilization and correspond to the numbers on the abscissa in fig. 23.3. Each scale line is equal to 1 mm. (Adapted from Marinos.[28])

changes in extractable amounts of gibberellins, auxin, and abscisic acid and changes in the growth rates of both pod and seeds, the growth of the pod apparently depending largely on the hormones supplied by the seeds. Growth of the ovules is initiated by the fertilization stimulus which activates the existing hormones and induces the synthesis of new ones.

To assess the full significance of seed formation one must realize that embryogenesis is a preparation for successful germination and the two phases of the life cycle should be viewed jointly.[12] During embryogenesis, the nutrient tissues outside the embryo and in the embryo itself synthesize and store large amounts of food materials. During germination, the same cells completely reverse the metabolic process and hydrolyze the stored nutrients. Some gene activation and deactivation must be involved in such a fundamental reversal in metabolism.

SEED COAT

The seed coat varies in structure in relation to the specific features of the ovule, such as number and thickness of integuments and pattern of vascularization, and to the developmental changes in the integuments during seed maturation. The seeds of angiosperms have mostly dry seed coats but some have fleshy appendages such as elaiosomes, or juicy layers such as the edible outer epidermis of the seed coat in pomegranate (*Punica*). Fleshy seed coats are common in the gymnosperms and have been recorded in Paleozoic pteridosperms. As was pointed out in chapter 22, seeds with fleshy investments are considered to be phylogenetically primitive and to have given way to seeds with reduced fleshy, edible parts attractive to animals (aril, caruncle, elaiosome) and then to dry seeds which, however, may be enclosed in a fleshy fruit wall.

The seed coat serves for protection of the enclosed embryo but the specific aspects of this protection are varied and complex, and some seed coats appear to be involved in controlling germination by restricting it to periods and conditions most favorable for seedling growth.[56] The inhibition of germination may be based on a high degree of impermeability of the seed coat to water, to oxygen, or to both. These effects can be traced to cuticular layers and their distribution. Phenolic compounds also appear to contribute to the impermeability of seed coats to water.[27] In some species, the seed coat seems to offer mechanical resistance to the growing embryo (*Fraxinus, Rosa, Crataegus*).

The break in dormancy of seeds depends on a balance between growth inhibitors and growth promoters. Among the numerous plant-produced germination inhibitors some are localized in the fruit wall or seed coat.[3] The stimulating effect of removal of the seed coat or of the seed coat and associated coverings (as the pericarp and other floret parts in grasses) on seed germination supports the view that the investments of the seed are one of the sources of germination inhibitors.[13,44]

The seed coats of some species have features that assist in seed dispersal.[16] The testa of seeds that are eaten by animals and man may be resistant to digestive processes and permit the seed to pass through the intestinal tract without damage. Seeds, the epidermis of which consists of mucilaginous cells and, therefore, swells and becomes sticky in contact with water, may adhere to animals and be carried to favorable locations or, on the contrary, may adhere to the soil and thus resist being carried to unfavorable locations by wind or rain. Mucilage may protect the germinating seed by preventing its desiccation, or control the germination process by excluding the passage of oxygen when moisture is in excess and could damage the emerging seedling. Seeds may be adapted to dispersal by associations with suitable structures derived from other floral parts or be dispersed by mechanisms evolved in the fruit.[16] In the following, several examples of seed coats are described.

RICINUS

The seed of *Ricinus* (Euphorbiaceae) develops from an ovule with two integuments (bitegmic) and a large nucellus (crassinucellate; fig. 23.5,*A*). A vascular bundle extends through the funiculus into the chalazal region and branches in the inner integument forming here a ring of vascular tissue as seen in transections (fig. 23.5,*B*). The embryo sac is deeply embedded in the nucellus, which is completely enclosed by the integuments. Both integuments form the micropyle. An outgrowth from the placenta, the *obturator,* slightly protrudes into the micropyle. The changes occurring after fertilization are the following.[50] In the outer integument, the outer epidermal cells extend tangentially and deposit material responsible for the brown color of the seed. The inner epidermal cells become columnar but the mesophyll between the two epidermal layers is crushed. A caruncle develops near the micropyle and

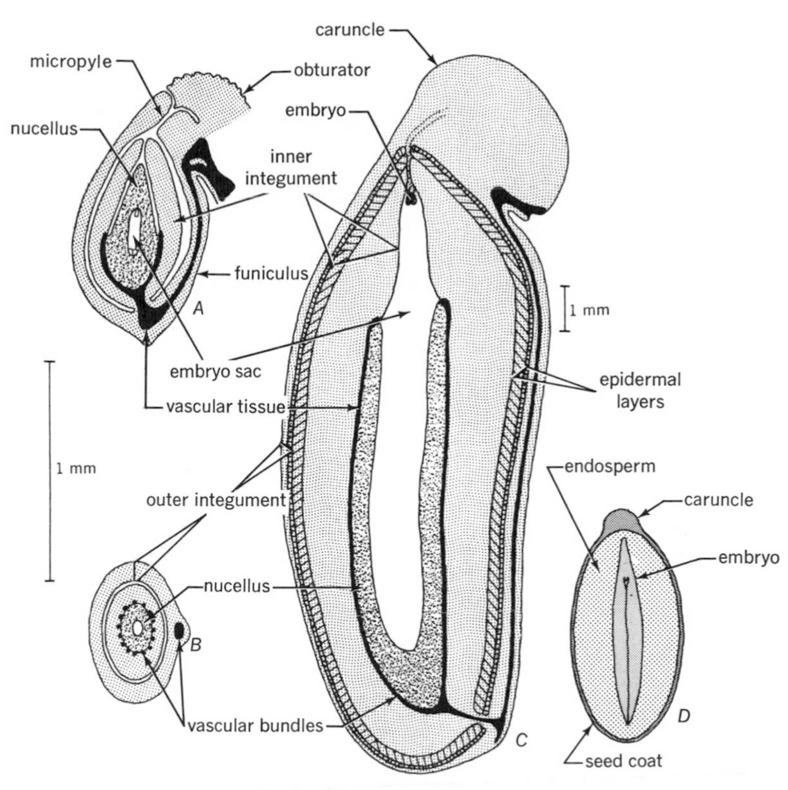

Figure 23.5 Seed of *Ricinus communis* (Euphorbiaceae). *A*, longitudinal section of ovule with a mature embryo sac. *B*, transection of ovule similar to that in *A*. *C*, longitudinal section of immature seed. *D*, longitudinal section of mature seed. (*A–C*, adapted from Singh.[50])

pushes the obturator toward the placenta. The cells of the outer epidermis of the inner integument elongate radially, assume a palisade arrangement, and become sclerified. The wide mesophyll and the inner epidermis of the inner integument are crushed. During the development of endosperm and embryo the nucellar cells in the micropylar region enlarge and vacuolate, those in the chalazal region undergo many divisions. This meristematic activity is largely responsible for the growth of the seed. The micropylar nucellar tissue, except a small remnant to which the embryo is attached, becomes crushed while the chalazal tissue is still present around the enlarged embryo sac (fig. 23.5,*C*). As the endosperm develops, the nucellar tissue is pushed laterally and becomes a layer of crushed cells lining the seed coat. In the mature state, the endosperm occupies the bulk of the seed and the embryo extends through the entire length of the seed (fig. 23.5,*D*)

BRASSICA AND SINAPIS

The bitegmic ovules of the Brassicaceae have rather thick integuments. The outer has two to five cell layers, the inner up to ten. In many species, the epidermal cells of the outer integument become almost filled with

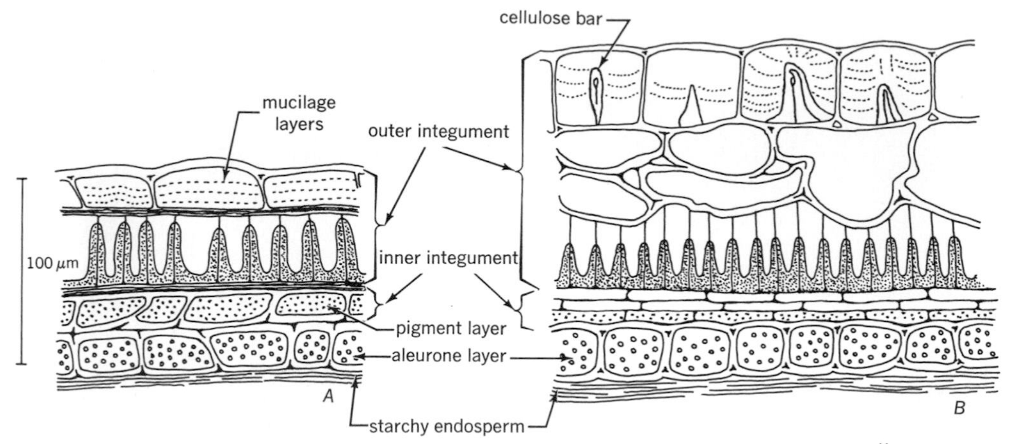

Figure 23.6 Cross sections of seed coat and aleurone layer of *Brassica* (*A*), and *Sinapis* (*B*). (After Z. Černohorský, *Graines des Crucifères de Bohême,* Opera Botanica Cechica, Vol. 5. 92 pp. 1947.)

mucilaginous material which appears in layers[53] (fig. 23.6). The mucilage, which consists of pectin and cellulose, swells when it comes in contact with water and, in some species, bursts the outer wall of the cell and forms a gelatinous film over the surface of the seed.[13] According to electron-microscope observations of the seed of *Plantago ovata,* dictyosomes appear to be concerned with the production of mucilage and its release via vesicles outside the plasmalemma.[19] In *Brassica* seed, a heavy cellulosic deposit on the inner tangential wall of the epidermis may form radially oriented bars, one in each cell (fig. 23.6,*B*). If subepidermal parenchyma is present in the outer integument, it develops thickened walls (fig. 23.6,*B*) or becomes crushed (fig. 23.6,*A*). The inner epidermis of the outer integument, the palisade layer,[55] is the strongest layer in most species, for its cells develop lignified thickenings on the radial and inner tangential walls (fig. 23.6). These cells are structurally most distinctive and most useful for systematic diagnosis. The inner integument dies and is compressed. In some species the inner epidermis of this integument becomes the pigment layer. The walls of the palisade layer also may be pigmented after the cells die.[13]

CUCURBITA

The seed of the Cucurbitaceae (fig. 23.7,*A*) develops from a bitegmic crassinucellate ovule but the inner integument disintegrates early in seed differentiation. The seed coat is thus derived from the outer integument.[49] The inner integument is narrow, the outer is wide, having 8 to 14 cell layers depending on the species. After fertilization, the outer integument further increases in thickness by periclinal divisions and differentiates into several distinct layers (fig. 23.8,*A*). The layer that remains the outermost after the periclinal divisions are completed—now the epidermis of the seed coat—consists of radially elongated cells (100 to 600 μm long) which have more or less thickened nonlignified walls (collapsed in fig. 23.8,*A*). The compact hypodermis, variable in thickness, consists of small cells with moderately thick lignified walls, the thickening forming a reticulum. Beneath the hypodermis is a layer of scleren-

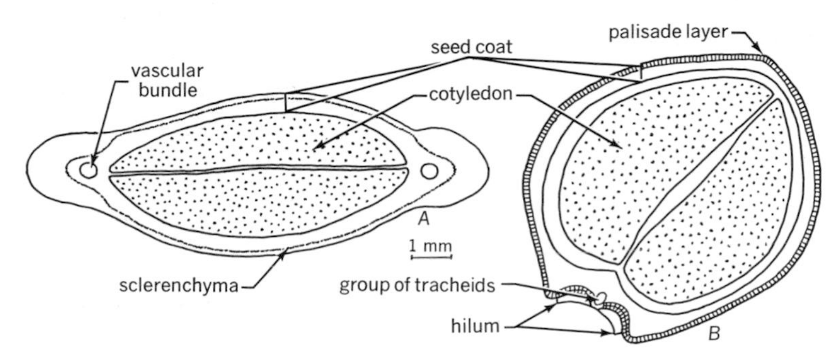

Figure 23.7 Diagrams of cross sections of seeds of pumpkin, *Cucurbita* (*A*), and bean, *Phaseolus* (*B*).

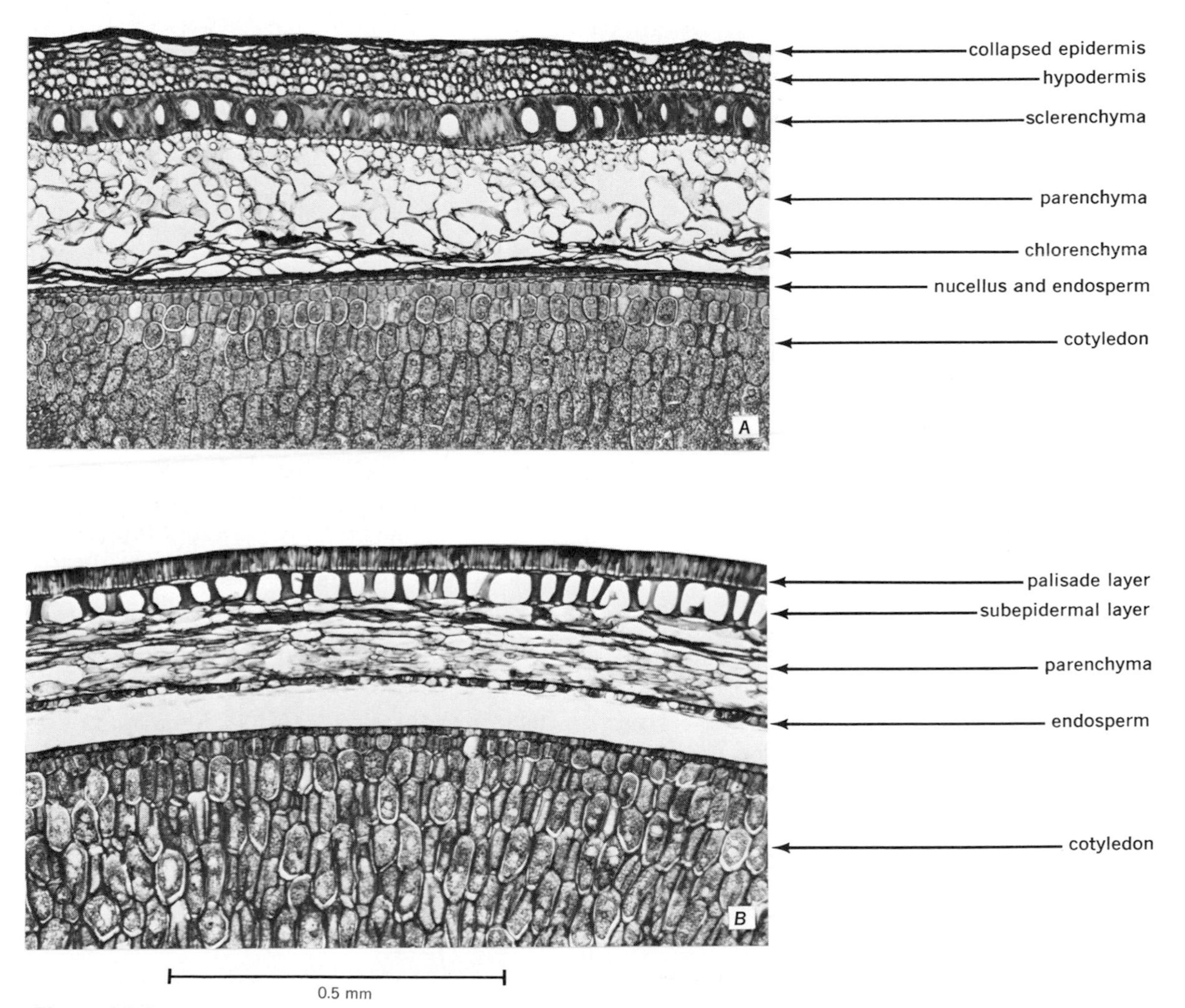

Figure 23.8 Cross sections of seed coats and parts of embryos of pumpkin, *Cucurbita* (*A*), and soybean, *Glycine* (*B*). (*B*, slide courtesy of A. T. Guard.)

chyma composed of slightly branched sclereids with thick lignified walls. A lacunate parenchyma (aerenchyma) with branched cells occurs beneath the sclerenchyma. Directly beneath the latter, the cells are small and compactly arranged. The aerenchyma is succeeded by tangentially stretched thin-walled chlorenchyma cells, the innermost of which constitute the inner epidermis of the outer integument. Some nucellar and endospermal cells occur next to the embryo. The nucellus is represented by two to four layers in the mature seed and its epidermis has a cuticle. In dry seeds the chlorenchyma usually separates from the seed coat as a thin green membrane and, together with the inner

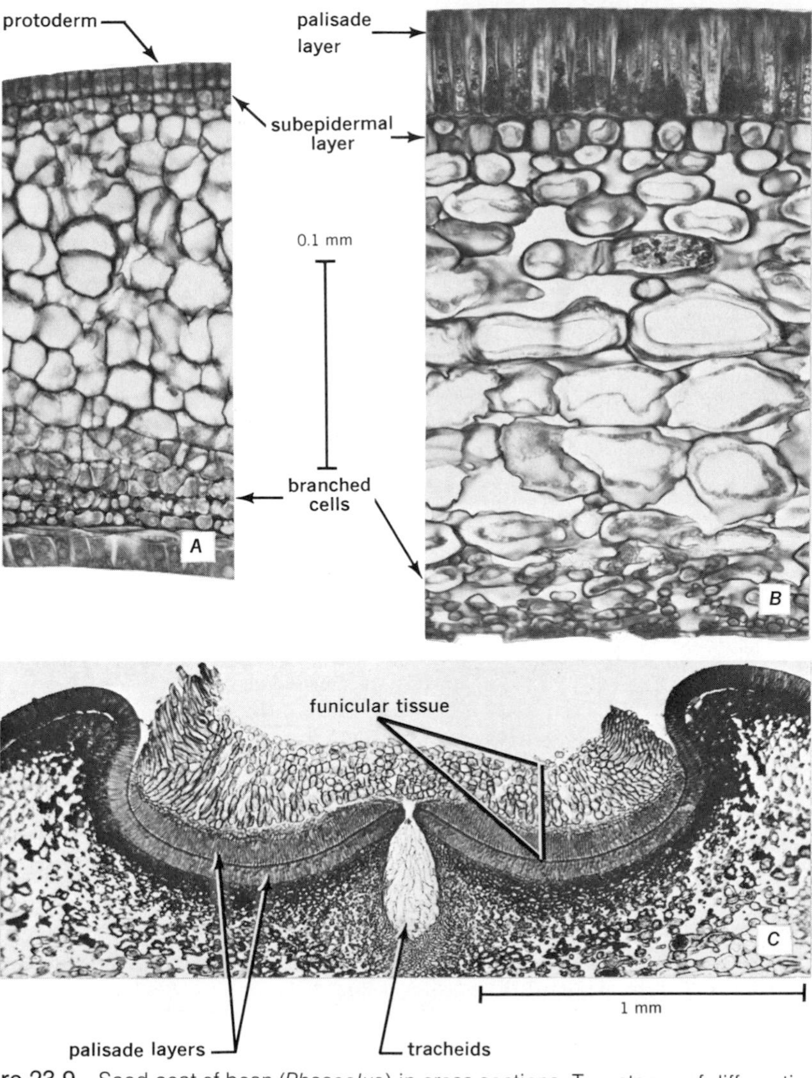

Figure 23.9 Seed coat of bean (*Phaseolus*) in cross sections. Two stages of differentiation: *A*, young; *B*,*C*, almost mature; *C*, hilum region.

epidermis, closely envelops the embryo and the remains of the inner integument, nucellus, and endosperm.

GLYCINE AND PHASEOLUS

The seed of Fabaceae (fig. 23.7,*B*) differentiates from an ovule with two integuments. The inner of the two integuments disappears during ontogeny of the seed, whereas the outer integument differentiates into a variety of distinct layers. The outermost layer, the epidermis, remains uniseriate and develops into a palisade layer of sclereids (macrosclereids; chapter 6) with unevenly thickened walls (figs. 23.8,*B,* and 23.9). Two palisade layers occur in the hilum region. The outer of these is derived from the funiculus (fig. 23.9,*C*). The cells of the subepidermal layer differentiate into the so-called columnar cells, also termed pillar cells, hourglass cells, osteosclereids, or lagenosclereids, depending on the distribution of wall thickenings and shape of cells (figs. 23.8,*B;* 23.9,*B;* and 23.10). The deeper tissue is a lacunate parenchyma with large tangentially elongated cells in the outer part and smaller much branched cells in the inner part (fig.

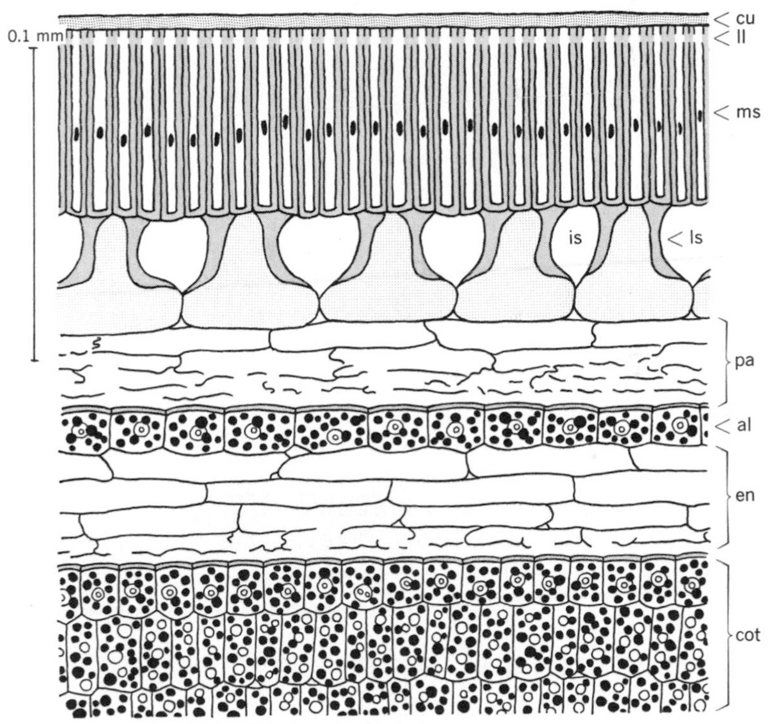

Figure 23.10 Partly diagrammatic drawing of transection of seed coat and associated tissues of *Crotalaria intermedia* (Fabaceae). Details are given from the top downward. Seed coat: *cu,* cuticle; *ll,* light line; *ms,* macrosclereids; *ls,* lagenosclereids (flask shaped) with intercellular spaces (*is*) among them; *pa,* parenchyma, most of it crushed. Associated tissues: *al,* aleurone endosperm; *en,* endosperm beneath aleurone layer; *cot,* cotyledon. Secondary walls, gray; storage protein, black dots; lipid globules, open circles. (Adapted from Miller.[31])

23.9,*B*). The subepidermal layer and the parenchyma beneath it have a common origin (fig. 23.9,*A*). The vascular system of many legume seeds is well developed. From the funiculus, the vascular bundle extends to the chalazal region where it branches. A compact group of tracheids of unknown function occurs in the hilum region (fig. 23.9,*C*).

The palisade layer attracts much attention because its structure in certain hard legume seeds is assumed to be causally connected with their high degree of impermeability. The so-called light line of the palisade cells (fig. 23.10) is thought to be the particularly impermeable region. The light-line effect results from the intense refraction in a restricted region in epidermal walls. In sections of seeds this region is oriented tangentially above the middle of a cell or close to its outer wall. The cell wall in the light-line region appears to be especially compact.[17]

The hard legume seeds achieve and maintain a very low percent moisture which is not affected by fluctuations in moisture content of the surrounding air. The attainment of this high degree of desiccation is ascribed to a combination of a highly impermeable testa with valvular action of the hilum.[20] The hilum acts like a hygroscopic valve: a fissure occurs along the groove of the hilum; this fissure opens when the seed is surrounded by dry air and closes when the outside air is moist. Thus, the entry of moisture is prevented, but loss of moisture is permitted.

NUTRIENT STORAGE TISSUES

Seeds storing food in the endosperm (fig. 23.12,*D,* and chapter 22; prevalent in the monocotyledons) or perisperm (Amaranthaceae, Chenopodiaceae, Polygonaceae) are said to be *albuminous,* those lacking these storage tissues, *exalbuminous* (fig. 23.7). The distinction is not absolute, however, for seeds usually show combinations of storage in the embryo and outside of it and the proportions of embryo volume to those of endosperm and perisperm are highly variable.

The endosperm is a tissue characteristic of angiosperms and is triploid when it is the product of three haploid nuclei—two polar nuclei and one sperm nucleus—fusing in the central cell of the embryo sac. Different taxa show variations in the number of polar nuclei fusing with the male gamete.[26]

Three methods of endosperm formation are recognized: *nuclear, cellular,* and *helobial* (after Helobieae, a monocotyledon taxon). In the nuclear endosperm, many nuclei are formed by free nuclear divisions. The endosperm may remain noncellular or cell walls may appear later (fig. 23.11). In the cellular endosperm, cell wall formation occurs beginning with the first mitosis (chapter 21) and continues as long as the endosperm is growing. In the helobial endosperm, the embryo sac is divided in two unequal cells the larger of which (the chalazal) usually develops in a noncellular manner, whereas the smaller micropylar cell shows variable behavior. The helobial type occurs mainly in monocotyledons.[51] The phylogenetic relation between the three types of endosperm formation is subject to divergent views.[52,59]

In the nuclear type of endosperm, the free nuclei commonly accumulate in the parietal layer of cytoplasm and the center of the cell is occupied by a large vacuole (fig. 23.11, *D*). In *Capsella,* however, the endosperm nuclei accumulate at two ends of the embryo sac, and the endosperm at the chalazal end becomes involved in digestion of nucellar cells.[45,46] In a number of plants, these nucellar cells enlarge greatly and become richly cytoplasmic before they break down in front of the advancing embryo sac (*Aesculus,*[24] *Allium,*[57] *Capsella*[45]). During their growth, the hypertro-

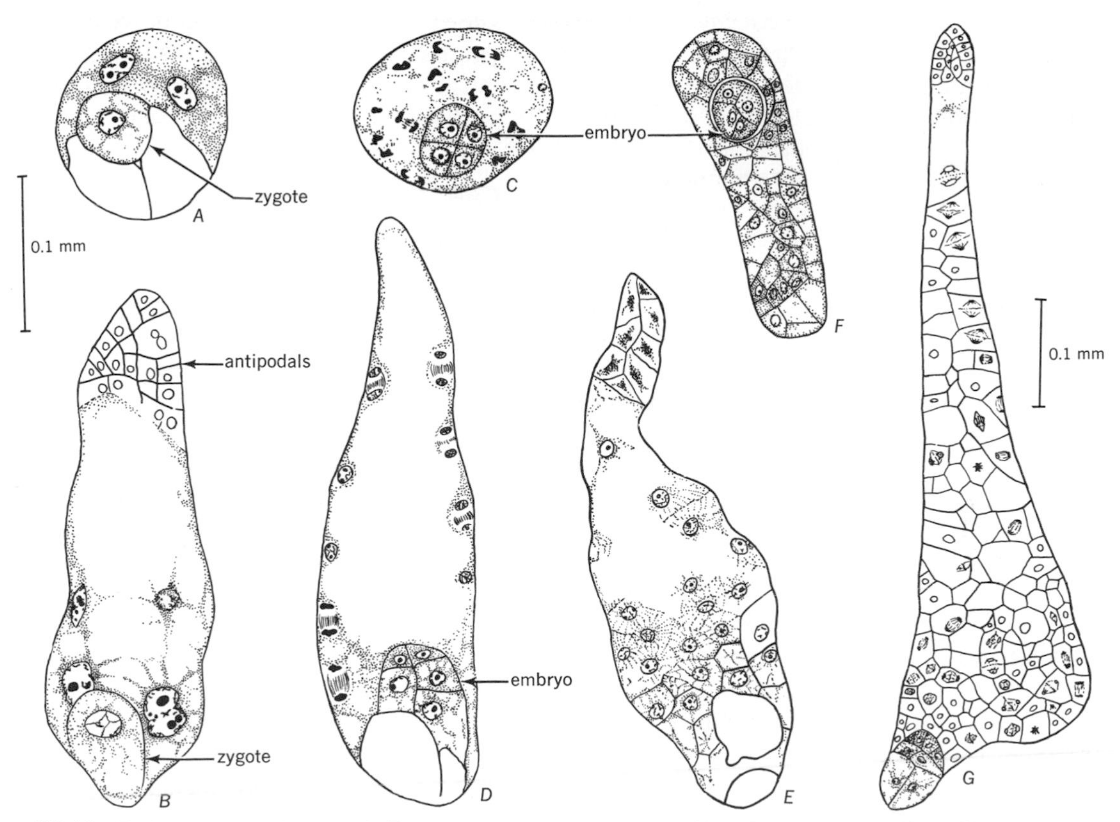

Figure 23.11 Endosperm development in *Zea mays* as seen in transverse (*A, C, F*), longitudinal (*B, D, G*), and oblique (*E*) sections of embryo sacs. *A, B,* zygote and a few endosperm nuclei 26 and 34 hours after pollination. *C, D,* a few-celled embryo and endosperm nuclei in division; passage from the 128- to the 256-free-nucleate stage 3 days after pollination. *E,* passage from free-nucleate to cellular stage 3.5 days after pollination; phragmoplasts between nuclei are associated with cell plates. *F, G,* cellular endosperm 4 days after pollination. (Redrawn from Randolph.[40])

phied nucellar cells, as observed in *Capsella,* stain for protein and nucleic acids and rapidly incorporate nucleic acid precursors, probably because the large nuclei characteristic of these cells are polyploid. During the disintegration of nucellar cells, large vacuoles in the adjacent endosperm are filled with vesicles and apparent cytoplasmic debris. The name "chalazal proliferating tissue" given to the hypertrophied cells of the nucellus is misleading because no proliferation of cells is involved in their specialization.

In *Eranthis hiemalis* (Ranunculaceae), the nuclear endosperm incorporates adjacent integumentary cells.[38] As the walls of the integument become disorganized, nuclei migrate into the endosperm and are broken into fragments. By the time the cell walls are formed in the endosperm the integumentary cell layers surrounding the embryo sac are reduced from eleven to seven. Just before the endosperm becomes cellular, a rest period occurs in the growth of the ovule and endosperm during which numerous autophagic vacuoles appear in the endosperm. These vacuoles sequester and digest cytoplasm and organelles.[9] The authors propose that possibly this process serves to rebuild the incorporated sporophytic cytoplasmic material into the gametophytic endosperm.

Cell wall formation in the endosperm begins while the nuclei are in the parietal location. Cellular endosperm first appears in the micropylar region, near the embryo (fig. 23.11,*F,G*). Commonly, phragmoplasts and cell plates are involved in the formation of cell walls (fig. 23.11,*E*). In *Stellaria media,* however, some endospermal walls grow freely in the manner of wall ingrowths.[37] This type of wall growth occurs in the micropylar region in addition to wall formation by means of cell plates. In *Pisum,* electron microscopy has revealed that endosperm cells deposit a wall layer next to the outer wall of the embryo so that the embryo-endosperm wall is a typical intercellular boundary composed of two primary wall layers and a middle lamella.[28]

The cellular endosperm gradually invades the central vacuolar space. Cell formation in the parietal region may be as orderly as in the cambium and at the end of these divisions the outermost layer becomes the aleurone layer.[13,36,40] In *Cocos nucifera,* the central cavity does not become filled with cells but retains a sap called coconut milk.[10] The endosperm in some taxa develops haustoria that invade various ovular tissues and serve as an additional device for the nutrition of the growing embryo.

The principal food materials stored in seeds are carbohydrates, proteins, and lipids. These compounds occur together but the relative amounts vary in seeds of different taxa.[54] On dry weight basis, seeds of cereals contain 70 to 80 percent of starch, those of peas and beans about 50 percent. Although the principal reserve material in *Zea* is starch, most of which is in the endosperm, the embryo contains around 50 percent oil. The seeds of rape and mustard (Brassicaceae) yield 40 percent oil and 30 percent protein, those of soybean (Fabaceae) 20 percent oil and 40 percent protein.

CARBOHYDRATES

Starch grains arise as single or multiple granules in amyloplasts. In cereal endosperm, large and small starch grains are often combined (fig. 23.12,*A*), the smaller appearing at later stages of seed development than the larger. Starch synthesis occurs in wheat endosperm when the peripheral layer is in the midst of cell formation so that, when dormancy sets in, the older inner cells have accumulated more starch than the younger outer cells.[15] In *Zea,* a still more pronounced gradient in starch accumulation extends from the upper endosperm to the lower.[22] At maturity, the starchy endosperm in Poaceae and certain other families is nonliving and the starch grains are free of the amyloplast envelope.[6,34]

During germination, starch grains become corroded on the surface as hydrolysis sets in. Phosphorylase activity is followed by those of α-amylase and maltase. In cereals, the breakdown of endosperm reserves depends on enzymes secreted by the aleurone layer. It is generally accepted that the synthesis of enzymes in the aleurone layer is stimulated by gibberellins (GA_3), for it can be evoked by exogenous GA_3 in an excised aleurone layer.[6] Cytochemical localization of phosphatase in barley *in vitro* and *in vivo* indicates that during imbibition of water by seeds enzyme activity increases with or without the addition of GA_3 and that the major fraction of the enzyme accumulates in the cell walls. In presence of GA_3, however, the wall is corroded and the enzyme is free to enter the endosperm. This observation suggests that GA_3 controls the release of the enzyme and that

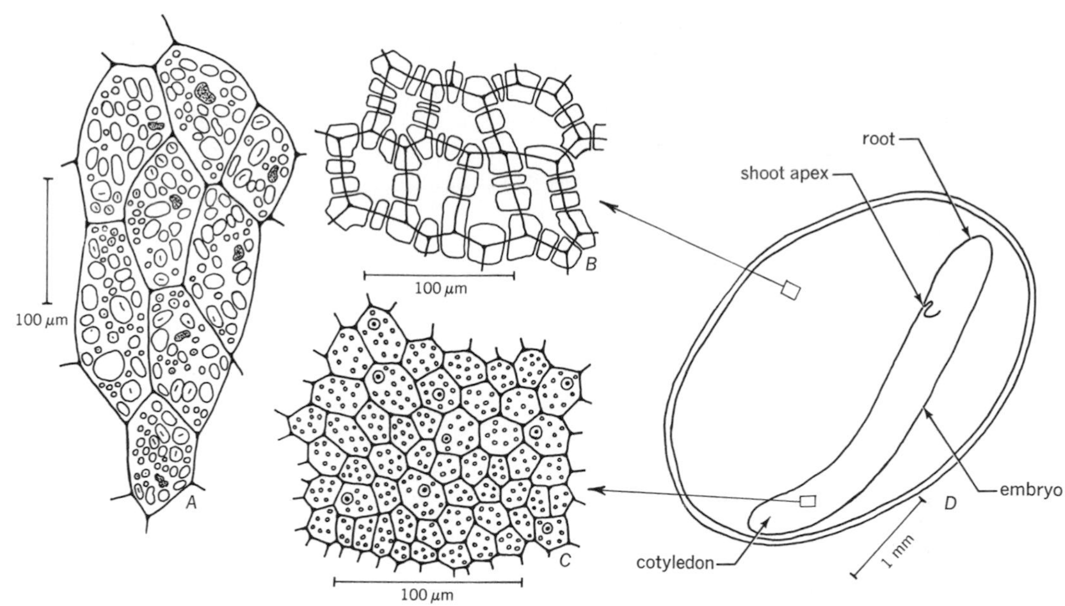

Figure 23.12 Endosperm and embryo. *A*, starchy endosperm of rye kernel. *B–D, Asparagus officinalis;* thick-walled endosperm (*B*), parenchyma of cotyledon with nuclei and stored nutrients (*C*), and longitudinal section of seed (*D*). (*B–D*, after W. W. Robbins and H. A. Borthwick, *Bot. Gaz.* 80:426–438, 1925.)

some of the enzyme synthesis may be independent of GA_3.[2]

The biochemical specialization of the outermost endosperm layer as an aleurone layer is not restricted to cereals (figs. 23.6 and 23.10). As observed in several Fabaceae,[41] carbohydrate breakdown in the endosperm cells is concurrent with digestion of protein bodies and corrosion of walls in the outermost endosperm layer. This layer thus appears to function as an aleurone layer in mobilizing its own protein reserves for synthesis of enzymes and secreting the latter through the corroded cell wall.

The glucose resulting from hydrolysis of starch in cereal endosperm is taken up by the scutellum, synthesized into polysaccharides, and transported to the growing seedling. The energy required in these processes is provided in the form of ATP by the actively multiplying mitochondria.[6]

In some seeds, carbohydrate reserves appear as thickened walls composed mostly of hemicelluloses giving mannose and other monosaccharides on hydrolysis.[30] This type of storage occurs in endosperm (*Asparagus*, fig. 23.12,*B–D; Coffea, Diospyros, Iris, Phoenix, Strychnos*) or cotyledons (*Impatiens, Lupinus, Primula, Tropaeolum*).

PROTEINS

Storage proteins are synthesized during seed development together with the reserve carbohydrates or lipids. Storage proteins must be distinguished from metabolic proteins that are concerned with normal cellular activities including the synthesis of storage proteins. As determined for representatives of the Fabaceae, storage proteins are globulins and are restricted to seeds.[32]

Storage proteins occur as discrete protein bodies (fig. 23.10). Those in the aleurone layer are called aleurone grains, but often the same term is used for protein bodies in any part of the seed. Protein bodies vary in composition. Some contain no inclusions in their homogeneous protein matrix, others include one or more of the following: globoids, in which insoluble salts of phytic acid are localized; protein crystalloids; protein-carbohydrate bodies; crystals of calcium oxalate.[1,21]

Ultrastructural studies indicate that reserve proteins accumulate in vacuoles (chapter 3) and when the process is completed the accumulations appear as bodies enveloped by a unit membrane, presumably the original tonoplast.[4] During germination, the protein becomes dissolved *in situ* and the protein bodies are replaced by vacuoles which may fuse into one large vacuole.[5,25,43] Reserve proteins are degraded to amino acids by proteinases and peptidases.[6]

LIPIDS

Lipids appear in the form of oil bodies in the cytoplasm of oil storing tissues as, for example, in cotyledons of white mustard, or in scutella of cereals. The delimitation of lipid bodies appears as an interface with the surrounding cytoplasm, possibly a monomolecular layer of phospholipid.[42] Lipid bodies are also referred to as lipid-containing vacuoles or as spherosomes and are then assumed to have a bounding unit membrane or half of a unit membrane.[47]

The seed lipid reserve is a triglyceride which is hydrolyzed *in situ* to glycerol and fatty acids by lipases. The fatty acids are utilized for synthesis of phospholipids and glycolipids, which are required as constituents of organelles, but most are converted to sugars and transported to the seedling body for growth. Microbodies of glyoxysome type are involved in the lipid oxidation pathway.[6]

REFERENCES

1. Altschul, A. M., L. Y. Yatsu, R. L. Ory, and E. M. Engleman. Seed proteins. *Ann. Rev. Plant Physiol.* 17:113–136. 1966.
2. Ashford, A. E., and J. V. Jacobsen. Cytochemical localization of phosphatase in barley aleurone cells: the pathway of gibberellic-acid-induced enzyme release. *Planta* 120:81–105. 1974.
3. Barton, L. V. Dormancy in seeds imposed by the seed coat. *Handb. Pflanzenphysiol.* 15:727–745. 1965.
4. Briarty, L. G., D. A. Coult, and D. Boulter. Protein bodies in developing seeds of *Vicia faba*. *J. Exp. Bot.* 20:358–372. 1970.
5. Briarty, L. G., D. A. Coult, and D. Boulter. Protein bodies of germinating seeds of *Vicia faba*. *J. Exp. Bot.* 21:513–524. 1970.
6. Ching, T. M. Metabolism of germinating seeds. Chap. 2. pp. 103–218. In: *Seed Biology*. T. T. Kozlowski, ed. Vol. 2. New York, Academic Press. 1972.
7. Chuang, T.-L., and L. R. Heckard. Seed coat morphology in *Cordylanthus* (Scrophulariaceae) and its taxonomic significance. *Amer. J. Bot.* 59:258–265. 1972.
8. Cooper, D. C. Embryology of *Pisum sativum*. *Bot. Gaz.* 100:123–132. 1938.
9. Cresti, M., E. Pacini, and G. Sarfatti. Ultrastructural studies on the autophagic vacuoles in *Eranthis hiemalis* endosperm. *J. Submicr. Cytol.* 4:33–44. 1972.
10. Cutter, V. M., Jr., K. S. Wilson, and B. Freeman. Nuclear behavior and cell formation in the developing endosperm of *Cocos nucifera*. *Amer. J. Bot.* 42:109–115. 1955.
11. Davis, G. L. *Systematic embryology of the angiosperms.* New York, John Wiley & Sons. 1966.

12. Dure, L. S. III. Seed formation. *Ann. Rev. Plant Physiol.* 26:259–278. 1975.
13. Edwards, M. M. Dormancy in seeds of charlock. I. Developmental anatomy of the seed. *Ann. Bot.* 19:575–582. 1968. II. The influence of the seed coat. *Ann. Bot.* 19:583–600. 1968. III. Occurrence and mode of action of an inhibitor associated with dormancy. *Ann. Bot.* 19:601–610. 1968.
14. Eeuwens, C. J., and W. W. Schwabe. Seed and pod wall development in *Pisum sativum,* L. in relation to extracted and applied hormones. *J. Exp. Bot.* 26:1–14. 1975.
15. Evers, A. D. Development of the endosperm of wheat. *Ann. Bot.* 34:547–555. 1970.
16. Fahn, A., and E. Werker. Anatomical mechanisms of seed dispersal. Chap. 4. pp. 151–221. In: *Seed Biology.* T. T. Kozlowski, ed. Vol. 1. New York, Academic Press. 1972.
17. Frey-Wyssling, A. *Die pflanzliche Zellwand.* Berlin, Springer-Verlag. 1959.
18. Gunn, C. R. Seed collecting and identification. Chap 2, pp. 55–143. In: *Seed Biology.* T. T. Kozlowski, ed. Vol 3. New York, Academic Press. 1972.
19. Hyde, B. B. Mucilage-producing cells in the seed coat of *Plantago ovata:* developmental fine structure. *Amer. J. Bot.* 57:1197–1206. 1970.
20. Hyde, E. O. C. The function of the hilum in some Papilionaceae in relation to ripening of the seed and the permeability of the testa. *Ann. Bot.* 18:241–256. 1954.
21. Jacobsen, J. V., R. B. Knox, and N. A. Pyliotis. The structure and composition of aleurone grains in the barley aleurone layer. *Planta* 101:189–209. 1971.
22. Kiesselbach, T. A. The structure and reproduction of corn. *Univ. Nebraska Coll. Agric. Res. Bull.* 161. 1949.
23. Kozlowski, T. T., and C. R. Gunn. Importance and characteristics of seeds. Chap. 1, pp. 1–20. In: *Seed Biology.* T. T. Kozlowski, ed. Vol 1. New York, Academic Press. 1972.
24. List, A., Jr., and F. C. Steward. The nucellus, embryo sac, endosperm, and embryo of *Aesculus* and their interdependence during growth. *Ann. Bot.* 29:1–15. 1965.
25. Lott, J. N. A., and C. M. Vollmer. Changes in the cotyledons of *Cucurbita maxima* during germination. IV. Protein bodies. *Protoplasma* 78:255–271. 1973.
26. Maheshwari, P. *An introduction to the embryology of angiosperms.* New York, McGraw-Hill. 1950.
27. Marbach, I., and A. M. Mayer. Permeability of seed coats to water as related to drying conditions and metabolism of phenolics. *Plant Physiol.* 54:817–820. 1974.
28. Marinos, N. G. Embryogenesis of the pea (*Pisum sativum*). I. The cytological environment of the developing embryo. *Protoplasma* 70:261–279. 1970.
29. Martin, A. C. The comparative internal morphology of seeds. *Amer. Midl. Nat.* 36:513–660. 1946.
30. Meier, H. On the structure of cell walls and cell wall mannans from ivory nuts and from dates. *Biochem. Biophys. Acta* 28:229–240. 1958.
31. Miller, R. H. *Crotalaria* seed morphology, anatomy, and identification. *USDA ARS Tech. Bull.* 1373. 73 p. 1967.
32. Millerd, A. Biochemistry of legume seed proteins. *Ann. Rev. Plant Physiol.* 26:53–72. 1975.
33. Mohana Rao, P. R. Seed anatomy in some Hamamelidaceae and phylogeny. *Phytomorphology* 24:113–139. 1974.
34. Müller, D. Tote Speichergewebe in lebenden Samen. *Planta* 33:721–727. 1943.
35. Netolitzky, F. Anatomie der Angio-

spermen-Samen. In: *Handbuch der Pflanzenanatomie*. K. Linsbauer, ed. Band 10. Lief. 14. Berlin, Gebrüder Borntraeger. 1926.

36. Neubauer, B. F. The development of the achene of *Polygonum pensylvanicum:* embryo, endosperm, and pericarp. *Amer. J. Bot.* 58:655–664. 1974.
37. Newcomb, W., and L. C. Fowke. The fine structure of the change from the free-nuclear to cellular condition in the endosperm of chickweed *Stellaria media*. *Bot. Gaz.* 134:236–241. 1973.
38. Pacini, E., M. Cresti, and G. Sarfatti. Incorporation of integumentary nuclei in *Eranthis hiemalis* endosperm and their disaggregation by the endoplasmic reticulum. *J. Submicr. Cytol.* 4:19–31. 1972.
39. Pate, J. S., and A. M. Flinn. Carbon and nitrogen transfer from vegetative organs to ripening seeds of the field pea (*Pisum arvense* L.) *J. Exp. Bot.* 24:1090–1099. 1973.
40. Randolph, L. F. Developmental morphology of the caryopsis in maize. *J. Agric. Res.* 53:881–916. 1936.
41. Reid, J. S. G., and H. Meier. The function of the aleurone layer during galactomannan mobilization in germinating seeds of fenugreek (*Trigonella foenum-graecum* L.), crimson clover (*Trifolium incarnatum* L.) and lucerne (*Medicago sativa* L.): a correlative biochemical and ultrastructural study. *Planta* 106:44–60. 1972.
42. Rest, J. A. , and J. G. Vaughan. The development of protein and oil bodies in the seed of *Sinapis alba* L. *Planta* 105:245–262. 1972.
43. Rost, T. L. The ultrastructure and physiology of protein bodies and lipids from hydrated dormant and nondormant embryos of *Setaria lutescens* (Gramineae). *Amer. J. Bot.* 59:607–616. 1972.
44. Rost, T. L. The morphology of germination in *Setaria lutescens* (Gramineae): the effects of covering structures and chemical inhibitors on dormant and non-dormant florets. *Ann. Bot.* 39:21–30. 1975.
45. Schulz, P., and W. A. Jensen. *Capsella* embryogenesis: the chalazal proliferating tissue. *J. Cell Sci.* 8:201–227. 1971.
46. Schulz, P., and W. A. Jensen. *Capsella* embryogenesis. The development of the free nuclear endosperm. *Protoplasma* 80:183–205. 1974.
47. Schwarzenbach, A. M. Observations on spherosomal membranes. *Cytobiologie* 4:145–147. 1971.
48. Singh, B. Development and structure of angiosperm seed. *Bull. Natl. Bot. Gard.* No. 89, 115 pp. 1964.
49. Singh, D., and A. S. R. Dathan. Structure and development of seed coat in Cucurbitaceae. VI. Seeds of *Cucurbita*. *Phytomorphology* 22:29–45. 1972.
50. Singh, R. P. Structure and development of seeds in Euphorbiaceae: *Ricinus communis* L. *Phytomorphology* 4:118–123. 1954.
51. Swamy, B. G. L., and K. V. Krishnamurthy. The helobial endosperm: a decennial review. *Phytomorphology* 23: 74–79. 1973.
52. Swamy, B. G. L., and N. Parameswaran. The helobial endosperm. *Biol. Rev.* 38:1–50. 1963.
53. Swarbrick, J. T. External mucilage production by the seeds of British plants. *Bot. J. Linn. Soc.* 64:157–162. 1971.
54. Vaughan, J. G. *The structure and utilization of oil seeds*. London, Chapman and Hall. 1970.
55. Vaughan, J. G., and J. M. Whitehouse. Seed structure and the taxonomy of the Cruciferae. *Bot. J. Linn. Soc.* 64:383–409. 1971.

56. Villiers, T. A. Seed dormancy. Chap. 3. pp. 219–281. In: *Seed Biology.* T. T. Kozlowski, ed. Vol. 2. New York, Academic Press. 1972.
57. Weber, E. Entwicklungsgeschichtliche Untersuchungen über die Gattung *Allium. Bot. Archiv.* 25:1–44. 1929.
58. Whiffin, T., and A. S. Tomb. The systematic significance of seed morphology in the neotropical capsular-fruited Melastomataceae. *Amer. J. Bot.* 59:411–422. 1972.
59. Wunderlich, R. Zur Frage der Phylogenie der Endospermtypen bei den Angiospermen. *Österr. Bot. Z.* 106:203–293. 1959.
60. Wunderlich, R. Some remarks on the taxonomic significance of the seed coat. *Phytomorphology* 17:301–311. 1967.

24

Embryo and Seedling

The embryo was introduced in chapter 2 as representing the first stage in the development of the sporophyte of the seed plant; and its origin from a fertilized egg (zygote) was described in chapter 21. The embryo was dealt with also as the essential part of the seed in chapter 23. The present chapter emphasizes the development of the embryo by the use of several examples and introduces embryo classification based on comparative embryogeny, including that induced experimentally. Brief reviews of germination and seedling structure relate the embryo to the adult plant. Only angiosperm embryos are discussed.

MATURE EMBRYO

The presence of a single cotyledon in the monocotyledons and of two cotyledons in the dicotyledons distinguish the two main taxa of angiosperms (fig. 24.1). In ontogeny, the cotyledons in dicotyledons arise as lateral organs at the apex of embryo axis and are positionally related to the apical meristem as are the foliage leaves of the shoot developing later. The relation of the single cotyledon in monocotyledons to the apical meristem of the embryo is less clear. According to the French embryologist Souèges,[48] the single cotyledon is terminal in origin, the shoot apex is lateral, and the whole monocotyledonous plant is a sympodium of lateral shoots. Developmental studies, however, suggest that the single cotyledon arises as a lateral organ but, by its growth, displaces the apical meristem to a seemingly lateral position.[13] This position of the emerging shoot may be the result of a delay in apical development, for the latter occurs only after the cotyledon attains considerable size.[55] The shoot apex is often late in being organized in the dicotyledons as well but is not displaced because the two cotyledons originate symmetrically.

Embryos vary in their relative volume and orientation in the seed (chapter 23), features that determine in part whether the embryo is upright, bent, or curved (fig. 24.2). The degree of development of the embryo varies

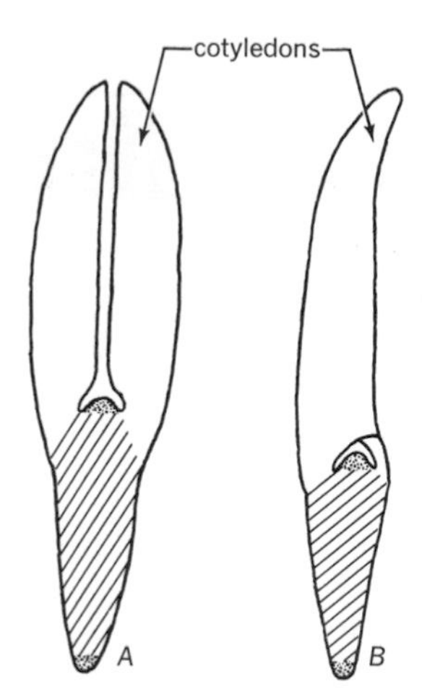

Figure 24.1 Diagrams of embryos of a dicotyledon (*A*) and a monocotyledon (*B*) in longitudinal sections. Shoot tip and root tip are stippled, hypocotyl-root axis hatched. (Adapted from Rauh.[35])

also. Some have longer, others shorter hypocotyls. The cotyledons may be thin (fig. 24.2,*C,D*) or fleshy as when they constitute the main site of food storage (e.g., embryos in Fabaceae; chapter 23). The epicotyl may be represented by its meristem only, or it may initiate one or more of the first true leaves and the corresponding nodes. The root also may be represented by its apical meristem or by a more extensive portion of the embryo axis (hypocotyl-root axis), that is, a radicle.

The procambium of the future vascular system appears in the embryo and is continuous between the axis and the cotyledon or cotyledons. In some species, initials of xylem and phloem are distinguishable in the embryo.[8] The young epidermis, or protoderm, constitutes the surface layer. The ground tissue is distinguished from the procambium by its less elongated and wider cells but both kinds of cells are filled with storage products in the mature dormant embryo.[32]

GRASS EMBRYO

The morphology of the grass embryo is so complex and so much a subject of controversy regarding the interpretation of its parts that it must be treated separately. A grass embryo reaches a relatively high degree of differentiation (fig. 24.3). Within the mature caryopsis, the embryo is appressed to the endosperm by its massive cotyledon, the *scutellum* (fig. 24.3,*A*). The embryo axis appears as though laterally attached to the scutellum. The lower part of the axis is a radicle bearing,

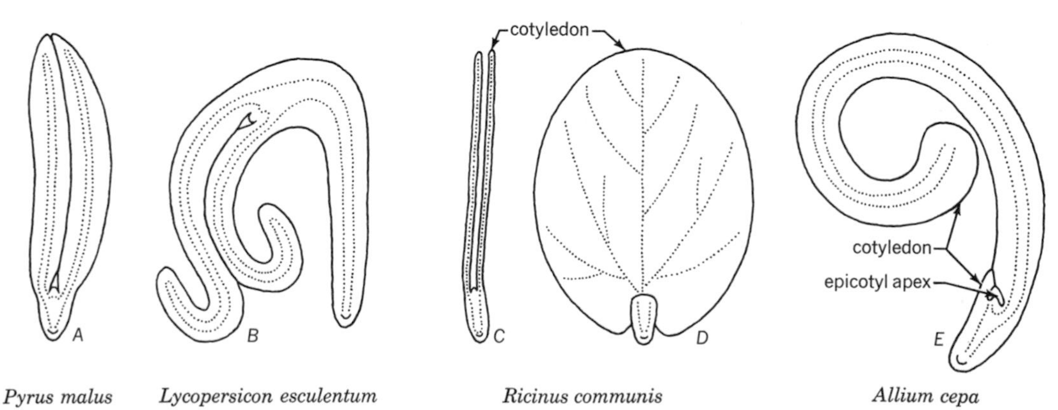

Figure 24.2 Diagrams of embryos in longitudinal sections perpendicular to the plane of cotyledons (*A–C, E*) and parallel with the same plane (*D*). Procambium outlined with dotted lines. (*A, C, D,* adapted from Rauh.[35])

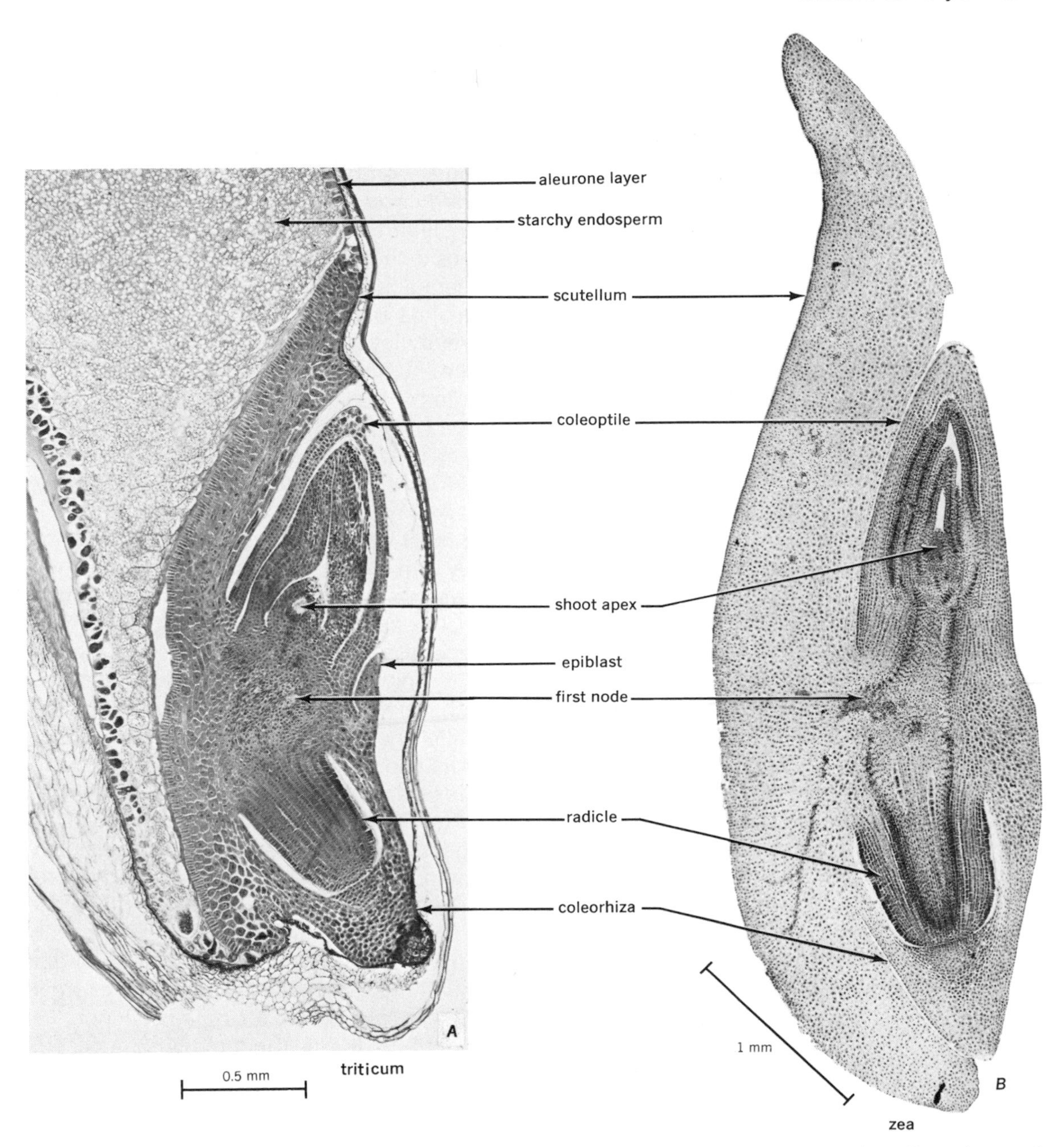

Figure 24.3 Embryos of Poaceae *Triticum* (*A*) and *Zea* (*B*), both in near median longitudinal sections. Some tissues of the caryopsis are present in *A*. Epiblast in *A*, none in *B*. The first node is the scufellar node.

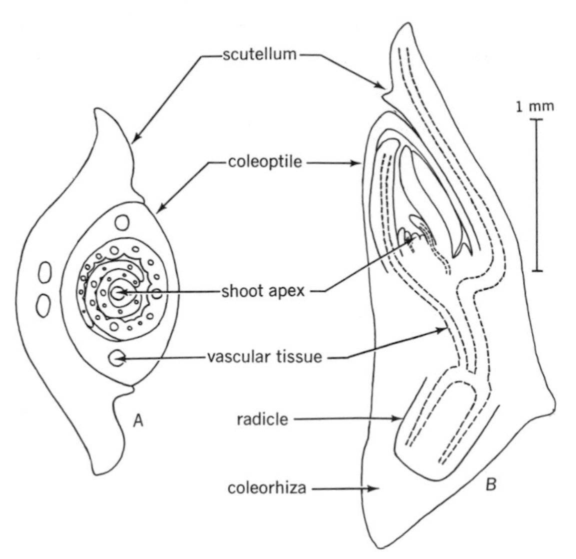

Figure 24.4 Transverse (*A*) and longitudinal (*B*) sections of embryos of barley (*Hordeum sativum*) from mature grains soaked for 24 hours. (Adapted from Merry.[22])

at its lower end, an apical meristem and a rootcap. The root and its rootcap are enclosed in the *coleorhiza* which, in the young embryo, is continuous with the rather bulky suspensor. Above the radicle is the scutellar node (consequently, there is no hypocotyl distinct from the radicle), then follows the epicotyl with several foliar primordia. The first and outermost of these is the *coleoptile.* The part of the axis between the scutellar node and the coleoptile is an internode. It is not regarded as an internode and is called *mesocotyl* by proponents of the view that the scutellum and the coleoptile constitute the cotyledon and its sheath.[10] In some Poaceae, a small outgrowth, the *epiblast,* occurs opposite the scutellum (fig. 24.3,*A*). A procambial system interconnects the shoot and the root (fig. 24.4). The procambium is elaborately branched in the scutellum.[1] Food reserves (lipid, protein bodies, starch) occur throughout the embryo including the scutellum.[38]

The scutellum is a shieldlike structure partly enclosing the embryo axis and the epicotyl (fig. 24.4). During germination, this organ is the source of the initial complement of hydrolytic enzymes causing the breakdown of the endosperm[2] and the probable source of the initial gibberellin stimulus.[51] The abaxial epidermis of the scutellum is a secretory epithelium releasing the enzymes and hormones to the endosperm. The wrinkles on the epidermal surface are sometimes referred to as glands but they are part of a continuous secretory layer. The scutellum also serves to absorb nutrients from the endosperm. As determined for germinating rice, glucose derived from the reserve starch in the endosperm is converted to sucrose in the scutellum which is then transported to the embryonic axis.[25]

In most grasses the coleoptile is a hollow cone with an opening near its apex which allows the shoot to emerge during germination. The opening is a slit through which the inner and outer epidermal layers are continuous.[28] Both layers have stomata, some of which serve as hydathodal pores. The coleoptile encloses the shoot with its leaf primordia and, because of its sensitivity to light, guides the shoot toward the surface of the soil after germination, protecting it on the way.

The different views expressed on the nature of the parts of the grass embryo are comprehensively reviewed by Brown.[5] It will suffice to cite here two interpretations to illustrate the problem. According to Guignard and Mestre,[11] the scutellum is a new type of organ that has evolved from a single cotyledon; the coleoptile is the first leaf modified as a protective structure and the mesocotyl is the first internode between the scutellum and the coleoptile; the coleorhiza is a degenerated primary root, whereas the normal root arising above the coleorhiza is adventitious; the epiblast is merely an extension of the coleorhiza. Brown[4] regards the scutellum as a cotyledon, the epiblast as an extension of the coleorhiza, the coleoptile and mesocotyl as innovations without counterparts, the coleorhiza as part of the base of the embryo remaining after the true primary root differentiates.

DEVELOPMENT OF EMBRYO

The egg and the zygote show cytological differentiation between the chalazal and micropylar poles so that the cells have an axial symmetry (chapter 21). The upper (chalazal) pole is the main seat of potentialities for further growth, for the embryo proper develops from cells formed at this pole.[7,55] The lower pole has essentially only a vegetative function. Growth at this pole produces the suspensor that anchors the embryo at the micropyle and also provides one of the mechanisms for the transfer of nutrients from the surrounding tissues to the embryo.

In their early stages of differentiation, embryos of the different angiosperm taxa have many features in common. The unicellular zygote becomes subdivided, usually first by a horizontal wall (chapter 21; fig. 24.9,*A*), then by a series of other horizontal, vertical, and oblique walls. Before the division progresses far enough to indicate differentiation into embryo proper and suspensor, the entire structure is often called *proembryo* (fig. 24.5,*A*). When, in dicotyledons, the emergence of the two cotyledons is in preparation, bilateral symmetry replaces the axial (fig. 24.5,*D*). Some embryologists regard this change in symmetry as the termination of the proembryo stage.[7] But the distinction between the embryo proper and the suspensor is detectable before the symmetry changes (fig. 24.5), and reference to bilateral symmetry would not help to delimit the proembryo stage in monocotyledons. Consequently, the term proembryo should not be defined too narrowly. If used at all, the term seems to serve best when it is applied to stages before the protoderm is initiated by periclinal divisions in the upper part of the young embryo (figs. 24.5,*A;* 24.7,*A;* 24.9,*A–F*).

The differentiation of the of the embryo from proebmryo is variable with regard to the organogenic destiny of the cells produced at the two poles of the zygote. In the following, the development of three kinds of embryo is described, that of a dicotyledon (*Capsella bursa-pastoris*), an amaryllid monocotyledon (*Allium cepa,* onion), and a poaceous monocotyledon (mainly *Poa annua* and *Hordeum vulgare*).

CAPSELLA EMBRYO

The inherently axial polarity of the early embryo is exhibited during the first division: a transverse wall divides the zygote into two unequal cells, a large basal cell and a smaller apical cell.[42] A series of transverse divisions in the basal cell give rise to a filamentous proembryo but the apical cell, which is the precursor of the embryo proper, divides by a vertical wall (fig.24.5,*A*). Another vertical division and subsequent transverse divisions convert the embryo proper into a globular structure consisting of two octants of cells. The upper octant is the source of cotyledons and the apical meristem of the epicotyl, whereas the lower octant develops into the hypocotyl. During the divisions beginning with the first and ending with the formation of the globular embryo, the size of cells decreases and the number of ribosomes and the associated density of protoplasts increase. The embryo attains 64 cells during the globular stage. It is still globular when periclinal divisions produce the protoderm and subsequent vertical divisions block out the procambium of the vascular cylinder, delimiting it

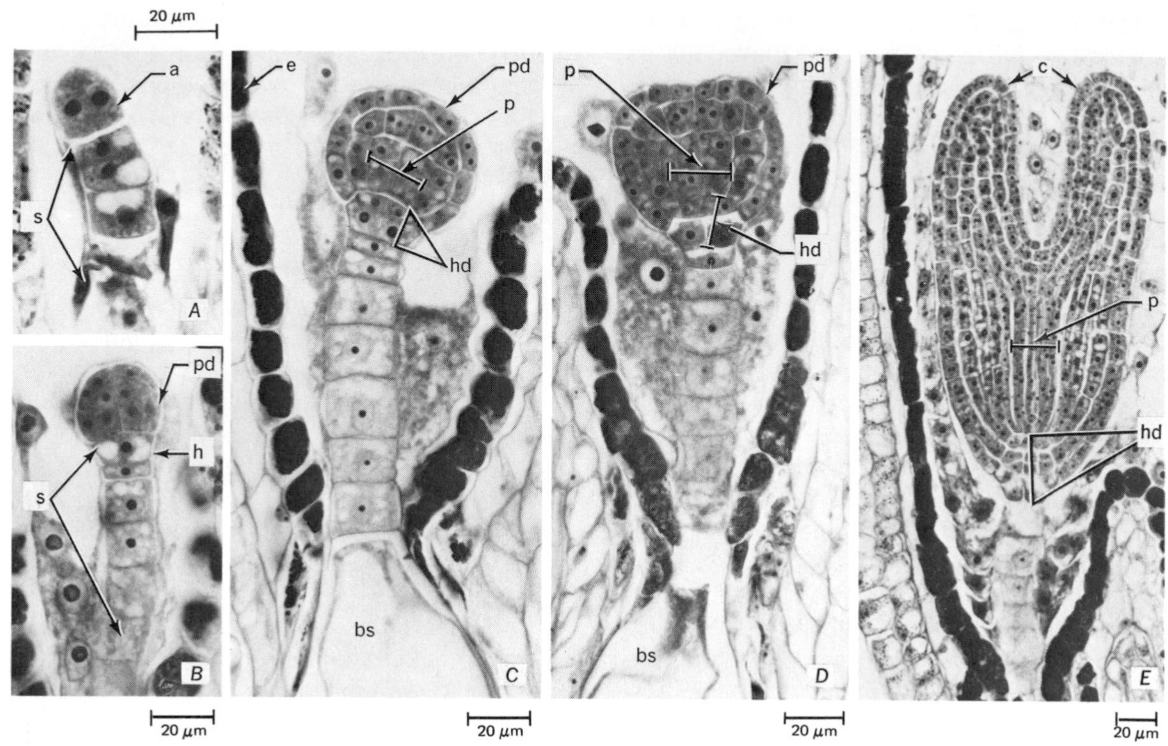

Figure 24.5 *Capsella bursa-pastoris* embryo in several stages of development shown in longitudinal sections of ovules. *A*, apical tier from which the embryo proper develops shows less vacuolated cells than the suspensor. *B*, globular embryo with early protoderm; the uppermost cell of suspensor is a hypophysis. *C*, globular embryo with protoderm and procambium; hypophysis has produced two tiers of derivatives. *D*, early heart-shaped embryo; beneath protoderm, anticlinal divisions have occurred at future cotyledonary sites. *E*, cotyledons partly developed. Details: *a*, apical tier; *bs*, basal cell of suspensor; *c*, cotyledon; *e*, endothelium; *h*, hypophysis; *hd*, derivatives of hypophysis; *p*, procambium; *pd*, protoderm; *s*, suspensor.

from the ground meristem of the cortex (fig. 24.5,*B*,*C*).

The flattening of the embryo introducing the bilateral symmetry occurs before cotyledonary growth is initiated (fig. 24.5,*D*). The embryo is now described as heart shaped. Periclinal and anticlinal divisions below the protoderm and anticlinal divisions in the latter bring about the elevation of the cotyledons. The embryo becomes green and its plastids develop stacked thylakoids.[42] The cotyledons and the now discernible hypocotyl elongate and the embryo is said to become torpedo shaped (figs. 24.5,*E* and 24.6,*B*). With continued growth, the embryo bends and the cotyledons reach the chalazal end of the embryo sac (fig. 24.6,*B–D*). The apical meristem between the cotyledons becomes a small mound and the procambium forms the core of the hypocotyl-root axis and extends into the cotyledons. During the growth of the embryo the endosperm passes from free-nuclear to the cellular state and the chalazal hypertrophied nucellar tissue is digested (fig. 24.6,*A*,*B*,*D*; chapter 23).

The derivatives of the basal cell of the two-celled embryo make a contribution to the embryo proper. The uppermost suspensor cell in the filamentous proembryo (fig. 24.5,*A*) divides transversely and the upper of the two cells becomes the *hypophysis* (fig. 24.5,*B*). Through a series of transverse and longitudinal divisions the hypophysis provides the initials of the cortex and completes the rootcap and its initial zone (figs. 24.5,*C–E*). Depending on the taxon, the apical meristem of the embryonic root may or may not assume the same cellular organization as the root of

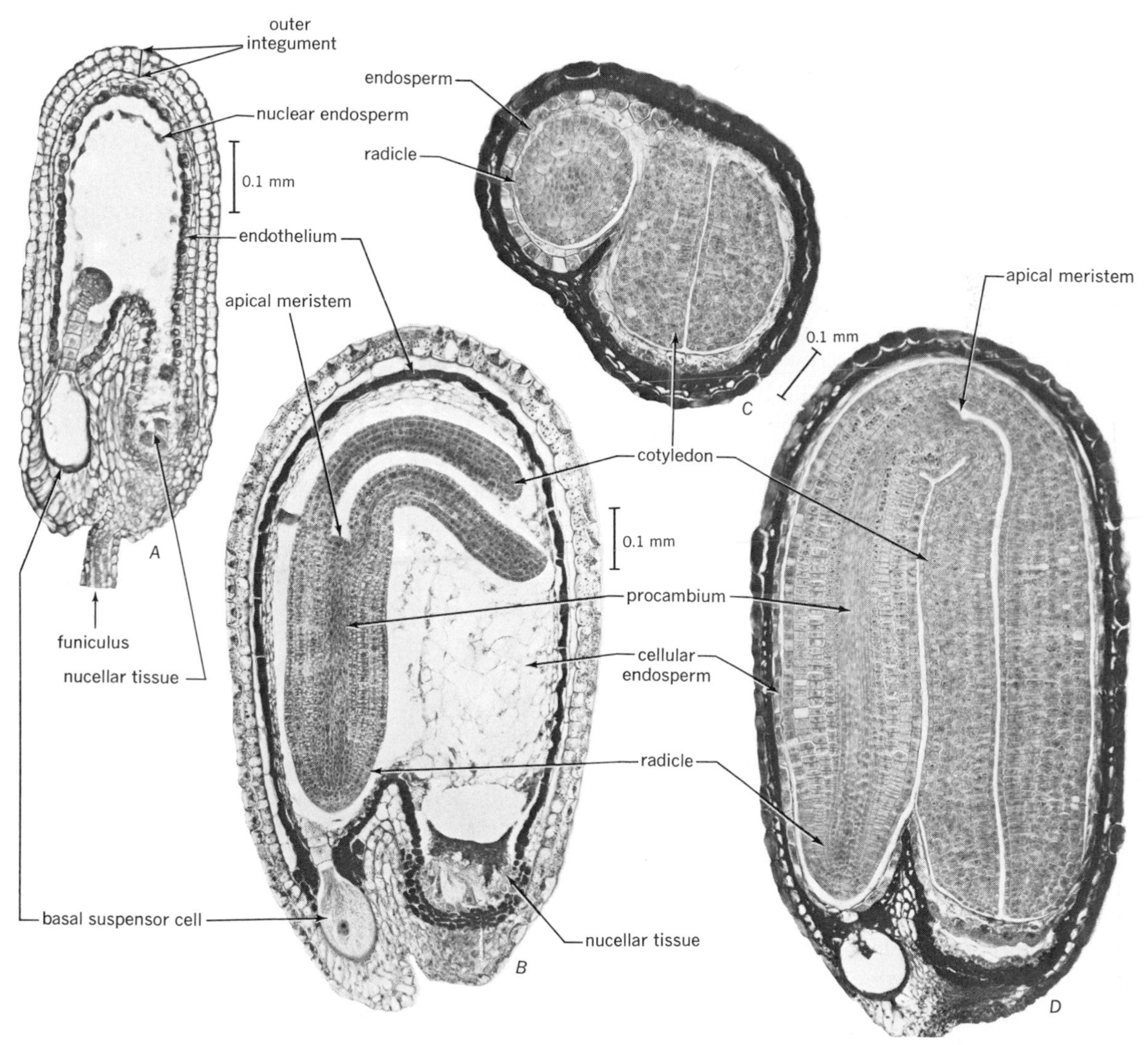

Figure 24.6 *Capsella bursa-pastoris* embryo in several stages of development shown in longitudinal (*A, B, D*) and transverse (*C*) sections. Complete profiles of ovules (*A, B*) and seeds (*C, D*). *A*, embryo in globular stage, endosperm in free-nuclear stage; beneath the outer integument, inner integument with endothelium, next to the embryo sac. *B*, embryo at stage of bending of cotyledons; cellular endosperm. *C, D*, mature embryo; much reduced endosperm.

the growing plant. In some dicotyledons, adventitious root primordia differentiate in the embryo hypocotyl.[50]

The long suspensor in *Capsella* remains threadlike except that its basal cell is much enlarged (fig. 24.6,*A*) and has other unique features. A rather similar combination of suspensor cells occurs in the embryo of *Stellaria media*[24] and, for convenience of description, the authors distinguish between "the basal cell" and "the chalazal suspensor cells." These designations are used in the following with reference to the *Capsella* suspensor.

Ultrastructural observations show that, in comparison with the embryo cells, the chalazal suspensor cells of *Capsella* are more vacuolate and contain more endoplasmic reticulum and dictyosomes, but have fewer ribosomes and less affinity for protein and nucleic acid stains.[43] The cells are connected with one another and with the basal cell by numerous plasmodesmata but no plasmodesmata occur in any suspensor wall facing the embryo sac lumen. The basal cell is highly vacuolate and has an extensive network of wall projections on the micropylar and adjacent lateral walls. The basal cell and the lower chalazal suspensor cells fuse with the embryo sac wall. During its enlargement, the basal cell crushes the integument. The chalazal suspensor cells begin to degenerate at the stage of heart-shaped embryo and are eventually crushed by the enlarging embryo (fig. 24.6,*D*). The basal cell survives longer. The ultrastructure of the suspensor, as observed in *Capsella* and *Stellaria,* supports the concept that this organ is a site of active metabolism and serves in the absorption and translocation of nutrients to the developing embryo.

ONION EMBRYO

Early divisions lead to formation of a club-shaped embryo (fig. 24.7,*A*). Later, the embryo becomes an almost spherical body on a thin suspensor (fig. 24.7,*B*,*C*). The cotyledon develops upward from the spherical body (fig. 24.7,*D*,*E*). A slight depression, or notch, on one side of the embryo reveals the site of the future apical meristem. The spatial relation between the notch and the cotyledon clearly illustrates the previously mentioned difficulty morphologists experience in deciding whether the single cotyledon in a monocotyledon is terminal or lateral.

The depression is shallow at first (fig. 24.7,*E*) but increases in depth as the tissue on its margin grows out. This marginal growth is a sheathlike extension from the cotyledon (fig. 24.8,*A*). The apical meristem arises as a small mound of embryonic cells at the bottom of the depression and initiates the first foliage leaf primordium. When the seed germinates, the first leaf emerges from the enclosure through a slit above the sheath (fig. 24.15). The apical meristem of the root and rootcap become organized at the base of the short hypocotyl, with the initials of the vascular cylinder, cortex, and rootcap forming three successive tiers (fig. 24.7,*E*). The protoderm is at first continuous with the rootcap initials, later with those of the cortex.[12]

The mature embryo has a protoderm, a somewhat vacuolated ground meristem, and a less vacuolated procambium. The latter extends from the root meristem to the base of the cotyledon where it widens out and forms a short branch directed toward the epicotyl and a long branch extending through the cotyledon (fig. 24.8,*B*).

GRASS EMBRYO

The development of the embryo in *Poa annua,* treated according to Souèges,[46] serves to illustrate the embryogeny in Poaceae and to give an example of the kind of analysis some embryologists carry out to correlate cell patterns in the early embryo with

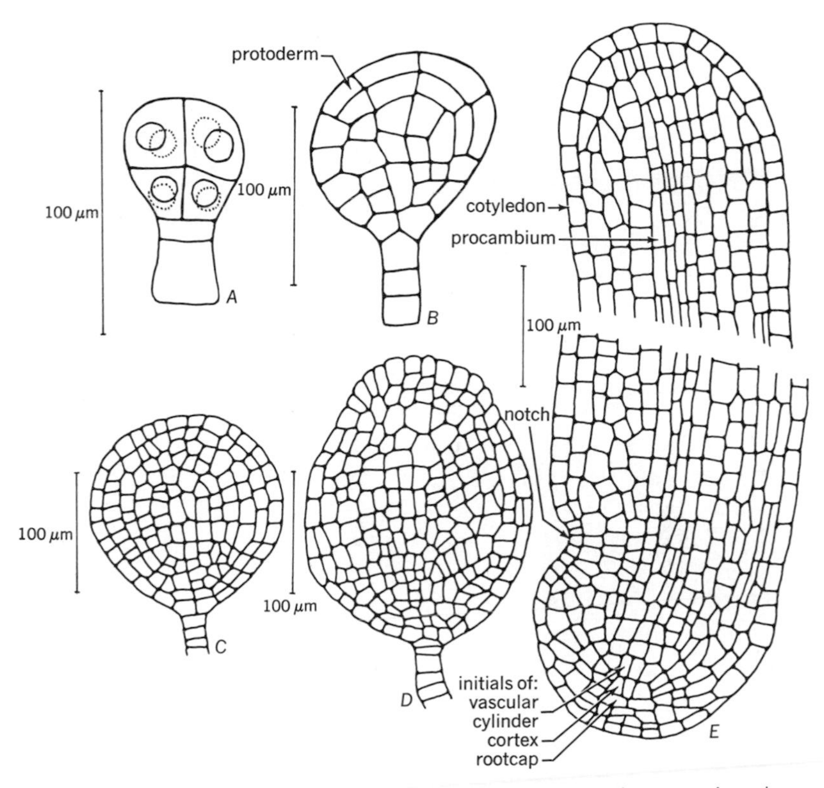

Figure 24.7 Embryo of *Allium cepa* (onion) in several stages of development. *A*, embryo body is distinct from suspensor. *B*, protoderm has been initiated in distal tiers. *C*, embryo body is spherical. *D*, beginning of elongation at the distal end: initiation of cotyledon. *E*, embryo still immature but with tissue regions blocked out; the notch is the future site of apical meristem of shoot.

organogenesis (fig. 24.9). A transverse division of the zygote (*A*) is followed by a vertical division in the apical cell and a transverse division in the basal cell, all resulting in a four-celled, three-tiered embryo (*B*). Subsequent divisions form four tiers (*C*) with tier 1 derived from the apical cell, tiers 2 to 4 from the basal cell. Cell division continues in each tier (*D–F*) but is not necessarily precise and consequently the cellular compostion of the tiers varies in different embryos, and the separation between adjacent tiers is not always equally clear.

According to the sequence depicted in figure 24.9, the protoderm is initiated in tier 1 (*G*), then its delimitation progresses basipetally (*H,I*). The upper part of the cotyledon (scutellum) arises from tier 1, the lower part from tier 2 (*J*). The coleoptile has a similar origin (*K*). The entire embryo axis and the epicotyl are derived from tier 2, the coleorhiza from tier 3, the rootcap and epiblast being part of the coleorhiza (*K,L*). Tier 4 produces the lower part of the embryo that corresponds to a suspensor in function.

In contrast to Souèges'[46] concept of grass embryogenesis, other authors find no evidence that the origin of organs in Poaceae

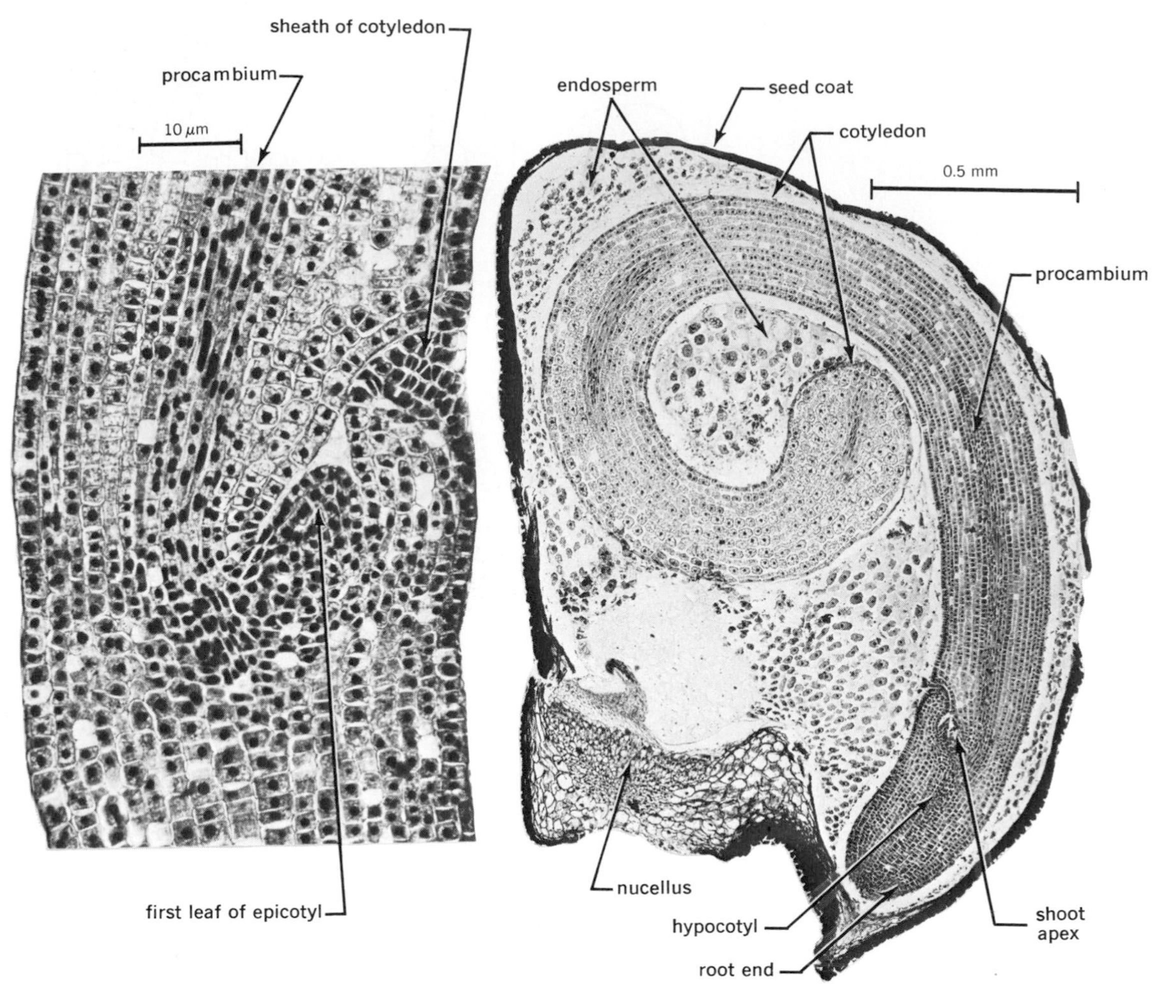

Figure 24.8 Mature embryo of *Allium cepa* (onion). *A*, median part of embryo with epicotyl sectioned through its first leaf and enclosed by the cotyledon. *B*, mature embryo within seed in longitudinal section. (*B*, from K. Esau, *Plant Anatomy*, 2nd ed. John Wiley & Sons, 1965.)

can be assigned to specific tiers in the young embryo (e.g., *Hordeum vulgare*,[22] *Zea mays*[34]).

Additional details on grass embryogeny appear in figure 24.10 with respect to *Zea mays*. About five days after fertilization the maize embryo becomes club shaped (*A*). The upper enlarged part gives rise to the main body of the embryo; the lower part is the suspensor. A ten-day-old embryo is elongated and thickened on one side because of growth of the scutellum (*B*). Opposite the scutellum on the embryo axis is the epicotyl apex. It becomes a small rounded prominence surrounded by a circular welt of tissue, the incipient coleoptile. As the coleoptile develops further (*C*), leaf primordia are initiated, and the growth of the epicotyl is reoriented from a

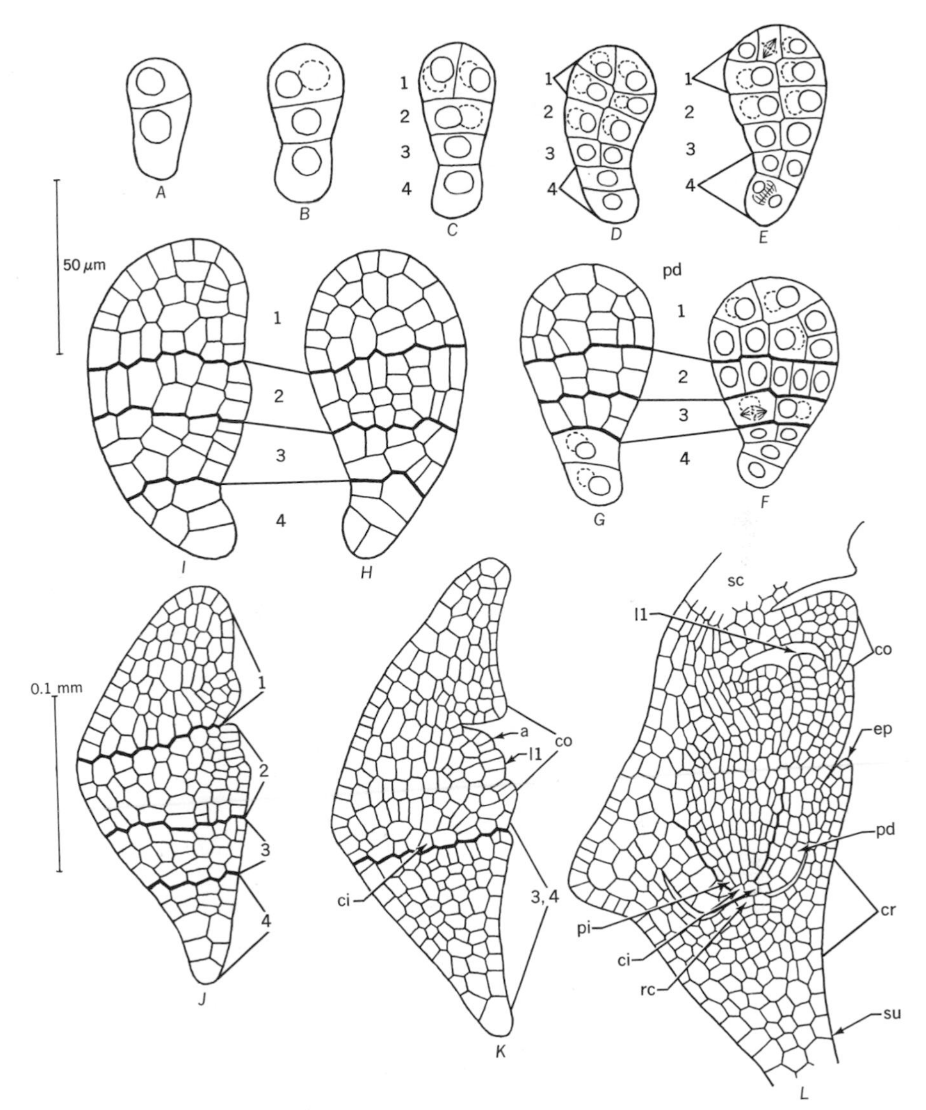

Figure 24.9 Embryo development in *Poa annua* (Poaceae). *A–C,* sequence in establishment of four tiers of cells. *D–H,* multiplication of cells in each of four tiers. *I–K,* continued cell multiplication and beginning of organ differentiation. *L,* growing foliar primordia and evidence of internal organization. Details: numbers, tiers of cells; *a,* apex of epicotyl; *ci,* cortical (periblem) initials; *co,* coleoptile; *cr,* coleorhiza; *ep,* epiblast; *l*1, first leaf; *pd,* protoderm; *pi,* vascular cylinder (plerome) initials; *rc,* rootcap initials; *sc,* scutellum; *su,* suspensor. (Adapted from Souèges.[46])

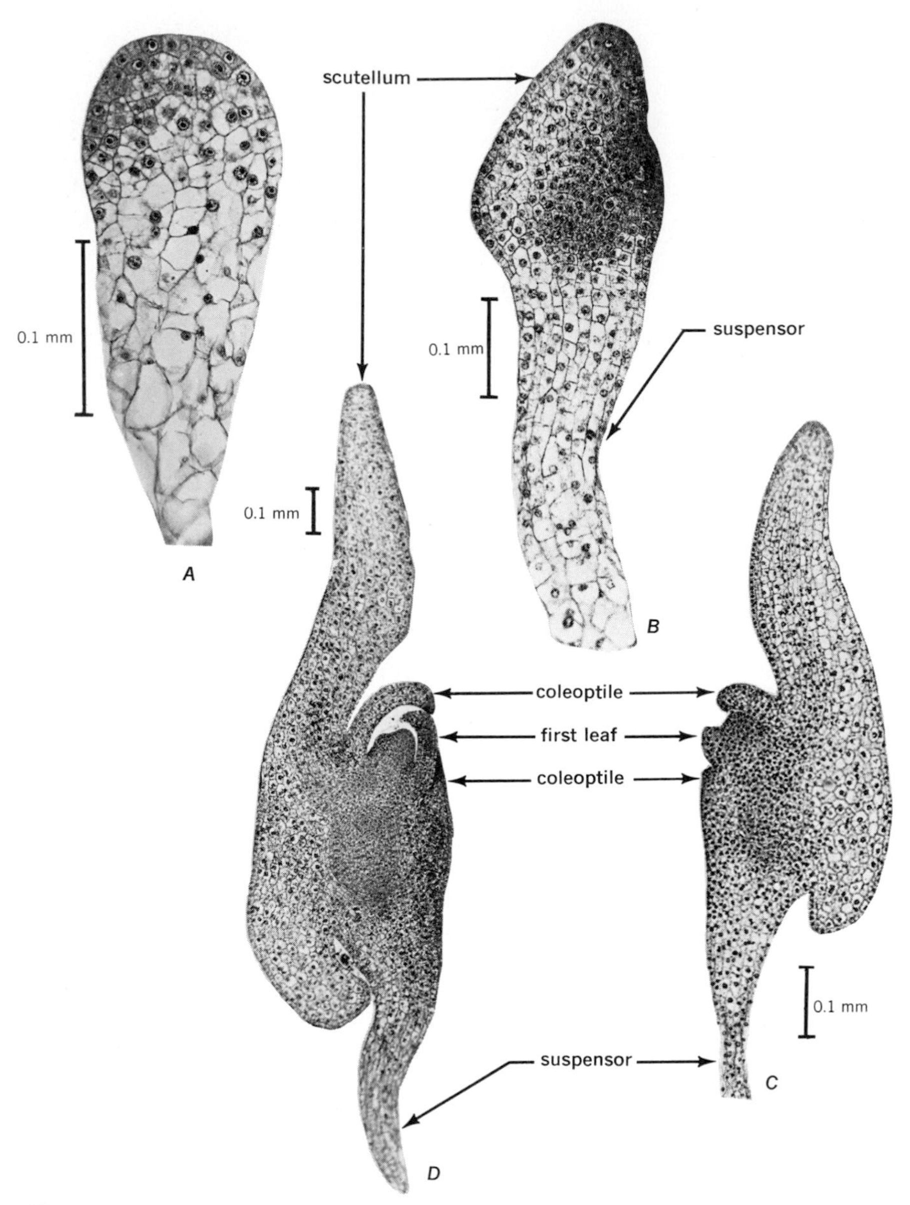

Figure 24.10 *Zea mays* (corn) embryo in several stages of development. The material was collected the following numbers of days after pollination: *A*, 5; *B*, 10; *C*, *D*, 15. (Courtesy of J. E. Sass. *D*, from J. E. Sass, *Botanical Microtechnique*, 3rd ed. The Iowa State College Press, 1958.)

lateral to a vertical direction (*D*). The scutellum enlarges and grows around the embryo axis (as shown for *Hordeum,* fig. 24.4,*A*) and eventually covers the scutellar node. The epiblast, as is found in the wheat embryo (fig. 24.3,*A*), develops rather late in embryogeny, after the scutellum has elongated considerably and the coleoptile has partly enclosed the apical meristem.

At the lower end of the embryo axis, above the suspensor, the radicle and rootcap are organized. The radicle is at first united with the coleorhiza tissue but becomes separated from it as the embryo matures (fig. 24.3,*B*). Above the scutellar node additional roots are initiated. These are called seminal adventitious roots. After germination, during the growth stage known as tillering, further adventitious roots develop at the nodes of the main and lateral shoots.

Ultrastructural studies on barley embryo (*Hordeum vulgare*[26,27]) deal with the early stages of development ending with the 20-celled embryo (figs. 24.11 and 24.12) that correspond with the stages of *Poa* embryo development in figure 24.9,*A–E.* The barley embryo does not develop a long, highly vacuolated suspensor: as other grass embryos, it undergoes little elongation in early growth. In the upper part of the embryo, the cleavage pattern is rather orderly through the octant stage, but is irregular at the base. The embryo is first attached to the nucellus by its basal cells but this connection is severed when the embryo reaches the 20-cell stage. Plasmodesmata are lacking in the walls of the zygote and in the outer walls of the embryo but are present between embryo cells.

The pyriform appearance of the zygote indicates axial polarity which, however, is not well expressed in the internal organization (fig. 24.11,*A*,*B*). The nucleus is located in the wider upper part but the micropylar end is as rich in cytoplasm as the chalazal end. The vacuoles are distributed at the periphery so that the organelles are crowded around the nucleus. The vacuoles seem to be somewhat more abundant at the base.

After several divisions the embryo becomes approximately cylindrical (fig. 24.11,*C*). Vacuoles are still peripheral but in the 13-celled embryo they are larger in the lower cells than in the octant region (fig. 24.12,*A*). Cell size decreases during the early divisions and consequently the 20-celled embryo is no larger than the somewhat younger embryo, and its nuclei are smaller than that in the zygote. The cytoplasm is less vacuolated at this stage and the distribution of organelles more uniform. The endosperm is still in the free-nuclear stage when the embryo consists of 20 cells. The inner layer of nucellus is lysed during early embryogeny.

DEVIATING KINDS OF EMBRYOGENY

Reduced and anomalous embryos not conforming in their development to the normal type discussed in the preceding pages are widely scattered in different taxa. Some occur in parasites and saprophytes. Embryos may deviate also in their origin. Some are formed apomictically, that is, by an asexual process as from an unfertilized egg cell (haploid parthenogenesis) or from some other cell of the gametophyte (haploid apogamy). If meiosis fails to occur and a diploid gametophyte is formed, there may be diploid parthenogenesis or diploid apogamy. Still other variants occur.[55] In a variety of ways, polyembryony may develop and result in seeds with more than one embryo.

Ontogeny of isolated embryos cultured in artificial media follows normal or abnormal paths depending on the physical and chemical factors in the environment.[33,55] In the origin of embryos from callus cells, or from isolated somatic cells, or from pollen grains,

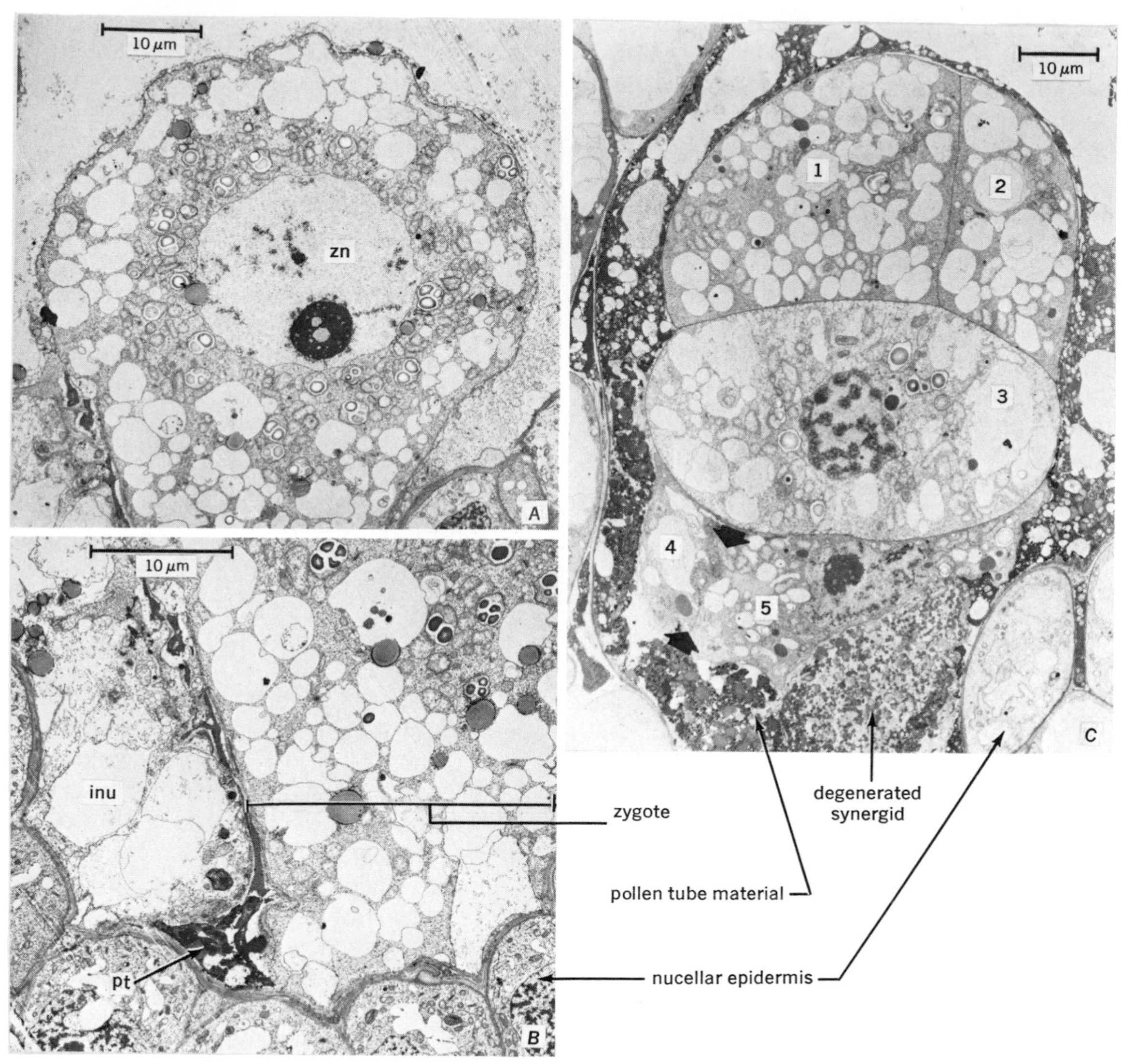

Figure 24.11 Ultrastructure of zygote (*A, B;* profile 50 × 70 μm) and 5-celled embryo (*C*) of barley (*Hordeum vulgare*). Longitudinal sections. Cells are numbered in *C.* Sections of cells 1, 2, 4 are considerably off median. Numerous vacuoles in cells, starch grains in plastids. Details: arrowheads in *C,* position of wall between suspensor cells 4 and 5; *inu,* inner, thin-walled nucellar cells; *pt,* pollen tube; *zn,* zygote nucleus. (From Norstog.[26])

more or less large complexes of meristematic cells (proembryonic mass) are formed before embryolike structures, *embryoids,* begin to develop.[13,16] This type of development resembles that of normal embryos in certain species having the so-called irregular or, more appropriately named, *delayed embryogeny.*[13] A close correspondence between adventitious and normal embryogeny is possible, however, as for example in the formation

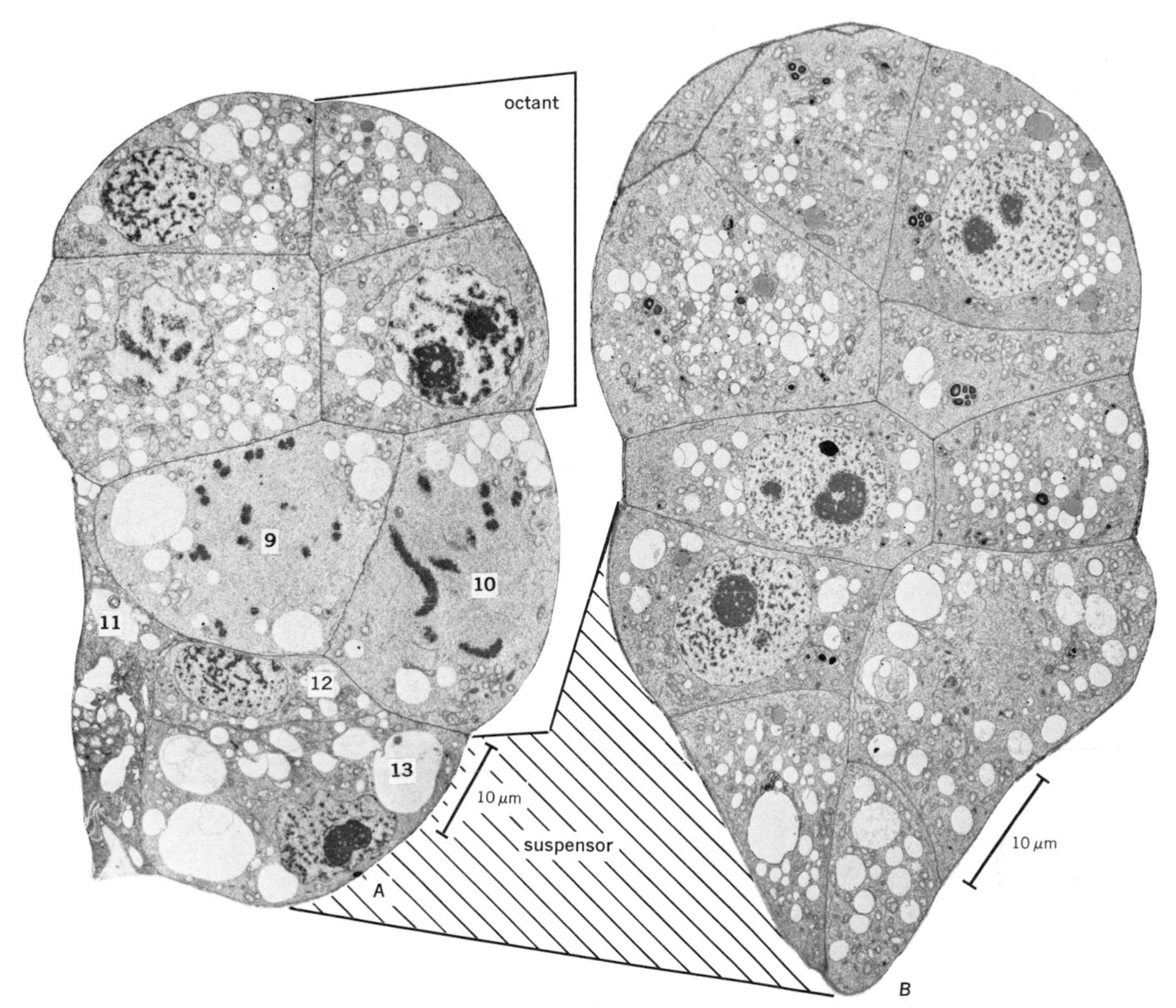

Figure 24.12 Ultrastructure of barley (*Hordeum vulgare*) embryo. Longitudinal sections of embryos composed of 13 (*A*) and 20 (*B*) cells. Actual sizes of embryo profiles: *A*, 45 × 84 μm; *B*, 44 × 82 μm. Cells 9 and 10 in *A* are in prophase. Vacuolation less prominent and organelle distribution more uniform than in the zygote. Irregular cell arrangement and imprecise delimitation of suspensor region. (From Norstog.[26])

of embryoids from epidermal cells of plantlets of *Ranunculus sceleratus* derived from callus cultures.[17]

CLASSIFICATION OF EMBRYOS

Embryos show a wide variation in the derivation of embryo parts from the tiers produced by the initial divisions in the zygote. Research in embryogeny that is concerned with the ontogenetic origin of embryo parts is widely used as a basis for establishing developmental categories of embryos. The major contributions in this direction were made by Souèges[47] and his disciples. Other frequently cited embryologists concerned with typology of embryos are Schnarf,[41] Johansen,[15] and Maheshwari.[20]

A survey of commonly listed categories of early dicotyledonous embryos are given in figure 24.13,*A–E*. (Monocotyledons are fitted into the same categories by comparison.) The foremost distinction among the embryo types is the relative amount of cellular material involved in the formation of embryo proper (shaded in fig. 24.13) as distinguished from the suspensor; that is, whether the embryo arises from the apical cell (*A*,*D*), from the apical cell and part of the basal cell (*B*,*E*), or only from part of the apical cell (*C*). For the second major separation the position of the first wall in the apical cell is used; that is, whether this wall is vertical (*A*,*B*) or horizontal (*C–E*). The two characters combined give five categories designated by the names of families selected to illustrate the particular type of embryogeny. The five categories are further subdivded on the basis of variations in later cell divisions. The resulting complex classification scheme is frequently criticized because of the many deviations from the type that may occur not only within a single large taxon but even in a species or a single plant.

One of the steps proposed to make the scheme simpler and to increase its usefulness is to abandon reference to the position of the first wall in the apical cell.[23] Another suggested change is the addition of the category with delayed embryogeny, that is, embryogeny preceded by the formation of a cell mass by relatively unordered divisions[13] (*Argemone, Fumaria, Tulipa*). The simplified reduced embryogeny series (fig. 24.13,I—III) has the advantage that it is less restrictive and also serves to unify normal embryogeny *in vivo* with adventitious embryogeny occuring

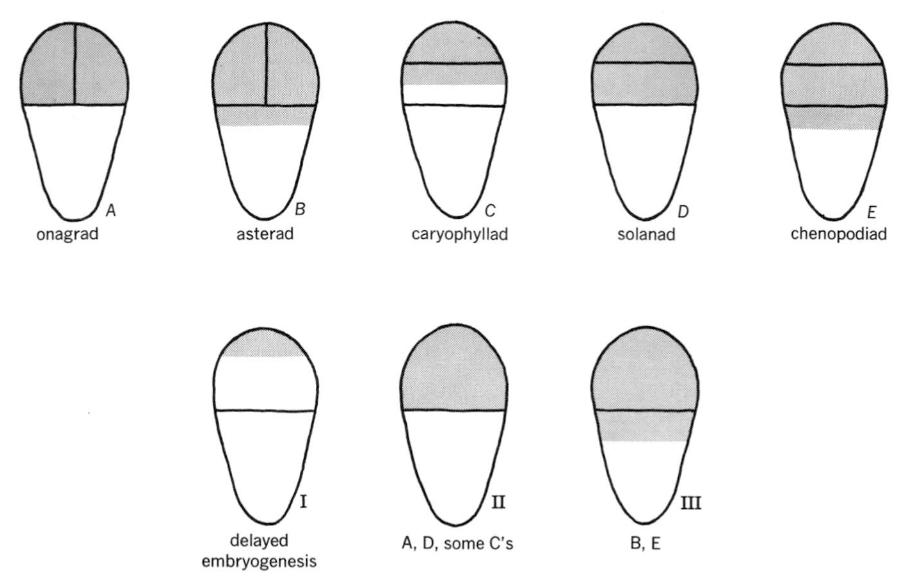

Figure 24.13 Diagrams illustrating classifications of embryos based on embryogeny. *A–E*, earlier classification emphasizing degree of participation of apical and basal cells in the construction of embryo proper (shaded areas) and position of first wall in the apical cell. I-III, later, simplified, classification. (Caryophyllad has affinities with I and II. Adapted from Haccius.[13])

in vivo and *in vitro,* for the development of adventitious embryos fits into the category designated as delayed embryogeny. In recognition of this unity, Haccius[13] proposes that the term embryoid be abandoned and the term embryo used for adventitious embryos of all kinds.

SEEDLING

Germination and seedling establishment

After a more or less prolonged dormancy caused by various internal factors, such as rudimentary or physiologically immature embryos, mechanically resistant or impermeable seed coats, or the presence of germination inhibitors, the seed germinates under appropriate environmental conditions. Germination is essentially a resumption of embryo growth after uptake of water, or imbibition.[54]

During imbibition, the water content of the seed rises, usually fast at first then slower, and the quiescent tissue becomes metabolically active. Enzymes already present are activated and new proteins with specific enzymic activities are synthesized for the digestion and utilization of the different kinds of stored materials (chapter 23). Cell extension and cell division are initiated and proceed according to a programmed pattern. This growth requires a continuous supply of water and nutrients. Before the embryo becomes a self-supporting seedling, it utilizes the food stored in the endosperm and the embryo itself.

The initial transformation of the dry embryo into an actively growing organism is associated with characteristic ultrastructural changes: progressive loss of reserve materials, an enhancement of the definition of cytoplasmic membranes that were rather obscure in the dry state, appearance of certain organelles and membrane systems that seemed to be absent in the dormant seed, and, finally, a start of cell division and differentiation.[49]

Some histochemical and ultrastructural work deals with changes in the cotyledons of germinating embryos. The cotyledons of *Cucurbita maxima,* which change from storage organs to greatly enlarged photosynthesizing leaves, show an incorporation of thymidine-2-^{14}C, and chloroplast development coincides with the time of this incorporation.[18] During the transition from fat degradation to photosynthesis in the cotyledons of *Helianthus annuus,* the enzyme patterns indicate that the glyoxysomal function in microbodies is replaced by peroxisomal function.[9] In the cotyledons of *Pisum sativum,* nuclei increase in size and degree of lobing.[44] In the scutellum of wheat, starch grains develop in the plastids, protein bodies are hydrolysed and the residual vacuoles coalesce, and the epithelial cells elongate to double their original length.[52]

The main event in differentiation of the embryo during germination is the beginning of development of conducting cells in the procambium. The earlier work on this stage of plant vascularization[8] has been expanded by studies of vascular differentiation in cotyledons of germinating pea seed,[45] in the scutellum of wheat before and after germination,[52] in the growing coleoptile of oat and wheat,[29,30,53] and in the growing embryo and young seedling in wheat.[52] The timing of vascular differentiation is related to various physiological events. Metabolism in the cotyledons is activated and controlled by stimuli from the embryonic axis and the movement of these stimuli appears to coincide with the establishment of vascular connection between the axis and the cotyledons.[45] At a certain postgermination stage of cotyledon development in *Cucurbita* mitochondrial respiration

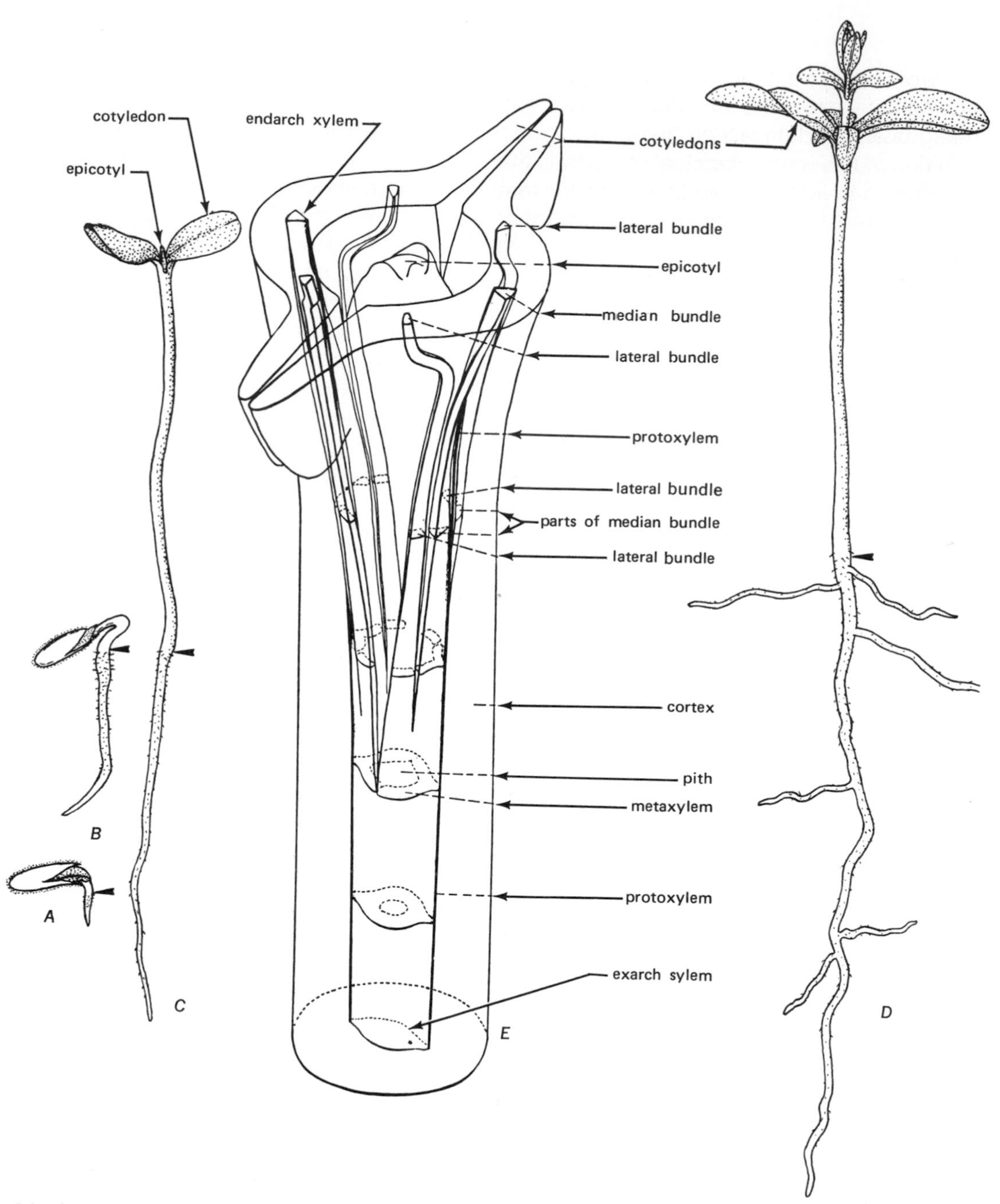

Figure 24.14 Seedling structure in flax (*Linum usitatissimum*). *A–D,* germinating seed and three stages of seedling development showing: growth of taproot (below arrowhead) and appearance of branch roots; elongation of hypocotyl (above arrowhead); unfolding of cotyledons and development of epicotyl. *E,* xylem system of the transition region through which root and cotyledons are connected. Phloem would be on outer periphery of xylem. (*A–D,* drawn by Alva D. Grant; *E,* from D. M. Crooks, *Bot. Gaz.* 95:209–239, 1933.)

activity is concentrated in the veins, possibly in relation to loading of the newly differentiated phloem with photosynthates.[19]

The morphological features of germination and seedling establishment are reviewed in a dicotyledon (*Linum,* flax) and several monocotyledons: onion, asparagus, and grasses. In both, monocotyledons and dicotyledons, germination may be *hypogeous,* with the cotyledons or cotyledon remaining under ground (*Pisum, Zea*), or *epigeous,* with the cotyledons or cotyledon carried above ground by the elongating embryo axis (*Ricinus, Allium*). If the cotyledons appearing above ground are thin, they become photosynthetically active (*Ricinus communis*). Fleshy cotyledons continue to supply the seedling with food until the supply is exhausted and the cotyledons die and are shed (*Phaseolus vulgaris*).

Linum usitatissimum (fig. 24.14) exemplifies a typical epigeous germination in a dicotyledon. The radicle emerges first (*A*) and becomes the main axis of the taproot system. The hypocotyl elongates at its base and is curved (*B*). The resulting tension causes the cotyledons and the enclosing seed coat to be pulled above ground. The hypocotyl straightens out, the seed coat is shed, and the cotyledons separate from one another and expand (*C*). The epicotyl apex starts producing leaves, nodes, and internodes. The first internodes remain short (*D*).

Allium cepa[35,40] (figs. 24.8 and 24.15) is a bulb forming monocotyledon with epigeous germination. An initial elongation of lower and middle regions of the cotyledon forces the root tip and the short hypocotyl to emerge through the seed coat at the micropyle (fig. 24.15,*A*). The cotyledon follows and becomes sharply curved (*B*). The point at the bend serves to penetrate the soil. The haustorial tip of the cotyledon remains within the seed and continues to draw upon the food reserve in the endosperm while the part emerging above ground becomes green and photosynthetically active. The side of the cotyledon still attached to the seed in the soil stops growing while the other side continues to elongate and becomes buckled and taut (*C*). The tension causes the shorter side and the attached seed to be pulled out of the soil. The cotyledon straightens out but the bend remains fixed (*D*). The testa is shed. The part of the cotyledon above the bend dries out from the tip down (*E*). In the meantime, adventitious roots emerge (*B–E*) and finally the first foliage leaf protrudes through the slit in the cotyledon (*E*). The positional relation of the epicotyl to the curvature in the cotyledon is not constant (compare figs. 24.8,*B* and 24.15,*B*).

In contrast to *Allium cepa,* the liliaceous monocotyledon, *Asparagus*[6,36] (fig. 24.16), shows hypogeous germination. The cotyledon elongates at its base and pushes the embryo axis from the seed (*A*). After this step, all the elongation occurs in the root and epicotyl (*B,C*). The latter breaks through the cotyledonary sheath and rises above ground but the cotyledon remains embedded in the endosperm (*D,E*). The epicotyl forms a succession of scales and axillary buds and the hypocotyl produces a short branched rhizome.

Germination in the Poaceae also is hypogeous, for the kernel with the enclosed scutellum remains below the surface of the ground (fig. 24.17). At the start of germination, the coleorhiza elongates and breaks through the pericarp, then the root breaks through the coleorhiza. At the other end of the kernel, the shoot covered by the coleoptile emerges. In *Zea,* this unit is pushed upward by the elongating first internode but in wheat, the shoot is raised mainly by growth of the second internode, that is, the internode above the coleoptilar node.[14]

In the Paniceae, the dispersal unit is derived from an entire floret so that the

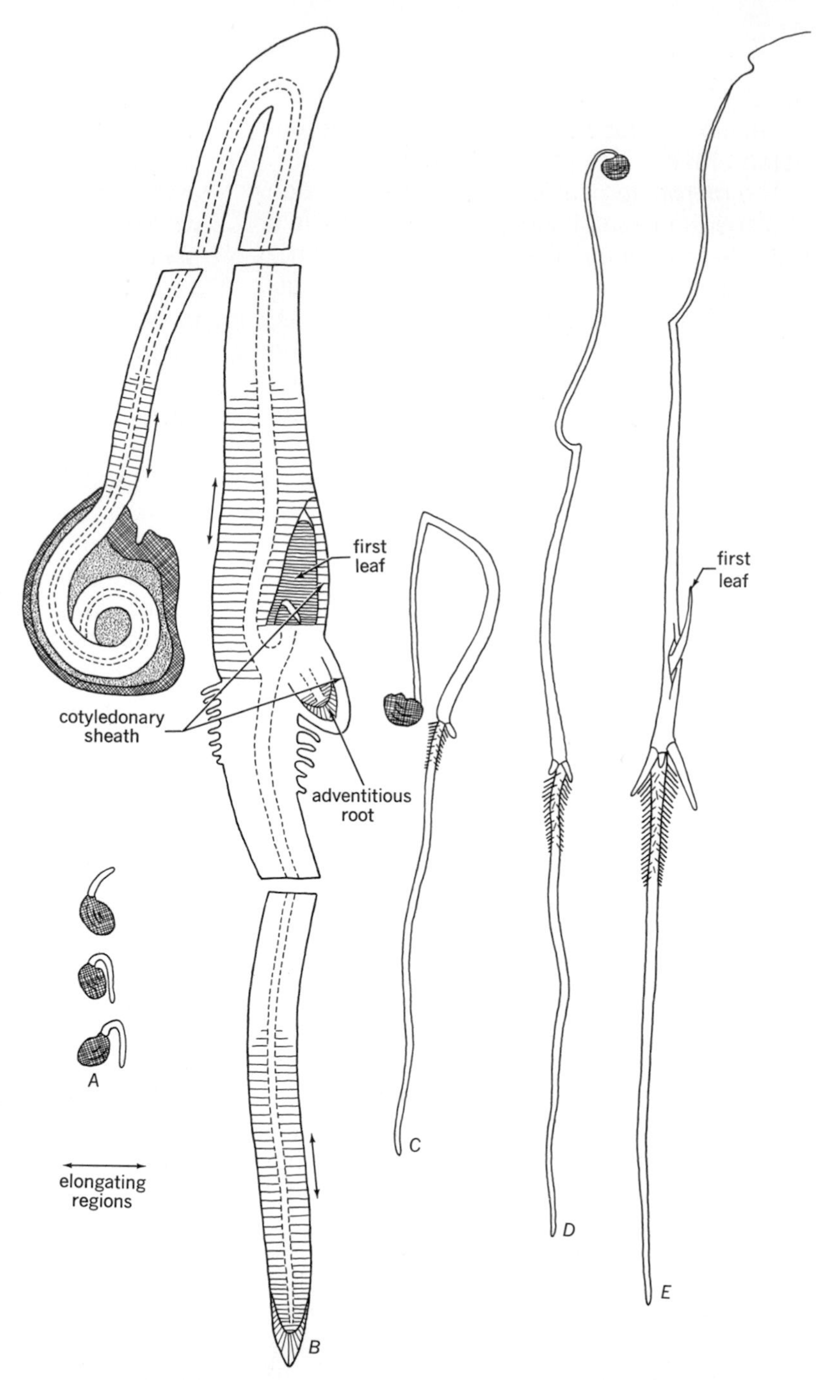

Figure 24.15 Seedling structure in onion (*Allium cepa*). *A*, germinating seed; *B–E*, stages in seedling development. *B*, early elongation of cotyledon and root; seed still under ground; *C*, elongation of cotyledon restricted to the side continuous with the root. *D*, cotyledonary tip and seed above ground. *E*, seed shed and cotyldon tip withered. Details: cross hatched, seed coat; stippled, endosperm. (Adapted from Sachs.[40])

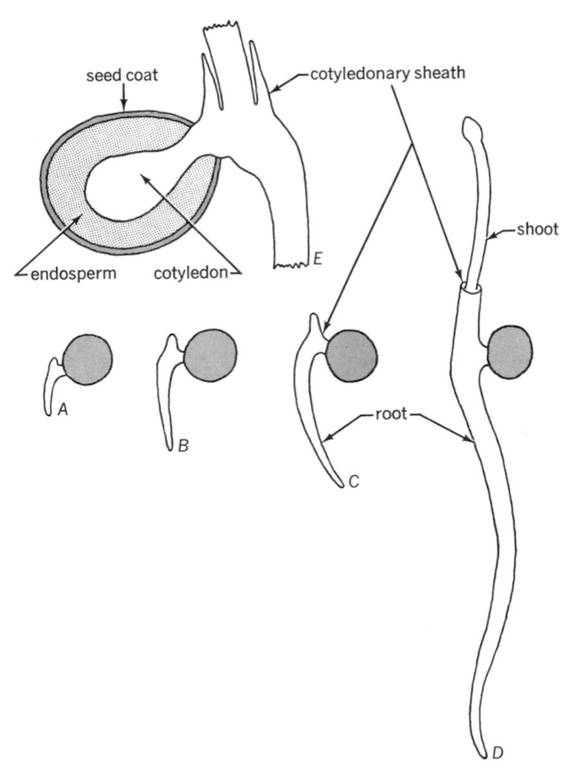

Figure 24.16 Early stages of *Asparagus medeoloides* seedling development. The seed and enclosed cotyledon remain under ground. The shoot has grown beyond the cotyledonary sheath in *D*. (Adapted from Courtot.[6])

caryopsis is enclosed in the lemma and palea of the floret. The lemma has a flaplike hinged structure, about 0.25^2 mm in area, near the micropyle. During germination, the coleorhiza emerges through this "germination lid".[37,39] At the distal end of the dispersal unit, the coleoptile with the enclosed embryonic shoot issues between the palea and the lemma. In the Zizanieae, the dispersal unit is a spikelet and the radicle must split the tough upper glume during germination. The mechanical resistance of the glumes was found to be one of the factors in the dormancy of the seed of *Themeda triandra*.[21]

After emerging, the coleoptile elongates until it reaches the soil surface. It then ceases to grow and the first foliage leaf of the epicotyl issues through the slit near the apex of the coleoptile. Adventitious roots of the first whorl, which are initiated in the embryo above the scutellar node, begin to elongate. Later, adventitious roots arise near six to ten higher nodes on the stem. These crown roots become the main root system of the plant. Buds arising in axils of lower leaves, each associated with adventitious roots, initiate the phenomenon of tillering. Inflorescence formation begins after the typical number of leaves are formed but for the most part are not yet unrolled.

The controversy regarding the nature of the parts of a grass embryo and the associated terminology is carried over into the descriptions of the grass seedling. Figure 24.17 illustrates the useful way to standardize the terminology for grass seedling parts suggested for the *Zea mays* seedling.[31]

Transition region

The embryo initiates the organization of the plant in the arrangement of its partly differentiated meristematic tissues: protoderm, procambium, and ground meristem. The embryo is an axiate structure with a root pole and a shoot pole. Polarity, which may be detected in the cytological organization of the egg, is also revealed in the development of the embryo and continues to be one of the dominant morphogenetic factors in the differentiation of the seedling. The effect of polarity is clearly expressed in changes in structure and physiological activities from one end of the seedling axis to the other. Hence, at the root pole of a seedling the root nature of the axis is manifested in the differentiation of the type of vascular cylinder characteristic of roots. In contrast, the shoot end of the plant shows the typical shoot organization of a higher plant characterized by the close relation between

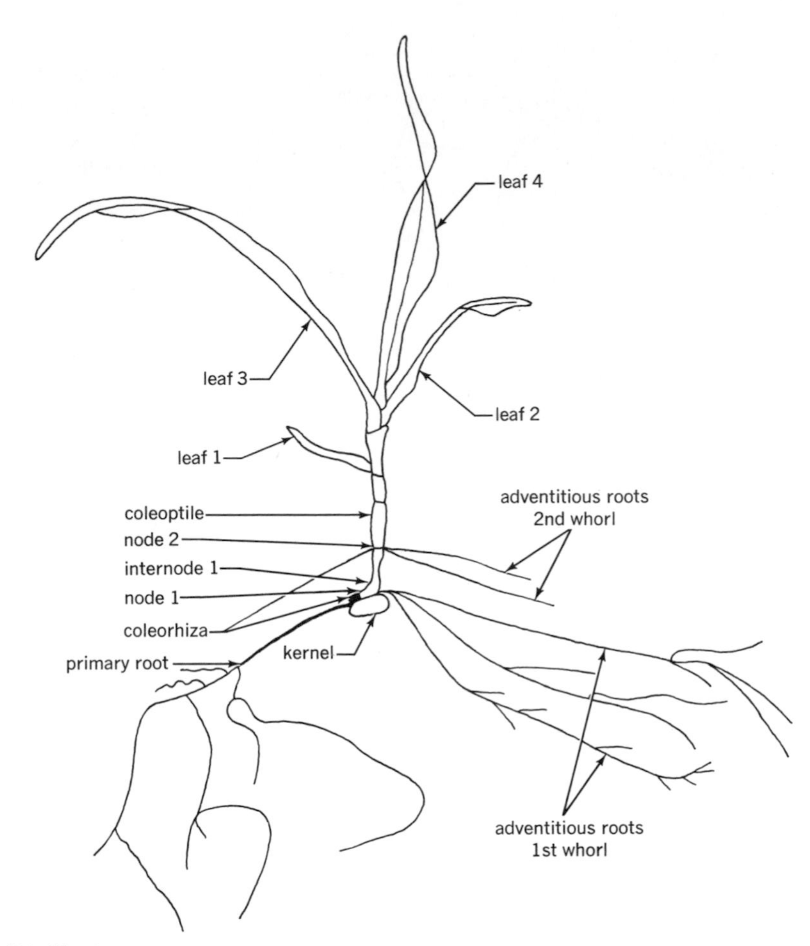

Figure 24.17 Seedling of *Zea mays*. Redrawn and labeled according to J. J. Onderdonk and J. W. Ketcheson, *Can. J. Plant Sci.* 52:1003–1006, 1973. Details: primary root, root that originated within the coleorhiza; first node, scutellar node; first internode, internode between scutellum and coleoptile; second node, coleoptilar node; first leaf, first leaf above the coleoptile.

the vascular system of the axis and the leaves borne on the axis (cotyledons and hypocotyl in a young seedling). As seen in a dicotyledonous seedling without internal phloem (fig. 24.14), the vascular tissue of the hypocotyl is separated above into strands that can be followed into the cotyledonary blades. These strands are leaf traces, cotyledonary traces in this instance.

Between the two levels, those of the shoot and the root, a connection exists between the cylindrical vascular system of the root and the system of strands of the upper hypocotyl (fig. 24.14, *E*). If one studies this connection level by level, beginning with the root for convenience, one gains the impression that root structure gradually changes into shoot structure. The compact cylinder of the root is re-

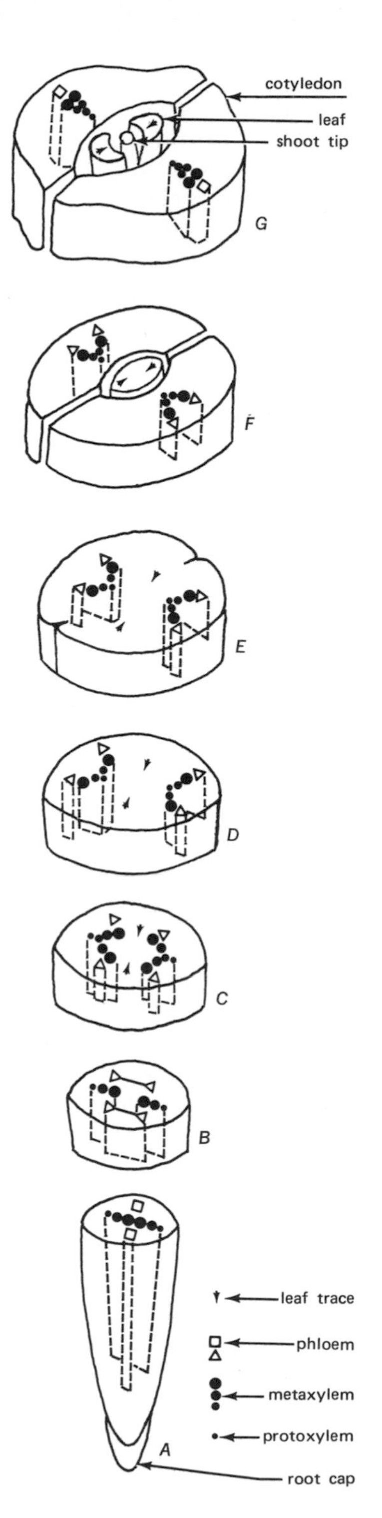

placed by a less compact one higher up. If no pith is present in the root, a pith may be evident at a higher level. Still higher, the vascular tissue appears to separate into two or more units, the cotyledonary traces.

In addition to these simple differences in the general form of the vascular system between one level and another, complex changes in the direction of differentiation of the xylem elements are observable at the successive levels of the seedling (fig. 24.18). The protoxylem elements in the root occur in peripheral positions in the vascular cylinder. The subsequently maturing metaxylem elements appear successively closer to the center. Thus, as seen in transection, the direction of maturation of the xylem elements is centripetal: the xylem is exarch (*A*). In the cotyledonary bundles (or first in the epicotyl in some plants) the order of xylem maturation is reversed. The protoxylem elements are located farthest from the periphery, and the metaxylem elements mature in the centrifugal direction: the xylem is endarch (*G*). The connection between the exarch xylem of the root and the endarch xylem of the shoot occurs through a part of the vascular system in which the relative position of the early and late xylem elements is intermediate between those in the shoot and the root (*B–F*).

The phloem differentiates centripetally at all levels but whereas in the root it has a radially alternate arrangement with the xylem (fig. 24.18,*A*), in the shoot it is associated with

Figure 24.18 Diagrams illustrating vascular connection between root and cotyledons in a dicotyledon seedling (*Beta vulgaris*). The root is diarch (*A*). The primary vascular system of the root is continuous with that of the cotyledons (*G*) through the hypocotyl. Between the two levels (root and cotyledons) the spatial relation between xylem and phloem and the direction of differentiation of the xylem are gradually changed. (Adapted from K. Esau, *Plant Anatomy,* 2nd ed. John Wiley & Sons, 1965.)

the xylem in the form of collateral vascular bundles (in plants without internal phloem; fig. 24.18,G). The reconciliation between the two types of arrangement occurs in connection with the exchange between exarchy and endarchy in the xylem.

The region of the seedling where the root and shoot systems are connected and where the structural details change from level to level in relation to the differences between the two systems is called the *transition region.* The change in pattern in the successive levels of the transition region is gradual and suggests, therefore, that graded influences from the root pole and the shoot pole are responsible for the development of the particular pattern. The developing seedling thus manifests the existence of gradients in its differentiation, another common morphogenetic factor.

The foregoing discussion has dealt with a relatively simple type of transition region in a dicotyledon. Many dicotyledons have more complex root and shoot connections, and in the monocotyledons the presence of only one cotyledon is associated with an asymmetrical structure of the transition region.[3,14] In the gymnosperms the frequent presence of more than two cotyledons contributes to the complexity of the transition region.[3]

REFERENCES

1. Avery, G. S., Jr. Comparative anatomy and morphology of embryos and seedlings of maize, oats, and wheat. *Bot. Gaz.* 89:1–39. 1930.
2. Briggs, D. E. Enzyme formation, cellular breakdown and the distribution of gibberellins in the endosperm of barley. *Planta* 108:351–358. 1972.
3. Boureau, E. *Anatomie végétale.* Vol. 1. Paris, Presses Universitaires de France. 1954.
4. Brown, W. V. The morphology of the grass embryo. *Phytomorphology* 10: 215–223. 1960.
5. Brown, W. V. The grass embryo—a rebuttal. *Phytomorphology* 15:274–284. 1965.
6. Courtot, Y. Recherches sur la morphologie, l'anatomie, la croissance et les movements révolutifs de deux Liliacées: *Bulbine annua* et *Asparagus medeoloides. Ann. Sci. Univ. Besançon. Sér. 3, Bot.* Fasc. 1:1–98, 1966.
7. Crété, P. Embryo. Chapter 7, pp. 171–220. In: *Recent advances in the embryology of angiosperms.* P. Maheshwari, ed. International Society of Plant Morphologists, University of Delhi. Ranchi, India, The Catholic Press. 1963.
8. Esau, K. *Vascular differentiation in plants.* New York, Holt, Rinehart and Winston. 1965.
9. Gerhardt, B. Untersuchungen zur Funktionsänderung der Microbodies in den Keimblättern von *Helianthus annuus* L. *Planta* 110:15–28. 1973.
10. Guignard, J.-L. Recherches sur l'embryogénie des Graminées; rapports des Graminées avec les autres Monocotylédones. *Ann. Sci. Nat., Bot. Sér.* 12. 2:491–610. 1961.
11. Guignard, J.-L., and J.-Ch. Mestre. L'embryon des Graminées. *Phytomorphology* 20:190–197. 1970.
12. Guttenberg, H. von. Der primäre Bau der Angiospermenwurzel. *Handbuch der Pflanzenanatomie.* Band 8. Teil 5. 1968.
13. Haccius, B. Zur derzeitigen Situation der Angiospermen-Embryologie. *Bot. Jb.* 91:309–329. 1971.
14. Hayward, H. E. *The structure of economic plants.* New York, Macmillan. 1938.
15. Johansen, D. A. *Plant embryology.* Waltham, Mass., Chronica Botanica Company. 1950.

16. Jones, L. H. Factors influencing embryogenesis in carrot cultures (*Daucus carota* L.). *Ann. Bot.* 38:1077–1088. 1974.
17. Konar, R. N., E. Thomas, and H. E. Street. Origin and structure of embryoids arising from epidermal cells of the stem of *Ranunculus sceleratus* L. *J. Cell Sci.* 11:77–93. 1972.
18. Lott, J. N. A. Changes in the cotyledons of *Cucurbita maxima* during germination. I. General characteristics. *Can. J. Bot.* 48:2227–2231. 1970. III. Plastids and chlorophylls. *Can. J. Bot.* 48:2259–2265. 1970.
19. Lott, J. N. A., and P. Castelfranco. Changes in the cotyledons of *Cucurbita maxima* during germination. II. Development of mitochondrial function. *Can. J. Bot.* 48:2233–2240. 1970.
20. Maheshwari, P. *An introduction to the embryology of angiosperms.* New York, McGraw-Hill. 1950.
21. Martin, C. C. The role of glumes and gibberellic acid in dormancy of *Themeda triandra* spikelets. *Physiol. Plant.* 33:171–176. 1975.
22. Merry, J. Studies on the embryo of *Hordeum sativum*—I. The development of the embryo. *Bull. Torrey Bot. Club* 68:585–598. 1941.
23. Mestre, J.-G. La signification phylogénétique de l'embryogénie. *Rev. Gén. Bot.* 74:273–322. 1967.
24. Newcomb, W., and L. C. Fowke. *Stellaria media* embryogenesis: the development and ultrastructure of the suspensor. *Can. J. Bot* 52:607–614. 1974.
25. Nomura, T., Y. Kono, and T. Akazawa. Enzymic mechanism of starch breakdown in germinating rice seeds. II. Scutellum as the site of sucrose synthesis. *Plant Physiol.* 44:765–769. 1969.
26. Norstog, K. Early development of the barley embryo: fine structure. *Amer. J. Bot.* 59:123–132. 1972.
27. Norstog, K. Nucellus during early embryogeny in barley: fine structure. *Bot. Gaz.* 135:97–103. 1974.
28. O'Brien, T. P., and K. V. Thimann. Histological studies of the coleoptile. I. Tissue and cell types in the coleoptile tip. *Amer. J. Bot.* 52:910–918. 1965.
29. O'Brien, T. P., and K. V. Thimann. Observations on the fine structure of the oat coleoptile. III. Correlated light and electron microscopy of the vascular tissues. *Protoplasma* 63:443–478. 1967.
30. O'Brien, T. P., S. Zee, and J. G. Swift. The occurrence of transfer cells in the vascular tissues of the coleoptilar node of wheat. *Aust. J. Biol. Sci.* 23:709–712. 1970.
31. Onderdonk, J. J., and J. W. Ketcheson. A standardization of terminology for the morphological description of corn seedlings. *Can. J. Plant Sci.* 52:1003–1006. 1972.
32. Paulson, R. E., and L. M. Srivastava. The fine structure of the embryo of *Lactuca sativa.* I. Dry embryo. *Can. J. Bot.* 46:1437–1445. 1968.
33. Raghavan, V. Nutrition, growth and morphogenesis of plant embryos. *Biol. Rev.* 41:1–58. 1966.
34. Randolph, L. F. Developmental morphology of the caryopsis in maize. *J. Agric. Res.* 53:881–916. 1936.
35. Rauh, W. *Morphologie der Nutzpflanzen.* Heidelberg, Quelle & Meyer. 1950.
36. Rivière, S. Les activités méristématiques durant l'ontogenèse d'une plantule de Monocotylédone à germination hypogée: l'*Asparagus officinalis* L. (Liliacées). *C. R. Acad. Sci. Paris D.* 277:293–296. 1973.
37. Rost, T. L. The morphology of germination in *Setaria lutescens* (Gramineae): the effects of covering structures and

chemical inhibitors on dormant and non-dormant florets. *Ann. Bot.* 39:21–30. 1975.

38. Rost, T. L., and N. R. Lersten. A synopsis and selected bibliography of grass caryopsis anatomy and fine structure. *Iowa State J. Res.* 48:47–87. 1973.
39. Rost, T. L., and A. D. Simper. The germination lid: a characteristic of the lemma in the Paniceae. *Madroño* 23:68–72. 1975.
40. Sachs, J. Ueber die Keimung des Samens von *Allium cepa. Bot. Z.* 21:57–62; 65–70. 1863.
41. Schnarf, K. *Vergleichende Embryologie der Angiospermen.* Berlin, Gebrüder Borntraeger. 1931.
42. Schulz, S. R., and W. A. Jensen. *Capsella* embryogenesis: the egg, zygote, and young embryo. *Amer. J. Bot.* 55:807–819. 1968. *Capsella* embryogenesis: the early embryo. *J. Ultrastruct. Res.* 22:376–392. 1968.
43. Schulz, S. P., and W. A. Jensen. *Capsella* embryogenesis: the suspensor and the basal cell. *Protoplasma* 67:138–163. 1969.
44. Smith, D. L. Nuclear changes in the cotyledons of *Pisum arvense* L. during germination. *Ann. Bot.* 35:511–521. 1971.
45. Smith, D. L., and A. M. Flinn. Histology and histochemistry of the cotyledons of *Pisum arvense* L. during germination. *Planta* 74:72–85. 1967.
46. Souèges, R. Embryogénie des Graminées. Développement de l'embryon chez le *Poa annua* L. *C. R. Acad. Sci. Paris* 178:860–862. 1924.
47. Souèges, R. *Embryogénie et classification.* Paris, Hermann. 1938, 1939, 1948, 1951.
48. Souèges, R. L'origine du cône végétatif de la tige et la question de la "terminalité" du cotylédon des Monocotylédones. *Ann. Sci. Nat., Bot. Sér.* 11. 15:1–20. 1954.
49. Srivastava, L. M., and R. E. Paulson. The fine structure of the embryo of *Lactuca sativa.* II. Changes during germination. *Can. J. Bot.* 46:1447–1453. 1968.
50. Steffen, K. Die Embryoentwicklung von *Impatiens glanduligera* Lindl. *Flora* 139:394–461. 1952.
51. Stoddart, J. L., H. Thomas, and A. Robertson. Protein synthesis patterns in barley embryos during germination. *Planta* 112:309–321. 1973.
52. Swift, J. G., and T. P. O'Brien. Vascularization of the scutellum of wheat. *Aust. J. Bot.* 18:45–53. 1970. Vascular differentiation in the wheat embryo. *Aust. J. Bot.* 19:63–71. 1971. The fine structure of wheat scutellum before germination. *Aust. J. Biol. Sci.* 25:9–22. 1972. The fine structure of the wheat scutellum during germination. *Aust. J. Biol. Sci.* 25:469–486. 1972.
53. Thimann, K. V., and T. P. O'Brien. Histological studies of the coleoptile. II. Comparative vascular anatomy of coleoptiles of *Avena* and *Triticum. Amer. J. Bot.* 52:918–923. 1965.
54. Torrey, J. G. *Development in flowering plants.* New York, Macmillan. 1967.
55. Wardlaw, C. W. *Embryogenesis in plants.* New York, John Wiley & Sons. 1955.

Glossary

This glossary defines selected terms used in the language of plant anatomy. A few cytological and morphological terms are also explained, mostly those that are apt to be forgotten after an introductory course in botany, or those that have several definitions. Although a uniformity in the use of terms is desirable, some terms are bound to change in meaning with the increase in our understanding of the phenomena they indicate. Definitions in this glossary that may deviate from those commonly given for the terms in question will inform the reader how these terms are used in the present book. The Latin and Greek origins of the terms may be looked up in *Jackson's Glossary* (reference in chapter 1).

Abaxial. Directed away from the axis. Opposite of *adaxial.*

Abscission. The shedding of leaves, flowers, fruits, or other plant parts, usually after formation of an *abscission zone.*

Abscission layer. In abscission zone; layer of cells the disjunction or breakdown of which causes the shedding of a plant part. Other term, *separation layer.*

Abscission zone. Zone at base of leaf, flower, fruit, or other plant part that contains an *abscission* (or *separation*) *layer* and a *protective layer,* both involved in the abscission of the plant part.

Accessory bud. A bud located above or on either side of the main axillary bud.

Accessory cell. See *subsidiary cell.*

Accessory parts in fruit. Parts not derived from the ovary but associated with it in fruit.

Accessory transfusion tissue. *Transfusion tissue* located within the mesophyll rather than associated with vascular bundle. In leaves of certain gymnosperms.

Acicular crystal. Needle-shaped crystal.

Acropetal development (or differentiation). Produced or becoming differentiated in a succession toward the apex of an organ. The opposite of *basipetal* but means the same as *basifugal.*

Actinomorphic. Refers to a flower that can be divided in two equal parts in more than one longitudinal plane; radially symmetrical or *regular flower.* Opposite of *zygomorphic.*

Actinostele. *Protostele* with star-shaped outline in transection.

Adaxial. Directed toward the axis. Opposite of *abaxial.*

Adaxial meristem. Meristematic tissue on the adaxial side of a young leaf that contributes to the increase in thickness of the petiole and midrib.

Adnation. In a flower; union of members of different whorls, as stamens and petals.

Adventitious. Refers to structures arising not at their usual sites, as roots originating on stems or leaves instead of on other roots, buds developing on leaves or roots instead of in leaf axils on shoots.

Aerenchyma. Parenchyma tissue containing particularly large intercellular spaces of *schizogenous, lysigenous,* or *rhexigenous* origin.

Aggregate fruit. A fruit developing from a single *gynoecium* (single flower) composed of separate carpels, as the strawberry or raspberry fruits.

Aggregate ray. In secondary vascular tissues; a

group of small rays arranged so as to appear to be one large ray.

Albedo. White tissue of the rind in citrus fruit.

Albuminous cells. In gymnosperm phloem; certain ray and phloem-parenchyma cells spatially and functionally associated with the sieve elements, thus resembling the companion cells of angiosperms but usually not originating from the same precursory cells as the sieve elements. Also called *Strasburger cells.*

Albuminous seed. A seed that contains endosperm in mature state.

Aleurone. Granules of protein (aleurone grains) present in seeds, usually restricted to the outermost layer, the *aleurone layer* of the endosperm. (Protein bodies is the preferred term for aleurone grains.)

Aleurone layer. Outermost layer of endosperm in cereals and many other taxa that contains protein bodies and enzymes concerned with endosperm digestion.

Aliform paratracheal parenchyma. In secondary xylem; vasicentric groups of axial parenchyma cells having tangential winglike extensions as seen in transections. See also *paratracheal parenchyma* and *vasicentric paratracheal parenchyma.*

Alternate pitting. In tracheary elements; pits in diagonal rows.

Amoeboid tapetum. In anther locules; tapetum assuming amoeboid form when it disintegrates during pollen wall development.

Amphicribral vascular bundle. Concentric vascular bundle in which the phloem surrounds the xylem.

Amphiphloic siphonostele. A stele in which the vascular system appears as a tube and has phloem both externally and internally from the xylem.

Amphivasal vascular bundle. Concentric vascular bundle in which the xylem surrounds the phloem.

Amyloplast. A colorless plastid (*leucoplast*) that forms starch grains.

Analogy. Means having the same function as, but a different phylogenetic origin than another entity.

Anastomosis. Refers to cells or strands of cells that are interconnected with one another as, for example, the veins in a leaf.

Anatomy. The study of structure.

Androecium. Collective term for the stamens in a flower of a seed plant; part of the flower in which male gametogenesis is initiated or also carried to completion.

Angiosperm. A plant taxon the seed of which is borne within a matured ovary (fruit).

Angstrom (originally Ångström). A unit of length equal to one tenth of a millimicron (mμ), or one tenth of a nanometer (nm). Symbol A or Å.

Angular collenchyma. A form of collenchyma in which the primary wall thickening is most prominent in the angles where several cells are joined.

Anisocytic stoma. A stomatal complex in which three subsidiary cells, one distinctly smaller than the other two, surround the stoma.

Anisotropic. Having different properties along different axes; optical anisotropy causes polarization and double refraction of light.

Annual ring. In secondary xylem; growth ring formed during one season. The term is deprecated because more than one growth ring may be formed during a single year.

Annular cell wall thickening. In tracheary elements of the xylem; secondary wall deposited in the form of rings.

Anomalous secondary growth. A term of convenience referring to types of secondary growth that differ from the more familiar ones.

Anomocytic stoma. A stoma without subsidiary cells.

Anther. The pollen-bearing part of the stamen.

Anthesis. The time of full expansion of flower; from development of receptive stigma to fertilization.

Anthocyanin. A water-soluble blue, purple, or red flavonoid pigment occurring in the cell sap.

Anticlinal. Commonly refers to orientation of cell wall or plane of cell division; perpendicular to the nearest surface. Opposite of *periclinal.*

Antipodals. Three or more cells located at the chalazal end of the mature embryo sac in angiosperms.

Aperture in pollen grain. A depressed region in the wall in which thick intine is covered by thin exine; the pollen grain emerges through the aperture.

Apex (pl. apices), or summit. Tip, topmost part, pointed end of anything. In shoot or root the tip containing the apical meristem.

Apical cell. The single cell that occupies the distal position in an apical meristem of root or shoot and is usually interpreted as the initial cell in the apical meristem.

Apical meristem. A group of meristematic cells at the apex of root or shoot that by cell division produce the precursors of the primary tissues of root or shoot; may be *vegetative,* initiating vegetative tissues and organs, or *reproductive,* initiating reproductive tissues and organs.

Apocarpy. Condition in flower characterized by lack of union of carpels (free carpels).

Apomixis. Reproduction by embryo that is formed without meiosis and/or fusion of gametes.

Apotracheal parenchyma. In secondary xylem; axial parenchyma typically independent of the vessels (pores). Includes *boundary* (*initial* or *terminal*), *banded,* and *diffuse* apotracheal parenchyma.

Apposition. Growth of cell wall by successive deposition of wall material, layer upon layer. Opposite of *intussusception.*

Areole. A small area of mesophyll in a leaf delimited by intersecting veins.

Aril. A fleshy outgrowth enveloping the seed and usually arising at the base of the ovule.

Articulated laticifer. Laticifer composed of more than one cell with common walls intact or partly or entirely removed; anastomosing or nonanastomosing. *Compound laticifer.*

Aspirated pit. In gymnosperm wood; bordered pit in which the pit membrane is laterally displaced and the torus blocks the aperture.

Astrosclereid. A branched, or ramified, type of sclereid.

Atactostele. A stele with the vascular bundles as though scattered within the ground tissue.

Axial organ. Root, stem, inflorescence, or flower axis without its appendages.

Axial parenchyma. Parenchyma cells in the axial system of secondary vascular tissues; as contrasted with ray parenchyma cells.

Axial system. All secondary vascular cells derived from the fusiform cambial initials and oriented with their longest diameter parallel with the main axis of stem or root. Other terms: *vertical system* and *longitudinal system.*

Axial tracheid. Tracheid in the axial system of secondary xylem; as contrasted with ray tracheid.

Axil. The upper angle between a stem and a twig or a leaf.

Axillary bud. Bud in the axil of a leaf.

Axillary meristem. Meristem located in the axil of a leaf and giving rise to an axillary bud.

Banded apotracheal parenchyma. In secondary xylem; axial parenchyma in concentric bands as seen in transection, typically independent of vessels (pores). See also *apotracheal parenchyma.*

Bark. A nontechnical term applied to all tissues outside the vascular cambium or the xylem; in older trees may be divided into dead outer bark and living inner bark, which consists of secondary phloem. See also *rhytidome.*

Bars of Sanio. See *crassulae.*

Basifugal development. See *acropetal development.*

Basipetal development (or differentiation). Produced or becoming differentiated in a succession toward the base of an organ. The opposite of *acropetal* and *basifugal.*

Bast fiber. Originally phloem fiber, now any extraxylary fiber.

Bicollateral vascular bundle. A bundle having phloem on two sides of the xylem.

Bifacial leaf. A leaf with palisade parenchyma on one side of the blade and spongy parenchyma on the other. A *dorsiventral leaf.* Conceived ontogenetically, a leaf that develops continuously from the original leaf primordium apex and includes tissues from both adaxial and abaxial sides of the primordium. Compare with *unifacial* leaf.

Bilateral symmetry. Refers to a flower having two corresponding or complementary sides and which thus can be divided, by a single longitudinal plane through the floral axis, in two halves that are mirror images of one another. Contrasted with *radial symmetry.*

Biseriate ray. A ray in secondary vascular tissue two cells in width.

Blind pit. A pit without a complementary pit in an

adjacent wall, which may face a lumen of a cell or an intercellular space.

Bordered pit. A pit in which the secondary wall overarches the pit membrane.

Bordered pit-pair. An intercellular pairing of bordered pits.

Boundary apotracheal parenchyma. In secondary xylem; axial parenchyma cells occurring either singly or in a layer at the beginning of a season's growth (*initial*) or at the close of one (*terminal*). See also *apotracheal parenchyma.*

Brachysclereid. A short, roughly isodiametric sclereid, resembling a parenchyma cell in shape; a stone cell.

Branch gap. In the nodal region of a stem; a region of parenchyma in the vascular cylinder of the stem located where the branch traces are bent toward the branch. Usually confluent with the gap of the leaf subtending the branch.

Branch root. A root arising from another, older root; also called secondary root if the older root is the primary root, or taproot.

Branch traces. Vascular bundles connecting the vascular tissue of the branch and that of the main stem. They are leaf traces of the first leaves (prophylls) on the branch.

Branched pit. See *ramiform pit.*

Bulliform cell. An enlarged epidermal cell present, with other similar cells, in longitudinal rows in leaves of grasses. Also called *motor cell* because of its presumed participation in the mechanism of rolling and unrolling of leaves.

Bundle cap. Sclerenchyma or collenchymatous parenchyma appearing like a cap on the xylem and/or phloem side of a vascular bundle as seen in transection.

Bundle sheath. Layer or layers of cells enclosing a vascular bundle in a leaf; may consist of parenchyma or sclerenchyma.

Bundle sheath extension. A plate of ground tissue extending from a bundle sheath to the epidermis in a leaf; may be present on one or on both sides of the bundle and may consist of parenchyma or sclerenchyma.

Callose. A polysaccharide, β-1,3 glucan, yielding glucose on hydrolysis. Common wall constituent in the sieve areas of sieve elements; also develops rapidly in reaction to injury in sieve elements and parenchyma cells.

Callus. A tissue composed of large thin-walled cells developing as a result of injury, as in wound healing or grafting, and in tissue culture. (The use of callus for accumulations of callose on sieve areas is deprecated.)

Callus tissue. See *callus.*

Calyptrogen. In root apex; meristem giving rise to the rootcap independently of the initials of cortex and central cylinder.

Calyx. A collective term for the sepals.

Cambial initials. Cells so localized in the vascular cambium or phellogen that their periclinal divisions can contribute cells either to the outside or to the inside of the axis; in vascular cambium, classified into *fusiform initials* (source of axial cells of xylem and phloem) and *ray initials* (source of the ray cells).

Cambium. A meristem with products of periclinal divisions commonly contributed in two directions and arranged in radial files. Term preferably applied only to the two lateral meristems, the *vascular cambium* and the *cork cambium,* or *phellogen.*

Carpel. Leaflike organ in angiosperms producing one or more ovules; a constituent of the *gynoecium.*

Cartilaginous. Like cartilage; a firm elastic material or tissue, translucent in color.

Caruncle. A fleshy protuberance near the hilum of a seed.

Casparian strip, or band. A bandlike wall formation within primary walls that contains suberin and lignin; typical of endodermal cells in roots, in which it occurs in radial and transverse anticlinal walls.

Cataphylls. Leaves inserted at low levels of plant or shoot, as bud scales, rhizome scales, and others. Contrasted with *hypsophylls.*

Cauline. Belonging to the stem or arising from it.

Caulis. Stem.

Cell. Structural and physiological unit of a living organism. The plant cell consists of protoplast and cell wall; in nonliving state, of cell wall only, or cell wall and some nonliving inclusions.

Cell membrane. A translation of the German Zellmembran which refers to the cell wall in that language.

Cell plate. A partition appearing at telophase between the two nuclei formed during mitosis (and some meioses) and indicating the early stage of the division of a cell (*cytokinesis*) by means of a new cell wall; is formed in the *phragmoplast.*

Cell wall. A more or less rigid membrane enclosing the protoplast of a cell and, in higher plants, composed of cellulose and other organic and inorganic substances.

Cellulose. A polysaccharide, β-1,4 glucan—the main component of cell walls in most plants; consists of long chainlike molecules the basic units of which are anhydrous glucose residues of the formula $C_6H_{10}O_5$.

Central cylinder. A term of convenience applied to the vascular tissues and associated ground tissue in stem and root. Refers to the same part of stem and root that is designated *stele.*

Central mother cells. Rather large vacuolated cells in subsurface position in apical meristem of shoot in gymnosperms.

Centric mesophyll. A modification of isobilateral mesophyll in which the adaxial and abaxial palisade layers form a continuous layer; found in narrow or cylindrical leaves.

Centrifugal development. Produced or developing successively farther away from the center.

Centripetal development. Produced or developing successively closer to the center.

Chalaza. Region in the ovule opposite the micropyle where the integuments and the nucellus merge with the funiculus.

Chimera. A plant consisting of a combination of tissues of different genetic composition. In periclinal chimera, cells of different composition are arranged in periclinal layers.

Chlorenchyma. Parenchyma tissue containing chloroplasts; leaf mesophyll and other green parenchyma.

Chloroplast. A chlorophyll-containing plastid with thylakoids organized into grana and frets, or stroma thylakoids, and embedded in a stroma.

Chromoplast. A plastid containing pigments other than chlorophyll, usually yellow and orange carotenoid pigments.

Cicatrice. The scar left by a wound or by the separation of one plant part from another (as a leaf from a stem) and characterized by substances protecting the exposed surface.

Circular bordered pit. A bordered pit with circular aperture.

Cisterna (pl. cisternae). A flattened, saclike membranous compartment as in endoplasmic reticulum, dictyosome, or thylakoid.

Cladophyll. A branch resembling a foliage leaf.

Closed vascular bundle. A bundle forming no vascular cambium.

Closed venation. Leaf venation characterized by anastomosing veins.

Closing layer. One of the compact layers of cells formed periodically in alternation with the loose filling tissue in a lenticel.

Coenocyte. An aggregation of protoplasmic units; a multinucleate structure; sometimes applied to multinucleate cells in seed plants.

Cohesion. In a flower; union of members of the same whorl, as sepals with sepals and petals with petals.

Coleoptile. The sheath enclosing the epicotyl in the embryo of Poaceae; sometimes interpreted as the first leaf of the epicotyl.

Coleorhiza. The sheath enclosing the radicle of the embryo in Poaceae.

Collateral vascular bundle. A bundle having phloem only on one side of the xylem usually the abaxial side.

Collenchyma. A supporting tissue composed of more or less elongated living cells with unevenly thickened nonlignified primary walls. Common in regions of primary growth in stems and leaves.

Colleter. A multicellular appendage or a multicellular trichome producing a sticky secretion. Found on buds of many woody species.

Columella. The central part of a rootcap in which the cells are arranged in longitudinal files.

Commissural vascular bundle. A small bundle interconnecting larger parallel bundles as in leaves of grasses.

Companion cell. A parenchyma cell in the phloem of an agiosperm associated with a sieve tube member and originating jointly with the latter from the same mother cell.

Compitum. A region in the style of a syncarpous gynoecium where stylar canals are joined into one cavity and where the pollen tubes can change their direction of growth from one carpel to another.

Complementary tissue. See *filling tissue.*

Complete flower. A flower having all types of floral parts: sepals, petals, stamens, and carpels, or tepals, stamens, and carpels.

Compound laticifer. See *articulated laticifer.*

Compound middle lamella. A collective term applied to two primary walls and middle lamella; usually used when the true middle lamella is not distinguishable from the primary walls. May also include the earliest secondary wall layers.

Compound sieve plate. A sieve plate composed of several sieve areas in either scalariform or reticulate arrangement.

Compression wood. The reaction wood in conifers which is formed on the lower sides of branches and leaning or crooked stems and characterized by dense structure, strong lignification, and certain other features.

Concentric vascular bundle. A vascular bundle with either the phloem surrounding the xylem (*amphicribral*) or the xylem surrounding the phloem (*amphivasal*).

Conducting tissue. See *vascular tissue.*

Confluent paratracheal parenchyma. In secondary xylem; coalesced aliform groups of axial parenchyma cells forming irregular tangential or diagonal bands, as seen in transection. See also *paratracheal parenchyma* and *alirfom paratracheal parenchyma.*

Conjunctive tissue. A secondary parenchymatic tissue interspersed with vascular tissue where the latter does not form a solid cylinder; as in monocotyledons and in dicotyledons with anomalous secondary growth.

Connate. Refers to united parts of the same whorl in a flower, as petals united into a corolla tube. See also *cohesion.*

Connective. The tissue between the two lobes of an anther.

Contact cell. An axial parenchyma or a ray cell physiologically associated with a tracheary element. Analogous to companion cell in the phloem. Also a cell next to a stoma.

Contractile root. A root that undergoes contraction some time during its development and thereby effects a change in position of the shoot with regard to the ground.

Convergent evolution. See *parallel evolution.*

Coordinated growth. Growth of cells in a manner that involves no separation of walls, as opposed to *intrusive growth;* also called *symplastic growth.*

Copal. A resinous substance exuding from various tropical trees and hardening in air into roundish or irregular pieces, colorless, yellowish, reddish, or brown.

Cork. See *phellem.*

Cork cambium. See *phellogen.*

Cork cell. A phellem cell derived from the phellogen, nonliving at maturity, and having suberized walls; protective in function because the walls are highly impervious to water.

Corolla. A collective term for the petals of a flower.

Corolla tube. The tubelike part of a corolla resulting from congenital or ontogenetic union of petals.

Corpus. The core in an apical meristem covered by the tunica and showing volume growth by divisions of cells in various planes.

Cortex. The primary ground tissue region between the vascular system and the epidermis in stem and root. Term also used with reference to peripheral region of a cell protoplast.

Cotyledonary trace. The leaf trace of cotyledon located within the hypocotyl and connecting the vascular system of the root with that of the cotyledon.

Crassulae (sing. crassula). Thickenings of intercellular material and primary wall along the upper and lower margins of a pit-pair in the tracheids of gymnosperms; also called *bars of Sanio.*

Cristae (sing. crista). Crestlike infoldings of the inner membrane in a mitochondrion.

Cross-field. A term of convenience for the rectangle formed by the walls of a ray cell against an axial tracheid; as seen in radial section of the secondary xylem of conifers.

Crystal sand. A mass of very fine free crystals.

Crystalloid. Protein crystal that is less angular than a mineral crystal and swells in water.

Cuticle. A layer of fatty material, *cutin,* rather impervious to water, located on the outer walls of epidermal cells.

Cuticularization. The process of formation of the cuticle.

Cutin. A complex fatty substance considerably impervious to water; present in plants as an impregnation of epidermal walls and as a separate layer, the *cuticle,* on the outer surface of the epidermis.

Cutinization. The process of impregnation with cutin.

Cyclosis. The streaming of cytoplasm in a cell.

Cystolith. A concretion of calcium carbonate on an outgrowth of a cell wall. Occurs in a cell called *lithocyst.*

Cytochimera. A plant with combination of tissues the cells of which have different numbers of chromosomes. See also *chimera.*

Cyto-histological zonation. Presence of regions in the apical meristem having distinctive cytological charateristics. The term is meant to imply that a cytological zonation results in a subdivision into distinguishable tissue regions. Should be replaced with *cytological zonation.*

Cytokinesis. The process of division of a cell as distinguished from the division of the nucleus, or *karyokinesis.*

Cytological zonation. See *cyto-histological zonation.*

Cytology. The science dealing with the cell.

Cytoplasm. In a strict definition, the visibly least differentiated part of the protoplasm of a cell that constitutes the groundmass enclosing all other components of the protoplast. Also called *hyaloplasm.*

Decussate. Arrangement of leaves in pairs which alternate with one another at right angles.

Dedifferentiation. A reversal in differentiation of a cell or tissue which is assumed to occur when a more or less completely differentiated cell resumes meristematic activity.

Dehiscence. The spontaneous opening of a structure, such as anther or fruit, permitting the escape of reproductive entities contained in the dehiscing structure.

Dendroid venation. A type of venation in which the minor veins do not form closed meshes about small areas of mesophyll.

Derivative. A cell produced by division of a meristematic cell in such a way that it enters the path of differentiation into a body cell; its sister cell may remain in the meristem.

Dermal tissue. See *dermal tissue system.*

Dermal tissue system. The outer covering tissue of a plant, epidermis or periderm.

Dermatogen. The meristem forming the epidermis and arising from independent initials in the apical meristem. One of the three histogens, *plerome, periblem,* and *dermatogen,* according to Hanstein.

Desmogen. Meristematic strand destined to differentiate into a vascular bundle. May be primary, that is, composed of procambial cells, or secondary, that is, derived from a cambium in plants in which the secondary body consists of vascular bundles embedded in secondary parenchyma tissue.

Desmotubule. Tubule (often appearing as a solid rod) connecting the two endoplasmic reticulum cisternae located at the two opposite ends of a plasmodesma.

Detached meristem. A meristem, with a potenial to give rise to an axillary bud, appearing detached from the apical meristem because of the vacuolation of intervening cells.

Determinate growth. Formation of a restricted number of lateral organs by an apical meristem; characteristic of a floral meristem.

Development. The change in form and complexity of an organism or part of an organism from its beginning to maturity; combined with growth.

Diacytic stoma. A stomatal complex in which one pair of subsidiary cells, with their common walls at right angles to the long axis of the guard cells, surrounds the stoma.

Diaphragms in pith. Transverse layers (diaphragms) of firm-walled cells alternating with regions of soft-walled cells which may collapse with age.

Diarch. Primary xylem of the root; having two protoxylem strands, or two protoxylem poles.

Dichotomous venation. A venation pattern in which the veins appear repeatedly branched into equal portions.

Dictyosome. A membrane system (also called organelle) composed of stacked cisternae each producing vesicles at the periphery; when highly active in this process a cisterna may appear netted. Also called *golgi body* and *golgi apparatus.*

Dictyostele. A stele in which large overlapping

leaf gaps dissect the vascular system into anastomosing strands, each with the phloem surrounding the xylem.

Differentiation. A physiological and morphological change occurring in a cell, a tissue, an organ, or a plant during development from a meristematic, or juvenile, stage to a mature, or adult, stage. Usually associated with an increase in specialization.

Diffuse apotracheal parenchyma. Axial parenchyma in secondary xylem occuring as single cells or as strands distributed irregularly among the fibers, as seen in transection. See also *apotracheal parenchyma.*

Diffuse porous wood. Secondary xylem in which the pores (vessels) are distributed fairly uniformly throughout a growth layer or change in size gradually from early to late wood.

Dilatation. Growth of parenchyma by cell division in pith, rays, or axial system in vascular tissues; causes the increase in circumference of bark in stem and root.

Distal. Farthest from the point of origin or attachment. Opposite of *proximal.*

Distichous. Arrangement of leaves in two vertical rows; two-ranked arrangement.

Dorsal. Equivalent to *abaxial* in botanical usage.

Dorsiventral leaf. Possessing distinct upper and lower sides. Term derived from the reference to the abaxial and adaxial sides of a leaf as dorsal and ventral, respectively. A *bifacial leaf.*

Double fertilization. The fusion of egg and sperm (resulting in a $2n$ fertilized egg, the zygote) and the simultaneous fusion of the second male gamete with the polar nuclei (resulting in a $3n$ primary endosperm nucleus). A unique characteristic of angiosperms.

Druse. A globular, compound, calcium-oxalate crystal with numerous crystals projecting from its surface.

Duct. An elongated space formed by separation of cells from one another (schizogenous origin), by dissolution of cells (lysigenous origin), or by a combination of the two processes (schizo-lysigenous origin); usually concerned with secretion.

Early wood. The wood formed in first part of a growth layer and characterized by a lower density and larger cells than the late wood. Term replaces *spring wood.*

Eccrinous secretion. The secretion leaves the cell as individual molecules passing through the plasmalemma and cell wall. Compare with *granulocrinous secretion.*

Ectodesma. See *teichode.*

Ectophloic siphonostele. A stele having a pith and one phloem region, externally to the xylem.

Egg apparatus. The egg cell and two synergids located at the micropylar end of the female gametophyte, or the embryo sac, in angiosperms.

Elaioplast. A leucoplast type of plastid forming and storing oil.

Elaiosome. An outgrowth on a seed or fruit storing oil and serving as food for ants.

Embryo sac. The female gametophyte of angiosperms, generally composed of seven cells: the egg cell, two synergids, and three (or more) antipodals (each with a single nucleus), and the central cell (with two nuclei).

Embryogeny (or embryogenesis). Formation of embryo.

Embryoid. An embryo, often indistinguishable from a normal one, developing not from an egg but from a somatic cell, often in tissue culture.

Enation. A term applied to outgrowths of the stem in certain primitive land plants. See also *microphyll.*

Enation theory. A theory that regards microphylls as simple enations in contrast to megaphylls, which are considered to have evolved from branch systems.

Endarch xylem. A xylem system in which the maturation of cells progresses centrifugally; that is, the oldest elements (protoxylem) are closest to the center of the axis. Typical of stems in seed plants; also of leaves, in which the protoxylem is on the adaxial side.

Endocarp. The innermost layer or layers of the pericarp.

Endodermis. The layer of ground tissue forming a sheath around the vascular region and having the casparian strip in its anticlinal walls; may have secondary walls later. It is the innermost layer of the cortex in roots and stems of seed plants.

Endodermoid. Resembling the endodermis.

Endogenous. Arising from a deep-seated tissue, as a lateral root.

Endomembrane system. Collective term for plasmalemma, tonoplast, endoplasmic reticulum, dictyosomes, and the nuclear envelope.

Endoplasmic reticulum. (Usually abbreviated to ER.) A system of membranes forming cisternoid or tubular compartments that permeate the cytoplasm. The cisternae appear like paired membranes in sectional profiles. The membranes may be coated with ribosomes (rough or granular ER) or be free of ribosomes (smooth or agranular ER).

Endosperm. The nutritive tissue formed within the embryo sac of angiosperms from the central cell containing the primary endosperm nucleus.

Endothecium. A wall layer in the anther, usually with secondary wall thickenings.

Endothelium. The innermost layer of the integument lining the embryo sac in some taxa. Also called *integumentary tapetum.*

Enucleate. Lacking a nucleus.

Epiblast. A small structure opposite the scutellum present in the embryo of some grasses.

Epiblem. Term used sometimes for the epidermis of the root. See also *rhizodermis.*

Epicarp. See *exocarp.*

Epicotyl. The shoot part of the embryo or seedling above the cotyledon or cotyledons consisting of an axis and leaf primordia. See also *plumule.*

Epidermis. The outer layer of cells in the plant body primary in origin. If it is multiseriate (*multiple epidermis*), only the outermost layer differentiates as the epidermis.

Epigeal. Said of seed germination when the cotyledon or cotyledons are raised above the surface of the ground. Opposite of *hypogeal.*

Epigyny. Floral structure characterized by the positioning of sepals, petals, and stamens above the ovary (*inferior ovary*). The tissue enclosing the ovary and bearing the other floral parts above it is interpreted as receptacular tissue in some flowers and as appendicular tissue (floral tube) in others.

Epipetalous stamen. A stamen adnate to the corolla.

Epithelium. A compact layer of cells, often secretory in function, covering a free surface or lining a cavity.

Epithem. Mesophyll of a hydathode concerned with secretion of water.

Ergastic substances. Passive products of protoplast such as starch grains, fat globules, crystals, and fluids; occur in cytoplasm, organelles, vacuoles, and cell walls.

Eukaryotic (also eucaryotic). Refers to organisms having membrane-bound nuclei, genetic material organized into chromosomes, and membrane-bound cytoplasmic organelles. Opposite of *prokaryotic.*

Eumeristem. Meristem composed of relatively small cells, approximately isodiametric in shape, compactly arranged, and having thin walls, a dense cytoplasm, and large nuclei; word means "true meristem."

Eustele. A stele in which the primary vascular tissue is arranged in strands around the pith; typical of gymnosperms and angiosperms.

Exalbuminous seed. A seed without endosperm in mature state.

Exarch xylem. A xylem system in which the maturation of cells progresses centripetally; that is, the oldest elements (protoxylem) are farthest from the center of the axis. Typical of roots in seed plants.

Exine. The outer wall of a spore or a pollen grain.

Exocarp. The outermost layer or layers of the pericarp. *Epicarp.*

Exodermis. The outer layer, one or more cells in depth, of the cortex in some roots; a type of *hypodermis,* the walls of which may be suberized and/or lignified.

Exogenous. Arising in superficial tissue, as an axillary bud.

External phloem. Primary phloem located externally to the primary xylem.

Extrafloral nectary. Nectary occurring on a plant part other than a flower. See also *nectary.*

Extraxylary fibers. Fibers in various tissue regions other than the xylem.

False annual ring. One of more than one growth layers formed in the secondary xylem during one growth season, as seen in transection.

Fascicle. A bundle.

Fascicular cambium. Vascular cambium originating from procambium within a vascular bundle, or fascicle.

Fascicular tissue system. The vascular tissue system.

Fertilization. The fusion of two gametes, especially of their nuclei, resulting in the formation of a diploid ($2n$) zygote.

Festucoid. Pertaining to the subfamily of grasses Festucoideae.

Fiber. An elongated, usually tapering sclerenchyma cell with a lignified or nonlignified secondary wall; may or may not have a living protoplast at maturity.

Fiber-sclereid. A sclerenchyma cell with characteristics intermediate between those of a fiber and a sclereid.

Fiber-tracheid. A fiberlike tracheid in the secondary xylem; commonly thick walled, with pointed ends and bordered pits that have lenticular to slitlike apertures.

Fibonacci series. Series of numbers formed by successive addition of the last two: 1, 2, 3, 5, 8, 13, 21, 34, etc. These numbers occur in phyllotaxes. The relation was formulated by Leonardo of Pisa, surnamed Fibonacci.

Fibril. See *macrofibril.* Also used to designate long, threadlike subcellular entities in the protoplast.

Fibrous root system. A root system composed of many roots similar in length and thickness; as in grasses and other monocotyledons.

Filament. The stalk supporting the anther in a stamen.

File meristem. See *rib meristem.*

Filiform. Threadlike.

Filiform apparatus. A complex of cell wall invaginations in a synergid cell similar to those in transfer cells.

Filiform sclereid. A much elongated slender sclereid resembling a fiber.

Filling tissue. The loose tissue formed by the lenticel phellogen toward the outside; may or may not be suberized. Also called *complementary tissue.*

Flank meristem. A misnomer used with reference to the peripheral region of an apical meristem. The use of the word flanks implies that the entity is two-sided. The term should be replaced with *peripheral meristem.*

Flavedo. Yellow-colored tissue of the citrus rind.

Floral apical meristem. See *apical meristem.*

Floral meristem. Floral apical meristem. See *apical meristem.*

Floral nectary. See *nectary.*

Floral tube. A tube or cup formed by the united bases of sepals, petals, and stamens, often in perigynous and epigynous flowers.

Florigen. A hypothetical hormone assumed to be concerned with the induction of flowering.

Follicle. A dry, dehiscent, many-seeded fruit derived from one carpel and splitting along one suture.

Fracture face. One of two inner surfaces of a membrane that is exposed when the membrane splits during the fracture process used in freeze etching. There are two fracture faces: the *P* face (on the protoplasmic half of the membrane), which is closest to the cytoplasm, nucleoplasm, plastid stroma, or mitochondrial matrix; the *E* face, which is closest to the extracellular, endoplasmic, and exoplasmic space (e.g., vacuole, dictyosome vesicle). (Definition by Branton, D. et al. Science 190:54,1975.)

Free nuclear division. Nuclear division (*karyokinesis*) that occurs without cell wall formation (*cytokinesis*); characteristic of endosperm in certain taxa.

Fruit wall. The outer part of the fruit derived either from the ovary wall (*pericarp*) or from the ovary wall plus accessory floral parts associated with ovary in fruit.

Fundamental tissue. See *ground tissue.*

Fundamental tissue system. See *ground tissue system.*

Funiculus. The stalk of the ovule.

Fusiform cell. An elongated cell tapering at the ends.

Fusiform initial. In vascular cambium; an elongated cell with approximately wedge-shaped ends that gives rise to the elements of the axial system in the secondary vascular tissues.

Gelatinous fiber. A fiber with little or no lignification and in which part of the secondary wall has a gelatinous appearance.

Genotype. The genetic constitution of an organism; contrasted with *phenotype.*

Geotropism. Growth the direction of which is determined by gravity.

Germination. The resumption of growth by the embryo in a seed; also beginning of growth of a spore, pollen grain, bud, or other structure.

Gland. A multicellular secretory structure.

Glandular hair. A trichome having a unicellular or multicellular head composed of secretory cells; usually borne on a stalk of nonglandular cells.

Gluten. Amorphous protein occurring in the starchy endosperm of cereals.

Glyoxysome. A microbody containing enzymes necessary for converting fats into carbohydrates.

Golgi apparatus. A collective term for all golgi bodies, or dictyosomes, in a given cell. Sometimes used for an individual dictyosme.

Golgi body. See *dictyosome.*

Grafting. A union of two plants part of one of which, the *scion,* is inserted on root or stem of the other, the *stock.*

Grana (sing. granum). Subunits of chloroplasts seen as green granules with the light microscope and as stacks of disk-shaped cisternae, the *thylakoids,* with the electron microscope; the chlorophylls and carotenoids associated with photosynthesis are located in the grana.

Granulocrinous secretion. The secretion passes an inner cytoplasmic membrane, usually that of a vesicle, and is extruded from the cell after the vesicle fuses with the plasmalemma and releases its contents to the outside. Compare with *eccrinous secretion.*

Ground meristem. A primary meristem, or meristematic tissue, derived from the apical meristem and giving rise to the ground tissues.

Ground tissue. Tissues other than the vascular tissues, the epidermis and the periderm; also called *fundamental tissue.*

Ground tissue system. The total complex of ground tissues of the plant.

Growth. Increase in size by cell division and/or cell enlargement.

Growth layer. A layer of secondary xylem or secondary phloem produced during a single growth period, which may extend through one season (*annual ring*) or part of one season (*false annual ring*) if more than one layer is formed in one season; also called growth increment.

Growth ring. A growth layer of secondary xylem or secondary phloem as seen in transection of stem or root; may be an *annual ring* or a *false annual ring.*

Guard cells. A pair of cells flanking the stomatal pore and causing the opening and closing of the pore by changes in turgor.

Gum. A nontechnical term applied to material resulting from breakdown of plant cells, mainly of their carbohydrates.

Gum duct. A duct that contains gum.

Gummosis. A symptom of a disease characterized by the formation of gum, which may accumulate in cavities or ducts or appear on the surface of the plant.

Guttation. Exudation from leaves of water derived from the xylem.

Gynoecium. Collective term for the carpels in an angiosperm flower; part of the flower in which female gametogenesis occurs.

Hadrom (or hadrome). The tracheary elements and the associated parenchymatic cells of the xylem tissue; the specifically supporting cells (fibers and sclereids) are excluded. (See also *leptom.*)

Half-bordered pit-pair. A pit-pair consisting of a bordered and a simple pit.

Haplocheilic stoma. Stomatal type in gymnosperms; subsidiary cells are not related to the guard cells ontogenetically.

Hardwood. Technical designation of the wood of dicotyledons.

Haustorium. A projection from a cell or tissue that acts as a penetrating and absorbing device. In parasitic angiosperms, a modified root capable of penetrating and absorbing materials from host tissues.

Heartwood. The inner layers of secondary xylem that have ceased to function in storage and conduction and in which reserve materials have been removed or converted into heartwood substances; generally darker colored than the functioning *sapwood.*

Helical cell wall thickening. In tracheary elements of the xylem; secondary wall deposited on the primary or secondary wall as a continuous helix; also referred to as *spiral cell wall thickening.*

Hemicelluloses. Polysaccharides more soluble and less ordered than the cellulose; common component of the cell wall matrix.

Heterocellular ray. A ray in secondary vascular tissues composed of cells of more than one form: in dicotyledons, of procumbent and square or upright cells; in conifers, of parenchyma cells and ray tracheids.

Heterogeneous ray tissue system. Rays in secondary vascular tissues all heterocellular, or homocellular and heterocellular rays are combined. Term not used for conifers.

Hilum. (1) The central part of a starch grain around which the layers of starch are arranged concentrically. (2) The scar left by the detached funiculus on a seed.

Histogen. Hanstein's term for a meristem in shoot or root tip that forms a definite tissue system in the plant body. Three histogens were recognized: *dermatogen, periblem,* and *plerome.* (See the definition of these terms.)

Histogen concept. Hanstein's concept stating that the three primary tissue systems in the plant, the epidermis, the cortex, and the vascular system with the associated ground tissue, originate from distinct meristems, the histogens, in the apical meristems. See *histogen.*

Histogenesis. The formation of tissues, hence, *histogenetic,* having to do with origin or formation of tissues.

Histogenetic. See *histogenesis.*

Homocellular ray. A ray in secondary vascular tissues composed of cells of one form only: in dicotyledons, of procumbent, or square, or upright cells; in conifers, of parenchyma cells only.

Homogeneous ray tissue system. Rays in secondary vascular tissues all homocellular, composed of procumbent cells only. Term not used for conifers.

Homology. Having the same phylogenetic, or evolutionary, origin but not necessarily the same structure and/or function.

Horizontal parenchyma. See *ray parenchyma.*

Horizontal system. See *ray system.*

Hormone. A chemical substance produced in one part of an organism and transported to another part in which it has a specific effect.

Hyaloplasm. See *cytoplasm.*

Hydathode. A structural modification of vascular and ground tissues, usually in a leaf, that permits the release of water through a pore in the epidermis; may be secretory in function. See *epithem.*

Hydrolysis. The disassembly of large molecules by the addition of water.

Hydromorphic. Refers to the structural features of *hydrophytes.*

Hydrophyte. A plant that requires a large supply of water and may grow partly or entirely submerged in water.

Hygromorphic. Synonym of *hydromorphic.*

Hyperplasia. Refers most commonly to an excessive multiplication of cells.

Hypertrophy. Refers most commonly to abnormal enlargement. Hypertrophy of a cell or its parts involves no cell division. Hypertrophy of an organ may involve both enlargment of cells and abnormal cell multiplication (*hyperplasia*).

Hypocotyl. Axial part of embryo or seedling located between the cotyledon or cotyledons and the radicle.

Hypocotyl-root axis. Axial part of embryo or seedling comprising the hypocotyl and the root meristem, or the radicle if one is present.

Hypodermis. A layer or layers of cells beneath the epidermis distinct from the underlying ground tissue cells.

Hypogeal. Said of seed germination when the cotyledon or cotyledons remain beneath the surface of the ground. Opposite of *epigeal.*

Hypogyny. Floral structure characterized by the positioning of sepals, petals, and stamens below the ovary (*superior ovary*).

Hypophysis. The uppermost cell of suspensor from which part of the root and rootcap in the embryo of angiosperms are derived.

Hypsopylls. Leaves inserted at high levels of a

plant, as floral bracts. Contrasted with *cataphylls.*

Idioblast. A cell in a tissue that markedly differs in form, size, or contents from other cells in the same tissue.

Imperfect flower. A flower lacking either stamens or carpels.

Included phloem. Secondary phloem included in the secondary xylem of certain dicotyledons. Term replaces *interxylary phloem.*

Increment. In growth, an addition to the plant body by the activity of a meristem.

Indehiscent. Not opening spontaneously; refers to a certain fruit type.

Indeterminate growth. Refers to apical meristem producing an unrestricted number of lateral organs; characteristic of vegetative apical meristem.

Inferior ovary. See *epigyny.*

Initial. (1) Cell in a meristem that by division gives rise to two cells one of which remains in the meristem, the other is added to the plant body. (2) Sometimes used to designate a cell in its earliest stage of specialization. More appropriate term for (2), *primordium.*

Initial apotracheal parenchyma. See *boundary apotracheal parenchyma.*

Inner bark. See *bark.*

Integument. Outer cell layers enveloping the nucellus of the angiosperm ovule and differentiating into the seed coat.

Integumentary tapetum. The deeply staining innermost integumentary epidermis lining the embryo sac in some taxa and apparently assisting in the nutrition of the embryo. Also called *endothelium.*

Intercalary growth. Growth by cell division that occurs some distance from the meristem in which the cells originated.

Intercalary meristem. Meristematic tissue derived from the apical meristem and continuing meristematic activity some distance from that meristem; may be intercalated between tissues that are no longer meristematic.

Intercellular space. A space between two or more cells in a tissue; may have *schizogenous, lysigenous, schizo-lysigenous,* or *rhexigenous* origin.

Intercellular substance. See *middle lamella.*

Interfascicular cambium. Vascular cambium arising between vascular bundles (fascicles) in the interfascicular parenchyma.

Interfascicular region. Tissue region located between the vascular bundles (fascicles) in a stem; also called *medullary* or *pith ray.*

Intermediary cell. A parenchyma cell in the phloem of a minor vein that serves as a connection between the photosynthetic tissue and the sieve tube system; may be a companion cell. May or may not have wall ingrowths. See *transfer cell.*

Internal phloem. The primary phloem located internally from the primary xylem. Term replaces *intraxylary phloem.*

Internode. The region between nodes of a stem.

Interpositional growth. See *intrusive growth.*

Intervascular pitting. The pitting between tracheary elements.

Interxylary cork. The cork that develops within the xylem tissue.

Interxylary phloem. See *included phloem.*

Intine. The inner wall layer of a pollen grain or spore.

Intraxylary phloem. See *internal phloem.*

Intrusive growth. A type of growth in which a growing cell intrudes between other cells that separate from each other along the middle lamella in front of the tip of the growing cell; also called *interpositional growth.*

Intussusception. Growth of cell wall by interpolation of new wall material within previously formed wall. Opposite of *apposition.*

Irregular flower. A flower in which one or more members of at least one whorl of the perianth differ in form from other members of the same whorl and which cannot be divided in two equal halves in more than one plane. See *bilateral symmetry.*

Isobilateral leaf. A leaf in which the palisade parenchyma occurs on both sides of the leaf. See also *unifacial leaf.*

Isobilateral mesophyll. See *isobilateral leaf.*

Isodiametric. Regular in form, with all diameters equally long.

Isolateral leaf. Synonym of *isobilateral leaf.*

Isotropic. Having the same properties along all axes. Optically isotropic material does not affect the light.

Karyokinesis. Division of a nucleus as distinguished from the division of the cell, or *cytokinesis.*

Lacuna (pl. lacunae). Space. Usually air space between cells, which may be *schizogenous, lysigenous, schizo-lysigenous* or *rhexigenous* in origin. Also used with reference to the *leaf gap.*

Lacunar collenchyma. A collenchyma characterized by intercellular spaces and cell wall thickenings facing the spaces.

Lamella. A thin plate or layer.

Lamellar collenchyma. A collenchyma in which cell wall thickenings are deposited mainly on tangential walls.

Lamina of leaf. Blade, or expanded part of the leaf.

Late wood. The secondary xylem formed in the later part of a growth layer; denser and composed of smaller cells than the early wood. Term replaces *summer wood.*

Lateral meristem. A meristem located parallel with the sides of the axis; refers to the *vascular cambium* and *phellogen,* or *cork cambium.*

Latex (pl. latices). A fluid, often milky, contained in laticifers; consists of a variety of organic and inorganic substances, often including rubber.

Laticifer. A cell or a cell series containing a characteristic fluid called latex.

Laticiferous cell. A nonarticulated, or simple, laticifer.

Laticiferous vessel. An articulated, or compound, laticifer in which the cell walls between contiguous cells are partly or completely removed.

Leaf buttress. A lateral protrusion below the apical meristem constituting the initial stage in the development of a leaf primordium.

Leaf fibers. Technical designation of fibers derived from monocotyledons, chiefly from their leaves.

Leaf gap. A region of parenchyma in the vascular cylinder of a stem located above the level where a leaf trace diverges toward the leaf. Also called *lacuna.* Involves no interruption of vascular connections.

Leaf sheath. The lower part of a leaf that invests the stem more or less completely.

Leaf trace. A vascular bundle in the stem extending between its connection with a leaf and that with another vascular unit in the stem; a leaf may have one or more leaf traces.

Lenticel. An isolated region in the periderm distinguished from the phellem in having intercellular spaces; the tissue may or may not be suberized.

Leptom (or leptome). The sieve elements and the associated parenchymatic cells of the phloem tissue; the supporting cells (fibers and sclereids) are excluded. (See also *hadrom.*)

Leucoplast. A colorless plastid.

Libriform fiber. A xylem fiber commonly with thick walls and simple pits; usually the longest cell in the tissue.

Light line. A continuous line parallel with the surface seen in sections of the epidermis of certain leguminous seeds; results from matching of strongly refractive regions in the cell walls of contiguous epidermal cells.

Lignification. The impregnation with lignin.

Lignin. An organic substance or mixture of substances of high carbon content derived from phenylpropane and distinct from carbohydrates. Associated with cellulose in the walls of many cells.

Lithocyst. A cell containing a *cystolith.*

Locule. The cavity within a sporangium containing the spores (e.g., pollen grains) or within an ovule containing the ovules.

Longitudinal parenchyma. See *axial parenchyma.*

Longitudinal system. In secondary vascular tissues. See *axial system.*

Lumen of a cell. The space bound by the cell wall.

Lutoids. Vesicles, also called vacuoles, in laticifers bound by a single membrane and containing a spectrum of hydrolytic enzymes capable of degrading most of the organic compounds in the cell.

Lysigenous. As applied to an intercellular space, originating by a dissolution of cells.

Lysis. A process of disintegration or degradation.

Lysosomal compartment. A region in the cell protoplast or cell wall where acid hydrolases, capable of digesting cytoplasmic constituents and metabolites, are localized. Bound by a single membrane in the protoplast and usually constituting the vacuolar system. Another term, *lytic compartment.*

Lysosome. An organelle bound by a single membrane and containing acid hydrolytic enzymes capable of breaking down proteins and other organic macromolecules; in plants, represented by vacuoles. See *lysosomal compartment.*

Lytic compartment. See *lysosomal compartment.*

Maceration. The artificial separation of cells of a tissue by causing a disintegration of the middle lamella.

Macrofibril. An aggregation of *microfibrils* in a cell wall visible with the light microscope.

Macrosclereid. Elongated sclereid with unevenly distributed secondary wall thickenings; common in seed epidermis of Fabaceae.

Malpighian cell. Synonym of *macrosclereid.*

Macrospore. See *megaspore.*

Mantle. The outer layers of the kind of apical meristem that shows a layered arrangement of cells.

Marginal growth. The growth along the margins of a leaf primordium that results in the formation of the blade.

Marginal initials. Cells along the margins of a growing leaf lamina that contribute cells to the protoderm.

Marginal meristem. The meristem along the margin of a leaf primordium concerned with marginal growth of the blade.

Mass meristem. A meristematic tissue in which the cells divide in various planes so that the tissue increases in volume.

Matrix. Generally refers to a medium in which something is embedded.

Mechanical tissue. See *supporting tissue.*

Medulla. Synonym for *pith.*

Medullary bundles. Vascular bundles located in the pith region.

Medullary ray. See *interfascicular region.*

Medullary sheath. See *perimedullary region.*

Megagametophyte. Female gametophyte in heterosporous plants; embryo sac within the ovule in angiosperms.

Megaphyll. Leaf in ferns and seed plants larger than the *microphyll* in lower taxa and having a leaf trace associated with a leaf gap.

Megasporangium. Sporangium in which megaspores are produced; nucellus of the ovule in angiosperms.

Megaspore. A haploid (1*n*) spore developing into a female gametophyte in heterosporous plants.

Megaspore mother cell. See *megasporocyte.*

Megasporocyte. A diploid (2*n*) cell that undergoes meiosis and produces four haploid (1*n*) megaspores; also called *megaspore mother cell.*

Megasporophyll. A leaflike organ producing megasporangia; carpel producing ovules in angiosperms.

Meristele. One of the vascular bundles of a dictyostele.

Meristem Tissue primarily concerned with protoplasmic synthesis and formation of new cells by division.

Meristematic cell. A cell synthesizing protoplasm and producing new cells by division; varies in form, size, wall thickness, and degree of vacuolation, but has only a primary cell wall.

Meristemoid. A cell or a group of cells constituting an active locus of meristematic activity in a tissue composed of somewhat older, differentiating cells.

Mesarch xylem. A xylem strand in which the protoxylem is in the center and the metaxylem differentiates both centripetally and centrifugally from the center.

Mesocarp. The middle layer of a pericarp.

Mesocotyl. The internode between the scutellar node and the coleoptile in the embryo and seedling of Poaceae.

Mesomorphic. Refers to structural features of *mesophytes.*

Mesophyll. The photosynthetic parenchyma of a leaf blade located between the two epidermal layers.

Mesophyte. A plant requiring an environment that is neither too wet nor too dry.

Mestome sheath. An endodermoid sheath of a vascular bundle; the inner of two sheaths in leaves of Poaceae, mainly those of the festucoid subfamily.

Metacutisation. Depostion of suberin lamellae in outer cells of root tips that cease to be active in growth and absorption at the end of seasonal growth. *Late suberization.*

Metaphloem. Part of the primary phloem that differentiates after the protophloem and before the secondary phloem, if any of the latter is formed in a given taxon.

Metaxylem. Part of the primary xylem that differentiates after the protoxylem and before the secondary xylem, if any of the latter is formed in a given taxon.

Micelles. The regions in cellulose microfibrils in which the cellulose molecules are arranged parallel to each other so that a crystalline lattice structure is present.

Microbody. Organelle bound by a single membrane and containing various enzymes except the hydrolytic. *Peroxisomes* and *glyoxysomes* are microbodies.

Microfibril. A threadlike component of the cell wall consisting of cellulose molecules and visible only with the electron microscope.

Microgametophyte. The male gametophyte of a heterosporous plant; pollen grain in a seed plant.

Micrometer. One thousandth millimeter; also called *micron.* Symbol μm.

Micromicron. One millionth micron. One hundredth angstrom. Symbol $\mu\mu$.

Micron. See *micrometer.*

Microphyll. Small leaf the trace of which is not associated with a leaf gap. Contrasted with *megaphyll.* See also *enation.*

Micropyle. The opening in the integuments of an ovule through which the pollen tube usually enters the embryo sac.

Microsporangium. Sporangium in which microspores are formed; anther locule and its walls in angiosperms.

Microspore. A haploid (1*n*) spore developing into a male gametophyte in heterosporous plants; the uninucleate pollen grain in seed plants.

Microspore mother cell. See *microsporocyte.*

Microsporocyte. A diploid (2*n*) cell that undergoes meiosis and forms four haploid (1*n*) microspores; also called *microspore mother cell* and, in seed plants, *pollen mother cell.*

Microsporophyll. Leaflike organ bearing microsporangia; in angiosperms, modified into stamen.

Microtubules. Nonmembranous tubules about 25 nanometers (250 angstroms) in diameter and of indefinite length. Located in the cytoplasm in a nondividing eukaryotic cell, usually near the cell wall, and form the meiotic or mitotic spindle and the phragmoplast in a dividing cell. Sometimes called *organelles.*

Middle lamella. Layer of intercellular material, chiefly pectic substances, cementing together the primary walls of contiguous cells.

Millimicron. See *nanometer.*

Mitochondrion (pl. mitochondria). Double membrane-bound cell organelle concerned with respiration; carries enzymes and is the major source of ATP in nonphotosynthetic cells.

Morphogenesis. Development of form; the sum of phenomena of development and differentiation of tissues and organs.

Morphology. The study of form and its development.

Mother cell. See *precursory cell.*

Motor cell. See *bulliform cell.*

Mucilage cell. Cell containing mucilages or gums or similar carbohydrate material characterized by the property of swelling in water.

Mucilage duct. A duct containg mucilage or gum or similar carbohydrate material. See also *duct.*

Multilacunar node. A node in stem having numerous gaps (and numerous leaf traces) related to one leaf.

Multiperforate perforation plate. In vessel member of the xylem; a perforation plate that has more than one perforation.

Multiple epidermis. A tissue two or more cell layers deep derived from the protoderm; only the outermost layer differentiates as a typical epidermis.

Multiple fruit. A fruit composed of several matured ovaries each produced in a separate flower.

Multiseriate ray. A ray in secondary vascular tissues that is few to many cells wide.

Mycorrhiza (pl. mycorrhizae). The symbiotic association of fungi and roots of higher plants.

May be *ectotrophic* (web of hyphae invests the root of the host) or *endotrophic* (the hypae are located within the root cells).

Myrosin cell. Cell containing glucosinolates ("mustard oil glucosides") and myrosinases, enzymes hydrolysing the glucosinolates. Occur in eleven dicotyledon families, the two largest of which are the Brasicaceae and Euphorbiaceae.

Nacré wall. See *nacreous wall.*

Nacreous wall. A nonlignified wall thickening that is often found in sieve elements and resembles a secondary wall when it attains considerable thickness; designation based on glistening appearance of the wall.

Nanometer. One millionth of a millimeter; symbol *nm*. Equal to one millimicron (mμ) or 10 angstroms (*A*).

Nectary. A multicellular glandular structure secreting a liquid containing organic substances including sugar. Occurs in flowers (*floral nectary*) and on vegetative plant parts (*extrafloral nectary*).

Netted venation. See *reticulate venation.*

Nodal diaphragm. A septum of tissue at the node of a stem extending across the otherwise hollow pith region.

Node. That part of the stem at which one or more leaves are attached; not sharply delimited anatomically.

Nodular end wall. The cell wall, at right angles to the longitudinal axis of a xylem parenchyma cell, that appears beaded because of deeply depressed pits.

Nodules. Enlargements on roots of plants,particularly in the Fabaceae, inhabited by nitrogen-fixing bacteria.

Nonarticulated laticifer. A simple laticifer consisting of a single, commonly multinucleate, cell; may be branched or unbranched.

Nonporous wood. Secondary xylem having no vessels.

Nonstoried cambium. Vascular cambium in which the fusiform initials and rays are not arranged in horizontal tiers on tangential surfaces. *Nonstratified cambium.*

Nonstoried wood. Secondary xylem in which the axial cells and rays are not arranged in horizontal tiers on tangential surfaces. *Nonstratified wood.*

Nonstratified cambium. See *nonstoried cambium.*

Nonstratified wood. See *nonstoried wood.*

Nucellus. Inner part of an ovule in which the embryo sac develops. Commonly considered to be equivalent to the megasporangium.

Nuclear envelope. The double membrane enclosing the nucleus of a cell.

Nucleolar organizer. A region on a certain chromosome concerned with the formation of nucleolus.

Nucleolus. Spherical body, composed mainly of RNA and protein, present in the nucleus of eukaryotic cells, one or more to a nucleus; site of synthesis of ribosomes.

Nucleus. In biology, organelle in a eukaryotic cell bound by a double membrane and containing the chromosomes, nucleoli, and nucleoplasm.

Obturator. An outgrowth of the placenta or of the lining of stylar canal that brings the pollen transmitting tissue close to the micropyle.

Ontogeny. The development of an organism, organ, tissue, or cell from inception to maturity.

Open vascular bundle. A bundle in which the procambium differentiates into vascular cambium after all primary vascular cells are formed.

Open venation. Leaf venation in which large veins end freely in the mesophyll instead of being connected by anastomoses with other veins.

Opposite pitting. Pits in tracheary elements disposed in horizontal pairs or in short horizontal rows.

Organ. A distinct and visibly differentiated part of a plant, such as root, stem, leaf, or part of a flower.

Organelle. A distinct body within the cytoplasm of a cell, specialized in function.

Orthostichy. A vertical line along which is attached a series of leaves or scales on an axis of a shoot or shootlike organ. Often incorrectly applied to a steep helix, or *parastichy.*

Osteosclereid. Bone-shaped sclereid having a columnar middle part and enlargements at both ends.

Outer bark. See *bark.*

Ovary. Lower part of a carpel (simple pistil) or of a

gynoecium composed of united carpels (compound pistil) containing the ovules and differentiating into the fruit.

Ovule. A structure in seed plants enclosing the female gametophyte and composed of the nucellus, one or two integuments, and funiculus; differentiates into the seed.

Paedomorphosis. Delay in evolutionary advance in some characteristics as compared with others resulting in a combination of juvenile and advanced characteristics in the same cell, tissue, or organ.

Palisade parenchyma. Leaf mesophyll parenchyma characterized by elongated form of cells and their arrangement with their long axes perpendicular to the surface of the leaf.

Panicoid. Pertaining to the subfamily of grasses Panicoideae.

Papilla (pl. papillae). A soft protuberance on an epidermal cell; a type of trichome.

Paracytic stoma. A stomatal complex in which one or more subsidiary cells flank the stoma parallel with the long axes of the guard cells.

Paradermal. Parallel with the epidermis. Refers specifically to a section made parallel with the surface of a flat organ such as a leaf; it is also a *tangential* section.

Parallel evolution. Evolution occurring in a similar direction in different taxa. Also called *convergent evolution.*

Parallel venation. Main veins in a leaf blade arranged approximately parallel to one another, although converging at base and apex of leaf.

Parastichy. A helix along which is attached a series of leaves or scales on an axis of a shoot or shootlike organ. See also *orthostichy.*

Paratracheal parenchyma. Axial parenchyma in secondary xylem associated with vessels and other tracheary elements. Includes *aliform, confluent,* and *vasicentric.*

Parenchyma. Tissue composed of parenchyma cells.

Parenchyma cell. Typically a not distinctly specialized cell with a nucleate protoplast concerned with one or more of the various physiological and biochemical activities in plants. Varies in size, form, and wall structure.

Parietal cytoplasm. Cytoplasm located next to the cell wall.

Parthenocarpy. The development of fruit without fertilization. The fruit is usually seedless.

Passage cell. Cell in exodermis or endodermis that remains thin walled when the associated cells develop thick secondary walls. Has casparian strip in endodermis.

Pectic substances. A group of complex carbohydrates, derivatives of polygalacturonic acid, occurring in plant cell walls; particularly abundant as a constituent of the middle lamella.

Pedicel. The stalk of an individual flower.

Peduncle. The stem of an inflorenscence.

Peltate hair. A trichome consisting of a discoid plate of cells borne on a stalk or attached directly to the basal foot cell.

Perfect flower. A flower having both carpels and stamens.

Perforation plate. Part of a wall of a vessel member that is perforated.

Perianth. Petals and sepals or tepals of a flower considered together.

Periblem. The meristem forming the cortex. One of the three histogens, *plerome, periblem,* and *dermatogen,* according to Hanstein.

Pericarp. Fruit wall developed from the ovary wall.

Periclinal. Commonly refers to orientation of cell wall or plane of cell division; parallel with the circumference or the nearest surface of an organ. Opposite of *anticlinal.* See also *tangential.*

Periclinal chimera. See *chimera.*

Pericycle. Part of ground tissue of the stele located between the phloem and the endodermis. In seed plants, regularly present in roots, absent in most stems.

Pericyclic fiber. See *perivascular fiber.*

Pericyclic sclerenchyma. See *perivascular sclerenchyma.*

Periderm. Secondary protective tissue that replaces the epidermis in stems and roots, rarely in other organs. Consists of *phellem* (cork), *phellogen* (cork cambium), and *phelloderm.*

Perigyny. Floral structure characterized by an extension above the receptacle resembling a cup and bearing the sepals, petals, and stamens. The cup may be receptacular or appendicular (floral tube) in origin.

Perimedullary region or zone. The peripheral region of the pith (medulla); also called *medullary sheath.*

Perinuclear space. Space between the two membranes forming the nuclear envelope.

Perisperm. Storage tissue in seed similar to endosperm but derived from the nucellus.

Perivascular fiber. A fiber located along the outer periphery of the vascular cylinder in the axis of a seed plant and not originating in the phloem. Alternate term, *pericyclic fiber.*

Perivascular sclerenchyma. Sclerenchyma located along the outer periphery of the vascular cylinder and not originating in the phloem. Alternate term, *pericyclic sclerenchyma.*

Peroxisome. A cell organelle of the type of microbody that is involved in glycolic acid metabolism associated with photosynthesis.

Petal. A unit of the corolla of a flower.

Petiole. Stalk of a leaf.

Phellem (cork). Protective tissue composed of nonliving cells with suberized walls and formed centrifugally by the phellogen (cork cambium) as part of the periderm. Replaces the epidermis in older stems and roots of many seed plants.

Phelloderm. A tissue resembling cortical parenchyma produced centripetally by the phellogen (cork cambium) as part of the periderm of stems and roots in seed plants.

Phellogen (cork cambium). A lateral meristem forming the periderm, a secondary protective tissue common in stems and roots of seed plants. Produces phellem (cork) centrifugally, phelloderm centripetally by periclinal divisions.

Phelloid cell. A cell within the phellem (cork) but distinct from the cork cell in having no suberin in its walls. May be a sclereid.

Phenotype. Physical appearance of an organism resulting from interaction between its *genotype* (genetic constitution) and the environment.

Phlobaphenes. Anhydrous derivatives of tannins. Amorphous yellow, red, or brown substances very conspicuous when present in cells.

Phloem. The principal food-conducting tissue of the vascular plant composed mainly of sieve elements, various kinds of parenchyma cells, fibers, and sclereids.

Phloem elements. Cells of the phloem tissue.

Phloem initial. A cambial cell on the phloem side of the cambial zone that is the source of one or more cells arising by periclinal divisions and differentiating into phloem elements with or without additional divisions in various planes. Sometimes called *phloem mother cell.*

Phloem mother cell. A cambial derivative that is the source of certain elements of the phloem tissue, such as, sieve element and its companion cells or phloem parenchyma cells forming a parenchyma strand. Used also in a wider sense synonymously with *phloem initial.*

Phloem parenchyma. Parenchyma cells located in the phloem. In secondary phloem refers to axial parenchyma.

Phloem ray. That part of a vascular ray which is located in the secondary phloem.

Phloic procambium. That part of procambium which differentiates into primary phloem.

Photoperiodism. Response to duration and timing of day and night expressed in the character of growth, development, and flowering in plants.

Photorespiration. The light-dependent production of glycolic acid in chloroplasts and its subsequent oxidation in peroxisomes.

Photosynthetic cell. A chloroplast-containing cell engaged in photosynthesis.

Phragmoplast. Fibrous structure (light microscope view) that arises between the daughter nuclei at telophase and within which the initial partition (*cell plate*), dividing the mother cell in two (*cytokinesis*), is formed. Appears at first as a spindle connected to the two nuclei, but later spreads laterally in the form of a ring. Consists of microtubules.

Phragmosome. Layer of cytoplasm formed across the cell where the nucleus becomes located and divides. The equatorial plane of the subsequently appearing phragmoplast coincides with the plane of the cytoplasmic layer. In electron microscopy, the term refers to enzyme-containing microbodies that are thought to be participating in cell plate formation.

Phyllode. Supposedly a flat, expanded petiole replacing the leaf blade in photosynthetic function.

Phyllotaxy (or phyllotaxis). The mode in which the leaves are arranged on the axis of a shoot.

Phylogeny. Evolutionary history of a species or larger taxon.

Phytochrome. A proteinaceous pigment occurring in the cytoplasm of green plants and serving as a photoreceptor for red-far-red light. Involved in timing certain processes such as dormancy, leaf formation, flowering, and seed germination.

Pinocytosis. Process of uptake of a substance by the protoplast through invaginations of the plasmalemma that are pinched off as vesicles (*pinocytotic vesicles*) the contents of which are incorporated into the cytoplasm.

Pinocytotic vesicles. See *pinocytosis.*

Pistil. Equivalent to the whole syncarpous gynoecium (compound pistil) or to a single carpel in an apocarpous gynoecium (simple pistil). Is composed of ovary, style, and stigma.

Pistillate. Refers to a flower with one or more carpels but no functional stamens.

Pit. A recess or cavity in the cell wall where the primary wall is not covered by secondary wall. Pitlike structures in the primary wall are designated *primordial pits, primary pits,* or *primary pit-fields.* A pit is usually a member of a pit-pair.

Pit aperture. Opening into the pit from the interior of the cell. If a pit canal is present in a bordered pit, two apertures are recognized, the *inner,* from the cell lumen into the canal, and the *outer,* from the canal into the pit cavity.

Pit canal. The passage from the cell lumen to the chamber of a bordered pit. (Simple pits in thick walls usually have canallike cavities.)

Pit cavity. The entire space within a pit from pit membrane to the cell lumen or to the outer pit aperture if a pit canal is present.

Pit-field. See *primary pit-field.*

Pit membrane. The part of the intercellular layer and primary cell wall that limits a pit cavity externally.

Pit-pair. Two complementary pits of two adjacent cells. Essential components are two *pit cavities* and the *pit membrane.*

Pith. Ground tissue in the center of a stem or root. Homology of pith in root and stem is uncertain.

Pith ray. See *interfascicular region.*

Placenta (pl. placentae). Region in the ovary where ovules originate and remain attached to maturity.

Placentation. Refers to the distribution of placentae in an ovary.

Plasma membrane. See *plasmalemma.*

Plasmalemma. Single membrane delimiting the cytoplasm next to the cell wall. A type of unit membrane. Also called *plasma membrane.*

Plasmodesma (pl. plasmodesmata). A connection of protoplasts of two contiguous cells through a canal in the cell wall. Consists of plasmalemma lining the canal, a central *desmotubule* connecting two endoplasmic reticulum cisternae located at the two opposite openings of the canal, and cytoplasm between the plasmalemma and the desmotubule.

Plastid. Organelle with a double membrane in the cytoplasm of many eukaryotes. May be concerned with photosynthesis (*chloroplast*) or starch storage (*amyloplast*), or contain yellow or orange pigments (*chromoplast*).

Plastochron (or plastochrone). The time interval between the inception of two successive repetitive events, as origin of leaf primordia, attainment of certain stage of development of a leaf, etc. Variable in length as measured in time units.

Plastoglobule. Globule in a plastid with lipid as the basic component.

Plate collenchyma. See *lamellar collenchyma.*

Plate meristem. A meristematic tissue consisting of parallel layers of cells dividing only anticlinally with reference to the wide surface of the tissue. Characteristic of ground meristem of plant parts that assume a flat form, as a leaf.

Plerome. The meristem forming the core of the axis composed of the primary vascular tissues and associated ground tissue, such as pith and interfascicular regions. One of the three histogens, *plerome, periblem,* and *dermatogen,* according to Hanstein.

Plicate mesophyll cell. A mesophyll cell with

folds or ridges of cell wall projecting into cell lumen.

Plumule. Embryonic shoot above the cotyledon or cotyledons in an embryo. See also *epicotyl.*

Polar nucleus. One of two nuclei in the central cell of a mature embryo sac. The two nuclei are derived from groups of nuclei at the two opposite poles of the eight-nucleate embryo sac.

Pollen. A collective term for pollen grains.

Pollen conducting tissue. See *transmitting tissue.*

Pollen grain. A microspore in a seed plant included in an elaborately structured wall (one cell). Also a germinated microspore having formed a microgametophyte, immature (two cells) or mature (three cells).

Pollen mother cell. See *microsporocyte.*

Pollen sac. A locule in the anther containing the pollen grains.

Pollen tube. A tubular cell extension formed by the germinating pollen grain; carries the male gametes into the ovule.

Pollination, The transfer of pollen from the anther to the receptive surface, stigma in angiosperms.

Polyarch. Primary xylem of root; having many protoxylem strands, or protoxylem poles.

Polyderm. A type of protective tissue in which suberized cells alternate with nonsuberized parenchyma cells and both kinds of cell have living protoplasts.

Polyembryony. Development of more than one embryo in a single seed.

Polymerization. Chemical union of monomers, such as glucose or nucleotides, resulting in the formation of polymers, such as starch, cellulose, or nucleic acid.

Polyribosome (or polysome). Aggregation of ribosomes apparently concerned with protein synthesis as a group.

Polysaccharide. A carbohydrate composed of many monosaccharide units joined in a chain, e.g., starch, cellulose.

Pore. A term of convenience for the cross section of a vessel in the secondary xylem.

Pore cluster. See *pore multiple.*

Pore multiple. In secondary xylem; a group of two or more pores (cross sections of vessels) crowded together and flattened along the surfaces of contact. *Radial pore multiple,* pores in radial file; *pore cluster,* irregular grouping.

Porous wood. Secondary xylem having vessels.

P-protein. Phloem protein. Found in cells of phloem of seed plants, most commonly in sieve elements. Formerly called slime.

Precursory cell. A cell giving rise to others by division. *Mother cell.*

Preprophase band. A ringlike band of microtubules delimiting the equatorial plane of the future mitotic spindle in cells preparing to divide.

Primary body (of plant). The part of the plant, or entire plant if no secondary growth occurs, that arises from the embryo and the apical meristems and their derivative meristematic tissues and is composed of primary tissues.

Primary cell wall. Version based on studies with the light microscope: cell wall formed chiefly while the cell is increasing in size. Version based on studies with the electron microscope: cell wall in which the cellulose microfibrils show various orientations—from random to more or less parallel—that may change considerably during the increase in size of the cell. The two versions do not necessarily coincide in delimiting primary from secondary wall.

Primary endosperm nucleus. Nucleus resulting from the fusion of the male gamete and two polar nuclei in the central cell of the embryo sac.

Primary growth. The growth of successively formed roots and vegetative and reproductive shoots from the time of their initiation by the apical meristems and until the completion of their expansion. Has its inception in the apical meristems and continues in their derivative meristems, protoderm, ground meristem, and procambium, as well as in the partly differentiated primary tissues.

Primary meristem. Often used for each of the three meristematic tissues derived from the apical meristem: protoderm, ground meristem, and procambium.

Primary phloem. Phloem tissue differentiating from procambium during primary growth and differentiation of a vascular plant. Commonly divided into the earlier *protophloem* and the later *metaphloem.* Not differentiated into axial and ray systems.

Primary phloem fibers. Fibers located on the outer periphery of the vascular region and originating in the primary phloem, usually the protophloem. Often called *pericyclic fibers.*

Primary pit. See *primary pit-field.*

Primary pit-field. A thin area of the primary cell wall and middle lamella within the limits of which one or more pit-pairs develop if a secondary wall is formed. Also called *primordial pit* and *primary pit.*

Primary plant body. See *primary body.*

Primary root. The taproot. Root developing in continuation of the radicle of the embryo.

Primary thickening meristem. A meristem derived from the apical meristem and responsible for the primary increase in thickness of the shoot axis. May appear as a distinct mantlelike zone. Often found in monocotyledons.

Primary tissues. Tissues derived from the embryo and the apical meristems.

Primary vascular tissues. Xylem and phloem differentiating from procambium during primary growth and differentiation of a vascular plant.

Primary wall. See *primary cell wall.*

Primary xylem. Xylem tissue differentiating from procambium during primary growth and differentiation of a vascular plant. Commonly divided into the earlier *protoxylem* and the later *metaxylem.* Not differentiated into axial and ray systems.

Primordial pit. See *primary pit-field.*

Primordium (pl. primordia). An organ, a cell, or an organized series of cells in their earliest stage of differentiation, e.g., leaf primordium, sclereid primordium, vessel member primordium.

Procambium. Primary meristem or meristematic tissue which differentiates into the primary vascular tissue. Also called *provascular tissue.*

Procumbent ray cell. In secondary vascular tissues; a ray cell having its longest axis in radial direction.

Prodesmogen. A meristem precursory to *desmogen* (*procambium*). The term has the same connotation as *residual meristem.*

Proembryo. Embryo in early stages of development, the stages before main body and suspensor become distinct or before the protoderm is initiated.

Prokaryotic (also procaryotic). Refers to organisms in which the nuclear material is not enclosed in a nuclear envelope and the genetic material is not organized into chromosomes and which lacks membrane-bound cytoplasmic organelles. Blue-green algae, bacteria, mycoplasmas. Opposite to *eukaryotic.*

Promeristem. The initiating cells and their most recent derivatives in an apical meristem. Also called *protomeristem.*

Prop roots. Adventitious roots developing above the soil level on the stem and serving as additional support of the plant axis.

Prophyll. The first or one of two first leaves on a lateral shoot.

Proplastid. A plastid in its earliest stages of development.

Protective layer. In abscission zone; layer of cells that, because of substances impregnating its walls, has a protective function in the scar formed by abscission of a leaf or other plant part.

Protoderm. Primary meristem or meristematic tissue giving rise to the epidermis; also epidermis in meristematic state. May or may not arise from independent initials in the apical meristem.

Protomeristem. See *promeristem.*

Protophloem. The first-formed elements of the phloem in a plant organ. First part of the primary phloem.

Protophloem poles. Term of convenience for loci of phloem elements that are the first to mature in the vascular system of a plant organ. Applied to views in cross sections.

Protoplasm. Living substance. Inclusive term for all living contents of a cell or an entire organism.

Protoplast. The organized living unit of a single cell including protoplasmic and nonprotoplasmic contents of a cell but excluding the cell wall.

Protostele. The simplest type of stele, containing a solid column of vascular tissue, with the phloem peripheral to the xylem.

Protoxylem. The first formed elements of the

xylem in a plant organ. First part of the primary xylem.

Protoxylem lacuna. Space surrounded by parenchyma cells in the protoxylem of a vascular bundle. Appears in some plants after the tracheary elements of protoxylem cease to function and are stretched and torn.

Protoxylem poles. Term of convenience for loci of xylem elements that are the first to mature in the vascular system of a plant organ. Applied to views in cross sections.

Provascular tissue. See *procambium.*

Proximal. Situated near the point of origin or attachment. Opposite of *distal.*

Pulvinus. An enlargement of the petiole of a leaf, or petiolule of a leaflet, at its base. A structure that has a role in the movements of a leaf or leaflet.

Quantasomes. Granules located on the inner surface of chloroplast membranes making up the thylakoids. Once thought to be functional units in photosynthesis.

Quarter-sawed oak. Oak wood sawed along a radial plane so that the radial surface showing the wide rays characteristic of this wood is exposed.

Quiescent center. The initial region in the apical meristem that has reached a state of relative inactivity; common in roots.

Radial parenchyma. See *ray parenchyma.*

Radial pore multiple. See *pore multiple.*

Radial section. A longitudinal section coinciding with a radius of a cylindrical body, such as stem.

Radial seriation. Arrangement of units, such as cells, in an orderly sequence in a radial direction. Characteristic of cambial derivatives.

Radial symmetry. Refers to a flower that can be divided in two equal parts in more than one longitudinal plane passing through the floral axis. Contrasted with *bilateral symmetry.*

Radial system. See *ray system.*

Radicle. The embryonic root. Forms the basal continuation of the hypocotyl in an embryo.

Ramified. Branched.

Ramiform pit. Pit that appears to be branched because it is formed by a coalescence of two or more simple pits during the increase in thickness of the secondary wall.

Raphe. A ridge along the body of the seed formed by the part of the funiculus that is adnate to the ovule (in an anatropous ovule).

Raphides. Needle-shaped crystals commonly occurring in bundles.

Ray. A panel of tissue variable in height and width, formed by the ray initials in the vascular cambium and extending radially in the secondary xylem and secondary phloem.

Ray initial. A meristematic ray cell in the vascular cambium that gives rise to ray cells of the secondary xylem and seconary phloem.

Ray parenchyma. Parenchyma cells of a ray in secondary vascular tissues. Contrasted with *axial parenchyma.*

Ray system. The total of all rays in the secondary vascular tissues. Also called *horizontal system* and *radial system.*

Ray tracheid. Tracheid in a ray. Found in the secondary xylem of certain conifers.

Reaction wood. Wood with more or less distinctive anatomical characteristics formed in parts of leaning or crooked stems and on lower (conifers) or upper (dicotyledons) sides of branches. See *compression wood* and *tension wood.*

Receptacle. The part of the flower stalk that bears the floral organs.

Redifferentiation. A reversal in differentiation in a cell or tissue and subsequent differentiation into another type of cell or tissue.

Regular flower. See *radial symmetry.*

Reproductive apical meristem. See *apical meristem.*

Residual meristem. Used in the sense of residuum of the least differentiated part of the apical meristem. A tissue that is relatively more highly meristematic than the associated differentiating tissues beneath the apical meristem. Gives rise to procambium and to interfascicular ground tissue.

Resin canal. See *resin duct.*

Resin duct. A duct of schizogenous origin lined with resin-secreting cells (*epithelial cells*) and containing resin.

Reticulate cell wall thickening. In tracheary elements of the xylem; secondary cell wall deposited on the primary so as to form a netlike pattern.

Reticulate perforation plate. In vessel member of the xylem; a type of mulitperforate plate in which the bars delimiting the perforations form a netlike pattern.

Reticulate sieve plate. A compound sieve plate with sieve areas arranged so as to form a netlike pattern.

Reticulate venation. Veins in leaf blade form an anastomosing system, the whole resembling a net. *Netted venation*

Reticulum. A net.

Retting. Freeing fiber bundles from other tissues by utilizing the action of microorganisms causing, in a suitable moist environment, the disintegration of the thin-walled cells surrounding the fibers.

Rexigenous. As applied to an intercellular space, orginating by rupture of cells.

Rhizodermis. Primary surface layer of the root. Use of the term implies that this layer is not homologous with the epidermis of the shoot. See also *epiblem.*

Rhytidome. A technical term for the outer bark, which consists of periderm and tissues isolated by it, namely cortical and phloem tissues.

Rib. An elongated protrusion as those along the large veins on the underside of a leaf.

Rib meristem. A meristematic tissue in which the cells divide perpendicular to the longitudinal axis of an organ and produce a complex of parallel, vertical files ("ribs") of cells. Particularly common in ground meristem of organs assuming a cylindrical form. Also called *file meristem.*

Ribosome. A cell component (sometimes called organelle) composed of protein and RNA and concerned with protein synthesis. Occurs in the cytoplasm, nucleus, plastids, and mitochondria.

Ring bark. A type of rhytidome resulting from the formation of successive periderms approximately concentrically around the axis.

Ring-porous wood. Secondary xylem in which the pores (vessels) of the early wood are distinctly larger than those of the late wood and form a well-defined zone or ring in a cross section of wood.

Rootcap. A thimblelike mass of cells covering the apical meristem of the root.

Root hair. A type of trichome on root epidermis that is a simple extension of an epidermal cell and is concerned with absorption of soil solution.

Sapwood. Outer part of the wood of stem or root containing living cells and reserves and in which conduction of water takes place. Generally lighter colored than the *heartwood.*

Scalariform cell wall thickening. In tracheary elements of the xylem; secondary wall deposited on the primary so as to form a ladderlike pattern. Similar to a helix of low pitch with the coils interconnected at intervals.

Scalariform perforation plate. In vessel member of the xylem; a type of multiperforate plate in which elongated perforations are arranged parallel to one another so that the cell wall bars between them form a ladderlike pattern.

Scalariform pitting. In tracheary elements of the xylem; elongated pits arranged parallel to one another so as to form a ladderlike pattern.

Scalariform-reticulate cell wall thickening. In tracheary elements of the xylem; secondary thickening intermediate between scalariform and reticulate.

Scalariform sieve plate. A compound sieve plate with elongated sieve areas arranged parallel to one another in a ladderlike pattern.

Scale bark. A type of rhytidome in which the sequent periderms develop as restricted overlapping strata, each cutting out a scalelike mass of tissue.

Scar tissue. Composed of necrosed cells injured by wounding and subjacent cells impregnated with protective substances. See also *cicatrice.*

Schizogenous. As applied to an intercellular space, originating by separation of cell walls along the middle lamella.

Schizo-lysigenous. As applied to an intercellular space, originating by a combination of two processes, separation and degradation of cell walls.

Sclereid. A sclerenchyma cell, varied in form, but typically not much elongated, and having thick lignified secondary walls with many pits.

Sclerenchyma. A tissue composed of sclerenchyma cells. Also a collective term for sclerenchyma cells in the plant body or plant organ. Includes *fibers, fiber-sclereids,* and *sclereids.*

Sclerenchyma cell. Cell variable in form and size and having more or less thick, often lignified, secondary walls. Belongs to the category of supporting cells and may or may not be devoid of protoplast at maturity.

Sclerification. The act of becoming changed into sclerenchyma, that is, developing secondary walls, with or without subsequent lignification.

Sclerotic parenchyma cell. A parenchyma cell that through deposition of a thick secondary wall becomes changed into a sclereid.

Scutellum (pl. scutella). The cotyledon in Poaceae embryo specialized for absorption of endosperm.

Secondary body. The part of the plant body that is added to the primary body by the activity of the lateral meristems, vascular cambium and phellogen. Consists of secondary vascular tissues and periderm.

Secondary cell wall. Version based on studies with the light microscope: cell wall deposited in some cells over the primary wall after the primary wall ceases to increase in surface. Version based on studies with the electron microscope: cell wall in which the cellulose microfibrils show a definite parallel orientation. The two versions do not necessarily coincide in delimiting secondary from primary wall.

Secondary endosperm nucleus. Nucleus resulting from fusion of two polar nuclei in the central cell of the embryo sac of angiosperms.

Secondary growth. In gymnosperms, most dicotyledons, and some monocotyledons. A type of growth characterized by an increase in thickness of stem and root and resulting from formation of secondary vascular tissues by the vascular cambium. Commonly supplemented by activity of the cork cambium (phellogen) forming periderm.

Secondary phloem. Phloem tissue formed by the vascular cambium during secondary growth in a vascular plant. Differentiated into axial and ray systems.

Secondary phloem fiber. A fiber located in the axial system of secondary phloem.

Secondary plant body. See *secondary body.*

Secondary root. See *branch root.*

Secondary thickening. Used for both deposition of secondary cell wall material and secondary increase in thickness of stems and roots.

Secondary tissues. Tissues produced by vascular cambium and phellogen during secondary growth.

Secondary vascular tissues. Vascular tissues (both xylem and phloem) formed by the vascular cambium during secondary growth in a vascular plant. Differentiated into axial and ray systems.

Secondary wall. See *secondary cell wall.*

Secondary xylem. Xylem tissue formed by the vascular cambium during secondary growth in a vascular plant. Differentiated into axial and ray systems.

Secretory cavity. Commonly refers to a space lysigenous in origin and containing a secretion derived from the cells that broke down in the formation of the cavity.

Secretory cell. A living cell specialized with regard to secretion or excretion of one or more, often organic, substances.

Secretory duct. Commonly refers to a duct schizogenous in origin and containing a secretion derived from the cells (epithelial cells) lining the duct. See *epithelium.*

Secretory hair. See *glandular hair.*

Secretory structure. Any of a great variety of structures, simple or complex, external or internal, that produces a secretion.

Seed coat. The outer coat of the seed derived from the integument or integuments. Also called *testa.*

Seminal adventitious root. Root initiated in the embryo on the hypocotyl or higher on the axis.

Sepal. A unit of the calyx.

Separation layer. See *abscission layer.*

Septate fiber. A fiber with thin transverse walls (septa), which are formed after the cell develops a secondary wall thickening.

Septum (pl. septa). A partition.

Sessile. Refers to a leaf lacking a petiole or a flower or a fruit lacking a pedicel.

Sexine. The outer layer of the exine of a pollen grain. Sculptured part of exine.

Sheath. A sheetlike structure enclosing or encircling another. Applied to tubular or enrolled part of an organ, such as a leaf sheath, and to a tissue layer surrounding a complex of another tissue, as a bundle sheath enclosing a vascular bundle.

Sheathing base. Applied to a leaf base that encircles the stem.

Shell zone. In axillary bud primordia; a zone of parallel curving layers of cells, the entire complex shell-like in form. A result of orderly cell division along the proximal limits of the primordium.

Sieve area. A pitlike area in the wall of a sieve element with pores commonly lined with callose and occupied by protoplasmic material that interconnects the protoplasts of contiguous sieve elements.

Sieve cell. A type of sieve element that has relatively undifferentiated sieve areas (with narrow pores), rather uniform in structure on all walls; that is, there are no sieve plates. Typical of gymnosperms and lower vascular plants.

Sieve element. The cell in the phloem tissue concerned with mainly longitudinal conduction of food materials. Classified into *sieve cell* and *sieve tube member.*

Sieve field. Old term for a relatively undifferentiated sieve area found on wall parts other than the sieve plates.

Sieve pitting. An arrangement of small pits in sievelike clusters.

Sieve plate. The part of the wall of a sieve element bearing one or more highly differentiated sieve areas. Typical of angiosperms.

Sieve tube. A series of sieve elements (sieve tube members) arranged end to end and interconnected through sieve plates.

Sieve tube element. See *sieve tube member.*

Sieve tube member. One of the series of cellular components of a sieve tube. It shows a more or less pronounced differentiation between sieve plates (wide pores) and lateral sieve areas (narrow pores). Also *sieve tube element* and the obsolete sieve tube segment.

Silica cell. Cell filled with silica, as in epidermis of grasses.

Simple laticifer. Laticifer that is a single cell. *Nonarticulated laticifer.*

Simple perforation plate. In vessel member of the xylem; a perforation plate with a single perforation.

Simple pit. A pit in which the cavity becomes wider, remains of constant width, or only gradually becomes narrower during the growth in thickness of the secondary wall, that is, toward the lumen of the cell.

Simple pit-pair. An intercelluar pairing of two simple pits.

Simple sieve plate. Sieve plate composed of one sieve area.

Siphonostele. A type of stele in which the vascular system appears in the form of a hollow cylinder; that is, pith is present.

Slime. See *P-protein.*

Slime body. An aggregation of P-protein.

Slime plug. An accumulation of P-protein on a sieve area, usually with extensions into the sieve area pores.

Softwood. Technical designation of the wood of conifers.

Solitary pore. A pore (cross section of a vessel in secondary xylem) surrounded by cells other than vessel members.

Specialization. Change in structure of a cell, a tissue, plant organ, or entire plant associated with a restriction of functions, potentialities, or adaptability to varying conditions. May result in greater efficiency with regard to certain specific functions. Some specializations are irreversible, others reversible.

Specialized. Refers to (1) organisms having special adaptations to a particular habitat or mode of life; (2) cells or tissues having a characteristic function distinguishing them from other cells or tissues, more generalized in their function.

Spherosomes. Spherical bodies in the cytoplasm containing mostly lipid. Limiting membrane is thought to be absent or consisting of a unit membrane or of half of a unit membrane.

Spindle fibers. Microtubules aggregated in a spindle-shaped complex extending from pole to pole in a cell with a dividing nucleus. Term refers to light microscope views.

Spiral cell wall thickening. See *helical cell wall thickening.*

Spongy parenchyma. Leaf mesophyll parenchyma with conspicuous intercellular spaces.

Sporophyll. A modified leaf or leaflike organ that bears sporangia. In angiosperms, refers to stamens and carpels.

Sporopollenin. The substance composing the outer wall, or exine, of the pollen grain or a spore. A cyclic alcohol highly resistant to decay.

Spring wood. See *early wood.*

Square ray cell. In secondary vascular tissues, a ray cell approximately square as seen in radial section. (Considered to be of the same morphological type as the *upright ray cell.*)

Stamen. Floral organ producing the pollen and usually composed of anther and filament. The stamens together constitute the *androecium.*

Staminate. Refers to a flower having stamens but no functional carpels.

Starch. An insoluble carbohydrate, the chief food storage substance of plants, composed of anhydrous glucose residues of the formula $C_6H_{10}O_5$ into which it easily breaks down.

Starch sheath. Applied to the innermost region (one or more cell layers) of the cortex when this region is characterized by conspicuous and rather stable accumulation of starch.

Stele (column). Conceived by Van Tieghem as a morphologic unit of the plant body comprising the vascular system and the associated ground tissue (pericycle, interfascicular regions, and pith). The *central cylinder* of the axis (stem and root).

Stellate. Star shaped.

Stereom (or stereome). Collective term for supporting tissue as contrasted with the conducting tissues *hadrom* and *leptom.*

Stigma. The region of the carpel, in many taxa at the apex of the style, that serves as a surface upon which the pollen germinates.

Stigmatoid tissue. A tissue cytologically and physiologically rather similar to the tissue of the stigma and serving as a path for the pollen tube in the style. Preferred designations, pollen conducting tissue and pollen *transmitting tissue.*

Stoma (pl. stomata). An opening in the epidermis of leaves and stems bordered by two guard cells and serving in gas exchange.

Stomatal complex. Stoma and associated epidermal cells that may be ontogenetically and/or physiologically related to the guard cells. Also called stomatal apparatus.

Stomatal crypt. A depression in the leaf, the epidermis of which bears stomata.

Stomium. A fissure or pore in the anther lobe through which the pollen is released. Its formation is a type of dehiscence.

Stone cell. See *brachysclereid.*

Storied cambium. Vascular cambium in which the fusiform initials and rays are arranged in horizontal tiers on tangential surfaces. *Stratified cambium.*

Storied cork. Protective tissue found in the monocotyledons. The suberized cells occur in radial files, each consisting of several cells all of which are derived from one cell.

Storied wood. Wood in which the axial cells and rays are arranged in horizontal tiers on tangential surfaces. (Rays alone may be storied.) *Stratified wood.*

Strasburger cells. See *albuminous cells.*

Stratified cambium. See *storied cambium.*

Stratified wood. See *storied wood.*

Striate venation. See *parallel venation.*

Stroma. The ground substance of plastids.

Stromacenter. Aggregated fibrils, each 85 angstroms in diameter and of uncertain length, found in the stroma of a chloroplast fixed with glutaraldehyde (or acrolein)-osmium tetroxide.

Style. Extension of the top of the ovary, usually columnar, through which the pollen tube grows.

Stylode. Stylar branch.

Styloid. An elongated crystal with pointed or square ends.

Subapical initial. A cell beneath the protoderm at the apex of a leaf primordium that appears to function as an initial of the interior tissue of the leaf. Questionable concept.

Suberin. Fatty substance in the cell wall of cork

tissue and in the casparian strip of the endodermis.

Suberization. Impregnation of the cell wall with suberin or deposition of suberin lamellae on the wall.

Submarginal initials. Cells beneath the protoderm along the margins of a growing leaf lamina that appear to contribute cells to the interior tissue of the leaf. Components of *marginal meristem* which is concerned with *marginal growth.*

Subsidiary cell. An epidermal cell associated with a stoma and at least morphologically distinguishable from the epidermal cells composing the groundmass of the tissue. Also called *accessory cell.*

Summer wood. See *late wood.*

Superior ovary. See *hypogyny.*

Supernumerary cambium layer. Vascular cambium originating in phloem or pericycle outside the regularly formed vascular cambium. Characteristic of some plants with anomalous type of secondary growth.

Supporting cell. See *supporting tissue.*

Supporting tissue. Refers to tissue composed of cells with more or less thickened walls, primary (collenchyma) or secondary (sclerenchyma), that adds strength to the plant body. Also called *mechanical tissue.*

Suspensor. An extension at the base of the embryo that anchors the embryo in the embryo sac and pushes it into the endosperm.

Symplastic growth. See *coordinated growth.*

Syncarpy. Condition in flower characterized by union of carpels.

Syndetocheilic. Stomatal type in gymnosperms; subsidiary cells (or their precursors) are derived from the same protodermal cell as the guard-cell mother cell.

Synergids. Two cells in the micropylar end of the embryo sac associated with the egg in the egg apparatus of angiosperms. Play a vital role in fertilization.

Tabular. Having the form of a tablet or slab.

Tangential. In the direction of the tangent; at right angles to the radius. May coincide with *periclinal.*

Tangential section. A longitudinal section cut at right angles to a radius. Applicable to cylindrical structures such as stem or root, but used also for leaf blades when the section is made parallel with the expanded surface. Substitute term for leaf, *paradermal.*

Tannin. General term for a heterogeneous group of phenol derivatives. Amorphous strongly astringent substance widely distributed in plants, and used in tanning, dyeing, and preparation of ink.

Tapetum. In anther, a layer of cells lining the locule and absorbed as the pollen grains mature. In ovule, integumentary epidermis next to the embryo sac; also called *endothelium.*

Taproot. The first, or primary, root of a plant forming a direct continuation of the radicle of the embryo.

Taproot system. A root system based on the taproot, which may have branches of various orders.

Tasche. In German, pocket. Covering of primordium of lateral root derived from the endodermis, as distinguished from the rootcap, which is derived from the pericycle.

Taxon (pl. taxa). Any one of the categories (species, genus, family, etc.) into which living organisms are classified.

Teichode. A linear space in the outer epidermal wall in which the fibrillar structure is more loose and open than elsewhere in the wall. Replaces the term *ectodesma.*

Telome. One of the distal branches of a dichotomized axis, a morphological unit in a primitive vascular plant.

Telome theory. A theory that regards the telomes as basic units from which the diverse types of leaves and sporophylls of the vascular plants have evolved.

Template. A pattern or mould guiding the formation of a negative or a complement. A term applied in biology to DNA duplication (template hypothesis).

Tension wood. Reaction wood in dicotyledons, formed on the upper sides of branches and leaning or crooked stems and characterized by lack of lignification and often by high content of gelatinous fibers.

Tepal. A member of the kind of floral perianth that is not differentiated into calyx and corolla.

Terminal apotracheal parenchyma. See *boundary apotracheal parenchyma.*

Testa. The seed coat.

Tetrarch. Primary xylem of root; having four protoxylem strands, or protoxylem poles.

Thylakoids. Saclike membranous structures (cisternae) in a chloroplast combined into stacks (grana) and present singly in the stroma (stroma thylakoids or frets) as interconnections between grana.

Tissue. Group of cells organized into a structural and functional unit. Component cells may be alike (simple tissue) or varied (complex tissue).

Tissue system. A tissue or tissues in a plant or plant organ structurally and functionally organized into a unit. Commonly three tissue systems are recognized, *dermal, vascular,* and *fundamental* (ground tissue system).

Tonoplast. A single cytoplasmic membrane bordering the vacuole. A kind of *unit membrane.*

Torus (pl. tori). The central thickened part of the pit membrane in a bordered pit consisting mainly of middle lamella and two primary walls. Typical of bordered pits in conifers and some other gymnosperms.

Trabecula (pl. trabeculae). A rodlike or spoolshaped part of a cell wall extending radially across the lumen of a cell. In initials and derivatives of vascular cambium in seed plants.

Trachea. Old term for xylem vessel implying a resemblance to an animal trachea.

Tracheary element. General term for a waterconducting cell, tracheid or vessel member.

Tracheid. A tracheary element of the xlyem that has no perforations, as contrasted with a vessel member. May occur in primary and in secondary xylem. May have any kind of secondary wall thickening found in tracheary elements.

Transection. *Transverse section.*

Transfer cell. Parenchyma cell with wall ingrowths (or invaginations) that increase the surface of the plasmalemma. Appears to be specialized for short-distance transfer of solutes. Cells without wall ingrowths may also function as transfer cells. See *intermediary cell.*

Transfusion tissue. In gynmosperm leaves, a tissue surrounding or otherwise associated with the vascular bundle and composed of tracheids and parenchyma cells. See also *accessory transfusion tissue.*

Transfusion tracheid. A tracheid in *transfusion tissue.*

Transition region. A region in the plant axis where root and shoot are united and which shows primary structural characteristics transitional between those of stem and root. Best exhibited in seedlings.

Transition zone. With reference to an apical meristem, a zone of orderly dividing cells disposed about the inner limit of the promeristem or, more specifically, of the group of central mother cells. Is transitional between the apical meristem and the subapical primary meristematic tissues.

Transitional cell. See *intermediary cell.*

Transmitting tissue, or pollen transmitting tissue. The tissue in the style of a flower through which the pollen tube grows between the stigma and the ovarian cavity. Also called pollenconducting tissue.

Transverse division (of cell). With reference to cell, division perpendicular to the longitudinal axis of the cell. With reference to plant part, division of the cell perpendicular to the long axis of the plant part.

Transverse section. A cross section. Section taken perpendicular to the longitudinal axis of an entity. Also called *transection.*

Traumatic resin duct. A resin duct developing in response to injury.

Triarch. Primary xylem of root; having three protoxylem strands, or three protoxylem poles.

Trichoblast. Commonly used for a cell in root epidermis that gives rise to a root hair.

Trichome. An outgrowth from the epidermis. Trichomes vary in size and complexity and include hairs, scales, and other structures and may be glandular.

Trichosclereid. A type of branched sclereid, usually with hairlike branches extending into intercellular spaces.

Trilacunar node. In a stem, a node with three leaf gaps related to one leaf.

Tropism. Refers to movement or growth in response to an external stimulus the site of which determines the direction of the movement or growth.

Tube cells. Elongated cells with lignified walls in the inner epidermis of pericarp of a caryopsis of Poaceae.

Tunica. Peripheral layer or layers in an apical meristem of a shoot with cells that divide in the anticlinal plane and thus contribute to the growth in surface of the meristem. Forms a mantle over the corpus.

Tunica-corpus concept. A concept of the organization of apical meristem of shoot according to which this meristem is differentiated into two regions distinguished by their method of growth: the peripheral, tunica, one or more layers of cells showing surface growth (anticlinal divisons); the interior, corpus, a mass of cells showing volume growth (divisions in various planes).

Two-trace unilacunar condition. Characteristic of a node in a stem in which two leaf traces pertaining to one leaf are associated with one leaf gap.

Tylose (pl. tyloses). In xylem, an outgrowth from a parenchyma cell (axial or one in a ray) through a pit cavity into a tracheary cell, partially or completely blocking the lumen of the latter. Growth is preceded by a deposition of a special wall layer on the side of the parenchyma cell that forms the wall of the tylose.

Tylosoid. An outgrowth resembling a tylose. Examples are outgrowths of parenchyma cells into sieve elements in phloem and of epithelial cells into intercellular resin ducts.

Undifferentiated. In ontogeny, still in a meristematic state or resembling meristematic structures. In a mature state, relatively unspecialized.

Unifacial leaf. A leaf having similar structure on both sides. Conceived ontogenetically, a leaf that develops from a growth center abaxial or adaxial to the initial leaf primordium apex and thus includes tissues only from the abaxial or adaxial side of the primordium. The validity of the ontogenetic concept is questionable. Compare with *bifacial leaf.*

Unilacunar node. In a stem, a node with one leaf gap related to one leaf. If two or more leaves are attached at such a node, each is associated with one gap.

Uniseriate ray. In secondary vascular tissues, ray one cell wide.

Unisexual. Usually refers to a flower lacking stamens or carpels. A perianth may be present or absent.

Unit membrane. A historical concept of basic membrane structure visualizing two layers of protein enclosing an inner layer of lipid, the three layers forming a unit. The term continues to be useful for describing sectioned membranes (profiles), exhibiting two dark lines separated by a clear space, as seen with the electron microscope.

Upright ray cell. In secondary vascular tissues, ray cell oriented axially (vertically in the axis) with its longest dimension.

Vacuolar membrane. See *tonoplast.*

Vacuolation. Ontogenetically, the development of vacuoles in a cell; in mature state, the presence of vacuoles in a cell.

Vacuole. Cavity within the cytoplasm filled with a watery fluid, the cell sap, and bound by a unit membrane, the tonoplast. Involved in uptake of water during germination and growth and maintenance of water in the cell. Also contains hydrolytic enzymes and has a lytic function.

Vacuome. Collective term for the total of all vacuoles in a cell, tissue, or plant.

Vascular. Refers to plant tissue or region consisting of or giving rise to conducting tissue, xylem and/or phloem.

Vascular bundle. A strandlike part of the vascular system composed of xylem and phloem.

Vascular cambium. Lateral meristem that forms the secondary vascular tissues, secondary phloem and secondary xylem, in stem and root. Is located between those two tissues and, by periclinal divisions, gives off cells toward both tissues.

Vascular cylinder. Vascular region of the axis. Term used synonymously with *stele* or *central cylinder* or in a more restricted sense excluding the pith.

Vascular meristem. General term applicable to *procambium* and *vascular cambium.*

Vascular ray. A ray in secondary xylem or secondary phloem.

Vascular system. The total of the vascular tissues in their specific arrangement in a plant or plant organ.

Vascular tissue. A general term referring to either or both vascular tissues, xylem and phloem.

Vasicentric paratracheal parenchyma. Axial parenchyma in secondary xylem forming complete sheaths around vessels. See also *paratracheal parenchyma.*

Vegetative apical meristem. See *apical meristem.*

Vein. A strand of vascular tissue in a flat organ, as a leaf. Hence, leaf venation.

Vein rib. In a leaf, ridge of ground tissue occurring along a larger vein, usually on the lower side of the leaf.

Velamen. A multiple epidermis covering the aerial roots of some tropical epiphytic orchids and aroids. Occurs in some terrestrial roots also.

Venation. The arrangement of veins in the leaf blade.

Vertical parenchyma. See *axial parenchyma.*

Vertical system. In secondary vascular tissues. See *axial system.*

Vessel. A tubelike series of vessel members the common walls of which have perforations.

Vessel element. See *vessel member.*

Vessel member. One of the cellular components of a vessel. Also *vessel element* and the obsolete vessel segment.

Vestured pit. Bordered pit with projections from the overhanging secondary wall on the side facing the cavity.

Wall. See *cell wall.*

Water vesicle. A type of trichome. An enlarged highly vacuolated epidermal cell.

Wood. Usually secondary xylem of gymnosperms and dicotyledons, but also applied to any other xylem.

Wound cork. See *wound peridem.*

Wound gum. Gum formed as a result of some injury. See also *gum.*

Wound periderm. Periderm formed in response to wounding or other injury.

Xeromorphic. Refers to structural features typical of xerophytes.

Xerophyte. A plant adapted to a dry habitat.

Xylary procambium. The part of procambium that differentiates into primary xylem. Also called xylic or xyloic procambium.

Xylem. The principal water-conducting tissue in vascular plants characterized by the presence of tracheary elements. The xylem may also serve as a supporting tissue, especially the secondary xylem (wood).

Xylem elements. Cells composing the xylem tissue.

Xylem fiber. A fiber of the xylem tissue. Two types are recognized in the secondary xylem, *fiber-tracheid* and *libriform fiber.*

Xylem initial. A cambial cell on the xylem side of the cambial zone that is the source of one or more cells arising by periclinal divisions and differentiating into xylem elements either with or without additional divisions in various planes. Sometimes called *xylem mother cell.*

Xylem mother cell. A cambial derivative that is the source of certain elements of the xylem, such as axial parenchyma cells forming a parenchyma strand. Used also in a wider sense synonymously with *xylem initial.*

Xylem ray. That part of a vascular ray which is located in the secondary xylem.

Xylotomy. The anatomy of xylem.

Zygomorphic. Irregular flower. See *bilateral symmerty.* Opposite of *actinomorphic.*

Index

Numbers in *italics* indicate illustrations located apart from the description of the subject in the illustrations.